4th International Conference on Science and Technology Applications in Climate Change

IOP Conference Series: Earth and Environmental Science
Volume 880

Online
1 – 2 July 2021

ISBN: 978-1-7138-4862-2
ISSN: 1755-1307

Printed from e-media with permission by:

Curran Associates, Inc.
57 Morehouse Lane
Red Hook, NY 12571

Some format issues inherent in the e-media version may also appear in this print version.

Institute of Physics
Dirac House, Temple Back
Bristol BS1 6BE UK

Phone: 44 1 17 929 7481
Fax: 44 1 17 920 0979

techtracking@iop.org

Additional copies of this publication are available from:

Curran Associates, Inc.
57 Morehouse Lane
Red Hook, NY 12571 USA
Phone: 845-758-0400
Fax: 845-758-2633
Email: curran@proceedings.com
Web: www.proceedings.com

TABLE OF CONTENTS

IOP Conf. Series: Earth and Environmental Science **880** (2021) 011001 doi:10.1088/1755-1315/880/1/011001

Preface

Following the success of 2014, 2016 and 2018 International Conference on Science & Technology Applications in Climate Change (STACLIM), the Institute of Climate Change (IPI), Universiti Kebangsaan Malaysia (UKM) is proud to extend our promotion of research and education for the advancement of climate change studies. The 2021 International Conference on Science & Technology Applications in Climate Change (STACLIM 2021) with the theme "Empowering Breakthroughs for Climate Change" is the fourth in the series of conferences organized by IPI. This year the conference was carried out in virtual form through the Webex platform (1 – 2 July 2021) due to the COVID-19 travel restriction. Through the virtual form, the science community is able to share their research findings in time.

The aim of this conference is to bring together researchers in fields of Environmental Science, Health Sustainability, Mathematics, Sustainable Energy, Economic Sustainability, Socio-Cultural Studies, Social Science, Atmospheric Science, and related fields, to present their research findings as well as create new opportunities for future research collaborations. This event is envisaged to witness active participation from various eminent environmental and earth scientists, engineers and students from academia, industry and government sectors for addressing complications associated with climate change and to draw forth novel and ground-breaking initiatives and solutions for climate resilience.

The plenary sessions in the main room was opened by three keynote speeches from leading experts including Prof. Dr. Joshua S. Fu from The University of Tennessee Knoxville, USA on "Integrated Engineering and Earth Systems Science as Solution to Climate Change: Lessons and Opportunities", Prof. Dr. Agamuthu Pariatamby from Jeffrey Cheah Institute on Southeast Asia, Malaysia on "Waste Management and Climate Change in Asia: Will it Mitigate or Aggravate?" and Distinguished Professor Dr. Biswajeet Pradhan from University of Technology Sydney, Australian "Geospatial Intelligence in Climate-Induced Hazard Modelling". As the keynote session were open for public registration, we had 119 participants joining the event. It was then followed by the oral presentation of 72 papers in 3 parallel breakout rooms. Each presenter was given up to 15 mins for presentation and Q&A sessions. There were additional of 13 non-presenters who joined in during the presentation session. Presenters and participants have attended the conference from their respective countries including Malaysia, Indonesia, Philippines, USA, China, Saudi Arabia, Iraq, Algeria, India, and Ukraine.

The conference went well with great support and synergy of the staff and personnel from Institute of Climate Change, UKM. To document and promulgate the research findings and ideas shared, we are very pleased to published the accepted abstracts and research papers of STACLIM 2021 in IOP Conference Series: Earth and Environmental Science (EES) (Online ISSN: 1755-1315).

Organizing Committee

PATRON

Prof. Dato' Dr. Mohd Ekhwan Toriman
Vice-Chancellor, Universiti Kebangsaan Malaysia

ADVISOR

Prof. Dr. Mohd. Nizam Mohd. Said
Director of Institute of Climate Change (IPI), UKM

CHAIRMAN

Associate Prof. Dr. Marlia Mohd Hanafiah
Head of Research Centre for Tropical Climate System (IKLIM), IPI, UKM

VICE CHAIRMAN

Sr. Associate Prof. Dr. Khairul Nizam Abdul Maulud
Head of Earth Observation Centre

SCIENTIFIC PROGRAMME COMMITTEE

Associate Prof. Dr. Rawshan Ara Begum (Lead)
Associate Prof. Dr. Liew Ju Neng
Associate Prof. Dr. Zaini Sakawi
Ts. Dr. Wan Shafrina Wan Mohd Jaafar
Dr. Ooi Chel Gee
Dr. Wee Hin Boo
Dr. Kuhan Chandru A/L Balasanthiran
Dr. Khairiatul Mardiana Jansar
Dr. Siti Aminah Bahari
Norzatul Ézzah Mohd Hasan
Nur Khairun Ayuni Jusoh Sukri
Noridawaty Mat Daud
Nurul Umairah Mohd Nizam
Md Iqbal Ansari

SECRETARIAT

Balqis Md. Desa (Lead)
Azizah Ahmad
Nur Zaimah Abd Wahid

TREASURER AND MARKETING

Muhamad Roszaini Roslan (Lead)
Nur Afiqah Mohd Zaidi
Syarina Md Sah

Siti Aisyah Mohamed

IT AND PUBLICITY

Khirunnisa Mazelan (Lead)
Noor Shuhaira Rejab
Khalisah Khairina
Akhtar Rasool

IT, LOGISTIC AND TECHNICAL COMITTEE

Mohd Azlan Shah Jaafar (Lead)
Ahmad Tarmizi Mohd Azmi
Siti Aisyah Mohamed
Noor Shuhaira Rejab
Mohd Sayuti Che Hamid

IOP Conf. Series: Earth and Environmental Science **880** (2021) 011002 doi:10.1088/1755-1315/880/1/011002

Peer review declaration (STACLIM 2021)

All papers published in this volume of IOP Conference Series: Earth and Environmental Science have been peer reviewed through processes administered by the Editors. Reviews were conducted by expert referees to the professional and scientific standards expected of a proceedings journal published by IOP Publishing.

- **Type of peer review:** Single-blind
  The review process is conducted in single-blind method that the reviewers have access to the name(s) of author(s) while author is not being informed on the name of the reviewer(s).

- **Conference submission management system:** EDAS Conference and Journal Management System
  Link: https://edas.info/index.php?c=28219

- **Number of submissions received:** 110 ***(Total number of papers)***

- **Number of submissions sent for review:** 72 ***(Final number of papers reviewed)***

- **Number of submissions accepted:** 60 ***(Final number of papers after acceptance)***

- **Acceptance Rate (Number of Submissions Accepted / Number of Submissions Received X 100):** 54.5%

- **Average number of reviews per paper:** 3.25
  Each paper have gone through at least two stages of review, namely abstract review, and full manuscript review, while papers with major and minor revision received additional 1 to 2 rounds of reviews.

- **Total number of reviewers involved:** 24

- **Any additional info on review process:**
  - First-stage of peer review was done on the submitted abstract; only those which are accepted will be prompted for full paper submission.
  - The full paper needed to achieve similarity index of less than 30% before it could be sent out for the second-stage rigorous peer review; only papers that have properly revised and addressed the comments from editors and reviewers could be accepted and proceeded for acceptance, hence publication in the IOP Conference series: Earth and Environmental Science.

- **Contact person for queries:**

Name: Dr Marlia Mohd Hanafiah
Affiliation: Institute of Climate Change (IPI), Universiti Kebangsaan Malaysia
Email : mhmarlia@ukm.edu.my

Name: Dr Rawshan Ara Begum
Affiliation: Institute of Climate Change (IPI), Universiti Kebangsaan Malaysia
Email : rawshan@ukm.edu.my

Name: Dr Maggie Chel Gee Ooi
Affiliation: Institute of Climate Change (IPI), Universiti Kebangsaan Malaysia
Email : chelgee.ooi@ukm.edu.my

Variations of precipitable water vapor from 2012 to 2019 over the Philippines using radiosondes

Y Rivera[1*], K C Capacete[1], S K Rodriguez[1], A R David[1] and E Macalalad[2]

[1] Mapúa Senior High School, Muralla St., Intramuros, Manila, Philippines
[2] Department of Physics, Mapúa University, Muralla St., Intramuros, Manila, Philippines

*Corresponding E-mail: riverayavin@gmail.com

Abstract. Precipitable Water Vapor (PWV) refers to the content of water vapor in the atmosphere which is significant in observing climate changes. The trends and variations of precipitable water vapor in Laoag, Legazpi, Mactan, and Puerto Princesa from 2012-2019, are presented through the use of radiosonde data derived from the database of the Integrated Global Radiosonde Archives (IGRA). These data were analyzed for possible patterns through a time series of its daily, monthly, and annual mean, together with a Lomb-Scargle periodogram, and Mann-Kendall test. The results observed varying trends and variability. Legazpi and Puerto Princesa with a minimum value of 20 mm, observed a gradual downward trend of PWV. Laoag and Mactan detected an upward trend of PWV with a minimum of 10 mm and 20 mm, respectively. It also showed an annual and bi-annual periodicity of PWV. Furthermore, all cities detected an increase of PWV during the wet months of May to September, while the dry months of October to April with slight variations over 8 years. In terms of seasonality, only Laoag observed a slightly different dry season, with January, February, and March experiencing around 5 mm less in monthly PWV variation compared to the other cities. The correlation of surface temperature and relative humidity of PWV observed an overall increasing trend while showing a general moderate positive correlation. This study can be used for future references for meteorologists for upcoming forecasting on the likelihood of different weather phenomena in the Philippines.

Keywords: Radiosondes, Variability, Trends, Precipitable Water Vapor

Track Name: Atmospheric Chemistry and Physics

1. Introduction

Water vapor is considered to be responsible for emitting around 60 % of the greenhouse effect which contributes to the global warming phenomenon. Precipitable water vapor (PWV) represents the content of water vapor in the atmosphere. It is defined as the height of liquid gathered from the top of the atmosphere by condensing its total in a column [1]. The characteristics and conditions of the atmosphere provided by PWV data are significant on predictions in meteorology, such as weather and climate forecasts [2]. Because of this, monitoring the variation of PWV is essential for observing climate variability [3].

PWV can be determined using various instruments. One of them are radiosondes. Radiosondes are meteorological instruments that have different sensors attached to a small radio transmitter. They are released into the sky with the use of helium-filled balloons. These sensors measure the vertical profile atmosphere as it ascends. This includes air temperature, pressure, atmospheric stability, humidity, wind speed, cosmic rays, and more [4]. Typically, radiosondes are launch twice a day (0000UT and 1200 UT). They usually reach up to an altitude of about 30 km. Thus, radiosondes are used to student much of the neutral gas in the atmosphere, including water vapor.

Several studies have shown the analysis of short-term variations in PWV using radiosondes and other techniques such as remote sensing. Macalalad and Macalalad [9] analyzed the variability of PWV in 2017 over Manila using Global Navigation Satellite System where it demonstrated the occurrences of dry and wet anomalies in certain months relative to the daily variability of rainfall [2]. In PWV data from GNSS was used to observe the atmospheric water vapor in Davao City, Philippines. Here, a threshold value of PWV was determined from historical data to potentially predict the probability of heavy and torrential rainfall occurrence in the future [5, 6]. A similar study investigated the variation of total PWV in climatic zones in Iran through the Moderate Resolution Imaging Spectroradiometer (MODIS). The researchers concluded that a notable increase of PW can be seen in the hot climatic areas in both summer and spring [7]. Similarly, Zhang *et al.*[8] discussed seasonal variation of PWV using the National Climatic Data Center and Integrated Global Radiosonde Archive (IGRA) database. It is discovered that warm seasons have larger PWV measurements (40 mm) compared to cold seasons (10 mm) denoting that there is an extensive rise of PWV above land in the past 10 years [8].

Other studies focused on the long-term changes or trends of PWV for a long period of time, using either radiosondes or remote sensing instruments. According to Wang *et al.* [9], the study analyzed global PW trends together with distinct spatial variations between land and ocean from 1988 to 2011. It is observed that there is an overall increase of PW in all datasets worldwide, while the ocean has a relatively larger spatial variability in the said period [9]. Also, Chen and Liu [3] studied PWV general trends worldwide from 1979-2014 using radiosonde data. The result showed that there is an overall increase in PWV trends in the past years and the possibility of a 13 % PWV addition in the atmosphere might be seen in the years to come [3].

In the past decade, there are inadequate work that studied water vapor trends over the Philippines using radiosonde data for a long period of time [10]. Therefore, this research aims to study possible trends of PWV using radiosonde data from 2012 to 2019 over Laoag, Legazpi, Mactan, and Puerto Princesa City. Spatial and temporal variation (interannual and seasonal) of PWV will also be studied.

2. Data

This study obtained PWV data calculated from radiosonde observations from IGRA (v2) from 2012-2019 from four stations in the Philippines, namely, Laoag, Legazpi, Mactan, and Puerto Princesa City. Figure 1 shows the station IDs and geographic information of each station. In this study, time series of PWV data is determined. From this, the Lomb-Scargle periodogram is calculated to estimate the main periodicity of the data set. This technique is mostly useful to calculate periodicities of unevenly spaced data set [11–13]. Monthly mean and standard deviation are also calculated to evaluate its average monthly variation over the 8-year period. This is correlated with monthly average values of surface meteorological parameters, such as relative humidity, and surface temperature. These are also obtained from the radiosonde profiles provided by IGRA. Here, correlation between these variables with PWV is explored.

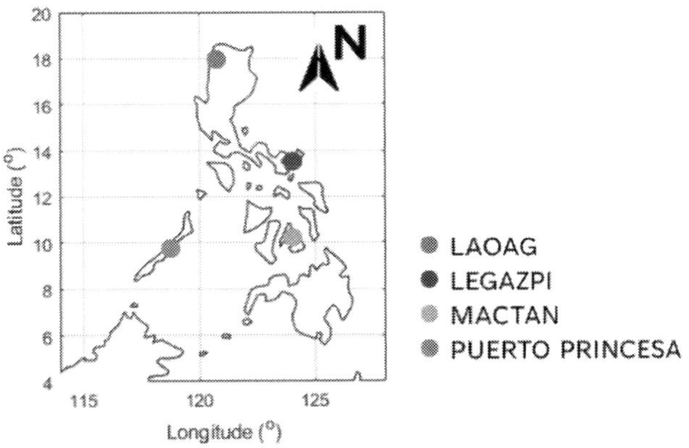

Figure 1. Stations used in this study.

3. Results and Discussion

Figure 2 shows the daily time series of PWV in the four cities from 2012-2019. Laoag saw an increase of PWV, observed in 2012, 2016, and 2017, which exceeded the 70 mm value. On the other hand, a decrease was observed in 2014, with the lowest PWV data in 2014 reaching below 10 mm. Mactan also observed a slightly increasing trend with PWV peaks surpassing 60 mm between 2014-2017. The daily PWV in Legazpi showed majority of data peaked above 60 mm, signifying rainfall within a certain period, while the lack of rainfall was observed in the PWV in 2014 and 2019 which reached below 20 mm. A slightly decreasing trend is observed due to numerous drops of the PWV measurements around the 20 mm level. PWV data in Puerto Princesa peaked in measurement above the 60 mm mark together with the decrease of PWV data at the beginning of each year of 2012-2019. In addition, PWV data slightly shows a downward trend which is seen in the dips of PWV, reaching 20 mm compared to its previous years. The four daily time series continue to present normal data, with only slight irregularities or surges in PWV. Finally, the data presented agree on the climate of their respective area, observing two different seasons in which PWV behaves according to either the dry or wet seasons from 2012 to 2019.

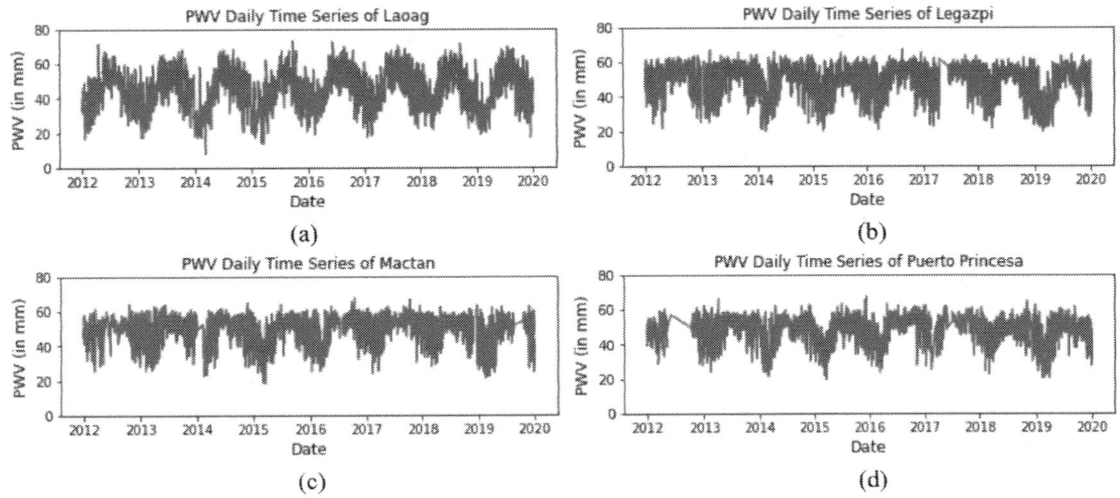

Figure 2. PWV daily time series from 2012-2019 at (a) Laoag, (b) Legaspi, (c) Mactan, and (d) Puerto Princesa.

Similar to the studies of Domingo and Macalalad [14], PWV shows recurring patterns without distinctive behaviors while showing a variation of PWV occurrences during pre-monsoon periods. High variability can indicate the occasional increase in moisture due to local weather events such as thunderstorms that may have contributed to the high variability of PWV during a particular season, which presents a relation to the results from the daily time series in Laoag. On the other hand, the indication of a PWV decrease after significant rainfall is caused by the occurrence of thunderstorms or typhoons [14].

The periodicity of PWV over Laoag, Legazpi, Mactan, and Puerto Princesa City which was shown through a Lomb-Scargle periodogram is shown in Figure 3. The behavior was observed to check the periodic variations or if the data occurs at intervals on unevenly sampled data. Here, Laoag, Legazpi, Mactan, and Puerto Princesa had peaks of about 365 days, with Laoag showing the strongest spectrum. This shows that PWV for these stations primarily vary annually. Secondary peaks are also observed at 182.5 days which indicates bi-annual periodicities over the four stations.

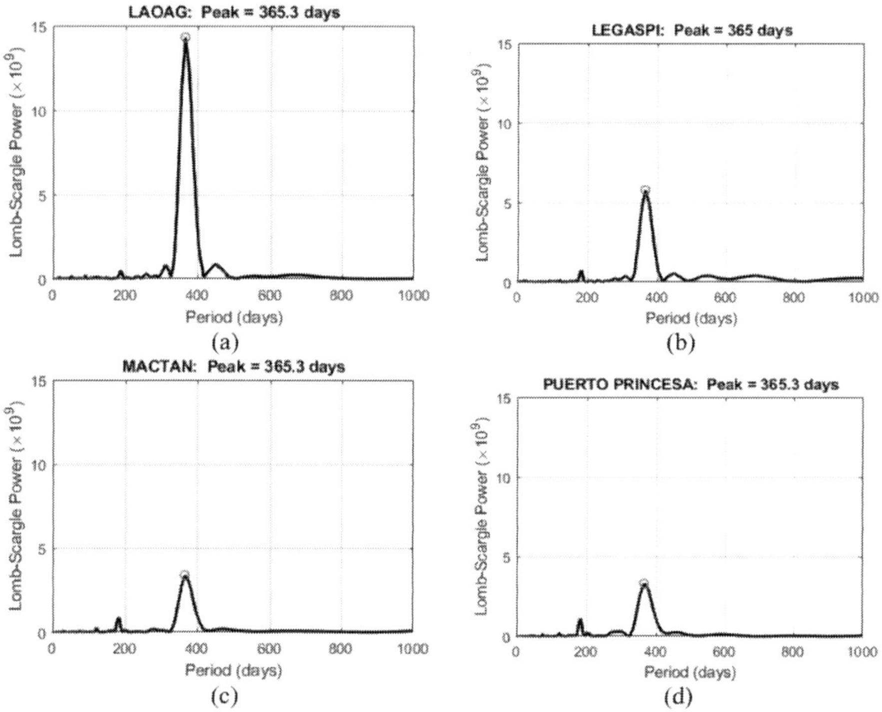

Figure 3. Lomb-Scargle periodogram (in arbitrary units) from 2012-2019 of (a) Laoag, (b) Legaspi, (c) Mactan, and (d) Puerto Princesa.

These results are consistent to Zhao *et al.* [15]. Here, a similar annual and bi-annual periodicities were observed from 4 global stations using Lomb-Scargle periodograms. However, in contrast with the annual observation of this research, the study of Zhao *et al.* [15] also observed seasonal periods in some areas [15]. This annual and semi-annual variability of PWV has also been observed in a series of stations over Turkey [16].

Figure 4 represented the annual mean of the four provinces using error bars provided by the standard deviation of each year. The annual mean PWV of Laoag from 2012-2013 and from 2016-2018 has shown concentrated data with PWV ranging from 40 mm-50 mm and years 2014-2015 and 2019 has indicated the values are more spread or it is likely less reliable with a PWV ranging from 40 mm-45 mm. While in Legaspi, we could see that there is a downfall despite having high PWV which ranges from 45 mm in 2012 to 50 mm in 2019. Mactan's annual mean has more likely concentrated data from 2013-2014 with a PWV data ranging from 50 mm-53 mm. An upward movement was observed from 2015-2017 with a range of 45 mm-50 mm and falling movement from 2018-2019 with a PWV from 50 mm-45 mm. Lastly, Puerto Princesa has started a downward movement that started from 2013 to 2015 and similar to 2018-2019 with a PWV ranging from 50 mm-45 mm. The only upward movement was observed in 2016-2017 with a low variability yet with the highest point of the PWV ranging from 50 mm-53 mm.

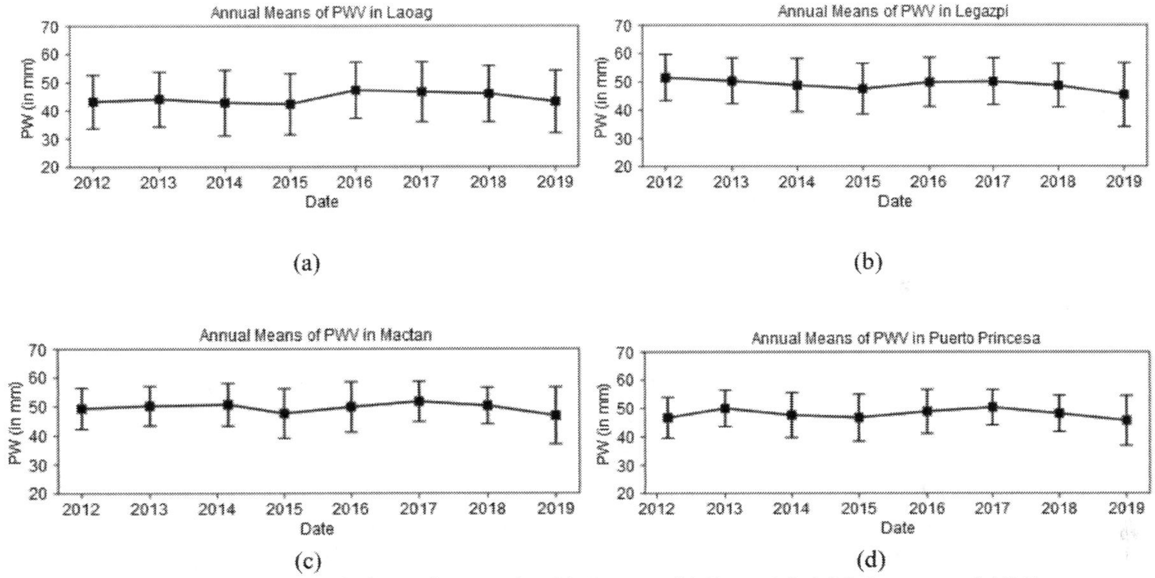

Figure 4. Annual mean variation of PWV for (a) Laoag, (b) Legaspi, (c) Mactan, and (d) Puerto Princesa. The error bars represent standard deviation.

The characteristics of the monthly variation with PWV is seen in Figure 5. This figure represents the results of how PWV from each month is seen in the 8-year observation. As observed, PWV rises from April to May which results in its peak in June and July and finally decreases as August ends. The approximate numerical values for each part of the rising and falling behavior of PWV data are 45-50 mm (i.e. rising; April-May), 55 mm (i.e. peak; June-July), and 50-40 mm (i.e. falling; August and succeeding months). Due to the high values of PWV in months from April to September, it could be concluded that the four cities have a cool climate during these months, particularly Legazpi. Mactan did not show a slight decrease in PWV after it peaked in April to September. On the other hand, Puerto Princesa has shown a great decrease compared to the other cities in August to September. In general, all the four cities had decreases in PWV values after the months wherein it peaked, yet these cities just varied in how much it lost compared to the maximum value of PWV.

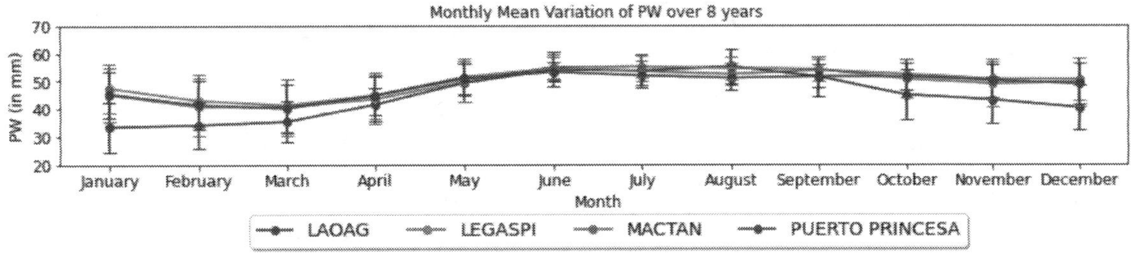

Figure 5. Monthly mean variation of PWV. The errors bars represent 1 standard deviation.

Similarly, the study of Macalalad and Macalalad (2019) used GNSS-derived PWV data over Manila in 2017. The monthly mean variation with error bars, shows a low variability in Manila despite having high PWV in the wet season, specifically in July with 3.9 mm and May with July 3.9 mm. However, this research concluded that it had a cool climate from April to September along with a rising action with high PWV [2]. In terms of climate as seen in the monthly variation of both studies, Laoag, Legazpi,

IOP Conf. Series: Earth and Environmental Science **880** (2021) 012001 doi:10.1088/1755-1315/880/1/012001

Mactan, Puerto Princesa, and Manila both have considered November to April as a dry season due to the lack of rainfall during these months. In comparison, the dry seasons of Manila generally observe less than 50 mm PWV measurements, while Laoag presents less than 45 mm. The dry season of Manila showed higher variability in April ranging from 6.2 mm to January with 9.9 mm. On the other hand, the cities of Legazpi, Mactan, and Puerto Princesa share nearly the same variability of PWV with the dry months of Manila.

The results of the Mann Kendall Test and Statistic was also computed, in which the PWV trend is achieved. The results noted that Kendall's τ coefficient will determine the behavior of the trend of PWV if it is increasing ($\tau > 0$), decreasing ($\tau < 0$), or monotonic ($\tau = 0$). Both Laoag and Mactan showed an increasing trend with the τ value being 0.6 and 0.2, respectively. On the other hand, Legazpi and Puerto Princesa presented a decreasing trend having a τ value of o –0.05 and –0.01, respectively.

Similarly, the study of Merrikphour and Rahimzadegan [7] used the Mann-Kendall Test to scrutinize the behavior of the trend of PWV. The only notable is that radiosondes were not the primary source for their PWV data as they utilized the Moderate Resolution Imaging Spectroradiometer (MODIS) of Western Iran. Their investigation and analysis on the spatial and temporal variations of PWV in Western Iran supported the results of the Mann-Kendall test [7]. Both studies apply the Mann-Kendall Test and Statistic as a secondary method in getting the behavior of a trend of data, which in the case of this research is the radiosonde data of PWV over the four cities of the Philippines.

Figure 6 shows the correlation and comparison of monthly PWV, surface temperature, and relative humidity from 2012-2019 for the four stations. Moreover, the approximate ranges of each variable are shown within the four correlation graphs above, namely, surface temperature (° C) having 30 as the maximum and 20 as the minimum; relative humidity (%) showing 70 as the lowest data to 95 as the highest, which ranges from 20 mm or 25 mm to 55 mm or 70 mm. Laoag shows a strong positive correlation between PWV and RH (R= +0.690) and it is expected in both parameters to describe the moisture content in the air, meanwhile, the strong positive correlation between PWV and temperature (R= +0.654) defines the formation of water vapor that is likely related to the [2]. However, Legaspi, Mactan and Puerto Princesa have weak-to-moderate positive correlation with RH with R-values ranged from 0.166 to 0.401. They also correlated with temperature whose R-values vary from 0.261 to 0.433. This indicates that at the monthly scale, PWV is affected by these variables in a multifaceted manner. Thus, it is recommended to correlate these using multi-variate correlation techniques to explore their complex relationship.

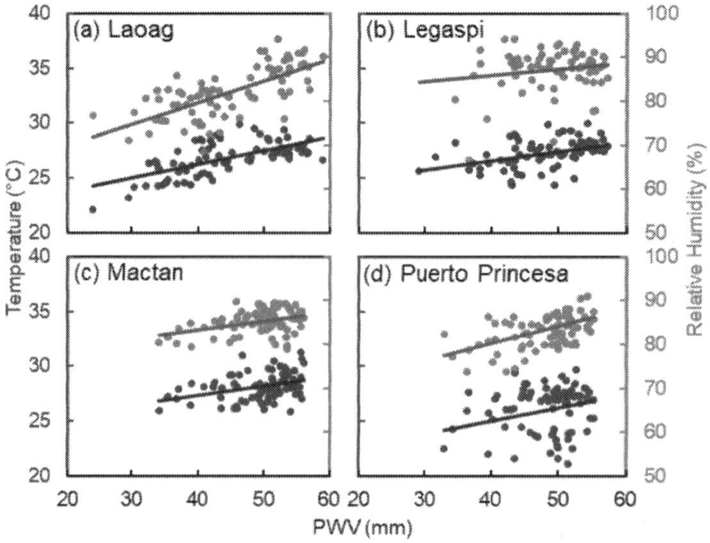

Figure 6. Correlation of Monthly PWV with Surface Temperature and Relative Humidity for (a) Laoag, (b) Legaspi, (c) Mactan, and (d) Puerto Princesa.

4. Summary and Recommendations

In this study, the general trend of daily PWV data derived from radiosonde observations from IGRA (v2) in Laoag has shown a range of peaks that reached 70 mm which signified rainfall, along with a measurement of PWV in 10 mm which presented the lack of rainfall. Similarly, radiosondes from Legazpi have shown the daily measurement of PWV that peaked above 60 mm and a minimum of 20 mm. Mactan showed an upward trend reaching up to 70 mm, and Puerto Princesa at approximately 60 mm. Mactan and Puerto Princesa shared a minimum of PWV data at the beginning of each year of 2011-2019 at 20 mm. The daily PWV has shown graphs with the upward or downward trend of PWV, as well as the monthly and annual mean variation with standard deviation error bars.

It has been shown that all four stations have annual and bi-annual periodicities as illustrated in their Lomb-Scargle periodograms. Other periodograms may also be explored in the future to extract (if any) smaller or larger scaled periodicities which may not be observed by the abovementioned technique. The Mann-Kendall test that was accomplished showed that Laoag and Mactan had an increasing trend throughout the study. While Legazpi and Puerto Princesa had a decreasing trend in the years 2012-2019.

In addition, the correlation of PWV with surface temperature and relative humidity showed a strong relationship in which these two variables (i.e., ST & RH) contribute greatly to the PWV output. The relationship of these three variables (i.e., PWV, ST, and RH) is shown through to the low temperatures that result in increased moisture in the air which contributed to the high production of PWV. It is also seen at times wherein the PWV is at a minimum, this means that the temperatures are high and the moisture in the air has decreased. With this, it is recommended to use multi-variable correlation approach to show that inter-relationship of theses parameters.

References

[1] Coster A J *et al* 1996 Measurements of Precipitable Water Vapor by GPS, Radiosondes, and a Microwave Water Vapor Radiometer *Proceedings of the 9th International Technical Meeting of the Satellite Division of The Institute of Navigation (ION GPS 1996)* (Kansas City, MO) 625–34

[2] Macalalad E P and Macalalad R V 2019 Variation of GNSS-derived Precipitable Water Vapor over Manila in 2017 *International Conference on Space Science and Communication, IconSpace* vol 2019-July (IEEE Computer Society) pp 235–8

[3] Chen B and Liu Z 2016 Global water vapor variability and trend from the latest 36year (1979 to 2014) data of ECMWF and NCEP reanalyses, radiosonde, GPS, and microwave satellite *J. Geophys. Res.* **121** 11442–62

[4] Fletcher S J 2017 *Data Assimilation for the Geosciences: From Theory to Application* (Amsterdam, Netherlands)

[5] Carnicer K M R, Castilla R M, Vallar E A and Galvez M C D 2017 On the variability of precipitable water vapor and its probabilistic modeling of precipitation occurrence in Davao city *Adv. Sci. Lett.* **23** 1379–83

[6] Carnicer K M R, Vallar E A and Galvez M C D 2020 On the Assimilation of GNSS PWV Measurements in Heavy to Torrential Rain Events In Davao City, Philippines *ARPN J. Eng. Appl. Sci.* **15** 2188–92

[7] Merrikhpour M H and Rahimzadegan M 2019 Analysis of temporal and spatial variations of total precipitable water vapor in western Iran using radiosonde and MODIS measurements *J. Appl. Remote Sens.* **13** 1

[8] Zhang Y, Xu J, Yang N and Lan P 2018 Variability and trends in global precipitable water vapor retrieved from COSMIC radio occultation and radiosonde observations *Atmosphere (Basel).* **9**

[9] Wang J, Dai A and Mears C 2016 Global water vapor trend from 1988 to 2011 and its diurnal asymmetry based on GPS, radiosonde, and microwave satellite measurements *J. Clim.* **29** 5205–22

[10] Ferreira A P, Nieto R and Gimeno L 2019 Completeness of radiosonde humidity observations based on the Integrated Global Radiosonde Archive *Earth Syst. Sci. Data* **11** 603–27

[11] Lomb N R 1976 Least-squares frequency analysis of unequally spaced data *Astrophys. Space Sci.* **39** 447–62

[12] Scargle J D 1982 Studies in astronomical time series analysis. II - Statistical aspects of spectral analysis of unevenly spaced data *Astrophys. J.* **263** 835

[13] VanderPlas J T 2018 Understanding the Lomb–Scargle Periodogram *Astrophys. J. Suppl. Ser.* **236** 16

[14] Domingo A L S and Macalalad E P 2020 Multi-year Precipitable Water Vapor Observations in East Manila, Philippines using Radiosonde and Global Navigation Satellite System ed Q Xiaosheng and O Amahmid *E3S Web Conf.* **199** 00002

[15] Zhao Q, Yao Y and Yao W 2019 Studies of precipitable water vapour characteristics on a global scale *Int. J. Remote Sens.* **40** 72–88

[16] Gurbuz G 2021 On variations of the decadal precipitable water vapor (PWV) over Turkey *Adv. Sp. Res.* **68** 292–300

Acknowledgments

The authors would like to express their utmost gratitude to the National Oceanic and Atmospheric Administration and DOST-PAGASA for providing the IGRA v2 radiosonde data used in this study.

Quantifying the joint distribution of drought indicators in Borneo fire-prone area

Mohamad Khoirun Najib[1*], Sri Nurdiati[1] and Ardhasena Sopaheluwakan[2]

[1] Department of Mathematics, Faculty of Mathematics and Natural Science, IPB University, Bogor 16680, Indonesia
[2] Center for Applied Climate Services, Agency for Meteorology, Climatology, and Geophysics, Jakarta 10720, Indonesia

*Corresponding E-mail: mkhoirun_najib@apps.ipb.ac.id

Abstract. Borneo island is prone to fire due to its large peat soil area. Fire activity in Borneo is associated with regional climate conditions, such as total precipitation, precipitation anomaly, and dry spells. Thus, knowing the relationship between drought indicators can provide preliminary knowledge in developing a fire risk model. Therefore, this study aims to quantify the copula-based joint distribution and to analyze the coincidence probability between drought indicators in Borneo fire-prone areas. From dependence analysis, we found that the average of 2 months of total precipitation (TP), monthly precipitation anomalies (PA), and the total of 3 months of dry spells (DS) provides a moderate correlation to hotspots in Borneo. The results show the probability of the dry-dry period is 26.63, 17.66, and 18.54 % for TP-DS, PA-DS, and TP-PA, respectively. All of these are higher than the probability of the wet-wet period, which is 25.01, 16.12, and 17.98 % for TP-DS, PA-DS, and TP-PA, respectively. Through the probability, the return period of TP-DS in the dry-dry situation 3.2 months/year, meaning the dry situation in total precipitation and dry spells that occur simultaneously could appear about 3 months in a year on average. Furthermore, the return period of PA-DS and TP-PA in the dry-dry situation is 2.12 and 2.22 months/year, respectively. Moreover, the probability of dry spells in dry conditions when given total precipitation in dry conditions is higher than given precipitation anomalies in dry conditions.

Keywords: Coincidence Probability, Copula, Drought, Dry Spells, Precipitation

Track Name: Atmospheric Chemistry and Physics

1. Introduction

Borneo island is prone to fire due to its large peat soil area, especially in Central Kalimantan Province, Indonesia [1]. In 2015, this province contained 30,400 km^2 peatland and experienced very high numbers of peatland hotspots which is 20,300 hotspots [2]. The conversion of peat forests to agricultural land has further exacerbated the problem by substantially increasing fuel loads and associated fire risk [3,4]. Fire activity in Borneo is sensitive to local climate conditions, such as total precipitation, precipitation anomaly, and dry spells (i.e., precipitation less than 1 mm day^{-1}) [5] and regional climate conditions such as El Nino-Southern Oscillation (ENSO). Madadgar *et al.* [6] define the fire risk model as the conditional probability of fire size given a specific climate condition [6]. Therefore, knowing the relationship between drought indicators is important as preliminary knowledge in developing a fire risk model.

The copula-based joint distribution can be employed to determine the relationship between the drought indicators. From the joint distribution, coincidence probability analysis [7] can investigate the frequency of the combination of two drought indicator periods, such as the frequency of dry-dry or wet-wet periods. Coincidence probability usually uses in hydrological frequency analysis [8,9]. However, very few experts use this coincidence probability analysis to identify coincidence between drought indicators conditions.

From the description above, this study aims to quantify the copula-based joint distribution and to analyze the coincidence probability between drought indicators in Borneo fire-prone areas. The study results can build knowledge about the relationship between drought indicators and become preliminary knowledge in developing a fire risk model in the future.

2. Materials and Methodology

2.1. Study area and data

The area in this study is limited to Borneo fire-prone areas selected based on clustering using the unsupervised learning method, which is a k-means algorithm [10]. The clustering uses hotspot data obtained from the National Institute of Aeronautics and Space of Indonesia (LAPAN). The data was derived from the Moderate Resolution Imaging Spectroradiometer (MODIS) sensors of the Terra and Aqua Satellites with a spatial resolution of 0.25°×0.25° from 2001-2020. We only use hotspot data with a confidence level of more than 80 % [5,11]. After defining the fire-prone area in Borneo, drought indicators such as total precipitation, precipitation anomalies, and dry spell were processed using converted precipitation data based on a satellite-derived product of SEMDP (The Space-based Weather and Climate Extremes Monitoring Demonstration Project). We use precipitation data based on Climate Prediction Center Morphing Technique (CMORPH) corrected for bias, known as CMORPH bias-corrected product (CMORPH CRT) [12]. The monthly total precipitation and precipitation anomaly data were processed using the CMORPH CRT monthly data. Meanwhile, the monthly dry spell data was processed using the CMORPH CRT daily data.

2.2. Dependence analysis

Kendall's rank correlation τ measures the degree of association between variables to capture any non-linear relationships, then τ has several desired properties over other rank correlation measures [13]. Each of the variables pair determines the statistically significant correlation [14].

Extremal dependence falls into two classes: asymptotic dependence or asymptotic independence [15]. Let (X,Y) is a pair of variables with distribution functions (F_X,F_Y), χ is the extremal dependence of (X,Y) [16]. Using transformed uniform (0,1) distributions, i.e. $U=F_X(X), V=F_Y(Y)$, the coefficient χ is given by

$$\chi(u)=2-\frac{\ln[P(U<u,V<v)]}{\ln[P(U<u)]} \tag{1}$$

where $\chi=\lim_{u\to1}\chi(u)$ for $0\le u\le1$ and $0\le\chi\le1$. If $\chi=0$ and $\chi=1$, the variable pair fall into asymptotically independent and perfect dependence, respectively. In the case of asymptotic independence $(\chi=0)$, the measure $\bar{\chi}$ uses to quantify the magnitude of the dependence of asymptotically independent variables

$$\bar{\chi}=\frac{2\ln[P(U>u)]}{\ln[P(U>u,V>u)]}-1 \tag{2}$$

where $\bar{\chi}=\lim_{u\to1}\bar{\chi}(u)$ for $-1\le\bar{\chi}\le1$. For positive association variables $\bar{\chi}=1$, for negative association variables $\bar{\chi}=-1$, and nearly independent variables $\bar{\chi}=0$ [14,16,17].

2.3. Copulas

For bivariate copula, let a 2-dimensional random vector X with marginal cumulative distribution function F_{X_1} and F_{X_2}, $F_{X_1}, F_{X_2}, \ldots, F_{X_m}$ in $\mathbb{R}$, i.e., non-decreasing, $F_{X_1}(-\infty)=0$ and $F_{X_1}(\infty)=1$. Sklar's theorem [18] says that the joint distribution F_X of this random vector is a function of its marginal distributions

$$F_X(x_1,x_2)=C(u_1,u_2) \quad (3)$$

where $C_X:[0,1]\times[0,1]\rightarrow[0,1]$ is a joint distribution function of the transformed random variables $u_i=F_{X_i}(x_i)\sim \text{uniform}(0,1)$ for $i=1,2$ [19,20]. We assume continuous and differentiable distribution functions F_{X_1} and F_{X_2}. Thus, C_X is unique and can be expressed by

$$C_X(u_1,u_2)=\int_0^{u_1}\int_0^{u_2} c_X(u_1',u_2')\, du_2'\, du_1' \quad (4)$$

where $u_i=F_{X_i}(x_i)$ for $i=1,2$. The function C_X is a copula function, while c_X is the corresponding copula density function. The important consequence of Sklar's theorem is that every joint probability density is also writable by the product of the marginal probability densities and the copula density.

$$f_X(x_1,x_2)=f_{X_1}\cdot f_{X_2}\cdot c_X(u_1,u_2) \quad (5)$$

There are many types of copula functions [7]. Different copulas differ in describing the dependence structures [21]. Eleven bivariate copulas are selected, including one-parameter and two-parameter copulas given in Table 1 [19, 21–28].

Table 1. The cumulative density functions and parameter ranges of selected bivariate copulas.

No	Copula	$C_X(u_1,u_2)$	Parameter Range
1	Normal	$F_{N(0,\Sigma)}(F^{-1}_{N(0,1)}(u_1),F^{-1}_{N(0,1)}(u_2))$	
2	Student-t	$F_{t(v,\Sigma)}(F^{-1}_{t(v)}(u_1),F^{-1}_{t(v)}(u_2))$	
3	Clayton	$(u_1^{-\theta}+u_2^{-\theta}-1)^{-1/\theta}$	$\theta>0$
4	Gumbel	$\exp[-(w_1^{\theta}+w_2^{\theta})^{1/\theta}]$, where $w_i=-\ln(u_i)$	$\theta\geq 1$
5	Frank	$-\frac{1}{\theta}\ln\left[1+\frac{w_1w_2}{(e^{-1}-1)}\right]$, where $w_i=e^{-\theta u_i}-1$	$\theta\neq 0$
6	Joe	$1-(1-w_1w_2)^{1/\theta}$, where $w_i=1-(1-u_i)^{\theta}$	$\theta\geq 1$
7	Galambos	$u_1u_2\exp\left[(w_1^{-\theta}+w_2^{-\theta})^{-1/\theta}\right]$, where $w_i=-\ln(u_i)$	$\theta\geq 0$
8	BB1	$\left[1+(w_1^{\delta}+w_2^{\delta})^{1/\delta}\right]^{-1/\theta}$, where $w_i=u_i^{-\theta}-1$	$\theta>0,\delta\geq 1$
9	BB6	$1-\{1-\exp[-(w_1^{\delta}+w_2^{\delta})^{1/\delta}]\}^{1/\theta}$, where $w_i=-\ln[1-(1-u_i)^{\theta}]$	$\theta\geq 1,\delta\geq 1$
10	BB7	$1-[1-(w_1^{-\delta}+w_2^{-\delta}-1)^{-1/\delta}]^{1/\theta}$, where $w_i=1-(1-u_i)^{\theta}$	$\theta\geq 1,\delta>0$
11	BB8	$\frac{1}{\theta}\left\{1-\left[1-\frac{w_1w_2}{1-(1-\delta)^{\theta}}\right]^{1/\delta}\right\}$, where $w_i=1-(1-\delta u_i)^{\theta}$	$\theta\geq 1,0<\delta\leq 1$

Note. *BB1 (Clayton-Gumbel), BB6 (Joe-Clayton), BB7 (Joe-Gumbel), and BB8 (Joe-Frank). Normal and student-t employ the linear correlation ρ as a parameter.*

2.4. Copula parameter estimation and goodness of fit

We use the inference of functions for margins (IFM) method to estimate the parameters of copulas [29]. The IFM method is a 2-step estimation method for the copula parameter θ. In the first step, estimate the parameters α_j (notated $\hat{\alpha}_j$) of each marginal distribution separately using X_1, X_2. In the second step, approximate the parameter θ by replacing α_j using $\hat{\alpha}_j$ in the log-likelihood function as

$$\hat{\theta} = \arg\max \sum_{t=1}^{n} \ln c_X \left[F_{X_1}(x_1^t; \hat{\alpha}_1), F_{X_2}(x_2^t; \hat{\alpha}_2); \theta \right] \quad (6)$$

where $\hat{\theta}$ is the estimated parameter for θ [7,30]. In this study, Kolmogorov–Smirnov error (D_n) [7], Root Mean Square Error (RMSE) [26], and Akaike's information criterion (AIC) [31] measure the goodness of fit of the joint distributions as follows:

Kolmogorov-Smirnov Error (D_n): $\max_{i=1,2,\ldots,n} |P_{Ei} - P_{Ti}|$ (7)

Root Mean Square Error (RMSE): $\sqrt{\frac{1}{n}\sum_{i=1}^{n}(P_{Ei} - P_{Ti})^2}$ (8)

Akaike's information criterion (AIC): $2k - 2\ln L$ (9)

where n is the sample size, k is the number of parameters of different distributions, L is the maximum likelihood function value of distributions, P_{Ei} and P_{Ti} are the empirical and theoretical frequency, respectively. Assuming the variables X and Y with the same length, Gringorten's formula can estimate the frequency of bivariate variables X and Y as follow

$$P_{Ei} = P(X \le x_i, Y \le y_i) = \frac{\#(X \le x_i, Y \le y_i) - 0.44}{n + 0.12} \quad (10)$$

where (x_i, y_i) is the combination of the i-th values of the increased order in the X and Y series.

2.5. Coincidence probability

The drought coincidence probability represents the frequency P in wetness, dryness, or normal condition, while the other indicator is wetness, dryness, or normal condition. To distinguish dry, normal, and wet conditions, tercile categories (i.e., below-normal, near-normal, and above-normal) are usually used [32]. Thus, this study uses the 1/3 and 2/3 quantiles as thresholds to define the condition of wetness and dryness in drought indicators conditions. As mentioned above, drought indicators include total precipitation, precipitation anomalies, and dry spells. For suitability, we used negative of total precipitation and negative of precipitation anomalies. Thus, for the drought indicator X, X_w and X_d were assigned the threshold quantiles of wetness $(P_w = 1/3)$ and dryness $(P_d = 2/3)$, respectively. Therefore, the degree of the wet-dry conditions is represented as wetness $(X < X_w)$, dryness $(X \ge X_d)$, and normalness $(X_w \le X < X_d)$. Thus, the wet-wet and dry-dry periods are given by

$$\begin{aligned} P_{ww} &= P(X < X_w, Y < Y_w) = C_X(u_w, v_w) \\ P_{dd} &= P(X \ge X_d, Y \ge Y_d) = 1 - u_d - v_d + C_X(u_d, v_d) \end{aligned} \quad (11)$$

where u and v are the marginal distribution of X and Y, while subscripts w, n, and d indicate wetness, normal, and dryness. For example, dd presents the dry–dry condition, and P_{dd} is the coincidence probability of this condition [7].

3. Results and Discussion

3.1. Identification of the Borneo fire-prone area using unsupervised learning

The unsupervised learning method, i.e., k-means clustering, mentioned above was used to identify the fire-prone areas in Borneo. K-means algorithm is a simple iteration method to partition a given dataset into *k* numbers of user-defined clusters [33]. This study selects 8-means clusters, and Figure 1 shows the results of the clustering process.

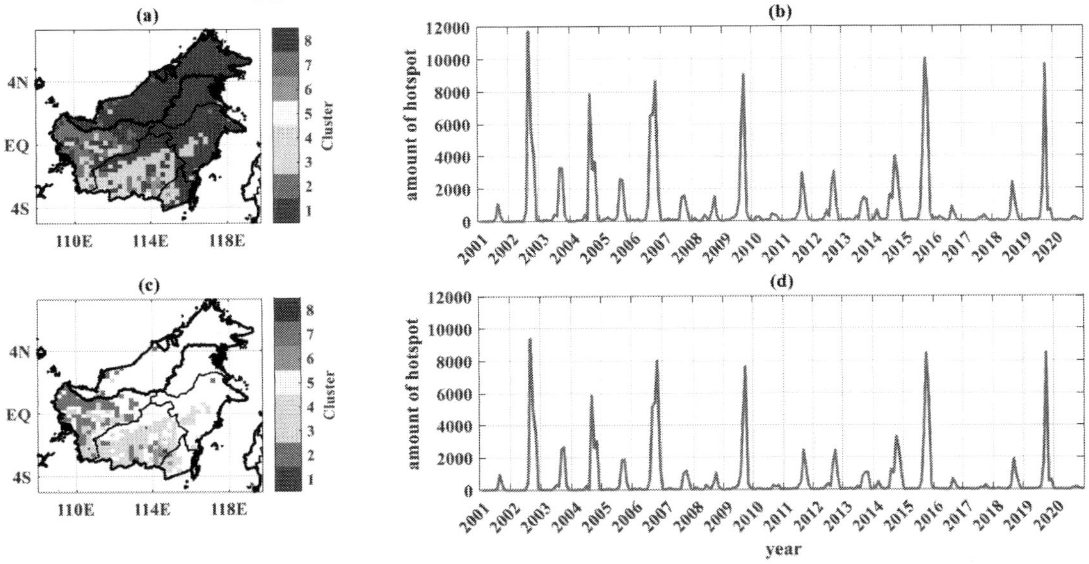

Figure 1. Clustering result of hotspots in Borneo in 2001-2020 using k-means clustering: (a) areas with all clusters, (b) total monthly hotspots of all clusters, (c) areas without cluster 1, and (d) total monthly hotspots without cluster 1.

Figure 1(a) shows the entire cluster area using the k-means clustering. Each color represents a different cluster. For example, cluster 1 (dark blue) is an area with a few hotspot characteristics [5]. From 969 grid points in Borneo, cluster 1 covers 679 grid points, about 70 % of Borneo Island, i.e., Malaysia, Brunei, and most East and North Kalimantan Provinces. In cluster 1, the maximum hotspot occurrence is only about 3.43 hotspots per grid point data on average, i.e., in 2002. Thus, we assume that cluster 1 is a low fire-prone area and removed from the study area, and the result area is given in Figure 1(c). After removing cluster 1 from the study area, the total monthly hotspots decreased slightly compared to the entire cluster but did not change the pattern, as seen in Figures 1(b) and 1(d).

3.2. Dependence analysis between drought indicators and total hotspot

After identifying the fire-prone areas, CMORPH precipitation data was extracted to derive the drought indicators in the Borneo fire-prone area from 2001-2020. After that, each data grid point aggregated on average. Thus, the time-series data represent the aggregate of drought conditions in Borneo fire-prone areas. Pearson, Kendall, and extremal dependencies mentioned above are employed to identify the association of each drought indicator to the total hotspots at several lag times, as seen in Figure 2.

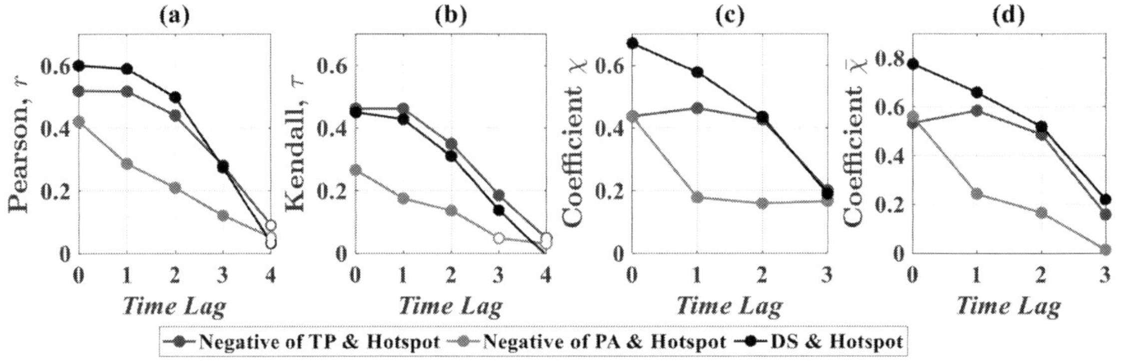

Figure 2. Assessment of the association between drought indicators and hotspots for a time lag of 0 to 3 months: (a) Pearson correlation, (b) Kendall correlation, (c) coefficient χ, and (d) coefficient $\bar{\chi}$.

Total precipitation and dry spells had moderate associations with hotspots ($\tau \geq 0.3$ and $r \geq 0.4$) at a time lag of 0 to 2 months and statistically significant correlations at a time lag of 3 months, although the correlation was small. In contrast, precipitation anomalies had a small association with hotspots ($\tau < 0.3$) but statistically significant correlations at a time lag of 0 to 2 months. Besides, the peaks in the correlations are zero lag at all drought indicators, representing that hotspot occurrence in fire-prone areas in Borneo is most influenced by drought indicators in the same month.

To ensure the association of drought indicators, especially at the extreme, Extremal dependencies are employed. Despite having a small Kendall-τ correlation, precipitation anomalies have a moderate positive association at extreme values at zero lag ($\chi > 0.4$). Meanwhile, dry spells have Extremal dependencies on lags 0, 1, and 2. A confusing thing occurs in total precipitation, where Extremal dependency on lag 2 is not high enough but has a moderate correlation. For differing with dry spells, this study assumes that total precipitation only has Extremal dependencies up to lag 1. Moreover, at lag 3, drought indicators are nearly independent at the extreme ($\bar{\chi} \to 0$). Consequently, the drought indicators we will use in this study are the average of 2 months total precipitation, monthly precipitation anomalies, and the total of 3 months of dry spells.

3.3. Copula parameter estimation and goodness of fit

As mentioned above, we use the inference of functions for the margins (IFM) method to estimate the copula parameter. We use generalized extreme value distribution to identify the univariate marginal distribution of drought indicators [7, 34]. Also, we use the Anderson-Darling test to test the feasibility of fitting these distributions to the data. A-D test returns a test decision for the null hypothesis that the data is from a population with a defined distribution. Table 2 shows the result of the marginal distribution fitting process. The statistical result shows that the chosen distribution all passed the Anderson-Darling test with a 5 % significance level.

Table 2. The result of the marginal distribution fitting process using generalized extreme value and lognormal distribution.

	Neg of Total Precipitation	Neg of Precipitation Anomaly	Dry Spells
Distribution	Generalized Extreme Value	Generalized Extreme Value	Generalized Extreme Value
A-D Test (p-value)	Do not reject (0.3474)	Do not reject (0.9732)	Do not reject (0.1853)

***Note.** Do not reject based on a 5 % significance level.*

The second step of the IFM method is estimating the copula parameter, as seen in Equation 6. Eleven selected bivariate copulas in Table 1 are constructed, and their parameters are estimated. Table 3 shows that Galambos, BB7, and BB7 copulas are the fittest model for bivariate joint distribution between Total

Precipitation-Dry Spells (TP-DS), Precipitation Anomalies-Dry Spells (PA-DS), and Total Precipitation-Precipitation Anomalies (TP-PA), respectively.

Table 3. The results of the copula parameter fitting process between TP-DS, PA-DS, and TP-PA.

Variable pair	Copula	θ	δ	KSE	RMSE	AIC
Neg of TP-Dry spells (TP-DS)	Galambos	2.2871	-	0.08321	0.030643	-352.45
Neg of PA-Dry spells (PA-DS)	BB7	1.4834	0.2161	0.05284	0.017313	-55.175
Neg of TP-Neg of PA (TP-PA)	BB7	1.5176	0.4640	0.05670	0.017263	-82.664

3.4. Tail dependence of the selected copula

Galambos copula is a type of extreme value copula family with a Pickands dependency function, i.e.,

$$A(t;\theta) = 1 - \left(t^{-\theta} + (1-t)^{-\theta}\right)^{-1/\theta} \tag{12}$$

where $t \in [0,1]$ and θ is the parameter of the Galambos copula [35]. In the context of extremes, it is natural to study the coefficient of upper and lower tail dependence [36]. For the bivariate case, the upper tail (λ_U) and lower tail (λ_L) dependency coefficients of extreme value copula are given by

$$\lambda_U = 2\left(1 - A(1/2)\right) \in [0,1] \quad \text{and} \quad \lambda_L = \begin{cases} 0, & A(1/2) > 1/2 \\ 1, & A(1/2) = 1/2 \end{cases} \tag{13}$$

Using the value of $\theta = 2.287$ from the fitting results in Table 4, we get $A(1/2) = 0.6307$. Therefore, the TP-DS variable pair has an upper tail dependency of 0.7385 and has no lower tail dependency.

Meanwhile, the PA-DS and TP-PA variable pairs both chose BB7 copula as the fittest model. BB7 copula is a 2-parameter copula that has upper and lower tail dependencies given by

$$\lambda_U = 2 - 2^{1/\theta} \quad \text{and} \quad \lambda_L = 2^{-1/\delta} \tag{14}$$

where $\theta \geq 1$ and $\delta > 0$ [37]. Using the fitting results in table 4, the PA-DS variable pair has an upper tail dependency of 0.4044 and has a lower tail dependency of 0.0405. Furthermore, the TP-PA variable pair has an upper tail dependency of 0.4211 and has a lower tail dependency of 0.2245.

3.5. Joint probability density function between drought indicators

The selected bivariate copula between TP-DS, PA-DS, and TP-PA are visualized using a contour plot of the joint probability density function and given in Figure 3. As mentioned in equation 5, the joint probability density function is the product of the marginal probability densities and the copula probability density function. The results show that the joint distribution between TP-DS, PA-DS, and TP-PA has an upper tail dependency. In this case, an upper tail dependency represents that the probability of dry-dry conditions is higher than wet-wet conditions. This is because this study area only covers fire-prone areas in Borneo (but not flood-prone areas) so that extreme wet events are not visualized in the joint distribution.

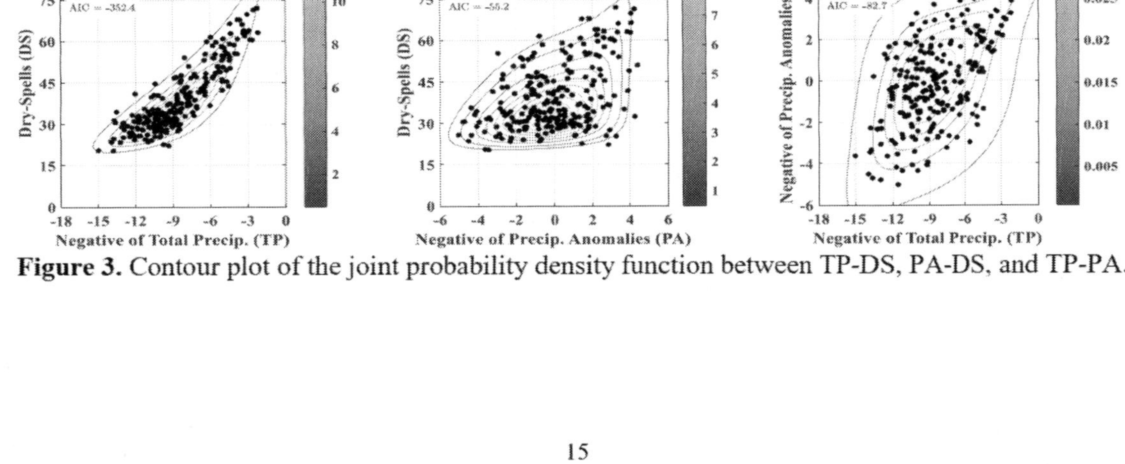

Figure 3. Contour plot of the joint probability density function between TP-DS, PA-DS, and TP-PA.

As mentioned above, this study uses the 1/3 and 2/3 quantiles as thresholds to define the condition of wetness and dryness in drought coincidence. Based on the result of the fitting process in Table 3, for wetness condition, we use $X_w = -9.8306$, $Y_w = -0.9513$, and $Z_w = 32.4272$ as a threshold for negative of total precipitation, negative of precipitation anomalies, and dry spells, respectively. Meanwhile, for dryness conditions, we use $X_d = -7.4105$, $Y_d = 0.9370$, and $Z_d = 41.1416$ as a threshold for negative of total precipitation, negative of precipitation anomalies, and dry spells, respectively. Using equation 11, the probability of the dry-dry period (P_{dd}) is 26.63, 17.66, and 18.54% for TP-DS, PA-DS, and TP-PA, respectively. All of these are higher than the probability of the wet-wet period (P_{ww}), which is 25.01, 16.12, and 17.98 % for TP-DS, PA-DS, and TP-PA, respectively, indicating the frequency of co-occurrence in dry conditions between the two drought indicators is higher than in wet conditions.

Through the probability, the return period of TP-DS in the dry-dry situation is 26.63 % or 1/3.75%, meaning the dry situation in total precipitation and dry spells that occur simultaneously appear every 3.75 months on average. By multiplying the return period by 12 months, the probability of dry conditions occurring in a year is 3.2 months/year, meaning the dry situation in total precipitation and dry spells that occur simultaneously could appear about 3 months in a year on average. Furthermore, the return period of PA-DS and TP-PA in the dry-dry situation is 2.12 and 2.22 months/year, respectively. From the coincidence probability, the conditional probability of dry spells in the dry situation given specific total precipitation in the dry situation is

$$P_{Z|X}(Z \geq Z_d \mid X \geq X_d) = \frac{P_{X,Z}(X \geq X_d, Z \geq Z_d)}{P_X(X \geq X_d)} = \frac{26.63\%}{1/3} = 79.89\%$$

and given specific precipitation anomaly in the dry situation is

$$P_{Z|Y}(Z \geq Z_d \mid Y \geq Y_d) = \frac{P_{Y,Z}(Y \geq Y_d, Z \geq Z_d)}{P_Y(Y \geq Y_d)} = \frac{17.66\%}{1/3} = 52.98\%$$

They implied that if given total precipitation in the dry situation (i.e., less than 7.41 mm day^{-1}), the probability of dry spells in the dry situation (i.e., more than 42.5 days) is 79.89 %. Otherwise, if given precipitation anomaly in the dry situation (i.e., less than -0.937 mm day^{-1}), the probability of dry spells in the dry situation (i.e., more than 42.5 days) is 52.98 %.

4. Conclusions

The drought indicator is an important variable that leads to many hotspots in Borneo. Therefore, knowing the relationship between drought indicators can provide preliminary knowledge in developing a fire risk model, which is the conditional probability of hotspots given a specific climate condition (such as drought indicators).

From dependence analysis, we found that the average of 2 months of total precipitation (TP), monthly precipitation anomalies (PA), and the total of 3 months of dry spells (DS) provides a moderate correlation to hotspots in Borneo. Based on these three variables, the bivariate copula was calculated. The results show the probability of the dry-dry period (P_{dd}) is 26.63 %, 17.66 %, and 18.54 % for TP-DS, PA-DS, and TP-PA, respectively. All of these are higher than the probability of the wet-wet period (P_{ww}), which is 25.01 %, 16.12 %, and 17.98 % for TP-DS, PA-DS, and TP-PA, respectively. Through the probability, the return period of TP-DS in the dry-dry situation is 3.2 months year^{-1}, meaning the dry situation in total precipitation and dry spells that occur simultaneously could appear about 3 months in a year on average. Furthermore, the return period of PA-DS and TP-PA in the dry-dry situation is 2.12 and 2.22 months year^{-1}, respectively. Moreover, the probability of dry spells in dry conditions when given total precipitation in the dry condition is higher than given precipitation anomalies in dry conditions.

References

[1] Yulianti N *et al* 2020 *ECOTROPHIC J. Ilmu Lingkung.* **14** (1): 62–73
[2] Miettinen J, Shi C, and Liew SC 2017 *Environ. Manage.* **60** (4): 747–57
[3] Wijedasa L S *et al* 2017 *Glob. Chang. Biol.* **23** (3): 977–82
[4] Santika T *et al* 2020 *Glob. Environ. Chang.* **64**: 102129
[5] Nurdiati S, Sopaheluwakan A, and Septiawan P 2021 *Agromet.* **35** (1): 1–10
[6] Madadgar S *et al* 2020 *Stoch. Environ. Res. Risk Assess.* **34** (12): 2023–31
[7] Wei X *et al* 2020 *Hydrol. Res.* **51** (5): 1120–35
[8] Yan B and Chen L 2013 *J. Hydrol.* **499**: 19–26
[9] Du H, Wang Y, Liu K, and Cheng L 2019 *Environ. Earth Sci.* **78** (7): 1–12
[10] Khairani N A and Sutoyo E 2020 *Int. J. Adv. Data Inf. Syst.* **1** (1): 9–16
[11] Ardiansyah M, Boer R, and Situmorang AP 2017 *IOP Conf. Ser. Earth Environ. Sci.* **54** (1): 012058
[12] Sun R, Yuan H, Liu X and Jiang X 2016 *J. Hydrol.* **536**: 302–19
[13] Li G, Peng H, Zhang J and Zhu L 2012 *Ann. Stat.* **40** (3): 1846–77
[14] Jane R, Cadavid L, Obeysekera J and Wahl T 2020 *Nat. Hazards Earth Syst. Sci.*: 1–30
[15] Ledford A W and Tawn J A 1997 *J. R. Stat. Soc. Ser. B Stat. Methodol.* **59** (2): 475–99
[16] Coles S, Heffernan J and Tawn J 1999 *Extrem.* **2** (4): 339–65
[17] Tilloy A, Malamud B, Winter H and Joly-Laugel A 2020 *Nat. Hazards Earth Syst. Sci.* **20** (8): 2091–117
[18] Sklar M 1959 *Publ. L'Institut Stat. L'Université Paris.* **8**: 229–31
[19] Schölzel C and Friederichs P 2008 *Nonlinear Process. Geophys.* **15** (5): 761–72
[20] Laux P et al. 2011 *Hydrol. Earth Syst. Sci.* **15** (7): 2401–19
[21] Li Z, Shao Q, Tian Q and Zhang L 2020 *Hydrol. Res.* **51** (5): 867–81
[22] Bezak N, Mikoš M and Šraj M 2014 *Water Resour. Manag.* **28** (8): 2195–212
[23] Orcel O, Sergent P and Ropert F 2020 *Nat. Hazards Earth Syst. Sci.* **21** (1): 239–60
[24] de Melo Mendes B and de Souza R 2004 *Int. Rev. Financ. Anal.* **13** (1): 27–45
[25] Zhang L and Singh VP 2007 *J. Hydrol. Eng.* **12** (4): 409–19
[26] Zhang L and Singh VP 2019 *Symmetric Archimedean Copulas* In *Copulas and their Applications in Water Resources Engineering* (Cambridge University Press) Chapter 4 pp 123–71
[27] Aldhufairi F A A, Samanthi R G M and Sepanski J H 2020 *Risks.* **8** (4): 106
[28] Buike A 2018 *Copula Modeling for World's Biggest Competitors* (Ph.D. Dissertation, Universiteit van Amsterdam)
[29] Joe H 2005 *J. Multivar. Anal.* **94** (2): 401–19
[30] Bouyé E *et al* 2000 *Copulas for Finance - A Reading Guide and Some Applications* (https://dx.doi.org/10.2139/ssrn.1032533)
[31] Tahroudi M N, Ramezani Y, De Michele C and Mirabbasi R 2020 *Hydrol. Res.* **51** (6): 1332–48
[32] De Andrade F M *et al* 2021 *Weather Forecast.* **36** (1): 265–84
[33] Wu X *et al* 2008 *Knowl. Inf. Syst.* **14** (1): 1–37
[34] Rinaldi A 2019 *Pros. Semin. Nas. Mat. dan Pendidik. Mat.* **2**: 21–38
[35] Budiarti R 2018 *Pengembangan Metode Pendugaan Minimum Distance pada Model Berbasis Copula Nilai Ekstrim: Aplikasi pada Pendugaan Value-at-Risk Portofolio* (Ph.D Dissertation, IPB University)
[36] Gudendorf G and Segers J 2010 *Extreme-Value Copulas* In *Copula theory and its applications* (Berlin, Heidelberg: Springer) pp 127–45
[37] Joe H, Li H and Nikoloulopoulos AK 2010 *J. Multivar. Anal.* **101** (1): 252–70

Acknowledgments

The authors would like to thank the Department of Mathematics, IPB University, and the Meteorological, Climatological, and Geophysical Agency of Indonesia for their support throughout this research.

MODIS-derived fire spatial and temporal distribution during haze season in Southeast Asia using empirical orthogonal function

Aulia Nisa'ul Khoir[1], Maggie Chel Gee Ooi[1*], Juneng Liew[2], Suradi[3], Andang Kurniawan[4] and Azka Ni'amillah[5]

[1] Institute of Climate Change, Universiti Kebangsaan Malaysia, 43600 UKM Bangi, Selangor, Malaysia
[2] Department of Earth Sciences and Environment, Universiti Kebangsaan Malaysia, 43600 UKM Bangi, Selangor, Malaysia
[3] Center for Applied Climate Information Services, Agency for Meteorology Climatology and Geophysics of Indonesia, Kemayoran, Jakarta, 10720, Indonesia
[4] Malang Climatological Station, Karangploso, Malang, 65152, Indonesia
[5] Yuvai Semaring Meteorological Station, Nunukan, 77456, Indonesia

*Corresponding E-mail: chelgee.ooi@ukm.edu.my

Abstract. One of the efforts to control the forest and land fire disasters which affect on the biomass burning haze is fire hotspots monitoring. Biomass burning haze in Southeast Asia (SEA) has become a recurring annual issue. This study aims to determine the spatial and temporal distribution of fire hotspots along SEA, so that it can serve as guidance for efforts to control them. The hotspot data used is derived from NASA's Fire Information for Resource Management System (FIRMS) MODIS sensors which is collected from 2001-2020. Spatial analysis of the re-gridded data shows the highest burning activities over SEA occurred in Feb-Apr, with >2000 fire events in the Indo-China area and >1000 fire events in Sumatra and Borneo. Empirical Orthogonal Function (EOF) was performed on monthly total hotspot data for 228 months for determining dominant patterns spatially and temporally. Based on the EOF analysis results, the three major modes have achieved a total variance of 71 %. The first mode (EOF1) explains 65 % of the total variance. The second (EOF2) and third (EOF3) modes account for 3.60 % and 2.97 % of the total variance respectively. The first and the third principal component identified high loadings over the Indo-China and Sumatra-Borneo regions respectively. Whereas the second principal component separates the fire areas into China and Indo-China region. Inter-annual pattern is dominant in the EOF1, while the inter-seasonal pattern is dominant in EOF2 and EOF3. ENSO, IOD, and MJO are factors that influence the pattern of the determined principal components. The result of this study provides general understanding on how the fire events varied over the past two decades in SEA.

Keywords: Fire hotspot, Biomass burning haze, Southeast Asia, Empirical Orthogonal Function, Principal Component Analysis

Track Name: Atmospheric Chemistry and Physics

1. Introduction

Widespread burning of forest which leads to transboundary haze has occurred almost every year in Southeast Asia (SEA). Transboundary haze causes various adverse effects in the countries that experience it, including health problems [1][2][3] and economic losses [4][5][6]. The high incidence of pollution due to transboundary haze has also led to an increase in household consumption of water and electricity in Singapore [7]. Recent studies reported that smoke haze occurrence in Malaysia has great negative consequences to human health and economic loss [8][9]. During the Southwest Monsoon period, the biomass burning event has greatly affected the concentration conditions of particulate matter in Southeast Asia [10]. The existence of various negative impacts called for the need to develop the efforts to mitigate the forest fire disaster in SEA. One of the first steps is to understand the condition of fire events that have occurred so far.

The distribution of the burning pollutants cannot be separated from the regional climate and weather characteristics, as in the study conducted in Northern China is also analyzed that various research sites exhibited different variability of aerosols based on meteorological, fuel type, elevation, and biomass burning sources [11]. Another research also studied on the connection of inter-annual climate variability, air quality, and biomass burning in Sumatra, Indonesia found a significant linear relationship between the visibility and ENSO signal as the visibility decrease occurred about 3 months earlier than peak ENSO [12].

This study aims to examine the principal temporal modes of fire activities and their spatial distribution in the Southeast Asia region using EOF to understand the role of the fire activities in the research region. The empirical orthogonal function (EOF) analysis is a method that is often used to study possible spatial modes of environmental variables and climatic data (i.e. patterns) and how they change with time. The fire activities data represented with the total of hotspots is also analyzed across the research area. The fire hotspot data is extracted from the Moderate Resolution Imaging Spectroradiometer (MODIS) data for the analysis of fire regimes in different areas [13]. Study of wildfires in Israel developed mapping methods of fire scars from MODIS imagery to examine the temporal and spatial patterns of wildfires in the 2000s [14]. MODIS hotspots are reliable to detect true burned areas, with only 1.8 % of them are not associated with actual burned patches, except for urban areas where very high commission errors were observed [15]. Spatial analysis to see hotspot distribution using MODIS data has been done in the previous paper [16][17][18]. The result of this study provides general understanding on how varied the fire events from MODIS data has been over the past two decades in SEA. The information of the spatial and temporal distribution of the fire hotspot events in the studied domain can become reference for forest fire disaster management and mitigation in SEA.

2. Data and Methods

2.1 Data

This research uses historical data on fire activities for the last 20 years, starting from January 2001 to December 2020. Fire activity data is obtained from NASA's Fire Information for Resource Management System (FIRMS). NASA FIRMS provides historical archive and near-real-time active fire products from the MODIS sensor which uses multiple channels to detect thermal anomalies [19]. The two satellites used are MOD14 (Terra) and MYD14 (Aqua) [20]. The thermal anomalies represent the center of a 1 km pixel that is corrected by the Terra and Aqua using thermal detection algorithm [20]. The daily data for fire activities used in the research are summed from Terra and Aqua satellites. Fire activity data are selected based on the confidence level of the data. Giglio, 2015 developed three confidence levels for hotspot information (Table 1). MODIS hotspots are very reliable to detect true burned areas [15]. As data validation process, it is necessary to filter the data before processing for the next level of analysis. A high level of confidence will result in fewer false alarms [21]. Hence, only data with a confidence level above 80% is used in this study. The research area focuses on the 90° E-130° E and 40° N-10° S domain that covers Southeast Asia and the southern China (Figure 1).

Table 1. Confidence levels of hotspot [20].

Range	Confidence Class
$0\ \% \leq C < 30\ \%$	Low
$30\ \% \leq C < 80\ \%$	Nominal
$80\ \% \leq C \leq 100\ \%$	High

2.2 Methods

Daily hotspot data in .csv (Comma Separated Values) format is regridded onto 0.5°x0.5° resolution using the R program into monthly data and stored in matrix form in the netCDF format. The monthly data is averaged over 20 years to see the spatial distribution of hotspots. The re-gridded data of fire activities is processed further using Empirical Orthogonal Function (EOF). EOF analysis is a multivariate analysis technique first introduced by [22] and developed independently by [23]. The main purpose of EOF analysis is to filter the large number of data variables to only a few representative variables, without changing most of the variance to describe the data [24].

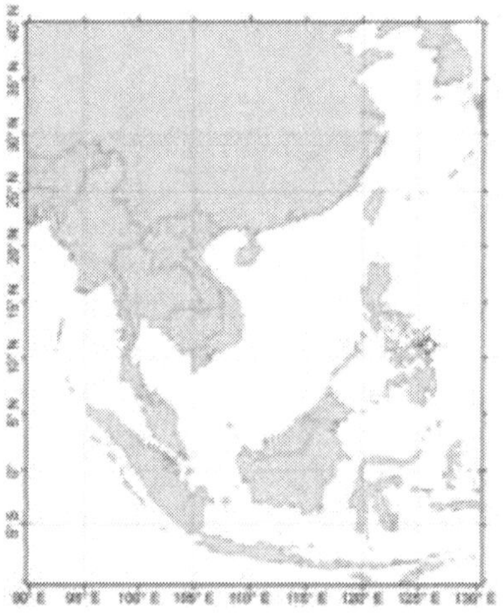

Figure 1. The study domain.

One way to minimize multicollinearity is spatial dimensions reduction or find new variables which are linear combinations of the original variables provided that the new variables were not correlated, namely EOF method based on the Eigenvalue Problem (EVP). This method is better known as Principal Component Analysis (PCA) method. The latter was used to analyze spatial and temporal variability of PM_{10} concentration across Malaysia [25]. The data were analyzed by means of the rotated principal component analysis and showed the variability of the PM_{10} concentration can be decomposed into four dominant modes in Malaysia. The most recent research on PCA to reduce variance from hotspots was carried out to identify two geographical regions for variance of biomass burning activities in Indo-China [26].

3. Result and Discussion

3.1 Fire Activities Distribution

The results of the spatial map of grid hotspot method with a resolution of 0.5°x0.5° is capable of explaining the distribution of the fire activities within the research area. Hotspots are mapped spatially and cumulatively over 20 years resulting in a varied distribution pattern. The result of MODIS satellite data processing with a confidence level of above 80 % with gridding methods generally shows that hotspots appear each month in different numbers and areas of the hotspot. The number of hotspots for

20 years monthly ranges from 0 up to more than 25,000 hotspots (Figure 2). This is related to the climate phenomenon which also varies in each territory.

Fire activities in January varied from a range of 50 to 5,000 total fire activities. The highest fire activities throughout the research area in January occurred in some parts of Indo-China region (Thailand, Laos, Cambodia, Vietnam, Myanmar). The value decreased gradually in the northern and southern part of Indo-China region. In February, fire activities increased, and it shows that the highest area of fire activities still located in the same location as the previous month. Total fire activities for this month fall in the range of 100 to 10,000 hotspots. Even though the number of burning activities has increased, it does not show a significant expansion of the burning activities area.

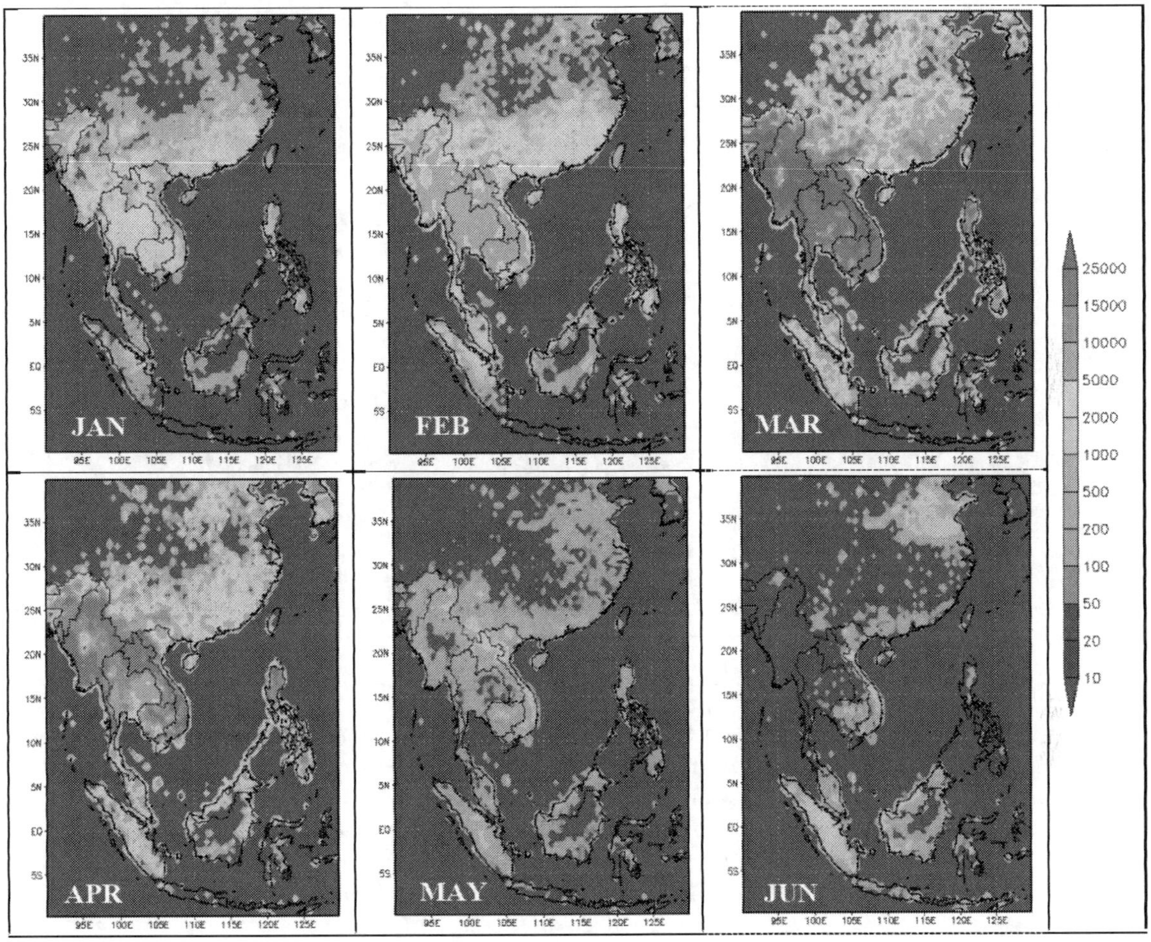

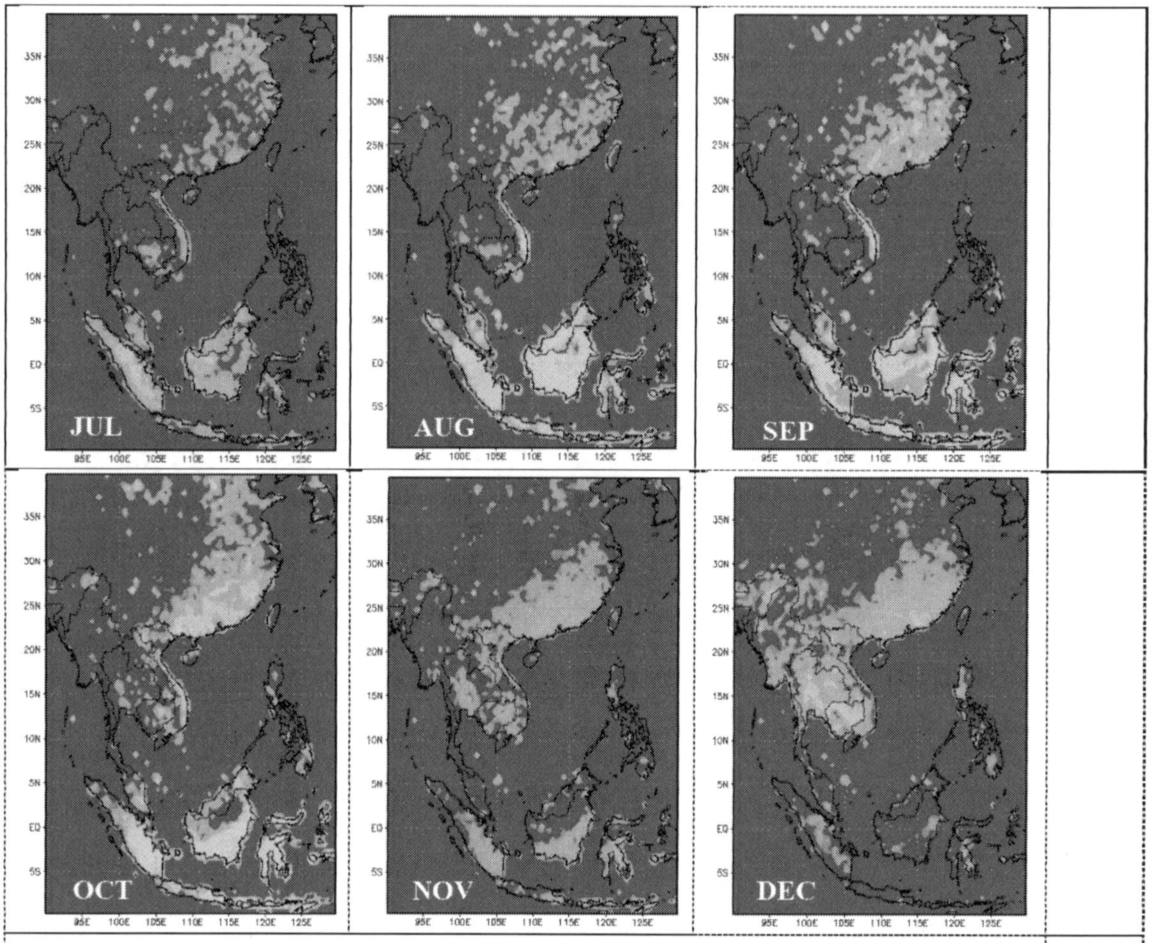

Figure 2. Total fire hotspot distribution.

March and April were the months with the worst fire activities as we can see from the monthly average throughout the study years, the number of hotspot occurrences is the highest. In April there were more than 15,000 fire activities, while in March there were more than 25,000 fire activities. The areas that had the highest number of burning activities were in Myanmar and Laos, with the number of hotspots over 25,000. In general, the area of Indo-China and the Philippine archipelago had the worst total of fire activities in both March and April. The area of the Malay Peninsula, a small part of Sumatra, and the coast of the Borneo Island also experienced an increase in the incidence of total fire activities reaching a range of total fire activities from 200 to 15,000. These are the months with the highest fire activities in the Malay Peninsula region. The highest hotspot occurrence value in the Malay Peninsula region is concentrated in the eastern part of the Malay Peninsula and gradually decreased to the west of the Peninsula area. The highest number of total hotspots in Sumatra Island is concentrated in the part of Sumatra which borders the Malacca Strait. The southeastern flank of Mainland China also has an increase in the number of total fire activities. The initial condition in January-February in southeast Mainland China only reached 5,000 fire activities, but it increased gradually and reached 15,000 fire activities in March and April.

A decrease in the occurrence of total fire activities in the research areas occurred in May and June, in addition to a decrease in total fire activities in the Indo-China region and the Philippine archipelago, there was also a reduction in the area affected. Fire activities that occur in the Indo-China region only occur in coastal areas, namely Vietnam and Cambodia. The difference compared to the previous months is that there has been a shift in areas that have high fire activities, which moved from the Indo-China region to the Sumatera region. In contrast to the decrease in area in the Indo-China and China regions, the total fire activities in the Borneo area began to expand in July and August. In July, the total fire activities that occurred in the research area were in the range of 50 to 5000, with the highest total fire activities occurring in Sumatra and the west part of Borneo. Fire activities in Java, Bali, Nusa Island, and Sulawesi began to emerge. In August, the area of fire activities in the Borneo area expanded and experienced an increase in total fire activities. The highest fire activities occurred in the Sumatera and Borneo regions with a total range of total fire activities reaching a range of 1,000 to 5,000. Fire activities on the Malay Peninsula ranged from 100 to 5,000 fire activities, while the occurrence of total fire activities in Indo-China was decreasing.

September is the peak of total fire activities in the regions of Borneo and Sumatra. The total range of fire activities occurs up to a range of 15,000. While the conditions of fire activities on the Malay Peninsula are in the same condition as the previous month, there have been occurrences of fire activities in eastern of China with a range of 100-2,000 total fire activities. Total fire activities were gradually reduced again in the Sumatera and Borneo region in October until December. In October, fire activities in China and the eastern coast of Indo-China began to re-emerge. The total value of its fire activities continued to increase from October to December in these areas and it expanded to Indo-China in December.

March and April are months with the highest number of hotspots, and it is located in the Indo-China region. The fire activities not only have an impact in the Indo-China region, but also have an impact on the surrounding area. Burning activities will create haze and how the haze will affect the surrounding environment depends on the conditions of the atmosphere.

Table 2. Statistic for the first three PC.

	PC1	PC2	PC3
Variance explained (%)	65.04 %	3.60 %	2.96 %
Cumulative variance (%)	65.04 %	68.64 %	71.6 %

3.2 EOF Spatial – Temporal Distribution

To see how the variation of the total hotspots in this area, EOF analysis was performed. The variances explained by the first three principal components are shown in Table 2. The first four principal components explained 71.60 % of the total variances. A large portion of the total variability was captured by the first principal component (PC1). PC1 explained the largest fraction of variance of approximately 65.04 %. The second mode (PC2) explaining 3.60 % of the total variance followed by the third principal component (PC3) explained 2.96 % of the total variance. Subsequent principal components with a lower fraction of explained variance were not considered for further analysis.

IOP Conf. Series: Earth and Environmental Science **880** (2021) 012003 doi:10.1088/1755-1315/880/1/012003

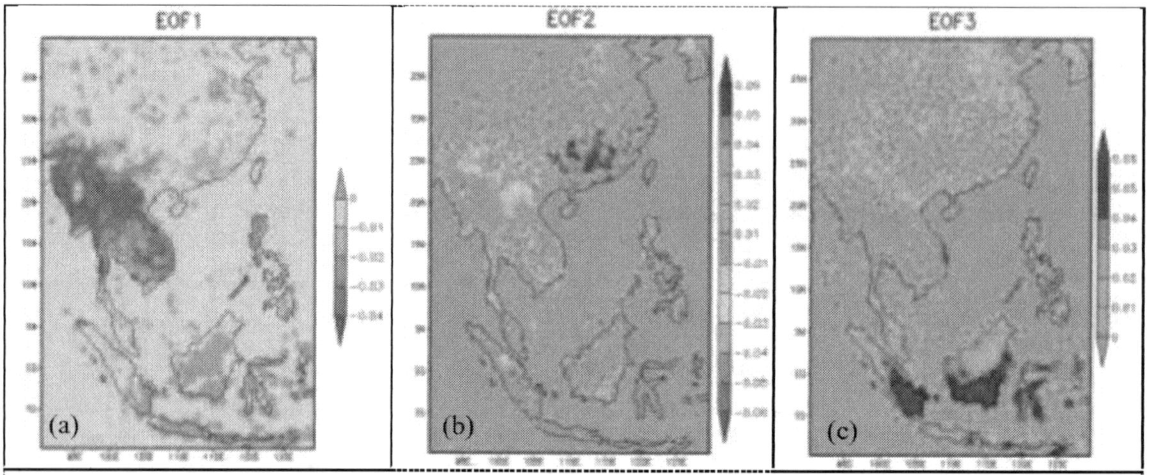

Figure 3. Principal mode of the empirical orthogonal function (EOF) analysis on the total of fire hotspot during 2001-2020.

Figure 3 shows the principal mode of the EOF. The principal component score of the EOF mode shows the value or contribution of each major component to each unit of observation. The principal component value can range between positive and negative. A high positive (negative) value means that the EOF mode makes a large contribution and has a positive (negative) effect on the unit of observation. Figure 3a shows the principal mode of total of fire hotspot for PC1 in the study area. The variability of total fire hotspots has a scale ranging from 0 to -0.04. The result identified high loadings over the Indo-China region and decreased gradually to the southern part of Indo-China. The EOF1 results indicate that Indo-China region has the larger contribution to variability of the total of fire hotspots during 2001-2020.

PC2 has a positive and negative variability of total fire hotspots with the range of the variability of total fire hotspots from -0.06 to 0.06 (Figure 3b). The second principal component separates the different fire areas, positive loadings is located in the southeast China and southeast part of the Borneo Island; and the opposite loadings is on the Indo-China region. The result in this mode shows the high variability of total fire hotspots with a positive contribution in the southeast part of the China region and southeast part of the Borneo Island; and negative contribution in the Indo-China region. Loadings in the southern part of Sumatra, southern part of Borneo, and the eastern coast of Vietnam are shown in PC3, while Indo-China showed weak loadings (Figure 3c). The variability of total fire hotspots has a positive scale ranging from 0 to 0.06. In EOF3 mode, these areas contribute the most to the variability of total fire hotspots.

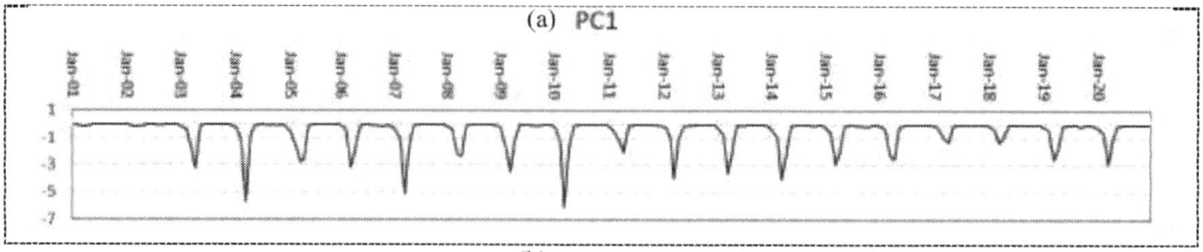

(b)

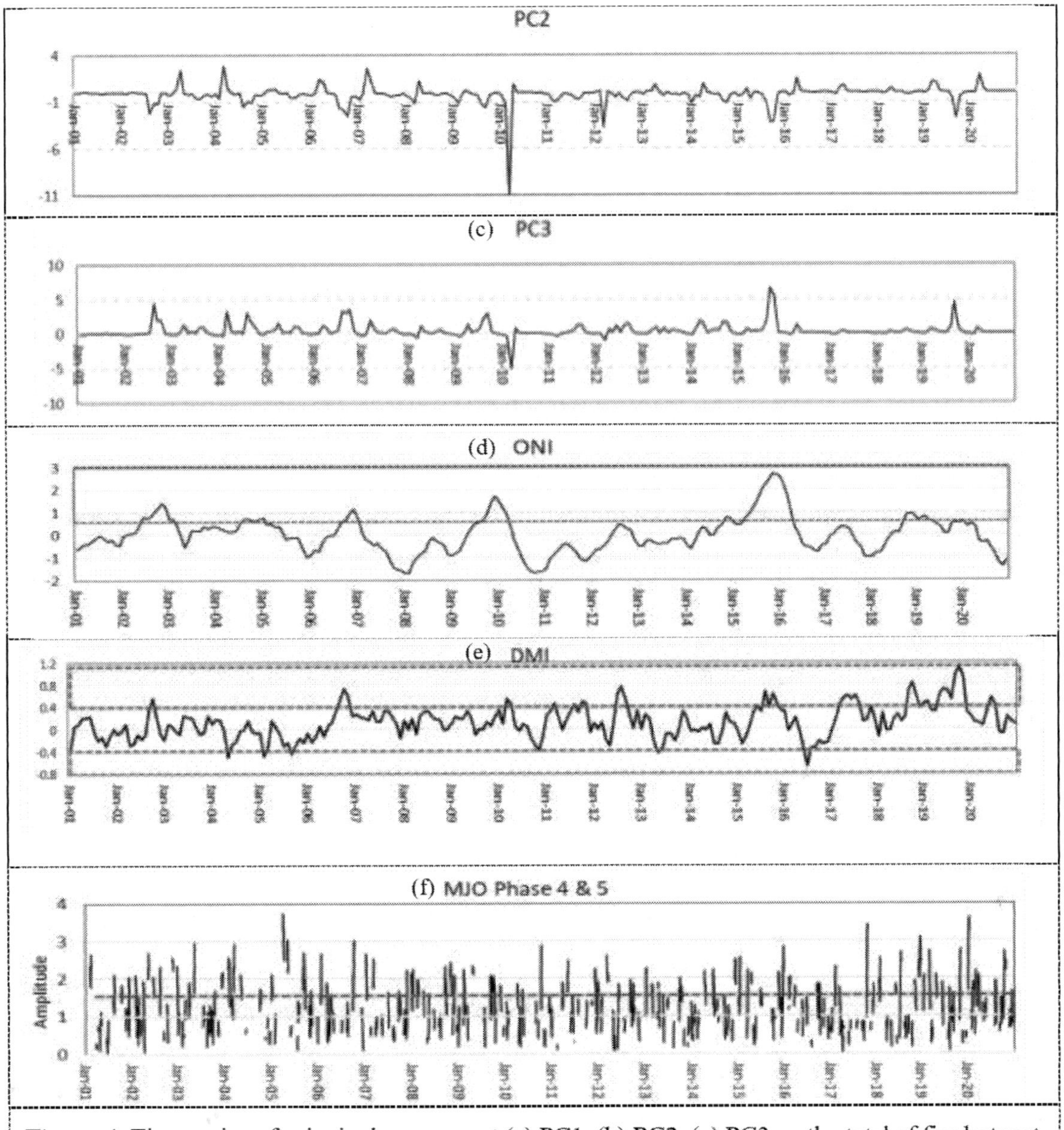

Figure 4. Time series of principal component (a) PC1, (b) PC2, (c) PC3 on the total of fire hotspot. (d) The Ocean Nino Index (ONI). Red dashed box represents El Nino events. (e) Dipole Mode Index (DMI). Red (blue) dashed box represents positive (negative) IOD. (f) Amplitude of the Madden-Julian Oscillation (MJO) in Phase 4 and 5. Yellow dashed line represents siginificant MJO and red dashed line represents strong MJO.

Figure 4 shows the principal component time series of PC1 that the PC1 region is dominated by negative variables throughout the years. In this first mode of principal component, an annual pattern occurs. Although it recurs every year in the cycle, the magnitude fluctuates every year. The peak always occurs at the beginning of the years and the strongest were in 2004, 2007, and 2010. To see what factors influenced the incidence of fire activities, it is necessary to look at the system that can give strong effect to the weather and climate over the Indo-China region.

The second and third PC give an inter-seasonal variability (figure 4b and 4c). There are positive and negative variations in PC2. However, similar to PC1, there was a high negative variable value for PC2 and PC3 in 2010. Looking at the PC2 area which the negative loading is also located in the Indo-China area, it can be said that the strong factor that influences PC1 also becomes an influence on PC2. PC3 was more dominant in having a high variable value in its positive value throughout the period except in 2010.

3.3 Hotspots and Weather Anomalies in Southeast Asia
The variability of fire activities is formed by similar factors that shape climate patterns. El Niño Southern Oscillation (ENSO) is known to dominate the inter-annual fire signal in Indo-China and Maritime Continent (MC). ENSO, as well as Indian Ocean Dipole (IOD), contributed as the inter-annual factor for the incidence of fire activities in Indonesia [27]. Warm phase ENSO conditions trigger the negative summertime precipitation anomalies in the MC. Fire activity increases dramatically with decreases of the precipitation. Likewise, IOD can be an indirect factor in fire activity, because of its effect on precipitation. During the strong ENSO years and a positive phase of the IOD, precipitation is less than usual in the Indo-China region. Conversely, during the La Niña years and a negative phase of the IOD, rainfall throughout the Indo-China Peninsula is above normal, except in parts of central Laos and northern Vietnam [28]. Likewise in the MC, the mixed diversity of ENSO and IOD events leads autumn droughts and is linked to frequent rampant forest wildfires [29].

Time series indices of ENSO and IOD; and the amplitude of MJO phases 4 and 5 are presented in the 4d-f graph to see the pattern similarity between the indices and PC value. By comparing the patterns, it can be seen whether the climate indices affect the PC value. On PC1 conditions, where the main contributors are in the Indo-China region, can be related to ENSO or IOD. The highest score for principal component on PC1 at the beginning of 2010 reached -6.10 when the ENSO moderate occurs in the end of 2009 until the beginning of 2010 (Figure 4d), coupling with the strong positive IOD (Figure 4e). Likewise at the end of 2002, early of 2007, 2016, and 2019, when the ENSO event occurred, there was also an increase in the signal of PC1. In early 2007, the score reached more than -5.00 which was influenced by the existence of a moderate El Nino scale that was active in those years and a positive IOD at the beginning of 2007. The signal fire was weakest from 2017 to 2018 within the study period, non-ENSO and non-IOD year occurred. Previous study reported that the increase in biomass burning activities over northern Indo-China is related to the intensification of the India-Burma Trough in the lower and middle troposphere as the result of the increase in north westerly wind behind the India-Burma Trough which could advect dryer air from the higher latitude to northern Indo-China. Dryer condition can help promote the local biomass burning activities. Furthermore, 2010 was a year with strong India-Burma Trough and SCSA so that the amount of biomass burning over northern Indo-China is greater than other years [26].

PC2 and PC3 have seasonal patterns during the study period. The peak of the PC of fire events occurs typically once or twice per season. PC2 is dominated by negative principal component. Furthermore, PC2 and PC3 in the year 2010 has a strong negative value as in PC1, means that of the El Nino-positive IOD years have increased the fire event occurrences in all cases. This is similar to the previous study where the combined effect of El Nino and positive IOD events are stronger compared to individual event, but depending on the season [30].

The time series of PC2 generally has two signal peaks, namely a positive signal at the beginning of the year and a negative signal at the end of the year. Recent research on the Inter Tropical Convergence Zone (ITCZ) mentioned that seasonal migration of ITCZ is related to the fire events in the MC [31]. Winter monsoon is formed by southern extent of ITCZ and lead the fire activity in Indo-China. Otherwise, northward extent of the ITCZ will form a Summer Monsoon and lead fire activity in the MC.

PC3 has a positive value of PC scale ranging from 0 to 0.06. The inter-seasonal pattern of the burning events that occurs can be caused by the MJO. If ENSO phase is the strongest indicator of fire events, the MJO often dictates when observed burning occurs in any given season [31]. El Nino events at the end of 2002, early of 2005, early of 2010, end of 2015, and early of 2019 caused an increase in positive signal on PC3. Sumatra and western Borneo are influenced by the phase of the MJO more than the other regions. Observed burning activity on the islands of Sumatra was more influenced by the phase and strength of the MJO [32]. Figure 4f shows the amplitude of the MJO during the study period. MJO is a tropical disturbance that propagates eastward along the tropics with a cycle of 30-60 days. The MJO phases which are related to climatic conditions in the Maritime Continent are MJO phases 4 and 5. A strong positive value for PC3 occurred in September of 2015 and a strong negative value in March of 2010. MJO entered phases 4 and 5 in September in 2015 with a strong MJO in the middle of the month, while MJO entered phases 4 and 5 in end of March 2010 with a strong MJO (Figure 4f). In August 2002 and September 2019, the principal component value was valued at more than 4.00, along with the entry of strong MJO into phases 4 and 5 that month (Figure 4f).

The highest variable value that occurred in 2015 can be affected by the ENSO activities as the PC3 region which is located in the southeast part of Sumatra and Borneo have known in many studies has correlation with ENSO activity [33][34][35]. PC scores give positive value above 3.00 in March 2004, August-October 2006, and September 2019 at the same time with strong MJO, while August 2004 and October 2015 occurred during non-MJOs. Knowing how the variability of hotspot events and the complexity of meteorology and climatology elements that can influence the variability of the hotspot need to be studied in more detail for each phenomenon. Furthermore, to see in more detail related to factors that affect hotspot occurrences, future study is suggested to use anomaly data of hotspots. By using the anomaly data, the influence factors are expected to be caught more clearly. Better understanding from the study with lead to mitigation measures which are important under a changing climate in the future.

4. Conclusion

The distribution of hotspot data for 2001-2020 in the study area is analyzed. The results shows that the distribution of hotspots varies each month over the Southeast Asia. EOF analysis conducted on monthly total hotspot data determines three dominant patterns spatially and temporally. The first mode (EOF1) explains 65 % of the total variance and identify high loadings over the Indo-China. The second (EOF2) and third (EOF3) EOF methods account for 3.60 % and 2.97 % of the total variance respectively. The second principal component separates the fire areas into China and Indo-China region, however the third principal component identify high loading in Sumatra-Borneo region. Annual pattern is dominant in the EOF1, while the inter-seasonal pattern is dominant in EOF2 and EOF3. Coupling of El Nino and positive IOD are factors that form the annual pattern on EOF1 which shows that loading is concentrated in the Indo-China region. Meanwhile, EOF2 and EOF3 that show the loading for MC, are more influenced by the presence of inter-seasonal factors such as MJO, although in certain years, the annual factor is stronger and still affects the incidence of fire events, as we know that MC climate system also linked with ENSO and IOD. Future studies need to discuss in more detail what driven factors influence the pattern of the determined principal components.

References

[1] Cheong K H *et al* 2019 Acute health impacts of the southeast asian transboundary haze problem - A review *Int. J. Environ. Res. Public Health* **16** 18

[2] Gao J *et al* 2017 Haze, public health and mitigation measures in China: A review of the current evidence for further policy response *Sci. Total Environ.* **578** 148–157

[3] Ho A F W *et al* 2019 Time-Stratified Case Crossover Study of the Association of Outdoor Ambient Air Pollution With the Risk of Acute Myocardial Infarction in the Context of Seasonal Exposure to the Southeast Asian Haze Problem *J. Am. Heart Assoc.* **8** 6 1–10

[4] Chen J *et al* 2019 Indirect economic impact incurred by haze pollution: An econometric and input-output joint model *Int. J. Environ. Res. Public Health* **16** 13

[5] Quan M and Shi Qiu Z *et al* 2013 An evaluation of the economic loss due to the heavy haze during January 2013 in China *China Environ. Sci* **33** 11 2087–2094

[6] Ouyang X *et al* 2019 Haze, health, and income: An integrated model for willingness to pay for haze mitigation in Shanghai, China *Energy Econ.* **84** 104535

[7] Agarwal S *et al* 2020 The impact of transboundary haze pollution on household utilities consumption *Energy Econ.* **85** 104591

[8] Othman J *et al* 2014 Transboundary smoke haze pollution in Malaysia: Inpatient health impacts and economic valuation *Environ. Pollut.* **189** 194–201
[9] Latif M T *et al* 2018 Impact of regional haze towards air quality in Malaysia: A review *Atmos. Environ.* **177** 28–44
[10] Radzi M *et al* 2004 Biomass burning as the main source of organic aerosol particulate matter in Malaysia during haze episodes *Chemosphere* **55** 8 1089–1095
[11] Wang S H *et al* 2015 Vertical distribution and columnar optical properties of springtime biomass- burning aerosols over Northern Indochina during 2014 7-SEAS Campaign *Aerosol Air Qual. Res.* **15** 5 2037–2050
[12] Wang Y *et al* 2004 Trends in atmospheric haze induced by peat fires in Sumatra Island, Indonesia and El Niño phenomenon from 1973 to 2003 *Geophys. Res. Lett.* **31** 4 1–4
[13] Chuvieco E *et al* 2008 Global characterization of fire activity: Toward defining fire regimes from Earth observation data *Glob. Chang. Biol.* **14** 7 1488–1502
[14] Levin N and Heimowitz A 2012 Mapping spatial and temporal patterns of Mediterranean wildfires from MODIS *Remote Sens. Environ.* **126** 12–26
[15] Hantson S *et al* 2013 Strengths and weaknesses of MODIS hotspots to characterize global fire occurrence *Remote Sens. Environ.* **131** 152–159
[16] Tansey K *et al* 2008 Relationship between MODIS fire hot spot count and burned area in a degraded tropical peat swamp forest in Central Kalimantan, Indonesia *J. Geophys. Res. Atmos.* **113** 23 1–8
[17] Yulianti N, Hayasaka H and Usup A 2012 Recent Forest and Peat Fire Trends in Indonesia The Latest Decade by MODIS Hotspot Data *Glob. Environ. Res.* **16** 1 105–116
[18] Reddy C S *et al* 2019 Identification and characterization of spatio-temporal hotspots of forest fires in South Asia *Environ. Monit. Assess.* **191**
[19] Justice C O *et al* 2002 The MODIS fire products *Remote Sens. Environ* **83** 1–2 pp. 244–262
[20] Giglio L, Descloitres J, Justice C O and Kaufman Y J 2003 An enhanced contextual fire detection algorithm for MODIS *Remote Sens. Environ.* **87** no. 2–3 pp. 273–282
[21] Giglio L 2015 *MODIS Collection 6 Active Fire Product User's Guide Revision A*. Department of Geographical science, University of Maryland
[22] Pearson K 1901 On Lines and Planes of Closest Fit to Systems of Points in Space*Philosophical Magazine* vol. **2** no. 11 pp. 559–572
[23] Harold H 1993 Analysis of a Complex of Stastistical Variables into Principal Components *J. Educ. Psychol.* vol. **24** no. 6 & 7 pp. 417–441 & 498–520
[24] Hannachi A, Jolliffe I T and Stephenson D B 2007 Empirical Orthogonal Functions and Related Techniques in Atmospheric Science: A Review *Int. J. Climatol.* pp. 1119–1152
[25] Juneng L, Latif M T, Tangang F T and Mansor H 2009 Spatio-temporal characteristics of PM10 concentration across Malaysia *Atmos. Environ.* vol. **43** no. 30 pp. 4584–4594
[26] Huang W R *et al* 2016 Interannual variation of springtime biomass burning in Indochina: Regional differences, associated atmospheric dynamical changes, and downwind impacts *J. Geophys. Res.* vol. **121** no. 17 pp. 10016–10028
[27] Field R D, Van Der Werf G R and Shen S S P 2009 Human amplification of drought-induced biomass burning in Indonesia since 1960 *Nat. Geosci.* vol. **2** no. 3 pp. 185–188
[28] Kim J S et al 2019 Identification of Hotspots of Rainfall Variation Sensitive to Indian Ocean Dipole Mode through Intentional Statistical Simulations *Hydrol. Earth Syst. Sci. Discuss.* no. June pp. 1–37
[29] Hu C *et al 2020* Mixed diversity of shifting IOD and El Niño dominates the location of Maritime Continent autumn drought *Natl. Sci. Rev.* vol. **7**, no. 7, pp. 1150–1153
[30] Amirudin A A *et al* 2020 Differential influences of teleconnections from the Indian and Pacific oceans on rainfall variability in Southeast Asia *Atmosphere (Basel).* vol. **11** no. 9 pp. 13–15
[31] Reid J S *et al* 2012 Multi-scale meteorological conceptual analysis of observed active fire hotspot activity and smoke optical depth in the Maritime Continent *Atmos. Chem. Phys.* vol. **12** no. 4, pp. 2117–2147
[32] Ramirez E M *et al* 2009 Relating Precipitation Phenomena with MODIS Detected Hot Spots in the Maritime Continent *Am. Geophys. Union*
[33] Islam M A *et al* 2018 Effects of El-Niño, Indian Ocean Dipole, and Madden-Julian Oscillation on surface air temperature and rainfall anomalies over Southeast Asia in 2015 *Atmosphere (Basel).* vol. **9** no. 9 pp. 1–14
[34] Rauniyar S P and Walsh K J E 2013 Influence of ENSO on the diurnal cycle of rainfall over the Maritime Continent and Australia *J. Clim.*, vol. **26** no. 4 pp. 1304–1321
[35] Chang C P, Wang Z, Ju J and Li T 2004 On the relationship between western maritime continent monsoon rainfall and ENSO during northern winter *J. Clim.* vol. **17** no. 3 pp. 665–672,

Acknowledgments

We gratefully acknowledge to Universiti Kebangsaan Malaysia for supporting and providing us with the funding (GGPM-2020-033). Furthermore, the authors appreciate the constructive comments from the anonymous reviewers.

Spatio-temporal of surface ozone (O_3) variations at urban and suburban sites in Sarawak region of Malaysia

H Mahidin[1,2], M T Latif[1]*, A Hamdan[3], J Salleh[3], D Dominick[4], M C G Ooi[5], L Juneng[1] and S Zawawi[6]

[1] Department of Earth Sciences and Environment, Faculty of Science and Technology, Universiti Kebangsaan Malaysia, 43600 UKM Bangi, Selangor, Malaysia
[2]Faculty of Applied Sciences, Universiti Teknologi MARA, Sarawak Branch, 94300 Kota Samarahan, Sarawak, Malaysia
[3]Department of Mathematical Sciences, Faculty of Computer and Mathematical Sciences, Universiti Teknologi MARA, Sarawak Branch, 94300 Kota Samarahan, Sarawak, Malaysia
[4]Centre for Atmospheric Chemistry, University of Wollongong, Wollongong, NSW2522, Australia
[5]Institute of Climate Change, Universiti Kebangsaan Malaysia, 43600 UKM Bangi, Selangor, Malaysia
[6]Faculty of Health Sciences, Universiti Teknologi MARA, Sarawak Branch, 94300 Kota Samarahan, Sarawak, Malaysia

*Corresponding E-mail: talib@ukm.edu.my

Abstract. Sarawak Region of Malaysia is currently experiencing a high demand for capital needs such as transformation forest to plantations, economic development, and improving transportation systems. Those land cover changes will increase primary pollutant emissions and trigger surface O_3 formation. Surface O_3 is a secondary pollutant and a significant greenhouse gas contributing to climate change and declining air quality. In this study, variations in surface O_3 concentrations at urban and suburban sites in Sarawak were explored using the Malaysian Department of Environment data spanning a two-year cycle (2018-2019). The primary aim of this study is to ascertain the variation of surface O_3 concentrations reported at four monitoring stations in Sarawak, namely Kuching (SQ1) (Urban), Sibu (SQ2) (Suburban), Bintulu (SQ3) (Suburban), and Miri (SQ4) (Suburban). The study also analysed the relationship between O_3 distribution and nitrogen oxides (NO and NO_2). The findings showed that O_3 concentrations observed in the region during the study period were lower than the maximum permissible value of 100 ppbv suggested by the Malaysian Ambient Air Quality Standard (2020). SQ4 (Miri) at suburban sites recorded the highest average surface O_3 concentrations with an hourly average and daily maximum O_3 concentration of 15.7 and 89.5 ppbv, respectively. Temperatures, UV exposure, and wind speed all impact the concentration of surface O_3 in Sarawak. In all stations, concentrations of O_3 were inversely linked with NO, NO_2, and relative humidity (RH). This research will assist the relevant agency in forecast, monitor, and mitigate the level of O_3 in the ambient environment, especially in the Sarawak Region.

Keywords: Surface ozone, Ozone precursors, Nitrogen oxides, Urban, Suburban.

Track Name: Atmospheric Chemistry and Physics

IOP Conf. Series: Earth and Environmental Science **880** (2021) 012004 doi:10.1088/1755-1315/880/1/012004

1. Introduction

The surface ozone (O_3) is one of the most important greenhouse gases (GHG), as it functions dually, depending on its elevation in the atmosphere[1]. O_3 plays a vital role in the chemistry of both the stratosphere and troposphere. Tropospheric O_3 is an important component of the troposphere that impacts air quality, atmospheric oxidation capabilities, and climate change [2]. Tropospheric O_3, often referred to as surface O_3 is an atmospheric pollutant that negatively impacts human health, contributes to respiratory problems and lung disorders such as asthma, and has a detrimental influence on plants, the environment, and agricultural productivity [3–7].

Surface O_3 is a significant contributor to air pollution in urban areas across the world [8]. The relationship between the production of O_3 in the presence of solar radiation and a complicated sequence of photochemical reactions involving two major O_3 precursors, nitrogen oxides NO_x ($NO_x = NO + NO_2$) and hydrocarbons such as volatile organic compounds (VOCs) [8,9]. The primary scientific concern associated with urban air pollution is these two main O_3 precursors [8]. It has been discovered that levels of O_3 are spread over various locations in major cities, as well as in suburban and industrial regions in Malaysia, with a particular focus on the present condition of air quality, the sources of O_3, and the long-term projections for the ambient air quality [10–14].

Consequently, it is imperative to locate and analyse surface O_3. Understanding the conditions that result in its formation and the role of climatic variables such as solar radiation, temperature, and wind direction can assist in a thorough understanding of photochemical processes and the relationship between O_3 concentration and weather variables. Thus, this study aims to determine the variation in surface O_3 concentrations recorded at four urban and suburban monitoring sites in Sarawak. Then, the relationship between O_3 and its precursors, nitrogen oxides NO_x ($NO_x = NO + NO_2$), and climatic variables is discussed, emphasising the challenges associated with ozone abatement.

2. Study Area and Methodology

2.1. Study area

The state of Sarawak is located on the northwest coast of Borneo and stretches throughout the whole length of the island and most of the South China Sea beaches. The state is 124,450 square kilometers in size. Its interior location has been recognised for the mountainous, dense rainforest that it encompasses; part of it is preserved as nature reserves. Figure 1 depicts the four monitoring stations within Sarawak's urban and suburban areas that were chosen for investigation.

2.2. Ozone and ozone precursors (NO, NO_2 and NO_x,)

The data source from the Department of Environment Malaysia (DOE) with measurements based on hourly concentrations of surface O_3 and oxides of nitrogen (NO, NO_2 and NO_x) between the period of January 2018 until December 2019 were used. In addition to air pollutant parameters, the stations also recorded temperature, relative humidity, solar radiation (UVb), wind speed, and wind direction. Surface O_3 measurements from 2018 to 2019 were measured using Ozone Analyzer (Model 49i). The continuous measurements of NO, NO_2, and NO_x were measured using Oxides of Nitrogen Analyzer (Model 42i) via chemiluminescence technique. The AIO2-Meteorology Sensor was used to detect wind speed, wind direction, temperature, and relative humidity, while the Photovoltaic Pyranometer was used to monitor solar radiation (UVb). Pakar Scieno TW (PSTW), a DOE-assigned entity, calibrates and manages all instruments.

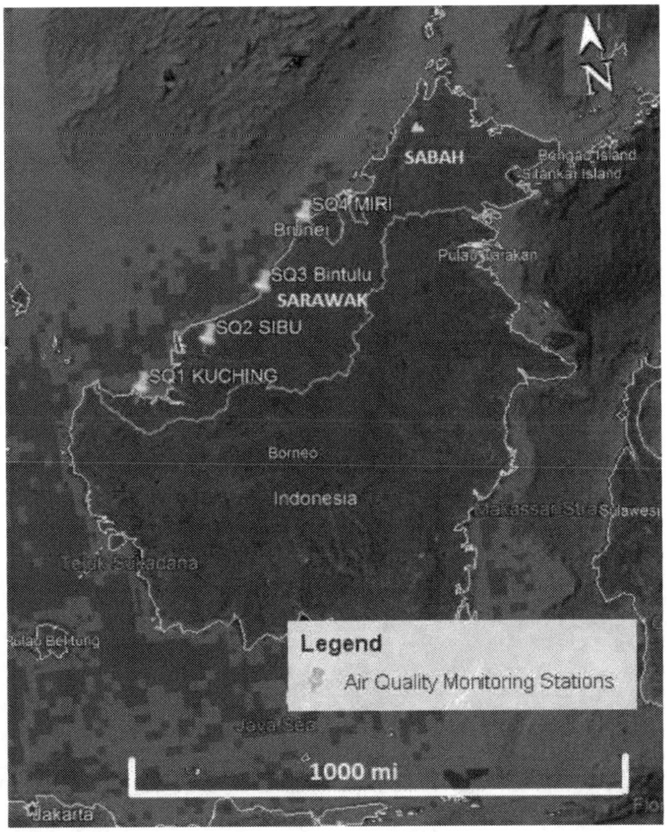

Figure 1. The location of four urban and suburban air quality monitoring sites of Sarawak.

2.3. Statistical data analysis
Descriptive statistics are used to provide data in a clear and easy-to-understand format. It has a descriptive statistic table and a boxplot created with R programming. Collecting, filtering, organising, summarising, and presenting data is also part of statistical analysis. The missing values for the data collected during the study period were between 4.69 % and 8.29 %. (Table 1). Missing values in the study period are not considered in the computation of statistics.

2.4. Polar Annulus
This polar plot is used to depict the temporal features of a pollutant concentration as a function of wind direction. It can show the monthly fluctuations in the concentration of surface O_3.

2.5. Diurnal Plot
It describes the characteristics, plot, and average of hourly values on a 24-hour scale. It may demonstrate the time series correlation and effect of O_3 precursors on surface ozone concentrations and the changes on a 24-hour timeframe.

2.6. Kolmogorov-Smirnov Test and Spearman Correlation Analysis
The Kolmogorov-Smirnov test used to decide if a sample comes from a population with a specific distribution. Kolmogorov-Smirnov test can be performed to assess the normality assumption.

Meanwhile, Spearman's rank correlation is a nonparametric measure of rank correlation in statistics (statistical dependence between the rankings of two variables). It determines how well a monotonic function can explain the relationship between two variables such as surface O_3 levels and their precursors and meteorological parameters.

3. Results and Discussion

3.1. Descriptive analysis

The descriptive statistics of hourly surface O_3 concentrations for urban and suburban sites from 2018 to 2019 are presented in Table 1. The hourly average concentrations of surface O_3 at the continuous air quality monitoring stations were recorded between 10.6 and 15.7 ppbv. SQ4 (Miri) at suburban sites recorded the highest average surface O_3 concentrations (M = 15.7 ppbv, SD = 10.2). This was followed by suburban site SQ3, Bintulu (M = 14.1 ppbv, SD = 10.6), urban sites SQ1, Kuching (M = 12.6 ppbv, SD = 9.65) and suburban site SQ3, Sibu M = 10.6 ppbv, SD = 8.08). The median values were within $\pm$ 2 ppbv of the average O_3 concentration, whereas the standard deviation ranged between 8.08 and 10.6 ppbv for all sites. The maximum O_3 concentration fell within the range of 54.3 to 89.5 ppbv recorded by the four stations. The varying local emissions of O_3 precursors from motor vehicles and industrial activities are assumed to be the primary factor for the difference in long-term hourly O_3 concentrations across all stations [12].

Table 1. Descriptive statistics for surface O_3 concentration between 2018 to 2019 in Sarawak Region (ASL: Altitude base on above sea level; U: Urban site; SU: Suburban site, SD: Standard Deviation, M: Median).

Location	Latitude	Longitude	Altitude (ASL) (m)	Station-Characteristic	Missing value (%)	Hourly [O_3] (ppbv)			
						Average	SD	M	Max
SQ1-Kuching	N 01° 33.734	E 110° 23.331	7	Kuching-U	5.86	12.6	9.65	11.2	69.7
SQ2-Sibu	N 02° 18.864	E 111° 49.915	9	Sibu-SU	8.29	10.6	8.08	8.96	54.3
SQ3-Bintulu	N 03° 10.625	E 113° 02.465	8	Bintulu-SU	6.19	14.1	10.6	12.6	71.9
SQ4-Miri	N 04° 25.481	E 114° 00.746	6	Miri-SU	4.69	15.7	10.2	14.9	89.5

3.2. Boxplot of surface O_3 concentration

The box plot in Figure 2 depicts the hourly O_3 concentration at study locations i.e., SQ1, SQ2, SQ3, and SQ4, from 2018 to 2019. In 2019, both SQ3 and SQ4 had increased in O_3 concentrations, as shown in a plot. The reason is most likely due to the proximity of the SQ4 and SQ3 monitoring stations to crowded residential districts and industrial areas. Compared to prior years, for SQ1, SQ2, and SQ3, the median O_3 concentration in SQ4 is the highest (16.01 ppbv) in 2019. Overall, from 2018 to 2019, all urban (SQ1) and suburban (SQ2, SQ3, and SQ4) stations exhibit an elevation in O_3 concentration. As a result, the O_3 concentration increased marginally but remained below the Malaysian Ambient Air Quality criterion of 100 parts per billion (ppb) for hourly O_3 concentration [15]. The O_3 concentration boxplot appears to skew to the right across all stations and years. Any skewness in the data indicates that the distribution is not normal.

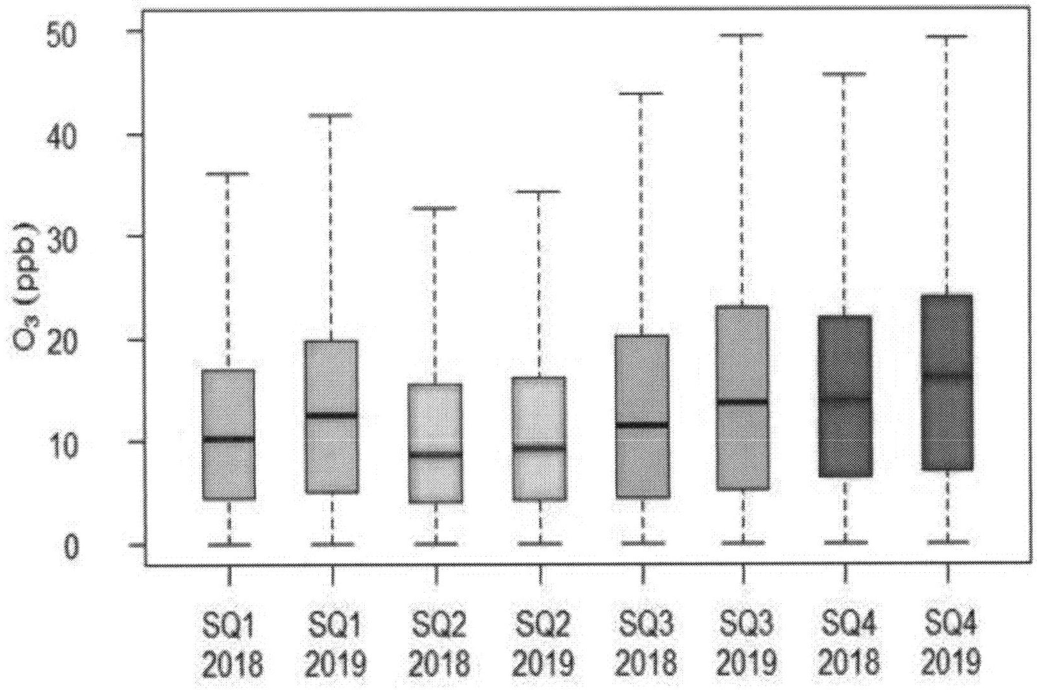

Figure 2. Boxplot of surface O_3 concentration at all sampling stations in 2018 and 2019.

3.3. Monthly variations of surface O_3 concentrations (Polar Annulus by season)

The surface O_3 concentration was analysed using average hourly data for every month to identify the monthly variation of O_3 concentration. Figure 3 presented the polar annulus of surface O_3 concentration for SQ1, SQ2, SQ3, and SQ4 by wind direction. The monthly plots show that the O_3 concentration in SQ1 was higher in Aug-Sept, influenced by north-westerly, north-easterly, and south-easterly winds. The O_3 concentration in SQ2 was higher in mid-Aug to mid-Sept by south-westerly, westerly and north-westerly winds. Meanwhile, SQ3 exhibits higher O_3 concentration, particularly in June and July-Sept, dominated by north-westerly winds. Lastly, SQ4 explicitly the higher O_3 concentration in mid-July-Sept with wind direction from the westerly winds.

Sarawak experienced the southwest (SW) monsoon from May to September, which is associated with a relatively dry period before the monsoon wind shifts to wet weather, which is more visible in abrupt spikes in rainfall levels. The current temperature and solar radiation rise due to this occurrence, which increases the rate of O_3 generation.

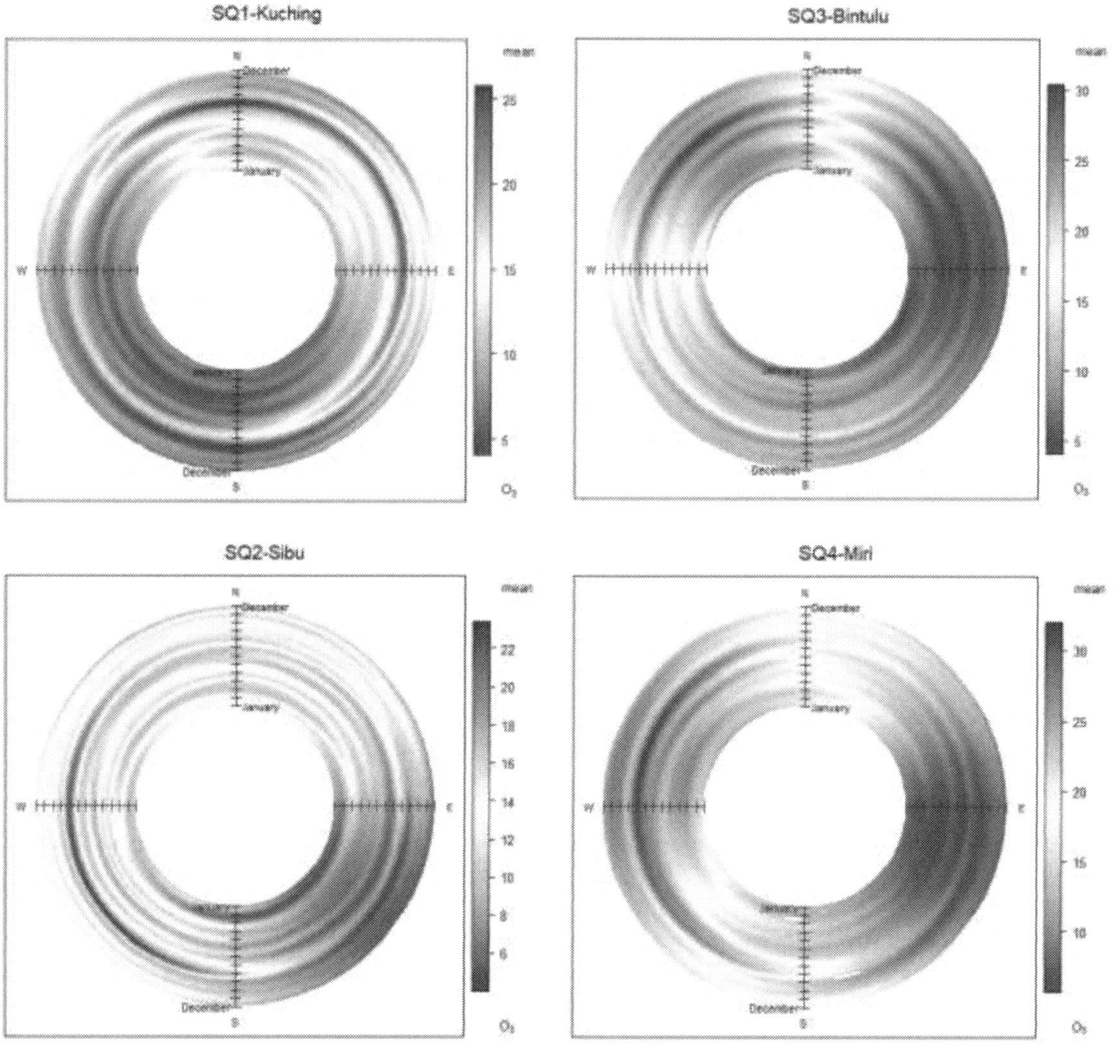

Figure 3. Polar annulus for mean hourly O_3 by month at all stations.

3.4. *Diurnal Fluctuation of Ozone and Its Precursor (NO, NO_2 and NO_x)*

Figure 4 (a-d) compares the diurnal variation of surface O_3 and its precursors, NO_2, NO, and NO_x. The diurnal plots of O_3 concentration for SQ1 (a), SQ2 (b), SQ3 (c), and SQ4 (d) were plotted according to the average hourly data set from 2018 to 2019. During daytime hours, the O_3 single peak was visible from all sites. O_3's diurnal fluctuation in urban and suburban areas is the primary feature of peak concentrations in the afternoon at about 1:00 p.m. As seen in Figure 4(c) and 4(d), the surface O_3 concentrations in suburban areas (SQ3 and SQ4) were higher than the mean concentration in urban areas (SQ1), and the maximum O_3 concentrations began to rise by 8:00 a.m. The O_3 peak occurred between 1:00 to 3:00 p.m. in suburban areas (SQ3 and SQ4), with mean peak values of 25.7 and 26.9 ppb, respectively, whereas suburban (SQ2) had the lowest mean peak value of 21.4 ppb.

Meanwhile, the O_3 peak occurred at about 2:00 p.m. in the urban location (SQ1), with a mean peak value of 23.8 ppb. This daily trend corresponds to the previous findings [11]. The concentration of surface O_3 was greatest during midday at S1 (Kuching) in the previous research, when the intensity of UVb was at its peak. A The high O_3 levels during the daytime may be attributed to incoming solar radiation, which encourages O_3 creation via photochemical reactions in the lower troposphere. In

IOP Conf. Series: Earth and Environmental Science **880** (2021) 012004 doi:10.1088/1755-1315/880/1/012004

contrast, the low O_3 levels during the night and early morning hours may be related to a lack of photochemical reactions, leading to a catalytic cycle for ozone loss. Urbanisation, industry, high-speed propulsion, and atmospheric emissions of a wide variety of active chemicals such as carbon dioxide and aerosol are significant factors in increasing O_3 concentration. Suburban regions (SQ3 and SQ4) were bordered by new townships with dense residential areas, urban infrastructure, and a significant volume of traffic, which may account for the elevated O_3 concentrations in both regions.

The O_3 formation is highly dependent on the precursor concentration and exhibits a distinct hourly variation. The diurnal pattern of precursor concentrations in this urban and suburban research region exhibited comparable bimodality observed in earlier research [11–13]. This bimodality is attributable to traffic emissions. The daily trends of each precursor pollutant (NO, NO_2, and NO_x) increased during the morning rush hours due to increasing traffic emissions and unstable atmospheric conditions. The resulting first peak of NO activity occurs between 7:00 to 9:00 a.m., and then it levels out and declines around 9:00 a.m., consistent with [11,16]. As is typically the case in polluted urban areas, NO reacts with O_3 and enters a "NO_X titrated state" [17]. This explains why the O_3 precursor concentrations at midday were low due to their conversion to a variety of secondary pollutants, including O_3, and the convective transmission of pollutants to higher elevations. The concentration of O_3 precursors started to rise again at 6:00 p.m. and gradually decreased after 11:00 p.m. However, as the diurnal plot indicated, it was unusual for SQ2; the second peak began early, approximately 1:00 p.m., with NO_x reaching its highest level of the day at 6:00 p.m., before rapidly decreasing after 7:00 p.m.

3.5. Kolmogorov-Smirnov Test and Spearman Rank Correlation Analysis

Part of the initial step in data analysis was to verify the normality assumption. A graphical method such as boxplots and histogram were used to assess the normality assumption. Further testing of normality using the Kolmogorov-Smirnov test revealed that the data in this study did not follow a normal distribution (sig. < 0.05). Since this data did not meet the normality assumption, the correlation between O_3 and its precursors and other meteorological parameters were assessed using non-parametric correlation (Spearman rank correlation).

Overall, the Spearman Rank correlation results for both years demonstrated that O_3 concentrations in all urban and suburban sites studied were negatively correlated with NO, NO_2, and RH. This conclusion is consistent with the fact that NO and NO_2 are known precursors of O_3, implying an increase in O_3 concentration relates to a decrease in these pollutants' levels. The direct association between relative humidity and precipitation or moist conditions contributed to the negative correlation between O_3 and relative humidity. The direct relationship between relative humidity and precipitation or wet circumstances explains why O_3 and relative humidity negatively correlate. This is consistent with occurrence of the heavy rain and thunderstorms that occur between November and January during the Northeast (NE) monsoon [18], resulting increases low temperatures, solar radiation (UVb) and wind speeds [14]. Increased relative humidity increased O_3 degradation due to reduced photochemical efficiency and influenced the chemistry between O_3 precursors. Additionally, there was a positive correlation between surface O_3 concentration and temperature, wind speed, wind direction, and solar radiation.

IOP Conf. Series: Earth and Environmental Science **880** (2021) 012004 doi:10.1088/1755-1315/880/1/012004

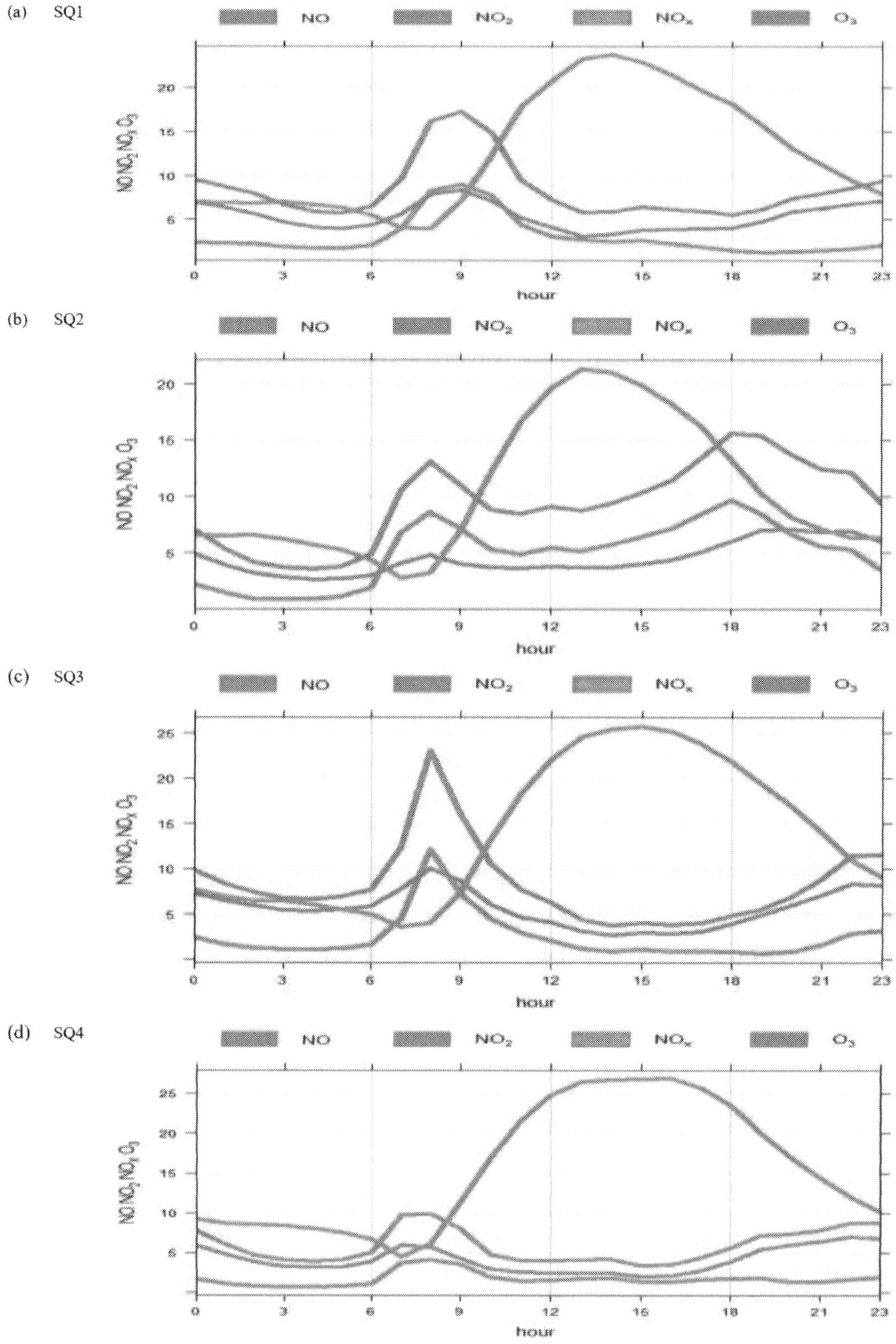

Figure 4. Diurnal variation of O_3 and its precursors.

4. Conclusion

Overall, the study demonstrates that all urban (SQ1) and suburban (SQ2, SQ3, and SQ4) stations show a rising level of O_3 from 2018 to 2019. However, the concentration of surface O_3 in the urban and suburban Sarawak region is still lower than the maximum permissible value of 100 ppbv suggested by the Malaysian Ambient Air Quality Standard (2020). SQ4, Miri at suburban sites recorded the highest average surface O_3 concentrations. Following that, suburban sites SQ3, Bintulu, urban sites SQ1, Kuching and suburban SQ2, Sibu. The concentration of surface O_3 in the Sarawak region is influenced by meteorological conditions, including temperature, UV radiation, and wind speed. During the SW monsoon season, the polar annulus has a greater O_3 level from July to September. The possibility that biomass burning may produce higher quantities of O_3 during the SW monsoon requires more study and analysis. This study implies that more research on surface O_3 and its precursors in Sarawak is necessary, particularly on surface O_3 to other O_3 precursors such as NO_x and VOC control strategies to determine the most effective solutions for comprehensive O_3 abatement. While the previous study highlighted various observational tendencies, little emphasis was made on the mitigating implications of these results in Malaysia's Borneo area. It is crucial to maintain O_3 precursor concentrations to sustain the O_3 concentration at or below the maximum limit. The challenges associated with O_3 mitigation could be tackled through the development of strategies and enactment of regulations such as strict NO_x emission limits for power plants and industrial combustion sources; enhanced vehicle inspection programmes in states; or the application of Internet of things technologies such as low-cost sensors to be installed in previously unmeasured urban and industrial areas.

References

[1] Sentian J *et al* 2020 Surface ozone variations at the great wall station, antarctica during austral summe. Adv Polar Sci. **31** (2) 92–102
[2] Wang T *et al* 2009 Increasing surface ozone concentrations in the background atmosphere of Southern China, 1994-2007 Atmos Chem Phys. **9** (16) 6217–27.
[3] Avnery S *et al* 2011 Global crop yield reductions due to surface ozone exposure: 1. Year 2000 crop production losses and economic damage Atmos Environ [Internet] **45** (13) 2284–96
[4] Feng Z *et al* 2019. Economic losses due to ozone impacts on human health, forest productivity and crop yield across China. Environ Int [Internet] **131** (May) 104966
[5] Pleijel H *et al* 2019 Ozone impact on wheat in Europe, Asia and North America – A comparison. Sci Total Environ [Internet]. 664 908–14
[6] Li Y *et al* 2020 Characteristics, source apportionment and health risks of ambient VOCs during high ozone period at an urban site in central plain, China. Chemosphere [Internet] 250 126283.
[7] Manisalidis I *et al* 2020 Environmental and Health Impacts of Air Pollution: A Review. Front Public Heal **8** (February) 1–13
[8] Izzah Mohd Hashim N *et al* 2019 Assessment of Surface Ozone Concentration in Northern Peninsular Malaysia IOP Conf Ser Mater Sci Eng. **551** (1)
[9] Wang T *et al* 2017 Ozone pollution in China: A review of concentrations, meteorological influences, chemical precursors, and effects. Sci Total Environ. 575 1582–96
[10] Sillman S 1999 The relation between ozone, NOx and hydrocarbons in urban and polluted rural environments *Dev Environ Sci.* **1** (C)1821–45
[11] Ahamad F *et al* 2020 Ozone trends from two decades of ground level observation in Malaysia. *Atmosphere (Basel)* **11** (7) 1–16
[12] Latif MT *et al* 2016 Seasonal and long term variations of surface ozone concentrations in Malaysian Borneo *Sci Total Environ.* 573 494–504
[13] Banan N *et al 2013* Characteristics of surface ozone concentrations at stations with different backgrounds in the Malaysian Peninsula *Aerosol Air Qual Res.* **13** (3) 1090–106
[14] Awang N R *et al 2016* Diurnal variations of ground-level ozone in three port cities in Malaysia. *Air Qual Atmos Heal* **9** (1) 25–39
[15] Dominick D *et al* 2012 Spatial assessment of air quality patterns in Malaysia using multivariate analysis *Atmos Environ.* **60** 172–81
[16] Department of Environment Malaysia (DOE) 2018 Laporan Kualiti Alam Sekeliling 2017 (Environmental Quality Report) Department of Environment Malaysia: Putrajaya, Malaysia.
[17] Kasparoglu S *et al* 2018 Spatial and temporal variation of O^3, NO and NO_2 concentrations at rural and urban sites in Marmara Region of Turkey. Atmos Pollut Res. **9** (6)1009–20

[18] Sa'adi Z *et al* 2017 Distributional changes in rainfall and river flow in Sarawak, Malaysia. *Asia-Pacific J Atmos Sci.* **53** (4) 489–500

Acknowledgments
The authors would like to thank the Department of Environment, Malaysia for providing air quality and meteorological dataset and Dana Kecemerlangan (DKCM 2020) UiTM Sarawak Branch 600-UiTMKS (RMU. 5/2) (21/2020/KCMT).

One-year monitoring of aromatic volatile organic compounds (VOCs) in urban areas of Malaysia

N S S L Hawari[1], M T Latif[1*], M Othman[2], N M Hanif[1], H H A Hamid[1], N I H Mustaffa[1], A A A Mohtar

[1]Department of Earth Sciences and Environment, Faculty of Science and Technology, Universiti Kebangsaan Malaysia, 43600 Bangi, Malaysia

[2]Institute for Environment and Development (LESTARI), Universiti Kebangsaan Malaysia, 43600, Bangi, Selangor, Malaysia

*Corresponding E-mail: talib@ukm.edu.my

Abstract. VOCs have been one of the important limiting factors of O_3 photochemical production in the urban tropical region. This study aims to determine the variation and composition of aromatic VOCs as well as their contribution to the O_3 formation. The hourly data of benzene, toluene, ethylbenzene and xylenes (BTEX) and ozone (O_3) were retrieved from January until December 2019 at three continuous air quality monitoring stations (S1, S2 and S3) operated by the Malaysian Department of Environment (DOE). Aromatic VOCs is detected by using an online gas chromatography analyser (GC 5000BTX). The annual average ΣBTEX concentrations ranged from 14.53 ± 12.43 μg m^{-3} to 25.04 ± 24.04 μg m^{-3} during the observation period, with toluene as the most dominant species with an average concentration of 10.65 ± 12.12 μg m^{-3}. The O_3 formation potential (OFP) was calculated and the results indicated that toluene is the highest contributor of O_3-forming potential, followed by m,p-xylene, o-xylene, ethylbenzene and benzene with the value of 127.8 μg m^{-3} (40.90 %), 103.0 μg m^{-3} (32.97 %), 48.66 μg m^{-3} (15.58 %), 26.17 μg m^{-3} (8.38 %) and 6.74 μg m^{-3} (2.15 %), respectively. Since the study of VOCs in Malaysia is quite limited, a more comprehensive study is currently underway to integrate the research on variation of VOCs in ambient air and its impact on the environment and human health.

Keywords: BTEX, Volatile Organic Compounds, Ozone Formation Potential, Urban areas

Track Name: Atmospheric Chemistry and Physics

1. Introduction

Volatile organic compounds (VOCs) have been one of the important limiting factors of ozone (O_3) photochemical production in the urban tropical region. Considerable efforts have been expended in recent times to understand the tropospheric O_3 formation mechanism, which is usually produced from photochemical reactions involving VOCs with various atmospheric radicals (i.e., OH radicals, HO_2 radicals, etc.) [1-4]. Anthropogenic VOCs emissions are complex and vary from city, with primary sources including automobile exhaust, gasoline volatilisation, solvent usage, industrial and domestic coal combustion and biomass burning, whereas additional sources are mainly affected by regional transport [5-7]. Anthropogenic VOCs, generally referred to as aromatic hydrocarbons, are the most frequently measured by previous studies. Aromatic VOC, particularly benzene, toluene, ethylbenzene and xylenes (BTEX) partitions is of great concern as it may cause a detrimental impact on human health [8-11]. The nature and extent of these health risk effects depend on the concentration levels of the species and the duration of their exposure to pose adverse health effects [12]. According to Ran et al. [13], atmospheric VOCs may exacerbate both respiratory and cardiovascular disease, particularly congestive heart failure (CHF) and chronic obstructive pulmonary disease (COPD).

BTEX is commonly considered as the key precursor to major air pollutants. Hot spot areas, typically urban areas, are ideal for studying BTEX emissions where the concentration levels in the atmosphere are significantly higher than the background remote areas [14, 15]. For instance, the urban air of Kuala Lumpur was observed to be slightly more polluted, resulting in health and air pollution issues [16]. Pollution levels in developing countries have dramatically increased in recent years due to a high volume of fresh potential emissions sources with large populations [17, 18].

Herein, this study aims to determine the variation of aromatic VOCs in selected cities in Malaysia and to evaluate its contribution to O_3 formation potential (OFP) in ambient air. It is hoped that the findings will help in designing effective mitigation strategies to improve the ambient air quality in Malaysia by providing information about VOCs variation, composition and their contribution. A more comprehensive study is currently underway to integrate the research on variation of VOCs in ambient air and its impact on the environment and human health.

2. Methodology

2.1. Study area

As shown in Figure 1, three monitoring stations were selected in this study: Cheras (S1), Shah Alam (S2) and Seremban (S3) stations. S1 (N 03° 6' 22.44", E 101° 43' 4.5") and S2 (N 03° 6' 16.9", E101°33'22.3") are located in the Klang Valley basin, whereas S3 (N 02° 43' 24.1", E 101° 58' 6.5") is situated approximately 67 km from Kuala Lumpur. All cities in this study are categorised as tropical rainforest climate region: hot and humid throughout the year. The elevation height of each station varied between 53 m and 79 m above sea level. The monitoring stations were selected based on the traffic conditions, proximity to the congested highways or streets and geographical distribution. S1 station represents a residential–urban environment and is near major roads and highways. For S2 station, it is a rapidly urbanising area and the most developed state in Malaysia. Meanwhile, the S3 station is the capital state of Negeri Sembilan, situated near the schools and resembles a residential setting.

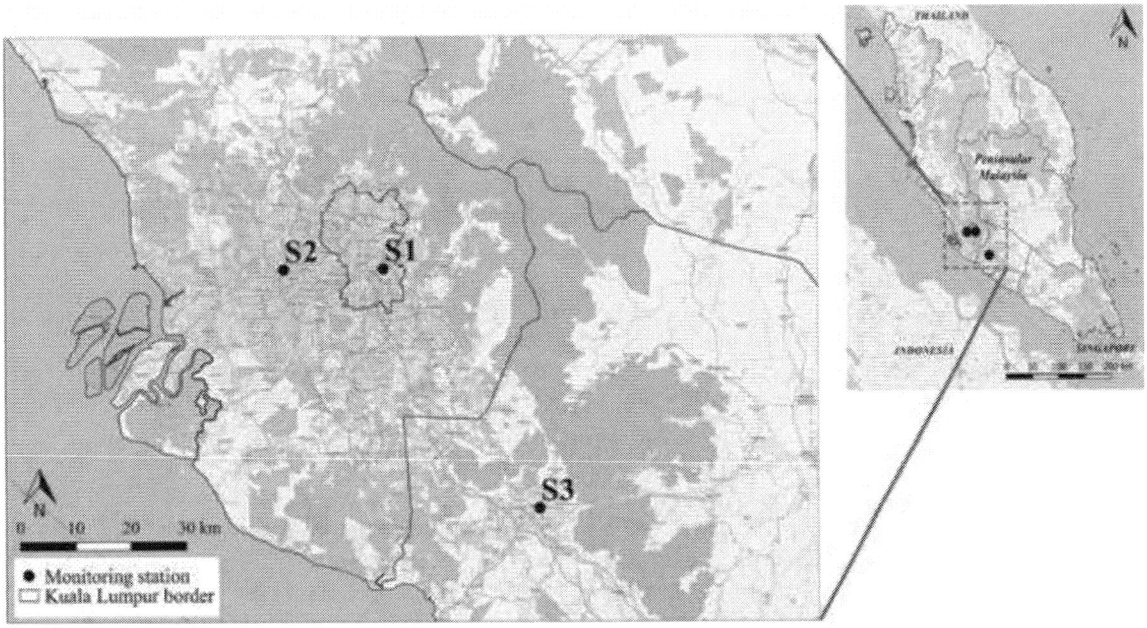

Figure 1. Map of study area showing S1 (Cheras), S2 (Shah Alam) and S3 (Seremban) as the monitoring stations.

2.2. Continuous sampling

The continuous sampling of aromatic VOCs and O_3 were carried out at three selected monitoring stations starting from January to December 2019. A compact online gas chromatography analyser (GC 5000BTX, AMA Instruments, Germany) is used to measure aromatic VOCs. The GC-5000BTX analyser is capable of capturing ambient aromatic VOCs concentration at a temporal resolution of 30 min. The device was installed with a capillary column AMASep (30 m x 0.32 mm ID, 1.5 μm film thickness) and a flame ionisation detector (FID) to detect VOCs parameters. In this study, five species of aromatic VOCs (benzene, toluene, ethylbenzene, m,p-xylene and o-xylene) were continuously collected on an hourly basis (local time). Prior to analysis, the temperature of the GC was set at 50°C for 3 min, 8° C/min to 130° C and was retained for 5 min. A volume of 30 mL of air will be pumped onto the enrichment tube and pre-concentrated the targeted VOCs before they are thermally desorbed into the GC system to detect BTEX species. Routine quality control (QC) was performed at single point calibration (10 ppb) of VOCs standard gas for the instrument validation performance.

2.3. Statistical data analyses

All statistical analyses were carried out using the Rstudio software - Openair package (version 2.8-3). In this study, the unit of the measurements was converted from part per billion (ppb) to micrograms per cubic metres ($\mu g\ m^{-3}$). The missing values (including data outliers) for each sampling station were estimated at 3.99 % (S3), 6.54 % (S2) and 10.33 % (S1). The statistical approach taken in this study omitted missing data values from the calculation to interpret the findings.

2.4. Ozone formation potential calculation

Briefly, calculating the O_3 formation potential (OFP) is one way to represent the relative contribution of VOC species towards the O_3 reactivity. The OFP was computed by multiplying the aromatic concentration with the maximum incremental reactivity (MIR). The MIR scale, comprehensively integrated by Carter [19], is a quantitative reactivity scale that expressed as the mass of additional O_3 formed per mass of VOC added to the emissions of various compounds under various atmospheric conditions. Herein, the updated coefficient value of the MIR scale is used to calculate the OFPs in this study [20, 21]. Every VOC species quantified a different MIR coefficient value; benzene had a value of 0.72 (g-O_3 g-VOCs-1), toluene had a value of 4.00 (g-O_3 g-$VOCs^{-1}$), ethylbenzene with 3.04 (g-O_3 g-$VOCs^{-1}$), m,p-xylene with 7.8 (g-O_3 g-$VOCs^{-1}$) and o-xylene with 7.64 (g-O_3 g-$VOCs^{-1}$). The O_3 potential can be calculated based on the following equation:

$$OFP\ (\mu g\ m^{-3}) = C_{VOC} \times MIR_i$$

Where;
C_{VOC}= Concentration of compound (µg m^{-3})
MIR_i= Maximum incremental reactivity coefficient value (g-O_3 g-$VOCs^{-1}$)

3. Results and Discussion

3.1. Descriptive data

Table 1 presents detailed results obtained from the preliminary analysis of BTEX concentration at the three monitoring stations. The annual average ΣBTEX concentration ranged from 14.53 ± 12.43 µg m^{-3} to 25.04 ± 24.04 µg m^{-3} during the observation period. Toluene was the most dominant species at all stations, with an average concentration of 10.65 ± 12.12 µg m^{-3}. In contrast, o-xylene had the lowest concentration at all stations with approximately 2.12 ± 2.02 µg m^{-3}. Moreover, the average level of benzene over the year was 3.12 ± 2.85 µg m^{-3}. The observed benzene in our study does not exceed the standard benzene value (5 µg m^{-3}) as suggested by the Central Pollution Control Board [22, 23]. Taken together, the ambient BTEX concentration was in the following order: o-xylene < ethylbenzene < benzene < m,p-xylene < toluene. A previous study in India demonstrated that the average concentration of BTEX was between 186 ± 107 µg m^{-3} and 456 ± 224 µg m^{-3}, with toluene accounting for 42-50 % and benzene accounting for 26-30 % of the total BTEX [24]. Other similar findings by Cruz et al. [25] stated that toluene was the most abundant species amongst the BTEX compounds with the average level ranging from 2.47 ± 1.49 µg m^{-3} to 3.18 ± 0.97 µg m^{-3} in the tropical urban area of Brazil. The high levels of toluene detected in ambient air can be explained by the fact that toluene is frequently added to gasoline to improve vehicle fuel efficiency. BTEX emissions in urban air are commonly caused by gasoline-powered automobile exhaust gas, evaporative emissions generated during the processing, storage, and distribution of gasoline and solvent usage. Intriguing findings by Hosaini et al. [26] identified the possible sources of anthropogenic VOCs in Kuala Lumpur city: motor vehicle exhaust or solvent (21-22 %), petrol pump or solvent usage (15 %) and industrial emissions (10 %).

Table 1. Annual average and standard deviation of BTEX concentration in µg m^{-3} (n=365).

Compound	S1	S2	S3	Average
Benzene	3.14 ± 3.07	3.28 ± 3.23	2.94 ± 2.24	3.12 ± 2.85
Toluene	12.08 ± 10.50	13.59 ± 20.88	6.27 ± 4.98	10.65 ± 12.12
Ethylbenzene	3.38 ± 3.49	3.64 ± 3.94	1.58 ± 1.56	2.87 ± 3.00
m,p-Xylene	3.93 ± 4.81	6.85 ± 8.18	2.43 ± 2.55	4.40 ± 5.18
o-Xylene	2.50 ± 2.16	2.57 ± 2.81	1.30 ± 1.09	2.12 ± 2.02
ΣBTEX	25.04 ± 24.04	29.93 ± 39.04	14.53 ± 12.43	23.17± 25.17

3.2. Monthly variation of BTEX

The variability of aromatic BTEX concentration for 12 months at the three monitoring stations is illustrated in Figure 2. The maximum average level of BTEX was reported at S2 station with a value of 61.12 µg m^{-3} in July and a minimum of 8.87 µg m^{-3} in February at S1 station. Contrary to expectations, the BTEX observed showed an unusual trend at the S2 station due to the influence of high levels of toluene and m,p-xylene compared to the other stations over the year. June-July undergoes dry season, which supposedly the BTEX concentration is at a minimum level during this month due to atmospheric dispersion, high intensity and temperatures [27]. This result may be explained by the fact that VOC can vary depending on monsoonal changes / seasonal differences, additional sources, relative reactivities, environmental conditions, etc. Another possible reason that caused the BTEX trend at S2 station was due to the characteristics of the studied areas, sampling points and sampling duration, city activities including emissions from industries, as well as differences in vehicle characteristics (eg. vehicle manufacturing quality and fuel composition). However, the study is limited by the lack of information on the monthly variation of VOC, particularly BTEX, in other tropical Asian cities.

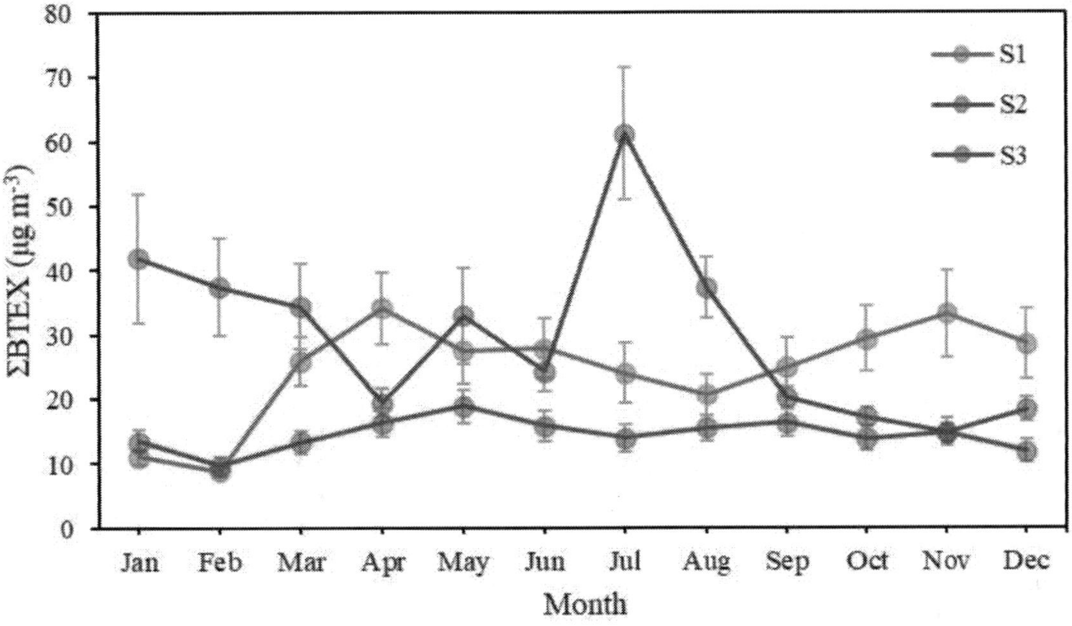

Figure 2. Monthly variation of BTEX levels according to each monitoring stations.

3.3. Ozone Formation Potential (OFP)

Based on Figure 3, the total OFP for the measured aromatic VOC at the sampling sites varied between 60.92 to 140.7 μg m^{-3}. World Health Organisation (WHO) has prescribed that the air quality guidelines for O_3 levels is below 120 μg m^{-3} [28, 29]. The leading contributors to O_3 formation during the sampling period are toluene, followed by m,p-xylene, o-xylene, ethylbenzene and benzene with the OFP value of 127.8 μg m^{-3} (40.90 %), 103.0 μg m^{-3} (32.97 %), 48.66 μg m^{-3} (15.58 %), 26.17 μg m^{-3} (8.38 %) and 6.74 μg m^{-3} (2.15 %), respectively. Although classified as the most toxic and carcinogenic species among C6-C9 aromatic VOCs [29-31], benzene had the lowest contribution to the O_3 formation mechanism throughout this research work. Overall, toluene and m,p-xylene significantly contributed to the level of O_3 in ambient air at all sampling points. These results were found to be comparable with a study conducted by Hosaini et al. [26] revealed that OFP in ambient Kuala Lumpur is ranked from benzene (6.50 μg m^{-3}) < ethylbenzene (32.8 μg m^{-3}) < toluene (135.9 μg m^{-3}) < o-xylene (139.9 μg m^{-3}) < m,p-xylene (145.8 μg m^{-3}). In Delhi, India, the total OFPs concentration was obtained at about 207.5 ± 123.4 μg m^{-3}, with toluene as the main contributor [32]. The major contribution of OFP in Hanoi, Vietnam [12] was the aromatic group (56.8 %): m,p-xylene (15.7 %), toluene (12.0 %), 1,2,4-trimethyl benzene (6.9 %), isopentane (6.3 %) and o-xylene (5.8 %).

The high OFP of VOCs clearly indicates the O_3 pollution issues in the future. For instance, previous research by [31] reported that aromatic VOC (59 %) dominated the high level of OFP and affected the local tropospheric O_3 in China, followed by alkenes and alkynes (26 %), alkanes (9 %) and carbonyl compounds (6 %). To sum up, the results of OFP indicated that aromatic VOCs significant impacts on surface O_3 in urban cities of Malaysia. It is well documented that O_3 in the lower troposphere is directly linked to its precursors, VOC. High ground-level O_3 were often seen in many hotspot cities, such as in Kuala Lumpur city. The O_3 episode events in Malaysia were mainly caused by enhanced precursors emission and meteorological conditions favourable to O_3 production (including higher temperature). This statement is proved with a recent case study by [33] revealed that the total concentration of O_3 episode days in Guanzhong Plain, China was significantly higher than non-episode days, with a value of 33.43 ± 13.64 ppb and 29.13 ± 14.31 ppb, respectively. In view of all that has been mentioned so far, determining the limiting factors of O_3 formation in the target regions is crucial for implementing regional O_3 reduction strategies.

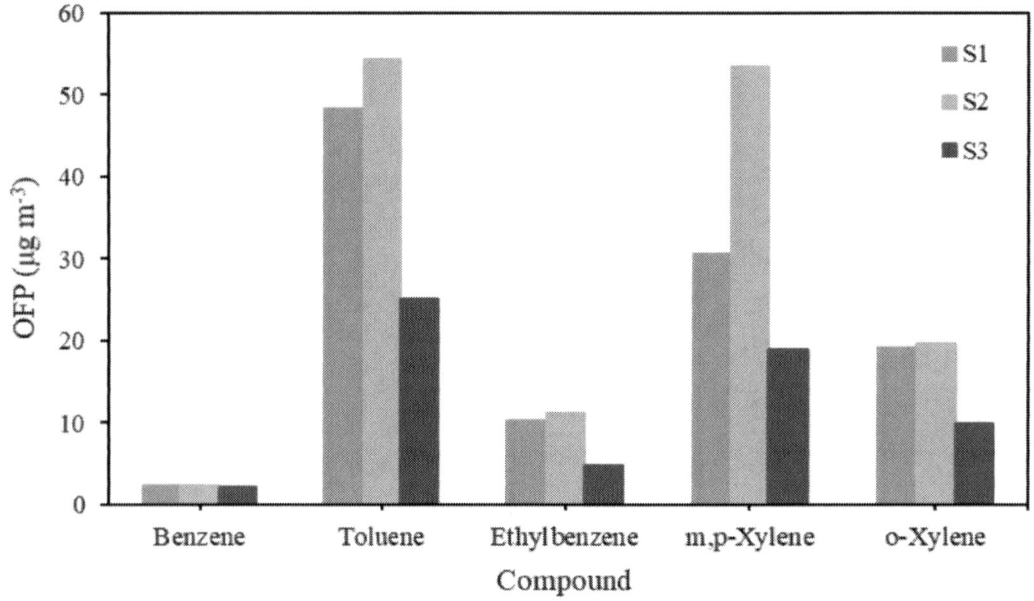

Figure 3. Aromatic VOC contributions to the total OFPs at S1, S2 and S3 stations.

4. Conclusion

Aromatic VOCs (BTEX) were continuously observed from January to December 2019. The present study was designed to determine the variation of aromatic VOCs in Malaysia's urban cities and their contribution to the O_3 formation mechanism. A notable finding from this study is that the annual average ΣBTEX concentration was approximately 23.17 ± 25.17 μg m^{-3}. All stations showed that toluene was the most dominant species with an average concentration of 10.65 ± 12.12 μg m^{-3}. The airborne BTEX concentration was in the following order: o-xylene < ethylbenzene < benzene < m,p-xylene < toluene. Additionally, toluene was also the highest contributors to O_3 formation in ambient air, followed by m,p-xylene, o-xylene, ethylbenzene and benzene. As VOCs play a crucial role in the O3 forming potential and photochemical oxidants associated with urban smog, it is important to determine the occurrence, sources, and behaviour of VOCs to establish effective air pollution abatement approaches. Therefore, these findings make several contributions to the current literature and provide new insight into the air quality level in Malaysia.

References

[1] Sillman S 1999 The relation between ozone, NOx and hydrocarbons in urban and polluted rural environments *Atmos. Environ.* **33** 1821–1845
[2] Poisson N, Kanakidou M, Bonsang B, Behmann T, Burrows J P, Fischer H, Gölz C, Harder H, Lewis A, Moortgat G K, Nunes T, Pio C A, Platt U, Sauer F, Schuster G, Seakins P, Senzig J, Seuwen R, Trapp D, Volz-Thomas A, Zenker T and Zitzelberger R 2001 The impact of natural non-methane hydrocarbon oxidation on the free radical and ozone budgets above a eucalyptus forest *Chemosphere - Glob. Chang. Biol.* **3** 353–366
[3] Tong H *et al* 2019 Radical formation by fine particulate matter associated with highly oxygenated molecules *Environ. Sci. Technol.* **53** 12506–12518
[4] Derwent R G, Jenkin M E, Saunders S M and Pilling M J 1998 Photochemical ozone creation potentials for organic compounds in northwest Europe calculated with a master chemical mechanism *Atmos. Environ.* **32** 2429-2441
[5] Tiwari V, Hanai Y and Masunaga S 2010 Ambient levels of volatile organic compounds in the vicinity of petrochemical industrial area of Yokohama, Japan *Air Qual. Atmos. Health* **3** 65–75
[6] Yadav R, Sahu L, Beig G and Jaaffrey S N A 2016 Role of long-range transport and local meteorology in seasonal variation of surface ozone and its precursors at an urban site of India *Atmos. Res.* 176–177
[7] Ohara T, Akimoto H, Kurokawa J, Horii N, Yamaji K, Yan X, *et al* 2007 An Asian emission inventory of anthropogenic emission sources for the period 1980–2020 *Atmos. Chem. Phys.* **7** 6843–6902
[8] Atkinson R and Arey J 2003 Gas-phase tropospheric chemistry of biogenic volatile organic compounds: a review *Atmos. Environ.* **37** 197–219
[9] Montero-Montoya R, Lopez-Vargas R and Arellano-Aguilar O 2018 Volatile organic compounds in air: sources, distribution, exposure and associated illnesses in children *Ann. Glob. Health* **84** 225–238
[10]Latif M T, Hamid H H A, Ahamad F, Khan M F, Nadzir M S M, Othman M *et al* 2019 BTEX compositions and its potential health impacts in Malaysia *Chemosphere* 2019 **237** 124451
[11]Pandey P and Yadav R 2018 A Review on Volatile organic compounds (VOCs) as environmental pollutants: fate and distribution *Int. J. Plant Environ* **4** 14-26
[12]Anchal G and Gupta N C 2018 A comprehensive study on spatio-temporal distribution, health risk assessment and ozone formation potential of BTEX emissions in ambient air of Delhi, India *Sci. Total Environ.* **659**
[13]Ran J, H Qiu, S Sun and L Tian 2018 Short-term effects of ambient benzene and TEX (toluene, ethylbenzene, and xylene combined) on cardiorespiratory mortality in Hong Kong *Environ. Int.* **117** 91–98
[14]Ly B T, Kajii Y, Nguyen T Y, Shoji K, Van D A, Do T N, Nghiem T D and Sakamoto Y 2020 Characteristics of roadside volatile organic compounds in an urban area dominated by gasoline vehicles, a case study in Hanoi *Chemosphere* **254** 126749
[15]Pandey P and Yadav R 2018 A review on volatile organic compounds (VOCs) as environmental pollutants: fate and distribution *Int. J. Plant Environ.* **4** 14–26
[16]Chin Y S J, De Pretto L, Thuppil V and Ashfold M J 2019 Public awareness and support for environmental protection-a focus on air pollution in peninsular Malaysia *PLoS One* **14** e0212206–e0212206
[17] Zhu X, Fan Z T, Wu X, Meng Q, Wang S W, Tang X *et al* 1994 Spatial variation of volatile organic compounds in a "Hot Spot" for air pollution *Atmos. Environ.* **42** 7329–7338
[18] Dehghani M, Fazlzadeh M, Sorooshian A, Tabatabaee H R, Miri M, Baghani A N *et al* 2018 Characteristics and health effects of BTEX in a hot spot for urban pollution *Ecotoxicol. Environ. Saf.* 155 133–143
[19] Carter W P L 1994 Development of ozone reactivity scales for volatile organic compounds *Air & Waste* **44** 881–99
[20]Carter W 2010 Updated maximum incremental reactivity scale and hydrocarbon bin reactivities for regulatory applications *California Air Resources Board* **1** 07–339
[21]Venecek M A, Carter W P L and Kleeman M J 2018 Updating the SAPRC Maximum incremental reactivity (MIR) scale for the United States from 1988 to 2010 2018 *J. Air Waste Manag. Assoc.* **12** 1301–1316
[22]Central Pollution Control Board 2009 *National Ambient Air Quality Standards*
[23]Sekar A, Varghese G K and Ravi Varma M K 2019 Analysis of benzene air quality standards, monitoring methods and concentrations in indoor and outdoor environment *Heliyon* **5** 02918
[24]Saxena P and Ghosh C 2012 A review of assessment of benzene, toluene, ethylbenzene and xylene (BTEX) concentration in urban atmosphere of Delhi *Int. J.Phys. Sci.* **7** 850–860
[25]Cruz L P S, Santos D F, dos Santos I F, Gomes Í V S, Santos A V S and Souza K S P P 2020 Exploratory analysis of the atmospheric levels of BTEX, criteria air pollutants and meteorological parameters in a tropical urban area in Northeastern Brazil *Microchem. J.* **152** 104265
[26]Hosaini P N, Khan M F, Mustaffa N I H, Amil N, Mohamad N, Jaafar S A *et al* 2017 Concentration and source apportionment of volatile organic compounds (VOCs) in the ambient air of Kuala Lumpur, Malaysia *Nat. Hazards* **85** 437–452
[27]Rad H D, Babaei A A, Goudarzi G, Angali K A, Ramezani Z and Mohammadi M M 2014 Levels and sources of BTEX in ambient air of Ahvaz metropolitan city *Air Qual. Atmos. Health* **7** 515–524
[28]World Health Organization 2006 *Air Quality Guidelines: Global Update 2005: Particulate Matter, Ozone, Nitrogen Dioxide, and Sulfur Dioxide*

[29]Hajizadeh Y, Mokhtari M, Faraji M, Mohammadi A, Nemati S, Ghanbari R, Abdolahnejad A, Fard A R F, Nikoonahad A, Jafari N and Miri M 2018 Trends of BTEX in the central urban area of Iran: A preliminary study of photochemical ozone pollution and health risk assessment *Atmos. Pollut. Res.* **9** 220–229
[30]Duerta-Davidson R, Courage C, Rushton L and Levy L 2001 Benzene in the environment: an assessment of the potential risks to health of the population *Occup. Environ. Med.* **58** 2–13
[31]Ran J, Qiu H, Sun S, Tian L, 2018 Short-term effects of ambient benzene and TEX (toluene, ethylbenzene, and xylene combined) on cardiorespiratory mortality in Hong Kong *Environ. Int.* **117** 91–98
[32]Skov H, Lindskog A, Palmgren F and Christensen C S 2001 An overview of commonly used methods for measuring benzene in ambient air *Atmos. Environ.* **35** S141–S148
[33]Hui L, Ma T, Gao Z, Gao J, Wang Z, Xue L, Liu H and Liu J 2021 Characteristics and sources of volatile organic compounds during high ozone episodes: a case study at a site in the eastern Guanzhong Plain, China *Chemosphere* **265** 129072

Acknowledgement
This research was supported by the Universiti Kebangsaan Malaysia (UKM) research grant DIP-2019-006 (Dana Impak Perdana) and in collaboration with the Department of Environment Malaysia (DOE).

Evaluation of regional simulation of surface ozone over Southeast Asia using ground-based observation at two existing Global Atmospheric Watch stations

Didin Agustian Permadi[1*] and Nguyen Thi Kim Oanh[2]

[1]Environmental Engineering, Institut Teknologi Nasional Bandung (ITENAS), Indonesia
[2]Environmental Engineering and Management, Asian Institute of Technology (AIT), Pathumthani, Thailand

*Corresponding E-mail: didin.agustian@gmail.com

Abstract. High level of ground level ozone concentrations was found in most of Southeast Asian (SEA) large cities and often exceeded the national ambient air quality standard. Ozone and PM_{10} are among of the critical air quality parameters that cause the unhealthy air quality index. Effort to mitigate ozone pollution is greatly complicated due to the photochemistry processes therefore photochemical smog modelling has been widely used. Surface ozone simulation in SEA was done using CHIMERE and weather research forecast (WRF) model. Emission inventory of ozone precursors was done for three countries in the domain, i.e. Indonesia, Thailand and Cambodia. Modelling performance evaluation for meteorological parameters and ozone at the SEA big cities was done in another study. This paper focused on the model evaluation conducted at the two remote sites represented by 2 (two) global atmospheric watch (GAW) remote stations of Bukit Kototabang (BKT) and Danum Valley (DNV). Evaluation result showed an overestimation of observed ozone in BKT while a contradictive result was seen in DNV station which was due to the ozone chemistry and inaccurate estimation of emissions (both anthropogenic and biogenic emission). The evaluation conducted at the remote sites was not even better than that conducted previously at the urban areas. Statistically, only mean normalized gross error and unpaired peak accuracy values that satisfy the criteria for surface ozone modelling suggesting major improvement required for ozone precursors emission inventory data.

Keywords: Ozone, GAW, WRF/Chimere, model performance, precursor

Track Name: Atmospheric Chemistry and Physics

1. Introduction

Surface ozone is a secondary pollutant that is known to cause detrimental effects on human health and reduction of agricultural yield [1]. Tropospheric ozone burden increased by 40 % from 1850 to the present day due to anthropogenic activities. Global satellite monitoring showed an increasing trend of ozone levels during the last decade [2]. Rapid development in Asia has been connected to the increase of surface ozone levels especially in the big cities which was affected by the local emission build up as well the favourable meteorological conditions [3]. Photochemical smog pollution abatement is greatly complicated in reducing NO_x or VOC due to the non-linear process in the photochemistry. In addition to the good data of monitoring, photochemical smog modelling has been widely used to understand the formation of ozone in an area as well as to simulate effects of precursor emission reduction on surface ozone concentration [4].

For Southeast Asia (SEA), limited studies are available for regional ozone simulation but mainly used the global emission inventory data with the coarse resolution. In addition, the maritime region of Indonesia was not included in the study domain. In this study, new emission inventory (EI) for three countries in SEA namely Thailand, Indonesia and Cambodia were developed [5]. Regional simulation of surface ozone was conducted for 1 year (2007) for SEA domain utilizing the newly developed EI. The evaluation of meteorological model performance and surface ozone modelling at the several SEA urban sites using the existing governmental air quality monitoring networks was done previously [6] suggesting a fair performance. This paper attempts to fill in gaps in evaluating regional model performance at the remote sites. Two existing Global Atmospheric Watch (GAW) stations at Bukit Kotatabang (BKT), Indonesia and Danum Valley (DNV), Malaysia were taken.

2. Methodology

Emission inventory (EI) is a foremost step prior to modelling task. The data for SEA were taken from our previous works including the emission of biomass open burning and fossil fuel combustion [5]. Estimation followed the framework of Atmospheric Brown Cloud Emission Inventory Manual (ABC EIM). Biogenic emission of isoprene, α-pinene, β-pinene, limonene, ocimene, and NO were calculated in the Model of Emissions of Gases and Aerosols from Nature (MEGAN) module provided in CHIMERE model. Data sources of EI and the coverage including species and sources are presented in Table 1. Initially, VOC speciation for anthropogenic emission was done using the available VOC source profile data (presented in Table 1).

Table 1. Emission data sources for the rest of countries in the domain for base year of 2007 (except Indonesia, Thailand, and Cambodia).

Sources	Species included									VOC Profile[a]
	SO_2	NOx	CO	VOC	PM_{10}	$PM_{2.5}$	BC	OC	CH_4	
Power generation	CGRER[a] [13]								EDGAR[b] [14]	[7], [12]
Industry										[7], [8], [11]
Residential										[7], [12]
Transportation										[7], [10]
Biomass open burning	Song *et al.* [15] and GFED3[c] [16]									[7], [9]
Biogenic emission	Calculated by MEGAN model in CHIMERE for 2007									Directly from MEGAN

Note: [a] – Center for Global and Regional Environmental Research, [b] - The Emissions Database for Global Atmospheric Research, and [c] – The Global Fire Emission Database.

For CHIMERE model, speciated VOC emission was initially converted to Statewide Air Pollution Research Center (SAPRC) using the methodology described in [17]. Further, the SAPRC speciated VOC

IOP Conf. Series: Earth and Environmental Science **880** (2021) 012006 doi:10.1088/1755-1315/880/1/012006

data were then converted to MELCHIOR (CHIMERE) using aggregation factors from [18] for totally 33 species. Ready input emission data were then prepared in netcdf format prior to CHIMERE simulation.

Simulation was done for 1 (one) year starting from 1st January – 31st December 2007. For this study monthly mean concentration of gases and aerosols for period of 1998-2002 for boundary conditions were taken from the global model simulation of LMDz-INCA while the initial condition used interpolation of global concentration fields from the same global model. Detail model set-up for both CHIMERE and WRF is presented in [5].

Bukit Kotatabang GAW station (BKT) is located in the Western part of Sumatra at 864 m above mean sea level (AMSL). This station is centralized at 0° 12′ 07″ (latitude) and 100° 19′ 05″ (longitude) which is surrounded by the tropical forest (see Figure 1). DNV is located 426 m AMSL which is centralized at the latitude of 4° 58‘ 53" North and the longitude of 117° 50‘ 37" East (see Figure 1). This site is situated within the 973 km^2 forest reserved areas managed by the Sabah Foundation.

For model evaluation and comparison at BKT, we selected the period of January-March and August-October 2007 to represent rainy and dry season, respectively. However, due to observation data completeness only January-March period was considered for DNV station. Statistical measures calculated were Mean Normalized Gross Error (MNGE), Unpaired Peak Accuracy (UPA), and Mean Normalized Bias Error (MNBE) with the criteria suggested by [14]. The definitions of all statistical measures used for the evaluation are presented in table 2 below.

Table 2. Definition of the evaluation statistical measures for ozone simulation.

Parameters	Formula	Suggested criteria
MNGE	$NGE = \frac{1}{N}\sum_{i=1}^{N}\frac{\lvert M_i - O_i\rvert}{O_i} 100$	Ozone: ≤ +35 %
UPA	$UPA = \frac{M_{max} - O_{max}}{O_{max}} 100$	Ozone: ≤ ±20 %
MNBE	$MNBE = \frac{1}{N}\sum_{i=1}^{N}\frac{M_i - O_i}{O_i}$	Ozone: ≤ ±15 %

Notes:
M-modelled value (model outputs for the first layer in different grids); O-observations; $\bar{O}$: mean- observation; N-number of observations.
Source: [18].

3. Results and Discussions

Figure 1 presents the spatial distribution of monthly average ozone of selected months (Jan, March, July and August) over the modelling domain for the year of 2007. In March, higher concentration of ozone was seen over the higher altitude as well as in August. In January and July, higher concentrations of ozone were seen in the lower altitude which seems to be affected by the seasonal weather pattern in the region.

Highest monthly average concentrations found in the domain for January, March and September were 66, 61 and 48 ppb. The maximum hourly concentrations for all months in the year ranged from 67-154 ppb. In January, ozone concentrations were seen high at Surabaya and Hanoi while in other months was seen also in other places such as South Sumatera, West Kalimantan and North Malaysia. Urban plumes from the SEA big cities were also well captured through the model simulation results. Northeast and Southwest monsoon patterns seemed to affect the ozone plume movement in the SEA. In January at the upper latitude, NE monsoon brought pollutants from the origin sources to the southwest direction while at the lower latitude (Indonesia) the plume moved to NE/E direction. In August and November at

IOP Conf. Series: Earth and Environmental Science **880** (2021) 012006 doi:10.1088/1755-1315/880/1/012006

the lower latitude the plume moved NW/W direction while in the upper latitude SW monsoon brought the pollutants to the NE direction.

Ozone concentrations at remote sites of the GAW station at two stations (BKT, Indonesia and DNV, Malaysia) were extracted from the model output and were included in the comparison. The scatter plots of hourly simulated ozone concentrations as compared to the measurements at the two stations are presented in figure 2. At BKT, the correlation coefficient derived from the hourly data presented in the scatter plot was 0.443 while at DNV it was slightly higher of 0.497. Even though better correlation coefficient was seen in DNV (maybe due to shorter period of data), evaluation the two stations showed somehow a contradictive result. Ozone simulation seems to be underestimated in DNV while overestimation was seen in BKT. Inaccurate EI especially the local emission sources might affect these discrepancies for both anthropogenic and biogenic emissions. In addition, complex photochemistry in the region especially under convective dominated meteorology could not be well captured by the models.

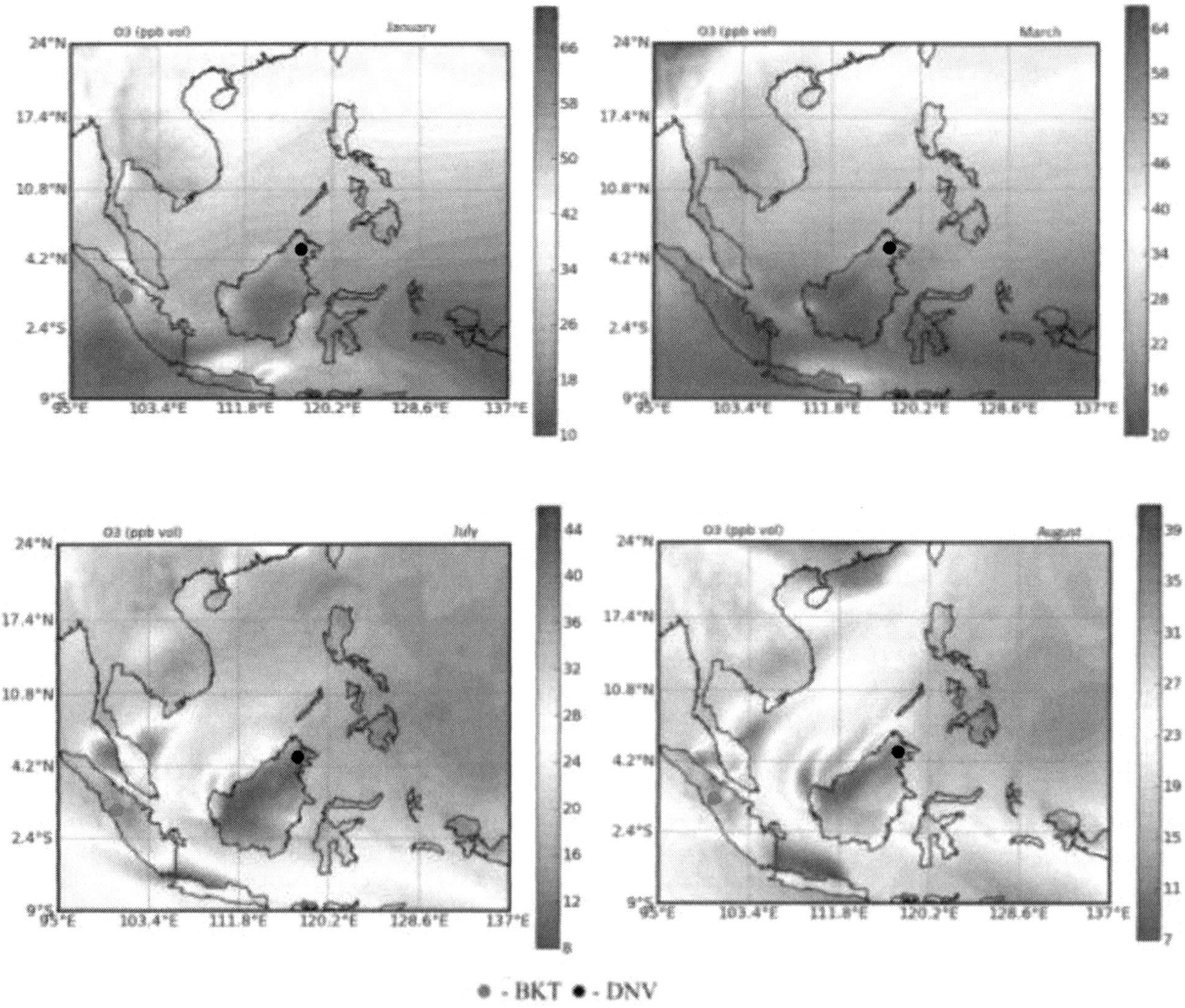

Figure 1. Spatial distribution of ozone for the month January, March, July and August 2007.

The next step is to calculate the statistical measures for model performance evaluation (presented in table 3) and the values were then compared to the suggested criteria. At BKT, two parameters (MNGE and UPA) satisfied the criteria while in DNV, only UPA satisfied the criteria. MNBE values at the two sites were well above the suggested criteria. Regional ozone simulation at the remote sites normally performs better than in the urban areas but perhaps inaccurate estimation of both biomass open burning and biogenic emission in Sumatra and Kalimantan would affect this moderate performance.

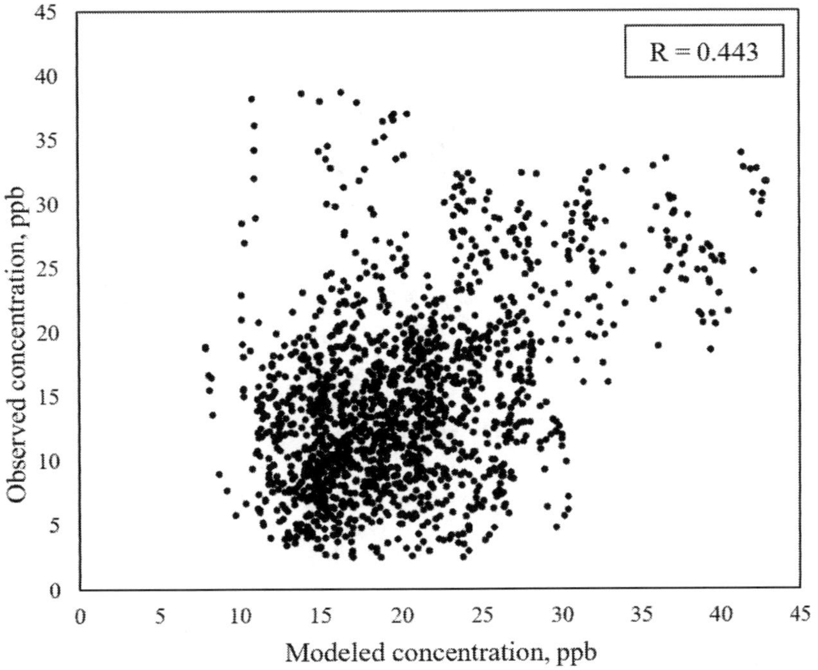

a) BKT station, Indonesia (period of Jan-March and Aug-Sept)

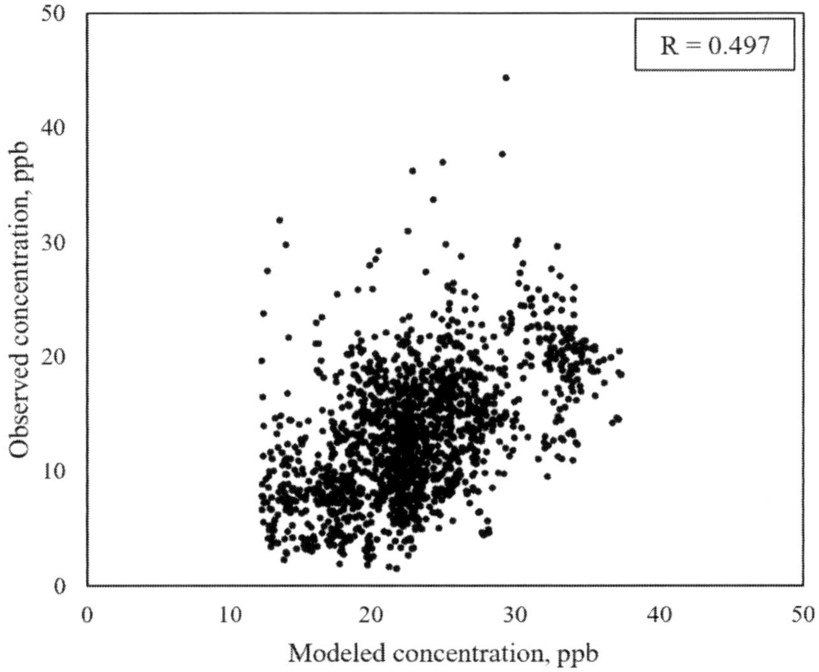

b) DNV station, Malaysia (period of Jan – March)

Figure 2. Scatter plots of modelled vs observed ozone concentrations at two GAW stations.

Table 3. Statistical measures analyzed for comparison between model output and observation.

No	Station name	Statistical measures		
		MNGE (%)	UPA (%)	MNBE (%)
1	BKT station, Sumatera, Indonesia	**34.9**	**14.7**	36.7
2	DNV station, Sarawak, Malaysia	52.3	**-15**	40.2

Note: Bolded values satisfy the criteria below
Criteria for ozone simulation [18]:
MNGE: ≤ +35%, UPA: ≤ ±20 %, and MNBE: ≤ ±15 %.

4. Conclusion

Photochemical smog modelling framework of WRF-CHIMERE was applied to simulate surface ozone over SEA. Complete modelling chain starting from emission inventory compilation for ozone precursor, VOC speciation, temporal and spatial distribution was done to prepare model ready input data. Simulated ozone over the higher altitude showed higher concentrations in Jan-Apr than those simulated during the period of May-Aug. Seasonal variation of ozone was successfully simulated that was affected by monthly variation of ozone precursor emission and meteorology. Monsoon circulation affected the ozone plume movement in the domain. In January at the upper latitude, NE monsoon brought pollutants from the origin sources to the southwest direction while in August SW monsoon brought the pollutants to the NE direction. In the same months at the lower latitude (Indonesia) the plume moved to NE/E direction while in August the plume moved NW/W direction. High ozone concentrations over the year were simulated over the equatorial zone. Modeling results were further evaluated using the ground observation available at the two GAW stations in BKT and DNV station. Model results seemed to overestimate the observation in BKT and a contradictive result was seen in DNV. However, at least two statistical parameters of MNGE and UPA satisfied the criteria for ozone simulation thus suggesting a moderate model performance as also seen in other big cities. Satellite data that observes the total ozone column can be used to evaluate the model performance especially for the spatial distribution. There are still rooms for improvement such as emission inventory updates, VOC speciation, temporal and spatial distribution of ozone precursor emissions.

References

[1] Young P J *et al* 2013 *Atmospheric Chemistry and Physics* **13** 203-260.
[2] Cooper O and Ziemke J 2015 *Bull. Amer. Meteor. Soc.*, in-preparation
[3] Granier C Bessagnet B Bond T *et al* 2011 *Climatic Change* **109** 163-190.
[4] Kim Oanh N T and Permadi D A 2009 *ILM Publications*, a trading division of ILM Holdings, Ltd.
[5] Permadi D A *et al* 2018 *Atmospheric Chemistry and Physics*. **18** 2725-2747.
[6] Permadi D A 2013 *Doctoral Dissertation*, Asian Institute of Technology (AIT), Thailand.
[7] Suwattiga P and Limpaseni W 2005 *Science Asia* **31** 395–401.
[8] Zhang B 2002 *Doctoral Dissertation* Asian Institute of Technology (AIT) Thailand.
[9] Andreae M O and Merlet P 2001 *Global Biogeochem. Cycles* **15** 955– 966.
[10] Ho S S H *et al* 2012 *Journal of the Air & Waste Management Association* **62** 221-234
[11] Yuan B Shao Min Lu and Wang B 2010 *Atmospheric Environment* **44** 1919-1926.
[12] Battye W and Harris J 2005 *Technical report: Improving Modeling Inventory Data: Speciation Profiles*, EC/R Incorporated, Chapel-Hill.
[13] Zhang Q *et al* 2009 *Atmos. Chem. Phys*. 9 5131–5153.
[14] Olivier J G 2002 Wilco BV Amersfoort, the Netherlands 58–88
[15] Song Y *et al* 2009 *Environ. Res. Lett.* **5** 014014
[16] Giglio L *et al* 2006 *Atmospheric Chemistry and Physics* **6** 11091–11141.
[17] Carter W P L 2008 *2nd International Conference on Atmospheric Chemical Mechanisms* Davis CA December 9-12
[18] Middleton P *et al* 1990 *Atmospheric Environment* **24** 1107–1133.
[19] Emery C A *et al* 2001 *Technical Report*: Retrieved from http://www.tceq.state.tx.us/assets/public/implementation/air/am/contracts/reports/mm/EnhancedMetModelingAndPerformanceEvaluation.pdf

Acknowledgments

Authors acknowledged the French Ministry of Foreign Affairs for the funding support through the SDCC-AIT/France partnerships. Atmospheric Brown Cloud Emission Inventory project team (AIT RRCAP) is also acknowledged to provide emission input data for the simulation.

Subsurface geological model of Malay basin using free air anomaly

Chia Jing Hwee[1], Farouq Jamil[1], Akhmal Sidek[1*], Zaidi Jaafar[1], Radzuan Junin[1], Shaziera Omar[1], Fadzlin Hasani[2], Aizuddin Supee[3] and Mohd Hariri Arifin[4]

[1] Petroleum Engineering Department, School of Chemical and Energy Engineering, Faculty of Engineering, Universiti Teknologi Malaysia, 81310 Johor Bahru, Johor.
[2] PETRONAS Research Sdn. Bhd., 43000 Bandar Baru Bangi, Selangor.
[3] Energy Management Group, School of Chemical and Energy Engineering, Faculty of Engineering, Universiti Teknologi Malaysia, 81310 Johor Bahru, Johor.
[4] Program Geologi, Pusat Sains Bumi dan Alam Sekitar, Fakulti Sains dan Teknologi, Universiti Kebangsaan Malaysia, 43600 Bangi, Selangor.

*Corresponding E-mail: akhmalsidek@utm.my

Abstract. The aim of gravity survey is to assist in the detection and delineation of subsurface geological features such as salt domes and faults. In this study, free air anomaly (FAA) data was adopted for mapping and modelling process to delineate subsurface geological features and basement depth in Malay Basin. FAA is the measured gravity anomaly after a free air correction is applied, and it is used for elevation correction. The data of FAA in this study is obtained from Earth Gravitational Model (EGM) 2008 released by the National Geospatial-Intelligence Agency (NGA)-EGM Development Team. Oasis Montaj software was used in the mapping and modelling process whereby the base map which constructed by the Oasis Montaj is used to form the FAA map of Malay Basin. Typically, the positive anomaly is associated with the high-density intrusion at the base of the crust, while in contrast (negative anomaly), it is related to the sedimentary basin in the upper crust. On top of that, the regional-residual anomaly, total horizontal derivative (THD) and 3D Euler Deconvolution enhanced maps were produced and interpreted to acquire comprehensive insight of subsurface geological features. To conclude, this study showed 5 % deviation as compared to previous reported works and the deepest basement depth encountered is 14.5 km.

Keywords: Free Air Anomaly, Malay Basin, 2D Gravity Model, 3D Euler Deconvolution

Track Name: Atmospheric Chemistry and Physics

1. Introduction

Applications of gravity method in oil and gas exploration are essential and beneficial whereby this method could significantly enhance any targets exploration which possess the density contrast at different depth such as salt domes, ore bodies, structure, and regional geology [1]. In addition, the basement rock depth data in the oil and gas industry could also be used to determine the depth of the basin as well as the maximum well drilling depth [2]. Generally, gravity method passively measures natural variation of the earth's gravity over a map area and correlate them with the variation of geological subsurface features. This method basically offers advantages such as relatively economical, non-invasive, non-destructive remote sensing and at the same time, capable in providing information related to the densities of subsurface rocks [3,4]. Fundamentally, gravity anomaly can be defined as small deviation from the average gravity force due to the density variation of subsurface rock. The average gravity force or commonly called as the standard gravity possess the value of 9.80665 m s^{-2} (equivalent to 980665 mGal) and it is measured as free fall acceleration of an object at above sea level [5]. There are several methods of acquisition in obtaining gravity anomaly which include airborne, shipborne, and inland. For this study, free air gravity anomaly or commonly named as free air anomaly (FAA) is used in the mapping and modelling, in which consider free air correction for the measured gravity anomaly for elevation correction [6] and the unit used to measure the FAA is mGal whereby 1 mGal is one over thousands of 1 cm s^{-2}.

The implementation of gravity method (FAA) in the oil and gas industry basically enhances the exploration of any targets specifically the targets with density contrast at different depth such as salt domes, ore bodies, structure, and regional geology [7]. This is due to the fact that the integration of gravimeter instrumentation provides more accurate FAA data as compared to the solely airborne (satellite) and marine acquisitions. Common practice exhibits that the FAA method is often applied in oil exploration involving salt due to the huge density difference of salt with the surrounding sediment, whereby it yields in positive value when in shallow and negative value when in deep [8]. On top of that, this FAA also capable to provide the information related to the densities of subsurface rocks. Inference about the distribution of underground strata can be made using the concept of different rock types possess different ranges of densities [9]. Furthermore, the FAA also could determine the faults as the faults often juxtapose rocks of differing density.

To our knowledge, previous scholars only reported the FAA on Malay Basin using the ERS-1 satellite data [10] and no FAA works has been performed on Malay Basin using the latest Earth Gravitational Model 2008 (EGM 2008) satellite data. Thus, it is essential to investigate the differences between them in terms of geological features and basement depth. For that purpose, the qualitative and quantitative interpretations of FAA data on Malay Basin using the EGM 2008 data were performed using Oasis Montaj software. Qualitative interpretations include regional anomaly, residual anomaly, total horizontal derivative (THD) and Euler Deconvolution, while quantitative interpretation include the 2D gravity crust model.

2. Geological Setting

Generally, the Malay Basin in Figure 1 (grey shaded area) is located at the North Peninsular Malaysia (completely offshore), South Cambodia, and Vietnam, and extend across the Gulf of Thailand and the South China Sea. The dimension of this basin is approximately 500 km long and 200 km width, and the total area is about 83,000 km^2 with broad range of FAA value is about -20 mGal to -30 mGal [11]. Arrows along the faults in Figure 1 (a) shows the initial movement on the Dungun Faults, Hinge-line Fault, and Tenggol Faults. Each dextral slips associated with the faults are due to its slight wrench depression [12]. The slip was reverted during the Late Oligocene to the Early of Miocene due to transpressional stress regime and compressional structures formation. The E-W trending anticlines (anticline that contain most hydrocarbons) was relocated up to 35 km as the Malay Basin is cross by major N-S faults [13]. Nonetheless, Paleogene wrench together with left-lateral movement of 30 to 35 km fault zone (also represents the axis of basin) form the E-W inclination of the large folds with Moho depth in Figure 1(b) is approximately ranging from 29 to 32 km [14].

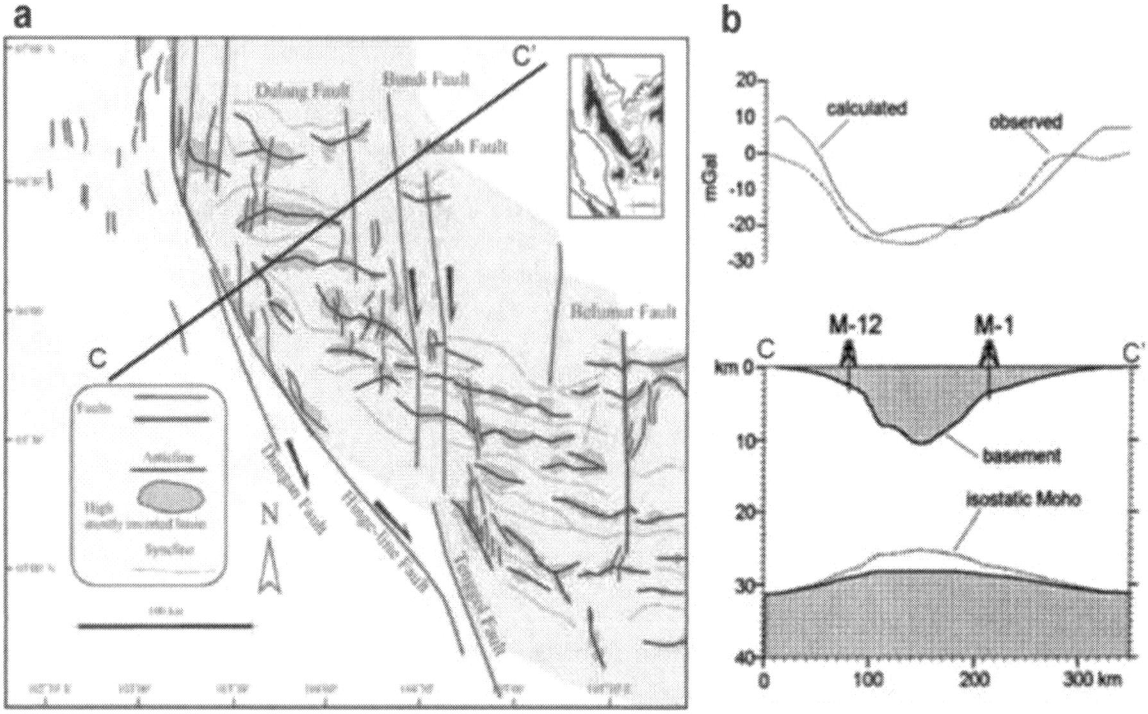

Figure 1. (a) Major structure of Malay Basin (after Tjia,[12]), (b) Gravity profile estimated Moho depth across northern part of Malay Basin (after Madon [22]).

3. Methodology

3.1. FAA Input Data

The FAA input data used in this study is the satellite airborne data obtained from EGM 2008: publicly published by the National Geospatial-Intelligence Agency (NGA). This gravitational model is basically complete up to spherical harmonic degree and order 2159,and contains additional coefficients extending to degree 2190 and order 2159 [15]. The data version used in this study is WGS84 (Version of EGM 2008 released in July, 2008) which includes the grids and programs for computing geoid undulations relative to WGS84 Ellipsoid.

3.2. FAA Extraction

Google Earth software was used to set the area of study whereby four place marks were added within the area of Malay Basin. The airborne path was generated within the area of investigation in Google Earth in order to obtain the FAA value corresponding to the coordinates. The FAA data (anomaly data) in the Google Earth are shown in different colours. The FAA value for each points of the airborne path can be obtained through the colour as each colour range represents different values. Next, the coordinate of each point of the airborne path are extracted by the airborne path tab and then saved in kml format before converted to Excel using the TCX Converter. These coordinates are basically in decimal degree form of ellipsoid system for longitude and latitude in Global Positioning System (GPS, WGS84). The extracted coordinates in the form of GPS system are then converted to Universal Transverse Mercator (UTM) due to the requirement of Oasis Montaj software, whereby the coordinate grid data must be in the form of unit meter, projection of UTM [16].

3.3. Qualitative Interpretation

In Oasis Montaj, the grid of FAA data was constructed using the minimum curvature technique. Base map was built corresponding to all the data coordinates. The grid then was imported to the base map to form the FAA map. The regional anomaly map was used to determine the deep crust especially the lower crust depth. The component of the long wavelength in the gravity field is contributed by the depth of body which exceeded the limit of exploration of hydrocarbon and minerals [17]. The regional anomaly map was produced by applying a low pass filter on the FAA map. The filtering was carried out by MAGMAP in Oasis Montaj through MAGMAP 1-step filtering. The aim of this filter is to observe the cut off wavelength at deeper area. The cut off wavelength basically eliminates all the wavelength which possess less value than the cut off value. Residual anomaly is the anomaly which represent the short wavelength and related to the presence of shallow structure. It is formed by eliminating the long wavelength in the isotactic compensation effect from the source of different boundaries between the mantle and earth's crust [18]. Principally, the residual anomaly is capable to observe the clearer local anomaly which could be associated with the shallow structures. In this study, the residual anomaly map was constructed using the regional anomaly map with low pass filter of MAGMAP in Oasis Montaj through MAGMAP 1-step filtering. Total horizontal derivative (THD) was used to increase the delineation of the boundary based on density from the gravity data, and it estimates the drastic changes of density in lateral.

THD are generally has low susceptibility to noise in the data and possess robust delineation either in shallow or deep situation. Horizontal gradient filter was applied to produce dx and dy grids. THD map then were produced from dx and dy grids by using the grid math. The relationship between the dx, dy and THD map is shown in Equation 1 [16]:

$$G_0 = \text{sqrt}\left((g_1 * g_1) + (g_2 * g_2)\right) \quad \text{(Equation 1)}$$

Where G_0 represents the THD, g_1 represents the dx grid and g_2 represents the dy grid.

The 3D Euler Deconvolution was applied to estimate the mineral depth in 3D map by using the Structural Index (SI) input as listed in table 1 and suitable window size [19]. Briefly, the generated Euler deconvolution map illustrates the boundaries for geology features based on the difference in gravity and estimate the anomaly depth or fault location in measured area. Conceptually, this SI is necessary to form the required result, while for the window size, it is required for the Euler deconvolution map construction. Window size used for analysis and interpretation should be the optimum size (not too big or too small) whereby it requires all the gravity value need to be included. In this study, seven window size were used which were 3x3, 5x5, 7x7, 9x9, 10x10, 13x13 and 15x15. The reason for the variation of window size was due to the fact that the small window size is capable to detect the shallow anomaly while the bigger size is capable to detect deeper area anomaly. On the other side, the 2D Crust Model was constructed using the GM-SYS method for subsurface structure illustration, and that model was basically based on the value of gravity data with the average density of basement for the rock is set to 2.76 g cm^{-3}, seawater-1.03 g cm^{-3} and average sedimentary rock-2.65 g cm^{-3} [20].

Table 1. Standard Euler Deconvolution (Structural Index).

SI	Gravity Field
0	Sill / Dyke / Ribbon / Step
1	Cylinder / Pipe
2	Sphere
3	NA

IOP Conf. Series: Earth and Environmental Science **880** (2021) 012007 doi:10.1088/1755-1315/880/1/012007

4. Results and Discussion

In this study, the analysis and interpretation of the Malay Basin has been made on the FAA map as shown in Figure 2. The qualitative enhanced map including the regional anomaly, residual anomaly, THD, and 3D Euler Deconvolution has been constructed using the Oasis Montaj software. These enhanced maps were then further analyzed and interpreted to estimate and to understand the major structure and faults of the Malay Basin. In addition, the 2D modelling has also been done by GM-SYS in Oasis Montaj for Malay Basin basement depth estimation.

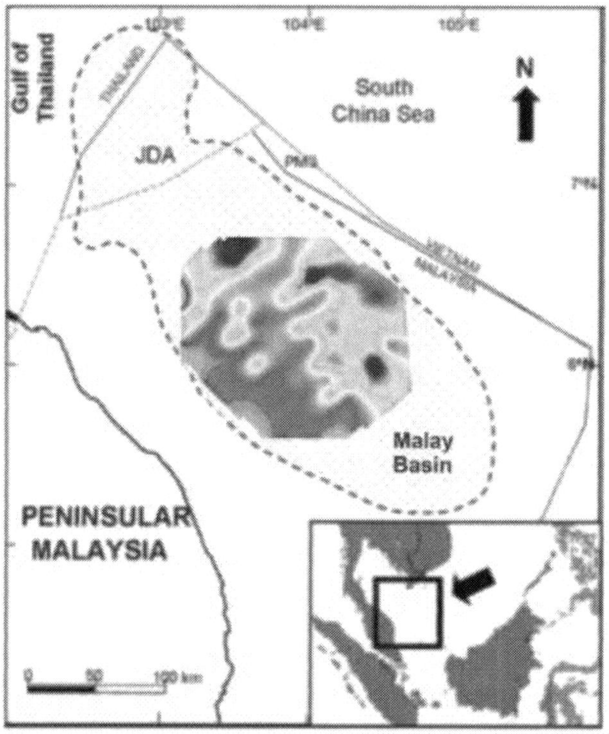

Figure 2. FAA map display in central part of Malay Basin.

4.1. Regional and Residual Anomaly Map

To determine the accurate cut off wavelength, several regional anomaly maps with different cut off wavelength values were constructed until the contour of the anomaly map became smooth. The cut off wavelength with the smooth contour line in the map will be the value of cut off wavelength for regional anomaly map. The value of cut off wavelength was determined as 125000 ground unit since the maps showed no changes from regional anomaly map with the cut off wavelength at 125000 (Figure 3a) and onwards. Hence, the cut off wavelength of 125000 is used for interpretation of residual anomaly within the study area (Figure 3b).

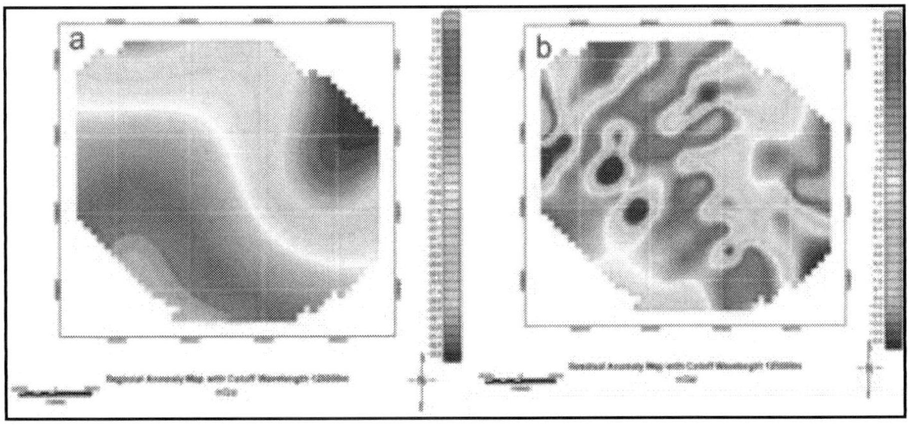

Figure 3. Regional anomaly map with (a) cut off wavelength at 125000 m and (b) residual anomaly map after cut off wavelength at 125000 m.

4.2. *THD and Euler Deconvolution Map*

Three THD maps were constructed which named as THD from FAA map (Figure 4a), THD from THD from residual anomaly map (Figure 4b). By observing the THD map generated from the FAA map both maps display several lineaments have similar dominant trend of anomaly reading in the direction to the reference lineament provided by Tjia [12]. These results are in agreement with the previous works same dominant trend in the central axis of Malay Basin which trending from northwest (NW) to southeast (SE)) [12] with some minor anomaly trends observed in the North-South direction.

All the three maps generally show similar major trend of the fault at the axial Malay fault zone, which trending from NW to SE. Furthermore, those maps also resemble major faults in Malay basin such as Dungun Fault, Hinge-line fault, Dulang Fault and Bundi Fault.

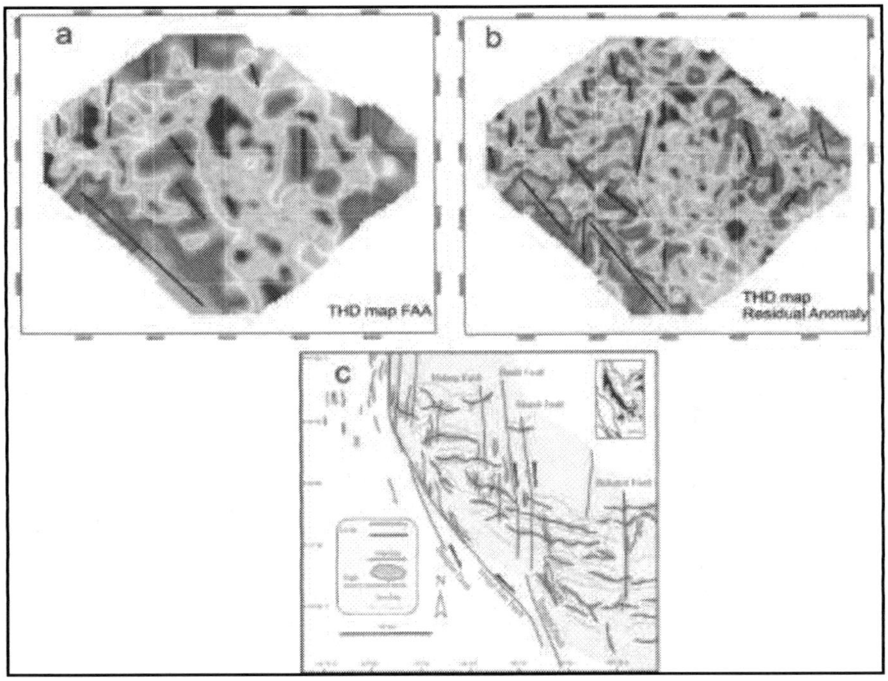

Figure 4. THD with lineament generated from (a) the FAA, (b) the residual anomaly, and (c) the reference lineament by Tjia [12].

The Euler Deconvolution map with the window size of 7x7 and 0 SI indicates that more anomalies (distributed in the whole map) observed for the depth of 2000-4000 m as shown in Figure 5. Meanwhile for the anomalies in the depth of 4000-6000 m and 6000-8000 m, all of them are distributed in the east direction and NW of the map, respectively.

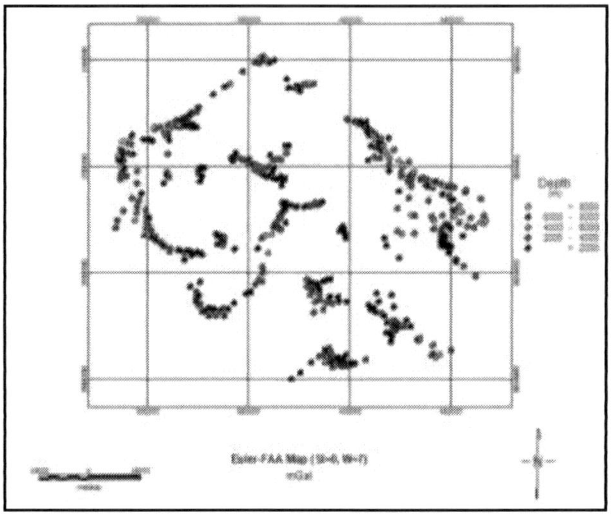

Figure 5. Euler Deconvolution map (SI=0, Window=7).

4.3. 2D Gravity Model

Figure 6 illustrates the 2D gravity profile of AA' directed from SW to NE which was modelled on the study area. The modelling shows the width of the profile is about 160 km and the range of depth encountered is 0-14.5 km with the deepest basement depth generated by GM-SYS is 14.5 km. The reference to built this 2D model was referred to Madon [21] with differentiate from observed and calculated error is acceptable below than 0.3.

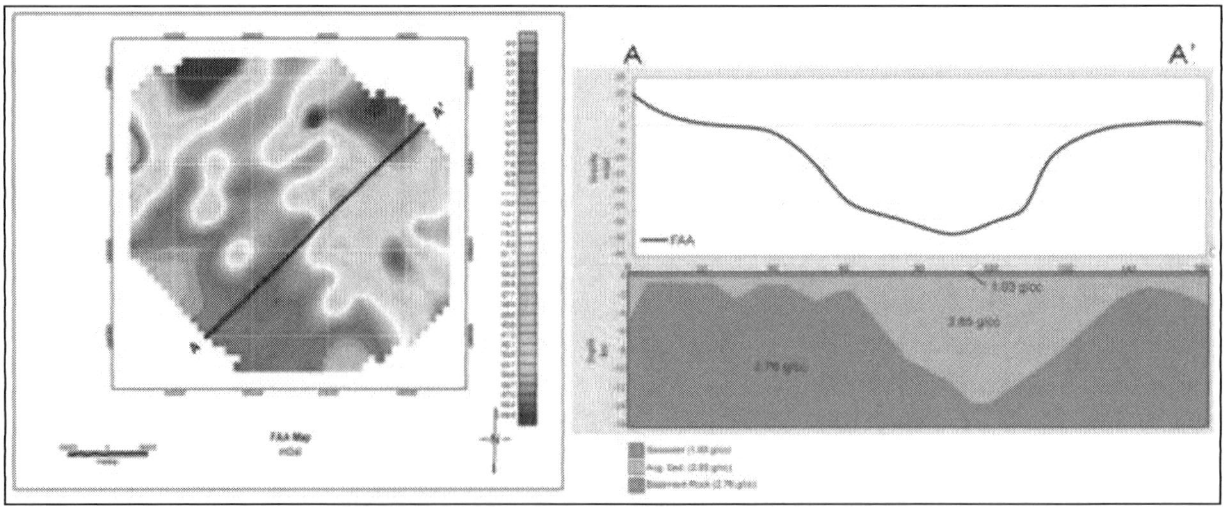

Figure 6. 2D gravity model for the central of Malay Basin.

5. Conclusions

Maximum anomalies were observed in the South (S) and SW direction while for the minimum anomalies, they could be seen in the East (E), NE and SE direction. Regional anomaly map shows the trend of NW-SE, while the THD maps detect the faults at similar locations as the geological map. From the study of Euler deconvolution, the most suitable window which produce efficient and clear anomalies was found to be 7x7. By comparing the Euler map with the geological structure map, similar faults at 2000-4000 m and 4000-6000 m were detected. The 2D gravity model reveals that the deepest basement depth found is 14.5 km which is about 5 % deviation from previous works using the the ERS-1 satellite data. For future works, it is suggested to match this FAA study with the seismic line and magnetic anomaly.

References

[1] Sidek M A M, Hamzah U and Junin R 201. Seismic Facies Analysis and Structural Interpretation of Deepwater NW Sabah. *Jurnal Teknologi* **75** (1)

[2] Maga D *et al* 2015 Fluid inclusions in quartz: implications for hydrocarbon charge, migration and reservoir diagenetic history of the Penyu Basin and Tenggol Arch, offshore Peninsular Malaysia.

[3] Rokhzadi H 2012 Separation of surface and deep geological structures by application of band pass filter and statistical comparison with other methods *Life Science Journal* **9** (3) 950-953

[4] Abdallah E E *et al* 2016 Integrated Interpretation of Remote Sensing and Satellite Gravity Data-Red Sea Sudan (Doctoral dissertation, Sudan University of Science & Technology).

[5] Petroškevičius *et al* 2014 An analysis of the equipotential surface waves of the Earth's gravity field *Journal of Vibroengineering* **16** (5) 2139-2148

[6] Märdla S *et al* 2014 Modelling the influence of terraced landforms to the Earth's gravity field In Gravity *Geoid and Height Systems* 157-162

[7] Barnes G and Barraud J 2012 Imaging geologic surfaces by inverting gravity gradient data with depth horizons *Geophysics* **77** (1) G1-G11

[8] Colombo D *et al* 2014 Exploration beyond seismic: The role of electromagnetics and gravity gradiometry in deep water subsalt plays of the Red Sea *Interpretation* **2** (3) SH33-SH53

[9] Madon M 2017. A brief review of gravity and magnetic data for Malaysia.

[10] Madon M B and Watts A B 1998. Gravity anomalies, subsidence history and the tectonic evolution of the Malay and Penyu Basins (offshore Peninsular Malaysia) *Basin Research* **10** (4) 375-392

[11] Tingay M R *et al* 2013 Evidence for overpressure generation by kerogen-to-gas maturation in the northern Malay Basin *AAPG bulletin* **97** (4) 639-672

[12] Tjia H D 2014 Wrench-Slip Reversals and Structural Inversions: Cenozoic Slide-Rule Tectonics in Sundaland *Indonesian Journal on Geoscience* **1** (1) 35-52

[13] Satti I A *et al* 2014 Analysis of overpressure mechanisms in a field of southwestern Malay Basin. In Offshore Technology Conference-Asia. Offshore Technology Conference.

[14] Morley C K 2013 Discussion of tectonic models for Cenozoic strike-slip fault-affected continental margins of mainland SE Asia *Journal of Asian Earth Sciences* 76 137-151

[15] Pavlis N K *et al* 2012 The development and evaluation of the Earth Gravitational Model 2008 (EGM2008) *Journal of geophysical research: solid earth* **117** (B4)

[16] Montaj G O 2013 Software for earth science mapping and processing *Geosoft Incorporation Ltd*

[17] Ali M Y *et al* 2017. Basement structure of the United Arab Emirates derived from an analysis of regional gravity and aeromagnetic database. Tectonophysics **712** 503-522

[18] Bassett D and Watts A B 2015 Gravity anomalies, crustal structure, and seismicity at subduction zones: 2. Interrelationships between fore-arc structure and seismogenic behavior. *Geochemistry Geophysics Geosystems* **16** (5) 1541-1576

[19] Reid A B and Thurston J B 2014 The structural index in gravity and magnetic interpretation: Errors, uses, and abuses *Geophysics* **79** (4) J61-J66

[20] Harder S H, Mauri S J and Marshall-Arrazola S 1993 Gravity modelling of extensional basins in Southeast Asia

[21] Madon M B 1998 Gravity anomalies, subsidence history and the tectonic evolution of the Malay and Penyu Basins (offshore Peninsular Malaysia) *Basin Research* **10** (4) 375-392

Acknowledgement

The authors are grateful to the UTM for interest and encouragement.

Review of potential astronomical optical observatory sites in the Philippines

Ruby Ann B Dela Cruz[1,2*] and Princess Tucio[1,2*]
[1]Center for Astronomy Research and Development, Rizal Technological University, Mandaluyong City, 1550, Philippines
[2]Department of Earth and Space Sciences, Rizal Technological University, Mandaluyong City, 1550, Philippines

*Corresponding E-mail: rbdcruz@rtu.edu.ph; pbtucio@rtu.edu.ph

Abstract: The Philippines is an archipelago of more than 7,600 islands with many geographical features such as mountain ranges, seas, and plains. The country has fairly developed cities that are now experiencing various kinds of pollution at different levels and intensities. Highly developed urban areas are brightly lit and are currently experiencing worsening light pollution. Despite this, the two biggest telescopes in the country are still installed in two observatories right in the country's capital city where light pollution is at its worst. This paper aims to identify several locations away from light pollution in the islands of Luzon, the biggest island in the Philippines, and the island of Palawan, where optical astronomical observatories may be built. The locations are chosen by studying the comparative light pollution scales in the various locations. The topography of the place and meteorological data such as temperature, humidity, and precipitation are likewise studied and considered as factors. Included in these factors is the fact that both the islands of Luzon and Palawan experience several typhoons every year, while earthquakes occur anywhere in the archipelago. For this reason, the observatories to be built must be designed to withstand Category 5 typhoons and strong magnitude 7 to 8 earthquakes. The results of the study identify the places on the island of Palawan and in northern Luzon which are good dark sites for the establishment of optical astronomy observatories. These locations can be declared as protected and developed as dark-sky sites dedicated to the study of Astronomy.

Keywords: Light Pollution, Optical Observatory, Astronomical Observatory

Track Name: Atmospheric Chemistry and Physics

1. Introduction

Astronomy as a field of science is important in the pursuit of knowledge regarding the universe around us. Yet, more importantly to a developing country, the field of Astronomy encourages developments in advanced technology that may also find application elsewhere. In the Philippines, this field is not prioritized or given much attention as institutions offering Astronomy as a dedicated program are difficult to find in the country.

The Philippines is a tropical country in Southeast Asia. In almost all months of the year, the archipelago is visited by typhoons which often spawn in the Pacific Ocean through some form in the South China Sea. For most of the year, the weather is hot and humid, relieved only by a few months from December to February when the weather is cool and pleasant especially in the northern regions. The archipelago is in the Pacific Ring of Fire and experiences frequent earthquakes and volcanic eruptions. All these phenomena should be considered when constructing observatories anywhere in the Philippines and it is not possible to meet every criterion as earlier mentioned.

This review was done to identify locations where astronomy facilities - specifically optical telescopes and observatories, can be placed. This is significant due to the country's relatively small size, which emphasizes finding such sites before they are taken up for other purposes. As it is right now, the Philippines has fairly developed cities and urban areas but also has plenty of relatively untouched land in Luzon alone that would make suitable observatory sites. In doing this, the identified locations can be protected and developed to increase their effectiveness in the purpose of astronomy.

The Philippines is a developing country with a tropical climate. Due to this, there are challenges to constructing an optical observatory in the country - still, its geography and rural areas afford its locations where the night sky was visible with minimal light pollution. This is in contrast to its major cities and capital region where light pollution drastically reduces the visibility of astronomical objects.

The vicinity of these observatories has to be preserved and protected from the encroachment of light brought about by establishments or agricultural practices which might cause light pollution. Government regulations must be made clear on this matter. The designation of observatories as protected preserves and may help in mitigating the effects or the progress of climate change.

2. Background of the Study

2.1 Astronomical Observatories

Observatories using optical telescopes use the light emitted or reflected by an astronomical object to produce images of the object. Due to this, they are affected by several factors that may obscure visible light from the observed object. These factors include the artificial brightness of the night sky, the atmospheric conditions of the area, and whether the view of the telescope is obstructed by physical objects such as geographical features.

A favorable location for an observatory, in terms of topography, would be where the surrounding land does not rise high enough to obstruct a significant portion of the night sky. Often, these locations are the tops or peaks of hills and mountains, or areas where the terrain is mostly flat such as deserts. Higher elevations are made even more advantageous, because light from an astronomical object travels through less atmosphere to reach a more elevated observatory, reducing atmospheric effects on the image [1]. Aside from distorting the image, atmospheric conditions can also reduce visibility in more direct ways - the biggest example being cloud cover. While clouds can simply block out objects from view, they can also reflect light from nearby sources of light pollution on the ground [2]. Not only that, but the atmosphere can also affect the equipment itself. Humidity can cause condensation, which can deteriorate reflective and refractive surfaces or even contribute to fungal growth [3]. Temperature can also affect equipment, as high temperatures can overheat electronics such as DSLRs and computers while cold temperatures cause the build-up of frost - both of these can limit observational periods [4].

Lastly, artificial night sky brightness, from sources of light pollution, can outshine much of the night sky. This is especially evident within the vicinity of cities where sources of light pollution are numerous and close together [5]. This makes placing optical observatories in the proximity of urban environments unfavorable as the bright night sky becomes a detriment to the visibility of astronomical objects. Most of the population is concentrated in urban areas, the biggest of which is the urban sprawl of Metropolitan Manila, which extends in a wide area from the province of Batangas south of Luzon to its central plains up to the Lingayen Gulf in the province of Pangasinan. This is where light pollution generally affects the night sky, the light emitted from urban areas brightens up the night sky making it non-conducive for astronomical observations.

Having said this, the Philippines has two major observatories - the Manila Observatory which is located at the Loyola Heights campus of the Ateneo de Manila University. The PAGASA Astronomical Observatory is sited inside the campus of the University of the Philippines, Diliman, Quezon [6]. Both are located in the National Capital Region, the most light-polluted region in the country. While these locations make them easily accessible, the conditions in these locations are detrimental to their effectiveness and purpose.

3. Methodology

The Philippines is a tropical country in Southeast Asia. It is on the western edge of the Pacific Ocean. In almost all months of the year, the archipelago is visited by typhoons which often spawn in the Pacific Ocean through some form in the South China Sea. These typhoons average about 20 per year. For most of the year, the weather is hot and humid, relieved only by a few months from December to February when the weather is cool and pleasant especially in the northern regions and during the rainy season brought by the southwest monsoon from May to be replaced by the northeast monsoon in September. The archipelago is in the Pacific Ring of Fire and experiences frequent earthquakes and volcanic eruptions. These earthquakes are often caused by shifting tectonic plates. The country also has a number of active volcanoes. All these phenomena should be considered when constructing observatories anywhere in the Philippines and it is not possible to meet every criterion as earlier mentioned. Despite all of these challenges, observatories can and should be built and there are locations that can offer the optimum advantages for an observatory.

Light pollution map is acquired from the Visible Infrared Imaging Radiometer Suite (VIIRS) 2019 Day Night Band [20] which has been propagated through the atmosphere using the radiative transfer code reported in [21]. The upward emission function and the radiance calibration were obtained using data from Sky Quality Meters [22] [23].

Topographical maps of Luzon were taken from National Mapping and Resource Information Authority (NAMRIA), a government agency that provides geospatial services to the public. The overlaid topographic map has a scale of 1:250,000 to decide the initial list of locations. These locations were chosen by their night sky brightness, proximity to light pollution, and highest area elevation. Furthermore, the average annual temperature, humidity, dew point, precipitation, and visibility of the locations from 2005-2015 were taken using the third-party website but data is from the PAGASA Science Garden Station. Each location's data were compared to each other to find the most suitable locations.

Using this method several sites around the northern areas of Luzon and the island of Palawan were located. The southeast of Luzon also had several potential sites but was considered too susceptible to typhoons compared to the other considered sites.

4. Results and Discussion

Table 1. Atmospheric data and the highest elevation of different locations.

Location	Elevation (m)	Yearly Precipitation (mm)	Humidity (%)	Dew Point (°C)	Visibility (km)	Mean Temperature (°C)
Catanduanes	781	209.9	84	25	15	28
Camarines Sur	921	212.4	84	25	15	28
Apayao	2344	145	77	24	19	29
Quezon	315	155.1	79	23	8	28
Sibuyan Island	1780	205.9	83	25	15	28
Palawan	1260	109.5	80	24	23	28
Averages	1233.5	172.967	81.167	24.33	15.833	28.167

IOP Conf. Series: Earth and Environmental Science **880** (2021) 012008 doi:10.1088/1755-1315/880/1/012008

Table 1 shows data taken from timeanddate.com and elevation was taken from topographical maps lifted from the government site of NAMRIA. The averages are listed in the lowest row, light green cells are above average while light red cells are below average. Mean Temperature is disregarded as there is no significant variation. The included locations were those initially chosen through their topography and night sky brightness.

a. Palawan

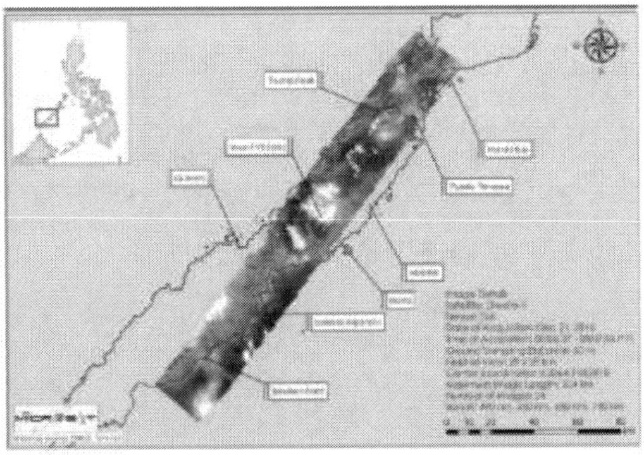

Figure 1. A mosaicked true-color SMI image of Palawan. © *PHL-Microsat.*

Palawan is an island province of the Philippines located in the MIMAROPA region. It lies between the South China Sea and the Sulu Sea, it stretches from Mindoro in the northeast to Borneo in the southwest. This island is said to be the largest province in the country with an area of jurisdiction measuring 450 kilometers (280 mi) long, and 50 kilometers (31 mi) wide. Professional astronomers travel to different places just to experience an astronomical phenomenon. These and many other people travel to different places just to chase eclipses as well as to gather in dark-sky sites for a few nights under stars and by attending 'star party', these people can enjoy the beauty of the stars and at the same time learn a lot about the sky. Natural phenomena like solar eclipses gather millions of nature-based tourists in many destinations that are appropriate for observations [24].

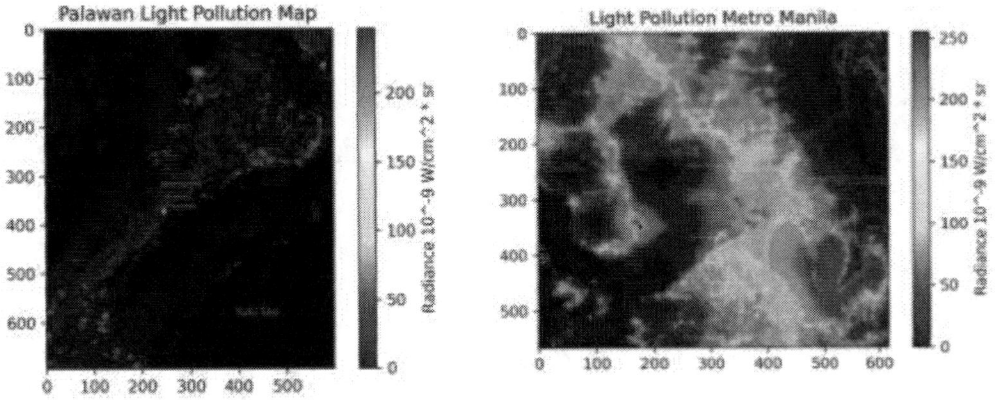

Figure 2. Comparison of Light Pollution Map of Palawan Island and Metro Manila Light Pollution Map.

As seen in the Figure 2, Palawan has much less light pollution compared to the mainland of Luzon, where in light increases night sky brightness, which causes celestial bodies with lower magnitudes to be less or not visible [19]. The possibility of observing celestial bodies is greatly affected also by climate and weather. High humidity can lead to dewing, reducing viewing or imaging time, strong winds can knock over equipment or blow debris, extreme cold can cause frost to form on optical/structural pieces, and extreme heat can affect the sensitivity of imaging equipment, as well as negatively impact equipment that require cooling [13]. Another factor that affects observation is cloud coverage. Aside from obscuring the sky, clouds can amplify sky luminance, increasing night sky brightness [2]. Palawan, as a good location is farthest away from the Pacific, where typhoons originate. It has relatively low annual precipitation at -109.5 mm, it is the lowest out of the initially chosen locations. Not only that, but it also has the most visibility at 23 km. This gives it very favorable conditions for placing an observatory, especially considering its topography as mountains such as Mt Mantalingahan provide high elevations upwards of 1 kilometer on which an optical observatory can be constructed. The main source of light pollution on the island is the city of Puerto Princesa, which is the most populated and urbanized city in Palawan. However, the magnitude of its artificial sky brightness is not as intense as in other areas like Manila. Multiple elevated locations are not in proximity to light-polluted areas that are suitable.

b. Apayao

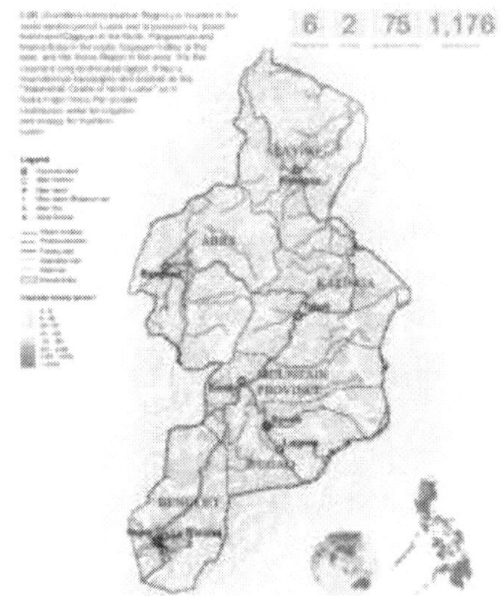

Figure 3. Cordillera Administrative Region Topography.

Cordillera Administrative region has a high mean elevation of 9,606 ft (2,928 m). Provinces such as Apayao, Mt. Province, Ifugao and Kalinga are within the third climatic type where seasons are not pronounced but are relatively dry from November to April and wet for the rest of the year.

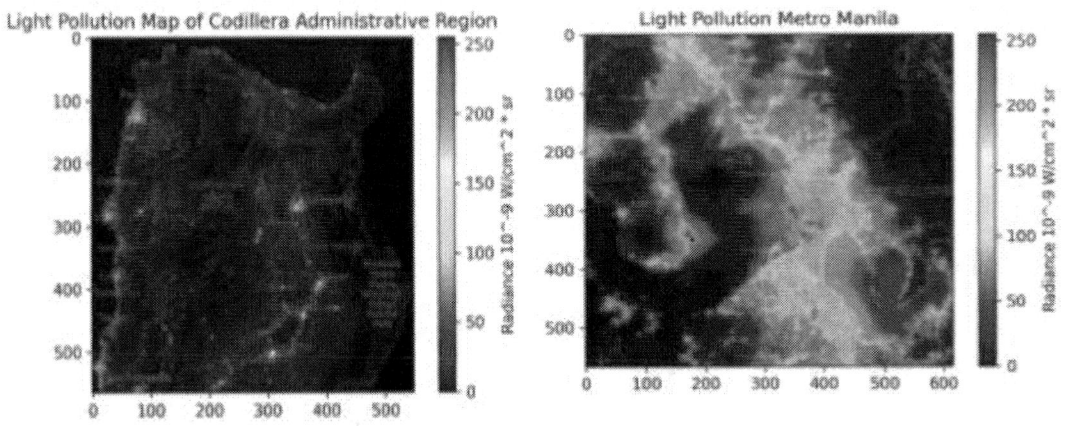

Figure 4. Light Pollution Map of Cordillera Administrative Region compared to Metro Manila.

Apayao has an area of 4,413.35 square kilometers (1,704.00 sq. mi). It is also the least densely-populated province in the Philippines. It is another of the locations chosen that was proven viable by the region's mountainous terrain providing elevations of 1,921 m (6,302 ft). It also has the second-lowest amount of yearly precipitation at 145mm as well as the second-highest visibility at 19 km. Not only that, but Apayao also has the lowest humidity out of the chosen locations – 77 %. Perhaps one caveat of the location, however, is its proximity to light-polluted areas such as Tuguegarao and Laoag, as well as the populated pockets of light present in the mountains of the region. However, the negative impacts of these on night sky visibility can be minimized by appropriate placement of the facility.

5. Summary of Findings

Several years have passed, the Philippines have established astronomy education and research. It may still be behind neighboring Southeast Asian countries, especially those who have established far way back and have established advanced space technologies. At that time, it was very evident that there was the scarcity of trained professional Filipino astronomers [12]. Observation is a part of studying astronomy, choosing a site location for an Astronomical Observatory is one of the first considerations of the entire process. Certainly, before any building takes place, a site must be chosen and confirmed, which requires site-scouting to be an essential first step in observatory planning. Often, the location will dictate certain parts of the telescope design.

Steps can be taken to prepare a facility for events such as typhoons and earthquakes, while it can also be built with the equipment to handle daily conditions such as high humidity. Equipment such as dehumidifiers can be used within the facility or heaters to keep the facility above the dew point. The structure of the facility can be built to withstand or minimize the damage caused by typhoons and earthquakes. Doing so would make the facility less costly to repair after these events.

Not only that, but sites suffering from bright night skies can be improved by implementing behaviors, ordinances, and technologies to reduce light pollution in the area. These can include using specific streetlights that minimize light being shone upward.

It should also be noted that due to the relatively short work period of this review, the researchers were not able to personally visit the sites for evaluation - this is more important to note for artificial night sky brightness. More accurate data may be taken by visiting the noted locations for first-hand accounts.

6. Conclusion

The Philippines has several viable locations within Luzon where an observatory may be constructed, mainly avoiding areas such as NCR where light pollution can be a major detriment to optical observation.

Although there are plenty of dark, elevated areas, as stated earlier, it is very difficult to find a location in the Philippines with the ideal atmospheric conditions due to the country's tropical climate. A prime example of this is the high humidity. Although Apayao had 77 % relative humidity, the lowest among the chosen locations, it is still high enough to be problematic for equipment [4].

Thus, as mentioned above, extra considerations should be taken when constructing a major observatory in the country regardless of the location. Not only that but selected locations viable for optical observations of the night sky should be protected from light pollution to retain their effectiveness as astronomy sites.

References

[1] ATNF. (2020). Australia Telescope National Facility 2020. https://www.atnf.csiro.au/the_atnf/index.html
[2] Kyba C C M, Ruhtz T, Fischer J and Hölker, F 2011 Cloud Coverage Acts as an Amplifier for Ecological Light Pollution in Urban Ecosystems *PLoS ONE* **6** 3 e17307
[3] Teitell L and Berk S 1952 Prevention of Mold Growth in Optical Instruments. Panama Canal Zone Exposure. *Industrial & Engineering Chemistry* **44** 5 1088–1095
[4] Bennett J 2018 What You Should Know About Humidity and Its Effects on Stargazing. Retrieved from https://www.stargazing.me.uk/how-does-humidity-affect-stargazing/
[5] Soardo P, Iacomussi P, Rossi G and Fellin L 2008 Compatibility of road lighting with star visibility. *Lighting Research & Technology* **40** 4 307–322
[6] Torres J R F 2004 Urban Astronomy in the Philippines. *Astronomy Education Review* **3** 1
[7] International Dark-Sky Reserve (n.d.). Light Pollution. Retrieved from: http://darksky.org/light-pollution/
[8] Falchi, F, Cinzano, P, Elvidge, C D, Keith, D M, & Haim, A 2011 Limiting the impact of light pollution on human health, environment and stellar visibility *Journal of Environmental Management* **92** 10 2714–2722
[9] Kyba C C M, Ruhtz T, Fischer J and Hölker F 2011 Cloud coverage acts as an amplifier for ecological light pollution in urban ecosystems *PLoS ONE* **6** 3
[10] Palawan topography Mosaic Map retrieved from https://blog.phl-microsat.upd.edu.ph/
[11]STAMINA4space. (2020, September 15). Contiguous image of Palawan area captured by Diwata-1. STAMINA4Space. https://stamina4space.upd.edu.ph/2017/03/31/contiguous-image-of-palawan-area-captured-by-diwata-1-779807042d2e/
[12] Kouwenhoven M B N, and Sese R M D 2012 Developing Astronomy Research and Education in the Philippines. International Astronomical Union
[13] Hudson K and Simstad T 2010 The share astronomy guide to observatory site selection. Neal Street Design Inc.
[14] Light pollution map. (2020, December 1). Light Pollution Map. https://www.lightpollutionmap.info/#zoom=6.89&lat=9.6824&lon=119.2334&layers=B0TFFFFFFFFFFFFFFF
[15]PAGASA. (2020). Tropical Cyclones. https://bagong.pagasa.dost.gov.ph/information/about-tropical-cyclone[16] Atmospheric Effects. (2021). Atmospheric Effects - Seeing. https://www.atnf.csiro.au/outreach/education/senior/astrophysics/atmosphere.html [17] NAMRIA | The Central Mapping Agency of the Government of the Philippines. (2021). Maps. https://www.namria.gov.ph/download.php#topo250Index
[18] The average annual temperature, humidity, dew point, precipitation, and visibility of the locations timeanddate.com
[19] Cinzano P, Falchi F, and Elvidge, C D 2001 The first World Atlas of the artificial night sky brightness. *Monthly Notices of the Royal Astronomical Society* **328** 3 689–707
[20] Cinzano P, and Falchi F 2012 Quantifying light pollution. *Journal of Quantitative Spectroscopy and Radiative Transfer* 139 13–20
[21] McCarthy J, Jacobson E, Kilduff T, Estes R, Levine P, Mills S, Elvidge C and Miller S 2013 On the potential to enhance the spatial resolution of the day-night band (DNB) channel of the visible and infrared imaging radiometer suite (VIIRS) for the second joint polar satellite system (JPSS-2) and beyond
[22] Duriscoe D, Luginbuhl C and Moore C 2007 Measuring Night-Sky Brightness with a Wide-Field CCD Camera PASP 119 192
[23] Cinzano P, Falchi F and Elvidge C 2001 The First World atlas of the artificial night sky brightness. Monthly Notices of the Royal Astronomical Society **328** 3 689-707
[24]The Editors of Sky & Telescope. (2020, April 20). The Observatories of. Sky & Telescope. https://skyandtelescope.org/astronomy-resources/the-observatories-of-sky-telescope/

Acknowledgments
This research is made possible with the assistance of Rafael Gamboa and Paolo Palacio of the Philippine Science High School. This work is supported by the Rizal Technological University - Center for Astronomy Research and Development (RTU-CARD) under the Science for Change (NICER) Program of the Philippine Council for Innovation, Energy, and Emerging Technology Research and Development (DOST-PCIEERD).

Earth's geomagnetic response to solar wind changes associated with solar events at low latitude regions at the TRE MAGDAS Station

R Umar[1*], S N A Syed Zafar[1], N H Sabri[2], M H Jusoh[3], A Yoshikawa[4], S Abe[4], T Uozumi[4]

[1] East Coast Environmental Research Institute, Universiti Sultan Zainal Abidin, 21300 Kuala Nerus, Terengganu, Malaysia
[2]Advanced Nano Materials (ANoMa) Research Group, Faculty of Science and Marine Environment, Universiti Malaysia Terengganu, 21030 Kuala Nerus, Terengganu.
[3]Applied Electromagnetic Research Group (AERG), Faculty of Electrical Engineering,
Universiti Teknologi Mara, 40450 Shah Alam, Selangor, Malaysia
[4] International Center for Space Weather Science and Education (ICSWSE), Kyushu University, Fukuoka, Japan

*Corresponding E-mail: roslan@unisza.edu.my

Abstract. The Sun's magnetic activity influences disturbances that perturb interplanetary space by producing large fluxes of energetic protons, triggering geomagnetic storms and affecting the ground geomagnetic field. The effect of two solar events, namely Coronal Mass Ejection (CME) and Coronal Holes, on geomagnetic indices (SYM/H), solar wind parameters and ground geomagnetic fields has provided magnetic ground data, which were extracted from the Terengganu (TRE, -4.21° N, 175.91° E) Magnetometer (MAGDAS) station, and investigated in this study. Results show that the physical dynamic mechanism in the Earth's magnetosphere is triggered by various solar wind parameters associated with CMEs and Coronal hole events during the minimum solar cycle of 24 at low latitudes. It is important to study solar wind-magnetosphere coupling because it has an impact on ground-based technological systems and human activities.

Keywords: Earth geomagnetic field, solar wind, geomagnetic indices, CME, coronal holes

Track Name: Atmospheric Chemistry and Physics

1. Introduction

Activities of the Sun, solar winds, magnetosphere, ionosphere, and thermosphere are defined as space weather that influences the performance of space-borne, human health and ground-based technologies [1]. Geomagnetic storms are caused by a variety of solar wind variables associated with solar activity, such as solar flares, coronal mass ejections (CMEs), coronal hole, high solar wind speed and solar energy particles. The increase in solar activity causes large fluxes of energetic protons that form a super geomagnetic storm [2]. CMEs is a solar phenomenon that contributes to the largest geomagnetic storm caused by high-density plasma when it is associated with the southward Bz component [4]. The formation of CMEs causes regional reconnections in the solar corona where light isotopes and plasmas in the solar corona are spread throughout the solar magnetic field [5]. Kilpua et al. (2019) found that high-speed CMEs are most likely preceded by geoeffective sheaths in the form of high-speed ejection sheaths that possess average high solar wind speeds and magnetic fields, and consequently, intense driving electric fields [6]. The occurrence of geomagnetic storms is mainly caused by CMEs driven by solar winds penetrating into Earth's magnetospheric system [7][8]. The coronal hole region has open magnetic fields that extend into interplanetary space; thus, high-speed solar winds can escape and interact with the magnetosphere resulting in magnetic storms. Recent studies on the coronal hole region have detected the geo-effectiveness of high wind speeds propagating in interplanetary space [9][10]. Meanwhile, observations have revealed the effect of solar winds and the orientation of interplanetary space on ground geomagnetic fields associated with solar activity. The ground horizontal component has a significant effect on the enhanced solar wind-magnetosphere coupling process. [11][12][13]. The main objective of this paper is to investigate the effects on solar wind parameters, geomagnetic storms and ground horizontal geomagnetic components caused by the occurrence of CMEs and coronal events during solar cycle of 24.

2. Methodology

This study investigated the effects of CMEs and coronal hole activity on solar wind changes, geomagnetic storms and Earth's horizontal geomagnetic field in low latitude regions. The geomagnetic storm that measures the symmetric part of the magnetospheric ring current activity at the equator is SYM/H. The Disturbance storm time (Dst) index and SYM/H are similar parameters, while the comparison of Dst Index and $SYMH^{-1}$ was only time interval. The Dst index was gauged every 3 hours and SYM H^{-1} was gauged per minute/data. Geomagnetic storms are classified according to weak (-30 < Dst< -50 nt), moderate (-50 <Dst< -100 nt), intense (-200 <Dst< -100 nt) and super storm (Dst < -200) stages. This study considered four solar wind parameters, such as solar wind speed (V), solar wind dynamic pressure (P), solar wind input energy (IE) and interplanetary magnetic field (IMF-Bz). Solar wind input energy is indicated as Epsilon (ε) and calculated using the original Akasufo parameter, as shown in Equation (1) [14]. Data concerning coronal hole and CME events were obtained from the SolarHam website (https://www.solarham.net/regions.htm), Solar and Heliospheric Observatory (SOHO) website (https://cdaw.gsfc.nasa.gov/CME_list/) and space weather live (https://www.spaceweatherlive.com/). Meanwhile, the SYM/H and solar wind parameters data were acquired from the OMNI website (https://omniweb.gsfc.nasa.gov/form/omni_min.html), which was provided by the Space Physics Data Facility (SPDF) based at NASA's Goddard Space Flight Center in Greenbelt, MD U.S.A. The data used were high resolution with a one-minute interval.

$$\text{Solar wind input energy} = \frac{4\pi}{\mu_o} VB^2 sin^4 \left(\frac{\theta}{2}\right) I_o 2 \text{ (erg/s)} \quad (1)$$

where, V= solar wind speed (kms^{-1}), B =magnitude of IMF (nT) and I_o are the 7 Earth radii. The polar angle of the IMF vector (Θ), projected onto the Y-Z plane, namely; [$\tan^{-1}$abs(B_y/B_z)] for B_z >0 and [180°-$\tan^{-1}$abs(B_y/B_z)] for Bz <0.

The horizontal geomagnetic field (H-component) was extracted from MAGDAS magnetometer with one-second data interval, which were then recorded and transferred to Japan. Data were collected from the MAGDAS magnetometer (MAGDAS-II) at TRE station, which is located at the East Coast Environmental Research Institute (ESERI) in Universiti Sultan Zainal Abidin, Terengganu, Malaysia. Details of selected MAGDAS magnetometer stations are listed in Table 1. The MAGDAS magnetometer is a ring core-type fluxgate magnetometer that measures three components in the geomagnetic field in the ground, namely the horizontal component (H), declination component (D) and the vertical component (Z).

Table 1. Details of selected MAGDAS magnetometer location.

Stations	Abbreviation	Geographic		Geomagnetic		Region
		Latitude (°)	Longitude (°)	Latitude (°)	Longitude (°)	
Terengganu	TRE	5.23	103.04	-4.21	175.91	Malaysia

3. Result and Discussion

This section presents an analysis of the dependence of solar wind changes on the Earth's horizontal geomagnetic field (H-component) at low latitude regions with data obtained from a selected MAGDAS magnetometer station, which is the TRE station. Figure 1 depicts the variety of solar wind changes and its effects on ground geomagnetic field (H-component) from 1st March to 30th March 2017. On 1st March 2017, the IMF-Bz, V, P, and IE responded to the occurrence of a moderate geomagnetic storm (SYM H^{-1} = -70 nT). High solar wind pressure (P =3-14 nPa) and high solar wind speed (V =400-700 km s^{-1}) carrying input energy (0.3-3.3 x10^{19} ergs) reacted to the moderate geomagnetic storm on 1st March 2017 in a strong southward Bz condition (5 to -7 nT). This moderate storm corresponded to a change in the H-component, which reached values of (4.150-4.142 x 10^4 nT) at the same time.

The weak geomagnetic storm (SYM/H = -30 nT) occurred in the middle of the 14th of March 2017 as a result of CME-associated coronal hole events on the 12th of March 2017. The weak storm was caused by by a weak solar wind pressure (P =2-3 nPa) with low solar wind speed (V=350-320kms^{-1}) in southward Bz orientation (0 to -5 nT) as well as low solar wind energy measuring (0-0.6 x10^{19} ergs) conveyed into Earth's magnetosphere. The influence of solar wind parameters and CME-associated coronal hole events on the H-component is not shown in Figure 1 due to instantaneous fluctuations during the same period caused by artificial noise.

In addition, it is noticeable that increases in solar wind speed (V =500-600 km s^{-1}), solar wind dynamic pressure (P =4-2 nPa) and solar wind input energy (1.5-0.5x10^{19} ergs) caused a weak geomagnetic storm (SYM/H = -50 nT) on 22th March 2017 with a weak southward Bz orientation (4 to -3 nT). As shown in Figure 2, the H-component decreased as the solar wind changed with a value of (4.148 - 4.140 x10^4 nT). In the early part of 28th March 2017, there was a moderate geomagnetic storm (SYM/H = -60 nT) caused by solar wind changes with a high solar wind speed (V=600-750kms^{-1}) associated with dynamic pressure (2-5 nPa), a strong southward Bz condition (10 to -10 nT) and a high energy transfer rate of (0-2 x10^{19} ergs). The analysis also indicated a moderate geomagnetic storm as well as change of the H-component on 28th March 2017, as shown in Figure 2 (f). The H-component is decreased with values varying (4.150-4.140 x10^4 nT). At the end of March, the Bz component showed a weak antiparallel orientation to Earth's magnetic field with values ranging from (0 to -6 nT), in response to a weak geomagnetic storm (SYM H^{-1} = -30 nT) on 30th March 2017. At the same time, there was a slight increasing trend in solar wind speed (V =580-620 km s^{-1}) and dynamic pressure (P =1-3 nPa) with solar wind input energy varying from (IE =0.3-0.6 x10^{19} ergs). As shown in Figures 1(e) and (f), the drop in SYM/H is equivalent to a very small decline in the H-component, which varied from (4.150 4.148 x10^4 nT).

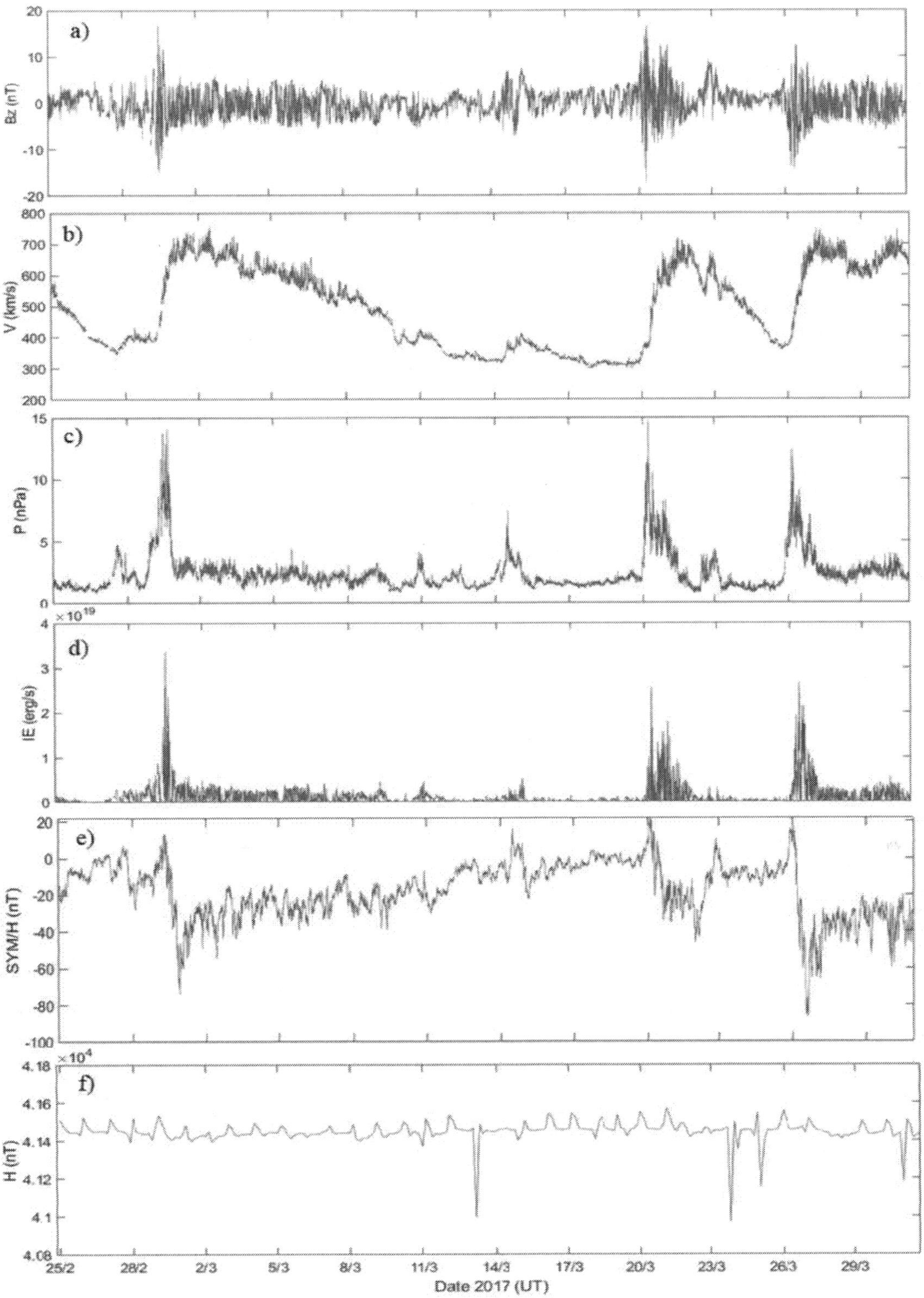

Figure 1. The variations in solar wind parameters a) IMF-Bz b) speed c) dynamic pressure d) solar wind input energy and geomagnetic storm e) SYM/H and ground geomagnetic field f) H-component of TRE magnetometer station.

Table 2. List of selected CMEs and Coronal hole.

Date	Solar activity			
	CME speed (km s^{-1})	Time (UT)	Coronal Hole	Time (UT)
28th February 2017	118	23:12:12	YES	10:26:17
12th March 2017	444	00:48:00	YES	10:24:10
20th March 2017	176	00:48:01	YES	10:25:53
27th March 2017	338	12:00:00	YES	10:40:29

According to Table 2, the association between CMEs and coronal hole events on 28th February 2017 responded to changes in solar winds, resulting in a weak geomagnetic storm on 1st March, 2017. Large amounts of plasma ejected from the CME event had driven the high solar wind to interact with the Earth's magnetic field, resulting in a geomagnetic disturbance [15]. This analysis shows that high dynamic pressure had compressed the Earth's magnetosphere with plasma ejections resulting in strong magnetic reconnection and moderate geomagnetic storms [16]. The plasma ejection is driven by a high solar wind speed in a southward Bz condition that transferred high input energy in the magnetosphere. These solar wind changes associated with CME and coronal hole events influenced the physical processes in the magnetosphere, causing a decrease in the H-component [17]. On 14th March 2017, a weak geomagnetic storm was also generated. According to the analysis, small coronal holes with weak solar wind pressure had contributed to low magnetosphere reconnection associated with southward Bz rotation, which caused low plasma densities and magnetic flux ejected into the inner magnetosphere. Tokumaru *et al.* [19] had supported this finding [18]. Figure 3 depicts the image of selected coronal hole events. However, disturbance in the ground caused by changes in solar winds was not shown due to artificial noise, which could be caused by man-made noise. Based on Figure 2, this is consistent with a short fluctuation in the H-component difference (diff-H) plot during the same period. The diff-H plot is used to estimate artificial noise on the ground in the ULF range.

The occurrence of a weak geomagnetic storm on 22th March, 2017 is indicated by a slow magnetosphere reconnection and low plasma injection process caused by the low solar wind's dynamic pressure in a weak southward Bz condition. At same time, it affected solar wind speed and conveyed a low energy density magnetic reconnection. Meanwhile, high solar wind changes associated with CME and coronal hole events on 27th March 2017 contributed to a moderate geomagnetic storm on 28th March 2017. This can be explained by high plasma densities caused by CME expansion and the expansion of high solar wind speed flow-out from coronal holes injected into the magnetosphere by solar wind pressure in strong southward Bz condition [19]. According to Abuninna *et al.* [21], northern and southern holes produced enhanced geomagnetic activity, as shown in Figure 2. The coronal hole area on 27th March 2017 was in the southern hemisphere [20]. The high magnetic reconnection in magnetosphere resulted in moderate storm, which was followed by a variation in the ground geomagnetic field with a decrease in the H-component. Finally, the analysis concluded that the wake of a weak storm on 30th March 2017 was triggered by the low effect of CME and coronal hole events on 27th March 2017, which contributed to a low acceleration of solar winds to Earth, resulting in a poor plasma environment in the magnetosphere [21]. Concurrently, the ground geomagnetic field had shown a decreasing trend in weak storms, indicating that the magnetosphere-ionosphere response was in action.

IOP Conf. Series: Earth and Environmental Science **880** (2021) 012009 doi:10.1088/1755-1315/880/1/012009

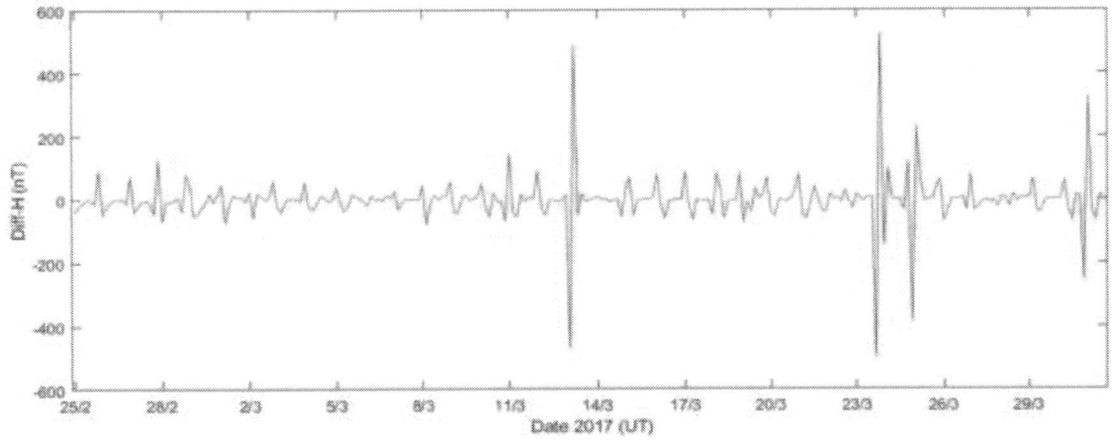

Figure 2. Differences in the H-component (Diff-H) plot at the TRE magnetometer station.

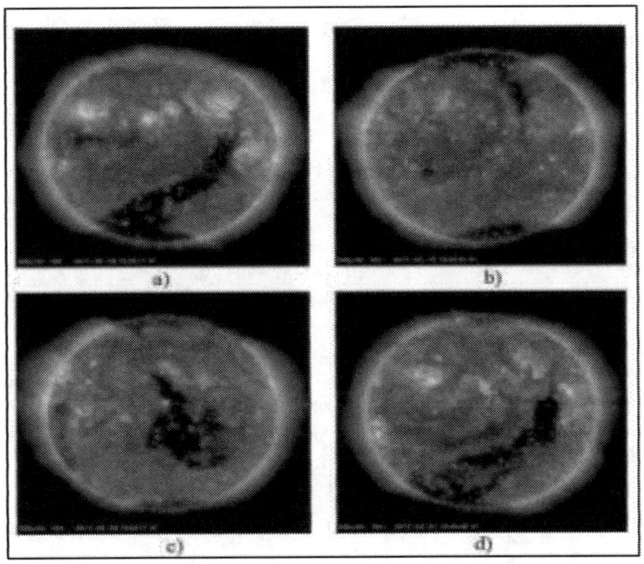

Figure 3. The selected image of coronal hole area obtained from SDO/AIA 193 on a) 28th February 2017 b)12th March 2017 c) 20th March 2017 d)27th March 2017.

4. Conclusion

The analysis in this study shows that CME and coronal hole events have an impact on the occurrence of weak and moderate geomagnetic storms during the minimum solar cycle of 24. The huge plasma cloud from the CMEs driven by high solar winds with compressed energy were ejected into the magnetosphere due to high solar wind pressure in a strong southward Bz condition, thus, generating a moderate geomagnetic storm on 1st March 2017 and 27th March 2017. In addition, the analysis suggested that the expansion of high solar wind speed flow-out from coronal holes in the southern hemisphere injected into the magnetosphere by high solar wind pressure also contributed to a moderate storm. In this study, a small coronal hole on 12th March 2017 had caused low solar wind speeds and weak solar wind pressure, which ejected fewer ions and particles into the magnetospheric system, implying a weak storm on 14th March 2017. The development of a weak storm on 22th March 2017 is explained by a weak southward Bz condition with weak solar wind pressure that caused lesser plasma densities ejected into the magnetosphere, resulting in a slow magnetic reconnection process. Following that, solar wind changes associated with CME and coronal hole events had revealed the response to geomagnetic storms and the current system that penetrate the ground, resulting in a reduction of H-component at the TRE station. It has been found that a high reduction in the H-component corresponds to a moderate storm occurrence compared to a decreased H-component during a weak storm. However, changes in the H-component during a weak storm on 14th March 2017 did not occur, resulting in a rapid fluctuation caused by artificial noise on the ground.

References

[1] Schwenn R 2006 Space weather: The solar perspective. *Living Reviews in Solar Physics* **3** (1) 1-72
[2] Siingh D *et al* 2011 Solar activity, lightning and climate. *Surveys in geophysics* **32** (6) 659
[3] Cliver E W 1994 Solar activity and geomagnetic storms: The first 40 years *Eos, Transactions American Geophysical Union* **75** (49) 569-575
[4] Gosling J T *et al* 1991 Geomagnetic activity associated with Earth passage of interplanetary shock disturbances and coronal mass ejections *Journal of Geophysical Research Space Physics* **96** (A5) 7831-7839
[5] Fu H S *et al* 2013 Energetic electron acceleration by unsteady magnetic reconnection *Nature Physics* **9** (7) 426-430
[6] Kilpua E K J *et al* 2019 Solar wind properties and geospace impact of coronal mass ejection-driven sheath regions: Variation and driver dependence *Space Weather* **17** (8) 1257-1280
[7] Singh A K Tonk A and Singh R 2014 Prospective of coronal mass ejections, solar flares and geomagnetic storms *Indian Journal of Physics* **88** (11) 1127-1133
[8] Mustajab F and Derouich M 2018 Geomagnetic response of interplanetary coronal mass ejections in the Earth's magnetosphere *Planetary and Space Science* 154 1-4
[9] Mordvinov A V and Yazev S A 2014 Reversals of the Sun's polar magnetic fields in relation to activity complexes and coronal holes *Solar Physics* **289** (6) 1971-1981
[10] Rotter T *et al* 2012 Relation between coronal hole areas on the Sun and the solar wind parameters at 1 AU *Solar Physics* **281** (2) 793-813
[11] Regi M De Lauretis M and Francia P 2015 Pc5 geomagnetic fluctuations in response to solar wind excitation and their relationship with relativistic electron fluxes in the outer radiation belt *Earth Planets and Space* **67** (1) 1-9
[12] Umar R *et al* 2019 Magnetic Data Acquisition System (MAGDAS) Malaysia: installation and preliminary data analysis at ESERI, UNISZA *Indian Journal of Physics* **93** (5) 553-564
[13] Syed Zafar S N A *et al* 2021 Behaviour of Geomagnetic storm, horizontal geomagnetic field and solar wind parameters during solar flare and CMEs event *Journal of Physics: Conference Series* **1768** (1) 012003
[14] Akasofu S I 1981 Energy coupling between the solar wind and the magnetosphere *Space Science Reviews* **28** (2) 121-190
[15] Eroglu E 2019 Modeling the superstorm in the 24th solar cycle *Earth Planets and Space* **71** (1) 1-12
[16] Rawat R Alex S and Lakhina G S 2010 Storm-time characteristics of intense geomagnetic storms (Dst≤− 200nT) at low-latitudes and associatedenergetics *Journal of atmospheric and solar-terrestrial physics* **72** (18) 1364-1371
[17] Syed Zafar S N A *et al* 2021 Effects of solar flares and coronal mass ejections on Earth's horizontal magnetic field and solar wind parameters during the minimum solar cycle 24. *Monthly Notices of the Royal Astronomical Society* **504** (3) 3812-3822
[18] Bianchi C and Meloni A 2007 Natural and man-made terrestrial electromagnetic noise: an outlook *Annals of geophysics*
[19] Tokumaru M *et al* 2017 Relation between coronal hole areas and solar wind speeds derived from interplanetary scintillation measurements *Solar Physics* **292** (3) 41
[20] Nakagawa Y Nozawa S and Shinbori A 2019 Relationship between the low-latitude coronal hole area, solar wind velocity, and geomagnetic activity during solar cycles 23 and 24 *Earth, Planets and Space* **71** (1)1-15
[21] Abunina M *et al* 2015 On the influence of the coronal hole latitude and polarity on the geomagnetic activity and cosmic ray variations. In Proceeding of Science *The 34th International Cosmic Ray Conference* 82

Acknowledgement

This study was supported by the grant FRGS/1/2019/STG02/UNISZA/02/3. Special thanks are dedicated to International Center for Space Weather Science and Education (ICSWSE) for providing MAGDAS instrumentation and assisting in the installation

IOP Conf. Series: Earth and Environmental Science **880** (2021) 012010 doi:10.1088/1755-1315/880/1/012010

A statistical analysis of the relationship between Pc4 and Pc5 ULF waves, solar winds and geomagnetic storms for predicting earthquake precursor signatures in low latitude regions

S N A Syed Zafar[1], Roslan Umar[1*], N H Sabri[2], M H Jusoh[3], A Yoshikawa[4], S Abe[4], T Uozumi[4]

[1] East Coast Environmental Research Institute, Universiti Sultan Zainal Abidin, 21300 Kuala Nerus, Terengganu, Malaysia
[2]Advanced Nano Materials (ANoMa) Research Group, Faculty of Science and Marine Environment, Universiti Malaysia Terengganu, 21030 Kuala Nerus, Terengganu
[3]Applied Electromagnetic Research Group (AERG), Faculty of Electrical Engineering, Universiti Teknologi Mara, 40450 Shah Alam, Selangor, Malaysia
[4] International Center for Space Weather Science and Education (ICSWSE), Kyushu University, Fukuoka, Japan

* Corresponding author: roslan@unisza.edu.my

Abstract. Short-term earthquake forecasting is impossible due to the seismometer's limited sensitivity in detecting the generation of micro-fractures prior to an earthquake. Therefore, there is a strong desire for a non-seismological approach, and one of the most established methods is geomagnetic disturbance observation. Previous research shows that disturbances in the ground geomagnetic field serves as a potential precursor for earthquake studies. It was discovered that electromagnetic waves (EM) in the Ultra-Low Frequency (ULF) range are a promising tool for studying the seismomagnetic effect of earthquake precursors. This study used a multiple regression approach to analyse the preliminary study on the relationship between Pc4 (6.7-22 mHz) and Pc5 (1.7-6.7 mHz) ULF magnetic pulsations, solar wind parameters and geomagnetic indices for predicting earthquake precursor signatures in low latitude regions. The ground geomagnetic field was collected from Davao station (7.00° N, 125.40° E), in the Philippines, which experiences nearby earthquake events (Magnitude <5.0, depth <100 km and epicentre distance from magnetometer station <100 km). The Pc5 ULF waves show the highest variance with four solar wind parameters, namely SWS, SWP, IMF-Bz, SIE and geomagnetic indices (SYM/H) prior to an earthquake event based on the regression model value of $R^2 = 0.1510$. Furthermore, the IMF-Bz, SWS, SWP, SWE, and SYM/H were found to be significantly correlated with Pc5 ULF geomagnetic pulsation. This Pc5 ULF magnetic pulsation behaviour in solar winds and geomagnetic storms establishes the possibility of using Pc5 to predict earthquakes.

Keywords: Geomagnetic field, magnetic pulsation, solar wind, earthquake precursor, regression

Track Name: Atmospheric Chemistry and Physics

1. Introduction

The ULF wave anomaly prior to large earthquakes has been intensively investigated during the past decade. Experimental results show that electromagnetic emissions have a frequency range when rocks are under stress, which means the pressure is triggered by a large current [1]. The earthquake precursor is most commonly analysed in a frequency range of f <10 Hz with skin depth less than 100 km and the epicentre distance from magnetometer being within 100 km [2]. Studies have found that ULF waves are generated due to the enhanced Sun-Earth coupling mechanism rather than some geological phenomena. The study shows that the Pc3-Pc5 wave range is a range that witnesses magnetized plasma wave propagation (Alfvén waves) in the magnetosphere and ionosphere [3][4]. The Pc3 magnetic pulsation appears to be very closely correlated to magnetic storms with seismic events [5]. Takla et al. (2018) discovered abnormal geomagnetic variations in a Pc3 range near the epicentre of some seismic events at the KTB station in Sumatra, Indonesia [6]. The significant change in polarization ratio analysis of the Pc4-Pc5 range before an earthquake event was revealed by several studies [7][8]. The Pc3-Pc5 range and occurrence with storms have shown a clear dependence associated with IMF and solar wind parameters [9][10]. Meanwhile, the relationship between Pc3-Pc5 magnetic pulsations and changes in the solar wind have been extensively studied in recent decades. The existence of Pc5 pulsations was triggered by fluctuations in solar wind dynamic pressure [11]. The Pc5 was found to be statistically correlated with solar wind speed in high latitude regions (Ji Kwon et al. 2019). The highest ULF power in the Pc5 range with high solar wind changes on the ground was observed in a declining solar cycle phase of 23 [12]. To determine and validate the ULF seismomagnetic phenomenon in solar winds and geomagnetic storms, this paper examined the possible ULF earthquake precursor signature with solar wind parameters and geomagnetic storms in low latitude regions. The ULF range (Pc4-Pc5) in solar winds and geomagnetic storms prior to an earthquake precursor was also examined.

2. Methodology

In order to investigate the relationship between geomagnetic pulsation and solar wind parameters associated with geomagnetic storms near equatorial (SYM/H) regions for earthquake precursor research, geomagnetic pulsation data near seismic events with a depth of less than 100 km, magnitude less than 5, and an epicentre distance less than 100 km from the magnetometer station were collected. The International Association of Geomagnetism and Aeronomy (IAGA) classified two magnetic pulsations in the ULF band, which were investigated based on solar wind parameters and geomagnetic storms by using Pc4 (6.7-22 mHz) and Pc5 (1.7-6.7 mHz) for earthquake precursor signature research. These geomagnetic pulsations were obtained and collected from the SuperMag website (http://supermag.jhuapl.edu/) that services the Davao magnetometer station (7.00° N, 125.40° E) located in the Philippines [13][14]. Figure 1 depicts the location of the Davao magnetometer station. The sample interval for these data was 1 s. The observation was analysed within two months of the data being collected, prior to the occurrence of earthquake events.

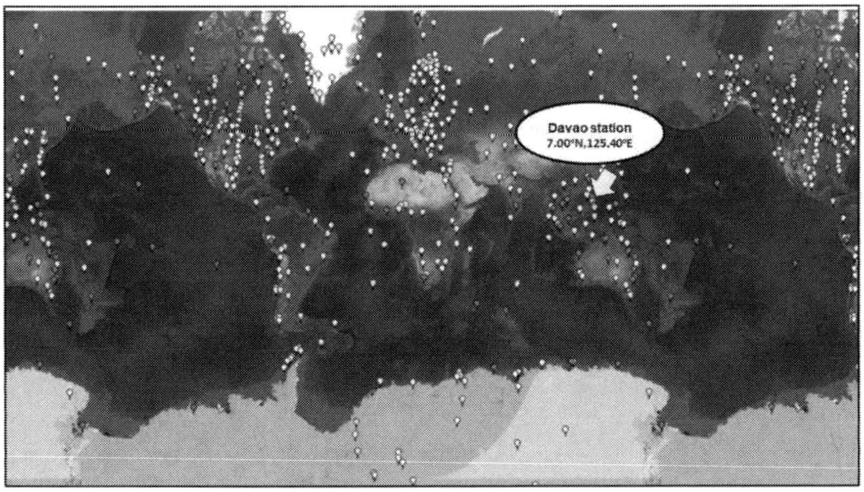

Figure 1. Location Davao magnetometer station (http://supermag.jhuapl.edu/).

This study selected four solar wind parameters, namely solar wind speed (SWS), solar wind pressure (SWP), IMF-Bz component (Bz), solar wind input energy (SWE) and symmetric part of the magnetospheric ring's current activity at the equator, which is (SYM/H). The sample interval for these data was 1-minute. Data were acquired from (https://omniweb.gsfc.nasa.gov/form/omni_min.html) , which were provided by Space Physics Data Facility (SPDF) based in NASA's Goddard Space Flight Centre in Greenbelt, MD U.S.A. Two months of data prior to the earthquake event were analysed with Pc4 and Pc5 ULF pulsation. The association of these parameters with Pc4 and Pc5 were determined for low latitude regions.

Multiple regression was used to examine the relationship of the strength between Pc4 and Pc5 magnetic pulsations, solar wind parameters and geomagnetic storms in low latitude regions for earthquake precursor signatures. The multivariate regression analysis was formulated according to the model equation below;

$$y = \beta_O + \beta_{1x_1} + \ldots \beta_{nx_n} + \varepsilon$$

Where, y is the dependent variable, x_1 is the independent variable, β_1 through β_n are the estimated regression coefficients, β_o is the value of Y when all independent variables (x_1 through x_n) are equal to zero and ε is the error. This equation model will be used for predicting the y variable as a linear function of the x variables [15]. In this study, there were two dependent variables, namely Pc4 and Pc5 ULF pulsations, which were analysed based on five independent variables, such as solar wind speed (SWS), solar wind dynamic pressure (SWP), IMF-Bz component (Bz), solar wind input energy (SIE) and magnetospheric ring current activity at the equator (SYM/H).

The multiple regression came out with the R^2 value, which could be used to estimate how a variation of the dependent variable changes as each independent variable change, as well as the strength of the variation of the dependent variable compared with selected independent variables. The R^2 coefficient (beta value) in the regression analysis determines the correlation between the independent variables, multicollinearity of independent variables (VIF-value) and significance (p-value), as discussed in the regression analysis [16]. Finally, performance of the regression model was determined. However, before performing the multivariate regression, the normality of assumptions for each variable was examined using the Shapiro-Wilk test. As the p-value >0.05 (alpha-value) for all variables, it was found to be normally distributed, thus, the null hypothesis was rejected because the null hypothesis of the observation was not normally distributed [17][18].

IOP Conf. Series: Earth and Environmental Science **880** (2021) 012010 doi:10.1088/1755-1315/880/1/012010

3. Result

The regression prediction model for Pc4 and Pc5 ULF geomagnetic pulsations with four solar wind parameters and geomagnetic indices was generated, as shown in equations (1) and (2). According to the results of the multiple regression analysis, the three solar wind parameters and geomagnetic indices discovered were significantly predictive using Pc4 ULF geomagnetic pulsation with $p<0.05$. The strength effect of each solar wind parameter and geomagnetic index on Pc4 variance is shown in Table 1 as a coefficient value. The coefficient values of five independent variables are as follows: (Bz = -0.0164, T = -10.28, $p <0.05$), (SWS = 0.0007, T = -21.65, $p <0.05$), (SWP = 0.0741, T = 31.65, $p<0.05$), (SIE = -0.0001, T = -1.16, $p >0.05$) and (SYM/H = -0.0038, T = -16.25, $p <0.05$). Based on the coefficient results, the variable with the most contribution was solar wind pressure, which had the highest coefficient value compared to other variables. This was followed by the IMF-Bz component, solar wind speed, geomagnetic indices (SYM/H) and solar wind input energy. In this analysis, the contribution of solar wind input energy to Pc4 was found to be insignificant, with a p-value greater than the significant value ($\alpha = 0.05$) and a very small coefficient contributing towards a small variance in Pc4. The VIF value in table 1 indicates how much of the variance is inflated in each coefficient. As a rule of thumb, a VIF value of 1 means it is not correlated, between 1 and 5 means it is moderately correlated, and more than 5 means highly correlated [19]. The result shows that the VIF value for solar wind parameters is more than 1, which means the association of solar wind parameters and geomagnetic indices is moderately colinear with each other in the model. The model degree explaining the variance of Pc4 ULF waves in R^2 value indicates how well Pc4 varies with respect to all solar wind parameters. Thus, the variance in Pc4 based on the four solar wind parameters and geomagnetic indices is described as $R^2 = 0.0482$, which corresponds to a 4.82 percent variance explained by four wind parameters and geomagnetic indices in low latitude regions. The R^2 for Pc4 in Table 2 shows that R^2-predicted is greater than R^2, indicating that the Pc4 prediction model is overfit [20][21]. In future studies with extended observation time, the unimportant independent variable should be excluded in order to fit the model and initiate the prediction model. Simultaneously, in this multiple regression analysis, the residual plot is used to verify whether the model is adequate to meet the assumptions of the analysis. The residual plot vs fitted value is used in this analysis. When the residual centre is zero, it indicates that the prediction model is correct on average. The residual plot of the Pc4 model shows a central tendency (Figure 2), despite the presence of a few outliers, indicating that this model cannot predict well.

$$\text{Pc4} = 1.1683 - 0.0164\,\text{Bz} + 0.0007\,\text{SWS} + 0.0740\,\text{SWP} - 0.0001\,\text{SIE} - 0.0038\,\text{SYM/H} \quad (1)$$

Table 1. Multiple Regression of Coefficient of 5 solar wind parameters and geomagnetic index for magnetic pulsation of Pc4.

Variables	Coefficients (Coef)	Standard error coef (SE coef)	T-value	p-value	Variation Inflation factor (VIF)
Constant	1.1683	0.0128	91.20	0.000	
Bz	-0.0164	0.0015	-10.98	0.000	2.05
SWS	0.0007	0.0000	21.65	0.000	1.27
SWP	0.0740	0.0023	31.65	0.000	1.36
SIE	-0.0001	0.0000	-1.16	0.244	1.72
SYM/H	-0.0038	0.0002	-16.25	0.000	1.32
			F (5, 48001) =485.76		

Table 2. Multiple regression result of Magnetic pulsation of Pc4.

R^2	R^2-adjusted (R^2-adj)	R^2-predicted (R^2-pred)
4.82 %	4.81 %	4.79 %

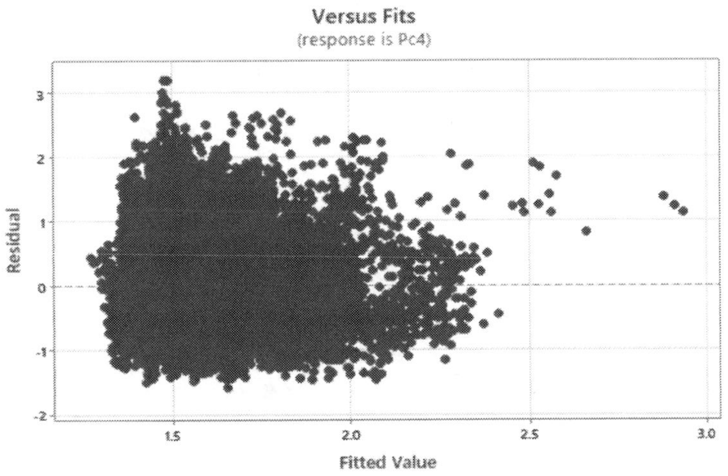

Figure 2. Residual versus fitted value plot of Pc4 with four solar wind parameters and geomagnetic storm (SYM/H).

The Pc5 prediction model with four solar wind parameters and geomagnetic indices was created based on Equation (2). Meanwhile, Table 3 shows the coefficient values for four solar wind parameters and geomagnetic indices with a Pc5 magnetic pulsation. These five independent variables' coefficient robustness (Bz = 0.0048, T = 2.82, p <0.05), (SWS = 0.0017, T = -43.80, p <0.05), (SWP = 0.1481, T =54.89, p <0.05), (SIE = -0.0001, T =9.79, p <0.05) and (SYM/H = -0.0046, T = -16.79, p <0.05) and it was found to be significantly predictive of Pc5 ULF magnetic pulsation. Solar wind pressure contributes the most to Pc5 variance, with a higher coefficient value than others. Sequence of the coefficient values is as follows: IMF-Bz component, solar wind speed, geomagnetic indices (SYM/H), and solar wind input energy. The solar wind input energy contributes very little to the variance of Pc5. The association between solar wind parameters and geomagnetic indices is moderately colinear, as indicated by the VIF values. Following that, a prediction model was created with $R^2 = 0.150$, which explains the variations of Pc5 with SWS, SWP, IMF-Bz, SIE, and SYM/H, implying that Pc5 has a 15.10 percent variance with selected solar wind parameters. As shown in Table 4, the R^2-predicted is greater than R^2, indicating a poor predictive ability. The residual plot of the Pc5 model reveals a central tendency with a few outliers, similar to that of the Pc4 model (see Figure 3).

According to the results of the analysis concerning the Pc4 and Pc5 prediction model, Pc5's magnetic pulsation has a higher variance in the IMF-Bz, solar wind speed, solar wind dynamic pressure, solar wind input energy components and geomagnetic indices (SYM/H) compared to Pc4. Pc5 has a higher R^2 value than Pc4, and its magnetic pulsation is better suited for prediction models with four solar wind parameters and geomagnetic indices as well as an earthquake precursor signature for earthquake depths of 100 km and an epicentre distance of 100 km. In addition, the regression model results show that Bz, SWS, SWP, SWE, and SYM/H have a significant influence on the ULF of geomagnetic pulsations (Pc5), which is useful for investigating earthquake precursor signatures in low latitude regions. According to Currie *et al.* [22], Sharma *et al.* (2012) [23] and Marchitelli *et al.* [24], there is a statistically significant correlation between solar wind changes and geomagnetic storms prior to the occurrence of an earthquake event with different correlation strengths at different latitudes [22][23][24]. Several previous studies have looked into the possibility of earthquake precursor signatures based on Pc4-Pc5 magnetic pulsations [25][26]. The solar wind-magnetosphere coupling mechanism results in the occurrence of geomagnetic storms, implying that charge carriers from plasma densities, driven by solar winds, are capable of flowing out into the lithosphere and emitting ULF waves. This may cause stress in the rock to increase, resulting in an earthquake [27]. Consequently, a preliminary study of the relationship strength of Pc4 and Pc5 geomagnetic pulsations with solar wind parameters and geomagnetic storm prior to predicting the earthquake precursor was carried out using a multiple regression approach. Moreover, the study also intended to determine the significance of solar parameters and geomagnetic storms used in the prediction of the Pc4 and Pc5 ULF magnetic pulsation and their impact on earthquake event preparation in low latitude regions.

$$Pc5=1.9250 + 0.0048\ Bz + 0.0017\ SWS + 0.1481\ SWP + 0.0001\ IE - 0.0046\ SYM/H \quad (2)$$

Table 3. Multiple Regression of Coefficient of four solar wind parameters and geomagnetic index for magnetic pulsation of Pc5.

Variables	Coefficients (Coef)	Standard error coef (SE coef)	T-value	p-value	Variation Inflation factor (VIF)
Constant	1.9250	0.0148	130.39	0.000	
Bz	0.0048	0.0017	2.82	0.005	2.05
SWS	0.0017	0.0003	43.80	0.000	1.27
SWP	0.1481	0.0027	54.89	0.000	1.36
SIE	0.0001	0.0000	9.79	0.000	1.72
SYM/H	-0.0046	0.0003	-16.79	0.000	1.32
			F (5, 48001) =1707.37		

Table 4. Multiple regression result of Magnetic pulsation of Pc5.

R^2	R^2- adjusted (R^2-adj)	R^2-predicted (R^2-pred)
15.10 %	15.09 %	15.07 %

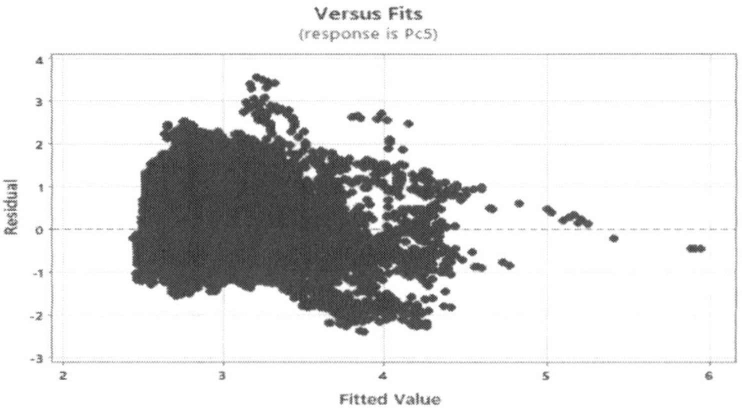

Figure 3. Residual versus fitted value plot of Pc5 with four solar wind parameters and geomagnetic indices (SYM/H).

4. Conclusion

This study concluded that the Pc5 ULF magnetic pulsation based on IMF-Bz, solar wind speed, solar wind dynamic pressure, solar wind input energy components and geomagnetic indices (SYM/H) has a greater effect than Pc4. Pc5 geomagnetic pulsation was found to be significantly correlated with the IMF-Bz, SWS, SWP, SWE, and SYM/H based on the generation of the regression model. This indicates that Pc5 is an appropriate wave band to study the possible earthquake precursor signature in ULF wave records by combining selected solar parameters and the SYM/H index. However, it appears that the Pc5 prediction model did not perform well and did not match model assumptions because R^2-predicted was greater than R^2 and a large residual with a few outliers was obtained. Future studies should fit and improve the model in order to realise the Pc5 prediction model for earthquake precursors at low latitude regions by eliminating irrelevant variables and using extended observation period data. In addition, the residual plot should be adjusted so that it corresponds with the prediction model's assumptions.

References

[1] Hattori K *et al* 2013 Investigation of ULF seismo-magnetic phenomena in Kanto, Japan during 2000–2010: case studies and statistical studies. *Surveys in Geophysics* **34** (3) 293-316
[2] Han P et al 2014 Statistical analysis of ULF seismomagnetic phenomena at Kakioka, Japan, during 2001–2010 *Journal of Geophysical Research: Space Physics* **119** (6) 4998-5011
[3] Silva G B, Padilha A L and Alves L R 2020 Latitudinal variation of Pc3–Pc5 geomagnetic pulsation amplitude across the dip equator in central South America. *In Annales Geophysicae* **38** (1) 35-49
[4] McPherron R L 2005 Magnetic pulsations: their sources and relation to solar wind and geomagnetic activity *Surveys in Geophysics* **26** (5) 545-592
[5] Masruri M F I *et al* 2018 Analysis of Preseismic Event Using Seismo-electromagnetics Data. *In 2018 2nd International Conference on Applied Electromagnetic Technology* 1-7 IEEE
[6] Takla E M *et al* 2018 Anomalous ultra-low frequency signals possibly linked with seismic activities in Sumatra, Indonesia NRIAG *Journal of Astronomy and Geophysics* **7** (2) 247-252
[7] Yusofa K A *et al* 2020 Normalized Polarization Ratio Analysis for ULF Precursor Detection of the 2009 M7, 6 Sumatra and 2015 M6 8 Honshu Earthquakes*E ngineering Journal SI* **3** (1) 35-41
[8] Hayakawa M *et al* 2000 ULF electromagnetic precursors for an earthquake at Biak, Indonesia on February 17, 1996, *Geophysical Research Letters* **27** (10) 1531-1534
[9] Hynönen R Tanskanen E I and Francia P 2020 Solar cycle evolution of ULF wave power in solar wind and on ground *Journal of Space Weather and Space Climate* **10** 43
[10] Filawati S 2021 Analysis of IMF Orientation and Pc5 Power During St. Patrick Geomagnetic Storm on 17 March 2015 *In Journal of Physics: Conference Series* **1805** (1) 012026
[11] Pranoto S C *et al* 2020 Ap c5 ULF waves during the geomagnetic storms on 7–8 September 2017 observed in the Indonesian region *In Journal of Physics: Conference Series* **1523** (1) 012025
[12] Kwon H J *et al* 2019 Characteristics of Pc5 activity at high latitudes stations in Antarctica *Journal of Atmospheric and Solar-Terrestrial Physics* **193** 105087
[13] Gjerloev J W 2012 The SuperMAG data processing technique *Journal of Geophysical Research: Space Physics* **117** (A9)
[14] Syed Zafar S N A *et al* 2021 Behaviour of Geomagnetic storm, horizontal geomagnetic field and solar wind parameters during solar flare and CMEs event *Journal of Physics: Conference Series, IOP Publishing* **1768** (1) 012003
[15] Uyanık G K and Güler N 2013 A study on multiple linear regression analysis *Procedia-Social and Behavioral Sciences* **106** 234-240
[16] Syed Zafar S N A *et al* 2019 Radio Frequency Interference on Nearby Radio Astronomical Lines: Relationship between Wind Speed and Radio Signal Strength Measured at East Coast of Peninsular Malaysia. *Sains Malaysiana* **48** (1) 183-189
[17] Sabri N H *et al* 2017 Correlation analysis of tropical rainforest climate effect on radio signal strength at KUSZA observatory, Terengganu. *Advanced Science Letters* **23** (2) 1268-1271
[18] Shapiro S S and Wilk M B 1965 An analysis of variance test for normality (complete samples). *Biometrika*, **52** (3/4) 591-611.
[19] Tranmer M and Elliot M 2008 Multiple linear regression. *The Cathie Marsh Centre for Census and Survey Research (CCSR)* **5** (5) 1-5
[20] Mat R *et al* 2018 The modelling of tropical weather effects on ultra-high frequency (UHF) radio signals using SmartPLS. *In IOP Conference Series: Materials Science and Engineering* 440
[21] Olive D J 2017 Multiple linear regression. In Linear regression 17-83
[22] Currie J L and Waters C L 2014 On the use of geomagnetic indices and ULF waves for earthquake precursor signatures *Journal of Geophysical Research: Space Physics* **119** (2) 992-1003
[23] Sharma A K, Patil A V and Haridas R N 2012 Investigation of ULF magnetic anomaly before moderate earthquakes. *Exploration Geophysics* **43** (1) 36-46

[24] Marchitelli V Harabaglia P Troise C and De Natale G 2020 On the correlation between solar activity and large earthquakes worldwide *Scientific reports* **10** (1) 1-10

[25] Yusof K A *et al* 2019 On effective ULF frequency ranges for geomagnetic earthquake precursor. *In Journal of Physics: Conference Series IOP Publishing* **1152** (1) 012033

[26] Ya'acob N *et al* 2011 Detecting the occurrence of ground ULF electromagnetic signal prior to earthquake events *In 2011 IEEE International RF & Microwave Conference* 54-57

[27] Athanasiou M A *et al* 2014 The ultra low frequency electromagnetic radiation observed in the topside ionosphere above boundaries of tectonic plates *Research in Geophysics* **4** (1)

Acknowledgement

This study was supported by the grant FRGS/1/2019/STG02/UNISZA/02/3. Special thanks are dedicated to Electromagnetic Research Group (EMRG) for their assistance in this work. Last but no least we want to acknowledge use of the data from SuperMAG and OMNIWeb.

Deriving offshore tidal datums using satellite altimetry around Malaysian seas

M H Hamden[1*], A H M Din[1,2] D D Wijaya[3] and M F Pa'suya[4]

[1] Geospatial Imaging and Information Research Group (GI2RG), Faculty of Built Environment and Surveying, Universiti Teknologi Malaysia, 81310 Johor Bahru, Johor, Malaysia
[2] Geoscience and Digital Earth Centre (INSTEG), Faculty of Built Environment and Surveying, Universiti Teknologi Malaysia, 81310 Johor Bahru, Johor, Malaysia
[3] Geodesy Research Division, Faculty of Earth Science and Technology, Institute of Technology Bandung, Jl. Ganesha 10, Bandung, Indonesia
[4] Environment and Climate Change Research Group, Faculty of Architecture, Planning & Surveying, Universiti Teknologi MARA, Perlis Branch, Arau Campus, 02600 Arau, Perlis, Malaysia

*Corresponding E-mail: mhanif87@live.utm.my; amihassan@utm.my

Abstract. Tidal datums are important for calculating spatial coordinates especially the elevation relative to mean sea level and also crucial for defining the state sovereignty boundaries over maritime areas. Normally, sea level was measured by tide gauges along the coastal for tidal datums computation. However, knowledge of tides is still restricted in coastal areas. Furthermore, tidal range at offshore was simply assumed to be similar as coastal due to the difficulties installing offshore tide gauges. The launching of satellite altimeter technologies with precise orbit determination since 1993 had provided significant accuracy of sea surface height (SSH) measurements. The observed SSH from satellite altimetry can be offered as tide gauge measurements at each location globally. This study aims to derive offshore tidal datums using satellite altimetry around Malaysian seas. SSH time series from TOPEX, Jason-1, Jason-2 and Geosat Follow On (GFO) were analysed using harmonic analysis approach to estimate harmonic constants. A minimum of 19 years tidal predictions were then performed using UTide software to determine Lowest Astronomical Tide (LAT) and Highest Astronomical Tide (HAT). These tidal datums were interpolated into regular 0.125° grids and were assessed with ten selected coastal tide gauges. The findings showed the Root Mean Square Error (RMSE) of spline interpolation yielded better accuracy, 25.5 cm (LAT_{MSL}) and 17.4 cm (HAT_{MSL}) as compared to the RMSE of Kriging interpolation, 31.8 cm (LAT_{MSL}) and 33.8 cm (HAT_{MSL}). In conclusion, deriving offshore tidal datums can serve as input data to unify marine database with coastal areas and also can support many marine applications.

Keywords: Tidal datums, Offshore, satellite altimetry, Lowest Astronomical Tide, Highest Astronomical Tide

Track Name: Coastal Management and Marine Ecosystem

1. Introduction

Tides are usually defined as the vertical periodic rise and fall of water on the surface of the earth that occurs twice in a little more than a day. According to [1], tides are also described as the periodic variation in the level surface of the ocean, inlets, bays, gulfs and estuaries resulting from the gravitational forces of the moon and the sun. Scientists have worked for the past two centuries to headway scientific knowledge of ocean tides [2]. The process of measuring tides is known as tidal observation by using an equipment called tide gauge station. Statistically averaging sea level measurements is the process of stabilising a fluctuating surface to serve as tidal datum [3]. There are many ways to average sea level which giving rise to a variety of tidal datums. Tidal datums are defined as a standard reference elevation used from which to reckon heights or depths in terms of a certain phase of the tide [4, 5, 6]. They are predominantly used to measure depth or water level as well as critical in defining spatial coordinates such as latitude, longitude and elevation with respect to Mean Sea Level (MSL). Not only that, they are also important as legal bodies for establishing the state sovereignty over maritime boundaries.

In determining the tidal datums, tides generally measured by tide gauges along the continental coastline which refer to coastal tide gauges. Although the long-term sea level observations can enhance the understanding of tides, the knowledge is still constrained in the vicinity of coastal areas. [2] stated that the tide measurements further away from coastal made by bottom pressure gauges is slow and less effective. In Malaysia, Department of Survey and Mapping Malaysia (DSMM) is the authority responsible for acquiring, processing, archiving and disseminating the long-term tidal data [7]. At present, there are eleven tide gauge stations along the Peninsular Malaysia coast (West Malaysia) and eight tide gauge stations along the coast of Sabah and Sarawak (East Coast). However, less discoverable on offshore tidal datums in this region is due to the difficulties installing the offshore tide gauges. There might be few offshore tide gauges being installed which could only restricted with the used by the offshore companies. In the past, the tidal range at the coastal was simply inferred to be similar to offshore and the tide phase was computed using shallow water wave theory [8].

Credit goes to an active remote sensing technique so called satellite altimetry, launched in 1970s that provided a comprehensive technique in measuring the global ocean. Satellite altimetry has been used by certain researchers to investigate ocean activities in the specific regions. For instance, [9] analyze the SSH from TOPEX series satellites of 19-year time series to extract the tidal harmonic constants in the Brazilian coast. It shows the Root Sum Square misfit (RSSmisfit) values are less than 12 cm in deep ocean. Besides, [2] has developed an empirical ocean tide models namely OSU12 models by utilizing an enhanced multi-mission satellite altimetry data from TOPEX, Jason-1/-2, Envisat and GFO based on a novel method via spatio-temporal combination, together with a robust estimation technique. The study shows substantial improvement which apparently in regions with high hydrodynamic variability. In Malaysia, [10] has studied on the derivation of tidal constituents from satellite altimetry for coastal vulnerability assessment. It used Mean Tidal Range parameter derived from tidal models where the tidal models were generated from TOPEX and Jason-1 data. The used of tidal models are projected to supplement the existing coastal management system in order to scrutinize the severity of coastal damages caused by sea level rise impacts, particularly in Malaysian coastal areas. Furthermore, Yahaya et al. [11] and Zulkifle et al. [12] have developed regional mean sea surface (MSS) models for Malaysian seas using multi-mission satellite altimetry data. This study generated MSS model by merging 11 years of repeated SSH observations from several satellite altimeter missions.

The improvement of precise orbit determination technique, instrumental and geophysical corrections have provided better accuracy of sea surface height (SSH) measurements since the launch of TOPEX/Poseidon mission [13]. The obtained SSH from satellite altimetry revisits at the similar point for each orbit cycle can be offered as tide gauge measurements at each location globally. Therefore, this study is an attempt to derive offshore tidal datums using satellite altimetry around Malaysian seas. SSH time series of TOPEX class (TOPEX, Jason-1 and Jason-2) and GFO missions were analysed using harmonic analysis approach to estimate the amplitude and phase of eight selected harmonic constants namely M2, S2, K1, O1, N2, K2, P1, Q1, MF, MM, SA and SSA. These harmonic constants were then used for tidal prediction of at least 19 years to determine the tidal datums. The tidal datums referred in this study are Lowest Astronomical Tide (LAT) and Highest Astronomical Tide (HAT) relative to MSL. Section 2 describes the materials and method, followed by results and discussion in section 3. Lastly, section 4 concludes the overall study.

2. Materials and method

2.1 Study Area and Datasets

Along track SSH from TOPEX class and GFO missions were extracted using Radar Altimeter Database System (RADS). The extracted SSH data spanning over 23 years for joint TOPEX class and 8 years for GFO mission as listed in table 1. TOPEX, Jason-1 and Jason-2 missions are denoted as TOPEX class because these missions are moving on the same orbit track. The study area was bounded within the geographical coordinates of 0° N - 9° N in latitude and 98° E - 121° E in longitude. Figure 1 illustrates the study area limited to the specified coordinates as well as the along track TOPEX class mission (blue) and GFO mission (green). The black triangles indicate the coastal DSMM tide gauge stations used to assess the estimated offshore tidal datum. Meanwhile, the red points are randomly selected for visualization of SSH time series.

Table 1. List of satellite altimetry data used.

Satellite Mission	Phase	Cycle	Period Time
TOPEX	A	001-364	January 1993 – August 2002
Jason-1	A	001-260	January 2002 – January 2009
Jason-2	A	001-303	July 2008 – October 2016
GFO	A	037-223	January 2000 – September 2008

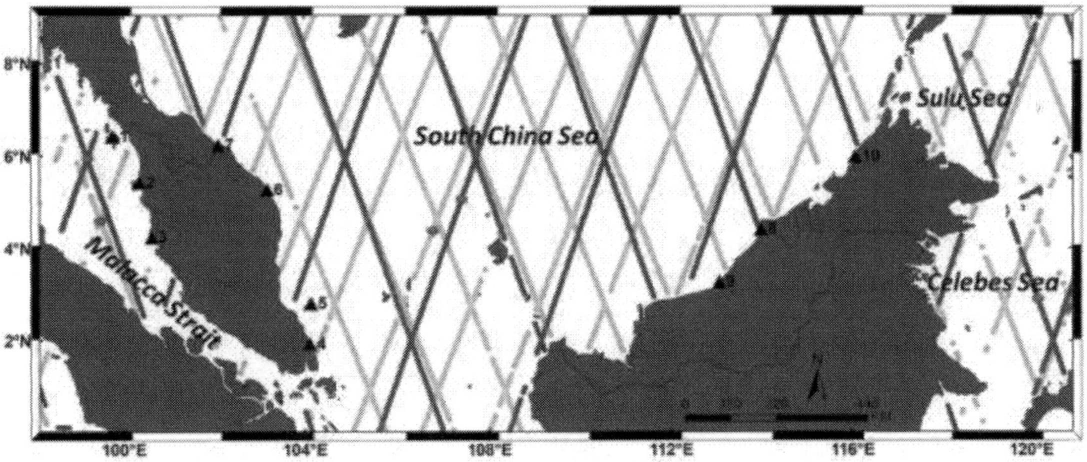

Figure 1. Along-track altimetry missions of TOPEX class (blue) and GFO (green) within the study area as well as the distribution of selected coastal DSMM tide gauge (black triangle).

2.2 Satellite Altimetry Processing and Formation of SSH Time Series

As aforementioned, all satellite altimetry used in this study were extracted using RADS server in Universiti Teknologi Malaysia (UTM), providing the latest information on orbits and geophysical corrections. The computation of SSH were processed by applying the preferred range and geophysical correction in the Malaysian region, removing the invalid data and also generating the corresponding refined corrections. SSH altimetry can be derived by using the equation (1) [10,14,15].

$$h_{ssh}^{obs} = H_{alt} - R_{obs} - (h_{dry} + h_{wet} + h_{ion} + h_{ssb} + h_{sol} + h_{pole} + h_{ocean} + h_{DAC} + \varepsilon) \quad (1)$$

Where, h_{ssh}^{obs} is the corrected sea surface height, H_{alt} is the satellite altitude, R_{obs} is the altimeter range measurement, $h_{dry}$is the dry tropospheric correction, h_{wet} is the wet tropospheric correction, h_{ion} is the ionospheric correction, h_{ssb} is the sea state bias correction, h_{sol} is the solid earth tide correction, h_{pole} is the pole tide correction, h_{ocean} is the ocean tide correction, h_{DAC} is the dynamic atmospheric correction and ε is the other errors induced in altimetry measurement. However, it is noted that $h_{ocean}$was excluded from being applied into the equation. This is to prevent the ocean tide signal to be eliminated as this signal is crucial in estimating harmonic constants.

2.2.1 Formation of SSH Time Series

This is done by extraction of SSH at each cycle of discrete points. Ground tracks of altimetry with repeated orbit missions generally do not coincide accurately with each other. Thus, collinear analysis is utilised to compute each SSH point of collinear tracks similar to the reference track. In this study, the collinear track of first cycle was treated as the reference track. Thus, from the reference track and corresponding collinear track, the formation SSH time series can be plotted as shown in figure 2.

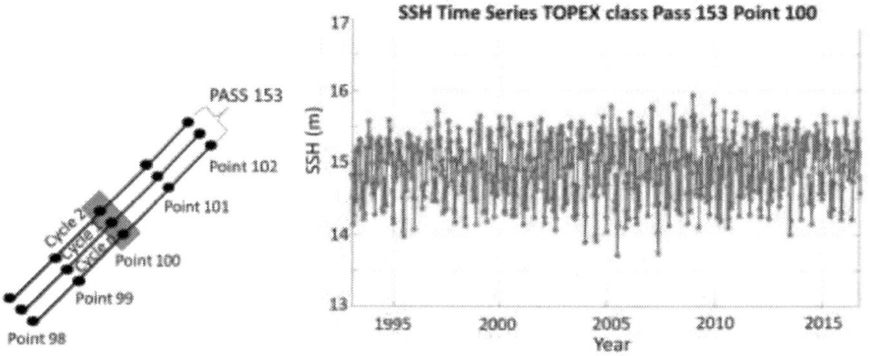

Figure 2. Schematic diagram on formation of SSH time series.

2.2.2 Tidal Aliasing.

TOPEX class and GFO missions have different repeated periods which are 9.9156 days and 17.0505 days, respectively. Thus, both missions suffer from tidal aliasing effect due to altimeter's long temporal sampling. According to Shannon sampling theorem, in order to completely rebuild the original signal, it must be sampled at least twice of the frequency, f_s . For instance, $f_N = 2f_s$, where f_N is the Nyquist frequency. A signal with period T_s can be completely rebuild if the samples are obtained at interval of less than $T_s/2$. If not, the signal of T_s are aliased to a longer period to T_a, which is aliased period [2,16]. The aliasing period can be computed by using the equation (2) [17].

$$T_a = \frac{2\pi\Delta s}{2\pi(f\Delta s - [f\Delta s + 0.5])} \quad (2)$$

Where, f is the frequency of tidal component, Δs is the period sample and the bracket [.] in the formula of $[f\Delta s + 0.5]$ is the fix function that return greatest integer less than argument. The information of actual period and aliased period from TOPEX class and GFO missions are tabulated in Table 2. The aliased period obtained from both missions are calculated by using equation (2). The consequence of the aliasing effect on tide signals detected from altimetry mission is that the length of each tidal constituent appears to be longer than its actual period. The aliasing period of each constituent is used in harmonic analysis to estimate amplitude and phase as well as for tidal prediction.

Table 2. Actual and Aliased Tidal Period for TOPEX class (9.9156 days) and GFO (17.0505 days).

Tidal Constituents	Actual Period (cph)	Actual Period (cpd)	TOPEX Aliased Period (cpd)	GFO Aliased Period (cpd)
M2	12.42	0.52	62.11	317.11
S2	12.00	0.50	58.74	168.82
K1	23.93	1.00	173.19	175.45
O1	25.82	1.08	45.71	112.95
N2	12.66	0.53	49.53	52.07
K2	11.97	0.50	86.60	87.72
P1	24.07	1.00	88.89	4467.14
Q1	26.87	1.12	69.36	74.05
MF	327.86	13.66	36.17	68.71
MM	661.31	27.55	27.55	44.73
SSA	4382.92	182.62	182.62	182.62
SA	8765.74	365.24	365.26	365.26

Notes: *cph = cycle per hour, cpd = cycle per day*

2.3 Tidal Analysis and Prediction

Harmonic analysis approach was adopted for tidal analysis and prediction in this study. This analysis is performed using MATLAB package 'Unified Tidal analysis and prediction' or UTide, developed by [18]. Based on [18], the UTide was built on the foundation of T_TIDE by [19], integrating concepts from [20] and [21]. SSH time series from TOPEX and GFO missions were analysed to estimate the selected tidal harmonic constants at each point of along-track. Then, these estimated tidal harmonic constants (amplitudes and phase lags) were put into tidal harmonic prediction equation in order to predict the tides. A tidal prediction can be computed by summing up the oscillating contributions of some number of tidal constituents. The formula for tidal analysis and prediction used in this study are expressed in the equation (3) [16, 18, 22]:

$$H(\varphi,\lambda,t) = Z_0(\varphi,\lambda) + \sum_{k=1}^{n} f_k\, A_k(\varphi,\lambda)\cos[\omega_k t + V_{0k} + u_k - G_k(\varphi,\lambda)] \quad (3)$$

Where, $H(\varphi,\lambda,t)$ is the sea level for predefined time (t) at discrete location (φ,λ) . $Z_0(\varphi,\lambda)$ is the mean sea level of analysed data and n represents the total number of tidal constituents used; A_k, ω_k and G_k represent the amplitude, frequency and delay phase of ith tidal constituents, respectively; V_{0k} is the astronomical argument, while f_k and u_k are nodal factor and nodal phase, respectively. Based on [23], LAT and HAT can be computed over a minimum of 19 years using harmonic constants estimated from at least 19 years of tidal data or other method known to produce reliable results. However, according to [24], using 19 years of tidal prediction to compute LAT and HAT are always reasonable. Thus, in this study, the tides were predicted for at least 19 years to determine the tidal datums.

2.4 Tidal Interpolation Method

The generation of tidal datums surfaces in this study involves two interpolation method namely Ordinary Kriging (OK) and a minimum curvature (MCr) technique called (regularised) spline. Both methods were used to interpolate the along-track altimetry data into a regular grid of 0.125°. Both were assessed to determine the best interpolation method for generation of tidal datums surfaces. Assessment was performed by comparing the interpolated points with the selected coastal tide gauges.

2.4.1 Ordinary Kriging

Kriging is a geostatistical interpolation method which is based on statistical models inclusive of autocorrelation. This method fits a mathematical function to specific points or all points within the specific radius to determine the output value of predicted locations. The formula is expressed in equation (4) [25, 26].

$$\hat{Z}(s_0) = \sum_{i=1}^{N} \lambda_i Z(s_i) \tag{4}$$

Where, $Z(s_i)$ is the measured value at the ith location; λ_i is an unknown weight for the measured value at the ith location; $\hat{Z}(s_0)$ is a predicted value and N is the total number of measured values.

2.4.2 Minimum Curvature Spline

This method is widely used in the earth sciences. The interpolated surface is equivalent to a thin, linearly elastic plate moving through each data with the minimum amount of bending [20]. The formula for spline interpolation is expressed in equation (5) [27, 28].

$$S(x,y) = T(x,y) + \sum_{j=1}^{N} \lambda_j R(r_j) \tag{5}$$

Where, N is the number of points, λ_j is the coefficient found by the solution of linear equations and r_j is the distance between point (x, y) and jth point. Since this study applied the regularised option, $T(x, y)$ and $R(r)$ were expressed in the equation (6) and (7).

$$T(x,y) = a_1 + a_2 x + a_3 y \tag{6}$$

Where, $a_1, a_2 \ldots$ are the coefficient found by the solution of linear equations.

$$R(r) = \frac{1}{2\pi}\left\{\frac{r^2}{4}\left[\ln\left(\frac{r}{2\tau}\right) + c - 1\right] + \tau^2\left[K_o\left(\frac{r}{\tau}\right) + c + \ln\left(\frac{r}{2\pi}\right)\right]\right\} \tag{7}$$

Where, r is the distance between the point and sample, τ^2 is the weight parameter, K_o is the modified Bessel function and c is the constant equal to 0.577215.

2.5 Statistical Data Assessment

The Root Mean Square Error (RMSE) is used to measure the quality of the results where it computes the difference between the predicted values and the true values. In this study, there are two statistical data assessment were performed. First, the RMSE was calculated to determine the reliability of the predicted SSH derived from satellite altimetry by comparing with the observed SSH. Second assessment was performed by calculating the RMSE values between the interpolated points from two interpolation techniques and the selected coastal tide gauges which can be expressed in equation (8).

$$RMSE = \sqrt{\frac{\sum (x_p - x_t)^2}{n}} \tag{8}$$

Where, x_p is the predicted values (i.e., interpolated points), x_t is the true value which is the coastal tide gauges, and n is the total number of samples. However, for the first assessment, x_p is indicated as predicted SSH values while x_t is the observed SSH values.

3. Results and Discussion

3.1. Time Series Modelling and Residuals

SSH time series of TOPEX class and GFO missions were modelled by applying the equation (3). These time series were predicted at each point of along-track altimetry as illustrated in figure 1. Each point from the along-track of predicted SSH time series were randomly selected (labelled as red dots in figure 1) at each Malaysian sea's regions namely the Malacca Straits, the South China Sea, the Sulu Sea, and the Celebes Sea. This is to visualise the modelled time series. It is noted that the point must be selected at the offshore area. Figure 3 shows the selected SSH modelled time series and its residuals from TOPEX class on the left column well as GFO mission on the right column. The observed and predicted SSH time series are plotted in blue and green lines, respectively. Meanwhile, the red plotted lines indicate as the residuals of SSH time series. These residuals were calculated by computing the differences between the predicted and observed SSH time series. Subsequently, the quality of the predicted SSH can be determined based on these residuals by using RMSE computation. It can be seen that the SSH time series data from TOPEX class is denser than GFO mission. This is because the repeated period of TOPEX class is shorter than GFO which is 9.9156 days and 17.0505 days, respectively. The highest RMSE value between observed and predicted SSH from TOPEX is at Malacca Straits which recorded 10.1cm, followed by Celebes Sea (9.8 cm), Sulu Sea (7.6 cm) and South China Sea (7.3 cm). This might be due that Malacca Straits is a closed sea area which the tidal characteristics would most likely have a large gradient. Nevertheless, the highest RMSE value from GFO mission is at Celebes Sea which recorded 10.9 cm, followed by Malacca Straits (7.9 cm), Sulu Sea (7.0 cm) and South China Sea (6.5 cm). Therefore, it can be inferred that the precision of tidal prediction from both missions were reasonable at offshore area since the RMSE is within 10.9 cm to 6.5 cm.

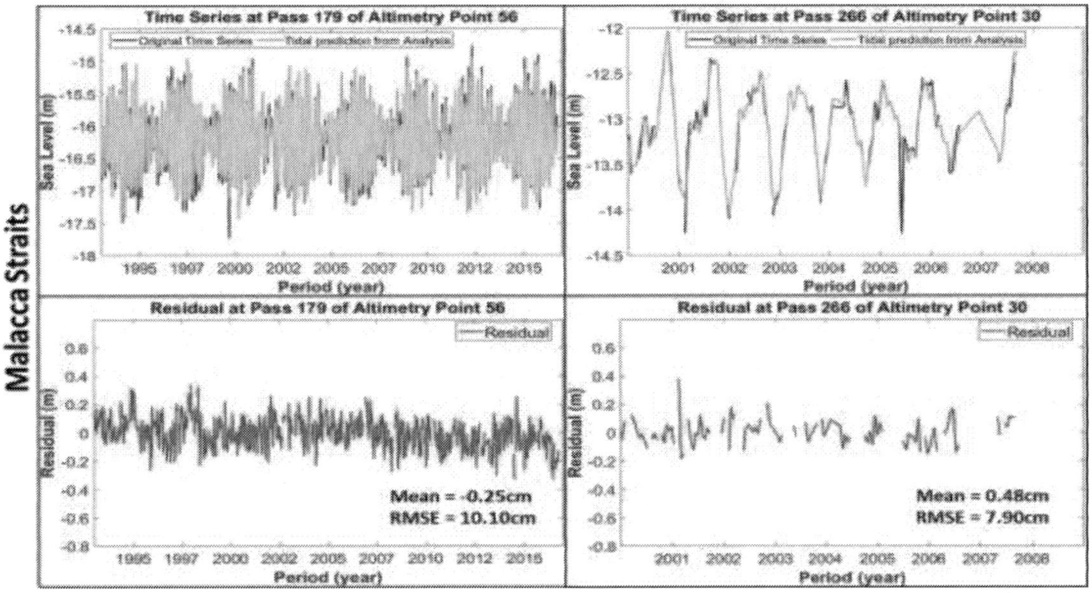

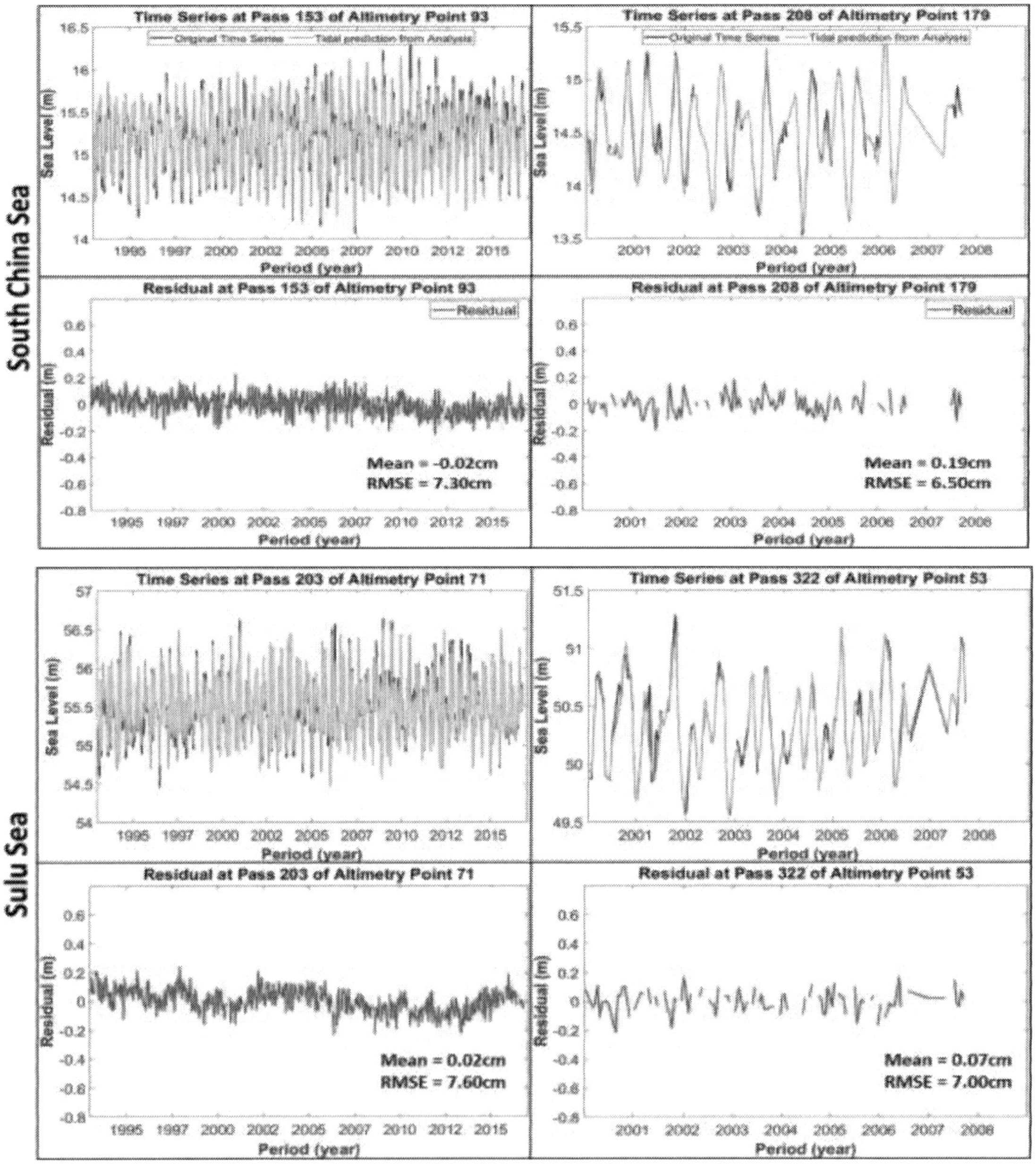
South China Sea
Time Series at Pass 153 of Altimetry Point 93
Original Time Series
Total prediction from Analysis
Sea Level (m)
Period (year)
Residual at Pass 153 of Altimetry Point 93
Residual
Residual (m)
Mean = -0.02cm
RMSE = 7.30cm
Time Series at Pass 208 of Altimetry Point 179
Residual at Pass 208 of Altimetry Point 179
Mean = 0.19cm
RMSE = 6.50cm
Sulu Sea
Time Series at Pass 203 of Altimetry Point 71
Residual at Pass 203 of Altimetry Point 71
Mean = 0.02cm
RMSE = 7.60cm
Time Series at Pass 322 of Altimetry Point 53
Residual at Pass 322 of Altimetry Point 53
Mean = 0.07cm
RMSE = 7.00cm

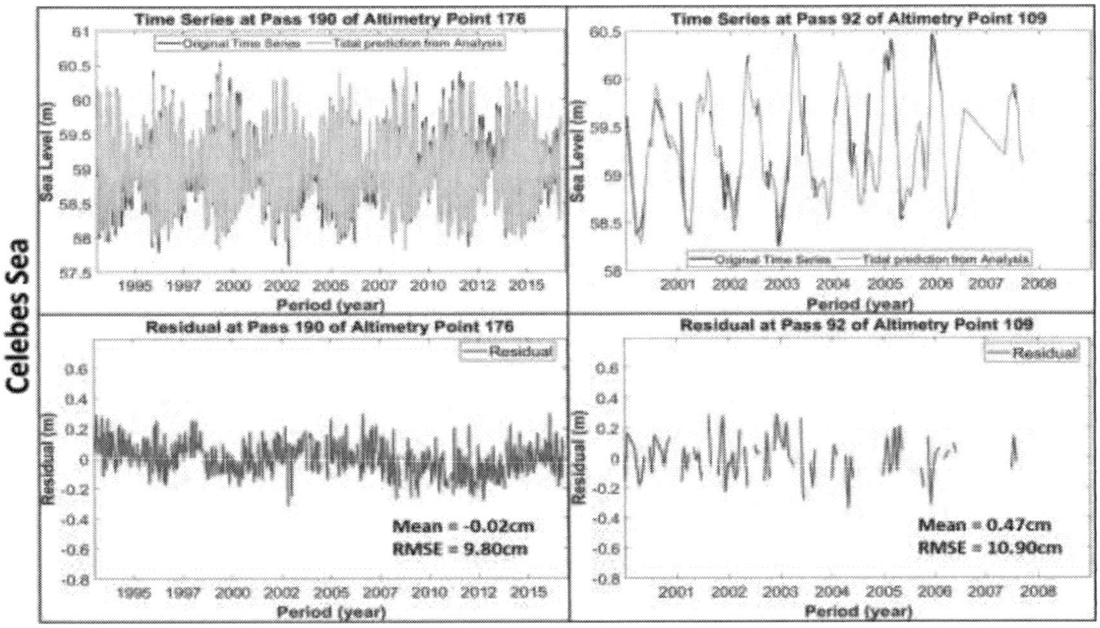

Figure 3. Observed (blue) and predicted (green) time series and residuals (red) of TOPEX class (left) and GFO (right) missions according to the areas (Malacca Straits, South China Sea, Sulu Sea and Celebes Sea).

3.2. LAT and HAT with respect to MSL

LAT and HAT can be defined as the lowest or (highest) water level which can be predicted to occur under average meteorological conditions as well as any combination of astronomical conditions. Both may be derived by the analysis of a number of years of tidal data or predictions which is normally 18.6 years to account for the full nodal cycle. In this study, the SSH time series from both missions were predicted for at least 19 years or more where predicted time series from TOPEX class is between 1993 until 2019. Meanwhile, predicted time series from GFO is between 2000 until 2019. The lowest and highest predicted tide indicate the LAT and HAT, respectively. Later, the derived LAT and HAT from satellite altimetry were interpolated and validated against the selected coastal DSMM tide gauges as distributed in Figure 1. Before the results could be compared with the tide gauges, the derived data need to be converted with respect to MSL. The computation of LAT and HAT with respect to MSL was depicted in Figure 4.

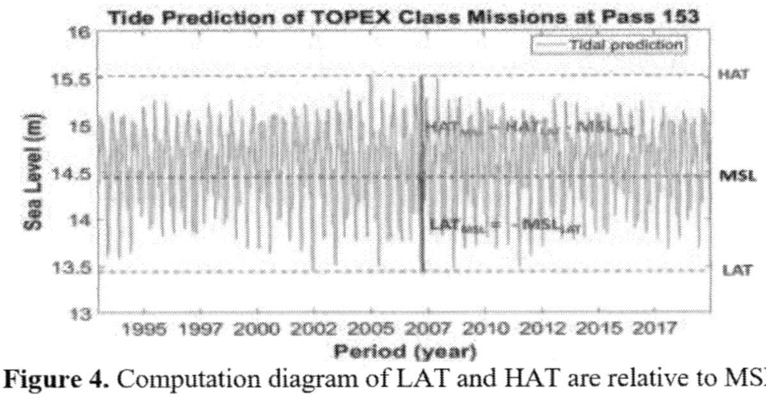

Figure 4. Computation diagram of LAT and HAT are relative to MSL.

Along-track satellite derived tidal datums from TOPEX class and GFO missions had to be merged together before being interpolated into a regular grid. Crossover offset was applied to the along-track

GFO mission to a TOPEX reference surface in order to minimise the orbital track errors and the discrepancy of satellite's orbit frame between two missions. The combination of TOPEX class and GFO along-track missions (LAT_{MSL} and HAT_{MSL}) were then interpolated into a regular grid of 0.125° by using ordinary kriging and minimum curvature spline method as shown in Figure 5. Based on Figure 5, it can be seen that the middle of the Malacca Straits has the greatest tidal range of LAT_{MSL} and HAT_{MSL} compared to other regions which recorded up to -2.6 m and 3.0 m, respectively. The tidal range of LAT_{MSL} and HAT_{MSL} in the middle of South China Sea, Sulu Sea, and Celebes Sea show the values within -0.9 m to -1.1 m (LAT_{MSL}) and the values within 0.9 m to 1.2 m (HAT_{MSL}). The greatest tidal range of LAT_{MSL} and HAT_{MSL} also can be seen at the southwest of East Malaysia in the South China Sea which recorded up to -2.1 m and 2.2 m, proportionately. Meanwhile, at the north-western part of South China Sea, near to the Gulf of Thailand depicted the lowest tidal range of LAT_{MSL} and HAT_{MSL} which recorded the values near to zero meter. Apparently, the tidal ranges are mostly larger near to the coastal areas compared to the offshore areas. It is visible at the coastal part of Celebes Sea as shown in Figure 5. The reason is due to the satellite altimetry data acquired near the coastlines are contaminated by the inclusion of land in the footprint signal or by the fact that the tide is on the ebb [2]. Moreover, the tidal datum models using ordinary kriging method generated smoother contour lines as compared to minimum curvature spline method. However, these gridded offshore tidal datum models were validated with the selected coastal tide gauges in order to identify the best interpolation method.

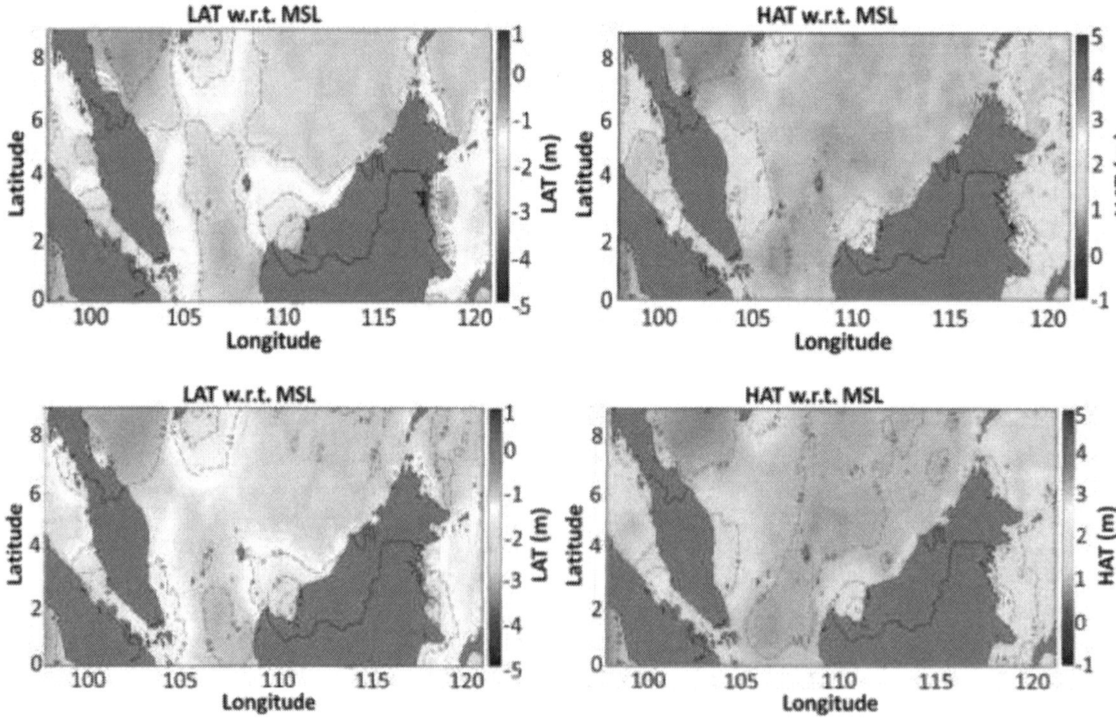

Figure 5. LAT_{MSL} and HAT_{MSL} models using ordinary kriging (Top) and minimum curvature spline (Bottom).

3.3. Statistical Assessment of Tidal Models

Generally, the best way to validate these offshore tidal datums are by comparing them with the offshore tide gauges. However, lack of deployment offshore tide gauges as well as difficulty in obtaining the offshore tidal data from the offshore authorities had hindered this validation method. Thus, this study adopted the statistical assessment of offshore tidal datums by validating the satellite altimetry derived offshore tidal datums models with ten selected DSMM tide gauges. The results of the assessment were described in Table 3 (LAT_{MSL}) and Table 4 (HAT_{MSL}).

Table 3. Statistical results between satellite derived offshore LAT_{MSL} (ordinary kriging and minimum curvature) and in situ tide gauge data (Units in meter).

Tide Gauge	Station	In Situ	Ordinary Kriging	Minimum Curvature	Ordinary Kriging – In Situ	Minimum Curvature – In Situ
1	P. Langkawi	-1.781	-1.758	-1.516	0.023	0.265
2	P. Pinang	-1.662	-1.117	-1.330	0.545	0.332
3	Lumut	-1.800	-1.293	-1.425	0.507	0.375
4	Tg. Sedili	-1.684	-1.389	-1.428	0.295	0.256
5	P. Tioman	-1.894	-1.584	-1.529	0.310	0.366
6	Cendering	-1.395	-1.210	-1.219	0.185	0.176
7	Geting	-0.717	-0.724	-0.919	-0.007	-0.202
8	Miri	-1.135	-1.212	-1.274	-0.077	-0.139
9	Bintulu	-1.389	-1.689	-1.528	-0.300	-0.139
10	K. Kinabalu	-1.223	-0.843	-1.089	0.380	0.134
				Mean	**0.186**	**0.142**
				STD	**0.258**	**0.211**
				RMSE	**0.318**	**0.255**

Table 4. Statistical results between satellite derived offshore HAT_{MSL} (ordinary kriging and minimum curvature) and in situ tide gauge data (Units in meter).

Tide Gauge	Station	In Situ	Ordinary Kriging	Minimum Curvature	Ordinary Kriging – In Situ	Minimum Curvature – In Situ
1	P. Langkawi	1.751	1.915	1.482	0.164	-0.269
2	P. Pinang	1.391	1.508	1.413	0.117	0.022
3	Lumut	1.578	1.449	1.458	-0.129	-0.120
4	Tg. Sedili	1.429	1.045	1.189	-0.384	-0.240
5	P. Tioman	1.762	1.609	1.559	-0.153	-0.203
6	Cendering	1.527	1.472	1.429	-0.055	-0.098
7	Geting	1.012	1.020	1.027	0.008	0.015
8	Miri	1.182	1.180	1.173	-0.002	-0.009
9	Bintulu	0.977	1.901	1.298	0.924	0.321
10	K. Kinabalu	1.213	0.974	1.141	-0.240	-0.072
				Mean	**0.025**	**-0.065**
				STD	**0.337**	**0.161**
				RMSE	**0.338**	**0.174**

For offshore tidal datum of LAT_{MSL}, the RMSE values obtained between the minimum curvature spline model and in situ data is smaller than the ordinary kriging model which recorded 25.5 cm and 31.8 cm, respectively. Meanwhile, the RMSE values of HAT_{MSL} between minimum curvature spline and in situ data is also smaller than the ordinary kriging model which yielded 17.4 cm and 33.8 cm, respectively. Thus, it can be inferred that the offshore tidal datums models generated using minimum curvature spline method has better agreement with coastal tide gauges compared to ordinary kriging models despite ordinary kriging produced smooth contour surfaces.

4. Conclusion

This paper generally studies on deriving the offshore tidal datums using SSH satellite altimetry data from TOPEX class and GFO missions around Malaysian seas bounded to the latitude of 0° N - 9° N and longitude of 98° E - 121° E. SSH time series of both missions were analysed by adopting harmonic analysis approach to estimate the selected tidal constituents. The estimated tidal constituents were then used to predict the tides at each of along-track altimetry time series points. The outcomes from this study illuminated that the predicted and observed tides at offshore areas have good precision which yielded the RMSE values within 6.5 cm to 10.9 cm. The models of LAT_{MSL} and HAT_{MSL} were generated by using two different interpolation methods namely ordinary kriging and minimum curvature (regularised) spline and were assessed with selected coastal tide gauges. The results showed the models (LAT_{MSL} and HAT_{MSL}) adopting regularised spline method have the smallest RMSE values which indicates the best interpolation method. Therefore, it can be concluded that the regularised spline method is the best interpolation method in this study to predict the offshore tidal datum compared to the ordinary kriging method.

In conclusion, this study indirectly can create an awareness towards Malaysian hydrographic society especially regarding the importance of satellite altimetry in supporting hydrographic survey practice. The encouraging results from this study are capable in establishing seamless vertical datum by integrating the tidal datums between coastal and offshore area. Other than that, deriving tidal datum is necessary to support in the establishment of marine boundaries as well as the requirement for conducting shoreline mapping.

References

[1] Rahibulsadri R *et al* 2014a Determination of Tidal Datum for Delineation of Littoral Zone for Marine Cadastre in Malaysia. In: Abdul Rahman A., Boguslawski P., Anton F., Said M., Omar K. (eds) *Geoinformation for Informed Decisions*. Lecture Notes in Geoinformation and Cartography. 219-230

[2] Fok H S 2012 Ocean Tides Modeling using Satellite Altimetry *Ohio State University Geodetic Science Report no.* **501**

[3] Hands E B 2005 Tidal Datums. In: Schwartz M.L. (eds) Encyclopedia of Coastal Science. *Encyclopedia of Earth Science Series*. Springer, Dordrecht.

[4] Gill S K and Schultz J R 2001 Tidal datums and their applications. *Silver Spring MD, NOAA, NOS Center for Operational Oceanographic Products and Services*, 102pp. & Appendix. (NOAA Special Publication NOS CO-OPS 1).

[5] Rahibulsadri R *et al* 2014b Tidal Datum Consistency for Marine Cadastre Littoral Zone Commencement in Malaysia *FIG Congress 2014 Engaging the Challenges – Enhancing the Relevance*, Kuala Lumpur, Malaysia, 16-21 June 2014

[6] Haigh I D 2016 Tidal Datum. In: Kennish M.J. (eds) *Encyclopedia of Estuaries. Encyclopedia of Earth Sciences Series.* Springer, Dordrecht.

[7] Din A H M *et al* 2017 Malaysian sea water level pattern derived from 19 years tidal data *J. Teknol* **79** 137–145

[8] Gill S K and Porter D L 1980 Theoretical Offshore Tide Range Derived from a Simple Defant Tidal Model Compared with Observed Offshore Tides *International Hydrographic Review*, Monaco LVII (**1**)

[9] Daher V B *et al* 2015. Extraction of tide constituents by harmonic analysis using altimetry satellite data in the Brazilian coast. *J. Atmos. Ocean. Technol* **32** 614–626

[10] Ainee A 2016 Derivation of Tidal Constituents from Satellite Altimetry Data for Coastal Vulnerability Assessment In Malaysia. *Master of Science Thesis* Universiti Teknologi Malaysia Skudai

[11] Yahaya N A Z *et al* 2016 *Mean sea surface (MSS) model determination for Malaysian seas using multi-mission satellite altimeter*. International Archives of the Photogrammetry *Remote Sensing and Spatial Information Sciences - ISPRS Archives* **42** 247–252

[12] Zulkifle M. I *et al* 2019 Determination of a localised mean sea surface model for Malaysian seas using multi-mission satellite altimeter *ASM Science Journal* **12** 81-89

[13] Din A H M *et al* 2014 Derivation of sea level anomaly based on the best range and geophysical corrections for malaysian seas using radar altimeter database system (RADS) *Jurnal Teknologi*, **71** 4 83-91

[14] Din A H M *et al* 2019 Sea level trend over Malaysian seas from multi-mission satellite altimetry and vertical land motion corrected tidal data. *Advances in Space Research* **63** 11 3452-3472

[15] Hamid A I A *et al* 2018 Contemporary sea level rise rates around Malaysia: Altimeter data optimization for assessing coastal impact *Journal of Asian Earth Sciences* **166** 247-259
[16] Pirooznia M et al., 2016 The Time Series Spectral Analysis of Satellite Altimetry and Coastal Tide Gauges and Tide Modeling in the Coast of Caspian Sea *Open J Mar Sci* **06** 258–269
[17] Lindsley R D 2013 Fitting Tidal Constituents to Altimeter Data
[18] Codiga D L 2011 Unified Tidal Analysis and Prediction Using the UTide Matlab Functions. Technical Report 2011-01. Graduate School of Oceanography University of Rhode Island Narraganset RI **59**
[19] Pawlowicz R *et al* 2002. Classical tidal harmonic analysis including error estimates in MATLAB using T_TIDE.Comput Geosci **28** 929–937
[20] Leffler K E and Jay D A 2009. Enhancing tidal harmonic analysis: Robust (hybrid L1/L2) solutions. *Cont Shelf Res* **29**78–88.
[21] Foreman M G G 1977 Manual for tidal heights analysis and prediction *Institute of Ocean Sciences, Pacific Marine Science Report* **97** 77-10
[22] Fu Y *et al* 2020 Evaluation of satellite-derived tidal constituents in the South China Sea by adopting the most suitable geophysical correction models *J. Oceanogr* **76** 183–196
[23] IHO 2018 Resolution on datums and benchmarks A2.5 3/1919. *International Hydrographic Organization*. https://www.iho. int/iho_pubs/misc/M3-E-AUGUST18.pdf
[24] Byun D S and Hart D E 2019 On Robust Multi-Year Tidal Prediction Using T_TIDE *Ocean Sci J* **54** 657–671
[25] Yilmaz H M 2007 The effect of interpolation method in surface definition: an experimental study *Earth Surf. Process, Landforms* **32** 1346-1361
[26] Environmental Systems Research Institute 2016a. Raster Interpolation toolset concepts: How Kriging works. Available at https://desktop.arcgis.com/en/arcmap/10.3/tools/3d-analyst-toolbox/how-kriging-works.htm# (Accessed 1 November, 2020).
[27] Franke R 1982. Smooth Interpolation of Scattered Data by Local Thin Plate Splines. *Computer and Mathematics with Applications* **8** 4 273-281
[28] Environmental Systems Research Institute 2016b. Raster Interpolation toolset concepts: How Spline works. Available at https://pro.arcgis.com/en/pro-app/latest/tool-reference/3d-analyst/how-spline-works.htm# (Accessed 1 November, 2020).

Acknowledgments
Highly acknowledged to the TU Delft, Altimetric LLC for providing the satellite altimetry data through Radar Altimeter Database System (RADS). Appreciation to the Department of Survey and Mapping Malaysia (DSMM) for providing coastal tide gauge data. This research is funded by the Ministry of Higher Education (MOHE) under the Fundamental Research Grant Scheme (FRGS) Fund, Reference Code: FRGS/1/2020/WAB05/UTM/02/1 (UTM Vote Number: R.J130000.7852.5F374).

The implementation and advancement of a regional geodetic vertical datum

N M Yazid[1], A H M Din[1,2]*, A H Omar[3], M F Pa'suya[4] and N M Abdullah[5]

[1] Geospatial Information and Imaging (GI2RG)
[2]Geoscience and Digital Earth Centre (INSTEG), Faculty of Built Environment and Surveying, Universiti Teknologi Malaysia, 81310 Skudai, Johor, Malaysia
[3]Geomatics Innovation Research Group (GnG), Faculty of Built Environment and Surveying, Universiti Teknologi Malaysia, 81310 Skudai, Johor, Malaysia
[4]Environment and Climate Change Research Group, Faculty of Architecture, Planning & Surveying, Universiti Teknologi MARA, Perlis Branch, Arau Campus, 02600 Arau, Perlis, Malaysia
[5]Faculty of Civil Engineering and Built Environment, Universiti Tun Hussein Onn Malaysia, 86400, Parit Raja, Johor, Malaysia

*Corresponding E-mail: amihassan@utm.my

Abstract. The essential parameter in computing three-dimensional coordinate system is the height or depth of the Earth's surface. It represents a particular reference surface that recognised as a vertical datum. The vertical datum is alienated into two foremost categories recognised as Mean Sea Level and Lowest Astronomical Tide. Different modifications approach, techniques and software programs are developed to determine vertical datum of a region with respect to geoid surface. This paper presents an effort to review and discuss the implementations and advancement of geodetic vertical datum based on geoid height reference surface. Hence, there are eight countries will be extracted and outlined in this paper consist of the United States of America, Australia, Taiwan, New Zealand, South Korea, Thailand, Philippines and Malaysia. An overview of geodetic vertical datum which implemented in these countries are summarised to support the future development of a regional vertical datum model. Then, the overview will also be utilised and analysed based on the essential elements and parameters for vertical datum model determination which include: data gathering, data input and analysis approach in order to develop a geodetic vertical datum model with good accuracy. These attempt and initiative are vital for the current and future implementation and advancement of geodetic vertical datum in the region of Malaysia across land and marine areas.

Keywords: Height, Vertical Datum, Mean Sea Level, Lowest Astronomical Tide, Geoid Model

Track Name: Coastal Management and Marine Ecosystem

1. Introduction

Vertical coordinates or heights of points are based on a coordinate surface known as vertical datum. The vertical datum of a height network is described as remitting the zero height to the Mean Sea Level

(MSL) referred to a particular period at a classified tide gauge in the prior 200 years. Traditionally, this reference height surface is implemented in countries to determine a regional vertical datum. Conventional spirit level is used to transfer the determined MSL value of the tide gauge measurement from the benchmark to another point for determining height differences that provides good accuracy [1]. Nevertheless, in a large area practically, it is costly and challenging particularly that require a huge point's number towards the need of height.

A seamless reference surface represents a continuous and time-invariance surface. Due to the irregular topographic surface of the Earth, the geodetic computations and determinations are difficult to be performed. To clarify this issue, geodesists utilise a flat and even mathematical reference surface to determine the irregular shape of the Earth's surface, hence, precisely explains as to determine a regional geoid height. [2] defines geoid as an equipotential surface of the Earth's gravity field closely approximates with mean sea level. Besides, geoid is also defined as the equipotential surface of the Earth's gravity field that best fits to the mean sea level [3] with a constant value of the gravity potential on the surface.

Global Navigation Satellite System (GNSS) is a revolution in this era of survey and mapping activities, which has turned out to be more necessitating than conventional levelling measurement [4]. The evolution of GNSS in providing a position with good accuracy up to mm level has been verified but there is lack of physical description. Thus, height transformation procedure is needed with the availability of high resolution and precise gravimetric geoid model. The GNSS ellipsoidal height is transformed to orthometric height (H) reference to geoid height by subtracting geoid heights (N) from the GNSS ellipsoidal heights (h).

GNSS levelling plays an important role in providing a good accuracy of geoid model. Currently, the height system has revolutionized by several countries through geoid modelling for determining orthometric height implementing GNSS levelling instead of conventional levelling. The initiation of new gravity from satellite gravity missions which provide an accurate Global Geopotential Model (GGMs) to be combined with the terrestrial gravity data either from land or marine area which the gravity data can also be integrated with several technique that require to be optimised properly with good geoid model accuracy.

Therefore, this paper affords to revise the techniques implemented towards the establishment of geodetic vertical datum implementation for a region. Six (6) countries are elected in this study in order to revise the geodetic vertical datum techniques and methods implements in Malaysia, United States of America, Australia, Taiwan, New Zealand, South Korea, Philippines and Thailand.

2. Geodetic Vertical Datum Implementation

There are numbers of vertical datum implemented for land and marine area in the determination of height or depth. However, the tidal observation is utilised to a reference datum, entirely [5]. Survey and mapping in Peninsular Malaysia is referred to Peninsular Geodetic Vertical Datum (PMGVD) via First Order Precise Levelling Network (FOPLN) as a vertical datum. Nevertheless, for charting activities, local Chart Datum (CD) via Lowest Astronomical Tide (LAT) are utilised. The knowledge about the discrepancy of height or depth at the coastal areas turn out to be crucial due to rapid development in a region [6]. The vertical network establishment in Peninsular Malaysia, Sabah and Sarawak are developed discretely. Besides, the levelling networks in Sabah and Sarawak are disjointed and referred to numbers of vertical datum [7].

Peninsular Malaysia Seamless Geoid Model 2014 (PMSGM2014) by [8] is the initial gravimetric geoid model in Peninsular Malaysia, which is computed using the Least Squares Modification of Stokes with Additive Corrections (LSMSA) or known as KTH approach. Nevertheless, this geoid model is developed by utilising terrestrial land and marine gravity without the utilisation of airborne gravity data. Thus, two types of geoid model is presented known as the gravimetric and hybrid (fitted) geoid models. The verification is performed by implementing GNSS-levelling points with 10 cm accuracy of the geoid model.

Nevertheless, an accuracy of a hybrid geoid model with a mean parametric surface corrector model established by [8] provide high Root Mean Square (RMS) of 7 cm as evaluated to the official geoid model over Peninsular Malaysia, MyGEOID. This is defying as the development of PMSGM2014 model without the utilisation of airborne gravity data with the gravity data problem related to the sparse of terrestrial gravity data that affected an accuracy of a gravimetric geoid model. Next in 2020, [9] has been developed, a latest height reference system modernisation in Peninsular Malaysia using LSMSA technique by integrating the airborne gravity data known as Peninsular Hybrid Geoid 2020 (PMHG2020). The hybrid geoid models were validated with GNSS levelling points and provided the optimum two parameters comprised of 50 km correlation length and 0.03 m with an accuracy, RMSE of 0.04716 m.

In United States of America, American Vertical Datum (GRAV-D) has been proposed by the National Geodetic Survey (NGS) in 2022 to redefine the vertical datum with the integration of airborne gravity, terrestrial gravity, latest satellite gravity and altimetry data from DTU13 known as XGEOID2020 [10]. The aim of this project is to achieve 2 cm accuracy level of the gravity-based vertical datum and comprised in the NGS 10-year plan.

The NGS has a distinctly mentioned mission to delineate, maintain and provide retrieve to the National Spatial Reference System (NSRS). The major fundamental of the NSRS is the height determinations with regard to the ellipsoidal height, orthometric height and dynamic height which any points or locations in the United States or its territories. Besides, NGS play a crucial role in of presenting the basic data and information for federal mapping activities in the nation.

The varies of the ellipsoidal heights over times in the United States have been estimating by NGS with the utilisation of GPS Continuously Operating Reference Stations (CORS). Moreover, the current regional gravity data has been surveyed and measured by NGS, then the gravimetric geoid model is developed. However, these attempts are inadequate at the maximum levels of accuracy, because of tremendously incongruent type of the available data and crustal motion.

The nation requires to handle a gravity survey due to the current regional gravity are insufficient for performing its requirement to the enormous NOAA mission. The high-resolution gravity survey needs to be conducted in recognition of the need to estimate the varies of heights which derived from estimated varies of gravity field information, particularly the geoid. Firstly, the improvement of existing gravity fields with dense spatial distribution, but with a short temporal span based on high resolution snapshot. Next, low-resolution movie to track the temporal changes to the gravity field on a broad scale which a re-occurring survey with very coarse spatial coverage and a long temporal span. Then, a terrestrial collaboration surveys to determine a regional gravity field data to the fine-scale regional determination of heights.

Hence, there are several issues need to be highlighted in a re-survey, for example, the data gaps, aged data, discontinuities, imbalance of spatial coverage, lack of data and knowledge with respect to gravity field varies over time and incompetence to retain the existing vertical datum realization. As a solution in order to establish a regional geoid model with good accuracy and track the varies through time, hence, a supplementary data in addition to the gravity survey will be vital. For example, the digital elevation models (DEM), geoid slopes from co-located (temporally and spatially) levelling and GPS surveys over large regions, rock densities of the largest mountain ranges of the United States, bathymetry of main lakes and near-shore regions of the United States, deflection of the vertical for testing upward continuation algorithms as required, gravity gradiometry and satellite derived gravity models.

Meanwhile, Australia implements a fitting gravimetric geoid or quasi-geoid models to GPS-levelling data and it is crucial in Australia due to distortions in the Australian Height Datum (AHD). Hence, Australia took an alternative to fit the Australian Gravimetric Quasigeoid 2017 (AGQG 2017) [11] model to a nationwide GPS-levelling dataset [12] to establish a model of the separation between the GDA2020 ellipsoid and the AHD. Hence, a direct transformation between ellipsoidal and AHD heights is permitted. This method is a combined gravimetric-geometric model or known as "hybrid quasigeoid model".

According to Kotsakis and Sideris [13], a combined gravimetric-geometric model are affected by errors in GPS ellipsoidal heights, heights estimations, levelling data and systematic errors in the geoid model. Consequently, the geometric component is determined and its complementary grid of improbability values utilising least squares prediction (LSP) [14]. The existing AUSGeoid model known as AUSGeoid2020 is established with respect to a combined gravimetric-geometric model be associated with a 1' by 1' grid of ambiguity values propagated from input data into the LSP-gridded surface. Thus, Australian geodetic datum is transformed from GDA94 to GDA2020 and referenced to International Terrestrial Reference Frame 2014 (ITRF2014) [15].The geometric component of AUSGeoid2020 was established by adopting least squares prediction (LSP) based on 7624 co-located GPS-levelling points across Australia. Hence, 40 points were determined to misfit AUSGeoid2020 by more than five standard deviations from the mean after handling a primary cross-validation test. Then, the points were removed and the geometric component redefined.

By referring to the outputs in [16], AUSGeoid2020 presents an approach to transform between GPS-derived ellipsoidal heights and AHD heights. Next, the differences of AHD height are computed. Pseudo-independent accessing of AUSGeoid 2020 has presented that it is capable to transform from ellipsoidal heights to AHD heights with ±27 mm, an absolute uncertainty. Thus, AUSGeoid 2020 is similar or enhanced as compared to conventional third-order levelling in Australia at distances beyond ~3 km.

The development of a high-resolution geoid model in Taiwan needs a substantial effort in gravity data collection and statistical techniques [17]. Taiwan is bounded by the Pacific Ocean to the east with deep trenches, the South China Sea to the south, the Taiwan Strait to the west and the East China Sea to the north. The topography of Taiwan is mostly rugged (up to 4000 m high), with flat regions only on its coastal plans. Land-based gravity surveys can only be performed along mountains treks and easy acssess area. Hence, a precise geoid model is required to overcome these issues.

In order to overcome these problems, three-dimensional, real-time cm-level positioning has been performed by utilising a continuous GNSS network. Next, the Kuroshio Current east of Taiwan and the surrounding seas perform large gradients in the dynamic ocean topography over Taiwan that can influence great variances in the vertical datums between Taiwan and its offshore islands. To end, Taiwan's ellipsoidal heights and most parts of Taiwan's offshore islands have been measured and surveyed by Light Detection and Ranging (LiDAR). A transformation method from ellipsoidal heights to orthometric heights necessitate a precise geoid model. The accuracy of a geoid model in low-lying areas and foothills play an importance role to weighing flooded zones and geohazards due to high sloping topography.

In geoid modelling, there are three categories of data implemented specified to terrestrial (land and near shore) gravity data such as terrestrial land gravity, airborne gravity and the offshore gravity data and described as follows:

1. Land gravity data

   This data set is classified into two division:

   a) Division 1: gravity data consisted over 1980 until 2003 at Taiwan's horizontal control points and first-order benchmarks

   b) Division 2: gravity data surveyed over 2004 until 2006 except gravity measurements based on horizontal control points [18].

2. Airborne gravity data

   Over 2004 until 2009, three airborne gravity measurements were accomplished to measure and survey gravity data over Taiwan [19] [20].

3. The offshore gravity data

   The gravity data were surveyed within 50 km to the five tide gauges and over waters 20 km offshore Taiwan. This survey is significance to improve vertical datum linking between Taiwan and islands, besides for the coastal geoid model improvement.

However, in order to fulfil the data gaps in the shipborne gravity data along the coastal area, the altimeter-derived sea surface heights (SSHs) was utilised to fulfil that occupied area. Then, the current approach is used to model a gravimetric geoid model that referred to gravity anomalies on the ground and the implementation of planar terrain corrections. Primarily, the height anomalies are determined resulted by transformation to geoidal heights. Hence, in the determination of a localised gravity fields information, a global geopotential model and a Digital Elevation Model (DEM) were needed. Next, a gravimetric geoid model in Taiwan is established which is implemented in the development of a hybrid geoid model using the computed geoid heights with respect to the remove-compute-restore technique. The details regarding to the procedure and step in the development of a new gravimetric-only and hybrid geoid models has been described in [17]. Based on the findings in [17], the combining of gravimetric geoid with the GPS and tidals records provides the errors identifications of about 40 to 50 cm in the existing geodetic vertical datums.

Next in 2009, New Zealand officially utilised a gravimetric quasi-geoid model as the primary national vertical datum known as NZVD 2009 model [21]. Previously, 13 various local vertical datum were adopted, each of a regional vertical datum were referenced to dissimilar tide-gauge with respect to localised mean sea level. Due to these issues, there were offsets of up to 0.4 m amongst the NZ local vertical datum. The NZGeoid2009 model was merged with the local vertical datum by computing the local offsets from the iteratively computed gravimetric quasigeoid model [22], [23], [24].

The gravity field data was implemented to determine NZGeoid 2009 for quasigeoid modelling. The terrestrial land gravity data largely comprise of ancient measurements with irregular accuracy and sparse distribution, commonly isolated to valley through rough terrain areas. Besides, satellite altimetry data provides large error in coastal areas, and the large survey vessels commonly for shipborne gravimetry surveys unable to navigate near to the coast. Hence, an appropriate set of gravity measurements with a consistent spatial distribution (seamlessly onshore and offshore) and consistent precision is needed in order to provide precise quasi-geoid modelling. Thus, as a solution, airborne gravimetry measurement gives large and good gravity fiefld coverage over challenging areas. It is an appropriate technique for the determination of regional geoid computations; hence, it will enhance the gravimetric quasi-geoid model [25].

The terrestrial, shipborne, altimeter and airborne gravity measurements were reduced to refined Bouguer gravity anomalies and then integrated onto a single grid at the topographic surface using least squares collocation (LSC). The gridded data were then transformed to estimate Molodensky gravity anomalies, by joining the Bouguer plate term. The modified Stokes integration [26] and remove-compute-restore technique was used in the determination of quasigeoid heights that implemented EIGEN-6C4 global gravity model as a reference field. The quasigeoid model was fitted to 1422 independent GPS-levelling-derived quasigeoid heights, consequent to removal of offsets between the dissimilar levelling datums in New Zealand. The outputs represent 4 mm more precise than EIGEN-6C4 and 14 mm more precise than NZGeoid 2009 based on the differences between the gravimetric quasigeoid and the GPS-levelling data. However, the largest-magnitude change in quasigeoid heights with 0.122 m subsequent to the contribution of airborne gravity data. Thus, the evaluation of a regional quasigeoid heights with GPS-levelling data were more precise for each set of integration parameters.

However, in 27 June 2016, New Zealand Vertical Datum 2016 (NZVD2016) substitutes NZVD2009 as the authorised vertical datum for New Zealand. This new vertical datum model is determined by the NZGeoid2016 geoid. The vertical datum values in the local vertical datum correlations grids differ with horizontal position. Where, it is slightly dissimilar from NZVD2009 that represents an average offset for the whole local vertical datum. NZVD2016 is constant with NZGD 2000. Thus, the normal-orthometric NZVD2016 heights can be converted to ellipsoidal NZGD2000 heights, conversely. NZVD2016 model is properly determined in the LINZ standard LINZS25009 (Standard for New Zealand Vertical Datum 2016) [27].

The National Geographic Information Institute (NGII) has established three Korean national geoid (KNGeoid) models recognised as KNGeoid13, KNGeoid14 and KNGeoid18 by utilising the Earth's gravity field model, satellite-altimetry-derived gravity data, land and ocean gravity data, airborne gravity data and digital elevation model (DEM) data. These geoid models are hybrid geoid models accomplished by NGII to the height measurement accuracy with respect to GNSS measurements.

Primarily, the KNGeoid13 model is developed by NGII for land area which developed based on gravity data from several national control points (unified control point, benchmark and triangulation point), airborne gravity data [28], DTU10 satellite altimetry data, EGM 2008, and 5m gridded topographic data. Then, in 2014, KNGeoid13 model was established by using the equivalent methods by combining the gravity data and shipborne gravity data from Korea Hydrographic and Oceanographic Agency (KHOA) that provides, 3.3 cm of residual's standard deviation [29]. The final procedure, the KNGeoid18 model was developed by implementing Experimental Gravity Field Model 2016 (XGM2016) with respect to GOCE gravity data [30] as the first sort of Earth Gravitational Model 2020 (EGM2020) with standard deviation of residual about 2.33 cm. Hence, EGM2008, EIGEN-6C4 and GECO model have been implemented as a Global Geopotential Model (GGM) with coefficients up to degree 2190. The accuracy of the geoid heights derived from the latest released Earth gravity model based on gravity data was validated with the geoid heights derived from the Global Navigation Satellite System (GNSS)-levelling, 1182 unified control points (UCPs) developed by the National Geographic Information Institute (NGII) over South Korea.

Hence, by referring to the output provided by [31], the geoid heights derived from the three high-degree global geopotential models (GGMs) and the three Korean National Geoid (KNGeoid) models represented regular distributions ranging from 17 to 33 m around South Korea. Next, the EGM2008 model gave a good residual accuracy with root-mean-square-error (RMSE) that relative to geoid heights accuracy. Thus, the EGM2008 model is suitable model than the GECO and EIGEN-6C4 for a region of South Korea and could be selected as an appropriate GGM for determining a gravimetric geoid model. Lastly, among the three KNGeoid models, the KNGeoid18 model was represented a good gravimetric geoid model accuracy over South Korea rather than the KNGeoid13 and KNGeoid14 models with respect to the GNSS/levelling-derived geoid heights.

Previously in Philippines, the determination of orthometric height and benchmarks was performed using Geodetic levelling, a tiresome procedure that obstruct the densification of benchmarks. 22851 benchmarks are established along major roads nationwide with a maximum divergence of 4.0 $mm\sqrt{K}$ between two level runs (1st order accuracy) during the Philippine Reference System 1992 (PRS92) in 2007. These level networks were linked to the respective reference tidal benchmarks to provide local mean sea level elevation. Hence, the primary step of determining a preliminary gravimetric geoid for the Philippines is through the Natural Resources Management Development Project (NRMDP) in year 1991. Land gravity data and altimetrically-derived anomalies at marine area and a reference global model from OSU89A as a reference model to degree and order 360 were utilised [32].

Then in 2014, a preliminary geoid model known as Philippine Geoid Model 2014 (PGM2014) has been performed with the technical assistance of National Space Institute-Denmark Technical University (DTU-space) and a financing from National Geospatial Intelligence Agency (NGA). The land gravity, airborne gravity, marine satellite altimetry and the latest satellite gravity data from the GOCE mission release 5 are performed with 0.30 m accuracy. The PGM2014 is determined by a set of Fortran routines known as *GRAVSOFT* system established by DTU-Space and Niels Bohr Institute, University of Copenhagen [33]. It conducts the base of major recent geoid computation projects like the joint Nordic "NKG" geoid models, conducted as joint geoid model computations of the Nordic and Baltic countries [34] under the auspices of the Nordic Commission for Geodesy (NKG), as well as the OSGM02 geoid model of the UK and Ireland [35] and numerous national geoid models performed from airborne surveys in current years such as Malaysia, Mongolia, Indonesia and many more.

Hence, a "remove-restore" technique was performed in determining the geoid with full 3-dimensional modelling going via the quasigeoid to the final geoid, where a spherical harmonic earth geopotential model (EGM/GOCE combination) is utilised as a base. The EGM08 model is encompassing GRACE satellite data which identifies the error spectrum of the EGM08 up to spherical harmonic degree 80 or so. The terrain part of the computations was referenced to RTM method, where topography is referred to a mean elevation level, and only residuals relative to this level is considered. The gravimetric geoid is determined using *spherical FFT* method with optimised kernels. This method is an alternative of Stokes integral in which there is an appropriate weighting of the long wavelengths from EGM08 and shorter wavelengths from the regional gravity data.

The gravimetric geoid is determined on a grid of 0.025°x 0.025° resolution covering the Kalayaan Island of West Philippine Sea using least squares collocation and Fast Fourier Transformation methods. Thus, the final input used in PGM 2014 computations are airborne gravity data, land gravity from NAMRIA, DTU10 global gravity anomalies from multi-mission satellite altimetry (the chosen open ocean area, away from the airborne gravity), SRTM 15" DEM data for the region and EGM08 and GOCE RL5 satellite data. A series of 190 GPS data in ITRF2005 levelling benchmarks is utilised to verify with the final computed geoid. These GPS data represented high errors relative to the geoid with high outliers in certain regions due to a combination of geodynamic effects, levelling and GPS errors.

For the improvement of a regional geoid model, several recommendations are suggested by [36]. Then, the PGM2014 was re-determined to the current Philippine Geoid Model (PGM2016) with an accuracy of 0.022 m, the accuracy is reduced compared to PGM2014 geoid model. Where, the additional land gravity stations (from 2017 until 2020) are combined with the similar airborne and satellite gravity data in order to re-determine a latest the Philippine geoid model.

The Royal Thai Survey Department (RTSD) collaborated with Chiang Mai University and Chulalongkorn University developed THAI12H regional geoid model [37] utilising 3979 land gravity points with as near to 5 cm accuracy in Bangkok Metropolitan Region. Nevertheless, Chao Phraya basin, the northern part of Thailand and another area represents the larger errors up to 30cm due to low intensity of gravity measurements performed before 1991. Besides, the outdated land relative gravimeters could influence the value of the gravity measurements. The establishment of a current geoid model for Thailand was performed in 2015 to develop a height modernisation system that connected the geoid model to the national real-time kinematic network (RTK GNSS network). This height modernisation system will represent elevation values significance for natural disaster management which spirit levelling unable to perform and other vertical positioning works [38].

TGM2017 was established in 2005 until 2007 [39] over the Short-Term National Water Strategy Plan, then published to public users in 2018. This model was developed based on the current gravity data sets of more than 10 000 terrestrial land and airborne gravity data over Thailand's territory by RTSD. Hence, TGM2017 model was determined from the airborne and terrestrial gravity data utilising remove-restore technique. Then, EGM2008 and DTU13 were utilised for the marine area and uncovered airborne gravimetric surveys for land area, respectively [40]. The gravimetric data were gridded utilising least-squares collocation within the study area as suggested by [41]. Next, the multi-band spherical Fast Fourier Transform (FFT) approach was utilised for determining and computing TGM2017 geoid model with the spatial resolution of one-arc minute regular grid around 1.8km.

For validating the TGM2017 geoid model, $N_{TGM.}$, 100 GNSS/levelling-derived geoid heights, N_{kolak} points was utilised using two procedures as elaborated in [38]. The values of geoid heights from the model were compared with those derived by the combination of Kolak-1915 orthometric (or leveled) heights and GNSS-based geodetic (ellipsoidal) heights. The validation of the new gravity measurements over Thailand enhanced the long- and medium-wavelengths of TGM2017 model by comparing with EGM2008 model at 360 and 2190 degree. Nevertheless, TGM2017 model was fitted to Kolak-1915 vertical datum, where EGM2008 was inconsistent. Then, the previous gravimetric geoid model known as THAI17G geoid model was utilised instead, to develop TGM2017 geoid model. Hence, the final validation represented the enhanced of TGM2017 geoid model than other geoid models with standard deviation of 4.9 cm.

IOP Conf. Series: Earth and Environmental Science **880** (2021) 012012 doi:10.1088/1755-1315/880/1/012012

3. Results and Discussions

Each country comprised of their own regional geodetic vertical datum developed based on gravity field information. The development of geodetic vertical datum is conducted based on selected and different approach as represented in Table 1. Hence, the geodetic vertical datum of Malaysia is developed by utilising Least Squares Modification of Stokes with additive corrections (LSMSA) approach. However, other countries such as United States of America, New Zealand, Thailand and Phillipines adopting Remove-Compute-Restore (RCR) approach in the development of their regional geoid model. Besides, Thailand and Philippines also utilised FFT method in a localised geoid computation. However, Australia represents a new technique called as the combined gravimetric-geometric model also recognised as hybrid quasi-geoid model. The combined gravimetric-geometric models are influenced by errors in GPS ellipsoidal heights, estimations in the determination of heights, errors in the levelling data and systematic errors in the geoid.

Thus, it can be figured out that airborne gravimetry provides a good data coverage especially over inaccessible areas like coastal areas and in rough topography. As discussed in previous section related to the geodetic vertical datum in New Zealand, the terrestrial data largely consist of historical measurements of varying accuracy with sparse distribution, commonly remote to valleys through areas of rough terrain areas. In addition to that, satellite altimetry data are unreliable in coastal areas [42] besides, the large survey vessels typically implemented for shipborne gravimetry also unable to navigate close to the coast. For the development of a precise quasigeoid modelling, the gravity measurements with a consistent spatial distribution (seamlessly onshore and offshore) and precision is needed with an adoption of a suitable gravity data for a regional geoid computation. Due to this reasons, airborne gravimetry appears well suited to account for the shortcomings of the existing gravity data in each country to enhance the gravimetric quasigeoid model [43]. Moreover, Global Geopotential Model (GGM) and marine gravity data from DTU are suggested to be utilised to cover up the lack of terrestrial gravity and airborne gravity data for land and marine area as performed by Thailand.

Table 1. The vertical datum applications in Malaysia, United States of America, Australia, Taiwan, New Zealand, South Korea, Philippines and Thailand.

Study Area	Data Gathering	Data Input	Analysis Approach
Peninsular Malaysia	Terrestrial land gravity data, Ship track marine gravity data from International Gravimetrics Bureau GGM and Satellite altimetry data from DTU10	Gravity data, geoid heights, othomertic heights and ellipsoidal heights	LSMSA and Fitted geoid
Peninsular Malaysia	Terrestrial gravity, airborne gravity, satellite gravity data from GGM, shipborne gravity, satellite altimetry from DTU13, GNSS levelling	Gravity data, geoid heights, othomertic heights and ellipsoidal heights	LSMSA Fitted geoid/hybrid geoid (fitted to PMGVD and MSL)
United States of America	Airborne gravity, latest satellite gravity models, Terrestrial gravity	Gravity data Timeline: from year 2014 to 2020 (NGS published annual experimental geoid (XGEOID) models.	R-C-R Fitted geoid
Australia	GPS ellipsoidal height, gravimetric quasi-geoid, altimeter-derived gravity	Gravimetric quasi-geoid, ellipsoidal height	A combined gravimetric geometric model known as hybrid quasi-geoid

	anomalies, levelling		model
Taiwan	Airborne gravity, shipborne gravity, terrestrial gravity and satellite altimetry data, 5 major tide gauges	Gravity data (land and marine), gravimetric geoid	Merging land and marine gravity using Least Squares Collocation (LSC) Fitted geoid
New Zealand	Land gravity, DEM, GPS levelling, shipborne gravity, satellite altimetry, scalar airborne gravity and GGMs	A one arc-minute resolution gravimetric quasigeoid model, EIGEIN-6C4 (GGM)	Modified Stokes integration using R-C-R
South Korea	Global Geopotential Models (EIGEN-6C4, EGM2008 and GECO), GNSS levelling, land gravity and DTU10 altimetry data	Gravity data	The combinations of gravity data from satellites and land gravity data (blank area of gravity data is filled with DTU10 altimetry data).
Philippines	Terrestrial land gravity, airborne gravity, DTU10 marine satellite altimetry, the newest satellite gravity data from GOCE mission release 5, DTM (15" SRTM) and GNSS levelling	Residual gravity anomalies, geoid residuals, gravity data, ellipsoidal heights, orthometric heights	• FFT to residual gravity anomalies • spherical FFT with optimized kernels (Stokes integral) • LSC • Fitted geoid • GRAVSOFT system • Remove-restore method
Thailand	Airborne gravity (land area and shorelines), terrestrial land gravity, GNSS levelling, DTU13 and EGM2008 marine gravity data	Orthometric heights, geoid heights	Remove-restore technique, LSC, FFT method

4. Conclusion

In a conclusion, the appropriate technique and approach for merging and integrating land and marine gravity data can be optimized which is crucial for developing the high-resolution gravity potential field especially for Malaysian region. The high-resolution of a regional gravity data is significance for establishing a regional geoid model (geodetic vertical datum) with good accuracy. Hence, multi-mission gravity data from terrestrial gravity data, airborne, shipborne, satellite altimetry and satellite gravity from GGM can be implemented to determine seamless vertical datum for a region. Where, each types of gravity data either land or marine area will provide their own gravity field accuracy depends on the types of measurements.

IOP Conf. Series: Earth and Environmental Science **880** (2021) 012012 doi:10.1088/1755-1315/880/1/012012

References

[1] Odumosu J O *et al* 2016 Empirical Geoid Modelling Using Classical Gravimetric Method FIG Working *W Issue* 8076
[2] Jekeli C 2016 Geometric Reference Systems in Geodesy Division of Geodesy and Geospatial *Science School of Earth Sciences Ohio State University*
[3] Sjoberg L E and Bagherbandi M 2017 Gravity Inversion and Integration *Springer International Publishing Switzerland*
[4] Amirrudin M A, Din A H M, Zulkifli N A, Amat M A C and Hamden M H 2020 Assessment of the accuracy and precision of myrtknet real-time services *Jurnal Teknologi* 831 93-103
[5] Din A H M, Abazu I C, Pa'Suya M F, Omar K M and Hamid A I A 2016 The impact of sea level rise on geodetic vertical datum of peninsular malaysia Paper presented at the International Archives of the Photogrammetry Remote Sensing and Spatial Information Sciences - *ISPRS Archives 424W1* 237-245
[6] Hassan M I and Rahmat NH 2016 The effect of coastline changes to local community's social economic *The International Archives of the Photogrammetry Remote Sensing and Spatial Information Sciences* Volume **XLII-4/W1**
[7] Ismail M K, Din A H M, Uti M N and Omar A H 2018 Establishment of new fitted geoid model in universiti teknologi malaysia Paper presented at the International Archives of the Photogrammetry *Remote Sensing and Spatial Information Sciences - ISPRS Archives* **424/W9** 27-33
[8] Sulaiman S 2016 Gravimetric Geoid Model Determination for Peninsular Malaysia Using Least Squares Modification of Stokes PhD Thesis Universiti Teknologi MARA
[9] Pa'suya MF 2020 New Modern Height Reference System for Peninsular Malaysia using Least Squares Modification of Stokes' Formula PhD Thesis Universiti Teknologi Malaysia
[10]NOAA 2020 Vertical Datums National Geodetic Survey Retrieved November 14 2020 from wwwngsnoaagov/datums/vertical/
[11] Featherstone W E, McCubbine J C, Brown N J, Claessens S J, Filmer M S and Kirby J F 2018a The first Australian gravimetric quasigeoid model with location-specific uncertainty estimates *J Geod* 922 149–168
[12] Featherstone W E, Brown N J, McCubbine J C and Filmer M S 2018b Description and release of Australian gravity field model testing data *Aust J Earth Sci* 651 1–7
[13] Kotsakis C and Sideris M G 1999 On the adjustment of combined GPS/levelling/geoid *networks J Geod* 738 412–421
[14] Mysen E 2014 On the computation of reliable formal uncertainties in the densification of GPS-levelling networks by least-squares collocation *J Geod* 8810 917–926
[15] Altamimi Z, Rebischung P, Métivier L and Xavier 2016 ITRF2014: a new reléase of the International Terrestrial Reference Frame modeling nonlinear station motions *J Geophys Res Solid Earth* 121B8 6109–6131
[16] Brown N J and McCubbine J C Featherstone W E Gowans N Woods A and BaranI 2018 AUSGeoid2020 combined gravimetric–geometric model: location-specific uncertainties and baseline-length-dependent error decorrelation *Journal of Geodesy* 9212 1457–1465
[17] Hwang C *et al* 2020 New gravimetric only and hybrid geoid models of Taiwan for height modernisation cross island datum connection and airborne LiDAR mapping *Journal of Geodesy 2020* **94** 83
[18] Hwang C, Wang CG and Hsiao YS 2003 Terrain correction computation using Gaussian guadrature *Comput* 2910 1259–1268
[19] Hwang C, Shih HC, Hsiao YS and Huang CH 2012 Airborne gravity surveys over Taiwan Island and Strait Kuroshio Current and South China Sea: comparison of GPS and gravity accuracies at different flight altitudes *Mar Geodesy* 353 287–305
[20] Hwang C *et al* 2014 New free-air and Bouguer gravity fields of Taiwan from multiple platforms and sensors Tectonophysics 611 83–93
[21] McCubbine J C *et al* 2018 The New Zealand gravimetric quasigeoid model 2017 that incorporates nationwide airborne gravimetry *Journal of Geodesy* 928 923-937
[22] Amos M J 2007 Quasigeoid modelling in New Zealand to unify multiple local vertical datums PhD thesis Curtin University of Technology Perth Australia
[23] Amos M J and Featherstone W E 2009 Unification of New Zealand's local vertical datums: iterative gravimetric quasigeoid computations *J Geod* 831 57–68
[24] Claessens S J, Hirt C, Amos M J, Featherstone W E and Kirby JF 2011 The NZGEOID2009 model of New Zealand *Surv Rev* 43319 2–15
[25] Hájková J 2011 Local geoid determination based on airborne gravity data Stud *Geophys Geod* 553 515–528
[26] Wong L and Gore R 1969 Accuracy of geoid heights from modified Stokes kernels *Geophys J Int* 181:81–91
[27] LINZ (2020) Land Information New Zealand. Retrieved June 24 2021 https://www.linz.govt.nz/
[28] Baek K, Lee J, Shin K, Kwon J H and Moon J 2014 Construction and precision verification of Korean national geoid model KNGeoid13 *In Proceedings of the 2014 Conference of the Korean Society of Surveying Geodest Photogrammetry and Cartography Seoul Korea 24-25 April 2014* 14 111-114
[29] Lee J and Kwon J H 2015 Construction and precision verification of Korean national geoid model KNGeoid14 *In Proceedings of the 2015 Conference of the Korean Society of Surveying Geodesy Photogrammetry and Cartography Masan Korea 23–24 April 2015* 177–179
[30] Pail R l *et al* Short note: The experimental geopotential model XGM2016 *J Geod* 2018 42 443–451
[31] Kim K B, Yun H S and Choi H J 2020 Accuracy Evaluation of Geoid Heights in the National Control Points of South Korea Using High-Degree Geopotential Model *Journal of Applied Sciences* 10 1466

[32] Gatchalian R, Forsberg R and Olesen A 2017 PGM2016: A new Geoid Model for the Philippines https://mycoordinates.org/pgm2016-a-new-geoid-model-for-the-philippines/
[33] Forsberg R C C T 2008 Overview manual for the GRAVSOFT Geodetic Gravity Field Modelling Programs Technical Report 2nd ed: DTU-Space
[34] Forsberg R D, Solheim and J Kaminskis 1996 Geoid of the Nordic and Baltic area from gravimetry and satellite altimetry Paper presented at the Symposium on Gravity Geoid and Marine Geodesy Tokyo 1996
[35] Forsberg R, Strykowski G, Iliffe J, Ziebart M , Cross P, Tscherning C C and Finch O 2002 OSGM02: A new geoid model of the British Isles Gravity and geoid 3
[36] Forsberg R 2003 Towards a cm-geoid for Malaysia Paper presented at the geoid computation workshop in Kuala Lumpur
[37] Dumrongchai P C, Wichienchareon and C Promtong 2012 Local Geoid Modeling for Thailand International Journal of Geoinformatics 84 15-26
[38] Dumrongchai P, and Duangdee N 2019 Evaluation of TGM2017 for Height System using GNSS/Leveling Data in Thailand 2019 International Transaction *Journal of Engineering Management & Applied Sciences & Technologies*
[39] Dumrongchai P and C Promtong 2017 Development of precise geoid model of Thailand Final Report the Royal Thai Survey Department Bangkok Thailand in Thai
[40] Andersen O B P, Knudsen and Stenseng L 2015 The DTU13 MSS Mean Sea Surface and MDT Mean Dynamic Topography from 20 Years of Satellite Altimetry *International Association of Geodesy Symposia Series* 144
[41] Moritz H 1980 Advance physical geodesy Herbert Wichman Verlag Karlsruhe
[42] Vignudelli S, Kostianoy A, Cipollini P and Benveniste J 2011 Coastal altimetry *Springer Berlin p* 566
[43] McCubbine J C *et al* 2018 The New Zealand gravimetric quasigeoid model 2017 that incorporates nationwide airborne gravimetry *Journal of Geodesy* 928 923- 937

Acknowledgments
Authors would like to acknowledge the Ministry of Higher Education (MOHE) under the Fundamental Research Grant Scheme (FRGS) Fund, Reference Code: FRGS/1/2020/WAB07/UTM/02/3 (UTM Vote Number: R. J130000.7852.5F304) for funding this study.

Wave runup and overtopping on smooth-slope NEXC block

Y A Benson[1,3*], A K Abd Wahab[1,2], I K Othman[1,2], M H Jamal[2], S Mat Desa[5], H L Lee[3] and M N Md Noh[4]

[1] Center For Coastal and Ocean Engineering (COEI), Universiti Teknologi Malaysia (UTM), Johore, Malaysia
[2] School of Civil Engineering, Engineering Faculty, Universiti Teknologi Malaysia (UTM), Johore, Malaysia
[3] Coastal Management and Oceanography Research Centre, National Water Research Institute of Malaysia (NAHRIM), Selangor, Malaysia
[4] Director General's Office, National Water Research Institute of Malaysia (NAHRIM), Selangor, Malaysia
[5] River Basin Research Centre, National Water Research Institute of Malaysia (NAHRIM), Selangor, Malaysia

*Corresponding E-mail: yannie@nahrim.gov.my

Abstract: Constant wave runup and overtopping during monsoon coupled with storm-surge events have poses threat to the coastal's community in flooding and land loss. The study was to further the research on the wave interaction issue using the modified NAHRIM Coastal Protection and Expansion (NEXC) block. The aim was to determine the significant relationship prediction model from the experiment variables due to water level changes. The study was conducted in 30 m long, 2 m height, and 1.5 m width of wave flume using gamma 3.30 of wave height JONSWAP spectrum under 1:15 and 1:8 mobile bed scenarios. Parameters were downscaled to 1:10 and based on Peninsular Malaysia's east coast hydrodynamics conditions. 36 different test scenarios were simulated every 20 minutes with three repetitions, enables 108 samples to be retrieved. Using statistical tools, correlation tests between the variables in the experiment results indicates wave runup, significant wave height and overtopping discharges are strongly correlated to the bed gradient and smooth-slope NEXC block. Changes in water level from shallow to deep, mild to steep mobile bed gradient with 30° to 60° block affect the relationship Hs-q decrease while $R_{u2\%}$-q positively increase. Overtopping was not directly affected by water level but positively affected on wave runup and negatively to significant wave height. The fitted relationship design model using a General Full Factorial method was verified with 0.338069 of standard error and 98.12 % of R-square. Finally, the significant relationship predictive model was obtained to have 26 interaction terms in the model successful.

Keywords: JONSWAP spectrum, mobile bed, smooth-slope block, wave flume, General Full Factorial

Track Name: Coastal Management and Marine Ecosystem

IOP Conf. Series: Earth and Environmental Science **880** (2021) 012013 doi:10.1088/1755-1315/880/1/012013

1. Introduction

The rising of the seawater level was expected to impact not only the coastal ecosystem but including the loss of functionality and biodiversity, coastal inundation and inland migration such as coastal land loss, increment possibility on wave interaction especially on the open long-span sandy beach exposed [1]. The higher sea levels will bring more intense characters of the sea level rise (SLR) [2]. This resulted in destructive forces of waves surge, leading to the penetration farther inland than before. Thus, it will cause the risk of groundwater pollution, saltwater intrusion, coastal erosion, property damages and also extinction of coastal flora and fauna [3-5, 22].

In terms of shoreline and existing coastal protection structure damages, the main factor that drives these effects was the propagation of aggressive wave interactions of wave runup and its overtopping [6, 7]. Nowadays these factors are considered as one of the major and various threats [15-18] that damaging beaches [31], coastal land loss and inundation due to the impact of rising seawater levels [21, 23]. the vulnerability potential is higher at low-lying, soil or sandy coastal areas with no protective coastal layer structure or no strong embankment [8-10, 14].

Lately, Malaysia coastal especially at Peninsular Malaysia has shown its hydrodynamic characteristic status as experiencing the unexpected extreme impact of sea level rise [11], leaving its east coastline the most vulnerable to wave regime [19, 20]. This is because the east coast of Peninsular Malaysia has a long span with open-wide sandy beach materials and exposed directly to the hydrodynamic regime from the South China Sea. Hence, wave interaction especially waves runup and its overtopping has become an important issue among researchers, scientists, local authorities, and young scholars. Finding the best solution in understanding its impacts on the management wise, social, economic and psychology was the main concern [35]. However, at the same time, it has eventually triggered an initiative to develop the innovative coastal structure dealing with the impacts, especially on the coastal flooding effects and expanding term.

In conjunction with research in wave interaction, the formula for wave runup, $R_{u2\%}$ (m) and overtopping discharges, q (l/s/m or m^3/s/m) [33] on sloping bed gradient and smooth slope coastal structure has been updated with its coefficient values, $c = 1.3$. The c value gives a slightly curved line on a log-linear graph, where the exponential distribution gives a straight line, with smooth plane and breaking (plunging) waves on the bed gradient as below:

$$\frac{q}{\sqrt{g \cdot H_{m0}^3}} = \frac{0.026}{\sqrt{tan\alpha}} \gamma_b \cdot \xi_{m-1,0} \cdot \exp[-\left(2.5 \frac{R_c}{\xi_{m-1,0} \cdot H_{m0} \cdot \gamma_b \cdot \gamma_f \cdot \gamma_\beta \cdot \gamma_v}\right)^{1.3}]$$

And with a maximum of slope as below formulae:

$$\frac{q}{\sqrt{g \cdot H_{m0}^3}} = 0.1035 \cdot \gamma_b \cdot \xi_{m-1,0} \cdot \exp[-\left(1.35 \frac{R_c}{H_{m0} \cdot \gamma_f \cdot \gamma_\beta \cdot \gamma^*}\right)^{1.3}]$$

These formulae will react if one the influence factor changes, such as γ_b is the influence factor for a berm, γ_f is for roughness elements on a slope, γ_β is for oblique wave attack and influence factor for a wall at the end of a slope,γ_v. Wave runup defined as the vertical difference between the highest point of wave run-up and the still water level (SWL) [33, 34] that will change with the beach and offshore wave properties. An exceeded by 2 % the number of incoming waves at the toe of the structure for sloping structures with gentle face bed gradient of range 1:8 to 1:15 is quoted as

$$\frac{R_{u2\%}}{H_{m0}} = 1.65 \cdot \gamma_b \cdot \gamma_f \cdot \gamma_\beta \cdot \xi_{m-1,0}$$

1.1 Issues, motivation and problem statement

Long-span and open wide effects of the east coastline of Peninsular Malaysia to the density wave regime from the South China Sea has posed the coastline in great threats. On the other hand, the tidal differences around the coastal area [12, 13] affects the intensity of the wave response towards the inland of the coastline or beach.

These issues have alarmed local authority and other parties to responding as it becomes a threat to all residents who inhabit the coastal areas especially during monsoon season with storm-surge coupled with high tide and SLR events [15, 26]. The effect may bring massive impact to resident's inhabit the coastal area, tourism, aquaculture, plantation area and coastal forest reserve area [25].

In 2016, NAHRIM as focal national excellence centre in coastal and its environment has taken the initiative to conduct detailed research and development, in producing an innovative coastal protection structure name as NAHRIM Coastal Protection and Expansion (NEXC) block that is expected to control the problem of coastal damages and at the same time may expand naturally of the impacted coastal land area. The results on the laboratory testing and one-year duration of block system installation with onsite performance analysis have proven its function to re-nourish the beach through the natural accretion process and effectively stabilize the impact of erosion [26, 27].

However, due to caused matter on wave interaction issue prevailed on the front of the block system, had not investigated. This has motivated further research using the modified NEXC block by using the irregular wave on the mobile bed. As applying mobile beds in the physical laboratory quite challenging, it is believed that added beach material in the research has become less interesting among the researchers.

A strong motivation for the study is aiming to determine the relationship statistically for optimal relationship prediction model in minimizing the overtopping discharges due to mobile bed and water level changes using the various angle of smooth-slope NEXC block with the irregular wave. Moreover, it is important to statistically analyze experiment data to assure reliability on the significant relationship towards a better understanding of the effects at the block area.

The main objective of this study is to examine and determine a significant relationship model of the smooth-slope NEXC blocks in affecting wave overtopping and runup due to water level changes in the sandy area.

2. Site conditions

The study had adopted the east Peninsular Malaysia coastal area as the downscale model in the physical laboratory and was conducted at the National Water Research Institute of Malaysia (NAHRIM) Laboratory, Selangor, Malaysia. The study area also has been choosing based on the coastal area was unique with the tidal types and location of wide-open facing to the South China Sea its long-span stretch coastline of sandy beach. Moreover, because of its strategic location, it received most of the active hydrodynamic activity throughout the year. The nearest and narrow part to the deepest bathymetry at -60 m depth and 30km away from the shoreline was located at latitude 4.86° and longitude 103.72° (Figure 1).

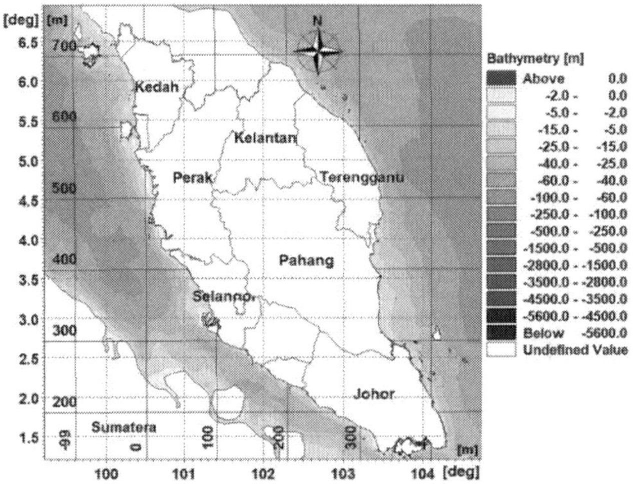

Figure 1. Bathymetry along the East coast of Peninsular Malaysia.

Peninsular Malaysia coastal area experiencing three types of tides. Based on their position, the east and southern part coastlines of the Peninsular experienced mixed tide prevailing diurnal and semidiurnal, while the major coastlines of the west experienced semidiurnal tide [13, 39, 40] (Figure 2). Furthermore, Peninsular Malaysia is mostly surrounded by the sea, which is affected by a wind-driven wave generated [29, 30]. This can be observed from the wave movement in Malaysia which follows the monsoon wind movements (Figure 3).

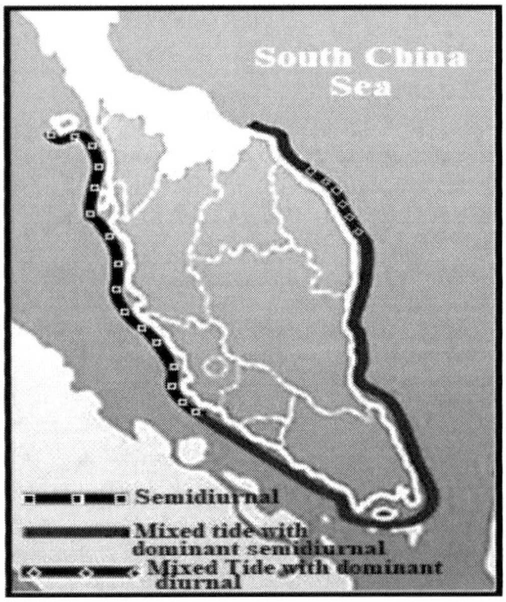

Figure 2. Types of the tide at Malaysia waters area [39].

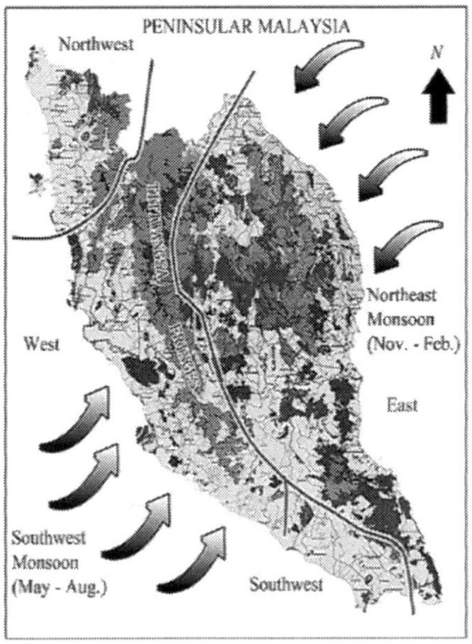

Figure 3. The Yearly Movement of Monsoon Winds and Waves in Malaysia [28].

3. Methodology and Experiment Setup

3.1 Geometry and Hydrodynamics parameters

Data records of 9 years wave from the offshore are analyzed using Extreme Value Analysis (EVA) for long-term times series statistical analysis to obtained 100 years return period of peak wave period, *Tp* and spectrum significant wave height, *Hs* in order downscale to laboratory use. The study has considering wave angles that perpendicular to the beach shoreline and suited it with the apparatus limits. The significant spectrum wave that perpendicular to the shoreline of the east coast of Peninsular Malaysia was from sector 61° to 90°.

The range of the spectrum wave height applied in the wave input was 0.1 m, 0.13 m, and 0.15 m using wave height of JONSWAP spectrum with gamma 3.30 were imposed on the block. Input parameters (Table 1) were downscaled to 1:10 and according to the hydrodynamic conditions using a combination of water depths of 0.9 m and 0.95 m with the 30°, 45° and 60° of the smooth-slope NEXC block's front-face angle (Figure 4).

Table 1. Input parameters used in the model.

Parameters	Values
Block sloping angle, α (°)	30, 45 and 60
Induce wave height, *Hmo* (m)	1.0, 1.3 and 1.5
Water Level, d (m)	0.9 and 0.95
Beach Slope, β (°)	4 and 7
Peak wave period, *Tp* (s)	2

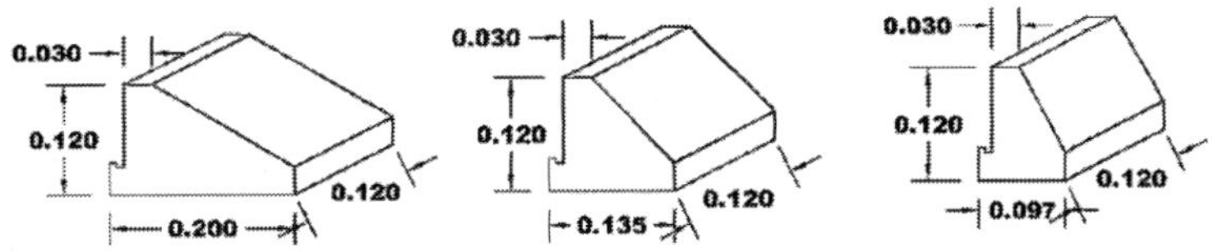

Figure 4. Three types of physical blocks used in the physical model.

3.2 Beach Profile and overtopping tank Setup

The beach profile setup was the most critical phase before filling up with the beach material and other model's accessories. A strong wood framework underneath the mobile bed is to reduce the usages on beach material of 100 µm of fine sand and bed gradient of the model are nicely compacted accordingly to gradient 1:15 and 1:8. Blocks were set at 13.5 m and 7.2 m of the horizontal length of the mobile bed to block's toe, Ls for both bed gradients respectively (Figure 5). The Beach foreshore in this study was assumed as one straight slope without any obstacle or berm. The wave overtopping tank was setup accordingly to the bed steepness with a 10 cm chute to guide the overtopping discharges flow into the tank [33, 34]. Extra water tank storage was prepared outside of the model which connects with the water pump from the overtopping tank.

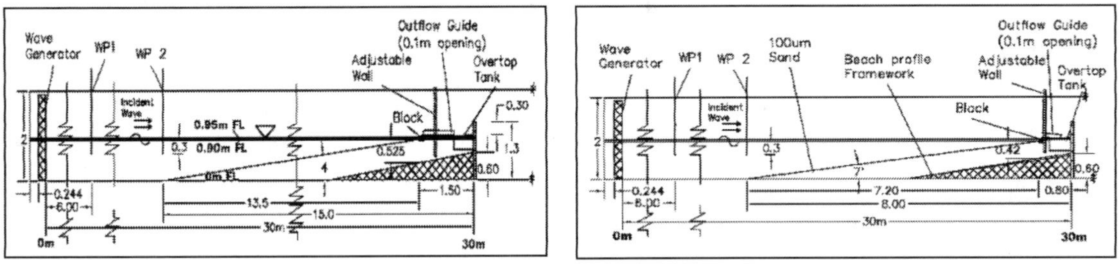

Figure 5. Schematic diagram of experimental wave in 30 m model with 1:5 (left) and 1:8 (right) beach gradient.

3.3 Calibrations

To ensure the success of the experiment, all the instruments were maintained checked and calibrated, especially the wave flume's paddle and wave probes. To define the minimum time simulation of mobile bed reaches its equilibrium in required model scale's size, the wave paddle was allowed to run for 0.5 to 1 hour with bed material but without other instruments and smooth-slope NEXC blocks.

The wave was the most challenging parameter to calibration as it needs extra attention and a precise method. The wave conditions were measured using 2 sets of twin wire resistance type on 30 cm wave probe's sensor and its capabilities, to read wave height up to 0.15 m. Both wave probes are connected to an installed computer with Hr Merlin software, which processed the signals into Hr-Daq software to derive the significant wave height, Hs, and means zero-crossing wave period, Tm from the wave energy spectrum. This study used the wave spectrum of JONSWAP which defined by wave height with Gamma 3.30.

Calibration on wave probes needs to calibrate at least 3 times repeated for each water level condition. This is to ensure signal transmitted from wave sequence to computer through probe device will eliminate any uncertainties in the process, with minimum calibration value, are achieved > 0.999. If the wave height and wave period are less than this value, then the setting needs to be adjusted and repeated until reached satisfactory calibration.

3.4 Data Analyzing

Statistical analysis was very important not only to assure the numerical accuracy and reliability of all statistical outputs but also used to provide solutions to complex problems. This can be applied to big data sets of many variables in finding a good pattern of the data. The results of statistical analysis have been significant for both in explanation a pattern and also for model prediction. It was one of the most convenient and easy tools in assisting to reach the goal of analyzing data using statistical software and its extension tools.

There are 4 parameters with 2 and 3 different levels this study has involved. Minitab Software has been chosen to achieve the most satisfactory of the statistical analysis data result. The tools to design the experiment was using General Full Factorial (GFF) method. The tools are quite amazing and well-known as one of the ideal tools in generating an empirical relationship model in practical to improving parameters [32] especially to interpret the response output of the experiment.

A correlation test was conducted using the Spearman rank method, r_s as it is appropriate in measuring the strength and direction of association between two ranked for both continuous and discrete ordinal variables [36, 37]. In theory, two method calculation will be applied according to the data, either it does have not (left) or has the tied ranks (right) as quoted below:

$$r_s = 1 - \frac{6\sum d_i^2}{n(n^2-1)} \quad \text{or} \quad r_s = \frac{\sum_i (x_i - \bar{x})(y_i - \bar{y})}{\sqrt{\sum_i (x_i - \bar{x})^2 \sum_i (y_i - \bar{y})^2}}$$

To choose between alternative models, Bayesian Information Criterion (BIC) is used to check, compare and determines objectively the best model, as it's the one that minimizes the considered information criterion to penalizes less the more complex model [38] in the search method that compares models sequentially. The term log-likelihood is giving everything aligned to their most favourable and p is the number of parameters considered in the model. BIC is defined as:

$$\text{BIC} = 2.(\textit{log-likelihood}) + (p.\log n)$$

To obtained a prediction value that might impact the future changes of the response, a robust fit regression model of the GFF method was used to analyze further variables of the correlated parameters and responses. As in this study, a linear trend is fitted to wave overtopping discharges of each parameter variable in an iteratively BIC technique. Then, the method will deal with solution determination and outliers' detection. Depending on the p-value and standard error, *S* value from the trend line, weights of measurements are adjusted accordingly and the trend line is then re-fitted. The process is repeated until the solution converges. As it is very rare in physic of the 3-ways interaction and the 4-ways interaction, and no this to allowed to have enough degree of freedom to do the analysis. So, any insignificant parameter has to set aside and refitting the model back. Randomized order in the block design of the variables has been applied in design tools to minimizing the effect of variability when it is associated with discrete data.

4. Results and Discussion

4.1 Defining and measured the results of the input parameter's response

Due to various not been previous physical laboratory doing comparisons of various smooth-slope NEXC block on different mobile bed gradients with irregular incident waves affecting response performance of each scenario block applied during experimental. The result of the experiment has found out the highest overtopping discharges, wave runup and significant wave height at bed toe was 8.50 l/s/m, 0.6 m and 0.025 m.

Figures 6a and 6b show the distribution data of 2 types of bed gradients in each figure. Indicated that the data in Figure 6a and 6b show the wave overtopping and significant wave height will decrease with increase in water level from shallow to deep, whenever mobile bed changes from mild to steep gradient with the smooth-slope block from 30° to 60°. However, Figure 6b interaction of wave runup and overtopping discharges show significant increases due to changes on these 3 parameters from lowest to the highest value.

Not much of the data distribution can be analysed using this visual and quantitative raw data. Thus, the further analysis method is required to understanding and better statistically analyze the relationship between the input parameters with the responses.

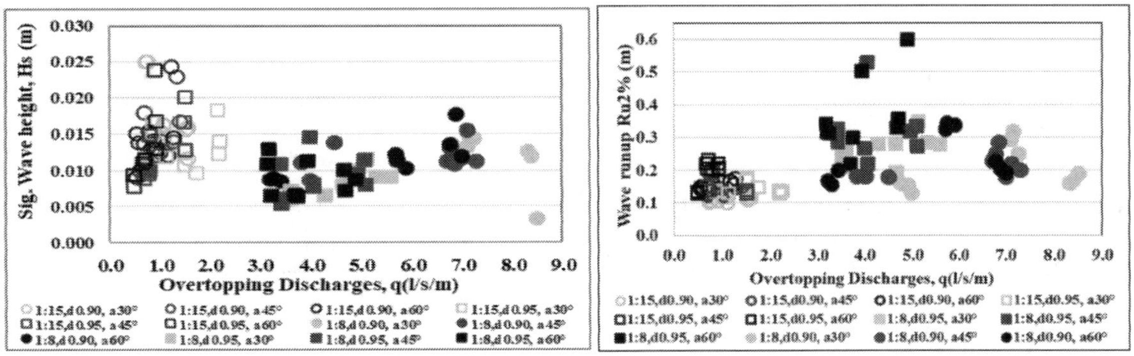

Figure 6. Distribution of significant wave height at bed toe (left) and wave runup (right) over the overtopping discharges with 1:15 and 1:8 beach gradient, the water level of 0.9 m and 0.95 m at 0.1 m, 0.13 m and 0.15 m wave height.

4.2 Significant wave height at the bed toe ($H_{s,bedtoe}$)

Using the correlation test, the confidence interval (CI) of the probability was set to 0.95 or 95 % that the true mean lies within the range. Table 2 indicates that most of the relationship between the $H_{s,bedtoe}$ with sample 2 are in a negative correlation with beach gradient (β_{bed}), water level (h_{water}) and smooth-slope NEXC block (α_{block}). While the relationship was positively correlated with the incident wave height (H_{mo}) with r_s equal to 0.334.

Table 2. Pairwise of Spearman rank's correlation on the $H_{s,bedtoe}$ with the input parameters.

Sample 1	Sample 2	N	Correlation, r_s	95 % CI for r_s	p-value
$H_{s,bedtoe}$ (m)	β_{bed} (°)	108	-0.475	(-0.616, -0.305)	0.000
$Hs,_{bedtoe}$ (m)	h_{water} (m)	108	-0.337	(-0.499, -0.153)	0.000
$Hs,_{bedtoe}$ (m)	α_{block} (°)	108	-0.029	(-0.217, 0.160)	0.762
$H_{s,bedtoe}$ (m)	H_{mo} (m)	108	0.334	(0.150, 0.496)	0.000

Even though, β_{bed} show the r_s value is -0.475, anyhow the value is less than 50 % of the strong r_s value. On the other hand, the sample with α_{block} is statistically not significant because its p-value (0.762), which is greater than the usual significance level of 0.05 and the r_s (-0.029) was weakly far from r_s = -1. As a result, it indicates that a negative significant linear relationship occurs in between the $H_{s,bedtoe}$ with the β_{bed} and d_{water}, while a positive linear relationship with H_{mo}.

4.3 Wave Runup ($R_{u2\%}$)

In the $R_{u2\%}$ relationship evaluation, clearly can be stated that the strongest relationships are with the β_{bed} with r_s value is 0.74, followed by 0.341 with the water level, 0.288 with the α_{block} and less correlated with the 0.158 of H_{mo}. However, the test indicates the relationship between $R_{u2\%}$ are with the α_{block} and h_{water}, which both have a p-value of 0.000, are strong significantly correlated.

The analysis results also clearly stated that there is a significant linear relationship in between the $R_{u2\%}$ with all three parameters of β_{bed}, h_{water}, and α_{block}, but not with the H_{mo}. Eventually, the analysis result has presently the $R_{u2\%}$ have a strongly significant linear relationship with the β_{bed} but significant with the h_{water}, and α_{block}.

Table 3. Table Pairwise of Spearman's rank correlation of the $R_{u2\%}$ with the input parameters.

Sample 1	Sample 2	N	Correlation, r_s	95 % CI for r_s	p-value
$R_{u2\%}$ (m)	β_{bed} (°)	108	0.740	(0.626, 0.823)	0.000
$R_{u2\%}$ (m)	h_{water} (m)	108	0.341	(0.157, 0.502)	0.000
$R_{u2\%}$ (m)	α_{block} (°)	108	0.288	(0.101, 0.456)	0.002
$R_{u2\%}$ (m)	H_{mo} (m)	108	0.158	(-0.033, 0.338)	0.102

4.4 Overtopping Discharges ($q_{overtop}$)

Analysis of the correlation in Table 4 indicates the $q_{overtop}$ have a positive correlation with the β_{bed} and the H_{mo} while showing a negative correlation with the d_{water} and the α_{block}. However, among these 4 set samples, the only d_{water} is not significantly correlated but strongly correlate with the β_{bed}, which the r_s (0.866) was close to the +1 with a p-value of 0.000.

Even though the relationship of $q_{overtop}$ with the H_{mo} and α_{block} is indicating r_s = 0.372 and -0.21, which this value was below 50 % of the r_s, these 2 samples are still significantly correlated. Thus, it is concluded that the $q_{overtop}$ have a strong positive linear relationship with the β_{bed} but a negative and positively significant linear relationship with α_{block} and H_{mo} respectively.

Table 4. Pairwise of Spearman rank's correlation on the $q_{overtop}$ with the experimental parameters.

Sample 1	Sample 2	N	Correlation, r_s	95 % CI for r_s	p-value
$q_{overtop}$ (l/s/m)	β_{bed} (°)	108	0.866	(0.798, 0.912)	0.000
$q_{overtopg}$ (l/s/m)	h_{water} (m)	108	-0.146	(-0.327, 0.045)	0.132

$q_{overtop}$ (l/s/m)	α_{block} (°)	108	-0.21	(-0.386, -0.020)	0.029
$q_{overtop}$ (l/s/m)	H_{mo} (m)	108	0.372	(0.191, 0.529)	0.000

4.5 Experiment output responses

To determine the stronger relationship between the 3 responses, all experiment output responses are evaluated. Analyze based on the matrix plot and the correlation value given, indicate that all responses are significant with each of the responses in the sample analyzed. However, the most significant correlate was $R_{u2\%}$ with the $H_{s,bedtoe}$ and with the $q_{overtop}$. Both show the p-value are 0.000 > 0.05 respectively.

However visually and analysed value shows that the strongest positive linear relationship is between the $R_{u2\%}$ with the $q_{overtopping}$, which the r_s value is 0.595. While the other two correlation analyses on responses: $R_{u2\%}$-$H_{s,bedtoe}$ and $q_{overtop}$-$H_{s,bedtoe}$, indicate have a negatively significant linear relationship, as both have r_s = -0.566 and -0.202 respectively. Hence the result strongly indicates that $q_{overtop}$ positively correlates and very dependent on $R_{u2\%}$ changes value. While $R_{u2\%}$ negatively correlate and dependent on the $H_{s,bedtoe}$ closely.

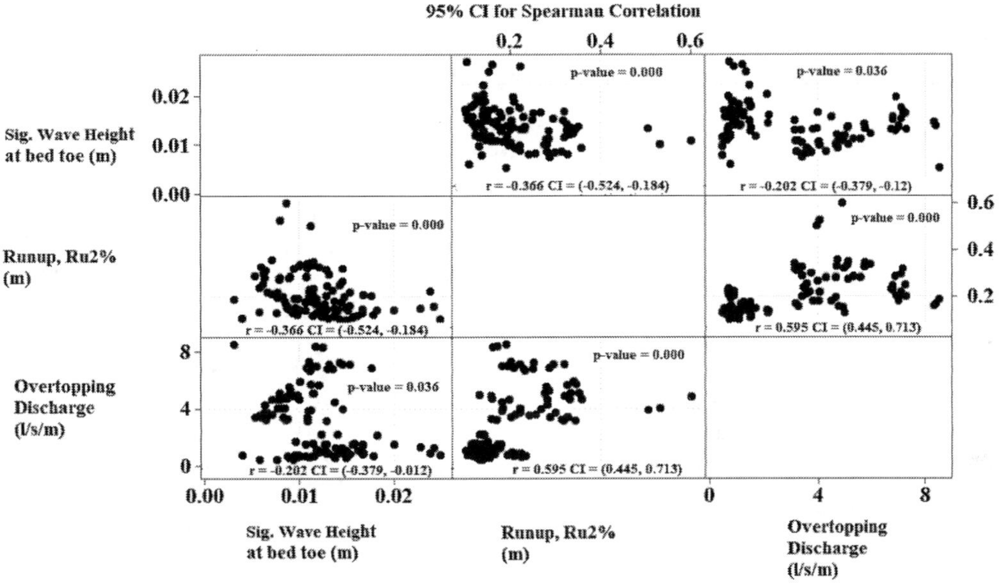

Figure 7. Presented the Spearman correlation's matrix plot with full correlation interval.

4.6 Significant relationship prediction model

The model was designed to develop a useful pre-experiment matrix setup of the physical model using the JONSWAP spectrum which will apply the parameters and its levels more than 2 on mobile bed material. The regression model used $H_{s,bedtoe}$ and $R_{u2\%}$ as a covariate and 95 % of the two-sided confident level for all intervals. To measure the strength of the difference between observed and expected values, a standardized residual was applied. The original design model reveals the model was 99.86 % in R-square (R-sq), S is 0.108253 and included both covariates in the model (Table 5).

Table 5. Summary of the optimum relationship prediction model.

Model	S	R-sq	R-sq(adj)	PRESS	R-sq(pred)	AICc	BIC
Before	0.108253	99.86 %	99.79 %	1.94273	99.67 %	-96.69	-37.97
After	0.338069	98.12 %	97.91 %	13.8863	97.63 %	89.39	120.38

However, to obtain the best model and using significant parameters in the model, the model was analysed based on the p-value terms and using stepwise with BIC. This means, in obtaining the fittest model, refitting and diagnostic tools need to eliminate the insignificant parameters. After refitted the model, clearly seen that no covariate is in the regression model and only strong significant parameters are included. The refitted design model shows the R-sq 98.12 % with S (0.338069) increase but still, less than 0.5, BIC increase from negative (-37.97) to positive (120.38) value and all coefficient term relationship have p-value less than 0.05 (Table 6), which indicate the model is considered as satisfied diagnostic and fitted. S value much have a big discrepancy as both the 3-ways and 4 ways interaction are set aside from the model makes design series value in this interaction terms have been ignored.

Figure 8 on the upper right of the Pareto chart indicate that all parameters interaction terms were critical in the model. Residual plotting also shows that no long tail in the pattern of the normal probability plot, fewer outliers and the residuals almost approximately follow a straight line. This verified the assumption of the residuals is consider are normally distributed. While residuals versus fits plot show but the points fall randomly on both sides of zero value, a slight curvature pattern in the points. However, it verifies the assumption of the residuals are randomly distributed and have almost constant variance. Additionally, the residuals in the observation order plot are fall randomly around the centre line in time order. Thus, an assumption that the residuals are independent of one another is verified.

Table 6. Coefficients and significant level of the interaction terms.

Term	Coef	SE Coef	95 % CI	T-Value	P-Value	VIF
Constant	3.0836	0.0325	(3.0190, 3.1482)	94.79	0.000	
β_{bed}						
0.0667	-2.0347	0.0325	(-2.0993, -1.9702)	-62.55	0.000	1.00
0.1250	2.0347	0.0325	(1.9702, 2.0993)	62.55	0.000	*
h_{water}						
0.90	0.4495	0.0325	(0.3849, 0.5140)	13.82	0.000	1.00
0.95	-0.4495	0.0325	(-0.5140, -0.3849)	-13.82	0.000	*
α_{block}						
0.57735	0.4203	0.0460	(0.3289, 0.5116)	9.14	0.000	1.33
1.00000	-0.0396	0.0460	(-0.1309, 0.0517)	-0.86	0.391	1.33
1.73205	-0.3807	0.0460	(-0.4720, -0.2893)	-8.27	0.000	*
H_{mo}						
0.10	-0.8662	0.0460	(-0.9575, -0.7749)	-18.83	0.000	1.33
0.13	0.0910	0.0460	(-0.0003, 0.1824)	1.98	0.051	1.33
0.15	0.7752	0.0460	(0.6839, 0.8665)	16.85	0.000	*
Bed gradient x h_{water}						
0.0667 0.90	-0.4858	0.0325	(-0.5504, -0.4212)	-14.93	0.000	1.00
0.0667 0.95	0.4858	0.0325	(0.4212, 0.5504)	14.93	0.000	*
0.1250 0.90	0.4858	0.0325	(0.4212, 0.5504)	14.93	0.000	*
0.1250 0.95	-0.4858	0.0325	(-0.5504, -0.4212)	-14.93	0.000	*
β_{bed} x H_{mo}						
0.0667 0.10	0.5084	0.0460	(0.4171, 0.5997)	11.05	0.000	1.33
0.0667 0.13	-0.1406	0.0460	(-0.2319, -0.0492)	-3.06	0.003	1.33
0.0667 0.15	-0.3678	0.0460	(-0.4592, -0.2765)	-8.00	0.000	*
0.1250 0.10	-0.5084	0.0460	(-0.5997, -0.4171)	-11.05	0.000	*
0.1250 0.13	0.1406	0.0460	(0.0492, 0.2319)	3.06	0.003	*
0.1250 0.15	0.3678	0.0460	(0.2765, 0.4592)	8.00	0.000	*
h_{water} x H_{mo}						
0.90 0.10	-0.2802	0.0460	(-0.3715, -0.1888)	-6.09	0.000	1.33
0.90 0.13	0.1577	0.0460	(0.0664, 0.2490)	3.43	0.001	1.33
0.90 0.15	0.1225	0.0460	(0.0311, 0.2138)	2.66	0.009	*
0.95 0.10	0.2802	0.0460	(0.1888, 0.3715)	6.09	0.000	*

0.95 0.13	-0.1577	0.0460	(-0.2490, -0.0664)	-3.43	0.001	*
0.95 0.15	-0.1225	0.0460	(-0.2138, -0.0311)	-2.66	0.009	*

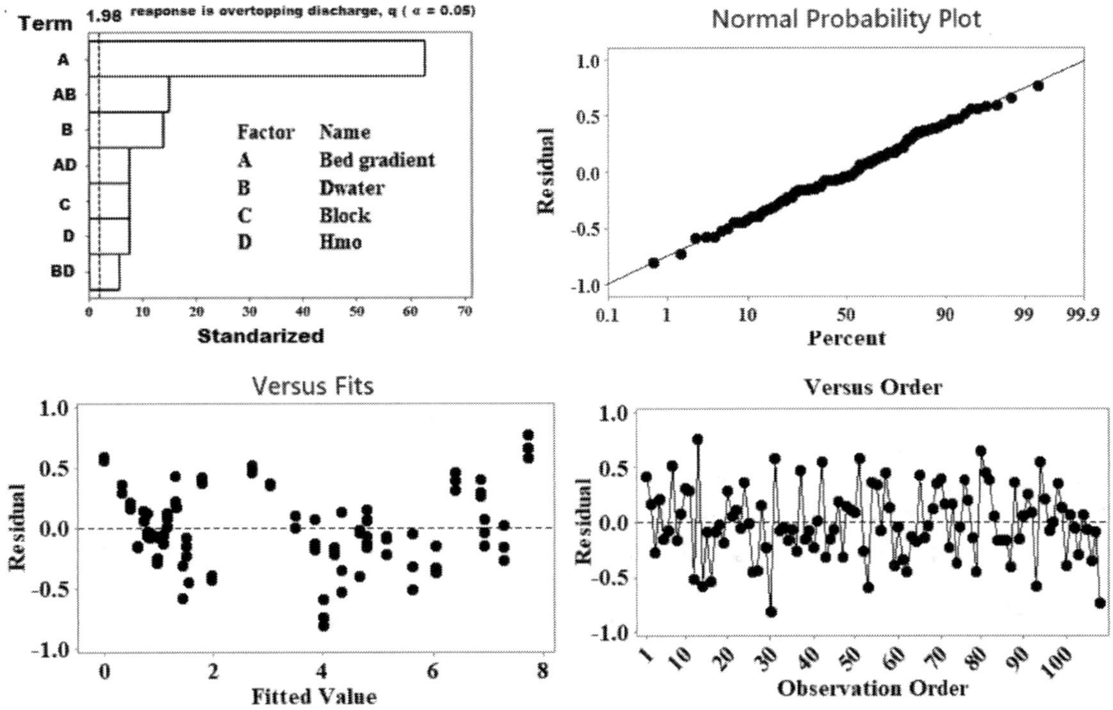

Figure 8. Pareto chart of the standardized effects (upper left) and residual plot for the fit regression model (normal probability plot, fitted value and observation order).

There are 10 linear terms, 37 interactions of 2-way terms, 43 interactions for 3-way terms and 48 interactions for 4-ways terms in the model before fitted and gone through the insignificant parameter's isolation. However, the fitted and significant relationship predictive model has obtained 10 linear terms and 16 interactions of 2-way term terms through the GFF method. This model is for direct predictive calculation on overtopping discharge on smooth-slope NEXC block using mobile bed due to water level changes and quote as follow:

Overtopping discharge, $q = 3.0836 - 2.0347\beta_{1:15} + 2.0347\beta_{1:8} + 0.4495h_{0.90} - 0.4495h_{0.95} + 0.4203\alpha_{30^\circ} - 0.0396\alpha_{45^\circ} - 0.3807\alpha^\circ - 0.8662H_{mo,0.10} + 0.0910H_{mo,0.13} + 0.7752H_{mo,0.15} - 0.4858\beta_{1:15},h_{0.90} + 0.4858\beta_{1:15},h_{0.95} + 0.4858\beta_{1:8},h_{0.90} - 0.4858\beta_{1:8},h_{0.95} + 0.5084\beta_{1:15},H_{mo,0.10} - 0.1406\ \beta_{1:15},H_{mo,0.13} - 0.3678\ \beta_{1:15},H_{mo,0.15} - 0.5084\beta_{1:8},H_{mo,0.10} + 0.1406\ \beta_{1:8},H_{mo,0.13} + 0.3678\beta_{1:8},H_{mo,0.15} - 0.2802h_{0.90},H_{mo,0.10} + 0.1577\ h_{0.90},H_{mo,0.13} + 0.1225h_{0.90},H_{mo,0.15} + 0.2802h_{0.95},H_{mo,0.10} - 0.1577h_{0.90},H_{mo,0.13} - 0.1225h_{0.90},H_{mo,0.15}$

5. Conclusion

The experimental result using 3 types of smooth-slope NEXC block on mobile bed using wave height JONSWAP spectrum due to water level changes have successfully obtained 36 different scenarios with 108 data at 30m long multi-functional wave flume at NAHRIM laboratory, Selangor, Malaysia. Experiment data have shown that both interactions of $q_{overtop}$-H_s will decrease but $R_{u2\%}$-$q_{overtop}$ significant increases with an increase in water level from shallow to deep, whenever mobile bed changes from mild to steep gradient with a smooth-slope block from 30° to 60°.

Correlation tests between the parameters and responses of the experiments indicate that $R_{u2\%}$, $H_{s,bedtoe}$ and $q_{overtop}$ are strongly correlated to the β_{bed} of the model. While both $R_{u2\%}$, and $q_{overtop}$ significant relationship with the α_{block}. Test on the responses output shown that $q_{overtop}$ dependently to the $R_{u2\%}$, meanwhile found out that $R_{u2\%}$ was dependant to the $H_{s,bedtoe}$. Finally, the fitted relationship design model with the most significant experimental variables has presented the optimal predictive model of $q_{overtop}$ has obtained 10 linear terms and 16 interactions of 2-way terms. However, this method needs to further analyze and fitted using the response optimizer method for a better model.

The analysis has also shown that the overtopping is not directly affected by water level but water level will be affected on wave runup and significant wave height. The optimal relationship prediction model of overtopping discharges was verified and satisfied to apply in experimental physical setup accordingly to the overtopping proportions required.

References

[1] IPCC 2019 Summary for Policymakers *IPCC Special Report on the Ocean and Cryosphere in a Changing Climate (SROCC)* [Pörtner H-O et al (eds)] 1-196
[2] Muis S *et al* 2016 A global reanalysis of storm surges and extreme sea levels *Nature Communications* **7** (11969) 12
[3] IPCC 2019 Chapter 4 Sea Level Rise and Implications for Low Lying Islands, Coasts and Communities *Summary for Policymakers- IPCC Special Report on the Ocean and Cryosphere in a Changing Climate (SROCC)* [Oppenheimer Met. al (eds)] 1-139
[4] Nicholls R J and Cazenave A 2010 Sea-level rise and its impact on coastal zones *Science* **328** (5985) 1517-1520
[5] USGCRP 2009 *Global Climate Change Impacts in the United States* "Climate Change Impacts by Sectors: Ecosystems" Karl T R Melillo J M and Peterson T C (eds) 1-196
[6] Shankar N J and Jayaratne M P R 2002 Wave run-up and overtopping on smooth and rough slopes of coastal structures *Ocean Engineering* **30** (2003) 221–238
[7] Kobayashi N 1997 Wave runup and overtopping on beaches and coastal structures *Research Report No CACR 97-09* 1-48
[8] Allsop W *et al* 2008 Direct Hazards From Wave Overtopping – The Forgotten Aspect Of Coastal Flood Risk Assessment? *Semantic Scholar* 1-11
[9] Le H T, Verhagen H J and Vrijling J K 2017 Damage to grass dikes due to wave overtopping *Natural Hazards* **86** 849–875
[10] Magar V *et al* 2010 Damage from Waves, Surges and Overtopping at Vertical Seawalls *Coast, Marine Structure and Breakwater* 144-157
11] IPCC 2014 Climate Change 2014: *Synthesis Report. Contribution of Working Groups I, II and III to the Fifth Assessment Report of the Intergovernmental Panel on Climate Change* [Core Writing Team, Pachauri R.K and Meyer L A (eds)] 151
[12] Mohd Yusoff N A, Ramli N L and Mohamed M R 2015 Investigation Of The Potential Harnessing Tidal Energy In Malaysia *ARPN Journal of Engineering and Applied Sciences* 9835-9841
[13] Md Din A H *et al* 2017 Malaysian Sea Water Level Pattern Derived From 19 Years Tidal Data *Jurnal Teknologi (Sciences & Egineering)* 79(5) 137-145
[14] Allsop W *et al* 2015 Wave overtopping at vertical and steep seawalls *Maritime Engineering* **158** (3) 103-114
[15] Yusof N S 2021 Faktor-Dan-Mitigasi-Hakisan-Pantai-Di-Terengganu *Perspektif 4/2020* 11
[16] Ngah H 2017 Hujan lebat, angin kencang di Pantai Timur *Berita Harian Online Nasional*
[17] Arrifin L 2020 Amaran waspada cuaca buruk di beberapa daerah Pantai Timur *Berita Harian*
[18] Bakar B 2016 6 pantai di Terengganu, Kelantan kritikal *Berita Harian*
[19] Senechal N *et al* 2011 Wave runup during extreme storm conditions *Journal Of Geophysical Research* **116** 1-13
[20] Hsu W T A Study on Coastal Flooding and Risk Assessment under Climate Change in the Mid-Western Coast of Taiwan *MDPI* **9**(6) 390
[21] Ehsan S *et al* 2019 Current and potential impacts of sea level rise in the coastal areas of Malaysia *IOP Conf. Series: Earth and Environmental Science* **228**
[22] Nicholls R J *et al* 2011 Sea level rise and its possible impacts given a 'beyond 4°C world' in the twenty-first century *Philosophical Transaction of The Royal Society A* **369** 161-181
[23] Faour G *et al* 2013 GIS-Based approach to the assessment of coastal vulnerability to sea level rise: case study on the Eastern Mediterranean *Journal of Surveying and Mapping Engineering* **1**(3) 41-48
[25] Nicholls R J 2004 Coastal flooding and wetland loss in the 21st century: Changes under the SRES climate and socio-economic scenarios *Global Environmental Change* **14**(1) 69-86
[26] Mohamed Rashidi A H *et al* 2017 Physical Hydraulic Modelling For The Development Of Innovative Coastal Protection Structure In A 2-D Wave Flume *Proceedings of the 37th IAHR World Congress* 3309-3317
[27] Mohamed Rashidi A H *et al* 2018 Beach Protection System Using Nahrim Coastal Erosion Protection And Beach Expansion (Nexc) Block At Pantai Batu 5, Port Dickson, Negeri Sembilan, Malaysia *Paper Proceedings: The 4th International Conference on Water Research (ICWR 2018)* 1-19
[28] Mahmud M R *et al* 2015 Assessment of Effective Seasonal Downscaling of TRMM Precipitation Data in Peninsular Malaysia *Remote Sensing* **7** 4092-4111

[29] Chu P C *et al* 2004 South China Sea Wind-Wave Characteristics. Part I: Validation of Wave watch-III Using TOPEX/Poseidon Data *Journal of Atmospheric and Oceanic Technology* **21** 1718-1733
[30] Aboobecker V M *et al* 2012 Wind wave variability in the South China Sea *Geophysical Research Abstracts* **14** 10508
[31] Elfrink B and Baldock T 2002 Hydrodynamics and sediment transport in the swash zone: A review and perspectives *Coastal Eng.* **45** 149–167
[32] MINITAB 2019 *MINITAB 19.2020.1(64-bit) (Version 19.2020.1) [Window]* Australia: Minitab Pty Ltd.
[33] EurOtop 2018 *Manual on wave overtopping of sea defences and related structures: An overtopping manual largely based on European research, but for worldwide application* [Van der Meer, J.W., Allsop, N.W.H., Bruce, T., De Rouck, J., Kortenhaus, A., Pullen, T., Schüttrumpf, H., Troch, P. and Zanuttigh, B.(eds.)] **2** 320
[34] Van der Meer J W *et al* 2016 *EurOtop: Manual on Wave Overtopping of Sea Defences and Related Sturctures : An Overtopping Manual Largely Based on European Research, but for Worldwide Application.* 2nd ed. 1-178
[35] Oppenheimer M *et al* 2019 Sea Level Rise and Implications for Low-Lying Islands, Coasts and Communities *IPCC Special Report on the Ocean and Cryosphere in a Changing Climate* [Pörtner H-O, Roberts D C, Masson-Delmotte V, Zhai P, Tignor M, Poloczanska E, Mintenbeck K, Alegría A, M Nicolai, Okem A, Petzold J, Rama B, Weyer N M (eds.)]. 321-445
[36] Kirch W 2008 Level of Measurement *Encyclopedia of Public Health* **2** 851–852
[37] Lehman A 2005 *Jmp For Basic Univariate And Multivariate Statistics: A Step-by-step Guide* 123
[38] Shao J 1993 Linear Model Selection by Cross-Validation *Journal of the American Statistical Association* **88**(422) 486–94
[39] Lee K S and Seng L Y 2009 Simulation Studies on the Electrical Power Potential Harnessed by Tidal Current Turbines *Journal of Energy and Environment* **1**(1) 18–23
[40] Mohd Yusoff N A, Ramli N L and Mohamed M R 2015 Investigation of The Potential Harnessing Tidal Energy In Malaysia ARPN Journal of Engineering and Applied Sciences 10(21) 9835-9841

Acknowledgements
Authors wishing to acknowledge National Water Research Institute of Malaysia (NAHRIM) in supporting the usage of a physical laboratory to carry out a physical model. Very appreciative of all NAHRIM technical staff's support and they are involving in conducting the 100m Multi-purpose Wave Flume. A huge thanks too for the great assistance and valuable encouragements from my supervisor, co-supervisor, colleagues and anonymous reviewers for their constructive comments and suggestions in improving this manuscript.

The influence of temperature and nutrient concentrations on growth rate, biomass, Chlorophyll-*a*, and biochemical compositions of *Tetraselmis suecica* (Chlorophyta)

A A Sas[1,2], A J Turki[3], A Affan[4]*, W A Al-Taisan[5], S K Das[1], S N Su[1] and Z C Cob[1]

[1]Department of Earth Sciences and Environment, Faculty of Science and Technology, Universiti Kebangsaan Malaysia, 43600 Bangi, Selangor, Malaysia
[2]Department of Marine Biology, Faculty of Environmental Science and Marine Biology, Hadhramout University, - Hadhramout, Yemen
[3] Department of Marine Chemistry, King Abdul-Aziz University, Jeddah, Saudi Arabia.
[4]Marine Biology Science, Faculty of Marine Biology, King Abdul-Aziz University, Jeddah, Saudi Arabia
[5]Department of Biology, College of Science, Imam Abdulrahman Bin Faisal University, Dammam, Saudi Arabia

*Correspondence E-mail: affan2007@gmail.com

Abstract. Microalgae have been got high attention due to its potentiality in aquaculture as live feed, and in industries as ingredients for pharmaceutical, cosmetics, and biofuel industries. The objective of this study was to know the optimum culture condition for profuse growth and biochemical composition of *Tetraselmis suecica* under two parameters: temperature (20° C, 25° C, and 30° C) and modified F/2 medium nutrients concentrations. Culture of group "A" 20°C was categorized as A1 (F/2 stock solution-A; 0.50 ml. L^{-1} and F/2 stock solution-B; 0.20 ml. L^{-1}), A2 (F/2 stock solution-A; 1. 00 ml L^{-1} and F/2 stock solution-B; 0.40 ml. L^{-1}) and A3 (F/2 stock solution-A;1.50 ml. L^{-1} and F/2 stock solution-B; 0.60 ml. L^{-1}). Cultures in 25° C and 30° C were also categorized as groups "B" 25° C (B1, B2, and B3) and "C" 30° C (C1, C2, and C3), respectively. The culture was done for 2 weeks with L:D cycle of 12:12 by using fluorescent light. The highest biomass production was 0.80, 0.64, and 0.45 gL^{-1} in C2, B3, and A3, respectively. Biochemical analysis showed that protein; 21.92, 20.83, and 18.68 %, lipid; 10.76, 9.42, and 11.71 %, carbohydrate; 38.51, 37.78, and 41.49 %, ash; 15.89, 15.61, and 13.7 %, and moisture; 14.26, 15.02 , and 14.42 % in biomass grown of "A", "B" and "C" culture group, respectively. From the study, it could be said that *T. suecic* is a eurythermal and mesotrophic habitant microalga which produce high protein and high carbohydrate in low and high temperature, respectively.

Keywords: Microalgae, media, culture condition, and biomass production

Track Name: Coastal Management and Marine Ecosystem

1. Introduction

Microalgae are microscopic photosynthetic organisms that are found in all aquatic environments. There are more than 40 species of microalgae with valuable properties that are used in aquaculture industries, and the algal biomass's lipid, carbohydrate, and protein composition determine its overall economic potential [1, 2]. Since they are a natural source of polyunsaturated fatty acids, they are a major food source for many aquatic organisms and the primary live feed component in marine hatchery operations [3]. In recent decades, Microalgae have been identified as a potential source of biofuel due to numerous advantages such as rapid growth potential, high biomass production rate, higher photosynthesis efficiency [4, 5], and can are grown in wastewater and on unproductive land [6]. Moreover, several chemical products and health foods have been mass-produced from microalgae, such as pigments, vitamins, long-chain unsaturated fatty acids, and antioxidants, for the nutraceutical industries [7].

Microalgae cultivation must be monitored and maintained to ensure optimal growth, high survival, biomass production, and enhancement of the biochemical composition due to their high metabolic flexibility in response to changes in environmental factors [8]. There is a continued need to establish factors that alter microalgal growth such as nutrients, light, temperature, salinity, and pH [9]. The growth rate of microalgae is influenced by the availability of nitrogen in the culture media, because it is a component of amino acid and chlorophyll [17]. Temperature is a fundamental environmental factor that strongly regulates microalgae growth. The temperature-growth range is ecologically important because it defines the temperature range in which microalgae can be metabolically active and determines their distribution [10]. Moreover, microalgal growth increases, and cell volume decrease with temperature [11]. There was no overall relationship between the percentage of microalgae biochemical compositions and temperature except ash (organic matter) [12], but it seems there is no consistent trend in the response of biochemical composition of microalgae to increases or decreases temperature [10].

Tetraselmis sp. are marine flagellate belong to *Chlorophyceae*, it grows as single, almost spherical, and motile cells with four equal flagella. They are used in relatively small amounts in aquaculture facilities as feed for shrimp larvae, rotifers, juvenile shellfish, and bivalve *molluscs* [13, 14]. This genus has also been discovered to have a wide range of antimicrobial activity [15]. In addition, its species have shown high potential as probiotics in fish [15]. Due of its high vitamin E content, *Tetraselmis sp.* have also been suggested as a source of vitamin E for human and animal consumption [16]. The temperature and nitrate concentration as the major causes of increasing cell abundance and growth rate of *Tetraselmis sp.* [17], *Tetraselmis suecica* is a microalgae commonly used in aquaculture and is thought to be an optimal source of long-chain polyunsaturated fatty acids (PUFAs), especially eicosapentaenoic acid (EPA) [18], as well as the possibility of using it to remove N and P for purify wastewater [19]. Different culture conditions, such as temperature, salinity, nutrient, and light, may influence microalgae growth and proximate composition. Thus, this study was aimed to investigate the effect of temperature, within the range 20° C, 25° C, and 30° C , and nutrient media concentrations on the growth and biochemical composition of *T. suecica* that isolated from coastal water.

2. Materials and methods

2.1 Culture medium preparation and Tetraselmis suecica isolation

2.1.1 Preparation modified F/2 medium (Guillard, 1975):

The modified F/2 culture medium stock solution was prepared as a concentrated solution, and for that natural seawater was filtered, thereafter, the different chemicals were diluted for making 'A' solution which was made with macronutrient and vitamins, and the solution 'B' was made with micronutrients most of them was trace metal that the microalgae need it in small amounts Table 1.

Table 1. Chemical composition of culture medium.

A- solution	Amount	B- solution	Amount
Chemicals	$g.L^{-1}$	Chemicals	$g.L^{-1}$
$NaNO_3$	290.00	$FeCl_3$. $6H_2O$	15.00
KNO_3	50.00	H_3BO_3	10.00
Ca $(NO_3)_2.4H_2O$	5.00	Co $(NO_3)_2$. $6H_2O$	2.00
Mg $(NO_3)_2$	5.00	$K_2Cr_2O_7$	0.50
K_2HPO_4	20.00	$CuSO_4.5H_2O$	2.00
KH_2PO_4	65.00	$MnSO_4.H_2O$	1.00
K_2SO_4	15.00	$ZnCl_2.6H_2O$	2.50
$MgSO_4.7H_2O$	5.00	Na_2-EDTA	5.50
Urea	2.50		
HNO_3	125.00 $ml.L^{-1}$		
H_3PO_4	15.00 $ml.L^{-1}$		
Vitamins			
Thiamin	10.00000		
D-Biotin	0.02500		
Cyanocobalamin	0.00001		

2.1.2 Isolation steps of Tetraselmis suecica

Tetraselmis suecica was isolated from coastal water and the isolation was done following the methods described by Affan et al [20]. In summary, a 1 mL diluted sample was transferred to a Sedgwick Rafter (S–R) counting chamber, then target species was isolated with a mouth-sucking micropipette (MSM) technique Affan et al [21]. To make MSM, the tip of a glass pasture pipette was heated over and drawn out, the MSM needle was positioned close to the *T.* suecica which was checked using an inverted microscope (MEIJI Techno- TC5100, JAPAN), and a gentle suction was applied to suck or picked up the single cell. Thereafter, *T. suecica* was transferred to a tissue culture multiwell plate for growing and subculture in autoclaved seawater which was enriched with F/2 medium nutrients. Then, the single species was again cultured in agar (Hi Media Laboratories, Ltd. India). To prepare agar culture medium, 2 % of agar [21] (w/v) and F/2 medium of 'A' and 'B' stock solution (0.5 mL of A- solution and 0.2 mL of B- solution) added and diluted, then autoclaved. *T. suecica* from the multiwall was transferred and grown in the agar plate. The agar plate was kept at a temperature of 25° C under fluorescent light (180 µmol photon.m^{-1} s^{-1}) on a 12:12 light:dark (L:D) cycle and the culture plate was checked every two days to observe the colony of *T. suecica.*

2.1.3 Determination of optimum growth conditions

For determination of the optimum growth condition the culture experiments were designed as follows; The culture was grown in three different temperatures (20° C, 25° C and 30° C) with three concentration of F/2 medium with fixed light intensity (as used during isolation) and salinity and pH of 38.00 psu and 8.00 psu, respectively of natural seawater. Each 1 L flask with 250 mL culture media was inoculated with approximately 20 cells ml^{-1} of *T. suecica.* The cultures were grown under fluorescent lights (180 µmol photons. m^{-2} s^{-1}) on a 12:12 light: dark (L: D) cycle for 2 weeks.

Table 2. The culture conditions temperature and nutrients concentrations.

Temperature	Solution "A" ml L^{-1}	Solution "B" ml L^{-1}
Group "A" 20° C	A1: 0.5, A2: 1, A3: 1.5	A1: 0.20, A2: 0.40, A3: 0.60
Group "B" 25° C	B1: 0.5, B2: 1, B3: 1.5	B1: 0.20, B2: 0.40, B3: 0.60
Group "C" 30° C	C1: 0.5, C2: 1, C3: 1.5	C1: 0.20, C2: 0.40, C3: 0.60

2.1.4 Estimation of dry biomass and cell abundance:

Each treatment was replicated twice. A 5 mL sample was collected from each culture flask on every two days and fixed with Lugol's iodine solution. Then, a 1 mL fixed sample was used for counting using with a Sedgwick-Rafter (S-R) counting chamber under an inverted microscope at 400x magnification. A 30 mL sample was collected for biomass estimation from each culture on same time the sample was collected for cell counting. The sample was filtered through pre weighed GF/C Whatman filter paper. A pre-weighed filter paper that was soaked in distilled water and dried at the same time as a blank. The biomass content filter paper was kept at 55° C in an oven, dried and weighed, and the dry weight biomass was calculated as gL^{-1}, which was plotted as a growth curve. The specific growth rate (l), defined as the increase in cell density per unit time (Pirt 1975) [23], and formulated as follows:

μ_{max} (day^{-1}) = ln (X_1/X_0) / $t_1 - t_0$

Where X_0 and X_1 are the biomass at the beginning (t_0) and the end (t_1) of the selected time interval between inoculation and maximum biomass production.

2.2 Estimation of Tetraselmis suecica chlorophyll a:

Chlorophyll *a* of *T. suecica* was estimated following the method described by Parsons et al [23], the short description of that respective method; the preparation of chlorophyll- *a* samples from 30 mL of culture aliquots which was filtered by standard vacuum filtration through a 0.45 µm pore size whatman GF/C filter paper. The filter paper was chopped up to extract the pigments, then placed in 10 mL of 90 % acetone in a 25 mL universal tube and stored at 4° C for 24 hours to allow time for the *T. suecica* pigments. After extracted period, the samples were removed from the freeze and allowed to warm at room temperature in a dark place. The acetone volume was restored to 10 mL, the solution was then centrifuged (ALC Multi Speed Refrigerated Centrifuge PK 121 R) for 5-10 min at 3000-5000 rpm. Finally, the chlorophyll- *a* concentration was determined using a Spectrophotometer (UV-VIS - mini 1240.), and was calculated using the formula described by Parsons et al [24],

The chlorophyll *a* ($\mu g.L^{-1}$) = $(27.6 * (A665 - A750) * s)/v$

A665= absorbance at 665 nm, A750= absorbance at 665 nm, s = solvent extract volume, ml, s = solvent extract volume, ml.

2.3 Biochemical Composition Analysis

The biomass from the best growth condition was selected for biochemical analysis, and the best growth condition was selected based on higher biomass production among the different culture conditions. The sample was harvested by centrifuged (ALC Multi speed, PK 121R, Korea) for 15 min at 4000 rpm Price et al [24] then preserved in -80° C temperature and finally freeze-dry powder for further study (Mini Lyotrap LTE, PE114732, Great Britain). The protein, lipid, carbohydrate, moisture and ash contents were

IOP Conf. Series: Earth and Environmental Science **880** (2021) 012014 doi:10.1088/1755-1315/880/1/012014

determined following the methods of the Association of Official Analysis Chemists AOAC (2005) [25], such as crude lipid was determined by Soxhlet extraction; crude protein, by the Kjeldahl method; ash, by calcinations in a furnace at 550° C (Jelrus P.D.Q-U.S.A.); and moisture by heating at 105° C for 20 h.

2.4 Statistical analysis

Statistical package for social sciences (SPSS 20.0) were applied for one way ANOVA test in the evaluation of the differences in the values followed by Tukey's tests. A P-value of <0.05 was considered as significant.

3. Results

There is no doubt that *Tetraselmis suecica* is useful in aquaculture, in the bioconversion of solar energy, and potentially useful as food for human consumption [26]. Therefore *T. suecica* was isolated and studied its growth response under conditions of temperature and nutrients (Table 2.).

3.1 Growth characteristics biochemical composition for T. suecica

3.1.1 Dry biomass production

The results showed the dry biomass of *T. suecica* varied from 0.33 to 0.45 g L^{-1} in "A" group and was highest in A3, followed by A2 and A1 (Figure. 1A). In the "B" group, the biomass production ranged from 0.51 to 0.64 gL^{-1}, with the highest value in B3, followed by B1 and B2 (Figure 1B). In the "C" group; the biomass production was varied from 0.62 to 0.80 gL^{-1}, and the highest was in C2, followed by C3 and C1 (Figure 1C). Maximum dry biomass was higher than that of all other culture conditions in C2.

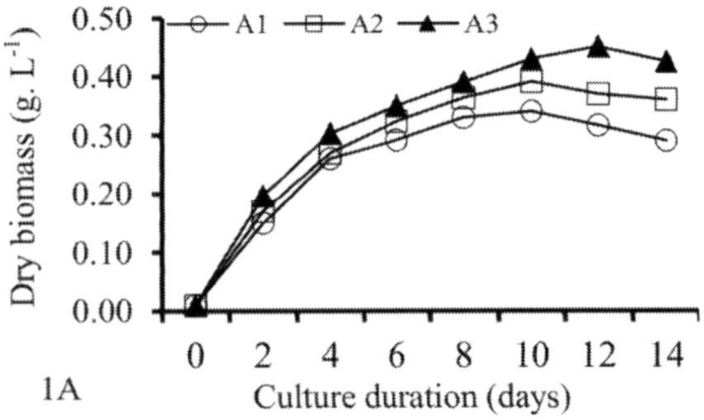

1A

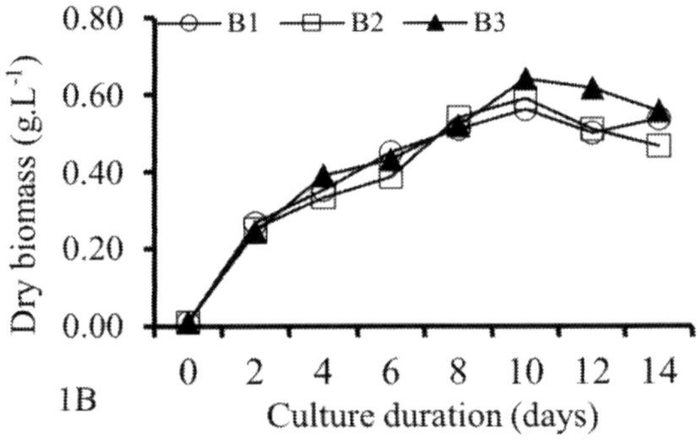

1B

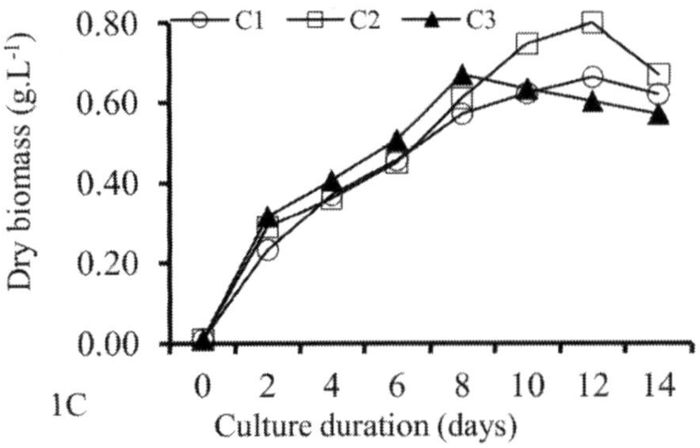

1C

Figures 1A, **1B**, and **1C** the dry biomass production of *T. suecica* at "A", "B", and "C" groups in F2 medium concentrations of A1, A2, A3, B1, B2, B3 and C1, C2, C3

3.1.2 Specific growth rate and cell abundance

The *T. suecica* grew under all culture conditions (temperature and F2 media) with different cell densities (represented by the column chart). The maximum specific growth rate (μ_{max} d^{-1}, represented by the line chart) was estimated as an informative way to ascertain microalgae activity, which can increase at exponential rates. The growth rate of microalgae varies greatly depending on the temperature, and the findings of our investigation into this phenomenon are presented in table 3. The highest mean values of specific growth rate (μ_{max} d^{-1}) of *T. suecica* was observed in culture at 25° C (0.41 ± 0.007 day^{-1}). The cell densities were the greatest in the 30° C water temperature condition (Figure. 2C), followed by 25° C (figure. 2B) and 20° C (figure. 2A). The highest μ_{max} occurred at 10th day after inoculation in all treatments except the 20°C water temperature condition, where, the highest μmax was 0.37d^{-1} and occurred on day 12 after being inoculated with A2 nutrient concentration. The highest cell density 5.51×10^5 cells mL^{-1} occurred at the μ_{max} of 0.36 d^{-1} on day 10th with A3 nutrient concentration in 20°C (Figure 2A). In "B" culture, the maximum cell density was found 5.38 ×10^5 cells mL^{-1} on day 10 with B2 nutrient concentrations conditions, while the highest μ_{max} was 0.42 d^{-1} in B3 culture (Figure 2B). Similarly, in "C" culture, the highest cell density was 5.72 ×105 cells mL^{-1} with the highest μ_{max} 0.37 d^{-1} on day 10th of culture (Figure 2C).

Table 3. Changes of mean specific growth rate (day^{-1}) of *T. suecica* under different temperatures and nutrients during the culture period.

Temperature groups (° C)	20	20	30
Tetraselmis suecica	0.36 ±0.008	0.41 ±0.007	0.35 ± 0.03

IOP Conf. Series: Earth and Environmental Science **880** (2021) 012014 doi:10.1088/1755-1315/880/1/012014

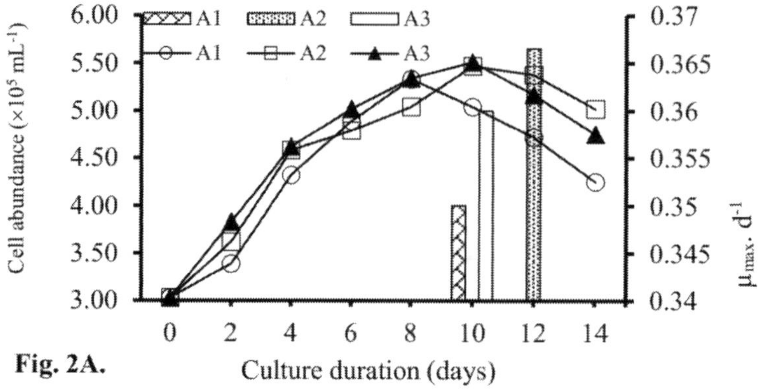

Fig. 2A.

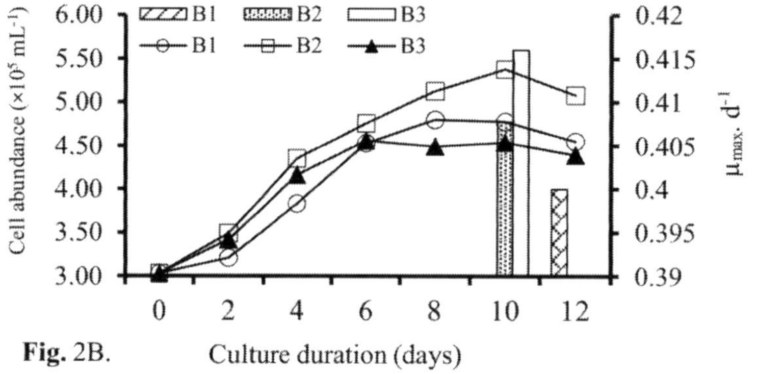

Fig. 2B.

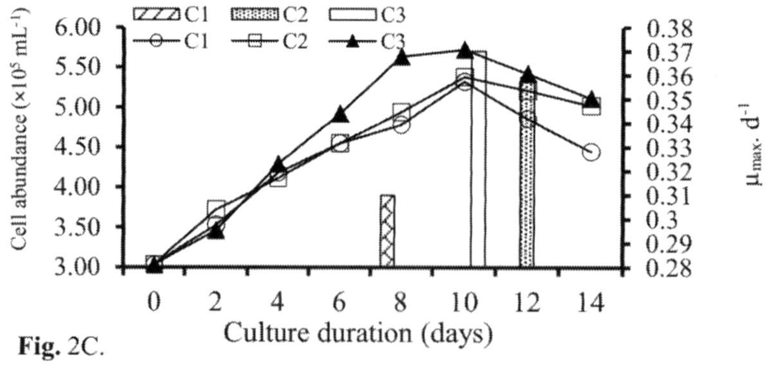

Fig. 2C.

Figure 2A, 2B, and 2C. Cell abundance (line chart) and specific growth rate (column chart) of *T. suecica* under different culture conditions.

3.1.3 Chlorophyll- a of Tetraselmis suecica

The chlorophyll-*a* concentration ranged from 0.23 to 3.10 mg L^{-1} and was the highest concentration in A1, followed by A3 in "A" culture group (Figure. 3A). In the "B" group, the chlorophyll a concentration varied from 0.25 to 3.41 mg L^{-1} with the highest in B2, followed by B3 (Fig. 3B). Similarly, in the "C" group, the chlorophyll- *a* concentration ranged from 0.24 to 3.94 mg L^{-1}, and the highest was in C2, followed by C3 (Figure 3C). Among nine culture conditions, the highest chlorophyll-*a* was found in C2 culture. Statistically, the results showed that there was no a significant difference of cell abundance and chlorophyll- *a* between nutrients concentrations ($P<0.05$) for all temperature treatments.

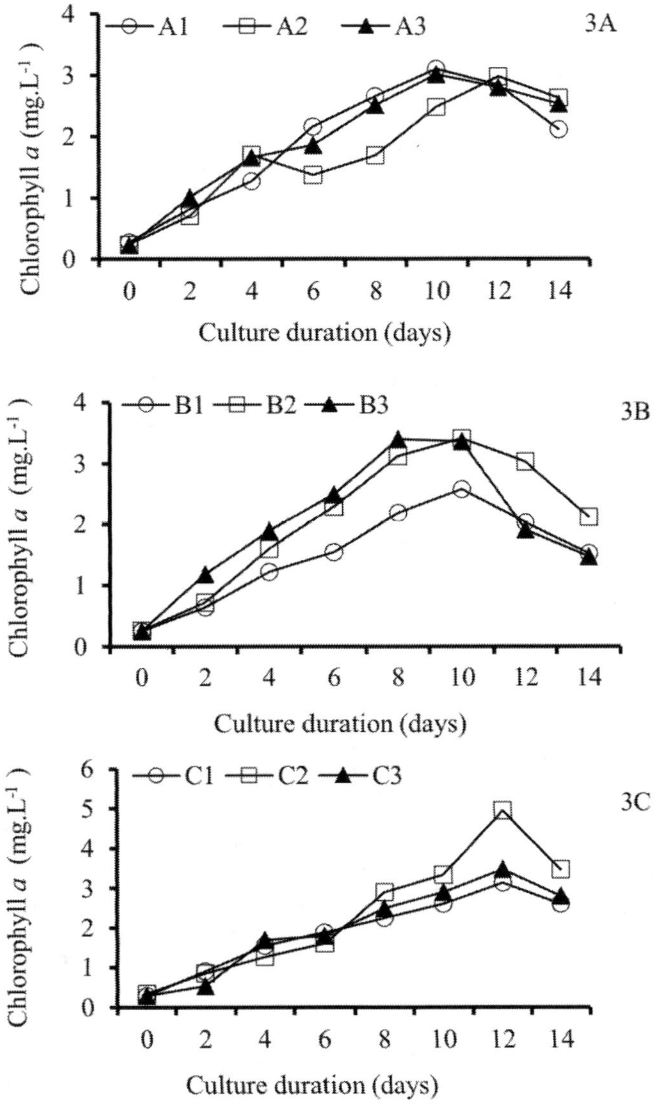

Figure 3A, 3B, and **3C** Chlorophyll *a* concentrations at “A”, “B”, and “C” groups in F2 medium concentrations of A1, A2, A3, B1, B2, B3 and C1, C2, C3 respectively.

3.1.4 Biochemical composition analysis

The biochemical composition of *T. suecica* determined as proteins, carbohydrates, total lipid, moisture, and ash in percentages of dried biomass of "A", "B", and "C" culture conditions. The *T. suecica* biochemical composition was as follows: protein; 21.92 %, 20.83 %, and 18.68 %, lipid; 10.76 %, 9.42 %, and 11.71%, carbohydrate; 38.51 %, 37.78 %, and 41.49%, ash; 15.89 %, 15.61 %, and 13.7%, and moisture; 14.26 %, 15.02 %, and 14.42 % in culture biomass grown in temperature of "A", "B", and "C", respectively (Figure. 4)

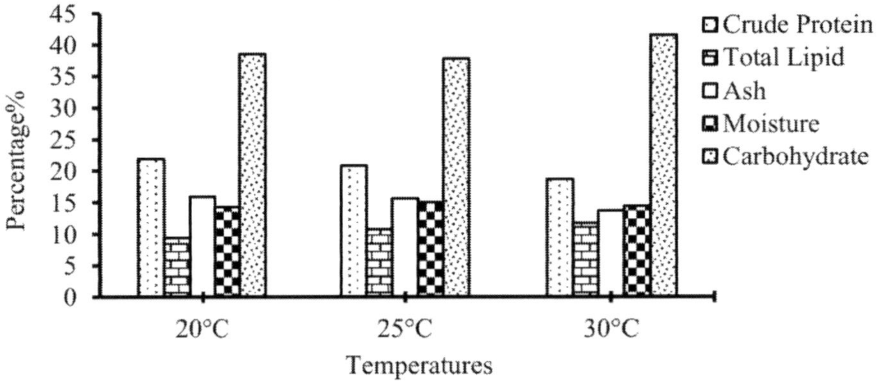

Figure 4. Biochemical compostions analysis at 20° C, 25° C and 30°bC.

4. Discussion

4.1 Effect of temperature and nutrients on growth

Tetraselmis sp., is one of the species isolated from coastal waters that is characteristic of mid-latitude areas with significant annual temperature fluctuations and has been widely used in aquaculture around the world [27]. In this study, isolation of a unialgal strain of *Tetraselmis suecica* (Chlorophyceae) from the direct seawater sample. The combination of MSM and agar culture techniques for the isolation of green algae was more effective and less time-consuming than serial dilution and agar plate culture isolation technique or direct pipetting. This species grew well in all culture conditions with remarkably highest cell abundance and chlorophyll-*a* concentration in C3 and C2 culture respectively. Therefore, it can be said that this species is euryhaline and mesotrophic and it makes bloom or vigorous growth at the temperature of 30° C. This result is in line with the previous study by Rukminasari et al [17], where their results indicate that the highest growth rate of *Tetraselmis sp.* was found at 30° C with double nitrate of Conway medium. The microalgal growth rate is expected to increase with temperature within range, and then rapidly decline above a critical temperature [28]. In this study, *T. suecica* reached the highest peak of growth at the temperature of 30° C with nutrients concentrations of C culture on day 10 of cultivation. This finding supported by previous study by Kim *et al*. [44] who found that the highest cell concentration of *Tetraselmis sp.* showed after 10 days. *T. suecia* produced the highest chlorophyll-*a* value of 3.94 mg L^{-1} at C2. Previous research conducted by Ginting et al [29] measured the values of chlorophyll-*a* of *Tetraselmis sp.* was 48 µg mL^{-1} and 40.88 µg mL^{-1}, and Sani et al [30] found total chlorophyll between 3.65-19.20 mgg^{-1} from dried biomass. The decrease in chlorophyll *a* content of *T. suecica* was most pronounced in the nutrient-stressed cultures [31], these results were in agreement with our findings of chlorophyll- *a* content. In addition, these results explain there is an ideal temperature range in which a specific microalgal strain shows the best growth performance [32].

4.2 Effect temperature and nutrients on biochemical compositions
The temperature plays an important role in controlling the physiological and biochemical processes in algae and their response to temperature changes varied with species [10]. High growth temperature has been linked to a substantial reduction in protein content, as well as increases in lipid and carbohydrate content [33, 34]. In our results, the lipid contents were low, due to the presence of significant amounts of ash [37]. There were no significant differences in the average crude protein content between different temperatures. The biochemical composition of microalgae often responds strongly to nutrient stress [35]. Under some conditions, total lipid content varies between microalgae species, ranging from very low 1 % to 75 % but it can exceed to very high 90 % of dry weight [36]. The biochemical composition of microalgal species (even in the same culture medium) can vary across culture conditions. A previous study reported the protein content was 70.24 %, 68.01 %, 68.67 %, 64.58 % and 62.81 % while the carbohydrate content was 9.88 %, 11.68 %, 12.72 %, 15.57 % and 19.63 % when *Spirulina maxima* was cultured in the same medium at 20° C, 25° C, 30° C, 35° C and 40° C [38]. Our result also showed low protein and high carbohydrate content at high temperature. Lipid content did not show any trend to increase or decrease with nutrient concentration and temperature rises and falls. Zheng et al [39] reported that plants adapt to temperature change is to decrease the degree of unsaturation of membrane lipids under high temperature and increase it under low temperature. Oliveira *et al.* [34] also found lipid content of 6.22 % and 6.22 % in *Spirulina maxima,* and 7.24 % and 6.32 % in *Spirulina platensis* when those species were culture at temperature of 20° C and 25° C. The lipid content of *T. suecica* was low at high temperatures which are similar with previously mentioned studied, and according to the results obtained the carbohydrate was high under all conditions of temperature, the cause is this species is one of the microalgae that has been found to accumulate large quantities of carbohydrates [37]. The lipid contents were low, due to the presence of significant amounts of fibers and ash. However, the lipid content of *T. suecica* in this study showed a minor increase with increasing temperature. This was in agreement with the previous report that the lipid content of *T. subcordiformis* increased with increasing temperature [40]. The ash content of *T. suecica* in this study was higher at 20° C being similar to the values obtained for the same species at the same temperature [39] and in other species such as *Tetraselmis sp.* and *T. chuii.* [37].

Several of the different descriptions of temperature effects on the growth and biochemical compositions of *Tetraselmis spp.* have been reported. At 20° C *T. subcordiformis* had the highest lipid content (22.25 %) [38], which corresponded to its optimal growth temperature. On the other hand, temperatures for optimal growth rates and highest biomass production of *T. chuii* were ranged from 20° C to 30° C [42]. Our study registered the carbohydrate contents were the highest value followed by protein and lipid these results agree with a previous similar study carried out on *T. suecica* cultured under 21° C and F/2 medium [43].

5. Conclusion

It's crucial to know how microalgae respond to growth and adapt to temperature changes and nutrients concentration. Based on the results of the study, it was concluded that the effect of temperature and nutrients on optimal growth and biochemical content of *Tetraselmis sp.* depends on the varied species, and could be increasing or decreasing by manipulating the culture conditions (temperature and media concentrations). *Tetraselmis suecica* showed maximum growth rates, cell density, biomass, and chlorophyll-*a* in higher temperatures 30° C and a 1, 1.5 ml L^{-1} of solution A and 0.40 ml L^{-1}, 0.60 ml L^{-1} of solution nutrients concentration of F2 media. An increase in the carbohydrate and lipid contents has been observed as temperatures rise, while the inverse was observed with protein content.

References

[1] Pulz O and Gross W 2004 *Applied microbiology and biotechnology* **65** (6) pp.635-648
[2] Williams P and Laurens L 2010 *Energy Environ. Sci.* 3 554e590
[3] Makridis P *et al* 2006 *Aquaculture* **255** (1-4) pp.76-8
[4] Yangmin G and Mulan J 2011 *Biotechnol Lett* 33 1269-84
[5] Wahidin S *et al* 2013 *Bioresour Technol* 129 7-11
[6] Katiyar R *et al* 2017 *Renew. Sustain. Energy Rev* 72 1083–93
[7] Pal D *et al* 2011 *Applied microbiology and biotechnology* **90** (4) 1429-1441
[8] Renaud *et al* 2002 *Aquaculture* **211** (1-4) pp.195-214
[9] Arau´jo S Garcia V 2005 *Aquaculture* **246** 1-4 405-412
[10] Teoh M *et al* 2010 *Malaysian Journal of Science* **29** (2) pp.82-97
[11] Sayegh F and Montagnes D 2011 *Bioresource technology*, **102**(3), pp.3007-3013
[12] Renaud S *et al* 2002 *Aquaculture* **211** (1-4) pp.195-214
[13] Muller-Feuga A *et al* 2003 Uses of microalgae in aquaculture. In: Stottrup, J.G., McEvoy, L. A. (Eds.), Live Feeds in Marine Aquaculture. Blackwell, *Oxford* pp. 253– 299
[14] Khatoon H *et al* 2014 *International Biodeterioration & Biodegradation* 95 11-18
[15] Zittelli *G* et al 2006 *Aquaculture* **261** (3) 932-943
[16] Irianto A and Austin B 2002 *J. Fish Dis.* 25 633–642
[17] Rukminasari N *et al* 2019 In *IOP Conference Series: Earth and Environmental Science* Vol. **370** No. 1 p. 012059
[18] Fabregas J *et al* 2001 *Marine biotechnology* **3** (3) 256-263
[19] Afshari A *et al* 2012 *Journal of Aquatic Animals and Fisheries* Vol. **2** No.8
[20] Affan A *et al* 2006 *Algae* 21 141–248
[21] Affan A *et al* 2009 *Journal of Phycology* 45 1405–1415
[22] Pirt S 1975 *John Wiley and Sons* **22** (3) 621-621
[23] Parsons R *et al* 1984 *Journal of the Fisheries Board of Canada* 18 1001-1016
[24] Price C *et al* 1974 *The Biological Bulletin* **147** (1) 136-145
[25] AOAC 2005 Official method of Analysis. 18th Edition, Association of Officiating Analytical Chemists, *Washington DC*, Method 935.14 and 992
[26] Fabregas J *et al* 1984 *Aquaculture* **42** (3-4) 207-215
[27] Huertas I *et al* 2011 *Proceedings of the Royal Society B: Biological Sciences* **278** (1724) 3534-3543
[28]Montagnes D *et al* 2003 *Aquat. Microbiol. Ecol.* 32 307– 313
[29] Ginting N *et al* 2018 *Bulletin Oceanografi Marina* **7** (2) 91-97
[30] Sani R *et al* 2014 *Journal of Food and Agroindustries*, **2** (2) 121-126
[31] Goiris K *et al* 2015 *Algal Research* **7** 51-57
[32] Venkataraman L 997 *Journal of Applied Phycology* 9 295–29
[33] Tomaselli L *et al* 1988 *Effects of temperature on growth and bichemical composition in Spirulina platensis strain M2*. In T. (Eds.), *Algal Biotechnology* 303 – 314
[34] Oliveira M *et al* 1999 *Aquaculture International* 7 261 – 275
[35] Goiris K *et al* 2015 *Algal Research* 7 51-57
[36]Spolaore P *et al* 2006 *Journal of bioscience and bioengineering* **101** (2) 87-96
[37] Pereira H *at al* 2019 *Molecules* **24** (17) 3192
[38] Oliveira M *et al* 1999 *Aquaculture International*. 7 261–275
[39] Zheng G *et al 2011Plant, cell & environment* **34** (9) 1431-1442
[40] Wei L *et al 2015 Journal of Oceanology and Limnology* **33** (1) 99-106
}41] Tulli F *et al* 2012 *Journal of aquatic food product technology* **21** (3) 188-197
[42] Chen S *et al* 2012 *Botanical Studies* **53** (1)
[43] Sharawy Z *et al* 2020 *Aquaculture Research* **51** (6) 2216-2228
[44] Kim G *et al* 2016 *Algae* **31** (3) 257-266

Acknowledgement

Our highest gratitude and appreciation to the staff of Marine Science laboratory, Faculty of Science and Technology, UKM during this research period. This research was supported by Fundamental Research Grant (FRGS/1/2017/STG03/ UKM/02/5) MOHE Malaysia, and UKM research fund (project no. GGPM-2020-020).

Submerging paddy cultivation area due to coastal flooding in Kuala Kedah, Malaysia

Samera Samsuddin Sah[1,2], Khairul Nizam Abdul Maulud[1,3*], Nurul A'idah Abd Rahim[3,4], Othman A. Karim[1], Suraya Sharil[1]

[1]Department of Civil Engineering, Faculty of Engineering and Built Environment, Universiti Kebangsaan Malaysia, Bangi 43600, Selangor, Malaysia
[2]Faculty of Chemical Engineering Technology, Universiti Malaysia Perlis, 02600 Arau, Perlis, Malaysia.
[3]Earth Observation Centre, Institute of Climate Change, Universiti Kebangsaan Malaysia, Bangi 43600, Selangor, Malaysia
[4]Coastal Management and Oceanography Research Centre, National Water Research Institute of Malaysia, Lot 5377, Jalan Putra Permai, 43300 Seri Kembangan, Selangor, Malaysia

*Corresponding E-mail: knam@ukm.edu.my

Abstract. Kuala Kedah is a coastal area where the majority of the community are paddy farmers and fishermen. Almost the entire coastal area is used as a paddy cultivation area. However, this area faces the threat of seawater intrusion into land due to climate change driven sea-level rise. The rising seawater has affected the surrounding area, not only in terms of crop yields but also property and livelihood to the locals. Therefore, this study is designed to detect and analyze the progress of seawater on land at the Kuala Kedah coastal area using a hydrodynamic approach. Mike 21 software was used to simulate the hydrodynamic effects on 2 segments (NA and SA) in this study area by considering two conditions namely Condition 1 (K1) and Condition 2 (K2) which are respectively with and without coastal protection structure. However, this structure was only built along the 2.5 km shoreline in the NA segment and not in the SA segment. The findings show that the coastal protection structure in K2 is effective in reducing 50 % of the impact of sea level rise in year 2100 at NA segment, while only 10 % at SA segment. Therefore, the construction of these structures permanently should be given consideration by local authorities in planning future development to ensure lowland areas are protected from coastal floods.

Keywords: Climate Change, Inundation, Modelling, Saltwater Encroachment, Seawater Intrusion.

Track Name: Coastal Management and Marine Ecosystem

1. Introduction

Malaysia is a country located in Southeast Asia with a coastline of 8840 km [1] .Malaysia's coastline is rich in coastal resources and has a wealth of natural biodiversity [2]. As a record, 70 percent of the Malaysia's population lived in coastal zone. The majority of communities in Malaysia's coastal zone consist of fishermen and farmers, most of whom are senior citizens aged between 40 and 65 years [3] who have been permanent residents for more than 15 years in the area. However, these groups were affected by the sea level rise due to climate change. According to previous researches [4–6], the surface temperature and sea level in Malaysia were increasing by time to time, as well as the frequency of extreme events such as flooding, high rainfall and heatwaves.

As reported by Abdul Hamid [7], Malaysia will be expected to experience floods due to rising sea levels between 2070 and 2100 if the annual temperature increase is around 4° C due to continuously high gas emissions. Therefore, low-lying areas along the northeast and west coasts of the Peninsula Malaysia including the northwest coast were forecasted to face the problem of coastal flooding [8]. One of the affected areas on the west coast is at the mouth of the Kedah River where it is an agricultural area that covers 40 % of land use in the coastal zone.

The coastal hydrodynamic study approach is used in monitoring and predicting coastal conditions based on the numerical model approach as part of the physical model test [9]. Hydrodynamic models are important tools for estuarine and coastal management and have been used in water quality studies, sediment transport, and predicting the effects of different climate scenarios on estuaries and coastal waters. Each of these models plays an important role in flood prediction, pollutant modeling and changes in estuarine and coastal morphology [10, 11].

In Kedah coastal zone, the main agricultural activity is paddy cultivation which is also facing the threat of flooding due to a rising sea level of about 0.5 m in the year 2100 [8, 12]. Subsequently, [12] also reported that the farmers, especially in the Kuala Kedah coastal area, had to bear losses of more than 75 percent due to the encroachment of seawater into paddy fields since 2016. The presence of sea water in this paddy field causes the crop to become stunted and eventually die due to salt stress [13–15]. After the seawater intrusion incident in year 2016, as recorded JPS Malaysia began to design and build a revetment along 2.5 km of coastline, only in the NA segment with a height of 3.2 m from the Mean Sea Level (MSL) to protect this low-lying area.

Therefore, the hydrodynamic simulation technique using MIKE 21 is a suitable approach to predict and forecast the seawater advancement in terms of distance and total inundated area in land [16]. This numerical modeling technique will consider the impact of sea level rise to 2100 in this study. Hence, this study was designed and focused on evaluating and assessing the impact of sea level rise on the advancement of seawater to land as well as the potential affected using Mike 21 for two different conditions with and without coastal protection structures, namely K1 and K2, and finally, produced the inundation risk maps that can be used to assist the local authorities in future coastal area development.

2. Materials and Methods

Generally, this study involved the Mike 21 software which is using a numerical model approach. This software is well-known hydrodynamic modeling using flexible mesh. Hydrodynamic module namely Mike 21 HD Flow Model was used to predict the coastal flooding along 55 km Kuala Kedah coastline, approximately and presented in the form of inundation map for the year 2020, 2050 and 2100. Figure 1 shows the flow of the hydrodynamic modelling process using Mike 21 FM HD.

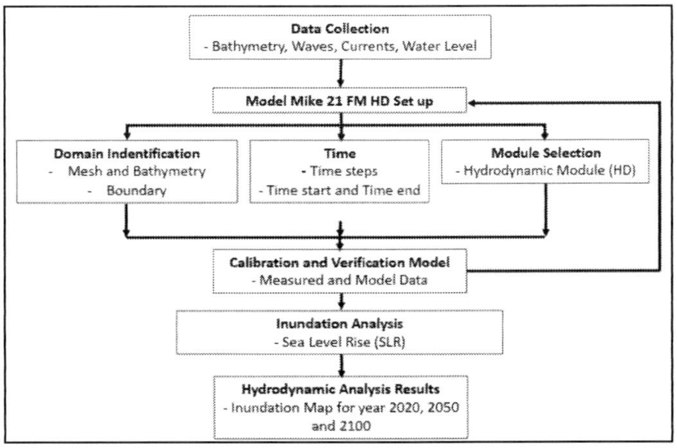

Figure 1. Flow process of Mike 21 FM HD.

The hydrodynamic parameters used in this study are bathymetry, waves, current speed, current direction and water level. The first step in this modeling, the identification of the domain and boundary was made by considering the location of the study area. The ocean area was divided into four polygons with different grid resolution as listed in table 1. Each polygon is designed based on the distance from the nearest area to the farthest from the shoreline as shown in Figure 2. Afterward, this mesh was uploaded into the Mike 21 FM HD and ready to process. There are three outputs produced after completing this model which are water level, current speed and current direction.

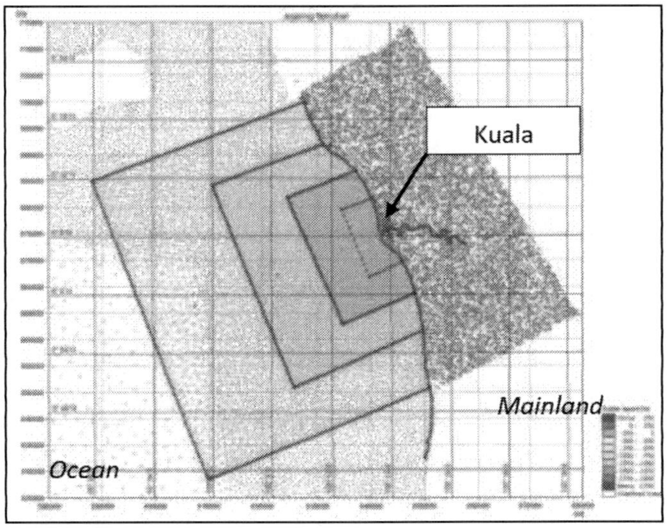

Figure 2. Flexible mesh using mesh generator.

Table 1. Grid resolution for area segment.

No	Area Segment	Distance from shoreline (m)	Grid Resolution (m)
1	Nearest area	5,000	100
2	Near area	13,000	200
3	Far area	26,000	600
4	Farthest area	40,000	1800

Next step, the water level, current speed and current direction results were calibrated and verified by comparing the modelled result with in-situ measurement data. The modelled result should meet the tolerance requirement as stipulated by the Department of Irrigation and Drainage (DID), Malaysia as stated in Table 2. The evaluation of this modelled result was done using Root-Mean-Square Error (RMSE) approach as shown in the formula below,

$$\sqrt{RMSE = \frac{1}{N}\sum_{1}^{n}(y-\acute{y})^2{}_i}$$

where n is the number of samples, y is the measured value, and $\acute{y}$ is the modelled value.

Table 2. Tolerance value for hydrodynamic study [17].

Parameter	Tolerance value
Water Level	Less than 10 %
Current Speed	Less than 20 %
Current Direction	Less than 20°

The final step, the coastal flooding due to sea level rise (SLR) for the year 2020, 2050 and 2100 were done using inundation analysis and mapped using ArcGIS software. Pulau Langkawi station was used as a reference point to analyse the SLR at the study area. In addition, only RCP 8.5 was chosen because it gives the most critical SLR value in the flooding analysis in this hydrodynamic model. This flooding analysis is divided into two conditions namely Condition 1 for analysis without coastal protection structure for northern part (NA) and southern part (SA) segment, and Condition 2 for analysis with coastal protection structure.

3. Results and Discussions

Hydrodynamic modeling in this study involves an area along 80 km of coastline from Kuala Perlis, Perlis to Yan, Kedah. The calibration and verification process were performed to ensure that the hydrodynamic modeling carried out were accurate and precise based on the actual conditions in the field. Subsequently, validate the findings from this modeling.

Figure 3. shows the calibration and verification graphs for the observed water level data and the modeling results. The blue line indicates the data observed by NAHRIM. Meanwhile, the black line is the result of this modeling. Therefore, the water level calibration value was determined based on the RMSE is 6.74 % which complies with the requirements of the guidelines issued by DID.

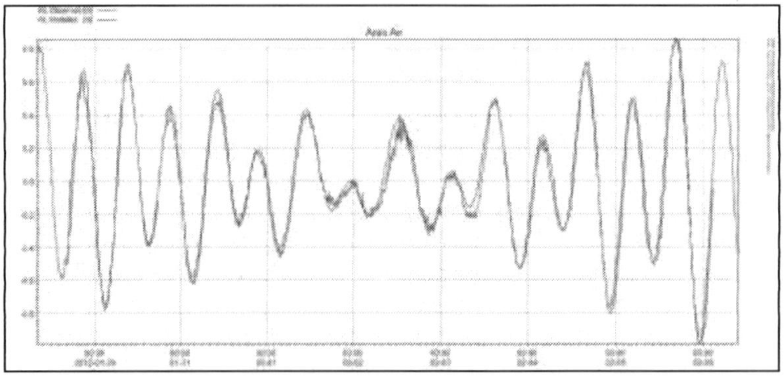

Figure 3. Calibration graph for water level parameters.

Next, figure 4 and figure 5 show the calibration graphs for the current speed and current direction. The RMSE value for current speed is 11.43 % while for current direction is 18.7°. These two parameters also meet the requirements as stated in the DID guidelines which are not more than 20% and 20°, respectively for speed and current direction.

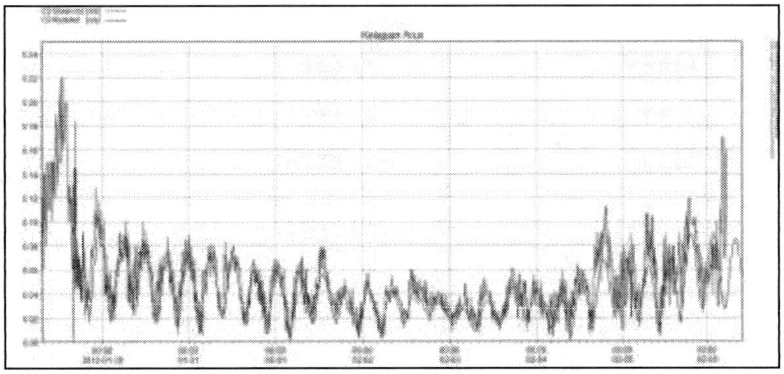

Figure 4. Calibration graph for current speed parameters.

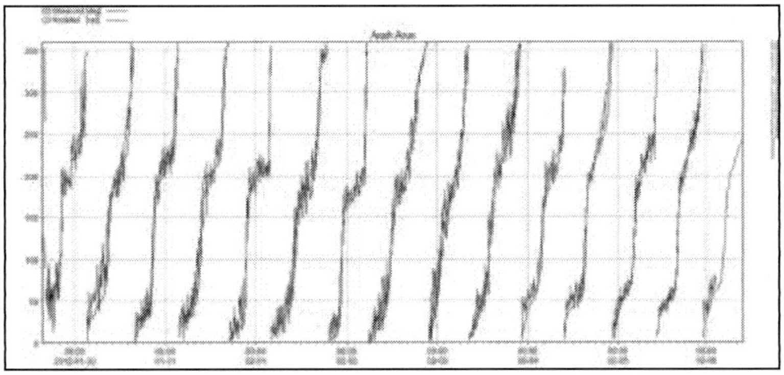

Figure 5. Calibration graph for current direction parameters.

Then, figure 6(a) shows the forecast condition of the areas affected by sea level rise in 2012, 2020, 2050 and 2100 for K1, while, Figure 6 (b) illustrates the affected area in 2020, 2050 and 2100 for K2. Visually, the SA segment is more affected than the NA segment from year to year for both K1 and K2. These simulation results show that more severe floods are expected to occur in the Kedah river estuary and low-lying zones. This is aligned with the findings by previous researchers [18,19].

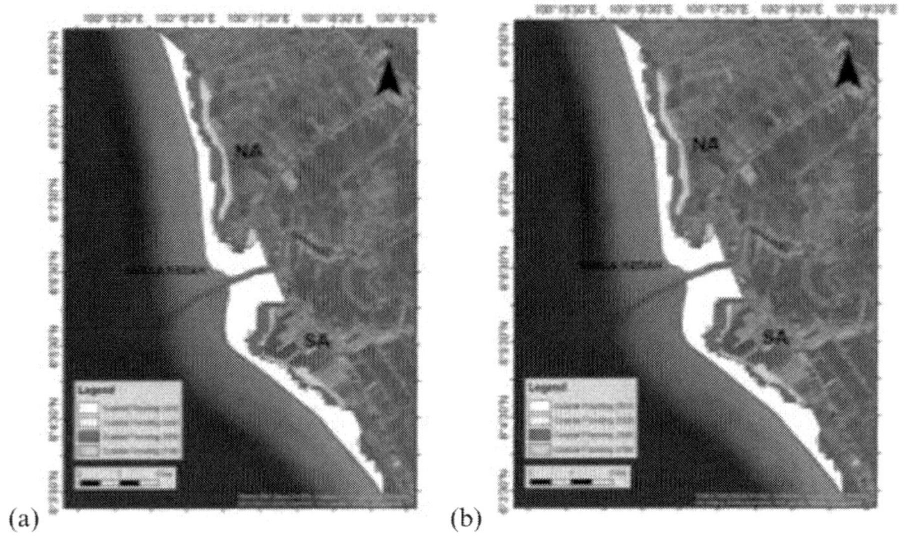

Figure 6. Coastal flooding for (a) K1 and (b) K2.

Furthermore, in 2100, 469.0 hectares and 498.9 hectares are forecast to be inundated in the absence of coastal protection structures, respectively for the NA and SA segments as illustrates in figure 7. But, if coastal protection structures are included in this flood analysis, the affected area is expected to be reduced by up to 50 %, even though this structure are only built in the NA segment.

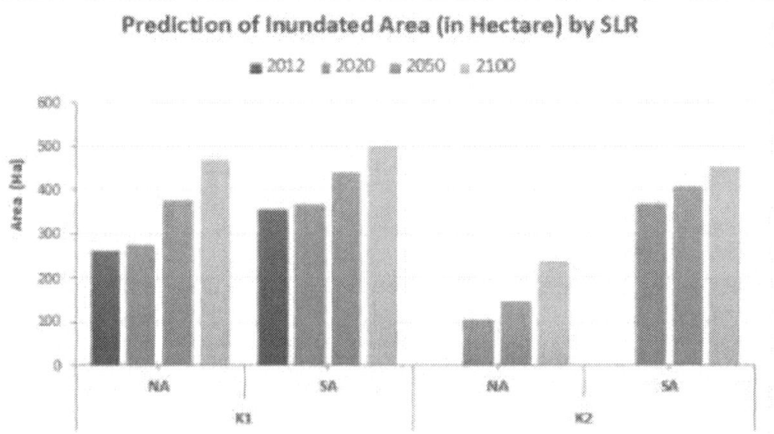

Figure 7. Total affected area for K1 and K2.

While, Table 3 shows the total areas affected by SLR for those years. In the year 2012, for K1, the land area will be affected up to a distance of 643.7 m and 788.5 m from the shoreline and continues to increase up to 926.5 m and 1212.6 m in the year 2100, respectively for the NA and SA segments. However, these values are seen to decrease for K2 for both segments.

Table 3. Seawater advancement distance for highest progress.

	K1 (meter)		K2 (meter)	
	NA	SA	NA	SA
2012	643.7	788.5	-	-
2020	752.7	810.6	586.4	717.5
2050	869.2	1123.8	690.1	879.8
2100	926.5	1212.6	759.9	960

Overall, the total area affected for the NA segment for K2 is reduced by more than 50 % compared to K1. However, the SA segment for K2 is only reduced by 10 % because of the absence of coastal protection structure at this segment. According to non-parametric statistical analysis namely Friedman Test, Chi-square for is equal to 4.0 which is greater than 3.84146 and the p-value is 0.046 less than 0.05 which can conclude that there is significant different between K1 and K2. This result was aligned with findings by [20–22] which found that the coastal protection structure able to absorb high wave energy and reduce its impact as well as preventing wave overtopping and flooding during heavy storm events.

4. Conclusion

Usually, coastal areas are areas filled with activities involving the sea and land such as fishing, farming, tourism and so on. The coastal area of Kuala Kedah, Kedah which is located on the west coast of Peninsular Malaysia is an agricultural area of paddy fields. However, this low-lying zone has the potential to face coastal flooding due to rising sea levels. This study was conducted to evaluate the impact of SLRs on potential coastal flooding using a hydrodynamic model namely Mike 21 FM HD. According to the results, coastal flooding would be the worst if there were no coastal protection structures to protect lowland areas. With the structure of coastal protection in NA segment, the rate of advance of seawater into the mainland and the affected area can be reduced by up to 50 %. These results can be input and reference to assist local authorities in planning the development of this area in the future.

References

[1] Selamat S N et al 2019 Multi method analysis for identifying the shoreline erosion during northeast monsoon season. *J Sustain Sci Manag*
[2] Mokhtar A *et al* 2020 Planning the Malaysian Coastline – Integrated Shoreline Management Plan. Apac 2019 1169–76
[3] Mohamed Shaffril HA et al 2015 The coastal community awareness towards the climate change in Malaysia *Int J Clim Chang Strateg Manag* **7** 516–33
[4] Daniel Tang K H 2019 Climate change in Malaysia: Trends, contributors, impacts, mitigation and adaptations *Sci Total Environ* **650** 1858–71
[5] NAHRIM 2017 Impact of Climate Change Sea Level Rise Projections
[6] Suparta W, Mohd Yatim AN 2019 Characterization of Heat Waves: A Case Study for Peninsular Malaysia *Geogr Tech* **14** 1460–155
[7] Abdul Hamid Z 2018 Climate change and human health 2018
[8] Ghazali N H M and Awang N A 2018 Impact of Sea Level Rise and Tsunami on Coastal Areas of North - West Peninsular Malaysia *Irrig Drain*
[9] Belibassakis K A and Karathanasi F E 2017 Modelling nearshore hydrodynamics and circulation under the impact of high waves at the coast of Varkiza in Saronic-Athens Gulf *Oceanologia* **59** 350–64
[10] Mardani N *et al* 2020 Improving the accuracy of hydrodynamic model predictions using lagrangian calibration. *Water (Switzerland)* **12**
[11] Vieira B F V *et al* 2020 Hydrodynamics and morphodynamics performance assessment of three coastal protection structures *J Mar Sci Eng* **8**
[12] Samsuddin Sah S *et al* 2020 Impact of Saltwater Intrusion on Paddy Growth in Kuala Kedah, Malaysia. *J Sustain Sci Manag*
[13] Tran T *et al* 2016 Improvement of Salinity Stress Tolerance in Rice: Challenges and Opportunities *Agronomy* **6** 54
[14] Kotagiri D & Kolluru VC 2017 Effect of salinity stress on the morphology & physiology of five different coleus species *Biomed Pharmacol J* **10** 1639–49

[15] Reddy I N B L *et al* 2017 Salt Tolerance in Rice: Focus on Mechanisms and Approaches. *Rice Sci* **24** 123–44
[16] Jia P *et al* 2018. Simulation of the effect of an oil refining project on the water environment using the MIKE 21 model *Phys Chem Earth* **103** 91–100
[17] Jabatan Pengairan dan Saliran Malaysia. Guidelines for Preparation of Coastal Engineering Hydraulic Study and Impact Evaluation 2013
[18] Mohd F A *et al* 2013 Assessment of coastal inundation of low lying areas due to sea level rise Assessment of coastal inundation of low lying areas due to sea level rise *IOP Conf Ser Earth Environ Sci* **169**
[19] Faour *et al* 2013 GIS-Based Approach to the Assessment of Coastal Vulnerability to Sea Level Rise: Case Study on the Eastern Mediterranean *J Surv Mapp Eng* **1** 41–8
[20] Zufayri Zulfakar M S *et al* 2020 The effect of coastal protections on the shoreline evolution at Kuala Nerus, Terengganu (Malaysia). *J Sustain Sci Manag* **15** 71–85
[21] Grases A *et al* 2020 Coastal flooding and erosion under a changing climate: Implications at a low-lying coast (ebro delta) *Water (Switzerland)* 12
[22] Jahangirzadeh A *et al* 2012 Effects of construction of coastal structure on ecosystem *World Acad Sci Eng Technol* **65** 663–74

Acknowledgments

The authors would like to thank the National Hydraulic Research Institute of Malaysia (NAHRIM) for the relevant data and would like to thank UKM for funding this study through research grant DIP-2020-030.

The effects of climate change on flood hazards in Kelantan River Basin Malaysia

Tze Huey Tam[1], Muhammad Zulkarnain Abd Rahman[1,*], Sobri Harun[2], Sophal Try[3], Ismaila Usman Kaoje[1], Mohd Radhie Mohd Salleh[1], and Kimberly Vivian Abai[1]

[1]TropicalMap Research Group, Faculty of Built Environment and Surveying, Universiti Teknologi Malaysia, 81310 UTM, Johor Bahru, Johor, Malaysia
[2] Department of Hydraulic and Hydrology, Universiti Teknologi Malaysia, 81310 UTM, Johor Bahru, Johor, Malaysia
[3] Disaster Prevention Research Institute, Kyoto University, Japan.

*Corresponding E-mail: mdzulkarnain@utm.my

Abstract. Climate change has had a significant impact on the hydrological cycle, causing changes in precipitation patterns in both frequency and magnitude. The aim of this study is to assess the effect of climate change on flood hazards in Kelantan River Basin, Malaysia. A distributed hydrological model called Rainfall-Runoff-Inundation (RRI) simulates floods under current and future climate scenarios. The Climate Change Factor (CCF) is a tool for forecasting future climate scenarios. The storm used in this analysis had 50-year and 100-year recurrence intervals every 24 hours (ARI). The finding shows that the streamflow in Guillemard station will increase in the future for both the 50- and 100-year ARI. The streamflow increased to 10329 m^3/s from 8434.9 m^3/s in the current state and to 11220.2 m^3/s from 9157.4 m^3/s in the 50- and 100-year ARI, respectively. In both cases, the 100-year ARI flood magnitude is significantly less than the 50-year ARI flood extent (current and future). However, the flood depth in several towns located downstream of the Kelantan River Basin is more significant for the 100-year ARI than for the 50-year ARI for both cases. The study's findings would be helpful to relevant agencies and government departments understand the current and potential flood hazard situation in the study area and assist them in developing effective mitigation strategies for future flood hazards.

Keywords: Climate Change, Flood Hazard, RRI, Kelantan River Basin, Climate Change Factor

Track Name: Land, Water, Forests and Food Security

1. Introduction

Human activities have had a significant impact on the Earth's energy budget by releasing anthropogenic greenhouse gases such as carbon dioxide (CO_2), methane (CH_4), and nitrous oxide (N_2O) into the atmosphere over the last few centuries. As a consequence, the global mean surface air temperatures over land and oceans increased. From 1880 to 2012, the global averaged combined land and ocean surface temperature increased by 0.85 [0.65 to 1.06]° C [1]. The Intergovernmental Panel on Climate Change (IPCC) AR5 report, on the other hand, reported an increase in global average temperature from 1.0° C in the lowest emission scenario to 3.7° C in the highest emission scenario by 2100 [2]. As a result, changing precipitation or melting snow and ice are altering hydrological systems in many regions [1]. Dore [3] examined global precipitation patterns using observed data and concluded that climate change is causing increased variability in precipitation across the globe, with wet and dry areas becoming wetter and drier, respectively. His findings also revealed increased precipitation in high latitudes (Northern Hemisphere), decreased precipitation in China, Australia, a small Pacific island, and increased variability in equatorial regions. Also, there is likely increase in extreme precipitation under a future climate in some areas in Europe [4], in Southeast Asia [5, 6], and in central America [7]. Tabari [8] analysed the relationship of future changes in extreme precipitation and flood events with water availability. The results showed that an intensification of extreme precipitation and flood events over all climate regions which increases as water availability increase from wet to dry regions. Also, there is an increase in the intensification of extreme precipitation and flood with the seasonal cycle of water availability.

Several studies have found an increasing precipitation trend in Malaysia [9-12], which has increased the number of extreme events in 21st century, particularly floods [9]. In Malaysia, floods are generally classified into two types: flash flood and monsoon flood [13]. The latter occurrs almost every year and has a severe impact along east coast of Peninsular Malaysia (the affected states included Kelantan, Terengganu, Pahang). Furthermore, Tan *et al.* [11] concluded that the monthly precipitation in Kelantan River Basin increases during the wet season. As a result, it is possible to predict where more floods will occur in the future. To date, only a few studies on flood hazard in Malaysia under climate change condition have been published [14, 15].

General Circulation Models (GCMs) are used to project future climatic variables to predict the likelihood of increased flood risk from global warming [16]. Various studied used GCMs model have been conducted in Malaysia. For example, [17] and [18] identified the best GCM models to conduct climate change study in Malaysia. [11, 19, 20] assessed the effects of climate change on streamflow in different basins. [10] examined the future precipitation patterns under different climate scenarios. However, only limited studies used Climate Change Factor (CCF) for projecting flood hazard [15]. CCF is define as a ratio of the design rainfall for the future period to the control period [21].

The main objective of this study to assess the effects of climate change on flood hazard in Kelantan River Basin under 50-year and 100-year return period using CCF and the Rainfall-Runoff-Inundation (RRI) model.

2. Material and Methods

2.1. Study Area

In Malaysia, the Kelantan River Basin is the most vulnerable to flooding. In December 2014, a most disastrous flood occurred, resulting in an estimated loss of 2.8 billion ringgit (685 million USD) with 151,072 victims, and ten deaths [22]. [23] conducted a flood hazard assessment for this extreme event in the Kelantan River Basin downstream. According to the findings, Tanah Merah had the highest flood depth of 5.9 m among the affected towns, followed by Pasir Mas, which had 3 m depth.

Figure 1 shows the study area. The Kelantan River Basin is located between 101.4° E and 102.7° E, and 4.6° N and 6.0° N, with an elevation range of 2 to 2174 m. The catchment area of the basin is about 13031.6 km^2. During the Northeast and Southwest Monsoons, the basin receives about 1530 mm year^{-1} and 993 mm year^{-1}, respectively [24].

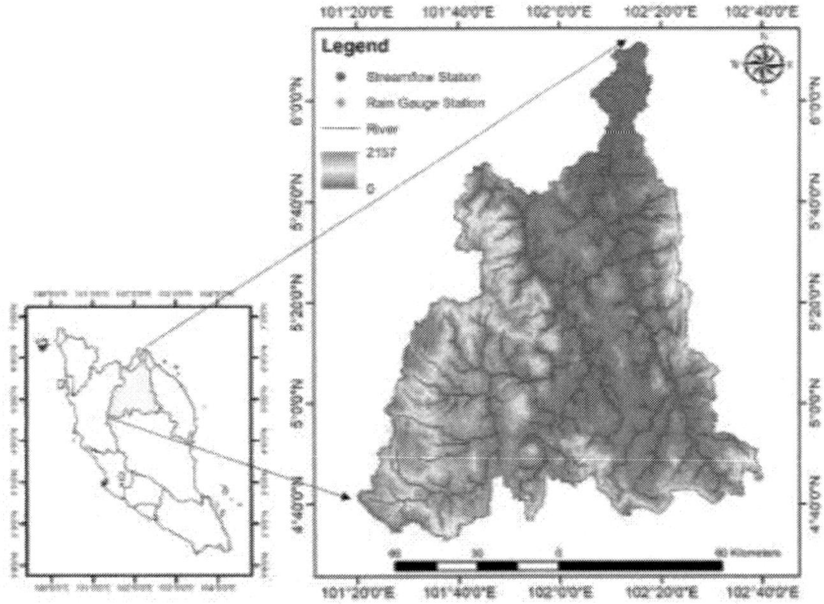

Figure 1. Kelantan River Basin.

2.2. *RRI model*

The RRI is a 2D fully distributed hydrological model that simulates rainfall-runoff and flood level simultaneously (Figure 2) [25]. At a pixel is river channel, this model presumes slope and river are at same pixel cell. 2D and 1D diffusive equations are used to calculate the flow on the slope pixel and the flow in a channel, respectively. This model simulates lateral subsurface flow, vertical infiltration flow and surface flow for describing the processes of rainfall-runoff-inundation. The lateral subsurface flow consists of saturated subsurface and surface flow. The Green-Ampt equation is responsible for vertical infiltration flow. The flow interaction of river channel and land is estimated based on different overflowing formula. The RRI model is a standalone product and its output, i.e., discharge, water level, and inundation can be displayed easily. In addition, the maximum flood inundation can be saved into ASCII format for further analysis, such as flood hazard mapping or identification of elements-at-risk, using with GIS platform additional data.

This study used the same model parameters as the previous paper [22] to simulate flood. They used various Near-Real-Time (NRT) Satellite Rainfall Products (SRPs) to simulate an extreme flood event in December 2014.

2.3. *Design Storm*

The Intensity Duration Frequency (IDF) of every rain gauge station were used to calculate the 24-hr design rainfall intensity. The empirical IDF equation is written in Equation 1.

$$i = \frac{\lambda T^k}{(d+\theta)^\eta} \quad (1)$$

where,
i = average rainfall intensity (mm/hr),
T = return period,
d = storm duration

The value of these parameters (i.e., λ, k, θ, and η) of every stations can refer to [21].

Once the design rainfall intensity is calculated, the future design rainfall intensity can be obtained by multiply with the coefficient of climate change factor (CCF). One can refer to [21] to gain more information about the CCF. Table 1 show the 50-year and 100-year rainfall depth of every station for current and future scenarios.

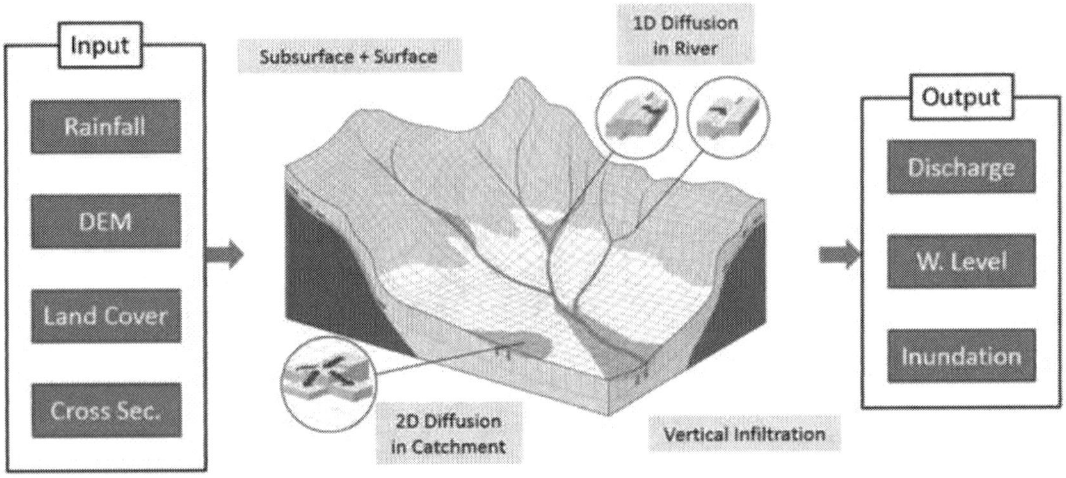

Figure 2. Schematic diagram of Rainfall-Runoff-Inundation (RRI) Model.

Table 1. Future and current 24-hour rainfall depth of 50-year and 100-year

	Current		Future		Increase	
	50 ARI	100 ARI	50 ARI	100 ARI	50 ARI	100 ARI
Balai Polis Bertam	214.4	240.9	293.7	337.2	37 %	40 %
Dabong	270.2	308.0	424.2	495.9	57 %	61 %
Brook	175.9	196.4	255.1	294.6	45 %	50 %
Gob	201.0	226.9	259.3	297.3	29 %	31 %
Gua Musang	200.2	222.9	238.2	269.7	19 %	21 %
Kuala Krai	392.7	460.9	518.4	617.6	32 %	34 %
Machang	405.4	471.9	559.5	665.4	38 %	41 %
Aring	313.5	356.9	417.0	485.4	33 %	36 %
Jeli	395.1	452.6	525.5	615.6	33 %	36 %
Lalok	295.7	338.9	375.5	437.2	27 %	29 %

3. Results and Analysis

3.1. Future rainfall depth changes

Table 1 depicts the 24-hour rainfall depth in current and future scenarios for 50-year and 100-year return periods. With a margin of at least 20 %, future rainfall depth exceeds current rainfall depth. Furthermore, for 50-year and 100-year return periods, the range increased from 19 % to 57 % and from 21 % to 61 %, respectively. Dabong and Gua Musang experienced the most remarkable and least significant percentage increases, respectively. Machang had the greatest rainfall depth of the stations, with 665.4 mm in the future and 471.9 mm in the current 100-year return period. On the other hand, Brook had the lowest rainfall depth in both the current and future scenarios, at 196.4 mm and 294.6 mm, respectively. The increased depth of rainfall implies that the future flood hazard in the Kelantan River Basin will worsen.

IOP Conf. Series: Earth and Environmental Science **880** (2021) 012016 doi:10.1088/1755-1315/880/1/012016

3.2. Impact of climate change on peak discharge

Figure 3 depicts the peak discharge for the 50-year and 100-year return periods in the current and future scenarios. The simulated streamflow collected based on Guillemard station. The 50-year peak discharge increased to 10,329 m^3 s^{-1} from 8,434.9 m^3 s^{-1}. At the same time, the 100-year peak discharge increased to 11,220.2 m^3 s^{-1} from 9,157.4 m^3 s^{-1}. The increase of peak discharge in the future is expected since the future rainfall depth increased too. These findings are consistent with the climate change impacts on water resources reported by [11].

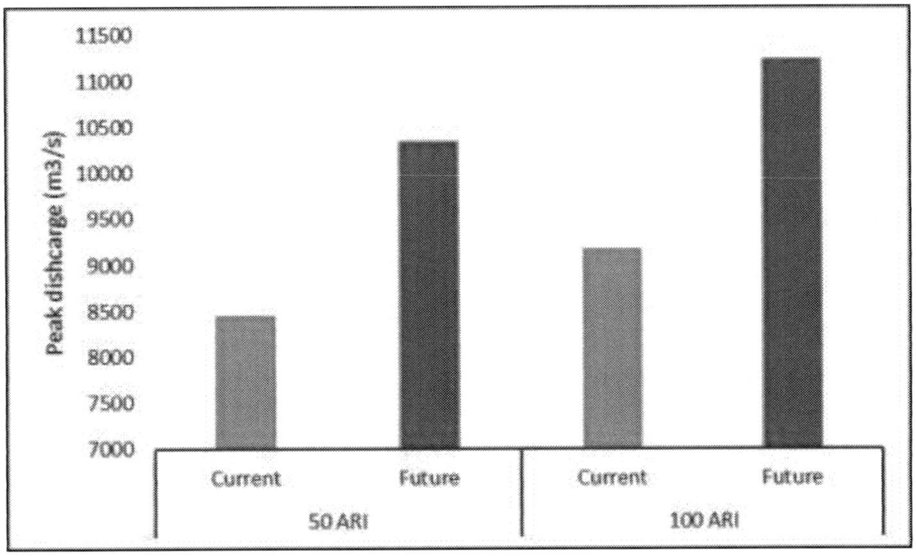

Figure 3. Comparison of current and future peak discharge for 50-year and 100-year.

3.3. Impact of climate change on flood depth

Figure 4 shows the current and projected flood depth in the affected towns downstream of the Kelantan River Basin. Overall, flood depths are expected to rise in the future for 50-year and 100-year. Tanah Merah had the significant flood depth among the towns, followed by Pasir Mas, Kadok, and Wakaf Baru. The rest of the towns, however, had similar flood depths. Tanah Merah had a projected flood depth of 3.85 m in a 100-years event, followed by Pasir Mas at 2.09 m, Kadok at 1.43 m and Wakaf Baru at 1.14 m. The remaining towns have a flood depth of less than 1 m. However, the highest 100-year flood depth in Tanah Merah in the future is less than the flood event in December 2014, as illustrated in Figure 4. The only explanation is that the digital elevation model (DEM) used in this study has a coarser spatial resolution of 0.9 km.

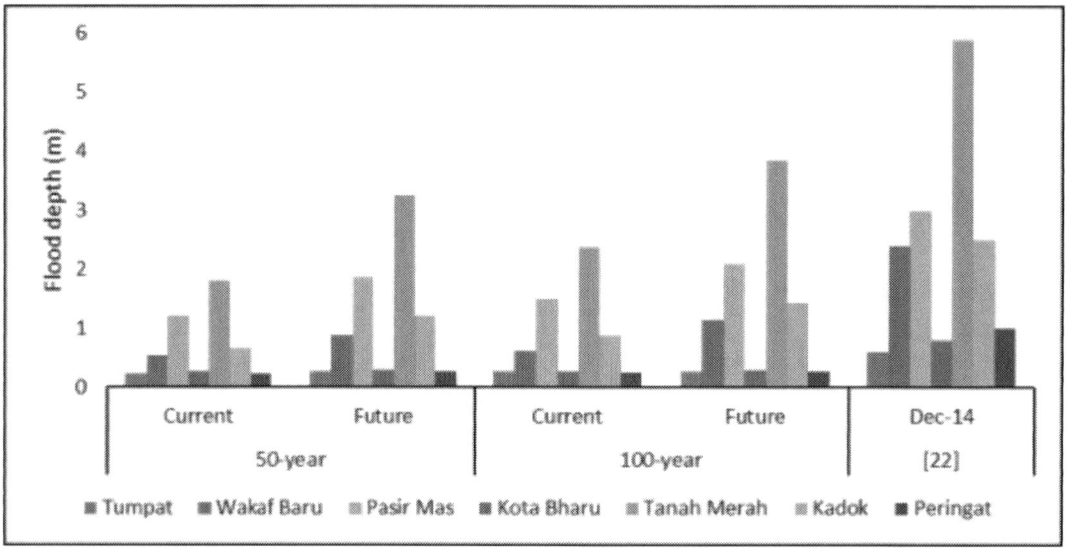

Figure 4. Current and future flood depth.

3.4. Impact of climate change on flood hazard

Figure 5 depicts the 50-year and 100-year flood hazard in the Kelantan River Basin downstream. The flood hazard classification is based on flood depth adopted from a previous study [26]. As shown in Figure 5, the flood hazard in the future appears to be increased compared to the current scenario, particularly in Tanah Merah. Furthermore, there are noticeable changes in the hazard level on the east and northwest sides of the basin. As a result, the total inundated area increased in the future. Table 2 displays detailed flood hazard information for each town. According to Table 2, only Tanah Merah will have a high level of flood hazard in the future, up from the current moderate level. In the current and future scenarios, the rest of the towns had levels that were moderate or lower than moderate.

IOP Conf. Series: Earth and Environmental Science **880** (2021) 012016 doi:10.1088/1755-1315/880/1/012016

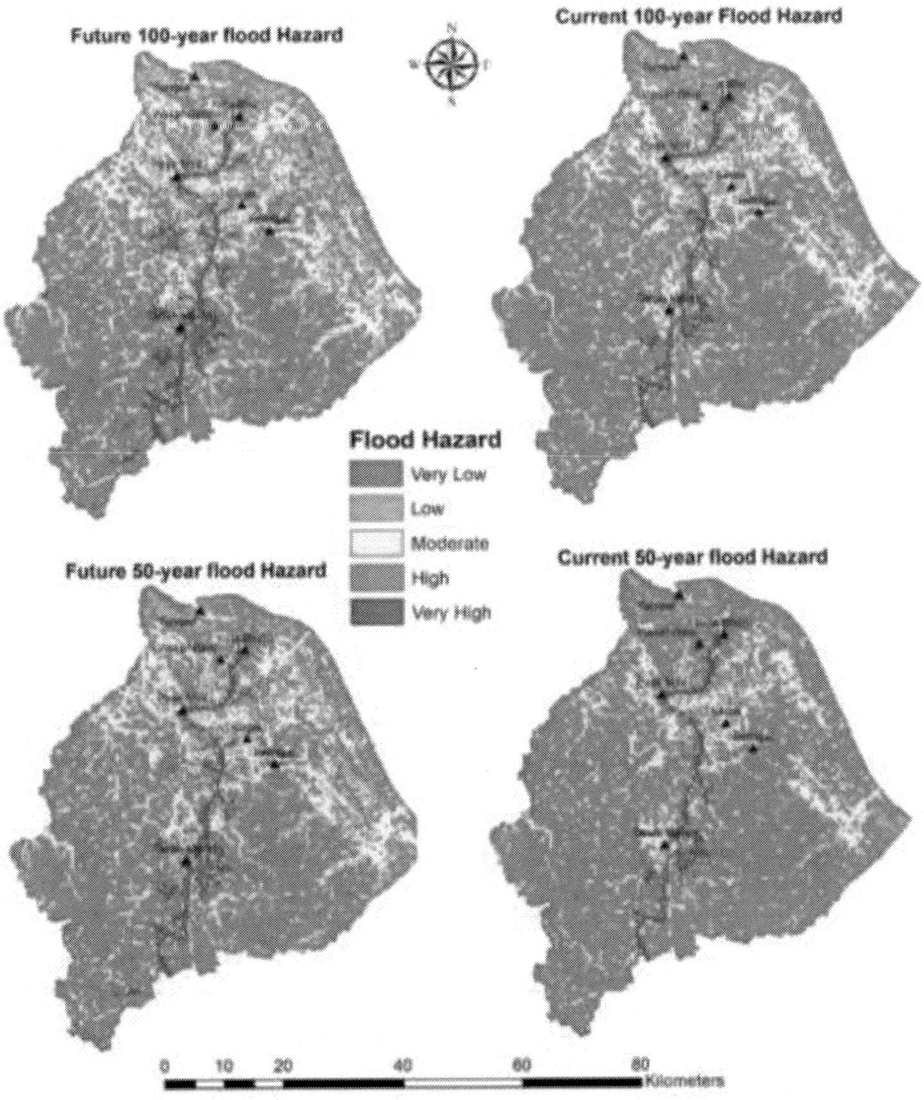

Figure 5. Flood Hazard in downstream of Kelantan River Basin.

Table 2. Degree of future and current flood hazard in the affected towns.

	50-year		100-year	
	Current	Future	Current	Future
Tumpat	Very Low	Very Low	Very Low	Very Low
Wakaf Baru	Low	Low	Low	Moderate
Kota Bharu	Very Low	Very Low	Very Low	Very Low
Pasir Mas	Moderate	Moderate	Moderate	Moderate
Kadok	Low	Moderate	Low	Moderate
Peringat	Very Low	Very Low	Very Low	Very Low
Tanah Merah	Moderate	High	Moderate	High

4. Conclusion

The effects of climate change on flood hazard in the Kelantan River Basin were discussed in this paper. The climate change factor (CCF) was used to forecast future rainfall depth over 50 and 100 years. The rainfall depth was used to drive the RRI model, which generated flood simulations and assessed flood hazards under various scenarios. According to the findings of this study, climate change poses a significant risk to the Kelantan River Basin. As a result, river basin management that seeks to reduce the risk of flooding must consider the impact of future climate change.

References

[1] IPCC 2014 Climate Change Synthesis Report. Contribution of Working Groups I, II and III to the Fifth Assessment Report of the Intergovernmental Panel on Climate Change. IPCC, Geneva, Switzerland
[2] IPCC 2018 Global Warming of 1.5°C.An IPCC Special Report on the impacts of global warming of 1.5°C above pre-industrial levels and related global greenhouse gas emission pathways, in the context of strengthening the global response to the threat of climate change, sustainable development, and efforts to eradicate poverty
[3] Dore M H I 2005 Climate change and changes in global precipitation patterns: What do we know? *Environment International* **31** (8) 1167-81.
[4] Madsen H *et al* 2014 Review of trend analysis and climate change projections of extreme precipitation and floods in Europe *Journal of Hydrology* **519** (PD) 3634-50
[5] Tangang F *et al* 2018 Future changes in annual precipitation extremes over Southeast Asia under global warming of 2 C *APN Science Bulletin* 8 (1)
[6] Raghavan S V *et al* 2018 Evaluations of NASA NEX-GDDP data over Southeast Asia: present and future climates *Climatic Change* **148** (4) 503-18.
[7] Imbach P *et al* 2018 Future climate change scenarios in Central America at high spatial resolution *PLOS ONE* **13** (4) e0193570.
[8] Tabari H 2020 Climate change impact on flood and extreme precipitation increases with water availability *Scientific Reports* **10** (1)
[9] Tang K H D 2019 Climate change in Malaysia: Trends, contributors, impacts, mitigation and adaptations *Science of The Total Environment* 650 1858-71
[10] Hussain M *et al* 2017 Projected changes in temperature and precipitation in Sarawak state of Malaysia for selected CMIP5 climate scenarios *International Journal of Sustainable Development and Planning* **12** (8) 1299-311
[11] Tan M L *et al* 2017 Climate change impacts under CMIP5 RCP scenarios on water resources of the Kelantan River Basin, Malaysia *Atmospheric Research* 189:1-10
[12] Amin I M Z b M *et al* 2019 Impacts of Climate Change on the Hydro-Climate of Peninsular Malaysia *Water* **11** (9) 1798
[13] DID 2009 Volume 1 – Flood Management.
[14] Aziz N A A *et al* 2016, editors. Effect of Climate Change to Flood Inundation Areas in Bertam Catchment, Pahang *Springer Singapore*
[15] Winarta B *et al* 2019 editors. Flood Inundation Assessment under Climate Change Scenarios in Kuantan River Basin, Malaysia *IOP Conference Series: Earth and Environmental Science*
[16] Mishra B K *et al* 2018 Assessment of future flood inundations under climate and land use change scenarios in the Ciliwung River Basin, Jakarta. *Journal of Flood Risk Management* 11 S1105-S15.
[17] Salman S A *et al* 2020 Selection of CMIP5 general circulation model outputs of precipitation for peninsular Malaysia *Hydrology Research* **51** (4) 781-98
[18] Noor M *et al* 2019 Selection of CMIP5 multi-model ensemble for the projection of spatial and temporal variability of rainfall in peninsular Malaysia *Theoretical and Applied Climatology* **138** (1) 999-1012.
[19] Ismail H *et al* 2020 Assessment of climate change impact on future streamflow at Bernam river basin Malaysia *IOP Conference Series: Earth and Environmental Science*
[20] Adib M N M *et al* 2020 Projected Streamflow in the Kurau River Basin of Western Malaysia under Future Climate Scenarios *Scientific Reports* **10** (1) 8336
[21] NAHRIM 2013 Technical Guide: Estimation of Future Design Rainstorm under the Climate Change Scenario in Peninsular Malaysia
[22] Yahaya N S *et al* 2015 The December 2014 flood in Kelantan: A post-event perspective
[23] Tam T H *et al* 2019 Application of Satellite Rainfall Products for Flood Inundation Modelling in Kelantan River Basin, Malaysia *Hydrology* **6** (4) 95.
[24] Wong C L *et al* 2016 Rainfall characteristics and regionalization in Peninsular Malaysia based on a high resolution gridded data set *Water* **8** (11) 500.
[25] Sayama T *et al* 2012 Rainfall–runoff–inundation analysis of the 2010 Pakistan flood in the Kabul River basin *Hydrological Sciences Journal* **57** (2) 298-312
[26] Usman Kaoje I et al 2020 Physical flood vulnerability assessment of buildings in Kota Bharu, Malaysia: an indicator based approach *International Journal of Disaster Resilience in the Built Environment*

Acknowledgments
The authors would like to express their gratitude to the Institut Penyelidikan Air Kebangsaan Malaysia (NAHRIM) for providing the hydrological data used in this study. Takahiro Sayama deserves special recognition for developing the RRI model as a public domain model. Finally, the authors would like to express their gratitude to Universiti Teknologi Malaysia (UTM) for financial support under grant GUP Q.J130000.3852.19J74.

IOP Conf. Series: Earth and Environmental Science **880** (2021) 012017 doi:10.1088/1755-1315/880/1/012017

Water treatment process using conventional and advanced methods: A comparative study of Malaysia and selected countries

N H Pakharuddin[1*], M N Fazly[1], S H Ahmad Sukari[1], K Tho[1] and W F H Zamri[1,2,*]

[1] Department of Mechanical and Manufacturing Engineering, Faculty of Engineering and Built Environment, Universiti Kebangsaan Malaysia, 43600 UKM Bangi, Selangor, Malaysia
[2] Institute of IR 4.0 (IIR4.0), Universiti Kebangsaan Malaysia, 43600 UKM Bangi, Selangor, Malaysia

*Corresponding E-mail: a162904@siswa.ukm.edu.my; wfathul.hakim@ukm.edu.my

Abstract. Water treatment is the process of removing all those substances, whether biological, chemical, or physical, that are potentially harmful to the water supply for human and domestic use. This treatment helps to produce water that is safe, palatable, clear, colorless, and odorless. The basic steps of water treatment include coagulation, precipitation, filtration, and disinfection. Water treatment before supplying water to consumers is essential to improve water quality to create a sustainable life. Water treatment can eliminate potential or certain harmful substances in the water to prevent the consumption of contaminated water sources that can cause potential health problems. Therefore, it is important to establish a water treatment facility with sufficient capacity to remove pollutants according to standards before being supplied to consumers. In this study, the focus of the discussion is on the use of river water as a source of water for consumers in Japan, Australia, Canada, and Malaysia after a water treatment process. This paper reviews the recent progresses of water treatment process using both conventional and advanced methods. A brief discussion on the water quality index of each country's rivers is presented. Several potential applications of Industrial Revolution 4.0 technology in the water treatment process are discussed. Adoption of the industrial revolution of technology in water treatment may provide many benefits to this field and excavate more potential improvement. This paper will deliver a scientific and technical overview and useful information to scientists, engineers, and stakeholders who work in this field.

Keywords: Water Treatment; River; Water Quality; Conventional Treatment; IR4.0 Technology

Track Name: Land, Water, Forests and Food Security

1. Introduction

Water scarcity and demand in clean water sources globally has influenced all parts of human existence which creates the biggest impacts on least developed nations and rural communities. The United

Nations' Sustainable Development Goal 6 (SDG 6) aims to ensure the accessibility and management of water and sanitation for all, including an end to open defecation, by 2030 [1]. It is known that due to rapid advancement of industrialization will results in destruction of ecosystem and biodiversity loss. As mentioned by World Health Organization (WHO, 2017), it is expected that undersupply of water will displace 700 million people by 2030, while desertification will put the livelihood of one billion people living in 100 countries across the world at risk by 2050 [1]. To fulfil the objectives of the SDG 6, the High-Level Panel on Water has developed innovative approaches to solve global water scarcity since recent years traditional financing solutions and technologies have proven to be insufficient in addressing these challenges. Concerns in the water industry were raised by several important issues that made a setback for modern technology of water treatment process to be applied in the world. These issues are also connected to the gap research of this study for instance, climate change, capacity of building infrastructure, lack of funding, improper and insufficient training.

In water treatment process, composition of precipitation in the rivers played a major role to define whether a treatment process is efficient in producing a good quality of water [2]. Due to changing of rising sea levels, unpredictable rainfall, saltwater intrusion would likely affect the change of precipitation patterns [2]. Not to mention, natural disasters such as earthquakes, storms happening in Japan, forest deforestation in Australia, floods in Malaysia can affect infrastructure of Water Treatment Plants (WTPs) in large-scale system and small-scale system [3]. Next, lack of funding for advanced water treatment process became a major barrier among countries. This is because, interests in technological innovations in water and disinfection remain insufficient [4]. The water innovations area is yet in its beginning phase of understanding the potential outcomes that 4IR can bring for real-life applications to help in the creation of smart and genuine organizations, networks, urban areas, and countries across the globe. Therefore, technologies should be combined and lead by development in organizations, plans of actions, and proper financing system for this to be a fruitful undertaking.

As mentioned before, lack of expertise between these countries could lead to a stagnant progress for advancement of water treatment process. In Malaysia, WTPs and wastewater business are in desperate need of a set up institutional program to train a huge number of staffs [5]. It was understood that the water industry (especially WTPs) is denied of formal conventional knowledge-based training for two decades since the privatization of Institut Kerja Raya Malaysia (IKRAM) in 1996 which paused many training programs in the water sector at central level [5]. Without formal centralized training institutions like IKRAM, the staffs will be dependent on in-house training which can limit newer expertise and adaptions towards technology.

Further into this paper, Water Quality Index (WQIs) of every country will be discussed as the quality of water, is an important aspect of water management system way before the age of 4IR [6]. Contaminations in the water such as pathogens, harmful chemicals can affect the overall properties of water system [7]. Every country has its own WQIs to ensure clean water consumption to avoid any acute and chronic health effects. However, sampling and testing of water source used to be done manually which is time consuming and absence of staffs attending can lead to incorrect data of WQIs. Hence, modification technological advances will be identified under the aspects of the Fourth Industrial Revolution (4IR) between Malaysia, Canada, Japan, and Australia.

In this fourth technological wave in industry, cyber physical systems can interact with one another using artificial intelligence (AI), Big Data, Drones, and the Internet of Things (IoT), etc. The 4IR mentioned, provide a progressive method of organization, production and distribution based on digital transformation and authorization that can erase limits between physical object [8]. In hindsight, the implementation of 4IR in water treatment process can contribute a lot to economic growth, employment, and sustainable development in governance of a nation. In this paper, the scope chosen for the source of water in every country is from river. Since 98 % of water came from river in Malaysia whilst it is 92 % for Canada [9]. In terms of common industry with Malaysia such as transportation, travel, and fisheries 78 % summed up the water source from river in Australia [10]. On the other hand, 11 major rivers in Japan have become the nation source of water (88 %) [25].

2. Water Treatment Process: Conventional vs Non-Conventional

There are two types of water treatment, namely conventional water treatment and unconventional water treatment. Conventional water treatment uses a combination of physical, chemical, and biological processes and operations. Preliminary, main, secondary, and tertiary and/or advanced water treatment are all words that are used to describe various levels of treatment in order of increasing treatment level. Non-conventional method water treatment is simpler than the conventional method of water treatment [11-12]. Non-conventional technologies have lower environmental impacts and reduce contaminant loads at lower costs than conventional treatments. Compared to the conventional method, the non-conventional method uses more advance equipment and technology [12-13]. The use of technology depends on the quality of the water source. Non-conventional will be used if and only the conventional water treatment is no longer feasible due to factors such as extreme water contamination. Figure 1 below shows the flow of processes that involve in typical conventional water treatment.

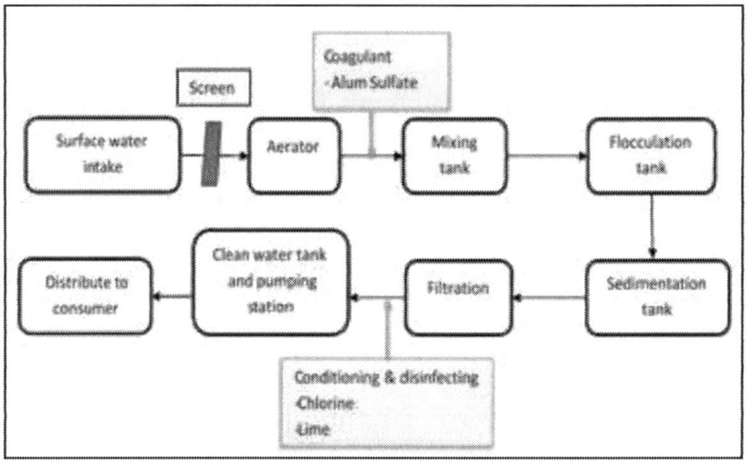

Figure 1. The typical conventional water treatment [11-13]

Before water is distributed to an urban community, it is treated to ensure safety and pleasant to drink. Of the great variety of water treatment processes, those mentioned later are by far the most applied [11-14]. Steps of water treatment;

a. Coagulation, Flocculation and Sedimentation
While many particles will gradually settle out from water over time, a process called sedimentation, some will not. To cause slow or non-settling particles to settle out more readily, a soluble chemical or mixture of chemicals is added to the water. Such chemicals are called coagulants and the process is called coagulation [11]. Coagulants react with the particles in the water, forming larger particles called flocs, which settle rapidly and can be removed as sludge [11-12]. Flocs can also be effectively removed by passing the water through a filter, either directly or after sedimentation. The process is controlled so that the coagulant chemicals are removed along with the contaminants [11].

b. Filtration
One of the oldest and simplest processes used to treat water is to pass it through a bed of fine particles, typically sand [12 -13]. Sand filtration will usually remove fine suspended solids and some other particles such as larger microorganisms. Sand filtration is even more efficient when the water being treated passes through the sand filter very slowly, although this requires large areas of land, not normally available in cities [12]. Filtration techniques have changed with the development of modern

plastics. The result is a new range of filter materials and methods to treat water for urban and industrial purposes [12].

c. Disinfection
While coagulation, flocculation on and filtration can remove quite a large amount of organic material and larger microorganisms from raw water, there are some important pathogens that are not eliminated during these treatments [13]. Disinfection kills harmful microorganisms that may be present in the water supply and prevents them from regrowing in the distribution systems. Without disinfection, waterborne diseases become an increased threat. Chlorine is the most widely used disinfectant for drinking water [12-13]. It is cheap, easy to use, effective at low dose levels against a wide range of infectious microorganisms and has a long history of safe use around the world [13].

2.1 Water Treatment Process in Malaysia
As previously mentioned, there are two types of WTP. Malaysia used both water treatment methods although the majority of water treatment plant still used conventional water treatment process for surface water [14-15]. For instance, water treatment process flow at Sungai Dua Water Treatment Plant (WTP) in which includes three types of water treatment technologies, mainly Sedimentation, Dissolved Air Floatation (DAF) and Lamella Clarifier [14]. There is a total of 5 treatment plants at Sungai Dua WTP. The plants consist of 2 sedimentation plants, 1 DAF Plant and 2 Lamella Clarifier Plants. The processes of water treatment are as Figure 2. In Malaysia, the only non-conventional WTP used is membrane technology. It is said there are only 3 WTP that use this technology in Malaysia [15-17]. The first plant with membrane technology in Malaysia using ultra-filtration process started operation in 2006 in Bukit Panchor, Pulau Pinang and then 2 other plants were built and started operation early 2008 in Selangor [15].

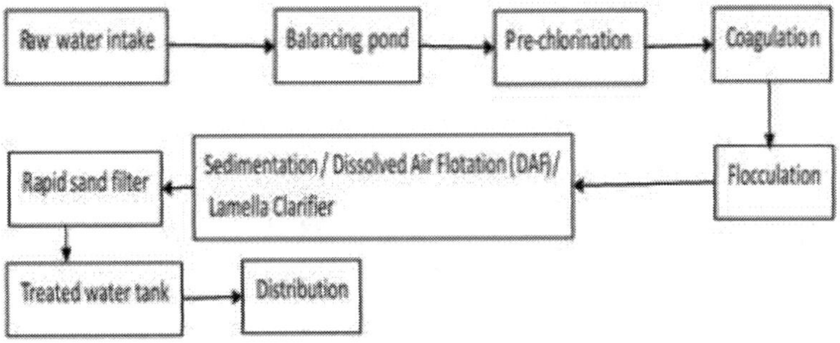

Figure 2. Water treatment processes of Sungai Dua WTP. [14-15]

In February 2013, the industrial-scale UF membrane drinking water treatment plant that located at Wakaf Bunut WTP, Kelantan was commissioned [16]. With a typical pore size of 0.02 m, ultrafiltration (UF) is a good barrier for bacteria and viruses and this method can be applied to the production of potable water [17]. Figure 3 shows the ultrafiltration method (UF) uses membrane modules as filtration media relative to conventional sand-using plant. A water treatment that uses ultrafiltration technology usually goes through simpler phases than conventional ones. Among the benefits of using UF technology is that it can generate high-quality clean water, lower operating costs, an easily upgradable system, and a compact system that reduces space.

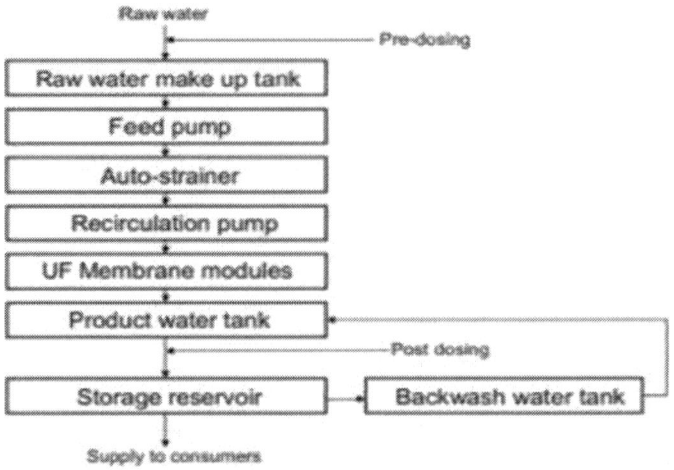

Figure 3. Water treatment process using UF Technology. Source: Bakhtiari et al. (2014) [15]

2.2 Water Treatment Process in Canada
In 2016, regional and municipal governments in Canada had owned over 1,200 Water Treatment Plants (WTP) [18]. With recent years, the number has increased to 4 % as industrial and water treatment construction projects have been active in Canada to fulfil the United Nations' Sustainable Development Goal 6 (SDG 6) [18]. For instance, the North End Sewage Treatment Plant (NEWPCC) Upgrade Project that is still in the schematic design phase and estimated to be completed in 2028 with $1.8 billion in total costs [18].

The raw water will flow from rivers in Canada to the Water Treatment Plant (WTP) across the country. Then, it is pumped to the WTP through the reservoirs and undergoes several steps of treatment to produce high-quality water. WTPs in Canada has added Ozonation and Ultraviolet (UV) Light in their water treatment process [18]. Liquid ozone is added to kill bacteria, and sodium bisulphite is added to remove excess ozone. On the other hand, disinfection of water will be done by exposing water to the UV light [18].

2.3 Water Treatment Process in Japan
In Japan, advanced water treatment only focused on ozone treatment and biological activated carbon treatment. For the ozone treatment process, minute bubbles are diffused slowly through the jade-tinted water by the tube at the bottom of the tank [19]. Ozone is a strong oxidizing agent that has the ability of disinfection of harmful microorganism. The bubbles will attach at the surface of microorganism and react with it. Thus, it can eliminate potential health risk without chemical means, so the water does not have unpleasant odors, nor does it taste of bold or disinfectant. Ozone purification process lasts 20 minutes and fine bubbles increase the efficiency to ensure more ozone dissolve in the water [19]. The process than entering the second stage of advanced water treatment by flowing the water into a filtration pond containing different grades of biological activated carbon. Organic matter and decomposition of microorganism propagated in activated carbon. The microporous carbon supports microorganisms that break down impurities and other by-products of ozone treatment. Active carbon is very effective at purifying water but need to replenish once every four years due to the reason bed gradually loses the ability to absorb organic material [19].

2.4 Water Treatment Process in Australia
In many pieces of Australia, surface water put away in repositories is the fundamental hotspot for civil water supply, making water supply helpless against dry spells; just a lot more modest offer comes from groundwater [20]. Non-traditional water sources, for example, seawater desalination, assume an

expanding job in Australia's water supply, with one desalination plant appointed to supply Perth and others being implicit Sydney, the Gold Coast, Melbourne, Adelaide, and another is intended to be worked at Port Augusta [20]. The utilization of recycled water- the non-consumable reuse of treated wastewater for water system of green spaces, fairways, farming yields or modern uses — is normal and expanding in Australia [20]. Among the 20 biggest water utilities in Australia, the biggest volume of reused water provided was by SA Water in Adelaide (25,047 ML or 29.6 % of sewage gathered), while the most minimal volume of reused water was by ACTEW in Canberra (2,104 ML or 7.4 % of sewage collected) [20].

WTPs in Australia has adopt Biological Activated Carbon Treatment (BAC) that is the biologically enhanced active carbon process which many water treatment industries considering using it for their utilities [21]. GAC (granular activated carbon) has long been used to remove dissolved organics from drinking water [21].

3. Comparison of Water Treatment Process

Generally, there are two different types of Water Treatment Process (WTP) that are conventional WTP and non-conventional WTP. The Table 1. will show the comparison of the conventional and non-conventional WTP between the countries. Coagulation, flocculation, sedimentation, and filtration are the most basic and commonly used conventional method in WTP. Hence, all the countries in the Table 1 using these methods in WTP. In Malaysia WTP use pulsatube clarifier and Actiflo' technology. Pulsatube clarifier technology is the technology that combining the process of flocculation and clarification in one area whereas Acttifflo' technology is the compact water clarification process in which raw water is flocculated with microsand and polymer in a draft tube reactor [22]. All countries except Japan use chlorination method to kill microorganisms in the water. Only Canada uses biological activated carbon as a medium for filtration while Malaysia, Japan and Australia used sand for filtration in conventional WTP. In Japan, membrane filtration technology is being used in the sedimentation method [19].

The only non-conventional WTP in Malaysia is membrane technology in the ultrafiltration (UF) system. Canada adopts ozonation technology and Ultraviolet (UV) light for their non- conventional WTP [18]. Ozonation is being used to kill bacteria by liquid ozone and Ultraviolet (UV) light are used during disinfection of water. Japan and Australia use biological activated carbon treatment to clarify water. To disinfect water, Japan uses ozone treatment process. The minute bubbles slowly disperse through the jade-tinted water through the tube at the bottom of the tank. This method makes the water does not have any unpleasant odor of disinfection. The advancement of technology might bring more aid, especially for Malaysia to counter any water disruption because the majority of water disruption Malaysia are caused by heavily polluted surface water supply.

Table 1. WTPs in Malaysia, Canada, Japan, and Australia

Country	Conventional WTP	Non- Conventional WTP
Malaysia	Coagulation; Flocculation; Sedimentation; Filtration; Chlorination; Dissolved AirFloatation (DAF); Pulsatube Clarifier Actiflo' Technology	Membrane Technology in Ultrafiltration (UF) system
Canada	Coagulation and Flocculation; DAF; Filtration; Chlorination; Fluorination	Ozonation Ultraviolet (UV) Light
Japan	Sedimentation (Membrane filtration technology); Sand Filtration	Ozone treatment; Biological activated carbon treatment
Australia	Coagulation; Flocculation; Sedimentation; Filtration Disinfection	Biological activated carbon treatment

3.1 Concept of Water Quality Index (WQI): WQI in Malaysia
In 1985, a study entitled "Development of Water Quality Criteria and Standards for Malaysia" was dispensed by the Malaysian government which aimed to develop standards for monitoring river water quality for domestic water use, fisheries and aquatic breeding, livestock drinking, recreation, and agricultural use [23]. The National Water Quality Standards (NWQS) are defined by six classes for river water classification supported the descending order of water quality. Table 2 below shows range of values of various water quality parameters for various classes of river waters [23].

Table 2. The classification and uses of river water quality in Malaysia. Source : Jabatan Alam Sekitar (2019) [23].

Class	Uses
Class I	Conservation of natural environment Water Supply I - Practically notreatment necessary Fishery I - Very sensitive aquaticspecies
Class IIA **Class IIB**	Water Supply II - Conventionaltreatment Fishery II - Sensitive aquatic speciesRecreational use body contact
Class III	Water Supply - III Extensive treatment required Fishery III - Common of economic value and tolerant species; livestockdrinking
Class IV	Irrigation
Class V	None of the above

The Water Quality Index (WQI) has been practiced for about 30 years. The WQI formula uses six parameters to see river water quality that is biochemical oxygen demand (BOD), chemical oxygen demand (COD), suspended solids (SS), ammonia-nitrogen (AN), and pH [23]. The WQI formula developed by Department of Environment Malaysia (DOE) is the idea for water quality assessment regarding pollution load and river water classification under the National Water Quality Standards (NWQS) [23].

In 2019, river water quality was assessed based on 8118 samples taken from 1353 manual monitoring stations covering 672 rivers in Malaysia [23]. Out of 672 rivers monitored, 408 (61 %) showed clean water quality, 205 (30 %) were slightly polluted and 59 (9 %) were polluted. Tabulated data in Jabatan Alam Sekitar report shows the average water quality that were monitored at Sungai Selangor in 11 stations including its parameters [23].

Based on the report From Jabatan Alam Sekitar Malaysia, 4 rivers are sampled for WQIs (Sungai Klang, Sungai Selangor, Sungai Pahang, Sungai Melaka), covering 64 stations. For Sungai Klang (13 stations) and Sungai Selangor (10 stations) are classified as Class III and Class II whereas Sungai Pahang (27 stations) and Sungai Melaka (14 stations) are Class II and Class III, respectively [23].

3.2 WQI in Canada
The Canadian Council of Ministers of the Environment (CCME) Water Quality Index (WQI) has been established since 2001, and itself act as a valuable tool for water resources management [24]. The CCME WQI model consists of three measures of variances from selected water quality objectives (Scope; Frequency; Amplitude). These three objectives will be combined to produce a value between 0 to 100 that represents the overall water quality. The water quality is categorized in five grades (excellent, good, fair, marginal, poor) [24]. With the scores from WQI, these rivers are classified into classes that are: Excellent as A; Good as B; Fair as C; Marginal as D; Poor as E. The water quality data are taken from 30 river sites across the province of Ontario, Canada for 5 years (2002-2008) [24]. Within that period, CCME WQI uses 7 or 10 parameters (ammonia, chloride, nitrate, phosphate, chromium, nickel, zinc, harness, pH, temperature), to be taken for sampling at these 30 river sites. It is

apparent from the data tabulated, in terms of station years; almost 65% (93v out of 144) fall in the same category that ranges from Class A and B [24].

3.3 WQI in Japan
Japan's rivers and lakes have moderately low levels of organic compounds and agricultural chemicals [25]. This phenomenon occurs only around rivers or streams, not in rural areas. Parameters for Japan WQI are pH, COD, NO_2, NO_3, Cl, E. coli, APE, and BPA [25]. In 2008, water samples from Hokkaido (15 Points), shown pH values ranged from 6.45 to 8.27, indicating alkaline properties due to the Hokkaido area's position as a reliable supplier of grain and livestock in Japan [25]. As a result, the rural water quality index is better than the urban water quality index. Furthermore, it was discovered that certain endocrine disruptors, such as APE and BPA, were present in water samples (Tokyo and Hokkaido) [25]. Effects of endocrine disruptors on the development and differentiation of humans and wildlife, especially aquatic wildlife. This suggests that environmental toxins, such as endocrine disrupting chemicals, should be re-examined as disrupting factors in differentiation and growth [25].

3.4 WQI in Australia
Water Quality Index (WQI) communicates the general water nature of a specific source at a specific time utilizing a 'solitary worth' in view of chosen water quality factors [26]. The WQI consolidates nine boundaries including temperature, disintegrated oxygen, pH, phosphate, nitrate, all out broke down solids, natural oxygen interest (Body) and fecal coliform [26]. The list is determined from Q worth and weight factor W, where Q demonstrates the degree of water quality comparative with any single boundary and the weight factor speaks to the general significance of the single boundary to the general water quality [26]. The generally water quality positioning measures falls under five classifications which are extremely terrible when WQI is < 25, awful when WQI is 26-50, moderate when WQI is 51-70, acceptable when WQI is 71-90 and excellent when in the scope of 91-100 [26]. The WQI for the port water quality during both high and low tides were reported as good except for Port Botany and Eden as stated as following [26]. The average value of WQI for Port Jackson (79.07), Port Botany (72.79), Port Kembia (76.78), Port Newcast (75.15), Port Yamba (77.67), and Port Eden (67.42) accordingly in the tabulated data [26].

4. Fourth Industrial Revolution (4IR) in Water Treatment

4.1 Malaysia: Iot (Internet of Things) and Drones
Air Selangor applied a smart river monitoring system (SRMS) that uses unmanned aerial vehicles (UAVs) or drones and low power wide area (LPWA) communication technology [27]. Thus, it also has an Intelligent Command Centre that provides real-time views of the water supply network using supervisory control and data acquisition (Scada) systems and remote sensor [27]. According to Mazlan Abbas, co-founder and CEO of IoT service provider Favoriot Sdn Bhd, installation of IoT sensors along the river and at intersection of major drains aid the law enforcement agencies to take prompt action once the sensors discover irregularities [27]. The sensor is connected to a central system that reflected real-time data in visual form. This sensor can inspect the river water quality and odor. Sensor may install at a specific are such as middle of riverbanks to monitor the condition of rover water before entering reservoirs. Such sensors have been installed at Sungai Klang [27]. According to The Star (2020), the total of RM2 million has been allocated by Selangor state government for 4 units military-grade high technology (DJI Matrice 300) drones to observe rivers in Selangor and guard against pollution of water resources in November 2020 [28]. The drones used for monitoring Klang River Basin, Selangor River and Sungai Langat [28]. Drones can catch image of the water samples and identify area of pollution including in remote area. Additionally, drones may provide supervision for illegal suspicious activities at night to prevent individuals pollute the water [28].

4.2. Canada: Iot (Internet of Things); Blockchains; Drones and remote sensing

A company in Canada called Aridea Solutions is performing a pilot project to help predict harmful water levels in water treatment facilities [29]. They invent the system called Plug and Sense Platform and collaborate with telecommunication firm. This system consists of sensor to measure water depth and chemistry composition elements [29]. The data collected is sent by 4G internet to communicate directly to Aridea's Terralytix Portal cloud platform. The project showed how the IoT can work efficiently and benefit water management while assuring cost reduction and removal of human error in the workflow when using automation [29].

Lake Winnipeg Foundation (LWF) through an RBC foundation grant launched Lake Winnipeg DataStream to collect data of water from lake Winnipeg [30]. All the sharing of data is through blockchains technology [30]. DataStream which leveraging the blockchains technology provides a higher level of security and transparency for managing data water in Canada. Everyone can check and know the latest authentic datasets over time. Blockchains technology have make authentication for LWF who constantly observe and protect the water of lake Winnipeg [30].

Toronto water is using drones to help conduct inspections in areas with difficult accessibility [31]. The fleet of aerial autonomous craft used is more advanced, with several drone units on hand named DJI Phantom 4, the Inspire 1 Pro, and the Mavic Air [31]. These drones will fly over major Toronto's waterway, such as Don River to inspect if there are any pipe blockages. With drones, the city water division has been able to discover decayed infrastructure [31]. Moreover, submarine drone is also used in the fleet (DeepTrekker DTG2) to be piloted into Toronto's larger pipes. This way, the need to hire a scuba diver can be avoided as inspection of pipe integrity can be made virtually instead [31].

4.3 Japan: Iot (Internet of Things) and Big data

The entire water distribution system in Japan, is control and monitored by online water network analysis to analyze its distributing pressure [32]. This would provide a real-time monitoring for the pump discharged pressure, valves openings based on historical setting. The system can prevent any losses in distributing the water by keeping up to date to piping network if there are any changes occur [32].

Besides IoT, Japan also is using Big Data for low emission sewage treatment control system [32]. This control system can reduce the release amount of any greenhouse gases by performing online calculation of a linear model of the water treatment process. Soft-sensor technology is the main components in this control system [32]. It has two function: to remove chemical oxygen demand and calculate volume of nitrous oxide gas. Monitoring and operating terminal would receive any measurement from soft sensor and giving a command based on operational setting through model prediction control module. The operational setting serves as an output to the distribution system to modify any changes [32].

4.4 Australia: Iot (Internet of Things) and Big data

Australia uses internet of things artificial intelligence to cut down the electricity usage and its accompanied cost in the water treatment plants [33]. A company, Melbourne water is trying to co-ordinates pump movement depending on the amount of treated water required on a given day in water treatment plant on a custom-developed platform [33]. Python, the platform determines optimal pump calibrations and sends them directly to the pump system without human intervention during its operation [33]. This program able to utilize historical data to determine the most energy efficient combinations of pumps and the associated speeds to run them at, to achieve the necessary flow rate and applies them in real time [33].

Australia using technology big data to collected and analyze discrete data on water resources in its country [34]. Australia uses big data to access water river quality by identified nutrient exceedances in accessed basin area. They used technology to identify and solve the problems of salinity and turbidity which was a major water quality issue for assessed basins in Australia [34]. Australia able to differentiate the significant gaps in water quality monitoring preclude a comprehensive assessment of all of Australia's river systems. Through big data technology, Australia able to collate and interpreted data for about 14,000 reaches across the more intensively used catchments. They examine the aquatic biota index using macro-invertebrates and a river environment index to acquire a balanced river ecosystem and usage of clean river water to other sectors [34].

5. Malaysia: Towards 4.0 IR Technology in Water Treatment

The summary of Fourth Industrial Revolution (4 IR) in Water Treatment can be shown in Table 3. As mentioned earlier that in Malaysia, water technology is still in its early stages and support. Aid from the government and NGOs would really start all possibilities towards technological innovations. Firstly, by providing courses related to Smart Water Technology in tertiary education to increase labor skills at the plants. Next, business related to water technology need to practice in creating strong ecosystem. This is because lack of supportive working environment can discourage adoption of new technology. In addition, key to the success of a project is getting a strong leadership from the beginning. Transformation leaders who involve primarily in every strategic planning and progress can ensure continuous supports from managements and staffs.

Table 3. Fourth Industrial Revolution (4IR) in Water Treatment.

Country	**4.0 IR Technology**				
	Block chain	IoT	Drones	VR	Big Data
Malaysia		✔	✔		
Canada	✔	✔	✔		
Japan		✔			✔
Australia		✔			✔

6. Conclusion

Every country has its own water method for advanced treatment but with same conventional treatment. Adoption of industrial revolution of technology in water treatment may provide many benefits to this field and excavate more potential improvement using cutting edge technology. Malaysia though has poorer water quality index due to limited on budget to kick start the major improvement projects compared to other three countries; which had successful using and showed improvement on water quality delivered. Malaysia has big potential to successfully implemented the technology in water treatment plant because they had set about to research and development in improving water treatment method by not limited to government's effort also by private company. In the nutshell, alternative ways that may bring Malaysia to the path taken by Canada, Japan and Australia are providing a strong ecosystem to support technology, promoting Smart Water Technology courses in tertiary education, and implementing strong leadership in management and administration.

References

[1] Division, S. 2020. Use of 4IR Technologies in Water and Sanitation in Latin America and the Caribbean Use of 4IR Technologies in Water and Sanitation in Latin America and the Caribbean (April).
[2] Akademi Sains Malaysia. 2013. Study on the Current Issues and Needs for Water Supply and Wastewater Management in Malaysia. *Journal of Chemical Information and Modeling, hlm.* **53**.
[3] Treacy J 2019. Drinking Water Treatment and Challenges in Developing Countries. The Relevance of Hygiene to Health in Developing Countries
[4] Results for Development Institute 2017 Challenge Funds and Innovation in the Water Sector: A Report to The High-Level Panel on Water, at https://sustainabledevelopment.un.org/content/documents/153732._HLPW_Final_Report_pdf_3.pdf 10
[5] Gofar N and Kassim A 2008. Mechanics of Rainfall Infiltration through Soil Slope.
[6] Kamaludin K H and Ismail W 2017 Water quality monitoring with internet of things (IoT) *IEEE Conference on Systems, Process and Control (ICSPC)*.

[7] Treacy J 2019 Drinking Water Treatment and Challenges in Developing Countries. The Relevance of Hygiene to Health in Developing Countries
[8] Rüßmann M, Lorenz M, Gerbert P, Waldner M, Justus J, Engel P and Harnisch M 2015 Industry 4.0: The Future of Productivity and Growth in Manufacturing Industries Boston Consulting Group Boston, MA, USA pp. 1–14.
[9] Canada E and C C 2020 Water Quantity in Canadian Rivers. Retrieved from www.canada.ca/en/environment-climate-change/services/environmental- indicators/water-quantity-canadian- rivers.html
[10] Jahan S and Strezov V 2017 Water quality assessment of Australian ports using water.
[11] World Health Organisation (WHO) 1996 Coagulation, Flocculation and Clarification Fact Sheet 2.13. Retrieved from http://www.who.int/water_sanitation_health/hygiene/emergencies/fs2_13.pdf
[12]World Health Organisation (WHO) 1996 Filration Fact Sheet 2.13. Retrieved from http://www.who.int/water_sanitation_health/hygiene/emergencies/fs2_13.pdf
[13] World Health Organisation (WHO) 2007 Water Treatment Session Objectives. Seminar Water Treatment.
[14] Thrumakunasagran V and Vadivelu V M 2014 Sungai Dua Water Treatment Plant. Network On Water Technology in Asia and Pacific 1–13.
[15] Bakhtiari A R *et al* 2014 Environment Asia 7 (1) 104–111
[16] Chew C M *et al* 2015 Malaysia's largest riverbank filtration and ultrafiltration systems for municipal drinking water production. *Water Practice and Technology* **10** (1) 59–65.
[17] Chew C M *et al* 2015 Practical performance analysis of an industrial-scale ultrafiltration membrane water treatment plant *Journal of the Taiwan Institute of Chemical Engineers* 46 132–139.
[18] City of Winnipeg 2009 Water and waste department. Retrieved from https://winnipeg.ca/waterandwaste/water/treatment/plant.stm
[19] Tadokoro H *et al* 2011. Smart water management and usage systems for society and environment. *Hitachi Review* **60** (3) 164–171.
[20] Report, F. 2012. Top-down efficiency review of SA Water (September).
[21] Rattier M, Reungoat J and Gernjak W 2012 Organic Micropollutant Removal by Biological Activated Carbon Filtration: A Review Urban Water Security Research Alliance Technical Report No. 53. *Urban Water Security Research Alliance Technical Report* (53) 45
[22] Nadya A *et al* 2015 Design and Development of New Debris Strainer in Water Treatment Plant. *Procedia Computer Science* **76** (Iris) 209–216
[23] Sekitar J A 2019 Laporan Kualiti Alam Sekeliling. Jabatan Alam Sekitar.
[24] Lumb A *et al* 2011 A Comparative Study of USA and Canadian Water Quality Index Models. *Water Quality, Exposure and Health* **3** (3–4) 203–216
[25] Kido M *et al* 2009. Comparison of general water quality of rivers in Indonesia and Japan. Environmental Monitoring and Assessment **156** (1–4) 317–329
[26] Jahan S and Strezov V 2017 Water quality assessment of Australian ports using water quality evaluation indices *PLoS ONE* **12** (12) 1–15
[27] New Straits Times 2020 The rise of digitalization in the water industry Retrieved 15 May 2021, from https://www.nst.com.my/news/nation/2020/12/650707/rise-digitalisation-water-industry
[28] The Star 2020 RM2 million drones to be used to catch polluters. https://www.thestar.com.my/news/nation/2020/10/20/rm2mil-drones-to-be-used-to-catch- polluters
[29] Libelium 2019 Smart water technology provides safe and clean water to indigenous communities in Canada. Retrieved from https://white-set4.rssing.com/chan-3948965/all_p41.html
[30] A Ross 2019 Ow technology id helping solve big water issues in Canada. Retrieved from https://discover.rbcroyalbank.com/how-technology-is-helping-solve-big-water-issues-in-canada/
[31] Jackson B 2018 Drones, digital twins, and connected sensors grids: How Toronto Water keeps the pipes flowing today and in the future. Retrieved from https://www.itbusiness.ca/news/drones-digital-twins-and-connected-sensor-grids-how-toronto-water-keeps-the-pipes-flowing-today-and-in-the-future/107604
[32] Tadokoro H *et al* 2011 Smart water management and usage systems for society and environment. *Hitachi Review* **60** (3) 164–171
[33] ITU News. 2018. How innovative technologies are transforming water management 2018. Retrieved from https://news.itu.int/innovative- technologies-transforming-water- management/
[34] Ghernaout *et al* 2018 Applying big data in water treatment industry: A new era of advance. *International Journal of Advanced and Applied Sciences* **5** (3) 89–97

Acknowledgments
The authors would like to thank the Geran Universiti Penyelidikan: GUP-2018-149, GP-2020-K016102, and GP-2019-K016102 for funding this work.

Effect of freeze-dried biofloc as A Dietary Supplement on Water Quality and Growth Performance of red Tilapia (Hybrid)

Tarq Binalshikh-Abubkr[1,2] and Marlia Mohd Hanafiah[1,3*]

[1] Department of Earth Sciences and Environment, Faculty of Science and Technology, Universiti Kebangsaan Malaysia, 43600 UKM Bangi, Selangor, Malaysia
[2] Department of Food Science and Technology, Faculty of Environmental Science and Marine Biology, Hadhramout University, P.O. Box 50512, Mukalla, Yemen
[3] Centre for Tropical Climate Change System, Institute of Climate Change, Universiti Kebangsaan Malaysia, 43600 UKM Bangi, Selangor, Malaysia

*Corresponding E-mail: mhmarlia@ukm.edu.my

Abstract. To investigate the effect of supplementation of dried biofloc produced by freeze-drying method on water quality and growth performance of red hybrid tilapia, two experimental diets (T0 treatment: 0 % biofloc; and T1 treatment: 4 % freeze-dried biofloc) were examined during 57 days of feeding trial. Diet T0 resulted in higher feed intake than T1 treatment ($P < 0.05$), while no significant differences were found between the two treatments in terms of weight gain, SGR, and FCR ($P > 0.05$). For water quality parameters, values of temperature, EC, TDS, ammonia and nitrate were significantly higher for T1 treatment ($P < 0.05$). While dissolved oxygen level was significantly lower for T1 treatment compared to T0 treatment ($P < 0.05$). Moreover, no significant differences were found for pH and salinity ($P > 0.05$). However, the two treatments showed water quality parameters levels within the range of standard levels of water quality for tilapia culture. In addition, proximate composition of experimental diets showed significantly higher lipid content for T0 treatment than T1 treatment ($P < 0.05$). While no significant differences were found between means of treatments for organic matter, protein, fiber, ash and energy contents ($P > 0.05$). Based on these results, a 4 % supplementation of freeze-dried biofloc can be successively included in formulated diets for red hybrid tilapia without any effects on growth in terms of weight gain, specific growth rate, feed conversion ratio and survival, and can result in acceptable water quality levels for red hybrid tilapia.

Keywords: Biofloc technology, red tilapia, water quality, growth performance, proximate composition

Track Name: Land, Water, Forests and Food Security

1. Introduction

In the recent years, feed expenses have become an urgent issue for aquaculture sector because of the high cost of feed ingredients and the continually increasing use of animal or plant origin (such as; fishmeal and soybean meal) as protein sources in aquafeeds to meet nutritional requirements for fish feed [1]. About 50 % of operating cost comes only from fish feeding [2]. Therefore, to maximize the benefit and to achieve the sustainable goals of this industry, other alternative, more economic protein sources, are needed to be used as a replacement for the partial or even total of traditional ingredients in the formulation of diets [3,4].

Recently, the use of several microbial biomasses as sources of protein has garnered more attention from researchers [5-9]. Among these, microbial flocs which known also as biofloc (BF) are widely used for culture animals and highly recommended to enhance production through provision of natural food in situ [10]. BF can be collected from biofloc technology systems as a wet meal to culture animals [11, 12] or as a dried meal after drying processes with high protein level [13-15]. BFT systems have many benefits that can increase the profitability of aquaculture industry and also can support the sustainability of this sector [16]. In that sense, BFT can keep land and water resources used for fish farming, and can reduce the environmental impact of aquaculture activities on the external water bodies [17]. The technology as well as the use of biofloc meal as a supplement or replacement of a part of feed ingredients is very promising and it can minimize feed costs [18].

As for the benefits associated with fish nutrition requirements, the nutritional value of biofloc has been investigated by many researchers, and BF showed protein content of 18-49, lipid content of 0.4-7.6, fiber content of 0.2-17.1, nitrogen free extraction (NFE) content of 16.5-36, ash content of 1.4-61, and moisture content of 7.8-10.1 g/100g dry weight [14-28]. For example, Crab et al. [29] reported protein content of 58 % in the bioflocs produced using glycerol with *Bacillus*. Biofloc can be a key to the sustainable aquaculture in future [16].

Biofloc can be successfully consumed by herbivorous and omnivorous fish species [16]. Fish consumption of BF depends on various factors such as the feeding habit of the species and its ability to eat the flakes and the suspended particles density [16]. Due to its omnivorous habit, tilapia can use BF as a food source, and BF can meet up to 50 % protein requirements of fish [30]. Red hybrid tilapia is one of potential fish species that can be cultured in freshwater and saline water and can tolerate wide range of water salinity [31-33].

The development of aquaculture sector and the increasing of fish production require information on the effect of feed ingredients, environmental conditions and other factors in the culture tank on the growth of fish [34]. Not only the growth, but also water quality can be affected by those factors [35]. For example, inedible part of the feed, metabolic waste products, and faces of fish will accumulate in culture tank, decompose, and decrease water quality parameters resulting in stressing the fish and production decrease [36]. Therefore, water quality parameters are the key factors which can limit the increase in production of fish farming and need to be managed properly for optimal standards of water quality for cultured animals [37].

Additionally, nutritional value of final product is also an important issue. Consumers prefer high quality, safe and more nutritious food products. Biofloc may bring higher profit of farmed fish if it applied as a replacement of commercial feed or served as a protein source in diets [15]. BF considered a high quality resource of protein, as a result of high level of protein in muscles of fish [29]. These advantages certainly should be more explored [38]. Although, bioflocs show an adequate protein and lipid content for use as an aquaculture feed, more research is required on their amino acid, fatty acid and mineral profile [39, 13].

Growth studies of cultured animals often rely on the proximate composition of experimental diets [12, 21], and fish composition as well. Knowledge of the proximate composition of fish and factors affecting that composition allows us to determine the nutrient transfer efficiency from the feed to the fish body, and also makes it possible to modify composition of the final product [40]. Thus, whole body composition can play an important role in improving the efficiency of formulated diets, reducing overall feed waste, and increase profitability of the culture operation [40]. The changes in body composition of tilapia fish determined by several researchers where carcass of the fish was analyzed at the beginning and end of the experiment [41-43].

Novel methods for integrating biofloc with the aquaculture feeding is required to meet nutritional needs of the cultured animals and maximize production, and also to minimize using traditional protein sources in feed [17, 28]. Therefore, to investigate the effect of biofloc as a dietary supplementation on growth and water quality of red hybrid tilapia reared in brackish water, supplement of 4 % (in dry weight) freeze-dried biofloc diet was evaluated and compared to a control diet without biofloc (100 % commercial feed).

2. Materials and Methods

2.1. Raw materials

2.1.1. Biofloc sampling

In this study, three indoor red hybrid tilapia biofloc system tanks at a commercial farm located in Sepang, Selangor, Malaysia were used for biofloc sampling. The biofloc was collected by using 40 μm plankton net then transported in a cold-store box with ice to laboratory at Universiti Kebangsaan Malaysia (UKM), Selangor, Malaysia. The collected biofloc was frozen immediately at – 80° C for two days then freeze-dried at a vacuum freeze-dryer for three days. After that freeze-dried biofloc was kept in zip-lock plastic bags and stored in a refrigerator at 4° C for later proximate analysis and also for experimental diet.

2.1.2. Tilapia fingerlings

Red hybrid tilapia fingerlings were purchased from a commercial fish farm located in Kuala Selangor, Malaysia. The fish were transported in specific plastic bags with oxygen to the laboratory (UKM). The bags including fish were dived into the aquarium water for 20 minutes to balance the water temperature. Then the fish were allowed to move slowly into the aquarium. The fish were not being fed for 24 h to avoiding any shock.

2.2. Experimental diets

Experimental diets; T0 a control diet (100 % commercial feed) without biofloc and T1 (4 % freeze-dried biofloc, in dry weight) were prepared to be examined for growth and water quality of red tilapia. All the dried ingredients were mixed with warm water to make dough. The dough was pressed through a pelletizer with 2 mm die, formulated in pellet form, and then dried at 60° C until moisture level of below10 % was reached. The dried diets were crushed to less than 2 mm size, then stored at 4 °C till use.

2.3. Experimental design and feeding trail

The experiment was conducted at the marine laboratory of Earth Sciences and Environment Department, Faculty of Science and Technology, UKM, Bangi, Selangor, Malaysia. An experimental design of 6 glass tanks (64.5×37×38) with proximately capacity of 90.5 L were used to rearing the fingerlings of tilapia. The tanks were provided with oxygen using a large air pump through a net of plastic pipes. Each tank was designed with an individual biology filter (1/6 of the tank size) and a small water pump (power 580 L hr^{-1}).

Before the start-up of the experiment, the fish were acclimated on the new environmental conditions at laboratory for a week. During the acclimation period the fish were fed a commercial feed in a proximate composition of; 12 % moisture, 15 % ash, 34 % protein, 6 % fat, and 7 % fiber as provided by the manufacturing (Red Eagle brand, Sri Edaran Hong Enterprise, Kelantan, Malaysia). After acclimation, 90 fishes were selected with average weight of 4 ± 0.068 g and distributed randomly into experimental tanks with a stocking density of 15 fingerlings per tank for eight weeks (57 days). Fish were fed three times per day at 10 am, 1 pm and 4 pm with a feeding ratio of 10 % of fish biomass weight per tank [44]. The feeding ratio was constantly adjusted throughout the rearing period and reduced to 8 % or 5 % depending on the weekly fish biomass gain [42,43,45]. Uneaten feed was siphoned and filtered for further calculations.

¼ of water was exchanged and the filter wool with fish wastes was removed from the primary stage of biology filter and replaced with a new one every week in each tank. After water exchanging salinity of 10 ppt was maintained to all experimental tanks by adding marine salt (Model; Aquarium Systems Instant Ocean, bucket 25 kg, France). The reason of using 10 ppt salinity is because that is the salinity of saline water, especially brackish water between 8 ppt [46] and 10 ppt [42], is recommended to improve the growth of tilapia.

2.4. Growth performance

Fish performance was evaluated by body weight gain (WG; g), feed conversion ratio (FCR), feed efficiency ratio (FER), feed intake per fish (FI; g), specific growth rate (SGR %), protein efficiency ratio (PER) and survival. These parameters were determined using methods by Lopez-Betancur et al., [43] and Prabu et al. [47] as following equations:

Weight gain WG (g) = Final Weight (g) – Initial Weight (g)

Specific growth rate: SGR= [(Ln final weight - Ln initial weight) / days] × 100

Feed conversion ratio: FCR = [total feed fed (g) / fish biomass increase (g)]

Feed efficiency ratio: FER = 1/FCR

Feed intake: FI = Total feed consumption per tank / number of fish per tank

Protein efficiency ratio: PER = biomass weight gain (g) / protein intake (g)

Survival = (Total number of fish survival / total number of fish stoked) x 100

2.5. Water quality parameters

Water quality parameters included; dissolved oxygen (DO), temperature, salinity, conductivity (EC), total dissolved solids (TDS) and pH were measured daily in the tanks of fish. A handheld multi-parameter water quality instrument (EcoSense®-EC300A, YSI, USA) was used for temperature, salinity, EC, and TDS. While DO was monitored continuously by a dissolved oxygen meter (YSI 5000) and a pH meter (LAQUA PH1100) was used for pH. A water sample of 500 ml was collected every week from each tank and filtered through dried and weighed Whatman 1822-047 GF/C filter paper under vacuum pressure and used for total ammonia-nitrogen (TA-N) and nitrate-nitrogen (NO_3-N) measurements according to standard methods [48].

2.6. Proximate composition

Proximate compositions of the experimental diets (control diet and 4 % freeze-dried biofloc diet) were analyzed in dry matter according to standard methods of AOAC [49]. The dried samples were ground by mortar and pestle and kept at 4° C for further analysis of moisture, ash, protein, lipid and fiber contents [50]. Samples were analyzed in triplicate for all proximate contents.

2.6.1. Moisture measurement

A known grams of flesh was pre-weighed (W1) on a crucible and place in a vacuum oven at 105° C for 24 hours. The sample was removed from the oven, cooled in a desiccator, and re-weighed (W2). Moisture percentage was calculated according to the formula:

Moisture (%) = (W1 – W2) ÷ W1 × 100

2.6.2. Total ash measurement

Porcelain crucible with dry sample from the moisture analysis was used for ash analysis incinerated in a muffle oven at 550° C for 4 hours. The crucible was removed from the muffle oven, cooled in a desiccator, and then weighed. Ash content was calculated according to the following formula:

Ash (%) = (ash weight ÷ sample weight) × 100

2.6.3. Protein measurement

Crude protein content was determined by kjeldahl method. The kjeldahl procedure basically divided into three parts: (1) Digestion, (2) Distillation and (3) Titration. The procedure used in this work measures the protein percentage as a nitrogen content of the dried sample. The protein content then was calculated assuming a ratio of protein to nitrogen for the specific food being analyzed ($N \times 6.25$). Calculation:

% Nitrogen = (value of HCl × 0.1 × 0.014 ÷ weight of sample) × 100

% of crude protein = % Nitrogen × Conversion factor

(Conversion factors for animal and plant origin are 6.25 & 5.90 respectively)

2.6.4. Lipid measurement
Crude lipid was determined by soxhlet extraction technique and was calculated as below:

% of crude lipid = (corrected weight of fat ÷ weight of sample) × 100

2.6.5. Crude fiber
Crude fiber was determined by fiber analyzer. The major components of natural dietary fiber are cellulose, hemicelluloses, lignin, and other non-starch plant polysaccharides such as pectin. Dietary fiber is often determined gravimetrically. In such a procedure, digestible carbohydrates, lipids, and proteins are selectively solubilized by chemicals or removed by enzyme-catalyzed hydrolysis. Then, non-solubilized and/or nondigested materials are collected by filtration, and the fiber residue is recovered, dried, and weighed. It is essential either that all digestible materials be removed from the sample so that only non-digestible polysaccharides remain or that the non-digestible residue be corrected for remaining digestible contaminants. Lipids are removed easily from the sample with organic solvents and generally do not pose analytical problems for the fiber analyst.

2.6.6. Total nitrogen free extract
The total nitrogen free extract (NFE) was measured as following:

NFE % = 100 – [moisture % + total ash % + crude protein % + crude lipid % + crude fiber %]

2.6.7. Gross energy
The gross energy was calculated by formula:

Gross energy (Kcal/100g) = [crude protein × 5.6] + [crude lipid × 9.44] + [crude fiber ×4.1] + [NFE × 4.1]

2.7. Data analysis
Statistical analysis was performed using statistical package SPSS version 22. Kolmogorov–Smirnov test for normality and Levene's test for homogeneity were used for variances. After the normality and homogeneity were met, significant differences of means among treatments were determined using analysis of variance (ANOVA) at 95 % confidence interval. The treatments were considered as independent variables (factors) and parameters of growth, water quality and proximate composition were considered as dependent variables (measured variables. The treatments were compared using one-way ANOVA followed by Tukey test ($P < 0.05$). Descriptive statistics were done for water quality variables to show maximum and minimum values for each treatment.

3. Results
Table 1 shows one-way ANOVA of the effect of experimental diets on the growth indicators of red tilapia. The Kolmogorov-Smirnov test indicated that the data came from a normal distribution. The Levene's test (Homogeneity of variance test) indicated that the data were homogeneous variance between the two groups of data. Treatments; T0 (0 % biofloc) and T1 (4 % freeze-dried biofloc) were considered as factors for ANOVA.

Statistically, no significant differences were recorded ($P > 0.05$) for growth indicators final weight (Wf), weight gain (WG), specific growth rate (SGR), feed conversion ratio (FCR), feed efficiency ratio (FER), protein efficiency ratio (PER) and survival (SG) between the two treatments. While the feed intake (FI) for T0 treatment (86.86 g) was significantly higher ($P < 0.05$) than T1 treatment (72.72 g) (Table 1).

Table 1. The effect of experimental diets on growth performance of red tilapia.

Growth indicators	Treatments	
	T0	T1
Initial weight (g fish^{-1})	4.05±0.04A	4.04±0.05A
Final weight (g fish^{-1})	20.98±0.24A	20.42±0.43A
Weight gain (g fish^{-1})	16.93±0.19A	16.48±0.46A
Initial biomass (g tank^{-1})	60.74±0.63A	60.60±0.68A
Final biomass (g tank^{-1})	272.99±24.07A	255.40±15.55A
Biomass gain (g tank^{-1})	212.25±23.43A	196.44±16.02A
SGR (%)	2.63±0.14A	2.57±0.12A
FCR	0.43±0.03A	0.44±0.03A
FER	2.34±0.14A	2.27±0.13A
FI (g)	86.86±5.79A	72.72±0.33B
PER	6.89±0.40A	6.67±0.37A
Survival (%)	86.67±6.67A	83.34±3.34A

Values are means of the experimental treatments ± SD, values followed with different capital letters in same row are significantly different ($P < 0.5$). SGR (%): Specific growth rate, FCR: Feed conversion ratio. FER: Feed efficiency ratio. FI: Feed intake (g/tank). PER: Protein efficiency ratio. T0: commercial feed 100 %; T1: freeze-dried biofloc 4 %.

In contrast, water quality parameters showed significant differences ($P < 0.05$) between means of the two treatments for temperature (T), dissolved oxygen (DO), electrical conductivity (EC), total dissolved solids (TDS), total ammonia-nitrogen (TA-N) and nitrate-nitrogen (NO_3-N). While no significant differences were recorded ($P > 0.05$) for pH and salinity (table 2).

In general, levels of T, DO, pH, salinity, EC, and TDS were ranged in 25.6-28.1, 7.46-10.68, 7.51-8.37, 8.98-10.35, 12.22-17.61, and 9.18-11.51, respectively for the two treatments (table 2). T1 treatment showed pH and salinity values (7.87 and 9.93, respectively) closer to T0 treatment (7.90 and 10.06, respectively). However, T1 treatment showed significantly higher ($P < 0.05$) temperature (26.27° C) than T0 treatment (26.22° C), and significantly lower ($P < 0.05$) DO level (9.03 ppm) than T0 treatment (9.15 ppm). Furthermore, EC and TDS levels (16.08 *m*S/cm and 10.76 ppm, respectively) tended to be significantly higher ($P < 0.05$) in T1 treatment compared to T0 treatment (15.85 *m*S cm^{-1} and 10.48 ppm, respectively).

Concentrations of total ammonia-nitrogen (TA-N) and nitrate-nitrogen (NO_3-N) ranged from 0.08-1.95 and 0.83-2.40 ppm, respectively. The concentrations of 1.26 ppm TA-N and 1.70 ppm NO_3-N were significantly higher ($P < 0.05$) for T1 treatment compared to 1.21 ppm TA-N and 1.58 ppm NO_3-N for T0 treatment (Table 2).

Table 2. Water quality parameters during the experiment period of 57 days (mean ± SD followed by minimum and maximum values).

Parameters	Treatments (means & range of values)	
	T0	T1
T (° C)	26.22±0.03A (25.6-28.1)	26.27±0.02B (25.7-28.1)
DO (ppm)	9.15±0.09A (7.61-10.68)	9.03±0.07B (7.46-10.50)
pH	7.90±0.02A (7.60-8.37)	7.87±0.02A (7.51-8.30)

Salinity (ppt)	10.06±0.08A (9.04-10.35)	9.93±0.07A (8.98-10.35)
EC (mS cm^{-1})	15.85±0.03A (12.62-17.61)	16.08±0.03B (12.22-17.54)
TDS (ppm)	10.48±0.05A (9.18-11.37)	10.76±0.06B (9.28-11.51)
TAN (ppm)	1.21±0.02A (0.08-1.95)	1.26±0.06B (0.09-1.92)
NO_3-N (ppm)	1.58±0.01A (0.83-2.20)	1.70±0.01B (0.90-2.40)

Values followed with different capital letters in same row are significantly different ($P < 0.5$). T0: commercial feed 100 %; T1: freeze-dried biofloc 4 %.

Results of proximate analyses indicated organic matter, moisture, crude protein, crude lipid, crude fiber, total nitrogen free extraction (NFE) and total ash contents ranged between 85.06 – 88.79 g/100g, 4.80 – 9.95 g/100g, 32.25 – 33.34 g/100g, 4.88-5.55 g/100g, 7.51-8.12 g/100g, 28.72-38.75 g/100g and 11.21-14.94 g/100g, respectively. Gross energy ranged between 387.59 – 418.79 Kcal/100g (Table 3).

Table 3. Proximate composition (%) of experimental diets (mean ± SD).

Proximate composition (%)	Diets	
	T0	T1
OM[1]	85.06±0.05A	88.79±0.41A
Moisture	9.95±0.01A	4.80±0.39B
Crude protein	33.34±1.29A	32.25±0.37A
Crude lipid	5.55±0.07A	4.88±0.04B
Crude fiber	7.51±0.40A	8.12±0.31A
Total NFE[2]	28.72±1.57A	38.75±0.85B
Total ash	14.94±0.05A	11.21±0.41A
Gross energy[3]	387.59±1.79A	418.79±2.92B

Values followed with different capital letters in same row are significantly different ($P < 0.5$).
1 Organic matter (OM) = 100 – ash
2 Nitrogen free extraction (NFE) = 100 – (crude protein + crude lipid + crude fiber + total ash + moisture)
3 Gross energy (Kcal/100g) = [crude protein × 5.6] + [crude lipid × 9.44] + [crude fiber ×4.1] + [NFE × 4.1]

To treatment showed contents of moisture and crude lipid (9.95 g/100g and 5.55 g/100g, respectively) significantly higher ($P < 0.05$) than T1 treatment (4.80 g/100g and 4.88, 38 g/100g, respectively), while total NFE of 28.72 g/100g and gross energy of 387.59 Kcal/100g significantly lower ($P < 0.05$) than T1 treatment (38.75 g/100g and 418.79 Kcal/100g) (Table 3). However, no significant differences were recorded ($P > 0.05$) between T0 treatment and T1 treatment in terms of organic matter, crude protein, crude fiber and total ash contents (Table 3).

4. Discussion

Growth performance of red tilapia in the present study did not affected by the method of dry (freeze-drying) in terms of final weight (*Wf*), weight gain (*WG*), specific growth rate (SGR), feed conversion ratio (FCR), feed efficiency ratio (FER), protein efficiency ratio (PER) and survival (SG) within all means of treatments. However, the feed intake (FI) for freeze-dried biofloc diet (T1 treatment) was affected by the 4 % supplement of freeze-dried biofloc, where the control treatment showed higher FI than another treatment.

FI can be calculated as total feed consumption per tank divided by number of fish per tank [46]. The total feed consumption is calculated by subtracting the amount of uneaten feed from the total feed fed in each tank. Uneaten feed could be resulting in poor water quality parameters in culture water [51] In this study, uneaten feed by fish in each tank was siphoned, filtered, dried, and then weighed to calculate the total consumed feed per tank during the 57 days of tilapia rearing. We found that the FI of T1 treatment was lower than the FI of T0 treatment (control). Therefore, the decrease in FI for the freeze-dried diets may be due to the increase of uneaten feed weight for the same treatment.

On the other hand, the drying method had clear effect on water quality parameter. From the results of this study, there were significant effects of using freeze-drying method on most of the water quality parameters, except for pH and salinity. Moreover, total ammonia and nitrate-nitrogen concentrations were higher for freeze-dried biofloc treatment compared to the control treatment.

In the present study, the values of DO, pH and EC might be influenced by the water temperature inside tanks. T1 treatment showed lower DO levels, but higher temperature compared to T0 treatment. In this respect, it is known that there is an inverse relation between temperature and the level of dissolved oxygen in water [52] Similar to Do levels, pH values can be influenced by temperature since there is also an inverse relationship between temperature and pH in water [53]. Although the pH values between the two treatments in the present study showed no significant differences, slightly increase was noted for T0 treatment, while T1 treatment was slightly decrease in pH. Meanwhile, the higher temperature in T1 treatment might cause increase in the EC values in the present study. Hayashi [54] reported that is a direct relationship between temperature and conductivity values in water. For example, increasing of 1°C in temperature causes increasing of 2 % in EC value [54]. Therefore, the decrease in DO level and the increase in EC values for T1 treatment in the present study may be due to the increase in temperature for the same treatment.

Nitrate is the final product of nitrification process of nitrogen cycle inside culture tank and is less toxic compared with other inorganic nitrogen compounds (nitrite). However, high levels of nitrate can negatively affect the growth of tilapia fish [50]. Unlike the nitrate, ammonia is the most toxic nitrogen compound which can cause high mortality for cultured fish [26].

In the present study, the use of freeze-dried biofloc as a supplement in the formulation of the experimental diet resulting in lower quality parameters in terms of T, DO, EC, TDS, total ammonia-nitrogen (TA-N) and nitrate-nitrogen (NO_3-N) for cultured water of red tilapia compared to the control tanks. Even though, the levels of these parameters are still within the range of standard levels for the cultured red tilapia [36,45].

Nutritionally, the control diet without biofloc (T0 treatment) showed higher content of crude lipid in the present study. While no differences in contents of organic matter, crude protein, crude fiber and total ash between the two treatments. However, freeze-dried biofloc diet showed better values of moisture, total NFE and gross energy. The increase in lipid value (4.88 g/100 g) for T1 treatment in this study can be explained by the low level of lipid in biofloc composition [18, 47].

5. Conclusions

Based on the growth performance results, freeze-dried biofloc diet provided good growth rates similar to control diet in terms of final weight (*Wf*), weight gain (*WG*), specific growth rate (SGR), feed conversion ratio (FCR), feed efficiency ratio (FER), protein efficiency ratio (PER) and survival (SG). However, the feed intake factor was lower for freeze-dried diet compared to the control. Although freeze-dried biofloc diet resulted in higher concentrations of TA-N and NO_3-N, these concentrations are still within the range of standard levels for the cultured red tilapia. This conclusion was also supported by the proximate composition of diets, since freeze-dried biofloc showed better values of moisture, total NFE and gross energy, and similar values of organic matter, crude protein, crude fiber and total ash compared to the control.

Overall, a 4 % supplementation of freeze-dried can be successively included in the formulated diets for red hybrid tilapia without any effects on growth in terms of weight gain, specific growth rate, feed conversion ratio and survival, and also can keep the water quality parameters within the acceptable range of red hybrid tilapia.

References

[1] Kuhn D D, Boardman G D, Lawrence A L, Marsh L and Flick G 2009 Microbial floc meal as a replacement ingredient for fish meal and soybean protein in shrimp feed. *Aquaculture*, **296**: 51-57.

[2] Yun H, Shahkar E, Hamidoghli A, Lee S, Won S and Bai S C 2017 Evaluation of dietary soybean meal as fish meal replacer for juvenile whiteleg shrimp, *Litopenaeus vannamei* reared in biofloc system. *International Aquatic Research*, **9** (1): 11–24.
[3] Himaja P H and Rajagopalasamy C B T 2016 Review on biofloc meal as an alternative ingredient in aquaculture feeds. *Journal of Aquaculture in the Tropics*; *New Dellhi*, **31** (3-4): 199-220.
[4] Neto J R S, Ribeiro F A S, Gonçalves A A and Emerenciano M G C 2019 Tilapia processing waste silage (TPWS): An alternative ingredient for *Litopenaeus vannamei* (Boone, 1931) diets in biofloc and clear-water systems. *Aquaculture and Fisheries*, **4** (5): 214–218.
[5] Hende S F D, Claessens L, Muylder E D, Boom N and Vervaeren H 2014 Microbial bacterial flocs originating from aquaculture wastewater treatment as diet ingredient for *Litopenaeus vannamei* (Boone). *Aquaculture Research*, 1-15.
[6] Martínez-Córdova L R, Martínez-Porchas M, Emerenciano M G C, Miranda-Baeza A and Gollas-Galva´n T 2016 From microbes to fish the next revolution in food production. *Critical Reviews in Biotechnology*, 1-9.
[7] Martinez-Porchas M and Vargas-Albores F 2017 Microbial metagenomics in aquaculture: a potential tool for a deeper insight into the activity. *Reviews in Aquaculture*, **9** (1): 42-56.
[8] Pikaar I, Matassa S, Rabaey K, Bodirsk B L, Popp A, Herrero M and Verstraete W 2017 Microbes and the Next Nitrogen Revolution. *Environ. Sci. Technol.*, **51** (13), 7297–7303.
[9] Glencross B D, Huyben D and Schrama J W 2020 The application of single-cell ingredients in aquaculture feeds – A review. *Fishes*, **5**, 22.
[10] D'Abramo L R, Silva J L and Kim T 2009 Optimization of crustacean quality through husbandry and adherence to post-harvest standards for processing. In book *'Shellfish Safety and Quality'*. Book Series: *Woodhead Publishing in Food Science Technology and Nutrition*. **167**: 339-358.
[11] Azim M E, Little D C and Bron J E 2008 Microbial protein production in activated suspension tanks manipulating the C:N ratio in feed and the implications for fish culture. *Bioresource Technology* **99** (9): 3590-3599.
[12] Emerenciano M, Ballester E L C, Cavalli R O, and Wasielesky W 2012 Biofloc technology application as a food source in a limited water exchange nursery system for pink shrimp *Farfantepenaeus brasiliensis* (Latreille, 1817). *Aquaculture Research*, **43**: 447-457.
[13] Shyne Anand P S, Kohli M P S, Kumar S, Sundaray J K, Roy S D, Venkateshwarlu G, Sinha A and Pailan G H 2014 Effect of dietary supplementation of biofloc on growth performance and digestive enzyme activities in *Penaeus monodon*. *Aquaculture*, **418-419**: 108-115.
[14] Khatoon H, Banerjee S, Yuan G T G, Haris N, Ikhwanuddin M, Ambak M A, and Endut A 2016 Biofloc as a potential natural feed for shrimp postlarvae. *International Biodeterioration & Biodegradation*, **113**: 304-309.
[15] Lee C, Kim S, Lim S and Lee K 2017 Supplemental effects of biofloc powder on growth performance, innate immunity, and disease resistance of Pacific white shrimp *Litopenaeus vannamei*. *Fisheries and Aquatic Sciences*, **20**: 15, 1-7.
[16] Crab R, Defoirdt T, Bossier P and Verstraete W 2012 Biofloc technology in aquaculture: Beneficial and future challenges. *Aquaculture*, **356-357**: 351–356.
[17] Bossier P and Ekasari J 2017 Biofloc technology application in aquaculture to support sustainable development goals. *Microbial Biotechnology* **10** (5): 1012-1016.
[18] Becerril-Cortés D, Monroy-Dosta M C, Coelho-Emerenciano M G, Castro-Mejía G, Cienfuegos-Martínez K and Lara-Andrade R 2017 Nutritional importance for aquaculture and ecological function of microorganisms that make up Biofloc, a review. *International Journal of Aquatic Science*, **8** (2): 69-77.
[19] Azim M E and Little D C 2008 The biofloc technology (BFT) in indoor tanks: water quality, biofloc composition, and growth and welfare of Nile tilapia (*Oreochromis niloticus*). *Aquaculture*, **283** (1): 29-35.
[20] Ekasari J, Angela D, HadiWaluyo S, Bachtiar T, Surawidjaja E H, Bossier P and Schryver P D 2014 The size of biofloc determines the nutritional composition and the nitrogen recovery by aquaculture animals. *Aquaculture*, **226**: 105-111.
[21] López-Elías J A, Moreno-Arias A, Miranda-Baeza A, Martínez-Córdova L R, Rivas-Vega M E and Márquez-Ríos E 2015 Proximate composition of bioflocs in culture systems containing hybrid red tilapia fed diets with varying levels of vegetable meal inclusion. *North American Journal of Aquaculture*, **77** (1): 102-109.
[22] Kuhn D D, Lawrence A L, Crockett J, Marsh L and Taylor D 2016 Evaluation of biflocs derived from confectionary food effluent water as a replacement feed ingredient for fishmeal or soy meal for shrimp. *Aquaculture*, **454**, 66-71.
[23] Liu W, Luo G, Tan H and Sun D 2016 Effects of sludge retention time on water quality and bioflocs yield, nutritional composition, apparent digestibility coefficients treating recirculating aquaculture system effluent in sequencing batch reactor. *Aquacultural Engineering*, **72-73**: 58-64.
[24] Wei Y F, Liao S and Wang A 2016 The effect of different carbon sources on the nutritional composition, microbial community and structure of bioflocs. *Aquaculture*, **465**: 88–93.
[25] Becerril-Cortés D, Monroy-Dosta M D C, Emerenciano M G C, Castro-Mejía G, Bermúdez B S B and Correa G V 2018 Effect on nutritional composition of produced bioflocs with different carbon sources (Molasses, coffee waste and rice bran) in Biofloc system. *International Journal of Fisheries and Aquatic Studies*; **6** (2): 541-547.
[26] Dauda A B *et al* 2018 Influence of carbon/nitrogen ratios on biofloc production and biochemical composition and subsequent effects on the growth, physiological status and disease resistance of African catfish (*Clarias gariepinus*) cultured in glycerol-based biofloc systems. *Aquaculture*, **483**: 120-130.
[27] Zhang Y, Yang Y, Chen X, Tan H and Luo G 2020 Ammonia nitrogen treatment capacity and nutrient composition of bioflocs with different particle sizes. *Journal of Fishery Sciences of China*, **27** (3): 295-306.

[28] Binalshikh-Abubkr T, Hanafiah M M and Das S K 2021 Proximate chemical composition of dried shrimp and tilapia waste bioflocs produced by two drying methods. *Journal of Marine Science and Engineering*, **9** (2), 193;
[29] Crab R, Chielens B, Wille M, Bossier P and Verstraete W 2010 The effect of different carbon sources on the nutritional value of bioflocs, a feed for *Macrobrachium rosenbergii* postlarvae. *Aquaculture Research*, **41**:559–567.
[30] Caldini N N, Cavalcante D H, Filho P R N R, and Sá M V C 2015 Feeding Nile tilapia with artificial diets and dried bioflocs biomass. *Animal Sciences, Maringá*, **37**(4): 335-341.
[31] Yue G H, Lin H R and Li J L 2016 Tilapia is the fish for next-generation aquaculture. *Int. J. Mar. Sci. Ocean Tech* **3** (1): 11-13.
[32] Ahmadi N, Baroiller J F, D'Cotta Carreras H and Morillon R 2016 Adaptation to Sali-nity. In: Torquebiau, E. (Ed.), Climate Change and Agriculture Worldwide. *Springer, Dordrecht*, pp. 45 – 58.
[33] FAO 2018 The state of world fisheries and aquaculture 2018, meeting the sustainable development goals. Rome. License: CCBY-NC-SA 3.0 IGO.
[34] Thorarensen H, Kubiriza G K and Imsland A K 2015 Experimental design and statistical analyses of fish growth studies. *Aquaculture* **448**: 483-490.
[35] Ariffin F D, Halim A A, Hanafiah M.M, Awang N, Othman M S, Azman, S A A and Bakri N S M 2019 The Effects of African Catfish, Clarias gariepinus Pond Farm's Effluent on Water Quality of Kesang River in Malacca, *Malaysia. Applied Ecology and Environmental Research*, **17** (2), 1531-1545.
[36] Putra I, Effendi I, Lukistyowati I and Tang U 2019 Growth and Survival Rate of Red Tilapia (*Oreochromis Sp.*) Cultivated in the Brackish Water Tank under Biofloc System. *Advances in Engineering Research*, **190**: 96-99.
[37] Nurhatijah N, Muchlisin Z A, Sarong M A and Supriatan A 2016 Application of biofloc to Maintain the water quality inculture system of the tiger prawn (*Peaneus monodon*). *AACL Biofluc*, **9** (4).
[38] Emerenciano M, Gaxiola G and Cuzon G 2013 Biofloc technology (BFT): A review for aquaculture application and animal food industry. In "Biomass Now - Cultivation and Utilization". InTech, 301-328.
[39] Ogello E O, Musa S M, Aura C M, Abwao J O and Munguti J M 2014 An appraisal of the feasibility of tilapia production in ponds using biofloc technology: A review. *International Journal of Aquatic Science*. **5** (1): 21-39.
[40] Chowdhury M A K and Bureau D P 2009 Predicting Body Composition of Nile Tilapia (*Oreochromis niloticus*). *Asian Fisheries Science*, **22**: 597-605
[41] Buyukcapar H M, Atalay A I and Kamalak A 2011 Growth performance of Nile tilapia (*Oreochromis niloticus*) fed with diets containing different level of hydrolysable and condensed tannin. *J. Agr. Sci. Tech.* **13**: 1045-1051.
[42] Durigon E G, Lazzari R, Uczay J, Lopes D L A, Jeronimo G T, Sgnaulin T and Emerenciano M G C 2020 Biofloc technology (BFT): Adjusting the levels of digestible protein and digestible energy in diets of Nile tilapia juveniles raised in brackish water. *Aquaculture and Fisheries*, **5** (1): 42-51.
[43] Lopez-Betancur D, Moreno I, Guerrero-Mendez C, Gomez-Melendez D, Macias M J and Olvera-Olvera C 2020 Effects of colored light on growth and nutritional composition of tilapia, and biofloc as a food source. *Appl. Sci.*, **10**, 362.
[44] Correa A S, Pinho S M, Molinari D, Pereira K R, Gutierrez S M, Monroy-Dosta M and Emerenciano 2019 Rearing of tilapia (*Oreochromis niloticus*) juveniles in a biofloc system employing periods of feed deprivation. *Journal of Applied Aquaculture* **32** (2): 1-18.
[45] Pradeep P J, Srijaya T C, Hassan A, Chatterji A K, Raghavan R, Withyachumnarnkul B and Jeffs A 2013 Growth performance of triploid red tilapia reared under laboratory conditions. *Journal of Applied Aquaculture*, **25**: 176-189.
[46] de Alvarenga E R et al., 2018 Moderate salinities enhance growth performance of Nile tilapia (*Oreochromis niloticus*) fingerlings in the biofloc system. *Aquaculture Research*, 1-8.
[47] Prabu E, Rajagopalsamy C B T, Ahilan B, Jeevagan I J M A and Renuhadevi M 2018 Effect of dietary supplementation of biofloc meal on growth and survival of GIFT tilapia. *Indian J. Fish.*, **65** (1): 65-70.
[48] APHA 2005 Standard methods for the examination of the water and wastewater. 22nd ed. *American Public Health Association*, Washington, D. C., USA.
[49] AOAC 2016 Official methods of analysis. 20th edition. Association of Official Analytical Chemists, Washington, D. C.
[50] Long L, Yang J, Li Y, Guan C and Wu F 2015 Effect of biofloc technology on growth, digestive enzyme activity, hematology, and immune response of genetically improved farmed tilapia (*Oreochromis niloticus*). *Aquaculture*, **448**: 135-141.
[51] Prangnell D I, Lupatsch I, Treece G D and Samocha T M 2019 Water Quality Management. In Sustainable Biofloc Systems for Marine Shrimp; Samocha, T.M., Ed.; *Academic Press, Elsevier Inc.*: London, UK, pp. 19–27.
[52] Summerfelt R C 1998 Water Quality Considerations for Aquaculture. *Aquac. Netw. Inf. Cent.*, **50**, 11–21.
[53] Hargreaves J A and C S Tucker 2004 Managing ammonia in fish ponds. *Southern Regional Aquaculture Center,* **4603**
[54] Hayashi M 2004 Temperature-Electrical Conductivity Relation of Water for Environmental Monitoring and Geophysical Data Inversion. *Environmental Monitoring and Assessment*, ***96***: 119-128.

A case study on water price calculation of key projects at Fenglinwan irrigation areas in JiangXi Province, China

Linlu Tan[1*], Xiaoming Zhu[2], Xiaoping Liu[1], Yiguo Wan[1], Arniza Fitri[3] and Siska Melinda[3]

[1]Jiangxi Academy of Water Science and Engineering, Nanchang, 330031, China

[2] College of Water Conservancy and Ecological Engineering, Nanchang Institute of Technology, Nanchang, 330099, China

[3]Department of Civil Engineering, Faculty of Engineering and Computer Science, Universitas Teknokrat Indonesia, Bandar Lampung, Indonesia

*Corresponding E-mail: 443062980@qq.com

Abstract. Reasonable water price plays an important role in the healthy operation of an irrigation area. This study evaluates the current operation and management of Fenglinwan irrigation area in JiangXi Province, including implementation of agricultural water price, employee salary, current comprehensive agricultural water price and investment of engineering facilities. The cost of agricultural water supply of key project in irrigation area and the total cost of irrigation area were calculated to measure the agricultural water price. The result shows that the cost which has been spent for operation and maintenance of irrigation area was about 3.76 million yuan per year and the total cost for irrigation area was about 5.33 million yuan per year. The excessive number of employees and low wage level have greatly affected the work enthusiasm of employees. The cost of engineering maintenance was too small to guarantee the good operation of irrigation and water conservancy projects. The current implementation of agricultural water price which is about 20 yuan per mu is lower than the evaluated operation and maintenance cost for agricultural water supply which is 71.13 yuan per mu. A part of that, the current comprehensive agricultural water price which is 49.20 yuan per mu is also lower than the whole cost for agricultural water supply which is about 100.87 yuan per mu. Neither the current implementation of agricultural water price nor the current comprehensive agricultural water price can meet the basic requirements of operation and management.

Keywords: Fenglinwan, irrigation areas, Jiangxi province, water supply cost, water price calculation

Track Name: Land, Water, Forests and Food Security

1. Introduction

Water is the most crucial thing for human life [1-5] and aquatic life [6-8] as well as for agriculture [9-10]. Reasonable setting price for agricultural water supply is an extremely important thing in the process of implementing the comprehensive reform of agricultural water price. It further can effectively exert the leverage of water prices, promote the transformation of agricultural water uses from extensive to be intensive, improve the enthusiasm of farmers to save water and maintain the farmland water conservancy facilities to continue healthy operation [11]. Fu *et al.* [12], Yu [13], Jia [14] and other researchers [15-18] proposed the calculation method of terminal water price in irrigation area. Li and Wang [19] analyzed the formation mechanism of agricultural water price from the evaluation of water price affordability of water users.

Besides, Li and Meng [20] analyzed the sensitivity of each component of agricultural water supply cost; while Du and Zhang [21] and Yuan [22] analyzed the price of agricultural water and the affordability of farmers. Then, Anhui Water Conservancy Accounting [23] investigated and studied the agricultural two-part water price in Pishihang Irrigation area. The direct management of some authority in large and medium irrigation areas in South China is the key project, but there are few studies focusing on the analysis and measurement of water price from the cost of key projects. This study evaluates the current operation and management of Fenglinwan Irrigation area; calculates the cost of agricultural water supply; and measures the water price in the irrigation area, so as to provide some reference for developing the formulation of water price of similar projects.

2. Overview of irrigation areas

Fenglinwan irrigation area is located in Guixi City, northeast of Jiangxi Province, with a designed irrigation area of about 75,400 mu and an actual irrigation area of about 52,800 mu. The total arable land area is 79,000 mu, of which 77,300 mu paddy field is a medium-sized irrigation area. The key project of Fenglinwan irrigation area consists of 3 medium-sized reservoirs and 1 main canal (7 main canals, 61 branch canals, more than 300 canal buildings). The total length of the main canal is 134.9 km, and the total length of the branch canal is 211.25 km (Figure 1).

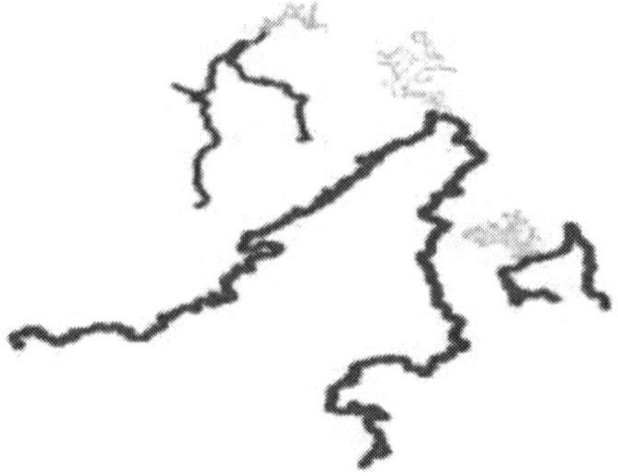

Figure 1. General layout of main and branch canals.

3. Evaluation of operational management

3.1 Execution of water price
According to Government of Guixi City, 2009 [24] about improvement of water fee collection in Guixi City, the standard of water charge for farmland irrigation is about 24 catties of rice/mu. In principle, water charge is mainly levied in kind. If farmers are willing to pay in cash, it is collected according to 20 yuan per mu. So far, the water uses for Fenglinwan irrigation area have not paid the water fees.

3.2 Personnel structure and salary level
A total of 32 public welfare posts and engineering operation management staff are in each water pipe unit. By the end of 2018, the actual number of posts in the three reservoir management committees was 41. By calculation, the average wage of 2016-2018 is 1,568,000 yuan per year, and the average wage level of personnel is 38,200 yuan per year/per person according to the actual number of people, which is far less than the average salary of 58,600 yuan of water conservancy practitioners in 2017 according to the 2018 Yingtan Statistical Yearbook [25-26].

3.3 Current comprehensive water price
So far, farmers in the Fenglinwan irrigation area have not paid the water fees. According to the policy requirements of the comprehensive reform of agricultural water price at the present stage, in order to meet the requirements of engineering operation, the way of charging water fee and government subsidy operation cost is adopted to implement the operation and maintenance of engineering. According to this requirement, combined with the actual situation of Fenglinwan irrigated district, the effective irrigation area of Fenglinwan irrigation area is about 52,800 mu. The water fee from 2016 to 2018 is 0 yuan, and the three-year irrigation water fund allocated by the government subsidy and maintenance fund is 2,597,900 yuan, which is converted into the comprehensive water price of 49.20 yuan per mu.

3.4 Engineering facilities
The average expenditure from 2016 to 2018 was 3,844,700 yuan, of which the engineering maintenance fund was 953,100 yuan, accounting for 24.79 % of the total expenditure. Due to the lack of funds, channel congestion, damage, poor operation of buildings, serious corrosion and other phenomena, seriously affecting the benign operation of irrigation projects.

4. Cost calculation for water supply

4.1 Cost Allocation
Fenglinwan Irrigation area is a medium-sized irrigation area with multiple hubs, simultaneous storage and diversion, and combination of reservoir and canal. The main function of the key project and sluice in the irrigation area is irrigation, without other functions. Therefore, in the cost calculation, the cost of the key project does not need to be shared; The main functions of the reservoir are flood control and irrigation, so cost allocation should be carried out when calculating direct wages and management costs. The reservoir capacity proportion method is used for allocation, and the allocation coefficients of Fenglinwan Reservoir, Daheyuan Reservoir, Taipingfan Reservoir and Fenglinwan Irrigation area are 0.85, 0.91, 0.83 and 0.86.

4.2 Water supply cost analysis

4.2.1 Employee compensation

Total approved wages for staff and workers were based on the average standard wages and number of staff and workers. The total employees for reservoirs management are 32 staffs (Table 1), including 10 staffs for Fenglinwan Reservoir Management Committee, 10 staffs for Taipingfan Reservoir Management Committee, and 10 staffs for Daheyuan Reservoir Management Committee. Based on Yingtan Statistical Yearbook [26], the average salary of water conservancy employees in 2017 was 58,623 yuan, and the monthly salary was 4,885.25 yuan.

Table 1. The calculation table of the number of employees to share the salary of the reservoir in Fenglinwan irrigation area.

Serial number	Reservoir name	Number of employees (people)	Apportionment factor	Apportionment number (people)
1	Fenglinwan	12	0.85	10.2
2	Daheyuan	10	0.91	9.1
3	Taipingfan	10	0.83	8.3
	Total	32		27.6

As shown in Table 1, the approved apportion number of Fenglinwan irrigation area is 27.6 people, so the employee salary is: average salary * approved number =5.8623*27.6= 1.618 million Yuan.

4.2.2 Direct material cost

Direct material costs include raw materials, raw water, auxiliary materials, spare parts, fuel, power and other direct materials consumed in the operation and production of water supply projects [24]. Fenglinwan irrigation area belongs to self-flow irrigation, there is no material cost in the actual operation process, so, the cost is calculated at 0 yuan.

4.2.3 Depreciation of fixed assets

Depreciation fee of fixed assets refers to the depreciation amount of water supply fixed assets calculated according to the prescribed depreciation method [25]. Medium-sized reservoirs in Fenglinwan irrigation area are depreciated according to the 40-year standard, irrigation channels are depreciated according to the 50-year standard, and sluices (small) are depreciated according to the 40-year standard. The residual rate is calculated at 5 %.

The annual depreciation is calculated by the net value of fixed assets * (1 − residual rate) * sharing coefficient / depreciation years, and the residual rate is calculated by 5 %. The fixed assets of the three reservoirs are 76,500,000 yuan, the average apportionment coefficient is 0.86, and the depreciation of fixed assets is 76,500,000*0.95*0.86/50=1,250,000 yuan. The capital used for water-saving supporting renovation in irrigation area is 1,684,790 yuan, and the depreciation of fixed assets is 16,847,900 * 0.95 / 50 = 320,100 yuan; that is, the total depreciation of fixed assets in Fenglinwan Irrigation area is 1,570,100.

4.2.4 Cost of major repairs

In principle, the cost of major repairs of fixed assets is calculated according to 1.4 % of the value of fixed assets after audit [26], considering the allocation coefficient of large and medium-sized reservoirs, so the cost of major repairs is 1,156,900 yuan.

4.2.5 Daily maintenance expenses

Calculation of maintenance cost for channel projects and channel system buildings is based on standard for maintenance quota of water conservancy projects and related provisions [26]. The key project of Fenglinwan irrigation area mainly undertakes agricultural irrigation, and basically has no other function.

In addition, for undertaking the water supply for irrigation, the reservoir also has flood control function. According to standard for water supply price based on water conservancy projects (Trial), the calculation of maintenance costs and agricultural water price should be shared according by the proportion of " Profitable inventory/ (Profitable inventory + flood control storage capacity)". The calculation results are shown in the Table 2 as follow.

Table 2. Daily maintenance cost calculation results table of Fenglinwan Irrigation area project.

Serial number	Project name	Actual cost (yuan)
First	Fenglinwan Irrigation key Project	385,060
Second	Taipingfan Irrigation key Project	209,652
Third	Daheyuan Irrigation key Project	121,191
Four	Water source (reservoir) project	470,104

In summary, according to the relevant standards, the daily maintenance costs for Fenglinwan Irrigation area is 1186000 yuan.

4.2.6 Other expenses

Other expenditures include engineering observation fees, temporary facility fees, and water quality testing fees which are actually incurred in the water supply project. There are no observation facilities in the irrigation area, and no temporary facility fees, mainly for water quality testing fees. The calculated cost of water quality testing is 48400 yuan per year.

4.3 Period cost analysis

Sales expenses refer to the various expenses incurred by agricultural water supply operators in the process of agricultural water supply sales. The sales expenses of Fenglinwan irrigation area should include depreciation fees, material consumption and other operating expenses of the water fee collection department. In order to ensure the full collection of water fees, without changing the current mode of water fee collection, it is still adopted by township, and 22.5 % is proposed as the principle of management fees. According to the current implementation of water price 20 yuan per mu, the effective irrigation area of Fenglinwan irrigation area is 52,800 mu, that is, the annual water charge is 1,056,000 yuan, and the sales cost is 237,600 yuan per year. Travel expenses, office expenses and transportation subsidies are calculated on the basis of the average value of the irrigation area for three years. According to statistics, the Fenglinwan Irrigation area has spent about 228,000 yuan per year in average from 2016 to 2018. The average sharing coefficient is taken as the basis of the calculation cost, that is, 228,000 * 0.86 = 196,100 yuan. By the end of 2018, the total number of employees in Fenglinwan Irrigation area is 41. The management cost of accounting is taken as the management cost of fixed staff, that is, 32*196,100/41=153,300 yuan. Fenglinwan Irrigation area has power generation income, with an average of 644,400 yuan per year from 2016 to 2018. Through the above analysis and operation cost accounting, the running-maintenance cost of Fenglinwan irrigation area is 3,755,800 yuan. Through the above analysis and operation cost accounting, the full cost of Fenglinwan Irrigation area is 5,325,900 yuan.

5. Total water price calculation

According to the calculation, the full cost of Fenglinwan Irrigation area is 5325900 yuan, and the effective irrigated area is 52,800 mu. The full cost water price =532.59/5.28=100.87 yuan per mu. According to the calculation, the running-maintenance cost of Fenglinwan Irrigation area is 5,325,900 yuan, and the effective irrigated area is 52,800 mu. The running-maintenance cost water price =375.58/5.28=71.13 yuan per mu.

6. Conclusions

The following conclusions and suggestions are drawn from the research and analysis of water price calculation in Fenglinwan Irrigation area:

(1) The institutional reform is not in place, and all the reservoir management committees have the phenomenon of overstaffing, especially the Fenglinwan Reservoir Management Committee has the largest number of overstaffing, which directly leads to the extremely low average wage level and greatly affects the enthusiasm of workers;

(2) Water fee is not collected in time, project maintenance funds are less, can't guarantee the benign operation of irrigation project;

(3) The current water price of 20 yuan per mu and the current comprehensive water price of 49.20 yuan mu^{-1} are both lower than the calculated running-maintenance cost water price of 71.13 yuan mu^{-1} and the full cost water price of 100.87 yuan per mu, which cannot meet the basic requirements of operation running-maintenance;

(4) It is suggested that the calculation cost should be reasonably considered when determining the water price of key projects in irrigation areas

References

[1] Fitri A, Hashim R, Abolfathi S, & Abdul Maulud K N 2019 Dynamics of sediment transport and erosion-deposition patterns in the locality of a detached low-crested breakwater on a cohesive coast *Water* **11** (8) pp 1721

[2] Fitri A, Hashim R, Song K I, and Motamedi S 2015. Evaluation of morphodynamic changes in the vicinity of low-crested breakwater on cohesive shore of Carey Island, Malaysia *Coastal Engineering Journal* **57** (04) pp 1550023.

[3] Fitri A, Hashim R, and Motamedi S 2017 Estimation and Validation of Nearshore Current at the Coast of Carey Island, Malaysia *Pertanika Journal of Science & Technology* **25** (3)

[4] Yao L, Li J, Shi S, and Fitri A 2019 Simulation Take-off Angle of a Ski Jump Energy Dissipater. In *IOP Conference Series: Earth and Environmental Science* IOP Publishing **365** (1), pp 012057

[5] Hashim R, Fitri A, Motamedi S, Hashim A M 2013 Modelling of Coastal Hydrodynamic Associated with Coastal Structures: A Reviews *Malaysian. Journal of Science* **32** pp 149-154

[6] Maulud K N A, Fitri A, Mohtar W H M W, Jaafar W S W M, Zuhairi N Z, and Kamarudin M K A 2021 A study of spatial and water quality index during dry and rainy seasons at Kelantan River Basin, Peninsular Malaysia *Arabian Journal of Geosciences* **14** (2) pp 1-19

[7] Fitri A, Maulud K N A, Pratiwi D, Phelia A, Rossi F and Zuhairi N Z 2020 Trend of Water Quality Status In Kelantan River Downstream, Peninsular Malaysia *Jurnal Rekayasa Sipil (JRS-Unand)* **16** (3) pp 178-184

[8] Fitri A, Maulud K N A, Rossi F, Dewantoro F, Harsanto P, and Zuhairi N Z 2021 Spatial and Temporal Distribution of Dissolved Oxygen and Suspended Sediment in Kelantan River Basin In *4th International Conference on Sustainable Innovation 2020–Technology, Engineering and Agriculture (ICoSITEA 2020)* Atlantis Press pp 51-54

[9] Fitri A, Yao L, and Sofawi B 2019 Evaluation of Mangrove Rehabilitation Project at Carey Island Coast, Peninsular Malaysia based on Long Term Geochemical Changes In *IOP Conference Series: Earth and Environmental Science* IOP Publishing **365** (1) pp 012055

[10] Hashim R, Roy C, Shamshirband S, Motamedi S, Fitri A, Petković D, and Song K I 2016 Estimation of wind-driven coastal waves near a mangrove forest using adaptive neuro-fuzzy inference system *Water resources management* **30** (7) pp 2391-2404

[11] Yuan J W 2016 Discussion on the Method of Determining Agricultural Terminal Water Price *Shandong Water Resources* **07** pp 61-62

[12] Fu T X, Deng H L, Xie H W and Chen Y. 2020 Calculation and Analysis of Agricultural Terminal Water Price in Ganfupingyuan Irrigation area *China Rural Water and Hydropower* **05** pp 10-12

[13] Yu G H 2012 *Research on Agricultural Irrigation Water Price in Irrigation area* Zhejiang University

[14] Jia Y 2014 Calculation of Water Supply Cost in Donglei Phase II Chouhuang Irrigation area *Assets and Finances in Administration and Institution* **27** pp 207-208

[15] Liu Y Y 2014 An Example Analysis of Water Price Cost Calculation in Taoqupo Irrigation area *Assets and Finances in Administration and Institution* **27** pp 217-218

[16] Qiu X L, Zhang P C 2019 Study on the Formation Mechanism of Agricultural Water Price in Ningxia Yanghuang Irrigation area *Agricultural Development & Equipments* **08** pp 69-70

[17] Fang D, Yu T and Zhou C X 2019 Study on the Mechanism of Agricultural Water Price Formation in Large and Medium-sized Irrigation areas in Yunnan Province *Water Resources Development Research* **19**(08) pp 24-29

[18] Ren J 2011 Water Price Calculation Method in Ningxia Yinhuang Water-lifting Irrigation area *Water Resources and Power* **29** (06) pp 140-142

[19] Li X R and Wang X B 2019 Analysis of Agricultural Water Price Formation Mechanism in Linquan Irrigation area - Taking High-efficiency Water-saving Irrigation Project as an Example *Journal of Nanchang Institute of Technology* **38** (03) pp 22-27

[20] Li J P and Meng W R 2020 Calculation of Agricultural Water Supply Cost and Sensitivity Analysis of Each Component - Taking Jiangkou Irrigation area as an Example *Agricultural Technology & Equipment* **07** pp 91-95

[21] Du L Y and Zhang X Q 2020 Research on Comprehensive Reform of Agricultural Water Price and Farmers' Affordability - Taking Subei Irrigation area in Jiangsu Province as an Example *Xinjiang State Farms Economy* **09** pp 9-17

[22] Yuan J H 2020 Cost Accounting of Agricultural Water Supply - A Case Study of Gangshi Irrigation area in Chongren County, Jiangxi Province *Agriculture and Technology* **40** (17) pp 59-63

[23] Anhui Water Conservancy Accounting 2013 *Investigation on Water Price of Two Departments of Agriculture in Pishihang Irrigation area of Anhui Province* **01** pp 20-23

[24] Ministry of Water Resources 2008 *Notice on the issuance of the Water Conservancy Project Water Supply Price Accounting Specification (Trial)* Gazette of the ministry of water resources of the people republic of China **01** pp 45-48

[25] Li H and Li J 2009 Water Conservancy Projects Should Extract Depreciation of Fixed Assets *Haihe Water Resources* **05** pp 109-110

[26] Anhui Water Conservancy Accounting 2008 *Specification for Water Supply Price Accounting for Water Conservancy Projects (Trial)* **01** pp 38-42

Acknowledgments

This research is supported by Research Project of Jiangxi Provincial Department of Water Resources (2018-007) and Jiangxi Provincial Department of Water Resources (201921BZKT07).

Numerical simulation of hydraulic optimization for regulating tank in pumping station

Xiaoming Zhu[1*], Sha Shi[1], Jing Si[1], Arniza Fitri[2], Dian Pratiwi[2] and Adelia Agustina[2]

[1] College of Water Conservancy and Ecological Engineering, Nanchang Institute of Technology, Nanchang, 330099, China
[2] Department of Civil Engineering, Faculty of Engineering and Computer Science, Universitas Teknokrat Indonesia, Bandar Lampung, Indonesia

*Corresponding E-mail: xmzhu@nit.edu.cn

Abstract. Based on the governing equations of steady incompressible fluid, renormalization group (RNG) turbulence model and SIMPLEC algorithm are used to calculate the steady flow field of regulating tank in the pumping station with six different geometries operating under same condition. The impacts of the layout schemes of guide walls for the flow field of the regulating tank are analyzed. The numerical results are verified by physical model experiment and good agreement is found. The results show that: 1) serious flow separation of side wall will occur in the regulating tank when the interval of diversion wall is 10 L; 2) the flow velocity in the regulating tank will be too low when the diversion wall spacing is 16 L; 3) the improvement of the flow pattern of the regulating tank is not obvious; and the project cost is increased when the excavation depth of the regulating tank is increased by 1 m; 4) the bottom velocity reached the non-silting velocity and the head loss of the regulating tank reducing nearly 1.2 m by using arrangement form of wide 21 L and narrow 10L of the guide walls, which provides a certain guarantee for the safe operation of the pumping station. The regulation tank layout scheme proposed in the paper can be applied to engineering practice.

Keywords: Guide wall, Numerical simulation, pumping station, Regulating tank

Track Name: Land, Water, Forests and Food Security

1. Introduction

Regulating tank is widely used in water supply and drainage system, and its application varies in different fields. As clear water supply is the main priority in our life [1-5], water tower and clean water tank in the water supply system; and storage tank in the building water supply system are broadly discussing [6-7]. All water tanks are categorized as regulating pool [7-9].

A pumping station in Shanghai does not belong to the hydraulic structure of water supply and drainage system, but it is an engineering design which includes regulating tank. There are two main functions of a regulating tank. Firstly, it is an important measure to ensure uninterrupted raw water within a certain period of time when carrying water from the Huangpu River. Since water usage may vary from day to night and from day to day, the regulating pool/tank can improve the safety of water supply. Secondly, the regulating pool/tank together with diversion wall can be set up to ensure no sediment deposition in the water transport buildings (it is because sediment and other impurities in the water can flow from the water source), so that, the flow of the channel reaches the non-silt flow rate.

In recent years, with the development of computer technology and computational fluid dynamics theory, hydrodynamic model and Computational Fluid Dynamics (CFD) model are becoming an important tool to understand the flow status of water [10-15]. COD model has been applied by many researchers and achieved many certain results [10-12]. Research on the regulating tank mainly focuses on the design [16-18], and does not consider the influence of the flow pattern inside the regulating tank. In order to ensure the safe and stable operation of the pumping station, a reasonable layout of the diversion wall is very important. This study tries to find a reasonable layout of the diversion wall through numerical test. Therefore, this study applied the computational fluid dynamics software ANSYS Fluent and RNG k-ε turbulence model to carry out CFD calculation on the whole model of a Shanghai pumping station with multiple diversion wall arrangement schemes under typical water level conditions. A reasonable diversion wall layout which flows pattern is optimal and hydraulic loss is minimum will be sought by comparing and analyzing the change of water flow pattern and hydraulic loss between different diversion wall layout schemes. This paper can provide some reference for the optimization design of similar project.

2. Materials and Methods

2.1 Calculation model and schemes

In this study, a pumping station in Shanghai is considered as the study area. The calculation model which includes the water distribution pipe, regulating tank, forebay, water inlet tank and water pump suction pipe as shown in Figure 1. In order to fully analyze the influence of the layout of diversion wall on the distribution of flow pattern and the change of head loss, six different layout forms of diversion wall are selected for calculation under typical water level conditions. Specifically, a number of diversion walls are arranged in the regulating pool to form a preliminary design scheme (scheme 1). Scheme 2 appropriately widens the flow channel on the basis of the scheme one. Scheme 3 adjusts the flow direction according to the scheme two. Scheme 4 enlarges the excavation depth base on the scheme three. Scheme 5 improves the flow pattern in the dead water area on the basis of the scheme 4, and scheme 6 further adjusts the size and spacing of diversion wall. The six layout plans of diversion walls are shown in Figure 2.

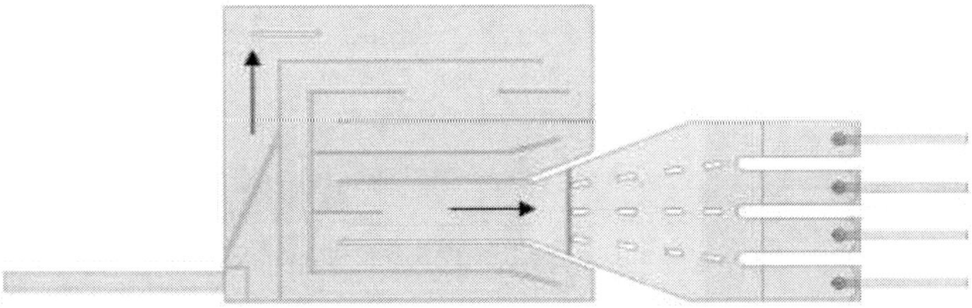

Figure 1. Geometric model of the pumping station.

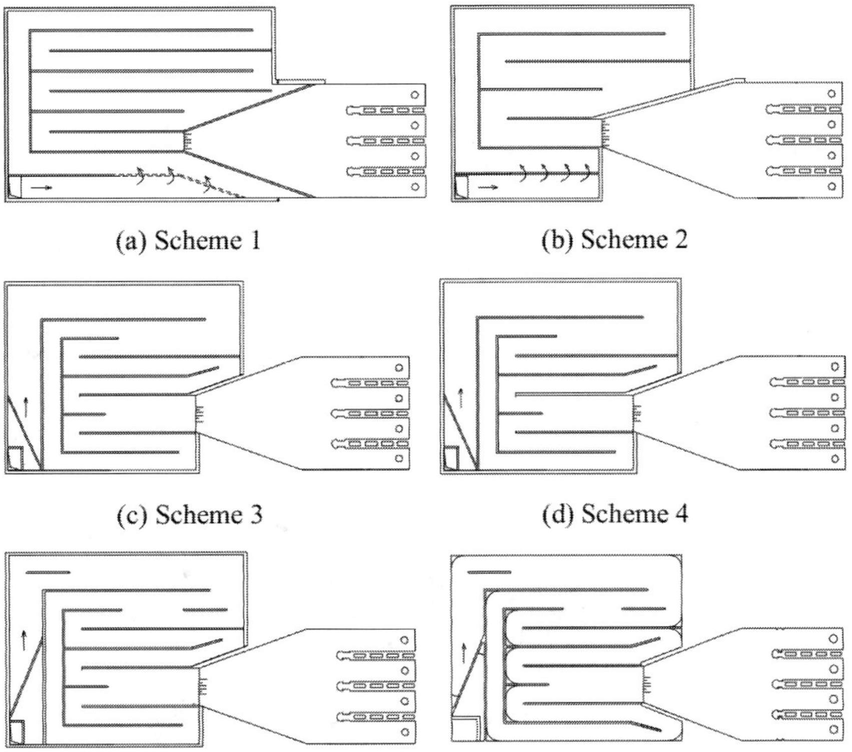

(a) Scheme 1 (b) Scheme 2

(c) Scheme 3 (d) Scheme 4

(e) Scheme 5 (f) Scheme 6

Figure 2. Six layout schemes of diversion walls

2.2 Calculation method and grid division

The flow in the regulating tank is turbulent and the flow is quite complex. However, under the assumption of continuous medium, the motion of fluid can be described by continuous equation and Navier-Stokes equation.

The equation of continuity:

$$\frac{\partial \boldsymbol{u}_i}{\partial x_i} = 0 \qquad (1)$$

And the equation of N-S:

$$\rho\frac{\partial \boldsymbol{u}_i}{\partial t}+\rho \boldsymbol{u}_j\frac{\partial \boldsymbol{u}_i}{\partial x_j}=\rho \boldsymbol{f}_i-\frac{\partial p}{\partial x_i}+\mu\frac{\partial^2 \boldsymbol{u}_i}{\partial x_j \partial x_j} \quad (2)$$

In the formula, $\boldsymbol{u}$ is the fluid velocity vector; ρ is the fluid density; f is the mass force; p is the pressure; μ is the turbulent viscosity.

It is generally considered that water flow is incompressible fluid and the heat exchange capacity is very small, so the energy conservation equation is not considered. The Reynolds averaged N-S equation is used to describe the turbulent motion, which is not close. It is necessary to introduce turbulence model for understanding the equations. In this paper, RNG k-ε model is used to simulate the flow characteristics in the computational domain.

The fluid domain mainly includes water distribution pipe, regulating tank, forebay, inlet channel and outlet pipe. In numerical calculation, grid is not only an indirect expression of geometric model, but also an important carrier of numerical calculation and analysis. The quality of grid directly affects the accuracy and efficiency of calculation, as well as the correctness and reliability of calculation results. There are many flow diversion walls in the whole channel of the regulating pool, and the geometric layout is complex. Considering the complexity of the layout of the regulating tank, the unstructured tetrahedral grid with strong geometric adaptability is divided by using the grid generation software ANSYS ICEM to locally encrypt the grid at the turning of the water flow, and the overall grid is shown in Figure 3. According to the analysis of grid independence, the total number of grid elements in the flow calculation domain of the whole midway pumping station is 9.91 million.

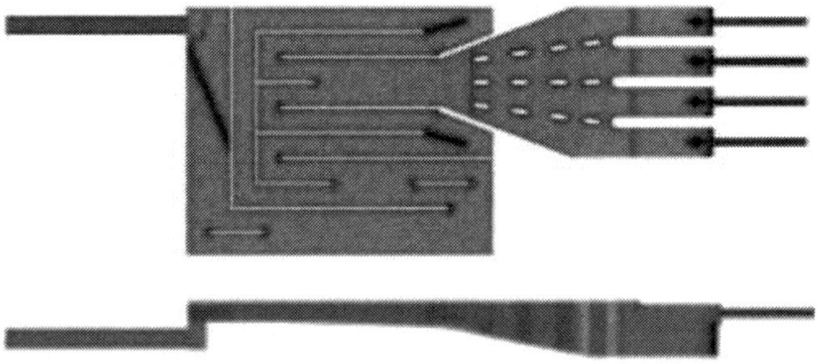

Figure 3. Mesh division of the pump station model.

2.3 Discretization of equations and boundary conditions

In the numerical simulation of a pumping station model in Shanghai, the water speed at the entrance area is adopted at the entrance boundary while the size and direction of the speeds are given. The outflow condition is adopted at the exit boundary, in other words, the gradient of each transport variable on the exit section is zero. For free/open water surface, it is assumed to be free stress boundary. The control equation is discretized by the finite volume method. Second order central difference scheme is adopted for diffusion term, and second order upwind scheme is adopted for convection term. The coupling of pressure and velocity adopts SIMPLEC algorithm.

3. Results and Discussions

3.1 Analysis of Flow Distribution

The geometric model of a pumping station in Shanghai as shown in Figure 1 is simulated using three-dimensional turbulence model. The calculated water level is 3.8m, and the starting combinations are 1# 2# and 4#. Figure 2 (a) to Figure 2 (f) are selected for the arrangement forms of the regulating tank. The spatial streamline distribution diagrams of the six schemes correspondingly are shown in Figure 4 (a) to Figure 4 (f). Based on Figure 4, we can demonstrate that there is an obvious side-wall flow phenomenon in Scheme 1, which can easily cause serious siltation (Figure 4 (a)). The water flow in the inlet flume at the front of the pump station is obviously skewed to one side, thus inducing the deterioration of the water flow pattern in the forebay of the pumping station. As shown in Figure 4 (b), with the increase of the wall channel width, the water flow pattern does not change substantially. The phenomenon of off-flow and backflow in the inner wall of the channel is more obvious, and even a small area of vortex appears. There is still serious flow deviation at the inlet of the forepool. In Figure 4 (c) and Figure 4 (d), the flow pattern in the regulating tank is roughly similar, and the flow pattern at the overflow weir is relatively good, followed by serious side-wall outflow. In Figure 4 (e), the flow pattern in the regulating tank is improved, and the initial flow pattern of the front-end channel in the front pool is good, but the flow pattern distribution becomes worse when it enters the forebay. In Figure 4 (f), the spatial streamline distribution is good, and the side-wall off-flow phenomenon is significantly improved, especially the water flow in the forebay is in the center, which is helpful to optimize the water flow pattern in the pump sump of the pumping station.

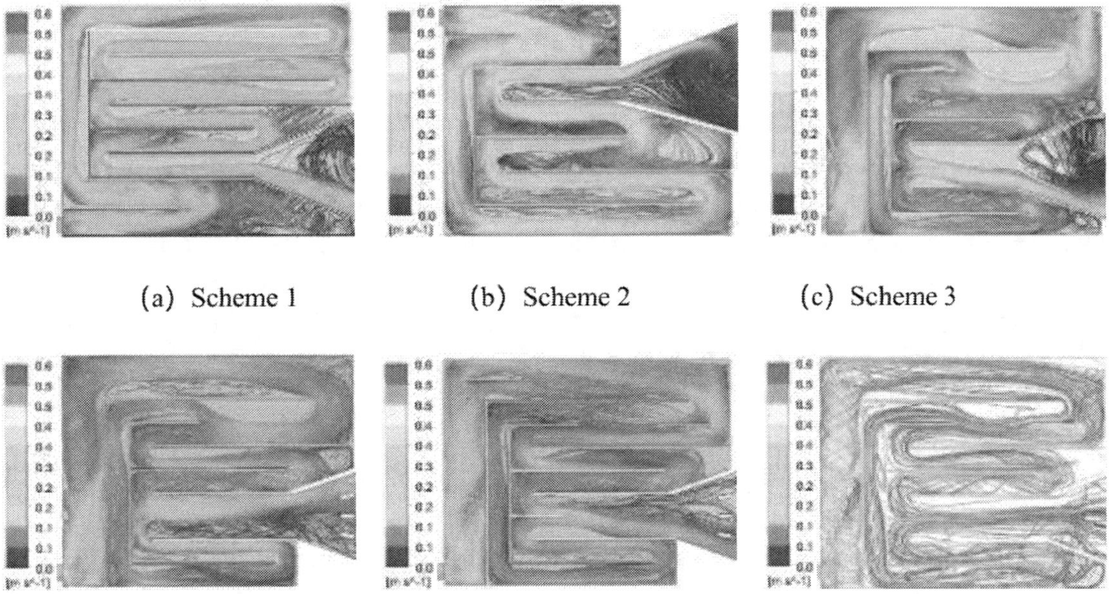

(a) Scheme 1 (b) Scheme 2 (c) Scheme 3

(d) Scheme 4 (e) Scheme 5 (f) Scheme 6

Figure 4. Spatial streamline distribution diagram of six schemes.

Figure 5. show diagrams of the velocity distributions at the bottom area for six schemes (a) to (f). It can be seen that the bottom velocity of waters in Figure 5 (a) to Figure 5 (e) reaches the non-silting velocity. However, there are various degrees of water bias in the front channel of the forepool, which is not conducive to the efficient and stable operation of the pumping station. In Figure 5 (f), the numerical results show that layout of Scheme 6 effectively homogenizes the flow distribution of the channel and improves the flow pattern of the regulating tank.

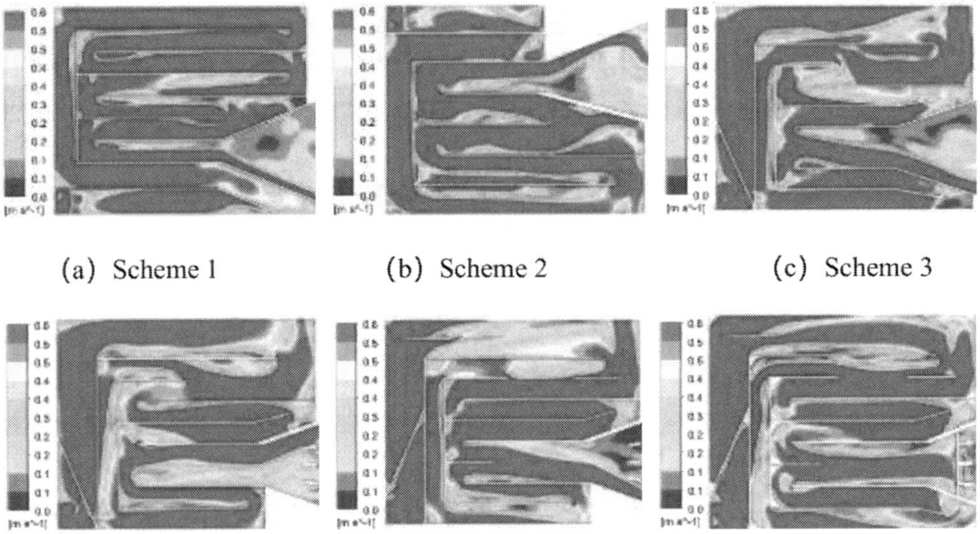

(a) Scheme 1 (b) Scheme 2 (c) Scheme 3

(d) Scheme 4 (e) Scheme 5 (f) Scheme 6

Figure 5. Cloud map of bottom velocity distribution of six schemes.

3.2 Analysis of Hydraulic Loss

In order to verify the accuracy of the numerical simulation study on the flow state of the regulating pond and the hydraulic loss of the inlet system, a physical model test is carried out for optimization scheme of the numerical simulation calculation as shown in Figure 6. The whole normal hydraulic model is used in the model test of the pumping station, which is designed according to the gravity similarity criterion. The model includes the inlet pipe, regulating tank, forebay, sump and outlet pipe. Considering the requirements of the model water flow in the resistance square region and the selection of the model pump, the model linear scale $\lambda_l=10$ is intended to be selected. According to the test and research requirements of the pumping station, the water pump with similar geometry and flow suitable for the simulation requirements is used as the test pump.

After investigation and comparison in many aspects, the 150ZLDB vertical mixed flow pump is proposed to be used as the test pump, and the water flow rate of the pump is adjusted through the gate valve on the outlet pipe, so as to meet the test requirements. The regulating tank of the prototype are made of reinforced concrete which roughness is 0.013~0.014, and corresponding roughness of the model is 0.009~0.010. The side wall of the model bottom plate is finely painted with pure cement, and its roughness can reach 0.010. The pool partition and the inlet pool are made of plexiglass and plastic plates, with a roughness of about 0.009, and the model meets the requirements of similar roughness.

Figure 6. Physical test model of the pumping station.

Physical model tests were carried out on the above six diversion wall layout schemes of regulating tank, and the head losses of the six-diversion wall physical models were collected, counted and converted. The head losses were compared with those collected in numerical simulation results, as shown in Figure 7. Based in Figure 7, both curves show a decreasing trend and tend to overlap, indicating that the numerical simulation results are reliable. Among the six schemes, scheme 6 has the lowest hydraulic loss, which is nearly 1.20 m lower than scheme 1, indicating that the reasonable layout of diversion wall can not only optimize the flow pattern, but also effectively reduce the hydraulic loss.

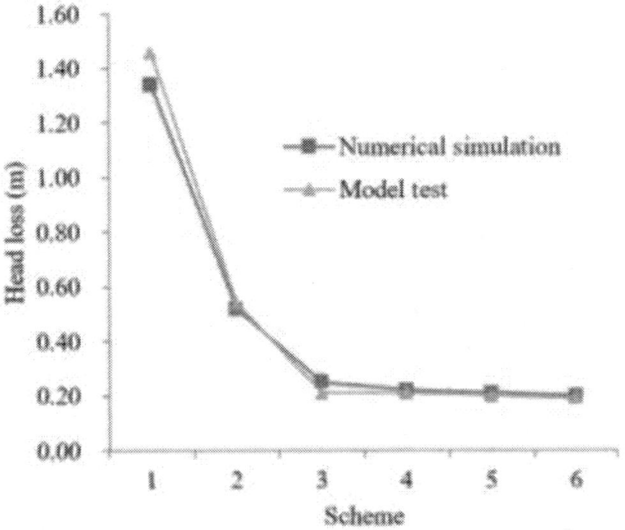

Figure 7. Comparison of head loss of each scheme in numerical model and physical model.

4. Conclusion

Based on the governing equation of steady incompressible fluid and the renormalization group turbulence model, this study calculates the three-dimensional flow in the regulating tank of the pumping station by using the SIMPLEC algorithm. Further, this study compares and analyzes the flow patterns of the regulating tank under different diversion wall for arrangement schemes. The calculation results show that optimizing the size spacing and layout direction of the diversion wall can effectively homogenize the flow distribution, improve the flow pattern, and reduce the hydraulic loss in the regulating tank. In this paper, scheme 6 is determined as the best diversion wall layout scheme, which can be a certain reference for engineering construction. This paper only for a pumping station in Shanghai. A reasonable arrangement of regulating pool diversion wall is provided, and the improvement measures for side-wall off-flow and flow deviation are relatively simple, and the search for better universal and diversified improvement measures remains to be studied.

References

[1] Maulud K N A, Fitri A, Mohtar W H M W, Jaafar W S W M, Zuhairi N Z, and Kamarudin M K A 2021 A study of spatial and water quality index during dry and rainy seasons at Kelantan River Basin, Peninsular Malaysia *Arabian Journal of Geosciences* **14** (2) pp 1-19

[2] Fitri A, Maulud K N A, Pratiwi D, Phelia A, Rossi F and Zuhairi N Z 2020 Trend of Water Quality Status In Kelantan River Downstream, Peninsular Malaysia *Jurnal Rekayasa Sipil (JRS-Unand)* **16** (3) pp 178-184

[3] Fitri A, Maulud K N A, Rossi F, Dewantoro F, Harsanto P, and Zuhairi N Z 2021 Spatial and Temporal Distribution of Dissolved Oxygen and Suspended Sediment in Kelantan River Basin In *4th International Conference on Sustainable Innovation 2020–Technology, Engineering and Agriculture (ICoSITEA 2020)* Atlantis Press pp 51-54

[4] Fitri A, Hashim R, and Motamedi S 2017 Estimation and Validation of Nearshore Current at the Coast of Carey Island, Malaysia *Pertanika Journal of Science & Technology 25*(3)

[5] Fitri A, Yao L, and Sofawi B 2019 Evaluation of Mangrove Rehabilitation Project at Carey Island Coast, Peninsular Malaysia based on Long Term Geochemical Changes *In IOP Conference Series: Earth and Environmental Science IOP Publishing* **365**(1) pp 012055

[6] Meng J, Zhang R and Meng J 2011 Probe into the function and design of the adjusting tank *Sci-Tech Inromation Development & Economy* **21**(12) pp 173-175.

[7] Wang L and Li H 2013 Design of equalization tank in Xihe raw water pumping station: Auxiliary project of middle route of south-to-north water divesion in Tianjin *China Water & Wastewater* **29** (16) 52-53

[8] Yao L, Li J, Shi S, and Fitri A 2019 Simulation Take-off Angle of a Ski Jump Energy Dissipater. *In IOP Conference Series: Earth and Environmental Science IOP Publishing* **365** (1) 012057

[9] Chen H, Yao L, and Fitri A 2019 The Influence Mechanism Research of Inflow Temperature in Different Time Scale on the Water Temperature Structure *In IOP Conference Series: Earth and Environmental Science IOP* Publishing **365** (1) 012058

[10] Cheng L, Liu C and Zhou J 2009 Numerical simulation on geometrical parameters for open pump sump *Transations of the Chinese Society for Agricultural Machinery* **40** (1) pp 50-55

[11] Yu Y and Cheng B 2012 CFD simulation and optimization on inflow patttern of diversion and intak pumping stations with side-inlets *Water Resources and Hydropower Engineering* **43** (2) pp 72-75

[12] Wang W 2013 *Numerical simulation of flow in open-style intake pool of irrigation and drainage combinated multi-function pumping stations* Yangzhou University

[13] Fitri A, Hashim R, Abolfathi S, & Abdul Maulud K N 2019 Dynamics of sediment transport and erosion-deposition patterns in the locality of a detached low-crested breakwater on a cohesive coast *Water* **11** (8) pp 1721

[14] Hashim R, Roy C, Shamshirband S, Motamedi S, Fitri A, Petković D, and Song K I 2016 Estimation of wind-driven coastal waves near a mangrove forest using adaptive neuro-fuzzy inference system *Water resources management* **30** (7) pp 2391-2404

[15] Hashim R, Fitri A, Motamedi S, Hashim A M 2013 Modelling of Coastal Hydrodynamic Associated with Coastal Structures: A Reviews *Malaysian Journal of Science* **32** pp 149-154

[16] Luo H and Han H 2015 Design principle and application of non-power regulating tank *Water&Wastewater Engineering* **41**(6) pp 40-42

[17] Han W, Li X and Yu C 2015 Discussion on design of adjusting tank for small industrial wastewater treatment station *Industrial water & Wastewater* **46** (2) pp 33-35

[18] Lou S, Yuan W and Zhang J 2014 Design of large-scale regional sewage transfer pump station and regulating pool engineering *Industrial water & Wastewater* **40** (11) pp 39-41.

Acknowledgments

This research is supported by the Science and Technology Research Project of the Department of Education in Jiangxi Province (Grant No. GJJ180953), the Belt and Road Special Foundation of the State Key Laboratory of Hydrology-Water Resources and Hydraulic Engineering (Grant No. 2018492011) and the Science and Technology Research Project of the Science and Technology Department of Jiangxi Province (Grant No. 20192ACBL20041).

River-flood forecasting methods: the context of the Kelantan River in Malaysia

M Abdul Majid[1*], M Hafidz Omar[2] M Salmi M Noorani [3]and F Abdul Razak[4]

[1,3&4] School of Mathematical Sciences, FST, Universiti Kebangsaan Malaysia
[2]Dept. of Mathematics & Statistics, KFUPM, Dhahran, Saudi Arabia

* Corresponding E-mail: abdulmajid_2000@yahoo.com

Abstract. River-flood forecasting is among the most important feasible non-structural approaches used in reducing economic losses and alleviating human sufferings. In spite of uncertainty in the forecasting of natural disasters, the current prevailing methods developed in many parts of the world in the recent history has made good progress to a great extent. The advancement is attributed mainly due to the availability of high-resolution weather data and the use of sophisticated computer modelling algorithms. However, it is desirable to conduct exploratory review studies to further improving the current state of affairs. The present paper reviews briefly the river-flood forecasting methods currently used worldwide with a specific focus in the context of the Kelantan River in Malaysia. Flooding in Malaysia is recurrent covering a large inhabited area compared with other natural disasters. Some of the popularly used methods in the literature such as statistical methods machine learning and methods based on chaos theory have been reviewed, The paper will also attempt to explore the future direction for research and development that might be useful specifically for dealing with the recurrent rivers flooding in Malaysia. A reasonably acceptable prediction of river streamflow is significantly important in disaster management and water resources management.

Keywords: Rivers flooding, forecasting methods, review, Kelantan River Malaysia

Track Name: Land, Water, forest and food security

1. Introduction

The Inter-Governmental Panel for Climate Change (IPCC) [1] and the United Nations Office for Disaster Risk Reduction, (UNISDR) [2] reports state that flooding is ranked as one of the most frequent, destructive and damaging natural disasters. The catastrophic disasters that have been reported between 1990 and 2019 make up about 90 % out of 1000 [3]. It further states that, because of these disasters, only in the year 2018 about 6500 have lost their lives reportedly and the livelihoods of over 57 million people have been affected with an economic loss of over US$ 140 billion worldwide. Floods record 43 % of most registered events between 1994 and 2013 consequently affecting about 2.5 billion people [4]. It is believed that such occurrences are on a steady rise [5] and in Malaysia this is happening mainly because of the annual monsoon rainfall. There is an increasing trend of several flooding in tropical regions. Malaysia's location near the equator mainly makes its climate hot and humid throughout the year. It has experienced several floods recorded as far back as late 1980s. The floods in Malaysia are a regular phenomenon and they occur during monsoon seasons. They affect the socio-economic structure immensely affecting a large inhabitants and disruption to normal life. In spite of several mitigation measures put in place by the government of Malaysia, the country has witnessed a long history of this natural disaster. No country is safe from extreme natural events. There is no foolproof method for completely avoiding it no matter how sophisticated and advanced tools and systems we develop for controlling it and the current measures are only partially successful.

Floods are classified as river floods, mud floods, dam floods and flash floods etc. A river flooding mainly causes due to heavy incessant rainfall over a short period exceeding the capacity of soil absorption. A flood is caused by above normal streamflow, leading to inundation of areas, which are normally not covered by water, as it is mainly restricted to abrupt change in high streamflow levels. The floods in Kelantan basin, occur in December frequently and with a high magnitude. Kelantan, northeastern region of Peninsular Malaysia, has an area of 15,099 km^2 and surrounded by Thailand, Perak, Pahang and Terengganu and a map of the river is given as figure 1 [6] and its characteristics in Table 1.

Table 1. Characteristics of Kelantan River.

Length	Catchment Area	Land use tributaries	Observation Station	Mountain (peak)
248 km	11,900 km^2	Galas River, Lebir River	Guillemard Bridge	Korbu 2183 m

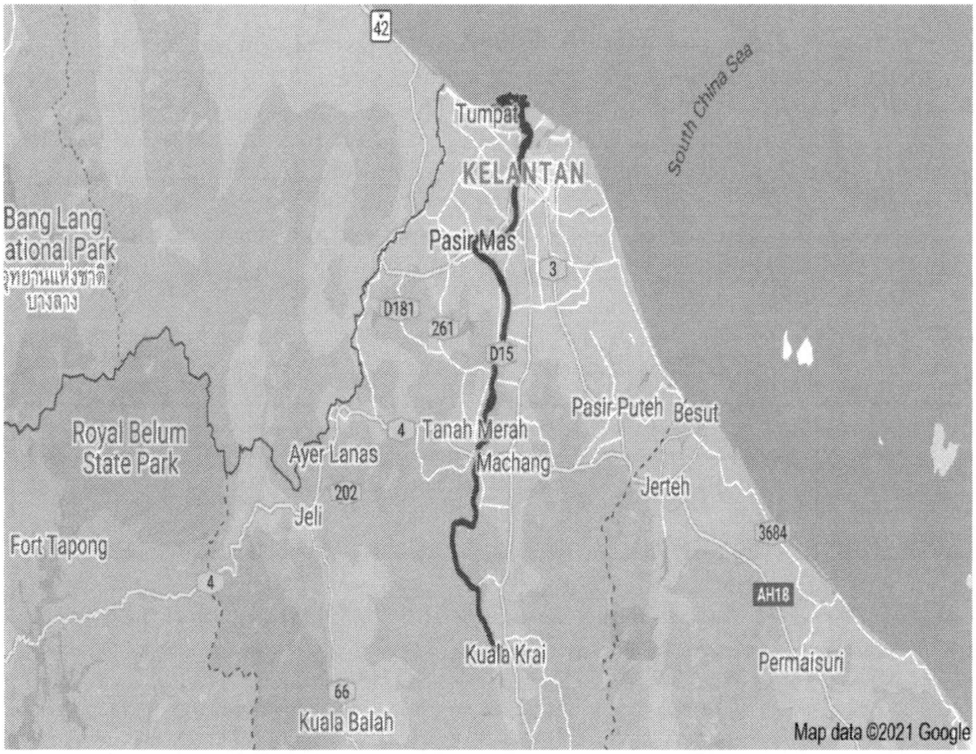

Figure 1. Map of Malaysia and location of Kelantan river.
https://www.google.com/search?q=kelantan+river+in+malaysia+map&oq=kelantan+river+in+malaysia+map&aqs=chrome..69i57j69i60l2.10496j0j7&sourceid=chrome&ie=UTF-8 [6]

River flow forecasting plays a vital role in controlling extreme hydrological events. Forecasting is an estimation and hence it is quite a tricky endeavor as it involves modelling errors, with a lot of uncertainty and nonlinearities involved in the process. The purpose of developing a flood early warning detection system is to reduce the risks and dangers and to prevent from the hazards of flooding. The real time precipitation data [7] and [8] and numerical prediction models [9] are considered two important factors in the development of flood early warning system. Moreover [10] has emphasized the importance of flood-disaster management actions and identified the development of an effective early warning system as an important milestone for the vulnerable world population by 2030.

The predictive hydrological and hydraulic modelling of the river flooding, though highly complex, but it is considered largely feasible with the current computational tools such as digital elevation models (DEM), and geographic information system (GIS), as they are believed to provide high accuracy in improving flood forecasting [11]. The non-structural early warning and alert systems can be developed based on forecasting of a river flow. In this study, a review of studies is undertaken on various forecasting methods and the focus is the Kelantan River only.

1.1. Models' Classification

The scientific studies of extreme hydrological events and natural disasters such as floods pave a long way in understanding the complex dynamics of these unusual occurrences and they are important to minimize the losses and maximize the possible benefits. It is well known that the complex behavior of natural systems cannot be accurately understood through laboratory experiments or modelling based on historical data. Nevertheless, scientists from diverse disciplines have been conducting numerous studies and there is an ever-increasing interest to have a proper control of such extreme events. Moradkhani and Sorooshian [12] describe a model as a simplified representation of real-world system. One important application of the river flow analysis and forecasts is the development of an early warning to floods.

Devia et al. [13] have classified hydrological models as empirical, conceptual and physically based. According to [14], the conceptual models take into consideration a priori structure of the model based on the perception of a hydrologist and the parameters are calibrated against the observed time series, while the physics-based models derive parameters from the field measurements and the identification of parameters must be done through inverse modelling approach. As non-structural measures for the mitigation of flooding, [15] proposed a classification of model structures and types that includes deterministic models as physics-based, conceptual and empirical and stochastic models as depicted in Figure 2. Table 2 and Table 3 use the classification as physical models and conceptual models following [16] for rivers in Malaysia.

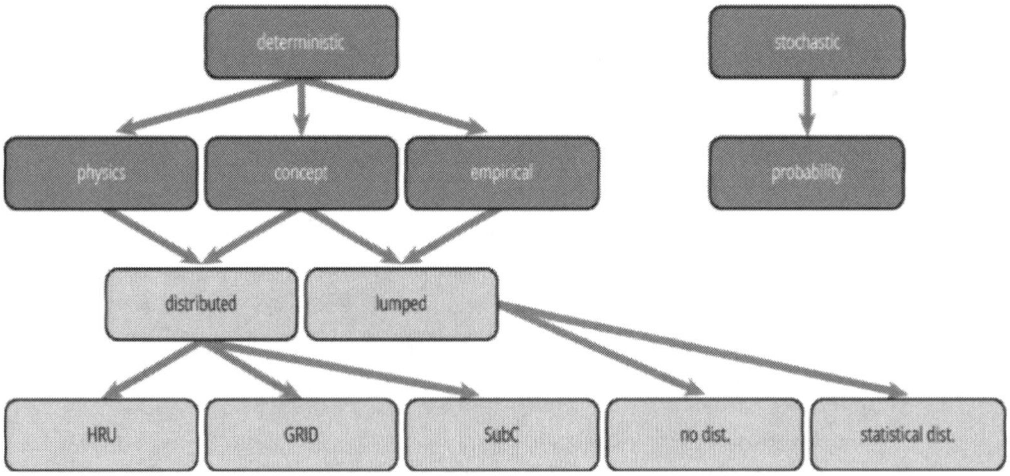

Figure 2: Classification of Model Structures and types [15]

2. Methodology

The resources that were used to search for the relevant articles for review have been accessed through the UKM's online library included Scopus, web of science, IEEE, Wiley, ProQuest, Science Direct and Taylor and Francis. These resources provided e-books, conference proceedings such as the proceedings of AICCE'19, AWAM international conference on civil engineering and ISFRAM 2014 Proceedings of the International Symposium on Flood Research and Management etc. apart from the Google Scholar and Springer Link. The search was restricted to flood forecasting methods in the context of Kelantan River during 1990 to 2021 only with a brief review of related studies in general worldwide.

The objective of this study is to review all the studies related to flood forecasting of Kelantan River, Malaysia with a brief overview of the studies conducted worldwide using methods in statistics, chaos theory and machine learning etc. The other objective is to highlight the study conducted by the authors [17] as a recent development. Finally, some future direction in research work will be provided.

3. Review of hydrological forecasting Studies

During the past few decades, the hydrological studies largely focused on the statistical methods such as ARIMA [18], SARIMA [19] including the river flow forecasting. The application of statistical methods usually establish linear and nonlinear relationships of variables using historical hydrological data. Apart from the statistical methods, several machine learning are also applied for river flood analysis and forecasting. The study of the dynamics of a river streamflow, inherently nonlinear in nature, requires the use of appropriate sophisticated methods. With the increased availability of hydrological data and the affordability of the computing capabilities, one of the popularly emerging approaches is the use of machine learning; rooted in artificial (machine) intelligence. [20] reported that these developments in data and computing could lead to changes in flood risk and impact assessment. The machine learning algorithms work on the principles of recognizing patterns in input data. These algorithms are currently used in forecasting of the nonlinear systems and there is a paradigm shift in modelling from physical based processes to the optimized data-driven modelling. Some of the known machine learning methods mentioned are Radial Basis Functions, Generalized Regression Neural Networks, Neighbor regression, CART regression trees, Bayesian Neural Network, Support Vector Regression and Gaussian Processes and Multi-Layer Perceptron, [21]. The ANN models work on historical data in contrast with the models requiring calculation of physical characteristics of a catchment. In spite of the ANN being a good potential candidate for forecasting over several statistical methods it also has some drawbacks. Some of the drawbacks include modeling overfitting error, gradient descent, slower rate of convergence, and high training data restriction. For a comparison between ANN and AR models, readers can refer to [22], [23] and [24], between ANNs and ARMA [25], [26] and [27]. The comparison revealed that the performance of ANN was superior to that of the statistical techniques. In a study by [28], the performance of SVR (support vector regression) model to predict long-term monthly flow discharge series was better than the performance of ARMA. The SVR also outperformed ANN, [29]. The short-term streamflow prediction by the phase space reconstruction method was reportedly better when compared with ANN, [30]. They have applied a combination of ARMA, KNN, ANN, MA-ANN, Phase Space Reconstruction and FNN to hydrologic time series prediction. The phase space reconstruction method and the recurrence analysis method derived from the chaos theory, based on the detection of a suitable embedding dimension and time delay, are also used as a forecasting method, [17].

Ttilford et al. [32] highlighted the need for a number of factors to consider when selecting a new flood forecasting solution for a catchment or site which includes the physical characteristics of the catchment and river flow; the quality and the availability of related data; the cost and time of development; technical and economic risks associated with the investment. Furthermore, the authors have proposed that the forecasting model must meet national targets for accuracy and lead-time.

3.1. Kelantan River flow analysis and forecasting

Kelantan historically has witnessed severe floods of varying magnitude, often times, affecting its adjoining states such as Terengganu and Pahang etc. The flooding event that occurred in late December of 2014 has particularly caused massive destruction and [33] have described it as the largest ever recorded in the history of Malaysia. According to [34] report, the flood of 2014 has been the most significant largest ever recorded and termed as "tsunami-like disaster" in the history of Kelantan.

According to [35], between 14-18, December 2014, the downstream of the Kelantan river was poured and inundated due to incessant heavy rainfall. This was followed by a second spell between 21-24, December 2014 and this time it was the upstream of the river. The soil saturation caused a rapid runoff and aggravated the flooding phenomenon and it was more severe in the Kuala Krai valley. The lives of the thousands of the residents were affected and with some fatalities. This clearly demands a thorough scientific understanding of the dangers and coming up with necessary measures to mitigate and control the dangers for the ever-growing population in the area through urbanization. Some of the possible non-structural and cost-effective measures are the development of an effective functional flood early warning system and the forecast modeling. [36] have applied stochastic flood forecasting model for Kelantan River.

Abdulkareem et al. [16] provides a comprehensive review study on the forecasting models in Malaysia. The report cited the studies focused on Kelantan River comprise about a quarter of the total studies conducted in Malaysia until 2018. The authors have classified models as physics-based and conceptual models. Our paper focuses only on the studies related to the Kelantan River due to the recurrent nature of flooding and its disastrous consequences. The lists of studies related to Kelantan River have been provided in Table 2 and Table 3. The modelling performance of the reviewed studies provided in tabular forms were mainly based on the values of R, R^2, NSE, RMSE, MAE and PBIAS among which RMSE is notably a popular common criterion.

Table 2. Physics-based Models for Kelantan River.

Model	**Research Paper**	**Model Performance criteria**	**Study**
HECHMS	[37]		To study the peak discharge variation and their factors related to discharge and runoff volume or different return period.
HECHMS	[38]		To look into the causes of flooding through the study of relationship between morphometric characteristics and the hydrological factors.
HECHMS	[39]	Coefficient of determination (R2), NSE, percentage bias (PBIAS), mean absolute error MAE	To study variations in peak flows
[HEC-HMS and Transient Rainfall Infiltration and Grid-Based Regional Slope Stability Analysis (TRIGRS)]	[40]		To assess the effect of Land use/land cover (LULU) change.
HECHMS	[41]	RMSE, MAE, NSE, R	To study the effects of environmental and ecological factors of watershed development and to assess flood prone areas during extreme storm events.
(HEC-HMS and Improved TRIGRS model)	[42]		To study the infiltration rate and governing equation related to land cover sensitivity maps and run off volume.
Integrated flood analysis system (IFAS)	[43]	Coefficient of efficiency (Ec) and peak discharge error (Ep)	To determine the applicability of IFAS model.

Model	Research Paper	Model Performance criteria	Study
HECHMS	[44]		To study the conditions and the type of land use and land cover on peak discharge and the volume of the runoff as well as the prominent factor resulting in maximum flows.
HECHMS	[45]	RMSE, R	To determine the influence of extreme rainfall on the performance of HEC-HMS model.

Table 3. Conceptual/Empirical Models for Kelantan River.

Model	Research Paper	Model Performance criteria	Study
Fuzzy logic	[46]	MAE, Relative Root Mean Square Error (RRMSE), NSE and R	To study the real time river water level forecasting as a flood early warning tool. .
Time Lag Forward Network	[47]	MSE, RMSE	To study forecasting of water level hourly.
Empirical model (ANN)	[48]	MSE	To forecast flood water level.
Empirical model [Numerical weather prediction (NWP)]	[49]	R	To effectively predict flood using NWP.
Empirical model [Support vector machine (SVM), frequency ratio (FR) and decision tree (DT)]	[50]		To study the flood susceptibility mapping with the comparison of the performances of the models used.
Empirical model [rule-based decision tree (DT), FR and logistic regression (LR)]	[51]	FR, LR	To compare the ability of DT and the combined use of FR and LR in flood susceptibility mapping
Conceptual model (Probability distributed model) Physically-based model (Kinematic wave model)	[52]		To produce a flood susceptibility map.
Empirical model (Logistic regression)	[53]		To produce flood susceptible map of flood prone areas with the aid of a statistical model and GIS.
Empirical model (Logistic regression)	[54]	Flood susceptibility index	To map out maximum flood prone area with the aid of multiple regression model using GIS techniques and remote sensing.
Neural network autoregressive exogenous input (NNARX) model	[55]	MSE	to predict floods and comparison with wavelet decomposition

Tables 2 & 3 provide lists of hydrological modeling methods that have been applied to various rivers of Malaysia highlighting the purpose of the study. Some of the studies have used only one method and others have used a combination of methods. One of the limitations that we found in all the studies that were focused on the Kelantan River forecasting is a very important aspect of the forecasting that is the floods early warning detection system.

Among several methods, Auto-regressive integrated moving average (ARIMA) model was reportedly used in only 8 % of the research studies, [16] in the context of Malaysia. Based on the review of previous studies of the use of forecasting models in hydrology, it can be seen that the method of recurrence plot and recurrence quantification analysis in combination with ARIMA has not been explored except the work carried out by the authors, [17]. Furthermore, we have not found any study in which ARIMA has been used as a forecasting method for Kelantan River. We believe that this is the second limitation observed, as a result, of the review of the related studies.

3.2. Recurrence analysis method and ARIMA method for Kelantan River flow forecasting

Firstly, an analysis of the time series data for the years 2000 to 2014 was carried out using recurrence plot, recurrence quantification and ARIMA to understand the dynamic behavior of the river flow and the indicators of early warning signs. The forecasting capabilities of these methods were also used including the highlights of the floods early warning detection indicators. [56] have put forth the concept of a recurrence plot as a qualitative visualization tool for detecting the recurrences in a dynamical system and later have provided an objective analysis through the technique known as recurrence quantification analysis. [58] have extensively applied the recurrence analysis method successfully for various dynamical systems. This technique was used for the analysis and forecasting of Kelantan river time-series data for 2000-2014 and for more details of the study, readers may refer to [17]. In addition, ARIMA model was also used for forecasting the flooding by using the year 2014 data as a holdout data. A comparison of the performance of forecasting models was also carried out between recurrence method and ARIMA. [17] provide details of the modeling information utilized. Models' performance based on RMSE revealed that ARIMA model predicted better result. The three phases are involved in this study, namely, the model-building phase, predictive phase, and finally the recurrence quantification analysis phase. The model-building phase used Kelantan daily river-flow data prior to 2014 while the predictive phase feature performance of the chosen models on the holdout 2014 daily river-flow data. Figure 3 and Figure 4 provide the time series plot of the stream flow data for the years 2000 to 2014 and the ARIMA model of the order (5, 0, 3) respectively. A closer examination of Figure 3 and 4 shows that the maximum flow for 2014 (i.e., the flooding event) does not fit the panel for Figure 4 as the main data that build the model for figure 4 are from 2000 to the end of 2013. The maximum for the data-building phase is not nearly as extreme as the maximum of 2014 that can be seen from Figure 5.

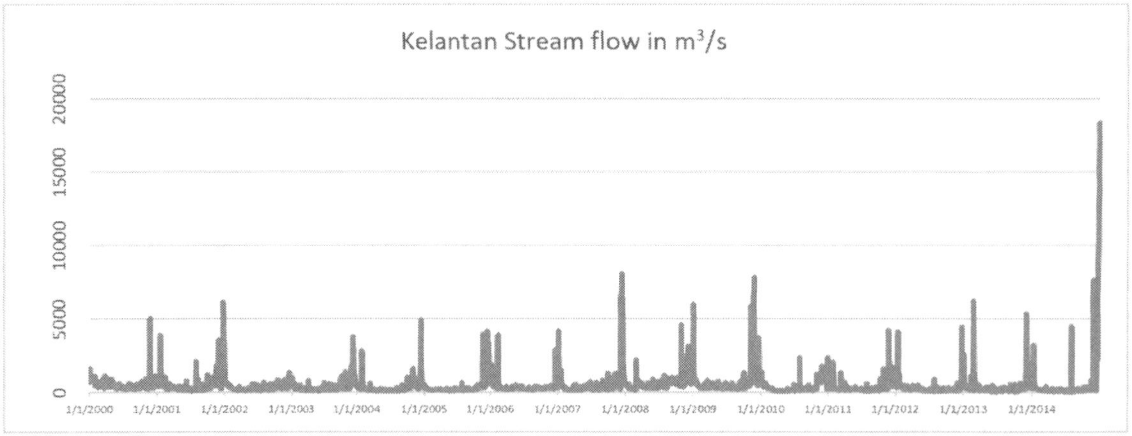

Figure 3. Kelantan River-flow data for all years including the 2014 hold-out data. [17]

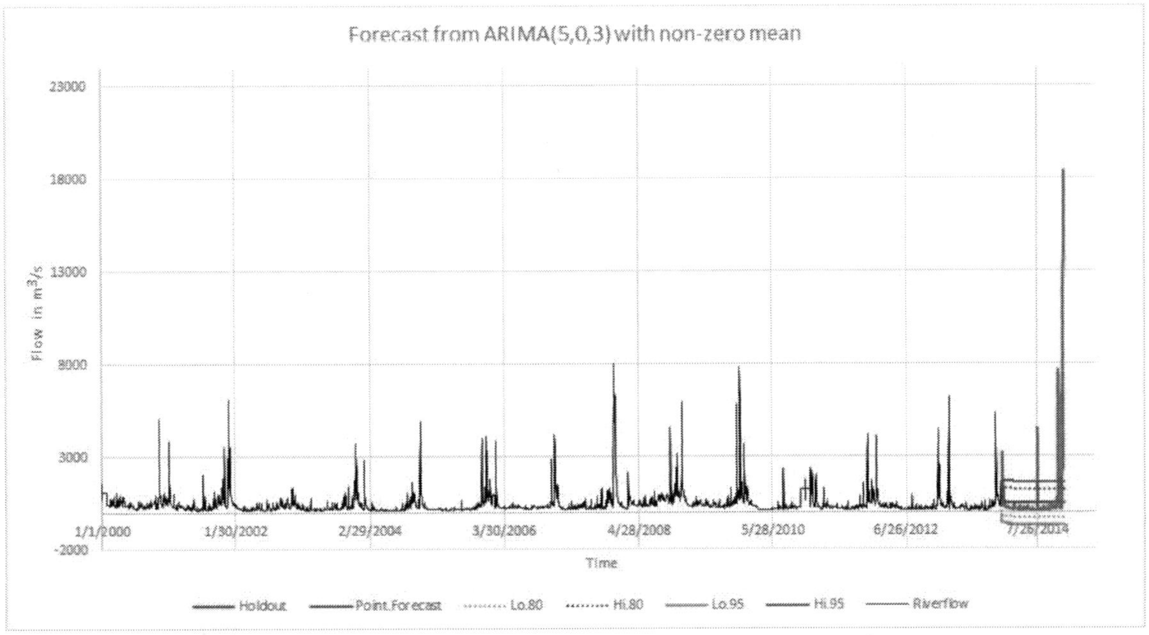

Figure 4. Kelantan River-flow time series data (in black), hold-out data (in red) represents 2014 data, and predicted forecast values (in blue), [17]

3.3. Floods early warning detection using ARIMA:

According to [32], hydrological forecasting models play a crucial role in the development of flood warning systems. The 15 year-time series data of the Kelantan River flow is depicted in Figure 4 from [17]. The figure also contains the year 2014 data highlighted in red color, the 80 % and the 95 % confidence bands in different shades of blue, and the predicted forecast values in blue color. It can be clearly observed that most of the 2014 time series data fall within 80 % (shorter) and 95 % (wider) confidence range except at the onset, latter half and at the end of 2014. The values outside the 80 % and 95 % confidence bands can be considered as an early warning indicator that the river flow is reaching aberrant volumes and may be dangerous. These bands can be incorporated into an early warning system so that populated areas surrounding rivers such as Kelantan River can take precautionary measures as soon as these bands are breached. As can be seen from Figure 5, this band is breached and has reached the alarming level especially when breaching the 95 % confidence bands at the beginning, the latter half and the end of the holdout year 2014. That is, in terms of the river flow in the holdout year 2014, the Kelantan River has experienced more than alarming river flow levels in the beginning and the latter half of the holdout year before the occurrence of the catastrophic flooding at the end of the year 2014. Figure 5 showed that the maximum river flow in December 2014 that caused the catastrophic flood is around 4 times higher than the peak river flow level that occurred at least 5 months earlier. RMSE criterion was used to test the performance of the ARIMA model and according to [16], a notable number of research studies in the context of Malaysia used it. The performance criterion used for both forecasting models, ARIMA and RQA, was RMSE. However, the RQA model also takes normalized mean square error (NMSE) into consideration in addition to RMSE.

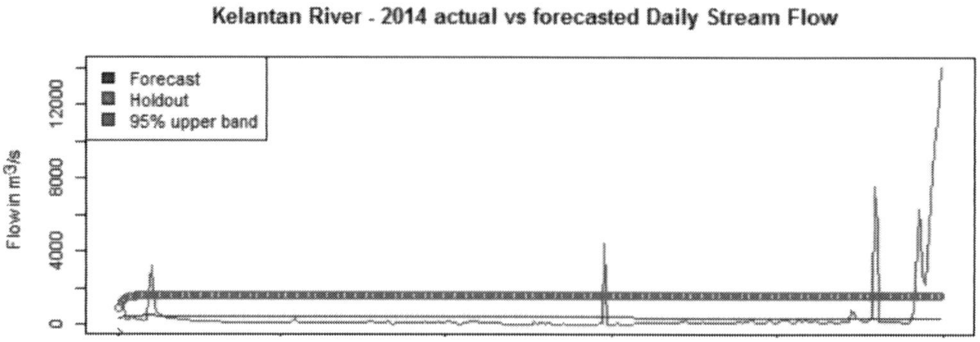

Figure 5. Kelantan River-flow time series hold-out data (in red), forecast (in blue) and 95 % upper confidence bands (in purple) for 2014. [17]

4. Conclusion

The present paper reviewed briefly the river floods forecasting methods currently used worldwide with a specific focus in the context of the Kelantan River in Malaysia. A summary of the forecasting methods such as machine learning, statistical and chaos theory has been provided with a focus on the methods used for the forecasting of Kelantan River. This review paper has also demonstrated the capability of the ARIMA method in the early warning detection. The values that fall outside the confidence bands can be considered as an early warning as indicated by the dangerous aberrant volumes of the river flow. These bands can be incorporated into an early warning system as potential predictors of an oncoming extreme event so that populated areas surrounding Kelantan River can take precautionary measures as soon as they are breached. The recurrence plot and recurrence quantification analysis is a novel and useful method especially for the analysis of the river flooding time series data. The authors have also used this technique for flood forecasting; however, its further rigorous and thorough work in this direction can be carried out in the future. This paper suggests the exploration of recurrence analysis method as a future direction for research and development in combination with other appropriate methods such as ARIMA and machine learning methods for dealing with the recurrent rivers flooding in Malaysia. Our future work will also focus on the probabilistic methods of forecasting river flow. A reasonably acceptable prediction of river streamflow is significantly important in disaster management and water resources management.

References

[1] IPCC 2020 Intergovernmental Panel on Climate Change
[2] UNISDR 2004 United Nations International Strategy for Disaster Reduction, Living with risk: A global review of disaster reduction initiatives
[3] HLPW 2018 High-Level Panel on Water Outcome Document: Making Every Drop Count, An Agenda for Action
[4] CRED- 2015 Canter for Research on the Epidemiology of Disasters Human Cost of Natural Disasters, A Global Perspective
[5] Kundzewicz Z W, Kanae S, Seneviratne S L, Handmer J, Nicholls N, Peduzzi P, Mechler R, Bouwer L M, Arnell N, Mach K, Muir-Wood R, Brakenridge R, Kron W, Benito G, Honda Y, Takahashi K, and Sherstyukov B 2014 Flood risk and climate change: global and regional perspectives, *Hydrolog Sci J* **59**, 1–28, https://doi.org/10.1080/02626667.2013.857411
[6] Map of Malaysia and location of Kelantan river https://www.google.com
[7] Billa L, Mansor S, and Mahmud A R 2004 Spatial information technology in flood early warning systems: An overview of theory, application and latest developments in Malaysia *Disaster Prevention and Management* **13**(5), 356-363
[8] Budhakooncharoen S 2004 Rainfall estimate for flood management using meteorological data from satellite imagery Ninth Biennial Conference on Engineering, Construction, and Operations in Challenging Environments, March 7-10,| League City, Houston, Texas, United States American Society of Civil Engineers
[9] Thielen-del Pozo J, Pappenberger F, Salomon P, Bogner K, Burek P, & de Roo A 2010 The state of the art of flood forecasting - Hydrological Ensemble Prediction Systems *EMS Annual Meeting Abstracts*, **7**(3/ 2016), 1
[10] SFDRR 2015-2030 Sendai Framework for Disaster Risk Reduction

[11] Modricka T M, Georgakakos K P, 2015 The character and causes of flash flood occurrence changes in mountainous small basins of Southern California under projected climatic change, *Journal of Hydrology Regional Studies* Volume **3**, 3/ 2015, Pages 312-336

[12] Moradkhani, H and Sorooshian, S, 2008 General review of rainfall-runoff modeling: model calibration, data assimilation, and uncertainty analysis *Hydrological modeling and the water cycle. Springer*. **291** p. ISBN 978-3-540-77842-4

[13] Devia, G K, Ganasri B P, and Dwarakish, G S 2015 A review on hydrological models *Aquatic Procedia* **4**, 1001–1007

[14] Wheater H S, Jakeman A J and Beven K J, 1993 Chapter 5 – Progress and directions in rainfall-runoff modelling, *In: Modelling change in environmental systems (Eds. A.J. Jakeman, M.B. Beck and M.J. McAleer), Wiley,* Chichester, UK, 101-132

[15] WMO 2013 World Metrological Organization, Expert Meetings on Intercomparison of Flood Forecasting Models, *Secretariat of the International Hydrological Programme of UNESCO and the Hydrology and Water Resources Programme,* Koblenz, Germany

[16] Abdulkareem J H P B, Sulaiman W N A 2018 Review of studies on hydrological modelling in Malaysia *Modelling Earth System. Environment.* **4,** 1577–1605 https://doi.org/10.1007/s40808-018-0509-y

[17] Abdul Majid M, Omar M. H, Noorani M S M 2021 Analysis and predictive validity of Kelantan River flow using RQA and time series analysis *Kuwait Journal of Science* **48** (1) pp. 25-36

[18] Kurunç A, Yürekli K, and Çevik O, 2005 Performance of two stochastic approaches for forecasting water quality and streamflow data from Yeşilirmak River Turkey Environmental Modelling and Software, **vol. 20** no. 9, pp. 1195–1200

[19] Valipour M, 2015 Long-term runoff study using SARIMA and ARIMA models in the United States, Meteorological Applications **vol. 22** no. 3, pp. 592–598

[20] Wagenaar D, Curran A, Balbi M, Bhardwaj A, Soden R, Hartato E, Mestav Sarica G, Ruangpan L, Molinario G, Lallemant D 2020 Invited perspectives: How machine learning will change flood risk and impact assessment *Natural Hazards and Earth System Sciences* **20**(4) 1149–1161

[21] Ahmed N K, Atiya A F, Gayar N E, El-Shishiny H 2010 An Empirical Comparison of Machine Learning Models for Time Series Forecasting *Econometric Reviews*;**29**(5–6): 594–621

[22] Raman H, Sunilkumar N, 1995 Multivariate modeling of water resources time series using artificial neural networks. *Hydrol. Sci. J.* **40** (2), 145–163

[23] Thirumalaiah K, Deo M C, 2000. Hydrological forecasting using neural networks *J. Hydrol. Eng.* **5** (2), 180–189

[24] Kisi O, 2003 River flow modeling using artificial neural networks *J. Hydrol. Eng.* **9** (1) 60–63

[25] Jain S K, Das A, Drivastava D K, 1999 Application of ANN for reservoir inflow prediction and operation *J. Water. Resour. Plann. Manage* **125** (5) 263–271

[26] Abraharty J R, See L, 2000 Comparing neural network and autoregressive moving average techniques for the provision of continuous river flow forecasts in two contrasting catchments *Hydrol.* Process. **14**, 2157–2172

[27] Castellano-Me'ndeza M, Gonza'lez-Manteigaa W, FebreroBande M, Manuel Prada-Sa'ncheza J, Lozano-Caldero'n R, 2004 Modeling of the monthly and daily behavior of the runoff of the Xallas river using Box–Jenkins and neural networks methods *J. Hydrol.* **296**, 38–58

[28] Lin J Y, Cheng C T, Chau K W, 2006 Using support vector machines for long-term discharge prediction *Hydrol. Sci. – J.* **51** (4), 599–611

[29] Dibike Y B, Velickov S , Solomatine D, Abbott M B 2001 Model induction with support vector machines: introduction and applications *J. Comput. Civil Eng.* **15** (3), 208–216

[30] B. Sivakumar, 2002 A phase-space reconstruction approach to prediction of suspended sediment concentration in rivers *Journal of Hydrology*, **258**,149-162

[31] Wu C L and Chau K W 2010 Data-driven models for monthly streamflow time series prediction Engineering Applications of Artificial Intelligence vol. 23 **no. 8** pp. 1350–1367

[32] Ttilford K A, Sene K, Khatibi R, 2007 Flood forecasting model selection, A structured approach *flood risk management in Europe, 401–416. Springer*

[33] Baharuddin K A, Abdull Wahab S F, Ab Rahman N H N, Mohamad A N, Kamauzaman H T, Noh A Y M, Abdul Majod M R 2015 The record-setting flood of 2014 in Kelantan: challenges and recommendations from an emergency medicine perspective and why the medical campus stood dry *Malaysian Journal of Medical Society*. **22**(2): 1–7

[34] Malay Mail Report 2014 https://www.malaymail.com/news/malaysia/2014/12/22/flood-victim-recounts-horrific-experience-says-night-felt-like-a-tsunami-at/806767

[35] Faizah C R, 2015 PhD Dissertation, Study on Early Forecasting of Flood through Historical Hydrologic Data Analysis and Numerical Simulation in Kelantan Watershed, Malaysia

[36] Osman S, Abdul Aziz N, Husaif N, Mohd Sidek L, Shakirah A, Hanum F, Basri H 2018 Application of stochastic flood forecasting model using Regression Method for Kelantan Catchment, *MATEC Web of Conferences*-**203**, 07001, ICCOEE-2018

[37] Abdulkareem J H, Sulaiman W N A, Pradhan B, Jamil N R 2018a Relationship between design floods and land use land cover (LULC) changes in a tropical complex catchment *Arab J Geosci* **11**:376

[38] Abdulkareem J H, Pradhan B, Sulaiman W N A, Jamil N R 2018b Quantification of runoff as influenced by morphometric characteristics in a rural complex catchment *Earth Syst Environ* **2**:145–162 https ://doi.org/10.1007/s4174 8-018-0043-0

[39] Adnan N A, Atkinson P M 2018 Disentangling the effects of longterm changes in precipitation and land use on hydrological response in a monsoonal catchment *Journal of Flood Risk Management* **11**:1063–1077

[40] Asmat A, Mansor S, Saadatkhah N, Adnan N A, Khuzaimah Z, 2016 Land use change effects on extreme flood in the Kelantan basin using hydrological model *ISFRAM 2015. Springer*, Singapore, pp 221–236
[41] Ghorbani K, Wayayok A, Abdullah AF 2016 Simulation of flood risk area in Kelantan watershed, Malaysia using numerical model. *Jurnal Teknolgy* **78**:51–57
[42] Saadatkhah N, Tehrani M H, Mansor S, Khuzaimah Z, Kassim A, Saadatkhah R 2016 Impact assessment of land cover changes on the runoff changes on the extreme flood events in the Kelantan River basin. *Arab Journal Geos*, **9**: 687
[43] Hafiz I, Sidek L M, Basri H, Fukami K, Hanapi M N, Livia L, Jaafar A S, 2014 Integrated flood analysis system (IFAS) for Kelantan River basin. *In: IEEE 2nd international symposium on telecommunication technologies*, pp 159–162
[44] Adnan N A, Zuraini, B 2014a Variation in hydrological responses estimation simulations due to land use changes. *In: International conference on civil, biological and environmental engineering,* Istanbul Turkey pp. 27–28
[45] Basarudin Z, Adnan N A, Latif A R A, Tahir W, Syafiqah N 2014 Event-based rainfall-runoff modelling of the Kelantan River basin *IOP Conf Ser Earth Environ Sci* **18**:120844
[46] Perera E D P, Lahat L 2014 Fuzzy logic based flood forecasting model for the Kelantan River basin, Malaysia *Journal of Hydro-Environ Research* *9* Issue 4 *Pages* 542-553
[47] Milad J, Lariyah S, Bte B H, Shakira J A, 2016 Flood Forecasting via Time Lag Forward Network; Kelantan, Malaysia *IOP Conference Series: Earth and Environmental Science,* Vol. **32**, 23–25
[48] Jaafar K, Ismail N, Tajjudin M, Adnan R, Hezri M, Rahiman F, 2016 Hidden neuron variation in multi-layer perceptron for flood water level prediction at Kusial station faculty of electrical engineering *In: 2016 IEEE 12th international colloquium on signal processing & its applications, 4–6 March, Melaka*, Malaysia, pp 4–6
[49] Mohtar I S A, Tahir W, Bakar S H A, Zuhari A Z M 2015 Use of numerical weather prediction model and visible weather satellite images for flood forecasting at Kelantan river basin. *ISFRAM 2014. Springer*, Singapore, pp 283–294.
[50] Tehrany MS, Pradhan B, Jebur MN, 2015 Flood susceptibility mapping using a novel ensemble weights-of-evidence and support vector machine models in GIS *Stochastic Environmental Research Risk Assessment* **512**: 332–343
[51] Tehrany M S, Pradhan B, Jebur M N 2013 Spatial prediction of flood susceptible areas using rule based decision tree (DT) and a novel ensemble bivariate and multivariate statistical models in GIS *Journal of Hydrology* **504**:69–79.
[52] Pradhan B, Youssef A M 2011 A 100-year maximum flood susceptibility mapping using integrated hydrological and hydrodynamic models: Kelantan River Corridor, Malaysia *Journal of Flood Risk Management* **4**:189–202.
[53] Pradhan B 2009 Flood susceptible mapping and risk area delineation using logistic regression, GIS and remote sensing *J Spat Hydrol* **9**:2
[54] Pradhan B, Shafiee M, Pirasteh S 2009 Maximum flood prone area map RADARSAT images and GIS: Kelantan River basin *Intern J Geomat* **5**:11–23
[55] Anuar M A S, Abdul Rahman R, Soh A C 2020 A Promising Wavelet Decomposition NNARX Model to Predict Flood: Application to Kelantan River Flood *International journal of electrical and computer engineering systems*, vol. **11**, issue 2 pg.89-99
[56] Eckmann J P, Kamphorst S O, Ruelle D. 1987. Recurrence plots of dynamical systems. *Europhysics Letters*, **4**, 973-977
[57] Webber C L, Jr, Zbilut J P 1994 Dynamical assessment of physiological systems and states using recurrence plot strategies. *Journal of Applied Physiology*, **76**, 965–973
[58] Marwan N, Romano M C Marco, Thiel J K, 2007 Recurrence plots for the analysis of complex systems, *Physics Reports* **438** 237 – 329

Acknowledgements
The first, third and fourth authors acknowledge the academic support from UKM. No financial support for this work was granted. The second author acknowledges the research facilities at KFUPM.

Numerical study on percolation and dam slope's stability of impermeable wall composed by clay and concrete for earth-rock dam

Xiaoming Zhu[1*], Linlu Tan[2], Jing Si[1], Sha Shi[1,3], Kai Yu[4] and Arniza Fitri[5]

[1]College of Water Conservancy and Ecological Engineering, Nanchang Institute of Technology, Nanchang, 330099, China

[2]Jiangxi Academy of Water Science and Engineering, Nanchang, 330029, China

[3]State Key Laboratory of Hydrology-Water Resources and Hydraulic Engineering, Hohai University, Nanjing 210098, China

[4]Yaohu Honors College, Nanchang Institute of Technology, Nanchang, 330099, China

[5]Department of Civil Engineering, Faculty of Engineering and Computer Science, Universitas Teknokrat Indonesia, Indonesia

*Corresponding E-mail: xmzhu@nit.edu.cn

Abstract. In order to study the percolation and dam slope's stability of an impermeable wall composed by clay and concrete for earth-rock dam, the steady seepage field of a reservoir in Jiangxi is numerically simulated by using finite element method based on the basic principles of saturated and unsaturated seepage under different ratios of clay to concrete. The results of the flow field of steady seepage are carried out and the dam slope's stability is analyzed. We found relatively minor impacts of concrete proportion on the discharge of seepage per unit width and the dam slope's stability. In this study, the impermeability performance of the impermeable wall is concerned. The results showed that impermeable effect of the impermeable wall has not obviously changed if the concrete proportion is below 0.43. When the proportion of concrete exceeds 0.43, the anti-seepage effect of the impermeable wall begins to change significantly. Besides, when the proportion of concrete reaches 1.0, the reduction of water level at front and behind impermeable wall, the discharge of seepage per unit width, and the safety factor of the Bishop method are all infinitely close to the value of the best anti-seepage performance and the most stable state of the dam slope.

Key words: Impermeable wall, Finite element analysis, Percolation, Dam slope's stability

Track Name: Land, Water, Forests and Food Security

1. Introduction

Several types of earth-rock structures have been applied in water resources managements [1-5]. As one type of dam structure, earth-rock dams are widely used since las few decades due to their advantages such as low investment, simple construction, fully use local materials and low requirements for topographic and geological conditions. Regarding all types of reservoirs in China which reaching up to 99 %, 30 % of dam collapse accidentally due to seepage and slope instability of earth-rock dams [6]. Impermeable wall is the key component of the whole earth-rock dam seepage-proof system, and its quality is directly related to the safety of the whole project. There are several studies on seepage and slope stability of earth-rock dam with concrete impermeable wall that have been done by local researches in China. Xin [7] studied the depth optimization of cut-off wall and its anti-seepage effect. Ni [8] and Wang *et al.* [9] studied the influence of water level rise and fall of reservoir on seepage of earth-rock dam and stability of dam slope. Gao *et al.* [10] studied seepage characteristics of imperfect cut-off wall in earth-rock dam reinforcement project. However, there is little research on seepage and slope stability of earth-rock dam with clay-bonded concrete cut-off wall. Therefore, this paper simulates the steady seepage field of earth-rock dam with clay-concrete cut-off wall by using finite element method based on the basic theory of saturated-unsaturated unstable seepage of porous media. To analyze the stability of dam slope under the condition of changed reservoir water level, the seepage parameters are compared for different clay-concrete ratio conditions and Bishop method is applied in this study. This study has beneficial to provide reference basis for seepage of earth-rock dam with clay-bonded concrete cut-off wall, stability of dam slope and designation of safe operation system of reservoir.

2. Material and Methods

2.1 Governing equation of unsaturated seepage flow

Frelund and Rahardjo [11] believe that the seepage in unsaturated soil conforms to Darcy's law and continuity equation. However, the permeability coefficient in unsaturated soil is no longer a constant, but a function of matrix suction, and the volume moisture content of soil changes with time. The seepage governing equations of unsaturated soil can be obtained by introducing Darcy's law into the seepage continuity equation as follows [12-13]:

$$\frac{k_x}{\gamma_w}\frac{\partial^2 \mu_w}{\partial x^2}+\frac{k_y}{\gamma_w}\frac{\partial^2 \mu_w}{\partial y^2}+\frac{\partial \theta_w}{\partial t}=0 \qquad (1)$$

In the formula (1): k_x, k_y are permeability coefficients in x and y direction respectively; γ_w is the volume weight of water; θ_w is the volumetric water content; t is the time.

2.2 Stability analysis equation of Bishop Method

The rigid body limit equilibrium method is used for the static stability calculation of dam slope. For the stable seepage period or the reservoir water level falling period, when the simplified Bishop method is used for the stability analysis, the formula of the effective stress method anti-sliding stability safety factor is:

$$K=\frac{\sum\left[c'\cdot b\cdot\sec\beta+\left(w_1+w_2\right)\sec\beta-\left(U-\gamma_w\cdot z\right)b\cdot\sec\beta\cdot tg\varphi'\right]/\left(1+tg\varphi' tg\beta/k\right)}{\sum\left(w_1+w_2\right)\sin\beta} \qquad (2)$$

IOP Conf. Series: Earth and Environmental Science **880** (2021) 012022 doi:10.1088/1755-1315/880/1/012022

For the reservoir water level falling period, when the simplified Bishop method is used for stability analysis, the formula for safety factor of anti-sliding stability of total stress method is as follows,

$$K=\frac{\sum\left[c_{cu}\cdot b\cdot\sec\beta+\left(W\sec\beta-U_i\cdot b\cdot\sec\beta\right)tg\varphi_{cu}\right]/\left(1+tg\varphi_{cu}tg\beta/k\right)}{\sum W\sin\beta} \quad (3)$$

In the formulas (2) and (3): *c'*, *Φ'* are the effective strength indexes; c_{cu}, φ_{cu} are the consolidated undrained total strength indexes; *b* is the width of soil strip; *w* is the solid weight of soil strip; w_l is the solid weight of strip above water level outside dam slope; w_l' is the sum of floating weight and solid weight of bars below the phreatic line above the water level outside the dam slope; w_2 is the distance between the water level outside the dam slope and the midpoint of the strip bottom; Z is the distance between the water level outside the dam slope and the midpoint of the strip bottom; u_i is the pore water pressure in the dam before the water level of the reservoir drops; *u* is the pore water pressure in the dam during the period of stable seepage or reservoir water level falling; *β* is the angle between the gravity line of the block and the radius passing through the midpoint of the bottom surface of the block; γ_w is the volume weight of water.

2.3 Determination of pore water pressure

The calculation of the pore water pressure of the dam body and dam foundation during the stable seepage period when the reservoir level is the design flood level and the check flood level can be determined according to the flow network calculated by seepage analysis, that is, the pore water pressure at any point on the same equipotential line in the flow network diagram is equal to the vertical head pressure from that point to the intersection of the equipotential line and the saturation line. When the effective stress method is used to calculate the upstream slope stability under the water level drops suddenly and the dissipation of pore water pressure is considered, the result of unsteady seepage period is used to calculate and the determination method of pore water pressure is the same as that of stable seepage period.

3. Results and Discussions

3.1 Reservoir overview

A reservoir in Jiangxi Province is a medium-sized reservoir, which is mainly used for irrigation, flood control, aquaculture and other comprehensive utilization and supplement of industrial and domestic water in the county. In the dam anti-seepage treatment scheme, the clay core wall is backfilled with the grab sleeve well and the anti-seepage wall is built with high-pressure jet grouting. The clay core wall is punched at the upper part and the effective thickness of the core wall is not less than 0.65 m. The anti-seepage wall is built with high-pressure jet grouting at the lower part. The thickness of the anti-seepage wall is $\delta \geq 0.20$ m, and the bottom of the anti-seepage wall extends into the bedrock 0.50m. The upstream normal water level is 58.90 m, the design flood level is 60.50 m, the check flood level is 62.40 m, and the dead water level is 48.40 m. The cross section of a dam in Jiangxi is shown in Figure 1.

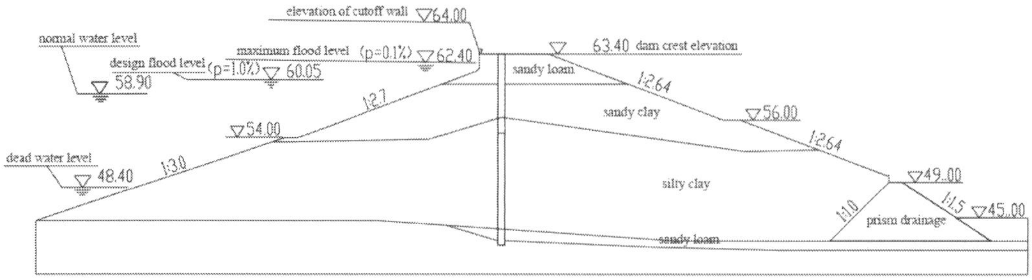

Figure 1. The cross section of a dam in Jiangxi.

Before the construction of clay core wall, the dry bulk density test with different times of ramming is carried out. The number of ramming is determined to be 14-16, the moisture content of soil material is all controlled at 22.7-28.7, and the dry bulk density detection is up to 1.47-1.55 g/cm^3. High pressure jet grouting seepage control wall, the slurry density control is 1.6g/cm^3, the pressure of high pressure water is 35 MPa, and the cement slurry pressure is 0.5 MPa.

3.2 Model design and Establishment

The cutoff wall of a reservoir dam in Jiangxi Province is a combination of clay and concrete. The upper part of the cutoff wall is clay core wall which is 8.9 m, and the lower part is concrete cutoff wall which is 12.1 m, and the total length is 21 m. In order to discuss and analyze the influence of impervious wall formed by the ratio of clay to concrete on the seepage and the stability of dam slope, eight working conditions of concrete accounting for the whole impervious wall height (hereinafter referred to as "the ratio of concrete") are selected for modeling, which are 0, 0.14, 0.29, 0.43, 0.58, 0.71, 0.83 and 1.0 respectively.

A two-dimensional seepage finite element model of a reservoir dam in Jiangxi Province is established according to the cross section of the dam (as shown in Figure 2). The seepage field under the condition of the change of reservoir water level is the result of the interaction of saturated seepage and unsaturated seepage. The selection of soil-water characteristic curve and permeability coefficient curve characterizing the unsaturated seepage characteristic is complex. At present, the soil-water characteristic curve and permeability coefficient curve are difficult to be obtained directly by theoretical analysis method, so they are estimated through corresponding mathematics model based on a small number of test values. SEEP/W module of Geo-Studio software can estimate the soil water characteristic curve of core wall and dam shell respectively by saturated volume water content (void ratio) of material, clay spline function and rockfill spline function provided by the module. At the same time, the module can estimate the permeability coefficient curve of the material through the saturated permeability coefficient of the corresponding material, the soil water characteristic curve and the Fredlund-Xing model provided by the module itself. Then, the seepage calculation results under various working conditions are transferred into slope/W module, and the Mohr-Coulomb criterion modified by Fredlund and Bishop method in limit equilibrium method are used to analyze the stability of upstream dam slope [7-8].

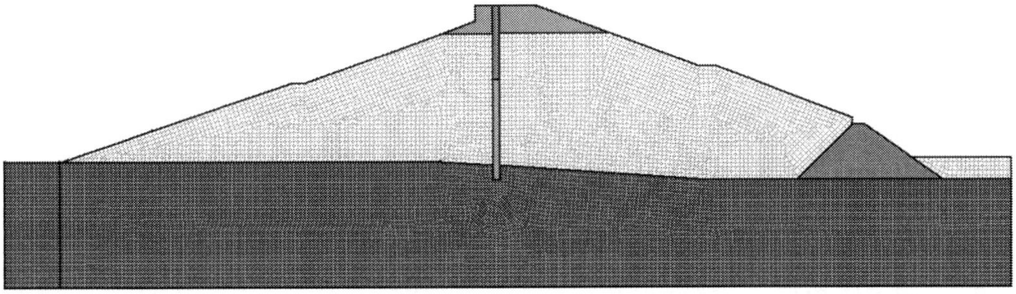

Figure 2. Two dimensional seepage finite element model of a reservoir dam in Jiangxi Province.

Taking the concrete ratio of 0.58 as an example, the seepage simulation results are shown in Figure 3. It can be seen from the simulation results that the maximum seepage gradient of the dam occurs in the impervious wall, and the gradient values are all within the allowable gradient range. The local seepage stability of the dam body is good, and the possibility of seepage failure is small. During the stable seepage period under various working conditions, the phreatic line escapes at the position close to the slope. The phreatic line is mainly lowered in the anti-seepage wall, and falls slowly in the dam body behind the anti-seepage wall. The anti-seepage performance of the dam is good. The distribution of the seepage calculation equipotential line of the dam conforms to the seepage law of the general concrete core wall.

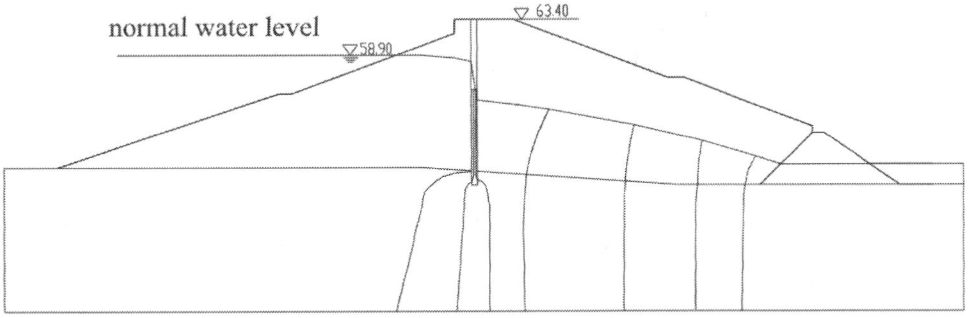

Figure 3. Calculation results of stable seepage when the concrete ratio is 0.58.

Taking the concrete ratio of 0.58 as an example, the dam slope stability simulation results are shown in Figure 4. Under the normal water level, design flood level and design flood level, the calculated safety factors *K* of Bishop method at the downstream of dam slope are 1.568, 1.552 and 1.532 respectively. The design code for roller compacted earth rock dam (SL274-2001) requires that the minimum safety factor of anti-sliding stability of dam slope is 1.30 when the dam is in normal operation condition. The calculation results show that the safety factor of anti-sliding stability of dam slope at the downstream of the dam meets the requirement of the codes.

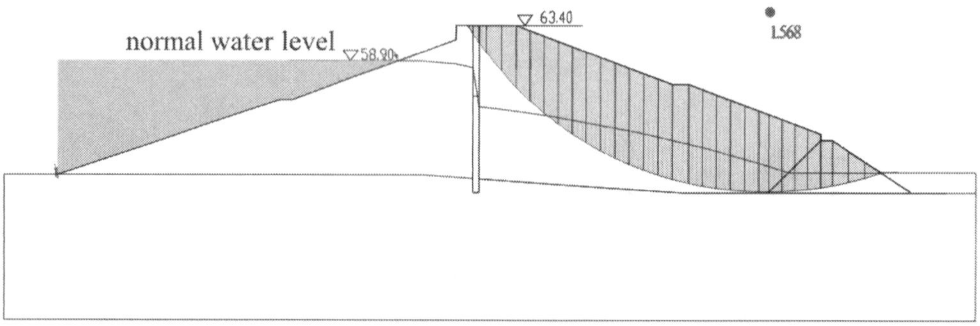

Figure 4. Downstream slope K=1.568 when the concrete ratio is 0.58.

3.3 Seepage and Dam slope stability

Three models of steady-state water level, including normal water level, design flood level and check flood level, were established under eight working conditions of concrete proportion of 0, 0.14, 0.29, 0.43, 0.58, 0.71, 0.83 and 1.0 respectively. The water level drop before and after the cutoff wall, single width seepage flow and safety factor of Bishop method were calculated under each working condition.

Figure 5 shows that the decrease of water level before and after the cutoff wall gradually increases with the increase of the ratio when the proportion of concrete is below 0.43, but the increase is not high. In terms of design flood level, the decrease of water level of concrete ratio of 0.43 is only 0.35 m more than that of concrete ratio of 0.29. The decrease of water level before and after the cutoff wall increases sharply and the increment increases gradually when the proportion of concrete increases from 0.43. Compared with the concrete ratio of 0.58, the water level drops 1.2 m more than the concrete ratio of 0.43. Compared with the ratio of concrete to 0.83, the water level decreased sharply to 2.1 m. At the same time, the water level drop before and after the cutoff wall is more than 5 m when the proportion of concrete is more than 0.58 and it shows that the performance of the cutoff wall is good. To sum up, the anti-seepage effect of the anti-seepage wall does not change significantly when the concrete proportion is less than 0.43 and the anti-seepage effect of the anti-seepage wall increases significantly when the concrete proportion is more than 0.43. As for the water level drawdown map before and after the cut-off wall, it is recommended to consider the cut-off wall with the concrete ratio of more than 0.58.

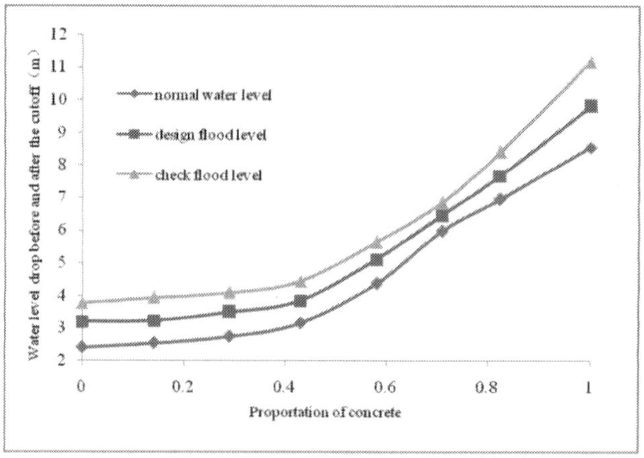

Figure 5. Water level drop before and after cutoff wall under various working conditions.

Figure 6 shows that the single width seepage flow under each stable water level is not high. With the increase of concrete proportion, the single width seepage flow is smaller and smaller and the difference is larger and larger. When the concrete proportion reaches above 0.71, the single width seepage flow is only half of that of the concrete ratio of 0. Simultaneously, the single width seepage flow decreases sharply to about 0.2 when the proportion of concrete reaches 1.0. Therefore, in terms of single wide seepage flow, the influence of concrete proportion is not too great. Under the condition that the conditions allow, it is suggested to consider the impervious wall with concrete proportion of more than 0.71.

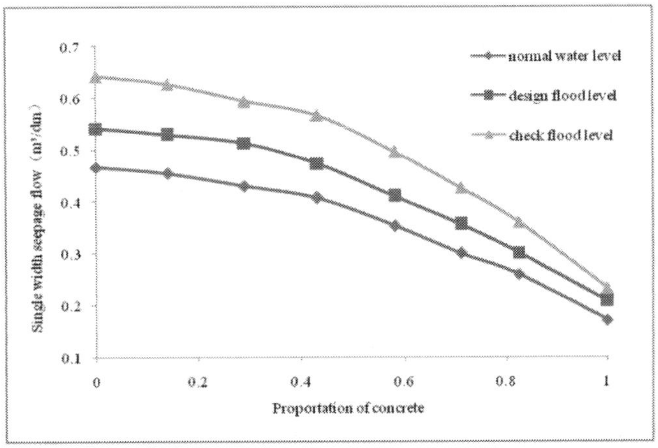

Figure 6. Single wide seepage flow under various working conditions.

As shown in Figure 7, the safety factor of Bishop method is greater than 1.30 under all working conditions. At the same time, with the increase of concrete proportion, the safety factor of Bishop method gradually increases, and the increase is originally larger. When the concrete proportion is 1.0, it increases sharply to about 1.6.

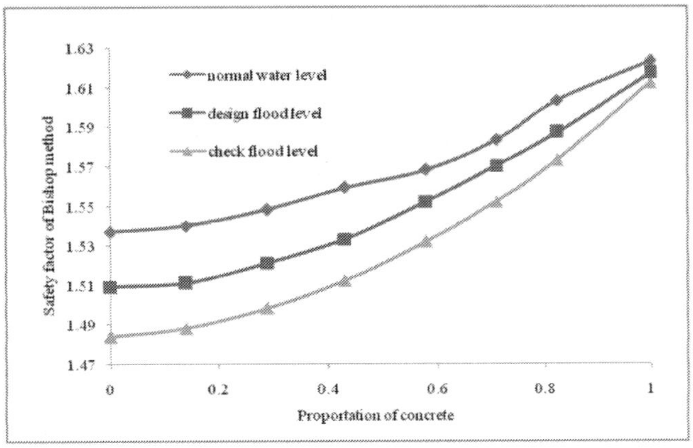

Figure 7. Safety factor of Bishop method under various working conditions.

4. Conclusions

(1) As for a reservoir in Jiangxi Province, when the proportion of concrete reaches 1.0, the water level drop before and after the anti-seepage wall, the single wide seepage flow and the safety factor value of Bishop method are all close to the best anti-seepage performance and the most stable state of the dam slope.

(2) It is suggested that the concrete anti-seepage core wall should be considered as much as possible when conditions permit. The anti-seepage effect of the anti-seepage wall increases significantly when the proportion of concrete is more than 0.43.

(3) The change of concrete proportion has little influence on the single wide seepage flow and the stability of dam slope. In terms of the anti-seepage performance of the anti-seepage wall, it is suggested to consider the anti-seepage wall with high concrete proportion.

References

[1] Fitri A, Hashim R, Abolfathi S, and Abdul Maulud K N 2019 Dynamics of sediment transport and erosion-deposition patterns in the locality of a detached low-crested breakwater on a cohesive coast *Water* **11** (8) pp 1721

[2] Fitri A, Hashim R, Song K I, and Motamedi S 2015 Evaluation of morphodynamic changes in the vicinity of low-crested breakwater on cohesive shore of Carey Island, Malaysia *Coastal Engineering Journal* **57**(04) pp 1550023

[3] Fitri A, Yao L, and Sofawi B 2019 Evaluation of Mangrove Rehabilitation Project at Carey Island Coast, Peninsular Malaysia based on Long Term Geochemical Changes. In *IOP Conference Series: Earth and Environmental Science* IOP Publishing Vol **365**(1) pp 012055)

[4] Fitri A and Yao L 2019 The impact of Parameter Changes of a Detached Breakwater on Coastal Morphodynamic at Cohesive Shore; A Simulation. In *IOP Conference Series: Earth and Environmental Science* IOP Publishing **365** (1) pp 012054

[5] Hashim R, Fitri A, Motamedi S, Hashim A M 2013 Modelling of Coastal Hydrodynamic Associated with Coastal Structures: A Reviews *Malaysian Journal of Science* **32** pp 149-154

[6] Gao J 2012 Study on the interaction of cut-off wall and earth-rock dam based on coupling of seepage and stress *Tianjin University*

[7] Xin X 2017 Study on optimization of anti-seepage wall depth and seepage control effect *Water Resources Power* **35** (12) pp 131-134

[8] Ni S and Zhang X 2016 Influence of ascending and descending water level on seepage field and stability of naban core earth-rock dam *Water Resources Power* **34** (3) pp 84-87

[9] Wang K, Xie L and Liu H 2018 Study on seepage field and slope stability of homogeneous earth-rock dam under action of drawdown of reservoir water level *Water Resources Power* **36** (08) pp 81-84

[10] Gao J and Yan Z 2017 Study on the permeability of cut-off wall with defects in the reinforcement projects of earth-rock dam *Yellow River* **39** (9) pp 125 – 134

[11] Fredlund D G and Rahardjo H 1997 *Saturated Soil mechanics* Trans. Binging: China Architecture and Building Press

[12] Fredlund D G, Xing A and Huang S 1994 Predicting the Permeability Function for Unsaturated Soils Using the Soil-water Characteristic Curve *Canadian Geotechnical Journa* **31** pp 533-546

[13] Geo-Slope 2001 *International Ltd. SEEP/W for Seepage Analysis* Canada: GEO-SLOPE International Ltd

Acknowledgments

This research is supported by the Science and Technology Research Project of the Department of

Education in Jiangxi Province (Grant No. GJJ180953), the Belt and Road Special Foundation of the State Key Laboratory of Hydrology-Water Resources and Hydraulic Engineering (Grant No. 2018492011) and the Science and Technology Research Project of the Science and Technology Department of Jiangxi Province (Grant No. 20192ACBL20041).

Evaluation of urban water security based on DPSIR model

Sha Shi[1*], Xueting Tao[2], Xiaona Chen[2], Hao Chen[1], Arniza Fitri[3] and Xiaojia Yang[2]

[1] School of Hydraulic and Ecological Engineering, Nanchang Institute of Technology, Nanchang 330099, China
[2]School of Yaohu, Nanchang Institute of Technology, Nanchang 330099, China
[3] Department of Civil Engineering, Faculty of Engineering and Computer Science, Universitas Teknokrat Indonesia, Bandar Lampung, Indonesia

*Corresponding E-mail: shisha2016@nit.edu.cn

Abstract. Assessment of urban water security is an important scientific basis for the sustainable utilization of urban water resources. This study established an evaluation model for urban water security based on DPSIR model, entropy method and comprehensive index method. The model was applied to Gong Qingcheng City (GQCC) in Jiangxi Province, China. An indicator system is developed based on four aspects including social development, water resources status, water pollution and water resources management, as well as summing up predecessors' experience. This study analysed per capita domestic water consumption; ratio of ecological water compensation; percentage of water conservancy and public infrastructure investment; and water consumption per ten thousand yuan of industrial output value; where ratio of urban sewage treatment are the main effects on urban water security. The results of single index evaluation show that the change of stress subsystem was most dramatic from 2013 to 2017, while others were relatively stable. This indicates that positive interventions were urgently needed to relieve pressure of the stress subsystem. The results of comprehensive index evaluation show that water security at GQCC was in a critical state. In this paper, the causal relationship between social development and water security was clarified while the main factors and other factors affecting urban water security was identified. Finally, a reference for the formulation of urban development planning based on water resources demand was provided.

Keywords: Comprehensive index method, DPSIR model, Entropy method, Evaluation, Urban water security.

Track Name: Land, Water, Forests and Food Security

1. Introduction

Urban water security includes water shortage, water pollution, water ecological degradation, water management and water disaster [1-10], etc. As a part of urban water space and ecological water environment, lakes are great significance for urban water security [14-15]. The conclusions of the scientific evaluation of urban water security are great scientific significance for urban development planning and improving the relationship between city and lake. However, there is no unified standard for the evaluation system of the urban water security at present, and the research is changing from qualitative to quantitative, from single evaluation to multi-method and multi-discipline comprehensive evaluation [16]. In this paper, DPSIR model framework and comprehensive evaluation method are used to evaluate the water security of lakeside cities. It has been done in order to solve the problems of difficult index selection and single index system, so that could improve the accuracy of evaluation results. Finally, the evaluation results can achieve multi-angle, objectivity, pertinence and applicability. DPSIR model framework is a conceptual model of evaluation index system widely used in environmental system [17]. It divides the natural system into five criterion layers (driving forces, Pressure, State, Impact and Responses, and under each criterion layer) and many indicators. Based on the interaction chain of "society-economy-nature", the analysis and evaluation are carried out. The DPSIR model framework fully selects evaluation indicators from five criterion layers to build the risk evaluation index database based on the DPSIR model framework. The selection of indicators is more comprehensive and the evaluation index system is more complete [18].

In this paper, Gong Qingcheng City (GQCC) of Jiangxi Province, China, is selected as the study area. By using DPSIR model framework to build index system, comprehensive evaluation method is adopted to obtain comprehensive index of water security. Urban water security of Gong Qingcheng City (GQCC) is then analysed, and therefore can provide reasonable suggestions for urban planning at Gong Qingcheng City.

2. Methods and Data

2.1 Index system construction

The selection of index is particularly important in the process of constructing the index system and it must be objective and reasonable. This study selects and analyses the index system of urban water security by using DPSIR conceptual model. The specific index system is shown in Table 1.

Table 1. Index system of urban water security.

Criteria	Alternatives	Unit	Attribute	Description
Drive	Annual growth rate of GDP	%	Positive	Represents the level of regional economic development
	Natural population growth rate	%	Negative	Represents the increase of population with the increase of water demand
	Water resources per unit area	10^4 m^3/km^2	Positive	Represents the finiteness of water resources
Pressure	Water consumption per 10,000 yuan of industrial output value	m^3/ (10^4 yuan)	Negative	Represents the pressure of regional socio-economic development on water resource consumption
	Agricultural water consumption rate	%	Negative	Represents the demand for water resources in agriculture
	Ratio of ecological water compensation	%	Negative	Represents the amount of water resources needed to maintain the ecological environment
	Per capita domestic water consumption	L/d	Negative	
	Discharge of wastewater of 10,000 yuan of output value	m^3/ (10^4 yuan)	Negative	Represents the pressure of water pollution
	COD concentration in wastewater	mg/L	Negative	Represents the content of organic substances in water
State	Per capita water resources	m^3/cap	Positive	Represents the consumption of water resources
	Utilization rate of water	%	Negative	Represents the consumption and

	resources			utilization level of water resources
	Full factor water quality compliance rate	%	Positive	Represents the degree to which water quality meets development and conservation needs
	GDP per capita	10^4 yuan	Positive	Represents the capacity of building water security
Impact	Rate of urbanization	%	Negative	Represents the influence of water resources shortage and pollution on regional development
	Green coverage rate in built-up areas	%	Positive	Represents the ecological environment quality of the region
Response	Ratio of urban sewage treatment	%	Positive	Represents the regulation level of water security
	Comprehensive utilization rate of industrial solid waste	%	Positive	
	Percentage of water conservancy and public infrastructure investment	%	Positive	Represents people's attention to water security
	Number of College Students per 10,000	No.	Positive	Represents the strength of water security awareness and sustainable development concept

2.2 Study area and data sources

Administrative Region of Gong Qingcheng City (GQCC) in Jiangxi Province is shown in Figure 1. The original data of water security evaluation in GQCC, Jiangxi Province from 2013 to 2017 are shown in Table 2. Data sources are Jiangxi Statistical Yearbook (2014-2018), Jiujiang Statistical Yearbook (2014-2018) and Jiangxi Water Resources Bulletin (2014-2018).

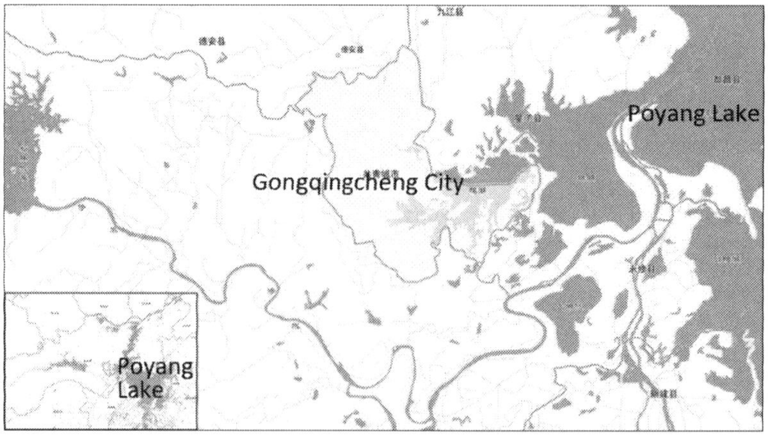

Figure 1. Administrative Region of GQCC in Jiangxi Province

Table 2. Primary data of water security evaluation in GQCC of Jiangxi province.

Criteria	Alternatives	2013	2014	2015	2016	2017
Drive	Annual growth rate of GDP	10.50	10.50	9.90	9.60	9.20
	Natural population growth rate	6.72	6.82	6.79	6.93	7.76
	Water resources per unit area	71.15	81.02	100.24	111.18	115.47
Pressure	Water consumption per 10,000 yuan of	114.97	116.71	105.37	82.44	74.45

	industrial output value					
	Agricultural water consumption rate	54.46	52.32	47.15	48.97	48.16
	Ratio of ecological water compensation	0.72	0.70	0.81	0.83	0.83
	Per capita domestic water consumption	158	158	158	158	157
	Discharge of wastewater of 10,000 yuan of output value	36.00	25.00	57.00	60.96	76.08
	COD concentration in wastewater	3.85	4.04	2.54	5.97	5.26
State	Per capita water resources	2796	3173	3910	4317	4460
	Utilization rate of water resources	19.59	16.97	12.46	10.91	10.49
	Full factor water quality compliance rate	68.90	63.80	80.40	80.40	87.50
	GDP per capita	8.01	8.89	9.32	7.58	8.85
Impact	Rate of urbanization	67.54	69.29	70.81	63.63	65.18
	Green coverage rate in built-up areas	49.52	48.74	47.84	47.65	42.29
Response	Ratio of urban sewage treatment	82.76	82.59	96.89	113.49	109.20
	Comprehensive utilization rate of industrial solid waste	47.93	100.00	100.00	100.97	125.93
	Percentage of water conservancy and public infrastructure investment	6.47	4.71	3.93	3.90	3.54
	Number of College Students per 10,000	166	173	184	191	190

2.3 Evaluation criteria

The key of evaluation work is the establishment of the standard value which directly affects the quality of evaluation results. There are 3 criteria to determine the standard value: (1) the standard value of the existing international or national standards, (2) the standard value is determined based on the development plans of the study areas or condition of environmental protection, (3) indicators without any standard reference should be determined according to previous experience, average value or peak value of the research area. On this basis, this study selects standard values based on the specific situation in GQCC, Jiangxi province as shown in Table 3.

Table 3. Evaluation standard values.

Criteria	Alternatives	Attribute	Standard value	Source
Drive	Annual growth rate of GDP	Positive	9.9	Five-year average value of Gongqingcheng
	Natural population growth rate	Negative	8	Refer to leading domestic cities
	Water resources per unit area	Positive	95.8	Five-year average value of Gongqingcheng
Pressure	Water consumption per 10,000 yuan of industrial output value	Negative	100	Five-year average value of Gongqingcheng
	Agricultural water consumption rate	Negative	55	The critical value
	Ratio of ecological water compensation	Negative	0.83	The highest value of Gongqingcheng in 5 years
	Per capita domestic water consumption	Negative	220	The highest value of Gongqingcheng in 5 years
	Discharge of wastewater of 10,000 yuan of output value	Negative	51	The highest value of Gongqingcheng in 5 years
	COD concentration in wastewater	Negative	4.3	The highest value of Gongqingcheng in 5 years
State	Per capita water resources	Positive	3700	Five-year average value of Gongqingcheng
	Utilization rate of water resources	Negative	40	Internationally recognized values
	Full factor water quality compliance rate	Positive	76.2	Five-year average value of

	GDP per capita	Positive	8.5	Gongqingcheng Five-year average value of Gongqingcheng
Impact	Rate of urbanization	Negative	65	Fifth census
	Green coverage rate in built-up areas	Positive	45	National ecological assessment standards
Response	Ratio of urban sewage treatment	Positive	95	National environmental protection model city assessment standards
	Comprehensive utilization rate of industrial solid waste	Positive	89.5	Energy conservation and emission reduction targets
	Percentage of water conservancy and public infrastructure investment	Positive	3	Extrapolating from developed countries
	Number of College Students per 10,000	Positive	100	Extrapolating from developed countries

2.4 Calculation of index weight

2.4.1 Standardization

Range method is used to standardize the primary data in order to avoid the impact caused by the difference of index in unit and dimension. Specifically, the indexes are divided into positive index and negative index. The formula is as follows:

Positive index：

$$A_{ij}=\frac{x_{ij}-min\ (x_{ij})}{max\ (x_{ij})\ -min\ (x_{ij})} \tag{1}$$

Negative index：

$$A_{ij}=\frac{max\ (x_{ij})\ -x_{ij}}{max\ (x_{ij})\ -min\ (x_{ij})} \tag{2}$$

Where, x_{ij} is the original data of the i-th evaluation object on the j-th evaluation index, A_{ij} is the standardized value of the i-th evaluation object on the j-th evaluation index, $A_{ij}\in[0,1]$.

2.4.2 Entropy method

Entropy method is a commonly weighting method. It analyses the weight according to the influence of the index on the system. For example, the smaller entropy of the indicator, the greater amount of information and the degree of variation to the system, which can play a greater role in the overall comprehensive evaluation.

First, formula (3) was used to determine the index proportion R_{ij}:

$$R_{ij}=A_{ij}/\sum_{j=1}^{m}A_{ij} \tag{3}$$

Then, calculate the entropy of the i-th index e_i:

$$e_i=-\frac{1}{lnn}\sum_{i=1}^{n}(R_{ij}-lnR_{ij}) \tag{4}$$

Further, calculate the difference coefficient of the i-th index g_i:

$$g_i = 1 - e_i \tag{5}$$

Finally, calculate the weight of i-th index ω_{ij}:

$$\omega_{ij} = g_i / \sum_{j=1}^{m} g_i \tag{6}$$

Where, i =1,2,3,...,n, j =1,2,3,...,m, R_{ij} is the proportion of the index value in all objects of the same index, e_i is the index j entropy value in the evaluation problem of m evaluation indicators and n evaluation objects, g_i is the difference coefficient of the i-th index, ω_{ij} is the entropy value of the j-th index.

The weight which is obtained from entropy method is shown in Table 4.

Table 4. Calculation results of entropy weight method.

Criteria	Alternatives	Entropy	Difference coefficient	Weight
Drive	Annual growth rate of GDP	0.8141	0.1859	0.0392
	Natural population growth rate	0.8688	0.1312	0.0277
	Water resources per unit area	0.8041	0.1959	0.0413
Pressure	Water consumption per 10,000 yuan of industrial output value	0.6737	0.3263	0.0688
	Agricultural water consumption rate	0.8239	0.1761	0.0371
	Ratio of ecological water compensation	0.5952	0.4048	0.0854
	Per capita domestic water consumption	0.1344	0.8656	0.1826
	Discharge of wastewater of 10,000 yuan of output value	0.8040	0.1960	0.0413
	COD concentration in wastewater	0.8015	0.1985	0.0419
State	Per capita water resources	0.8068	0.1932	0.0407
	Utilization rate of water resources	0.8216	0.1784	0.0376
	Full factor water quality compliance rate	0.8084	0.1916	0.0404
	GDP per capita	0.8163	0.1837	0.0387
Impact	Rate of urbanization	0.7929	0.2071	0.0437
	Green coverage rate in built-up areas	0.8668	0.1332	0.0281
Response	Ratio of urban sewage treatment	0.6820	0.3180	0.0671
	Comprehensive utilization rate of industrial solid waste	0.8621	0.1379	0.0291
	Percentage of water conservancy and public infrastructure investment	0.6638	0.3362	0.0709
	Number of College Students per 10,000	0.8182	0.1818	0.0383

2.5 The method of water security evaluation

Based on the single factor evaluation method, the comprehensive index method divides the evaluation object into multiple measurable indexes. After selecting the corresponding evaluation criteria for the indexes, the weighted evaluation of each index is carried out. Finally, the comprehensive score of the evaluation object is obtained, which is the water security index in this paper.

Let x_i be the actual value of the evaluation index, p_i be the safety index of index *i*, s_i be the standard index of index *i*, and $0 \le p_i \le 1$.

For the positive index, the safety value is the standard value:

$$p_i = \begin{cases} 1 & x_i \ge s_i \\ x_i/s_i & x_i < s_i \end{cases} \quad (7)$$

For the negative index, the unsafe value is the standard value:

$$p_i = \begin{cases} 0 & x_i \ge s_i \\ 1 - x_i/s_i & x_i < s_i \end{cases} \quad (8)$$

The comprehensive index model is shown in the following formula:

$$E = \sum_{i=1}^{n} W_i P\ (x_i) \quad (9)$$

Where, E is the comprehensive index of urban water security evaluation, n is the total number of indicators, W_i is the weight of each index, and , $P\ (x_i)$ is the security index of each index.

3. Results and Discussion

3.1 Results

According to the previous research, the status of water safety in GQCC, Jiangxi province can be evaluated by the value of *E* and ranked as corresponding five levels, as presented in Table 5. Table 6 presents the safety index in GQCC.

Table 5. The range of safety class

Safety class	Range	Gradient ramp
safe	$0.8 \le E \le 1$	
	$0.6 \le E < 0.8$	
critical	$0.4 \le E < 0.6$	
	$0.2 \le E < 0.4$	
dangerous	$0 \le E < 0.2$	

Table 6. The safety index in GQCC

Alternatives	Safety index				
	2013	2014	2015	2016	2017
Annual growth rate of GDP	1	1	1	0.97	0.93
Natural population growth rate	0.16	0.15	0.15	0.13	0.03
Water resources per unit area	0.74	0.85	1	1	1
Water consumption per 10,000 yuan of industrial output value	0	0	0	0.18	0.26
Agricultural water consumption rate	0.01	0.05	0.14	0.11	0.12
Ratio of ecological water compensation	0.13	0.16	0.02	0	0
Per capita domestic water consumption	0.28	0.28	0.28	0.28	0.29
Discharge of wastewater of 10,000 yuan of output value	0.29	0.51	0	0	0

COD concentration in wastewater	0.1	0.06	0.41	0	0
Per capita water resources	0.76	0.86	1	1	1
Utilization rate of water resources	0.51	0.58	0.69	0.73	0.74
Full factor water quality compliance rate	0.9	0.84	1	1	1
GDP per capita	0.94	1	1	0.89	1
Rate of urbanization	0	0	0	0.02	0
Green coverage rate in built-up areas	1	1	1	1	0.94
Ratio of urban sewage treatment	0.87	0.87	1	1	1
Comprehensive utilization rate of industrial solid waste	0.54	1	1	1	1
Percentage of water conservancy and public infrastructure investment	1	1	1	1	1
Number of College Students per 10,000	1	1	1	1	1
Annual growth rate of GDP	0.4872	0.5229	0.5430	0.5279	0.5349

3.2 Weighted Index Value

The values of water security index in GQCC, Jiangxi province are determined by using entropy weight method (EWM). There are 5 indicators with the weighted value of over 0.05, including water consumption per capita, ecological water replenishment ratio, proportion of investment in water conservancy and other infrastructure, industrial water consumption amount per unit output value of 10,000 yuan and urban disposal rate of sewage. It suggests that these 5 indicators pose a more significant impact on urban water security when other weighted indicators value are mostly around 0.04. The percentage of industrial solid wastes utilized, coverage rate of afforestation in developed area and the rate of natural increase (RNI) stay at the bottom of the list, with the weight values of 0.0291, 0.0281 and 0.0277 respectively, having a moderate impact on water security.

3.3 Trends of Criterion Layer

Based on the analysis of five subsystems in the evaluation criterion layer of urban water security, comparatively significant fluctuation can be detected in pressure and situation subsystem while driving force, response and influence subsystem remained stable.

As shown in Figure 2, during the five-year period, the weighted values of the driving force system were fluctuated between 0.07 and 0.08, similarly, the value for impact system stayed almost unchanged at 0.028. In addition, the response system was the most efficient in view of its gentle upward trend, reflecting that GQCC has made continuous efforts to improve and maintain the water security environment as well as popularize relevant knowledge. More dramatically, the weighted value of pressure system changed in the shape of "S", the data of state subsystem declined after the highest point at 2015 but shown an upward trend in general. In a conclusion, the evaluation value of response subsystem occupied the highest point, followed by the state, pressure driving force and influence system respectively.

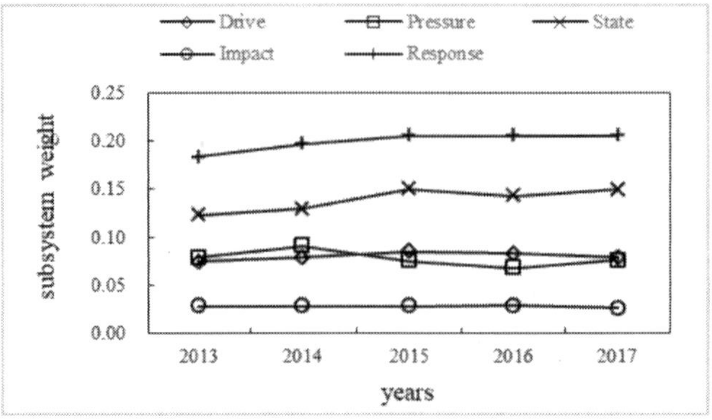

Figure 2. The evaluation value criterion layer.

3.4 Trends of Water Security

According to the formula (9), the comprehensive evaluation index of water security in GQCC from 2013 to 2017 can be calculated (see figure 3). During this five-year period, the trend of index values changed in the shape of "S", reaching the highest point at 2015, that is 0.543. To sum up, the water security index in GQCC is at a critical safety state, suggesting that water security situation is still not promising.

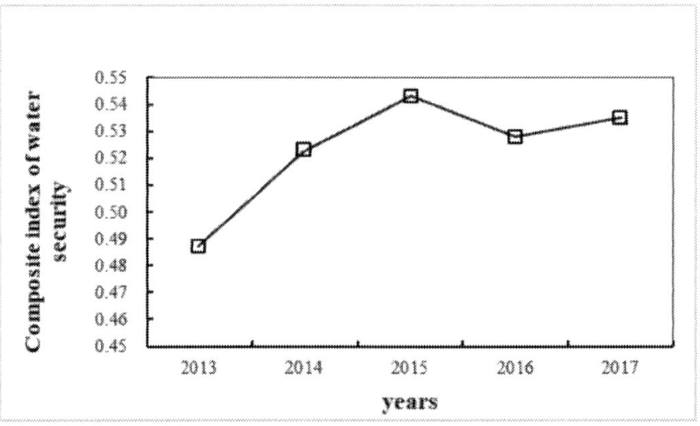

Figure 3. Composite index of water security in 2013-2017.

4. Conclusion

By the analysis of index values, some conclusions can be made that: (1) there are five main factors to urban water security which are water consumption per capita, ecological water replenishment ratio, proportion of investment in water conservancy and other infrastructure, industrial water consumption amount per unit output value of 10,000 yuan and urban disposal rate of sewage; (2) from 2013 to 2017, the change in pressure system was the most prominent while other systems were relatively stable, indicating that intervention measures are urgently needed to alleviate the pressure subsystem; (3) the water security index of GQCC is in a critical safety state. According to the above evaluation results, it can be known that the water security of GQCC requires more indicators to support the response subsystem for alleviating the threat of pressure subsystem and promoting the sustainable urban water security development.

Hence, we put forward some suggestions as following: (1) fully use the lakes in GQCC, strengthen the implementation of the water system connection project, improve the storage network system of unified and coordinated water resources allocation, and alleviate the water pressure of industry and residents; (2) strictly implement the management of water resources and promote the formulation and implementation of water-saving standards and water quotas in various fields, especially in the industrial sector; (3) By speeding up the construction of information infrastructure for water monitoring, especially paying attention to the standards of "quality" and "quantity" of industrial pollution discharge, we can minimize the harm of water pollution to the water environment and promote the transformation of water environment management guided by environmental quality; (4) GQCC need to optimize the spatial pattern and improve usage efficiency of urban space, and land, scientifically plan and construct new and old urban areas, reduce the shortage of water resources caused by excessive population density, and optimize the allocation of water resources.

References

[1] Robert C and Brear S 2016 *Urban Water Security* (independent consultant and environmental advisor, London, UK) chapter 2 pp25-34
[2] Maulud K N A, Fitri A, Mohtar W H M W, Jaafar W S W M, Zuhairi N Z, and Kamarudin M K A 2021 A study of spatial and water quality index during dry and rainy seasons at Kelantan River Basin, Peninsular Malaysia *Arabian Journal of Geosciences* **14**(2) pp 1-19
[3] Fitri A, Maulud K N A, Pratiwi D, Phelia A, Rossi F and Zuhairi N Z 2020 Trend of Water Quality Status In Kelantan River Downstream, Peninsular Malaysia *Jurnal Rekayasa Sipil (JRS-Unand)* **16** (3) pp 178-184
[4] Fitri A, Maulud K N A, Rossi F, Dewantoro F, Harsanto P, and Zuhairi N Z 2021 Spatial and Temporal Distribution of Dissolved Oxygen and Suspended Sediment in Kelantan River Basin In *4th International Conference on Sustainable Innovation 2020–Technology, Engineering and Agriculture (ICoSITEA 2020)* Atlantis Press pp 51-54
[5] Fitri A, Hashim R, Song K I, and Motamedi S 2015. Evaluation of morphodynamic changes in the vicinity of low-crested breakwater on cohesive shore of Carey Island, Malaysia *Coastal Engineering Journal* **57** (04) pp 1550023.
[6] Fitri A, Hashim R, Abolfathi S, & Abdul Maulud K N 2019 Dynamics of sediment transport and erosion-deposition patterns in the locality of a detached low-crested breakwater on a cohesive coast *Water* **11** (8) pp 1721
[7] Fitri A, Hashim R, and Motamedi S 2017 Estimation and Validation of Nearshore Current at the Coast of Carey Island, Malaysia *Pertanika Journal of Science & Technology* **25** (3)
[8] Fitri A, Yao L, and Sofawi B 2019 Evaluation of Mangrove Rehabilitation Project at Carey Island Coast, Peninsular Malaysia based on Long Term Geochemical Changes *In IOP Conference Series: Earth and Environmental Science IOP Publishing* **365** (1) pp 012055
[9] Fitri A, and Yao L 2019 The impact of Parameter Changes of a Detached Breakwater on Coastal Morphodynamic at Cohesive Shore; A Simulation *In IOP Conference Series: Earth and Environmental Science IOP Publishing* **365** (1) pp 012054
[10] Yao L, Li J, Shi S, and Fitri A 2019 Simulation Take-off Angle of a Ski Jump Energy Dissipater. *In IOP Conference Series: Earth and Environmental Science IOP Publishing* **365** (1) pp 012057
[11] Chen H, Yao L, and Fitri A 2019 The Influence Mechanism Research of Inflow Temperature in Different Time Scale on the Water Temperature Structure *In IOP Conference Series: Earth and Environmental Science IOP Publishing* **365** (1) 012058
[12] Hashim R, Roy C, Shamshirband S, Motamedi S, Fitri A, Petković D, and Song K I 2016 Estimation of wind-driven coastal waves near a mangrove forest using adaptive neuro-fuzzy inference system *Water resources management* **30** (7) pp 2391-2404
[13] Hashim R, Fitri A, Motamedi S, Hashim A M 2013 Modelling of Coastal Hydrodynamic Associated with Coastal Structures: A Reviews *Malaysian. Journal of Science* **32** pp 149-154
[14] Fitri A, Hasan Z A and Ghani A A 2011 Effectiveness of Aman Lake as flood retention ponds in flood mitigation effort: study case at USM Main Campus, Malaysia *3rd International Conference on Managing Rivers in 21st Century: Sustainable Solutions for Global Crisis of Flooding, Pollution and Water Scarcity (Rivers 2011)* Penang Malaysia
[15] Fitri A, Hasan Z A and Ghani A A 2011 Determining the Effectiveness of Harapan Lake as Flood Retention Pond in Flood Mitigation Effort In *Proceedings of 2011 4th International Conference on Environmental and Computer Science (ICECS 2011).*
[16] Hoekstra A Y, Buurman J, van Ginkel K C H 2018 *J. Urban water security: A review Environmental Research Letters* **13** (5) pp 053002
[17] Shikun S, Yubao W, Jing Liu *et. al.* 2016 Sustainability assessment of regional water resources under the DPSIR framework *Journal of Hydrology* pp1-9
[18] Hongmei X, Lahua J 2011 Study on regional water security assessment based on DPSIR model *Journal of Safety and Environment* **11** (1) pp 96-100

Acknowledgments

The authors would like to thank the Department of Education in Jiangxi Province (Grant No. GJJ180953) to financially support in completing this study. We also would like to thank the Belt and

Road Special Foundation of the State Key Laboratory of Hydrology-Water Resources and Hydraulic Engineering (Grant No. 2018492011) and the Science and Technology Research Project of the Science and Technology Department of Jiangxi Province (Grant No. 20192ACBL20041).

Future impacts of climate change on sediment influx rate in hydropower reservoir using SWAT

A Z Abdul Razad[1*], S H Shamsuddin1, A Setu[1] and L Mohd Sidek[2]

[1] TNB Research Sdn Bhd, No 1, Lorong Air Hitam, Kawasan Institusi Penyelidikan Bangi, 43000 Kajang, Selangor, Malaysia
[2] Civil Engineering Dept, UNITEN, Jalan IKRAM-UNITEN, 43000 Kajang, Selangor, Malaysia

*Corresponding E-mail: azwin.razad@tnb.com.my

Abstract. Climate change causes more frequent and intense rainfall events, leading to severe erosion in the catchment and sediment transferred into rivers and reservoirs. This study focus on long term sediment load in major rivers in Cameron Highlands and prediction of annual sediment inflow into Ringlet Reservoir from 2000 to 2030. Soil Water Assessment Tool (SWAT) is used as the simulation tool, utilising future gridded rainfall 2017 to 2030 under CCSM and future land use 2030. Future annual rainfall is minimum at 1551 mm (in 2030) and maximum at 3150 mm (in 2029). The future projected annual sediment load into Ringlet Reservoir from 2017 to 2030 is averaged at 354,013 m^3/year, ranging from 216,981 to 461,886 m^3/year. Comparing between the historical period of 2000 to 2016 and future projection (2017–2030), annual sediment load shows an increase of 12 %. To combat the increase sediment yield, catchment management such as erosion control plan, drainage and runoff control must be developed to minimise sediment yield and subsequent effect of high sediment load transport via rivers and drainage network.

Keywords: Climate change, land use, future rainfall, sediment yield, hydropower reservoir

Track Name: Land, Water, Forests and Food Security

1. Introduction

Average rate of global reservoir sedimentation is estimated to be 1 % [1] while Basson [2] estimated an average of 0.93 % sedimentation loss per year. A decrease in storage due to excessive sedimentation reduces the storage availability for hydropower generation, water supply, and irrigation. This directly reduces the revenue but increases the operation cost to remove sediment or clearing intake blockage. Flood storage reduction increasing the possibility of using spillways more frequently to divert floodwaters. This could increase the risk of flooding to downstream areas [3, 4]. Modern dam engineering standards have been modified and improved to ensure that dams are designed properly to an acceptable limit. Future development and impact of climate change would affect the hydrological regime and sediment generation, thus these must be incorporated in the design methodology.

Based on the study on impact of future climate change on the hydrologic regime and water resources of Peninsular Malaysia [5] both the historical and future simulations of the hydro-climate of Peninsular Malaysia were performed by downscaling the climate simulations of CGCM from 2000 to 2100. The results of these analyses indicated an expected increase in inter-annual and intra-seasonal variability with increased hydrologic extremes events (higher high flows, and lower low flows) in Kelantan, Pahang, Terengganu, and Kedah watersheds in the future. An increase up to 51 % in maximum monthly rainfall over the North East Coastal Region and 10 % increase of annual rainfall for Kelantan, Terengganu and Pahang are expected [6].

Soil and Water Assessment Tool (SWAT) is a physically-based catchment model which was developed to simulate the rainfall runoff process, erosion, sediment yield, and pollutant load resulting from climate change, land use change, and management plan in the catchment on the long term daily and sub-daily time scale [7, 8]. SWAT application is also dominant in Southeast Asia, primarily in land use impacts and climate change assessment on water yield, sediment, and nutrient transport with limitations on extreme events simulation [9]. SWAT computes runoff using the SCS curve number method, while erosion and sediment yield is determined based on USLE and MUSLE equations. Sediment load at the rivers is based on modified Bagnold's equation, of which excess sediment will be deposited in the channel [10]. SWAT needs extensive input data to meet the requirement for prediction but if data are lacking, then careful calibration and validation must be carried out.

This paper describes the process involved in simulating the future impacts of climate change and land use on annual inflow and sediment load into Ringlet Reservoir, a hydropower reservoir located in Cameron Highlands using SWAT model.

2. Study Area

Ringlet Reservoir is located in the state of Pahang. It is part of Cameron Highlands–Batang Padang Hydroelectric Scheme, which is used mainly as a regulating reservoir to generate electricity. Figure 1 illustrates the location of Ringlet Reservoir and its contributing catchment and major features of Cameron Highlands–Batang Padang Hydroelectric Scheme. Water impounded in Ringlet Reservoir is originated mainly from Sg Bertam, Sg Telom, Sg Habu, Sg Ringlet, Sg Kial, Sg Kodol and Sg Plauur in the catchment of Cameron Highlands. The catchment, which comprises an area of 180.6 km^2, has been facing rapid changes in land use since the 1960s due to highland agriculture and urbanisation. Increase in soil loss has increased the sediment load transported by the river network, causing an increase in sediment inflow into Ringlet Reservoir. Cameron Highlands has three active townships, namely Ringlet, Tanah Rata and Brinchang.

The elevation ranges from EL1068 m to the highest of EL2168 m above mean sea level with monthly temperature ranges from 14 °C to a maximum of 23 °C. The catchment is subjected to two rainy seasons; from April to May and from September to November with an average annual rainfall of 2400 mm to 2700 mm. Monthly rainfall ranges from minimum of 100 mm in January and maximum of 300 mm in October to November while evaporation is averaged at 1.8 mm/day [11].

Figure 1. Location of Ringlet Reservoir and its catchment area.

3. Materials and Methodology

Cameron Highlands has a good hydrological network involving several rainfall and stream flow stations and meteorological station, which were installed and maintained by Tenaga Nasional Berhad (TNB), Department of Irrigation and Drainage (DID) and Meteorological Department of Malaysia (MMD). Long term sediment discharge data from 1986 were used to derive the sediment rating curves, which was coupled with observed stream flow at Sg Bertam to generate continuous sediment load. Land use, soil type, soil map and topography were obtained from the Department of Agriculture (DOA), Food and Agriculture Organisation (FAO) and TNB Research respectively. Rainfall, evaporation, humidity, wind speed and solar radiation, were obtained from Tenaga Nasional Berhad (TNB), Department of Irrigation and Drainage (DID) and Meteorological Department (MMD). Table 1 summarises data used in this study.

An ensemble approach to project future climate is based on four emission scenarios of which the best possible is represented by SRES B1, the worst possible is SRES A1FI, the most likely scenario is SRES A1B, and the second worst possible is SRES A2 [12]. Therefore, future rainfall data under climate change scenario of CCSM3 SRES A1B (most likely) and CCSM3 SRES A2 (second worst possible) are obtained from NAHRIM.

Table 1. Input data in this study.

Data type	Year	Resolution	Source
Topography Digital Elevation Model (DEM)	2000	30 m horizontal resolution	SRTM
Soil Map	-	1:5,000,000	FAO
Land use map	2006, 2010, 2015 and 2030	NA	DOA, Majlis Daerah Cameron Highlands (MDCH)
Rainfall data	2000 to 2016,	Daily	TNB DID, MMD
Future rainfall data	2017 to 2030	Daily	NAHRIM
Streamflow data at Sg Bertam	2000 to 2016	Daily	TNB
Sediment data at Sg Bertam	1986 to 2016	Instantaneous	TNB

3.1 Land use analysis (2006 to 2030)

The land use changes in Cameron Highlands are mainly due to forest conversion to urbanization and agricultural activities. As shown in Figure 2, forest occupies an average of 65 % of the entire catchment from 2006 to projected future in 2030 while agricultural activities show increasing trend from 18 % in 2006 to 25 % in 2030. Urbanisation increase marginally by 1 % in the period of 24 years. Other land use activity such as tea, grassland, scrub forest and water body remain fairly constant throughout those years. This shows that there will be increase in runoff and sediment yield if the future rainfall follows the similar patterns of historical data.

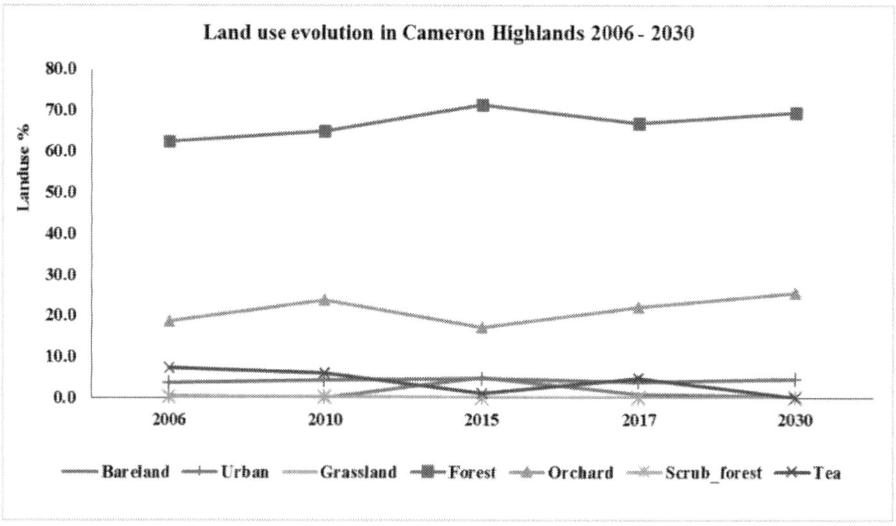

Figure 2. Long term land use change from 2006 to 2030.

3.2 Long term rainfall analysis (2000–2030)

By analysing historical and future rainfall under climate change scenario CCSM3 SRES A1B and CCSM3 SRES A2, average total monthly rainfall data from 10 stations from 2000 to 2030 (Figure 3) shows that there is decreasing trend of rainfall especially in the future period of 2016 to 2030, based on Mann-Kendall trend analysis ($p<0.1$) for both scenarios. However, CCSM3 SRES A1B shows higher extreme in 2028 with maximum average of 752 mm in Oct 2028 while minimum average of 10 mm in May 2021.

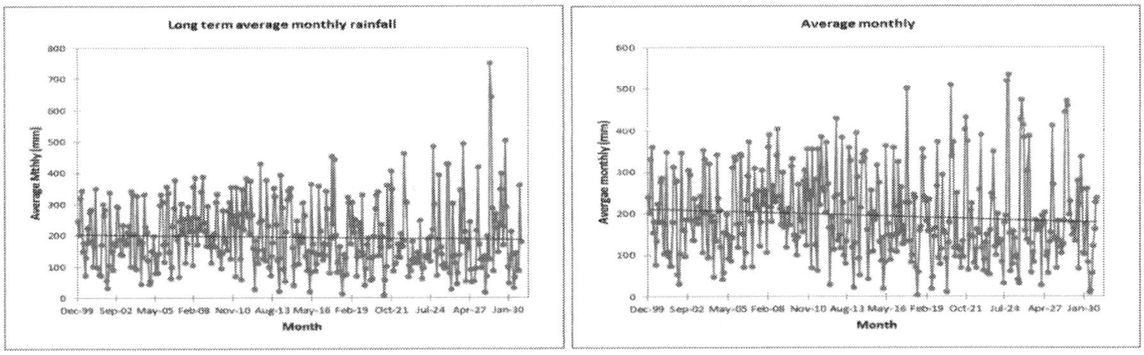

Figure 3. Long term average monthly rainfall in Cameron Highlands from 2000 to 2030 (Left: CCSM3 SRES A1B, Right: CCSM3 SRES A2).

Similar decreasing trend on annual rainfall from 2000 to 2030 is also observed under both scenarios of CCSM3 SRES A1B and CCSM3 SRES A2. Referring to Figure 4 and Figure 5, the extreme low occur in 2023 and 2030 respectively, while the extreme high occurs in 2029 for CCSM3 SRES A1B but none for CCSM3 SRES A2. Based on these trends, inflow and sediment load into the reservoir is expected to be greatly affected under scenario CCSM3 SRES A1B especially in the year 2029 and 2023.

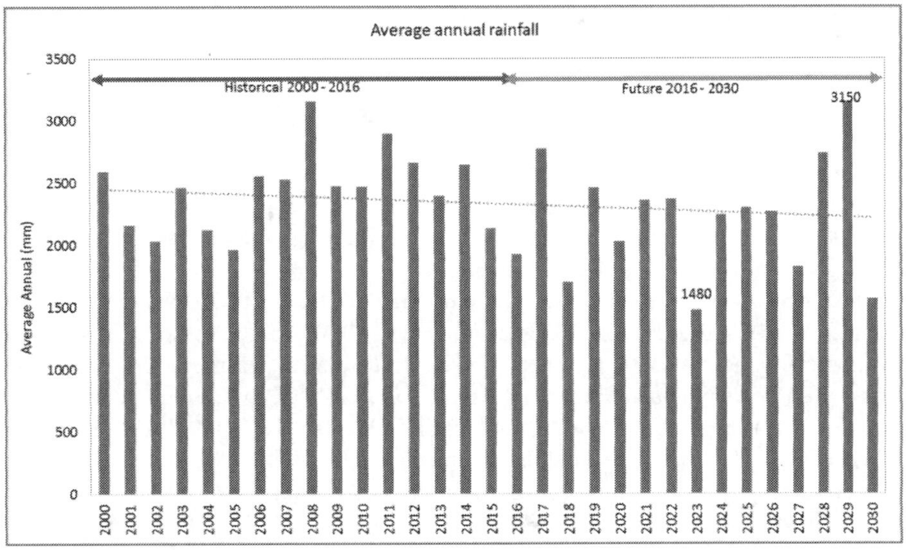

Figure 4. Long term average annual rainfall in Cameron Highlands from 2000 to 2030 under CCSM3 SRES A1B.

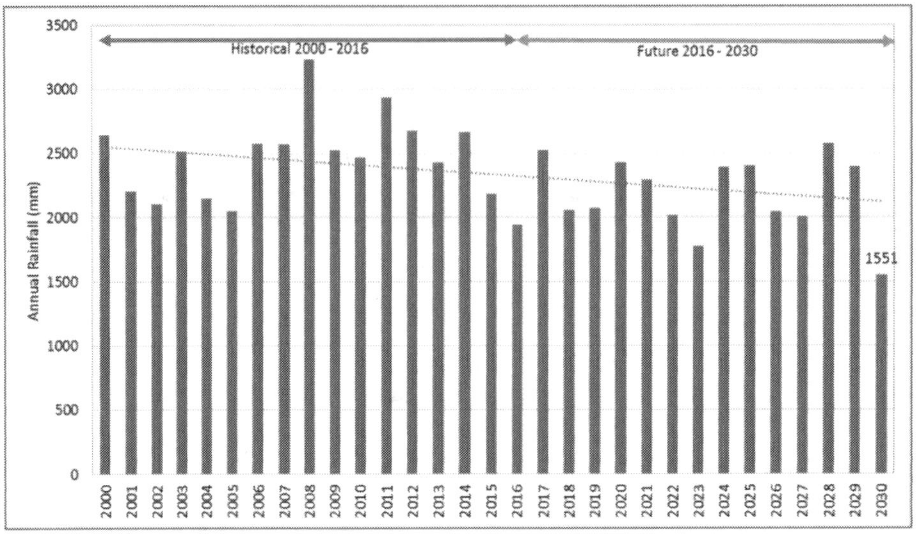

Figure 5. Long term average annual rainfall in Cameron Highlands from 2000 to 2030 under CCSM3 SRES A2

3.3 SWAT Model Setup, Calibration and Validation

The catchment and its reaches were both delineated in Arc SWAT using DEM and stream network layer, and overlaid with maps of soil, land use and slope to produce 52 reaches, 52 sub-basins and 305 HRUs. Land use maps for the year 2006, 2010, 2015 and 2030 were reclassified using SWAT land use code namely forest (FRST), orchard (ORCD), residential (URMD), rangeland (RNGE) and water body (WATR). Slopes were divided into three categories: 0-20 %, 20-35 % and >35 %. Weather data from 2000 to 2016 included humidity, temperature, solar radiation and wind speed were statistically analysed and input into the SWAT weather database. Daily future rainfall data under climate change impact from 2017 to 2030 was analysed as future scenario input.

Monthly streamflow, sediment load and sediment yield from 2001 to 2015 were simulated, of which observed streamflow and sediment load at Sg Bertam from 2001 to 2006 and 2010 to 2015 were used for model calibration and validation, respectively. Performance of model during calibration and validation was evaluated using Nash Sutcliffe efficiency index (NSE), volume percent bias (PBIAS) and ratio of RMSE (RSR). If the simulated results obtained NSE > 0.5, RSR < 0.7, PBIAS $< \pm 25$, the model is considered calibrated. Hence, the parameters can be used to simulate for other scenarios. In this study, the calibrated parameters that control hydrological process are CN2, SOL_AWC, GW_DELAY, OV_N, ALPHA_BF, GWQMN, SOL_BD, SOL_K, GW_REVAP, REVAPM and RCHRG_DP. High CN2 leads to more runoff and high peak flows. For sediment routing in the channels, the parameters used are PRF, SPEXP, SPCON, CH_EROD, CH_COV, while parameters of USLE_P, USLE_C, USLE_K, SLOPE, SLSOIL and LAT_SED are used for sediment yield from the catchment. Summary of research methodology is shown in Figure 6.

IOP Conf. Series: Earth and Environmental Science **880** (2021) 012024 doi:10.1088/1755-1315/880/1/012024

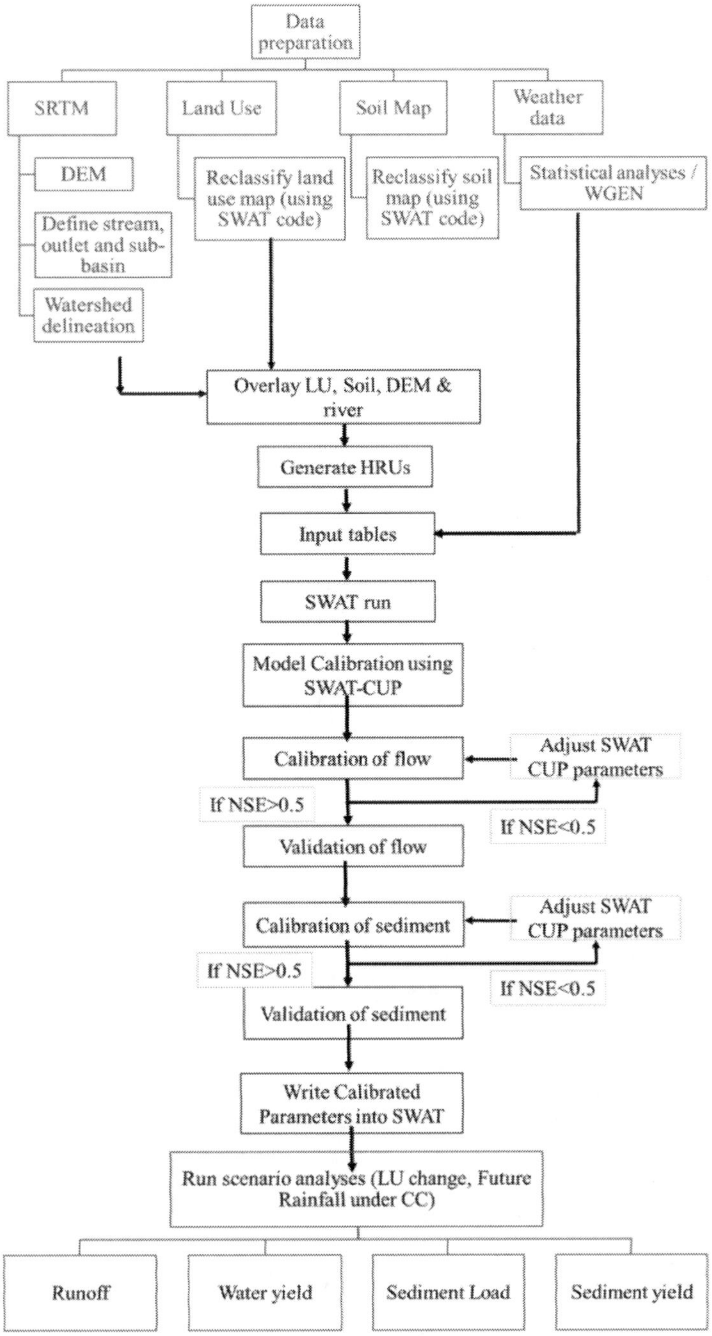

Figure 6. Methodology for simulation of sediment influx rate using SWAT.

4. Results and Discussion

SWAT delineated catchment of Cameron Highlands into 52 sub-basins (sub-catchments) and 52 reaches, assigned using numerical order from 1 to 52. Observed stream flow and sediment load data at Sg Bertam (at Robinson Falls) were used for calibration and validation. Total inflows and total sediment load into Ringlet Reservoir were accumulated stream flows from reach 18 (Sg Telom), reach 21 (Sg Kial), reach 46 (Sg Bertam) and reach 51 (Sg Ringlet).

4.1 Streamflow calibration and validation results

SWAT was able to simulate the rainfall runoff process and peak monthly flows although it overestimated the low flows that was less than 0.7 m^3/s during calibration and validation period as depicted in Figure 7 and Figure 8, respectively. The monthly flow was well simulated from 2001 to 2006. The monthly validation plot in Figure 6 showed good match between simulated and observed monthly flows from 2010 to 2015 and the results also responded well with rainfall distribution. Statistical performance of calibration and validation of streamflow show good performance with NSE > 0.65, $R^2 > 0.65$ and PBIAS < ±10% [13].

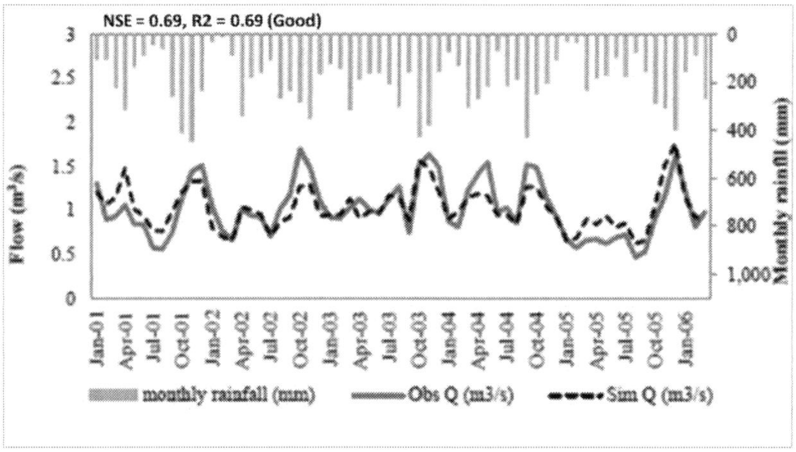

Figure 7. Monthly stream flow calibration plot from 2001 to 2006.

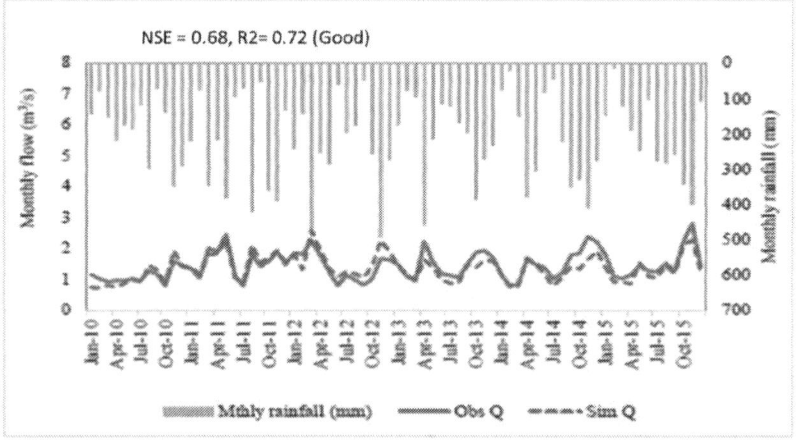

Figure 8. Monthly stream flow validation plot from 2010 to 2015.

4.2 Sediment load calibration and validation results

Observed sediment data at Sg Bertam from 2001 to 2006 and 2010 to 2015 were used for sediment load calibration and validation, respectively. From Figure 9 and Figure 10 below, SWAT simulated the sediment load better in the validation period, as compared to calibration. Peak matches well for period 2010 to 2015 as compared to 2001 to 2006. Statistical performance of the model marked by NSE > 0.5 and $R^2 > 0.5$ showed that sediment load is simulated satisfactorily.

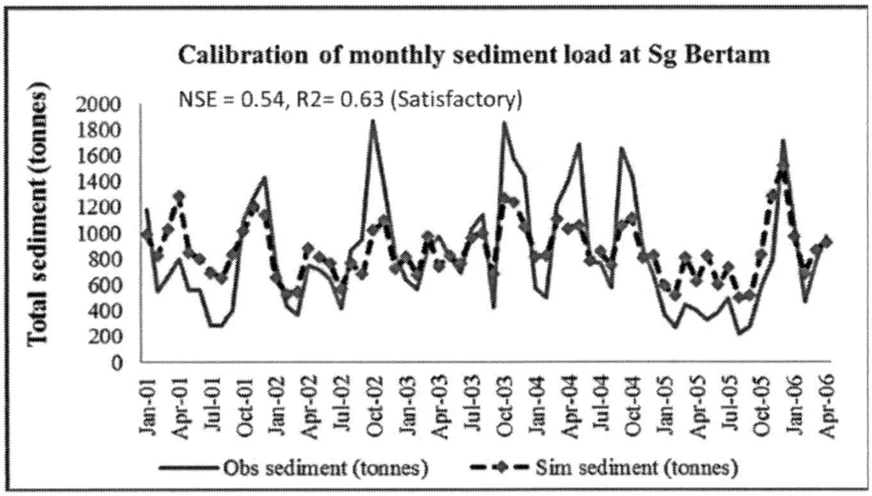

Figure 9. Monthly sediment load calibration plot from 2001 to 2006.

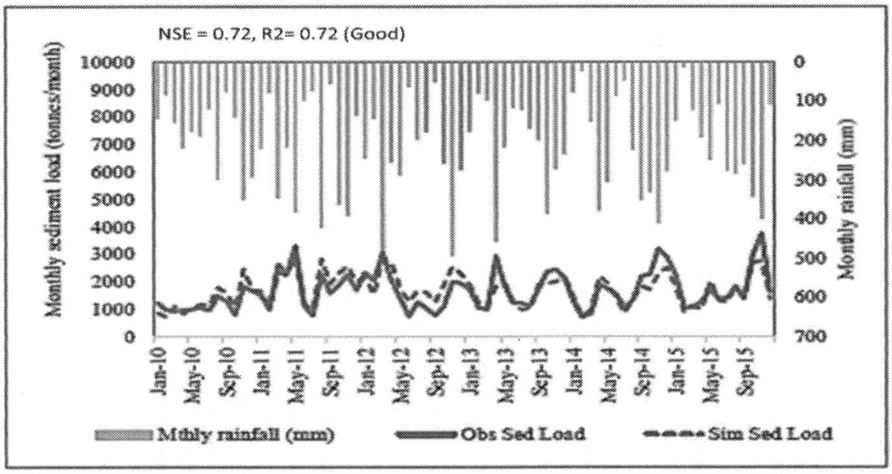

Figure 10. Monthly sediment load validation plot from 2010 to 2015.

From these calibration and validation results, SWAT is able to simulate the flows and sediment load at Sg Bertam to an acceptable accuracy. This enables the model to simulate and predict the variation of flows, sediment yields and sediment loads in other sub-catchment and rivers in Cameron Highlands, thus the impacts of land use on rainfall-runoff-sediment transport processes can be predicted. Using the calibrated parameters, SWAT was employed to simulate stream flow in all rivers in the catchment using historical rainfall from 2000 to 2016 and future rainfall future rainfall from CCSM3 SRES A2 of 2017 to 2030, under different land use from 2006, 2010, 2015 and 2030.

4.3 Annual inflow and annual sediment inflow under varying land use and climate change impact

SWAT simulated monthly stream flow of the major rivers such as Sg Telom, Sg Bertam Sg Ringlet and Sg Habu. They were summed up to determine the yearly variation of stream flow. Figure 11 illustrates the variation of annual inflow of Ringlet Reservoir based on the simulated flows in the sub-catchments. Using the historical rainfall from 2000 to 2016, future rainfall from CCSM3 SRES A2 and variation of land use from 2006, 2010, 2015 and 2030, the long term annual inflow from 2000 to 2030 ranged from 5.22 m^3/s to 10.2 m^3/s, with an average of 7.37 m^3/s. Based on the results of the Mann-Kendall tests ($p<0.05$), there is reducing trend on annual inflow from 2000 to 2030. From Figure 12, annual sediment load simulated from 2000 to 2030 showed an average of 548,257 tonne/year (or 338,430 m^3/year), with minimum and maximum of 311,644 tonne/year and 773,139 tonne/year respectively. Despite the annual rainfall is showing reducing trend under climate change scenario from 2017–2030, the annual sediment load shows higher average from 2017–2030 (354,013 m^3/year) as compared to 2000 to 2016 (316,210 m^3/year). Comparing between the historical and future projection, sediment load shows an increase of 12 %.

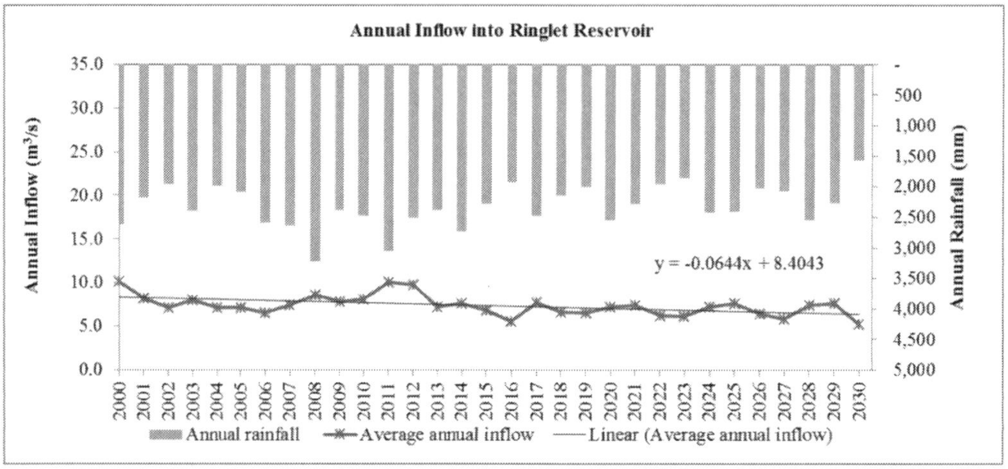

Figure 11. Long term predicted annual inflow into Ringlet Reservoir.

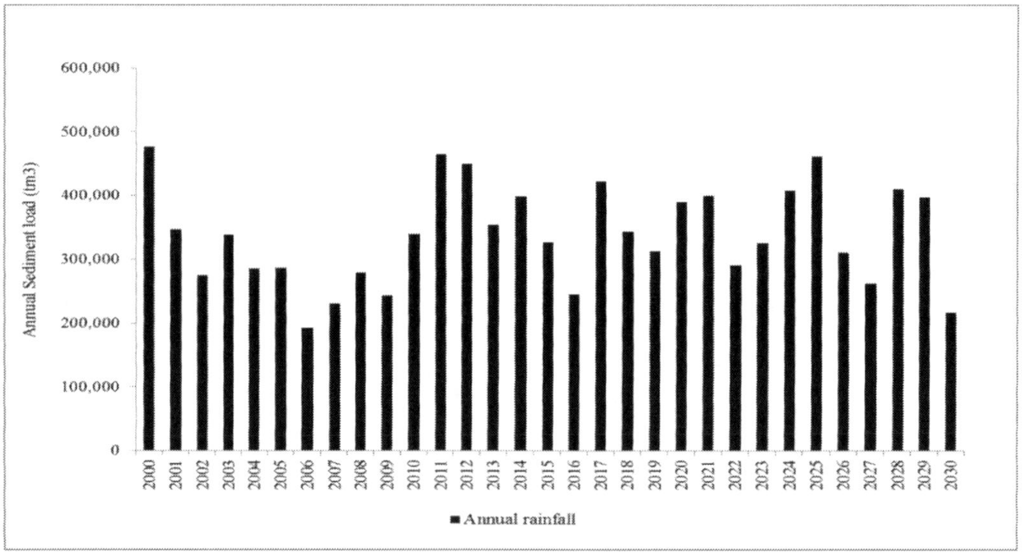

Figure 12. Long term predicted annual sediment load into Ringlet Reservoir.

5. Conclusion

Cameron Highlands is located in highland area, surrounded by active agricultural and commercial activities. The catchment receives high amount of rainfall throughout the year, with average of 2400mm to 2700mm. Hydrological process of rainfall runoff and sediment load are successfully modelled using SWAT as physically based hydrological model. Model runs from 1999 to 2015 with 2 years warm up performed well, achieving good calibration and validation results, with NSE > 0.5, R^2 > 0.5, RSR < 0.7 and PBIAS < ±10%. By varying the land use using land use map of 2006, 2010, 2015, 2030 and using future projected rainfall under CCSM3 SRES A2 scenario, SWAT simulates flow and sediment load on monthly variation. Long term annual inflow from 2000 to 2030 ranged from 5.22 m^3/s to 10.2 m^3/s, with an average of 7.37 m^3/s. Based on the results of the Mann-Kendall tests ($p<0.05$), there is reducing trend on annual inflow from 2000 to 2030. The future projected annual sediment load into Ringlet Reservoir from 2017 to 2030 is averaged at 354,013 m^3/year, ranging from 216,981 to 461,886 m^3/year. Comparing between the historical period of 2000 to 2016 and future projection (2017–2030), annual sediment load shows an increase of 12%. This shows that annual sediment load increases in future although the rainfall in future is on decreasing trend. Extreme high rainfall and extreme low rainfall are expected in the future, which could lead to the increase. To combat the increase in sediment yield, catchment management such as erosion control plan, drainage and runoff control must be implemented to minimise sediment yield and high sediment load transport via rivers and drainage network.

References

[1] Annandale G W 2001 Reservoir conservation and sediment management *Water week 2001* Water Bank Group (Washington, DC)

[2] Basson G 2008 Reservoir sedimentation—An overview of global sedimentation rates and predicted sediment deposition. *Oral Contribution to the International CHR Workshop—Expert Consultation: Erosion, Transport and Deposition of Sediments* (Berne) pp 74-79

[3] Biswas S S and Pani P 2015 Estimation of soil erosion using RUSLE and GIS techniques: a case study of Barakar River basin, Jharkhand, India *Modeling Earth Systems and Environment* **1**(4) 42. pp 1-13

[4] Department of Irrigation and Drainage Malaysia DID 2017 Malaysia Dam Safety Management Guidelines (MyDAMS)

[5] Shaaban A J Amin M Z M Chen Z Q and Ohara N 2011 Regional Modeling of Climate Change Impact on Peninsular Malaysia Water Resources *J. Hydrol. Eng.* 2011 **16** 1040–1049

[6] Amin M Z M Shaaban A J Ercan A Ishida K Kavvas M L Chen Z Q and Jang S 2017 Future climate change impact assessment of watershed scale hydrologic processes in Peninsular Malaysia by a regional climate model coupled with a physically-based hydrology model. *Sci. Total. Environ.* **575** 12–22

[7] Arnold J G Srinivasan R Muttiah R S and Williams J R 1998 Large area hydrologic modeling and assessment part I: Model development *J. Am. Water Resour. Assoc.* **34**(1) pp 73–89

[8] Mulla N H Krishna B M and Manoj Kumar B 2019 A Review on Water Quality Models: QUAL, WASP, BASINS, SWAT and AGNPS *International Journal of Scientific Research in Civil Engineering* **3**(10) pp 58–68

[9] Tan M L Gassman P W Srinivasan R Arnold J G and Yang X Y 2019 A Review of SWAT Studies in Southeast Asia: Applications, Challenges and Future Directions. *Water (Switzerland)* **11**(5) pp 1–25

[10] Arnold J G Kiniry J R Srinivasan R Williams J R Haney E B and Neitsch S L 2013 Soil & Water Assessment Tool: Input/output documentation *Version 2012 Texas Water Resources Institute TR-439 650* Retrieved from http://swat.tamu.edu/media/69296/SWAT-IO-Documentation-2012.pdf

[11] Abdul Razad A Z Sidek L M Jung K Rahman N F and Shamsuddin, S H 2020 Prediction of reservoir sedimentation using Soil Water Assessment Tool (SWAT) towards development of sustainable catchment management *IOP Conference Series: Material Science and Engineering* **736**

[12] Amin M Z M Ercan A Ishida K Kavvas M L Chen Z Q and Jang S H 2019 Impacts of climate change on the hydro-climate of peninsular Malaysia *Water (Switzerland)* **11**(9) pp1798

[13] Moriasi D N Arnold J G Van Liew M W Bingner R L Harmel R D and Veith T L 2007 Model Evaluation Guidelines for Systematic Quantification of Accuracy in Watershed Simulations *Transactions of the ASABE* **50**(3) Pp 885-900

Acknowledgments

Author would like to thank TNB Research and TNB, collaborators and personnel involved in the whole project implementation. The content of this paper is purely for research and academic purpose, which in any circumstances does not represent the view of TNB and its subsidiaries.

Characterization of aquifer system using electrical resistivity tomography (ERT) and induced polarisation (IP) techniques

N Akhtar[1], M S Mislan[2], M I Syakir[1*], M T Anees[3] and M S M Yusuff[1]

[1]School of Industrial Technology, Universiti Sains Malaysia, Pulau Pinang, Malaysia, 11800
[2]Groundwater Development, Puchong Selangor, Malaysia, 47100
[3]Department of Geology, Universiti of Malaya, Kuala Lumpur, Malaysia

[*]Corresponding E-mail: misyakir@usm.my

Abstract. Groundwater plays a major role as an alternative freshwater resource for irrigation and industrial purposes. This study aimed to characterize the subsurface of aquifer systems in TelukIntan district, Perak, Malaysia using Electrical Resistivity Tomography (ERT) and Induced Polarization (IP) methods. The horizontal profiling (TL1 and TL2) was conducted at length of 400 m. The estimated depth is 150 m below ground level (b.g.l.). An ABEM SAS 4000 Terrameter and ABEM LUND ES464 Imaging System were applied to create a resistivity pseudo-section using polar-dipole configuration. The collected geo-electrical data was interpreted using RESIST software with partial curve matching and computer iteration. ERT and IP survey profile results were validated with in-situ borehole data from borehole 2 (B2). Eleven samples of soil profile were collected at depth from 5.6 m to 61.2 m, with average percentage of sand, silt and clay are 93.77 %, 5.78 % and 0.02 %, respectively. The geology of subsurface settings is the key factor in determining the aquifer system characterized by interlayer sand-silt sequence indicating the saturation zone of aquifer underlain by shale at the bottom. Further study on hydraulics perspective is important to understand the overall capacity of the aquifer.

Keywords: Aquifer system; sub surface profiling; groundwater; electrical resistivity tomography (ERT); induced polarisation (IP)

Track Name: Atmospheric Chemistry and Physics

1. Introduction

Groundwater is a critical resource for the environmental system and human consumption [1]. The availability, quantity, and quality of groundwater mainly depend on aquifer characteristics, natural processes, human activities, and underlying geological formations [2,3]. Groundwater is stored either in alluvial formations (clay, silt, sand, gravel, sandstone, limestone, or unconsolidated substances, etc.) or hard rock formations such as igneous and metamorphic rocks. UNESCO has estimated that the worldwide water demand will increase by 20 % and up to 30 % (4600 km^3 per year) in 2050, particularly in three sectors namely, agriculture, household, and industry [4]. In Malaysia, large quantity of fresh water demand is increasing too i.e. for rice cultivation and drinking water [5,6]. Given the fact that Malaysia is water abundance region, the potential of groundwater resource is not fully utilised currently due to surface water remain as the major resources (97 %) for industries, agriculture, and other human consumption in Malaysia.

Previous studies also reported that surface water quality is decreasing in Malaysia due to a growing number of contamination sources such as palm oil mills, sewage treatment plants, irrigation practices and other industrial applications [6,7], hence it will limit the accessibility and availability of surface water particularly for irrigation purposes. In this context, the groundwater was identified as one of Malaysia's alternative water supplies [8]. Various methods have been performed to study the hydraulic properties of aquifer system i.e. nuclear magnetic resonance, multilevel slug tests, pumping tests, grain size distribution from core samples, and single-borehole measurements in completed boreholes, as flow meter measurements [9,10]. Furthermore, the geophysical information of the subsurface characteristic is useful as an alternative approach to wildcat drilling [11], hence provide better decision making in determine the best potential well for drilling.

The geophysical investigation is essential to avoid drilling abortive wells, and it leads to addresses failing boreholes problems effectively because they provide the knowledge of sub- aquifer boundaries and subsurface lithology properties of a specific location [12–14]. In fact, geophysical is often plagued with uncertainties and ambiguities in interpretation, and it provides a cost-effective and rapid way of obtaining continuous information from near-surface [15–17].

Electrical resistivity tomography (ERT) or (electrical resistivity imaging Induced polarization and (IP) methods have been mainly used in the geophysical survey to determine the subsurface profiles and resistivity of substances [18,19]. ERT and IP methods are based on a current flow by the subsurface medium which is comprised of fissures and pores of soils/rocks entirely or partly filled with water and layers of various resistive materials [11]. Resistivity is a parameter measured in the unit of ohm meter (Ω.m) which determines the resistivity of a particular substance to the electrical current flow. Electrical current is produced in the ground, the possible difference is measured, and resistivity can be determined.

ERT has been used for different purposes such as monitoring of pollutants, hydrogeology, volcano imaging, engineering geology (studying and tracking leakage in dams and embankments), agricultural fields, etc. [18]. In such cases, ERT is a suitable technique and helps to view the electrical characteristics of the subsurface features because groundwater and soil moisture are mostly electrically conductive [20]. The calculated resistivity through the ERT method is influenced mainly by the amount of water in the soil and by the concentration in porous media of dissolved solids [17]. ERT is limited to mapping the medium electrical resistivity, while the IP method provides the ability to distinguish specific physical characteristics of the subsurface which is known as chargeability.

The IP method has been developed especially for the prospecting of minerals [21]· IP survey is a helpful technique to map unconsolidated sediment layers of lithological material [22]. It also provides an essential parameter that can be utilized for the results interpretation by geophysical methods to resolve ambiguities. Nonetheless, due to the heterogeneity of rock and soils at the local scales in which the polarization process operates, the interpretation of IP parameters in subsurface formations may be complex [23].

The objective of this study is to characterize the aquifer system in the subsurface lithology of LabuKubung, TelukIntan district, Perak, Peninsular Malaysia. A combination of ERT and IP methods is used to investigate the subsurface profile and potential groundwater resources based on resistivity and chargeability values.

2. Materials and Methods

2.1. Study area

The study area is in the tropical region of Peninsula Malaysia, a part of the TelukIntan district in the Perak state; its name is LabuKubung, as shown in Figure 1. The approximate coordinates are 4° 7' 18.35" N and 101° 2' 0.67" E at an average elevation of 39 to 41 meters above mean sea level

IOP Conf. Series: Earth and Environmental Science **880** (2021) 012025 doi:10.1088/1755-1315/880/1/012025

(Figure 1). The general topography of the study area is naturally flat plain and it is surrounded by paddy fields and main roads [24]. This study area receives approximately 2450 mm rainfall per year. The average temperature between 23.5 °C (minimum) and33.5 °C (maximum) with average relative humidity varies from 64 % to 94 %. The dry season lies between June to August and the wet season lies from March-May and October-December.

2.2. Geology and hydrogeology

The geology of Peninsular Malaysia is distinguished through three longitudinal north-south belts such as the Eastern Belt, the Central Belt, and the Western Belt (Figure 2). These belts have been associated with the differences in stratigraphy, magmatism, structure, geological evolution, and geophysical signatures [25,26]. Perak state is situated in the Western belt and consists of continental sequences such as quaternary deposits, sandstone, siltstone, shale, metamorphic rocks, and unconsolidated formations, as well as ranged from Cambrian to Quaternary period [27,28]. Batchelor [29] has been described as the stratigraphy of the study area lies in the Gula formation of the Holocene age (480±120 year). Hassan [30] reported that this formation has been divided into five members based on localities such as Bagan Datoh, TelukIntan, MatangGelugor, Port Weld, and ParitBuntar. LabuKubung area is located in Bagan Datoh member, which is mainly composed of clay, silt, and sand deposits, as well as represents shallow marine, estuarine, and offshore environments [30].

The alluvial aquifers can be found long the coastal region of the Western Peninsular Belt with potential range from 30 to 50 m^3/d [24] while the central belt depicts hard rock aquifer potential. Alluvium soils are deposited from surface water which normally occurs in a discontinuous and irregular formation on the flood plain in this area [24]. Thus, due to low cementation between grains, most alluvial formations demonstrate high porosity [31,32]. Teluk Intan area consists about 15 m thickness of alluvium soils [24]. As well, studies showed that aquifers with meta-sedimentary, sedimentary and igneous rocks settings have the capacity for pumping with average well yields range from 10 to 40 m^3/h [33].

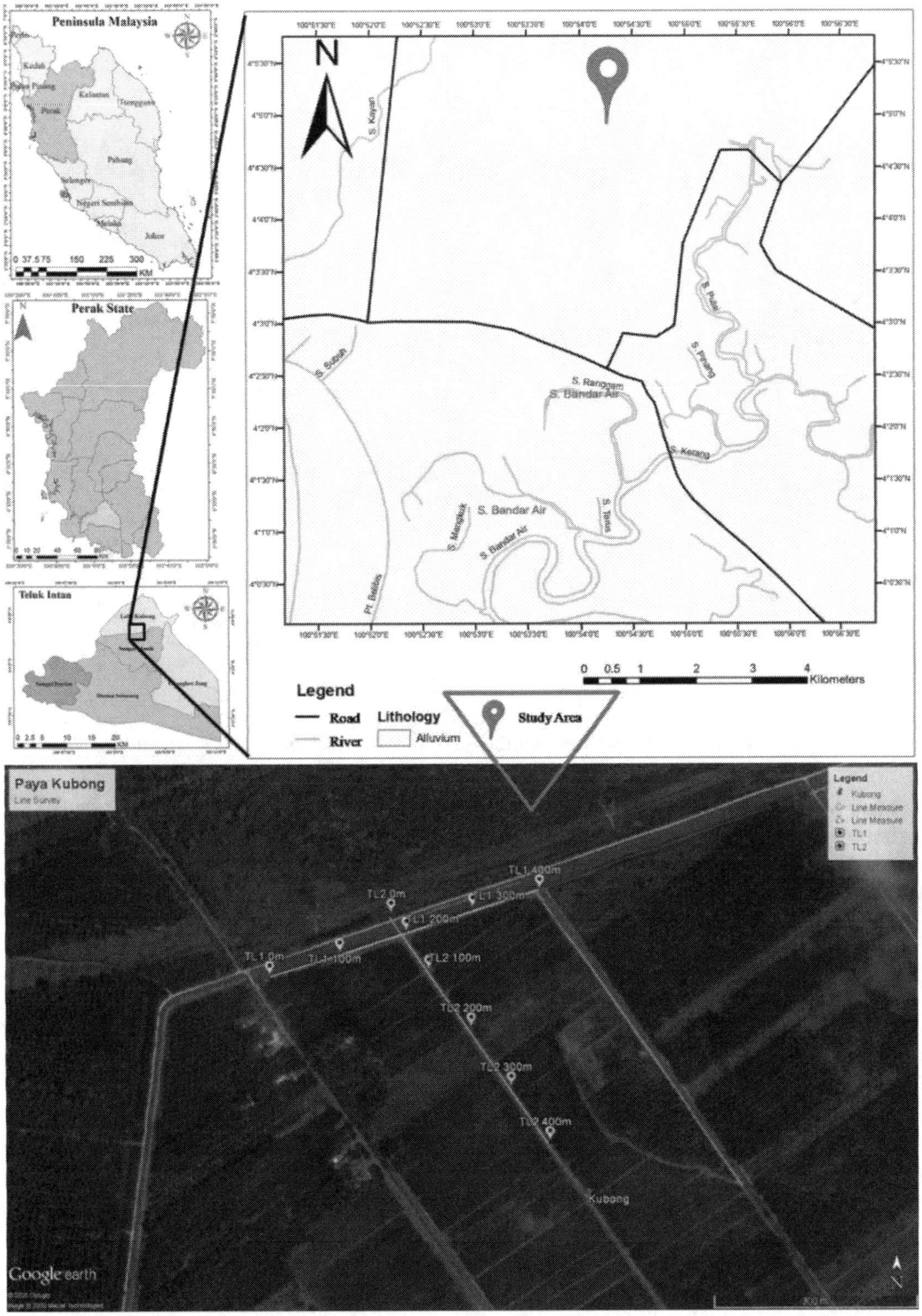

Figure 1. The map is showing the study area of TelukIntan, Perak, Malaysia.

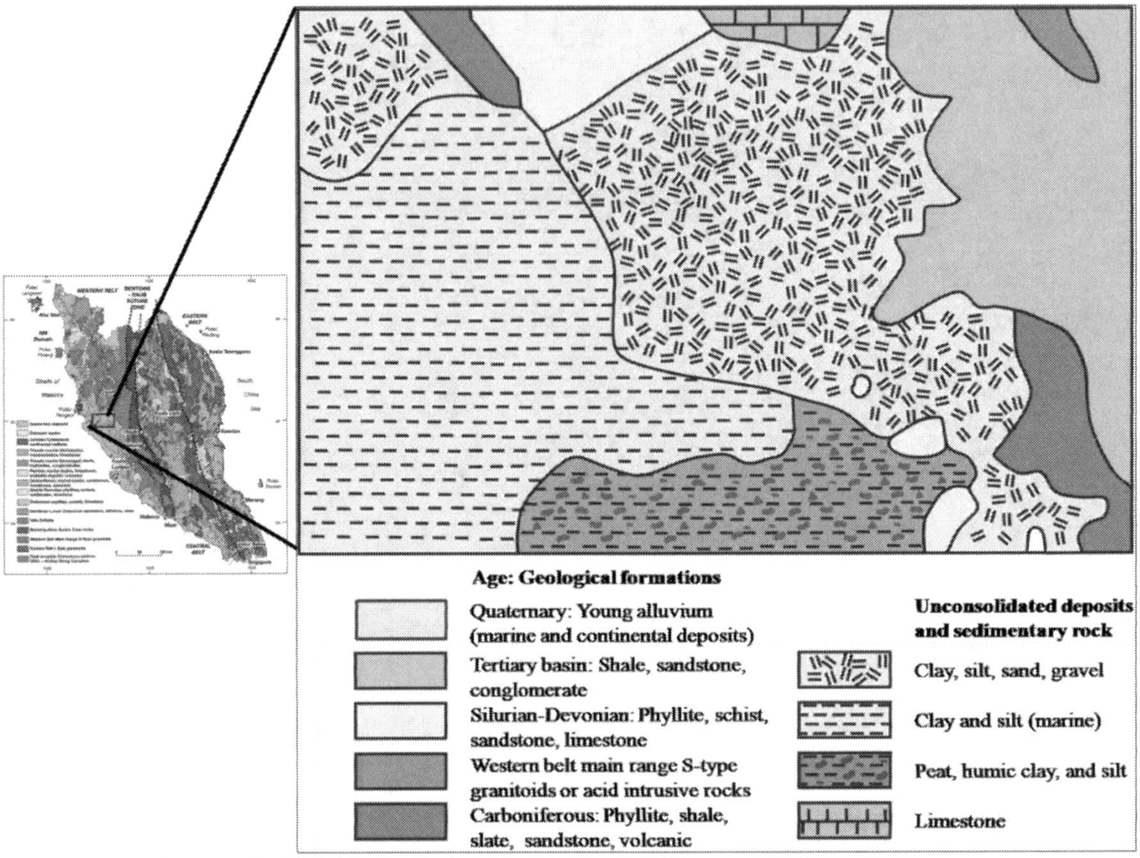

Figure 2: The map is showing the geological formation of the study area.

2.3. *Field experiments*

2.3.1. *ERT and IP surveys*

The two-dimensional (2D) ERT and IP surveys were used in this study. The 2D process is applied as a combination of assessment of both sounding and profiling that examines subsurface lithological objects by changing resistivity horizontally in order to obtain the information simultaneously regarding resistivity in a horizontal and vertical direction [14]. ERT is a cost-effective, rapid, and flexible method for obtaining essential information for groundwater exploration planning and the evaluation of shallow subsurface profiles. Further, ERT method investigates the ground resistivity distribution that can vary with the content of water and the lithological material [23]. The resistivity distribution in ground will provide depictions of subsurface heterogeneity or homogeneous. The subsurface is typically heterogeneous in the study area so that the resistivity observed is apparent for ERT surveys. The greater the resistivity contrast, the easier to detect and recognize the characteristics of the various materials. Moreover, inverse modelling (inversion) is applied to measure the true resistivity distribution to produce a model based on apparent data [34]. From a resistivity survey, a measured resistivity model is produced. By using RES2DInv software or similar, with inversion technique of robust or smoothness constraint, a calculated resistivity model is then produced. Both of these models will be compared of their correctness or similarity hence producing the inverted model (this will show a true resistivity value that can be compared to the geological resistivity table) and error shown means the differences between the measured and calculated model that the inverted model come out with [35]. The resistivity of earth materials is influenced by the metal content in the rock, porosity, clay content, permeability, and pore saturation degree. Thus, the resistivity value of various geological materials is shown in Table 1 and Table 2.

Table 1. Resistivity values of various types of waters [36].

Types of water	Resistivity (Ω.m)
Precipitation	30 – 1000
Surface water, in areas of igneous rock	30 – 500
Surface water, in areas of sedimentary rock	10 – 100
Groundwater, in areas of igneous rock	30 – 150
Groundwater, in areas of sedimentary rock	> 1
Sea water	≈ 2
Drinking water (max salt content 0.25%)	> 1.8
Water for irrigation and stock watering (max salt content 0.25%)	> 0.65

Table 2. Resistivity values of some types of soils [37].

Type of water	Resistivity (Ω.m)
Alluvium	10 – 800
Sand	60 – 1000
Clay	1 – 100
Groundwater	10 – 100
Sandstone	8 – 4 × 10
Shale	20 – 2 × 100
Limestone	50 – 4 × 100
Granite	5000 – 1000000

IP is the earth's ability to hold electric charge overtime or chargeability. The IP effect estimates the soil capability to be polarized when it is influenced by an electric field, which implies energy is reversibly stored in the soil during polarization. The IP measurements were conducted in the time domain for this research. IP surveys are performed by transfer of current to the ground, whereas the potential difference is estimated to evaluate the resistivity. The ground's potential decay is estimated again in windows after the current is shut off.

The chargeability measured the voltage decay curve, and the proportion of the residual voltage amplitude was determined with the DC potential immediately after the current shut-off. The higher the chargeability, the longer the charge is held; the chargeability decays over time, typically a few seconds but sometimes up to minutes, and will eventually disappear. IP is a resistivity derivative; however, the first measure of the resistivity and IP is calculated at the end of a resistivity cycle for IP measurements. Thus, the chargeability value for various geological materials is shown in Table 3.

Table 3. Chargeability value for different geological material [37].

Material type	Chargeability (m.sec)
Groundwater	0
Alluviums	1 - 4
Gravel	3 - 9
Schist	5 - 20
Sandstone	3 - 12
Argillite	3 - 10
Quartzite	5 - 12

2.3.2. Electrode arrangement and instruments

In the field measurement, the ERT technique was applied with the multi-electrode electrical imaging system such as ABEM Lund Imaging System using various layouts. Furthermore, it is an automated imaging system to process, collect, and display resistivity based on the variation in conductivity between the geological rock formation and the adjacent substances. This imaging system is a new technology recently developed to examine the underground properties of regions with complex geological formations. The use of other techniques, including resistivity soundings, is not productive. The resistivity imaging system defines the density, the electrodes' configuration, and the amount of data to be collected. Also, it can estimate a range of constant separation traverses by growing the electrode spacing. Because the separation increases the penetration of more depth, the contoured vertical image of apparent resistivity measured indicates vertical and lateral changes in apparent resistivity across the section. Moreover, ABEM Terrameter SAS (Signal Averaging System) 4000, electrode selector LUND ES464, and other equipment have been used to build up a resistivity pseudo-section, as shown in Figure 3.

Figure 3. This showed the equipments of ERT and IP surveys during the in-situ field survey.

A computerized device is used to select the active electrodes for each measurement automatically. The general principle of using an ERT device typically with the ABEM Terrameter with 62 electrodes for making a pseudo-section resistivity is given in Figure 4. Whereas SAS meter automatically records consecutive readings and displays the average results on a meter screen simultaneously. Moreover, the SAS-4000 Terrameter can work in three mechanisms: a) resistance measurement (mode of resistivity survey), b) voltage measurement mode, and c) induced polarization. Moreover, the direct current (DC) has been passed into the surface since DC may penetrate more in-depth than the alternative currency (AC). Also, the device includes four steel rods and two electric cables roller of 100 m (two rollers in each set). One pair has been used to transmit electric current into the ground, and the other to calculate the potential difference. Further, the voltage/current (V/I) ratio is automatically measured and recorded on the meter screen. The various types of electrode arrays are used to investigate the electrical resistivity, such as Schlumberger, Wenner, dipole-dipole, polar, pole-pole, and pole-dipole [38]. However, the pole-dipole configuration has been applied in this study to acquire the resistivity data (Figure 4). All measurements have been collected stacked twice in the field in order to assist the quality control as well as the flagging of erroneous data. During the survey, resistivity data were obtained from the polar-dipole array configuration with a remote-pole positioned at an equal "infinite" distance. Figure 4 below showed the equipment for the resistivity survey.

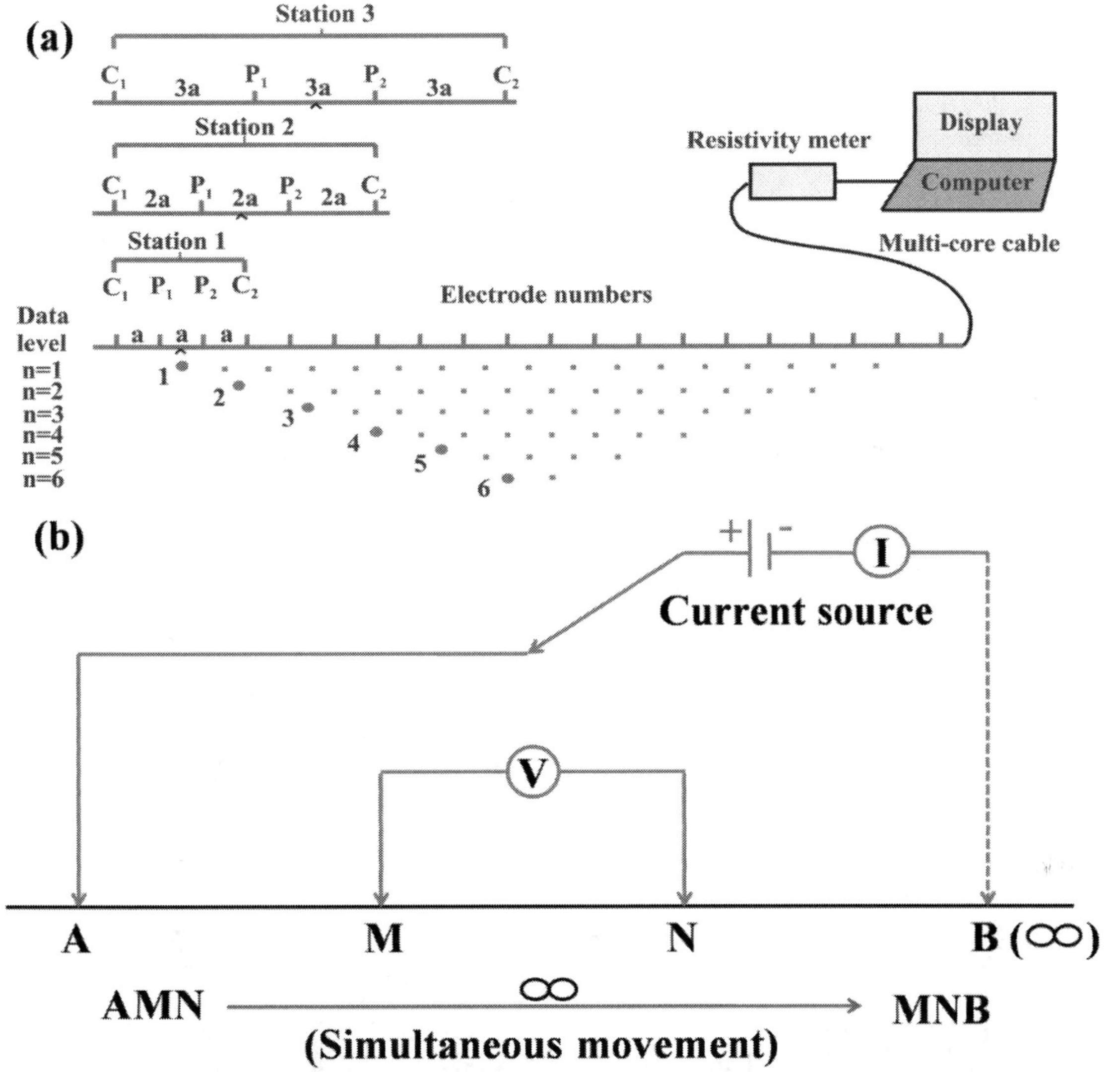

Figure 4. (a) The diagram shows the general measurements for pseudosection development (modified after Hasan et al. [39], with the current electrode C1, the remote electrode C2 at infinity (∞) and the potential electrodes P1 and P2, (b) the principal schematic for ERT and IP measurement using a pole dipole electrode configuration, while V is the voltage, I is current, A is the current electrode, B is the remote electrode at infinity (∞), and the potential electrodes denote M and N.

2.3.3. Resistivity data processing and acquisition

The present study has been performed for the data collection by applied 2 ERT and 2 IP survey lines, whereas the names given for survey lines as TL1 and TL2, as mentioned in figure 1. Further, 2 survey lines were conducted to the length of 400 m with an electrode interval of 5 m, as well as estimated depth of each survey were 150 m. The outer electrodes have been utilized for current and were placed on the earth's surface with 62 electrodes at a distance of 5, 10, 15, 20, 25, 30, 35, 40, 45, 50, 55, 60, 65, 70, 75, 80, 85, 90, 95, 100, 105, 110, 115, 120, 125, 130, 135, 140, 145, 150, 155, 160, 165, 170, 175, 180, 185, 190, 195, 200, 205, 210, 215, 220, 225, 230, 240, 245, 250, 260, 265, 270, 275, 280,

IOP Conf. Series: Earth and Environmental Science **880** (2021) 012025 doi:10.1088/1755-1315/880/1/012025

285,290,295, 300, 305, 310,315, 320, 325, 330, 335, 340,345, 350, 355, 360, 365, 370, 375, 380, 385, 390, 395, and 400.

The ERT and IP surveys data were obtained subsequently exported to RES2DINV compatible file formats in order to carry out the inversions with the software developed by Loke [40]. The 2D model used by the inversion process consists of a number of blocks having their distribution and size automatically generated by the program. This model is designed to evaluate the resistivity of blocks, creating an apparent resistivity pseudosection that agrees with the actual measurements, which is a function of the electrode array. The depth of the bottom row of blocks is set to be approximately equal to the equivalent depth of investigation. The thickness of the first layer is generally set to about 0.6 times the electrode spacing for the pole-dipole arrays. A correction is applied to the resistivity and IP data for surface topography effects during the initial loading of the files. The optimization approach effectively attempts to decrease the variance between the apparent resistivity values measured and calculated by adjusting the model block resistivity. A Root-Medium-Squared (RMS) error is used to measure this difference. The least potential RMS error model can often indicate large and unrealistic differences in model resistivity values. It may not be the "best" model from a geological perspective [41]. The most conservative technique is to select the model at the iteration, after which the RMS error is not substantially changed. The iteration RMS error was approximately 30.1 % in the ERT surveys and 6-9 RMS error in IP surveys for all inversions models. The inversion routine is used for the non-linear least-squares optimization technique and, most parameters, automatically fixed by the software.

2.3.4. Particle size analysis

In situ soil measurement was carried out with 11 soil samples from the pumped well collected in plastic boxes at a depth between 5.6 m to 61.2 m. However, nine samples were the soil samples, 1 sample was the peat, and the last sample was the bedrock as shale rock. Nine soil samples were brought back to the laboratory and dried in an oven at 40 °C for 24 hours. After dried, a Malvern Master Sizer Scirocco 2000 laser grain-size analyzer (Malvern Instruments, Malvern Company, UK) has been used for determining the particle size distributions of sediments. A soil particle size distribution was defined in terms of the calculation of percentage volume of sand, silt, and clay for the range of 2–2000 μm. The volumetric percentages were classified as clay (0-2 μm), silt (2–50 μm), and sand (50-2000 μm) based on the United States Department of Agriculture method [42].

3. Results

The geophysical survey was carried out using 2D ERT, IP surveys, and the parameter of the survey lines (TL1 and TL2) addressed in Figures 4 and 5 below. In the study area, previous studies indicated that the marine sediments describe the coastal region of TelukIntan with abundant mangrove pollen, fine to coarse sand, silt, clay, and intercalated peat layers [29,30]. The interstitial peat horizons show different phases of regression along with the environment of coastal lagoons and estuaries with sea water [12], [24], [44].

3.1. Survey line TL1

The 2D inverted resistivity model using pole-dipole array at TL1 site has been performed at a length of 400 m with an electrode interval of 5 m. The direction of the survey line TL1 was 225° N, with the starting outer electrode located at 4° 7' 22.09" N and 101° 1' 51.82" E. In contrast, the last electrode was positioned at 4° 7' 26.10" N and 101° 2' 4.41" E. TL1 line was started at a higher topographical point of 41 m and gradually decreased to the lowest point of 38 m. The resistivity profile demonstrates value range from 10-50 Ωm (Clays), 70-150 Ωm (Clay), 250-500 Ωm (alluvium and sands), and 1500-2000 Ωm (shale), suggesting the subsurface setting characterized by clays at the top followed by sand in the middle and shale as the bedrock (Figure 5).

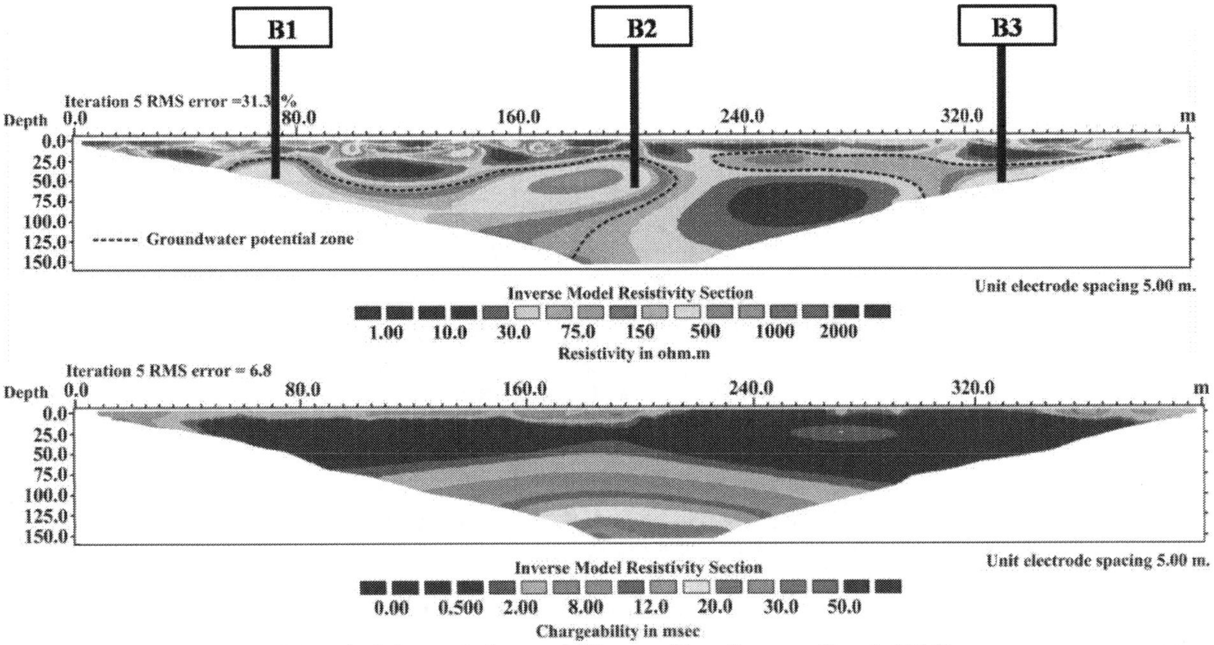

Figure 5. Resistivity and chargeability profile of survey line 1 (TL1)

This was confirmed by the particle size analysis, as shown in Table 4. 11 samples of soil profile were collected at depth from 5.6 m to 61.2 m (Figure 8), with average percentage of sand, silt and clay are 93.77 %, 5.78 % and 0.02 %, respectively.

Table 4. Particle size distribution of soil samples collected from Borehole 2 (B2)

Particle size (µm)	Alluvial formation	S1 (5.6 m)	S2 (11.2 m)	S3 (16.8 m)	S4 (22.4 m)	S5 (28 m)	S6 (36.6 m)	S7 (39.2 m)	S8 (50.4 m)	S9 (44.8 m)	S 10 (56.0 m)
		Volume in (%) of clay, silt and sand									
0 – 2	Clay	0	0	0	0.21	0	0	0	0	0	0
2 – 50	Silt	11.53	10.37	2.73	26.42	1.41	0	0.5	0.53	4.02	0.07
50 – 2000	Sand	87.99	89.63	97.27	73.38	98.58	100	95.51	99.49	95.96	99.95

3.2. Survey line TL2

The results of ERT and IP methods in survey line TL2 site using pole-dipole array has been performed at a length of 400 m with an electrode interval of 5 m. The direction of the survey line TL2 was 225° N, with the starting outer electrode located at 4° 7' 24.98" N and 101° 1' 57.47" E, as well as the last electrode, was located at 4° 7' 14.31" N and 101° 2' 4.85" E. The resistivity values observed between the 1 Ωm to 50 Ωm at a depth of 1-10m indicate the topsoil clay layer and the ERT values ranged between 50-1000 Ωm at a depth of 10-20 m, which was interpreted as the weathered layer, as shown in Figure 6. The hard layer considered at the minimum depth of 20 m below the earth surface, where the resistivity value is between 1000-5000 Ωm. On the other hands, IP results also showed the weathered and hard rock layers in this survey, whereas chargeability values observed between 2.5 to 30 msec at a depth of 0 to 100 m in this study area (Figure 6).

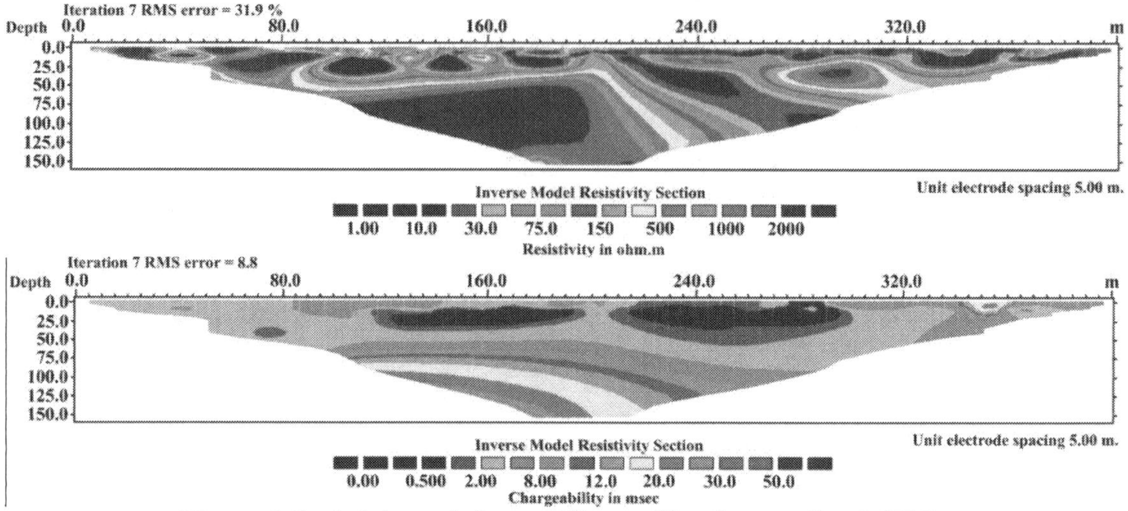

Figure 6. Resistivity and chargeability profile of survey line 2 (TL2)

4. Discussions

TL1 and TL2 survey lines provide better perspective about the subsurface characteristic of the studied area (Figure 5 and 6). TL1 line suggests three potential groundwater zones located between the distance of 50-160 m, 160-220 m, and 240-340 m at depth of 15 m, 15 m and 30 m b.g.l. (below ground level), respectively, as shown in Figure 5. However, B2 was selected for drilling due accessibility factor. Further, eleven samples of soil profile were collected at depth from 5.6 m to 61.2 m (Figure 8), with average percentage of sand, silt and clay are 93.77 %, 5.78 % and 0.02 %, respectively. The geology of subsurface settings (silt, clay, sand, some peat and shale) is the key factor in determining the aquifer system (Figure 7). The black-dashed line (TL1 survey line, Figure 5) demarcates the groundwater potential zones in the study area characterized by interlayer sand-silt sequence (Figures 7 and 8) indicating the saturation zone of aquifer [45] underlain by shale at the bottom [46]. Consequently, the interlayer sand-silt sequence or sand layers indicate the permeable beds or saturation zone may consider as an aquifer system [45], as shown in (Figure 7 and 8). Further study on hydraulics is important to understand the overall capacity of the aquifer.

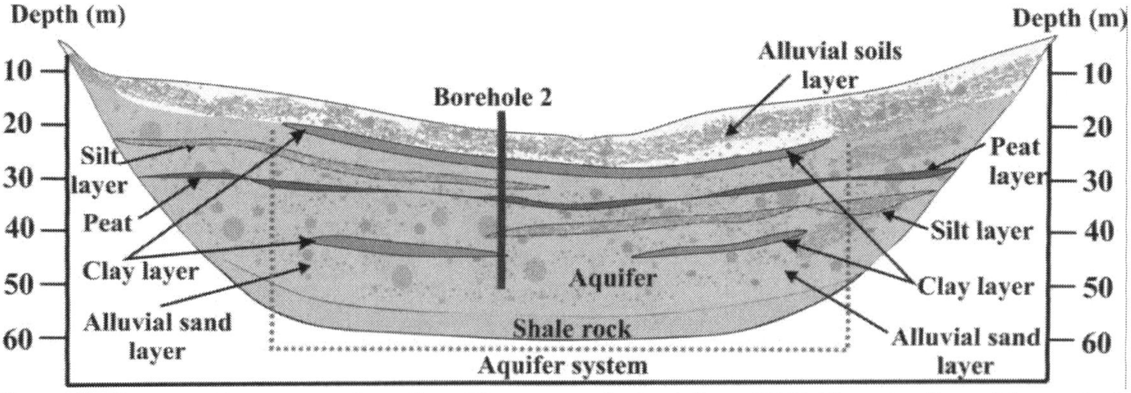

Figure 7. Cross-section perspective of aquifer system derived from ERT, IP and soil particle analysis results.

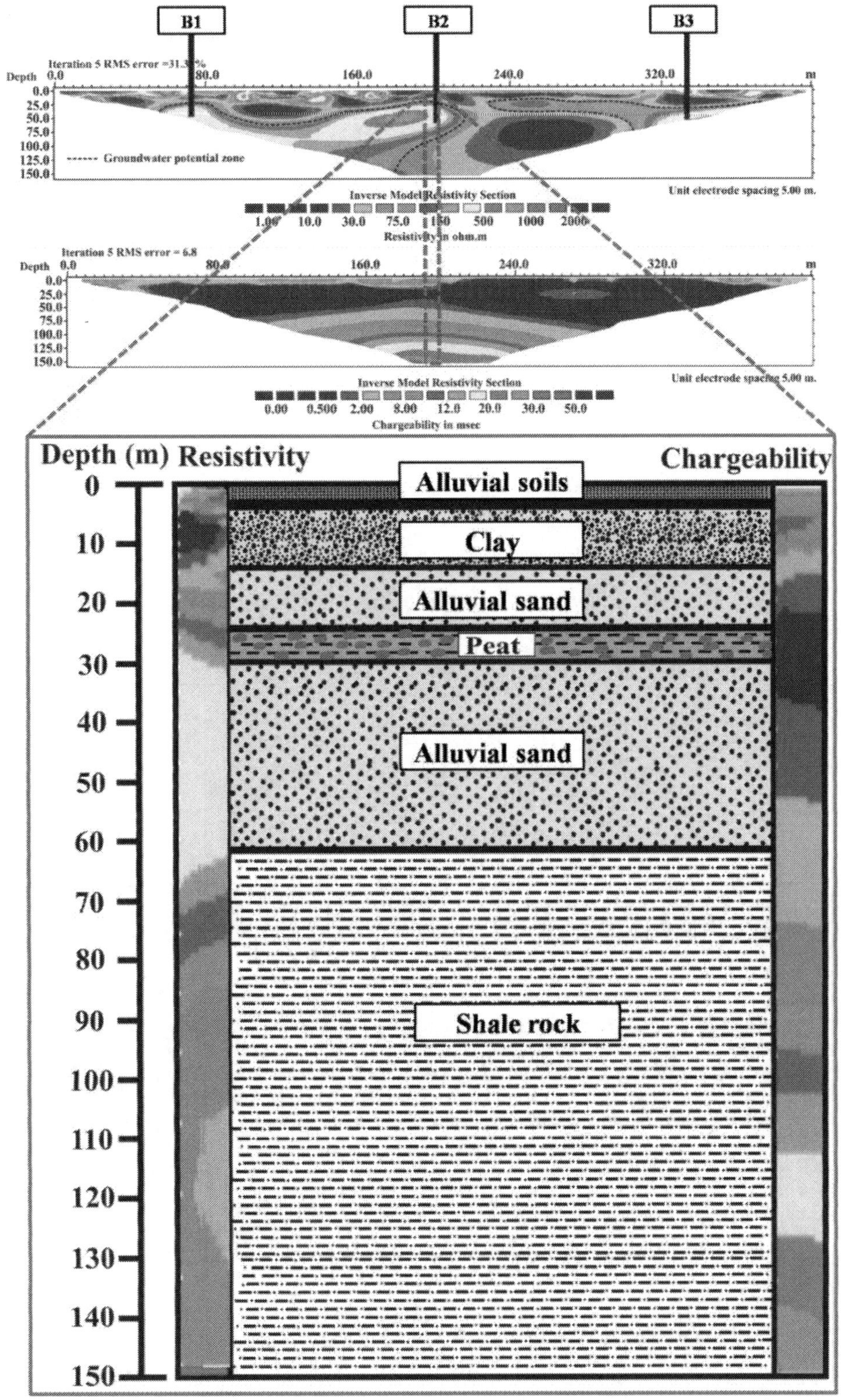

Figure 8. ERT and IP results showed the high potential groundwater system and geological formations (red dashed line) in survey line TL1 (for B2).

5. Conclusion

This study performed the geophysical survey using ERT and IP techniques to investigate the subsurface profile and identify the aquifer system in LabuKubung, TelukIntan, Perak. ERT and IP results using pole-dipole electrode configuration demonstrated the sub-surface lithology of the Quaternary period, which indicated the alluvial formations of sand, silt, clay, and peat, as well as the findings revealed the low values of resistivity (70-500 Ωm) and chargeability (0-1 msec.). Two surveys (TL1 and TL2) have been performed where three potential zones in the TL1 line were identified based on ERT and IP results i.e., 50-160 m, 160-220 m, and 240-340 m at a depth of 15 m, 15 m and 30 m.b.g.l., respectively. However, the subsurface profiling at TL2 indicates high resistivity and chargeability values suggesting a hard rock system. The geology of subsurface settings (silt, clay, sand, and shale) is the key factor in determining the aquifer system, characterized by interlayer sand-silt sequence indicating the saturation zone of aquifer underlain by the shale at the bottom.

References

[1] Rahman A *et al* 2020 Assessment of Hydrochemical Backgrounds and Threshold Values of Groundwater in a Part of Desert Area, Rajasthan, India. *Environ. Pollut* **266** 1–12.
[2] Priyanka B N *et al* **2018** Estimating Anisotropic Heterogeneous Hydraulic Conductivity and Dispersivity in a Layered Coastal Aquifer of Dakshina Kannada District, Karnataka. *J. Hydrol. 565* 302–317
[3] Akhtar N *et al* 2020 Multivariate Investigation of Heavy Metals in the Groundwater for Irrigation and Drinking in Garautha Tehsil, Jhansi District, India. *Anal. Lett* **53** (5) 774–794
[4] UNESCO *The United Nations World Water Development Report 2018 Nature-Based Solutions for Water* Paris 07 SP France
[5] NWRS 2011 Review of the National Water Resources Study (2000-2050) and Formualtion of National Water Resources Policy pp 1–480.
[6] Salam M A *et al* 2019 Water Quality Assessment of Perak River, Malaysia *Pollution* *5* (3) 637–648
[7] Huang Y F *et al* 2016 Quality of Water Resources in Malaysia. In *Research and Practices in Water Quality* IntechOpen Malaysia pp 65–94
[8] Chu L H 2004 Groundwater Utilization and Management in Malaysia. In *Thematic Session on Geo-environment of Delta and Groundwater Management in East and Southeast Asia* Reedman A Park Y S Eds 41st Coordination Committee for Geoscience Programmes in East and Southeast Asia (CCOP) Tsukuba Japan pp 83–93.
[9] Rathinasamy V *et al* 2018 Review on Rock Aquifers Researches Conducted in Malaysia. In *MATEC Web of Conferences* **250** pp 1–7
[10] Kaleris V K and Ziogas A I 2020 Using Electrical Resistivity Logs and Short Duration Pumping Tests to Estimate Hydraulic Conductivity Profiles. *J. Hydrol* **590** 1–11
[11] Ibuot J C *et al* 2019 Investigation of Litho-Textural Characteristics of Aquifer in Nkanu West Local Government Area of Enugu State, Southeastern Nigeria. *J. African Earth Sci* **153** 197–207
[12] Attwa M *et al* 2011 Evaluation of DC, FDEM and IP Resistivity Methods for Imaging Perched Saltwater and a Shallow Channel within Coastal Tidal Flat Sediments *J Appl Geophys* **75** (4) 656–670
[13] Anomohanran O 2015 Hydrogeophysical and Hydrogeological Investigations of Groundwater Resources in Delta Central, Nigeria *J Taibah Univ Sci* **9** (1) 57–68
[14] Rolia E & Sutjiningsih D 2018 Application of Geoelectric Method for Groundwater Exploration from Surface (A Literature Study). *In AIP Conference Proceedings 1977* Depok Indonesia pp 1–9
[15] Muchingami I *et al* 2012 Electrical Resistivity Survey for Groundwater Investigations and Shallow Subsurface Evaluation of the Basaltic-Greenstone Formation of the Urban Bulawayo Aquifer *Phys Chem Earth* **50** 44–51
[16] Kearey P *et al* 2002 *An Introduction to Geophysical Exploration*, 2nd ed.; Blackwell Scientific: *Oxford Publishing House*
[17] Riwayat A I *et al* 2018Application of Electrical Resistivity Method (ERM) in Groundwater Exploration. In *International Seminar on Mathematics and Physics in Sciences and Technology (ISMAP 2017) Journal of Physics: IOP Conferences Series 905* pp 1–9
[18] Ahmed A S *et al* 2019 Multiscale Induced Polarization Tomography in Hydrogeophysics: A New Approach *Adv Water Resour* **134** 10–28
[19] Blanchy G *et al* 2020 an Intuitive Open Source Software for Complex Geoelectrical Inversion/Modeling *Comput. Geosci* **137** 1–12
[20] Mehmood Q *et al* 2020 Optimizing Groundwater Quality Exploration for Irrigation Water Wells Using Geophysical Technique in Semi-Arid Irrigated Area of Pakistan *Groundw Sustain Dev* **11** (3) 1–8
[21] Commer M *et al* 2011 Three-Dimensional Induced Polarization Data Inversion for Complex Resistivity *Geophysics* **76** (3) 157–171
[22] Benoit S *et al* 2019 Characterization of Spatially Variable Riverbed Hydraulic Conductivity Using Electrical Resistivity Tomography and Induced Polarization *Hydrogeol J* **27** (1), 395–407

[23] Amaya A G *et al* 2016 Electrical Resistivity Tomography and Induced Polarization for Mapping the Subsurface of Alluvial Fans: A Case Study in Punata (Bolivia) *Geosci* **6** (51) 1–13
[24] Fauzie M B J **2013** Evaluation of Groundwater Potential for Irrigation at the Seberang Perak Aquifer, Malaysia, Universiti Putra Malaysia
[25] Metcalfe I 2013 Tectonic Evolution of the Malay Peninsula *J Asian Earth Sci* **76** 195–213
[26] Makoundi C *et al* 2014 Geology Geochemistry and Metallogenesis of the Selinsing Gold Deposit Central Malaysia *Gondwana Res* **26** (1) 241–261
[27] Pour A B & Hashim M 2015 Structural Mapping Using PALSAR Data in the Central Gold Belt, Peninsular Malaysia *Ore Geo. Rev* **64** (1) 13–22
[28] Sun C *et al* 2019 Genesis of the Selinsing Gold Deposit, Peninsular Malaysia: Constraints from Mineralogy, Geochemistry and in Situ Sulfur Isotope Compositions of Sulfides *Ore Geol Rev 113* 1–19
[29] Batchelor D A F 2015 Clarification of Stratigraphic Correlation and Dating of Late Cainozoic Alluvial Units in Peninsular Malaysia *Bull Geol So. Malaysia* **61** (12) 75–84
[30] Hassan K A **1990** Summary of the Quaternary Geology Investigations in Seberang Prai, Pulau Pinang and Kuala Kurau. *Bull. Geol. Soc. Malaysia* **26** (4) 47–53
[31] Fitts C R 2002 Physical Properties. In *Groundwater Science* London Academic press Elsevier Science LTD London pp 23–47
[32] Schwartz F W & Zhang H 2003 Introduction to Ground Water. In *Fundamentals of Groundwater* John Wiley & Sons New York p 115
[33] Sapari N *et al* 2011 Quantity and Quality of Groundwater in Fractured Metasedimentary Rocks of the West Coast of Peninsular Malaysia. *Sains Malaysiana* **40** (6) 537–542
[34] Dahlin T 2001 The Development of DC Resistivity Imaging Techniques *Comput Geosci* **27** (9) 1019–1029
[35] Loke M H *et al* 2003 A Comparison of Smooth and Blocky Inversion Methods in 2D Electrical Imaging Surveys *Explor Geophys* **34** (3) 182–187
[36] McNeill J D 1980 *Electrical Conductivity of Soils and Rocks. Geonics Limited*
[37] Saad R *et al* 2012 Groundwater Detection in Alluvium Using 2-D Electrical Resistivity Tomography (ERT) *Electron J Geotech Eng* **17** 369–376
[38] Malik A K A *et al* 2019 Comparison between Multiple Gradient and Pole Dipole Array Protocols for Groundwater Exploration in Quaternary Formation *Civ Eng Archit* **7** (6) 77–85
[39] Hasan M *et al* 2020 Geophysical Investigation of a Weathered Terrain for Groundwater Exploitation: A Case Study from Huidong County China *Explor Geophys* **5** 1–21
[40] Loke M H 2003 RES2DINV Manual: Rapid 2-D Resistivity & IP Inversion Using the Least-Squares Method: Geoelectrical Imaging 2-D & 3D GEOTOMO SOFTWARE; Universiti Sains Malaysia, Penang, Malaysia
[41] Gao Q *et al* 2018 Evaluation of a Weathered Rock Aquifer Using ERT Method in South Guangdong China *Water (Switzerland)* **10** (3) 1–22
[42] NRCS U 1993 *Soil Survey Division Staff (1993) Soil Survey Manual. Soil Conservation Service*
[43] SSDS Soil survey manual, no. 18. U.S. Govt. Print. Office, Washington, DC: Soil Survey Division Staff, United States Department of Agriculture. Handbook. No. 18, **1993**.
[44] Roslan N 2017 The Potential Susceptibility of Urban Hardrock Aquifers to Hydraulic and Contaminant Stresses : The Case of Shah Alam, Malaysia, The University of Birmingham
[45] Freeze R A & Cherry J A 1979 Physical Properties and Principles. In *Groundwater*; Prentive-hall, Englewood Cliffs, New Jersey: United States of America p 604
[46] Şen Z 2015 Basic Porous Medium Concepts. *In Practical and Applied Hydrogeology*; Elsevier Inc.: Istanbul Technical University, Istanbul, Turkey 43–97

IOP Conf. Series: Earth and Environmental Science **880** (2021) 012026 doi:10.1088/1755-1315/880/1/012026

Effectiveness of a groundsill structure in reducing scouring problem at Cimadur River, Banten Province

Arniza Fitri[1*], Li Yao[2,3], Dian Pratiwi[1], Arlina Phelia[1], Susarman[1], Fajar Dewantoro[1], Destiana Safitri[1], Khairul Nizam Abdul Maulud[4&5]

[1]Department of Civil Engineering, Faculty of Engineering and Computer Science, Universitas Teknokrat Indonesia, Bandar Lampung, Indonesia
[2]School of Water Conservancy and Ecological Engineering, Nanchang Institute of Technology, Nanchang, 330099, China.
[3]National and Provincial Joint Engineering Laboratory for the Hydraulic Engineering Safety and Efficient Utilization of Water Resources of Poyang Lake Basin, Nanchang, 330099, China
[4]Department of Civil Engineering, Faculty of Engineering & Built Environment, Universiti Kebangsaan Malaysia, 43600 UKM Bangi, Selangor, Malaysia
[5]Earth Observation Centre, Institute of Climate Change (IPI), Universiti Kebangsaan Malaysia, 43600 UKM Bangi, Selangor, Malaysia

*Corresponding E-mail: arniza@teknokrat.ac.id

Abstract. Groundsill is a transverse threshold structure that can be built on water body as one alternative in reducing the upstream water velocity and therefore could maintain the upstream sedimentation rates. Due to scouring problems faced in the Cimadur River, Banten Province, a groundsill structure has been constructed at one location in Cimadur River which is near to the Bantar Karang bridge structure. The scouring problem is worried to damage the abutment of the bridge structure. This study aims to investigate the effectiveness of the groundsill structure in reducing the scouring problem in Cimadur River, Banten Province. The effectiveness is determined by comparing the riverbed elevation around the Bantar Karang bridge structure, before and after 5 months construction of the groundsill structure. The results showed that the presence of the groundsill structure has increased the riverbed elevation especially around the abutment of the Bantar Karang bridge (approximately 25 cm in average after 5 months installation of the groundsill structure). It is therefore can slowly reduce the local scouring problem around Bantar Karang bridge. Besides, the results also showed that there is small reduction of riverbed elevation at the downstream area of the groundsill structure approximately about 10 cm in average. It can be concluded that the groundsill structure is quite effective in reducing the local scouring problem in Cimadur River, especially upstream area of their structure around Bantar Karang bridge.

Keywords: Cimadur River, Groundsill Structure, Scouring, Sedimentation

Track Name: Land, Water, Forests and Food Security

1. Introduction

River is one type of surface waters that continuously flow from upstream to downstream area [1-5], while bridge is a structure connecting areas separating by the river [6]. In water resources management, erosion or scouring problem in the water body is the most serious problem that need to be solved [7-13]. Scouring is an enlargement of a flow accompanied by transfer of material through the action of fluid motion. Scouring problem faced in the river can damage river structures and cause bad impact to the river environments.

Since the past decades, the scouring problem in the river especially near to the bridge foundation has become the major discussion by researchers [14-17]. One approach that can be applied in order to reduce the scouring problem in the river is by constructing a structure in the river that can be implemented in reducing the water flow velocities, such as a groundsill structure [18]. Groundsill structure is one alternative structure that can be built in the water body in order to reduce the water flow velocity and maintain sedimentation rates at the upstream area of its structure [18]. There were some researchers that have applied a groundsill structure as an effort to reduce scouring problem and protect the important structures in the river [18-21]. Generally, the groundsill structures were built up at downstream area of the threatened structure of being damage in the river. Recently, there is still lack studies discussed about the effectiveness of a groundsill structure in reducing scouring problem in the river. Therefore, this study aims to investigate the effectiveness of a gorundsill structure that has been built in the Cimadur River, Banten Province especially near to Bantar Karang bridge. The study can provide the important information for the water resources managers regarding scouring problem and existence of groundsill structure in the river.

2. Materials and Methods

2.1 Study Area

This study is carried out at the middle part of Cimadur River located at Lebak District, Banten Province. Geographically, Lebak district is positioned at 6° S to 7° S (latitude) and 105° 25" E to 106° 30" E (longitude) as shown in the Figure 1 below. There are scouring problems being faced in Cimadur River since the last decade. One location of the scouring problems in Cimadur River is near to Bantar Karang bridge (Figure 1). Continuous scouring problem is worried to breakdown the abutment structure of the Bantar Karang bridge, therefore, an effort in reduction of the scouring problem is urgent importance.

Topographic condition in the middle part of Cimadur River is characterized with steep slope where the riverbed elevations are approximately between 80 m and 400 m above means sea level (MSL). Steep slope of the riverbed in this area causes the high-speed velocities in the water flow and give high potential for the local scouring problem in the river. Based on field measurement using current meter, water velocities in the middle part of Cimadur River are approximately 2.64 m/s in average.

IOP Conf. Series: Earth and Environmental Science **880** (2021) 012026 doi:10.1088/1755-1315/880/1/012026

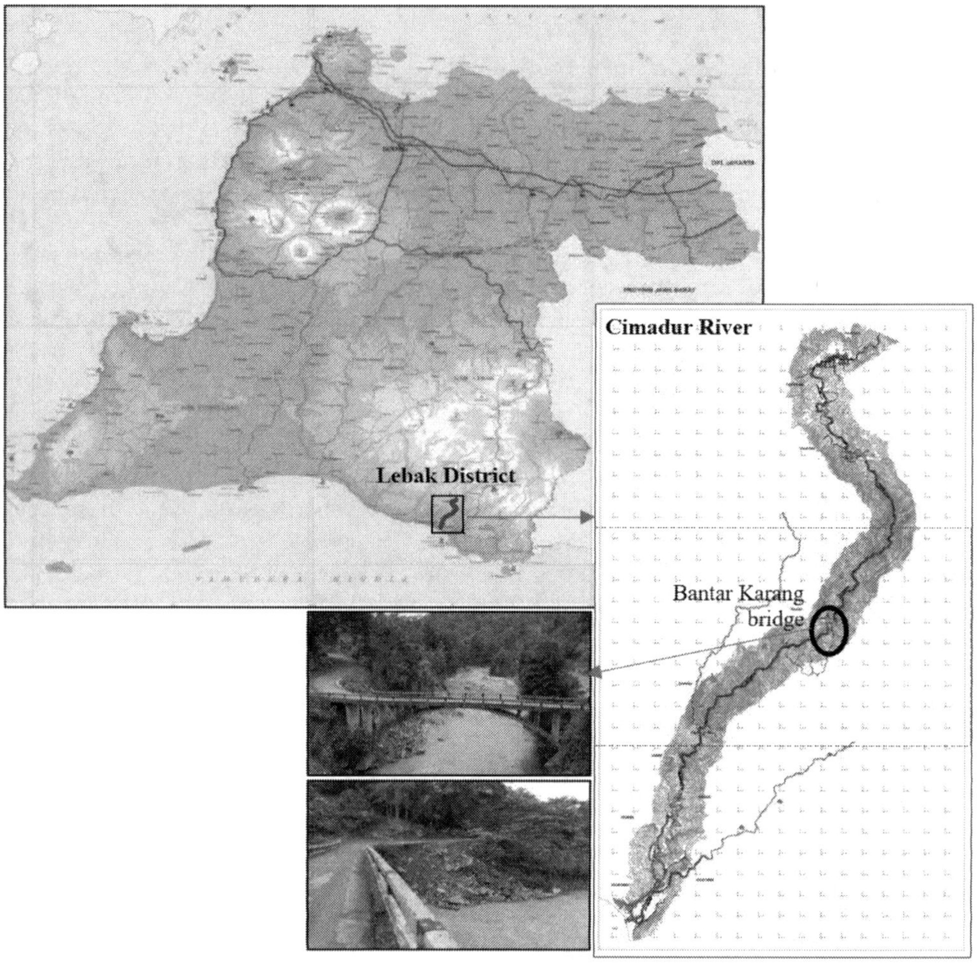

Figure 1. Location of the study area at Lebak district, Banten Province.

2.2 Construction record of the Groundsill Structure

In 2019, a groundsill structure has been constructed at one cross section in the middle part of Cimadur River especially near to Bantar Karang bridge. The position of the groundsill structure is about 90 m at downstream of the Bantar Karang bridge. The local scouring problem faced in Cimadur River around Bantar Karang bridge has encouraged an idea to construct the groundsill structure. Figure 2 show the detail description for the groundsill structure near to Bantar Karang bridge structure.

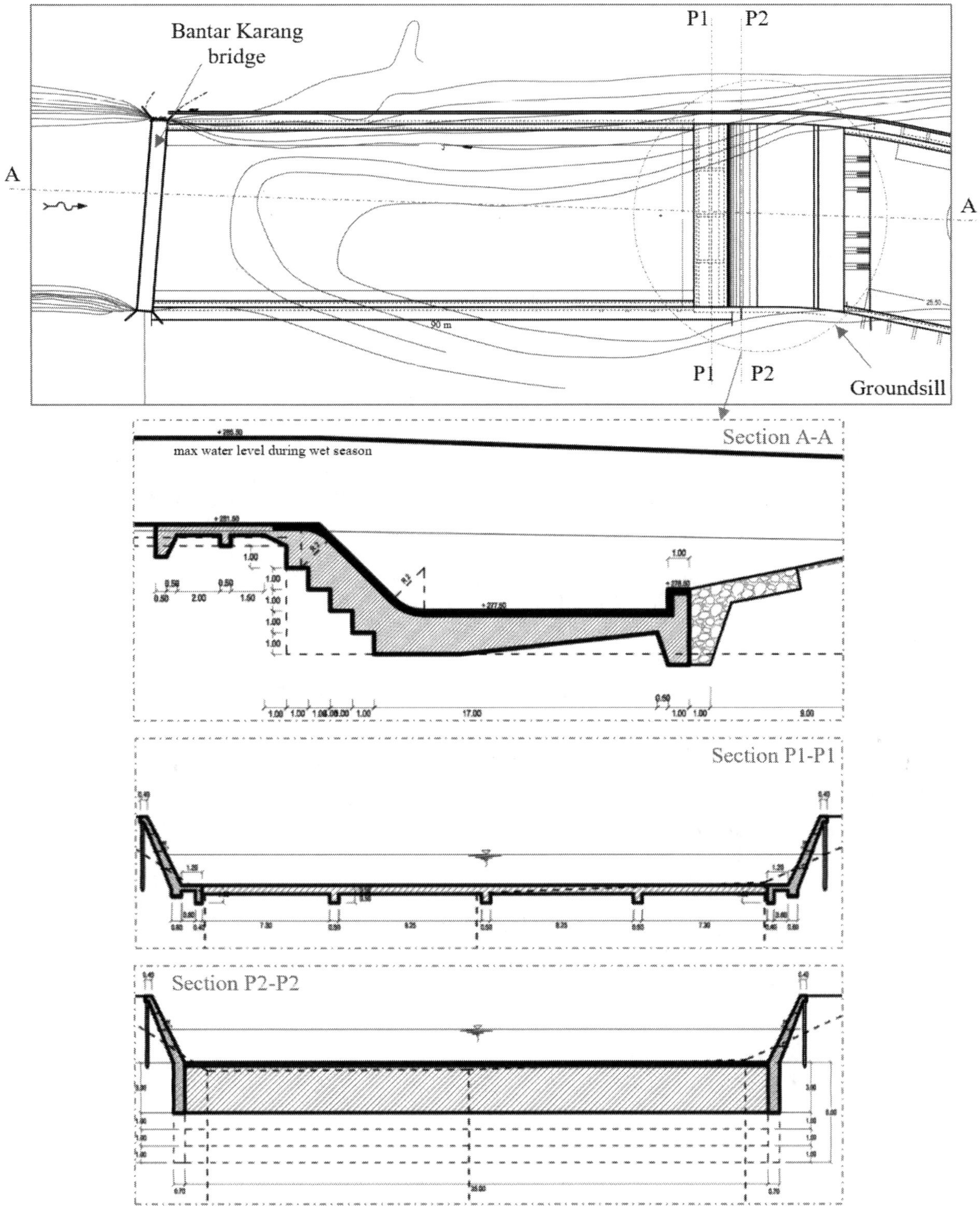

Figure 2. Detail description of the groundsill structure.

The groundsill structure has been designed based on 50 years return period of the flood discharges in Cimadur River. Dimensions of the groundsill structure are 35 m of the width and 7 m of the maximum height from the ground level. The maximum crest level of the groundsill structure is 281.5 m above mean sea level (MSL). The distance of the groundsill structure from the Bantar karang bridge is about 90 m positioned at downstream area of the bridge.

2.3 Data Collection

Riverbed elevations at a cross section (cross-section 1) in Cimadur River, Lebak district which is exactly located at 70 m upstream from the groundsill structure or 20 m downstream from the Bantar Karang bridge were monitored before and after 5 months construction of the groundsill structure. Besides, riverbed elevations at another cross section (cross-section 2) in Cimadur River, Lebak district located at 5 m downstream from the groundsill structure was also monitored (before and after 5 months construction of the groundsill structure). The riverbed elevations at cross-section 1 and cross-section 2 were monitored using theodolite equipment during normal water level condition at dry season.

2.4 Analyses of the Effectiveness of the Groundsill Structure in Reducing Scouring Problem

The effectiveness of the groundsill structure in reducing scouring problem around the Bantar Karang bridge, in Cimadur River is determined by comparing the riverbed elevations at both cross sections (cross-section1 and cross section 2) in Cimadur River as mentioned in sub-section 2.3. Increment of riverbed elevations in the cross-section area means that there is an increase of sediment accumulations around the cross-section area and therefore could reducing the local scouring problem near to the cross-section area. In contrast, reduction of the riverbed elevations in the cross-section area means that there is reduction of sediment accumulations around to cross-section area and therefore can cause increasing the local scouring problem at surrounding of the cross-section area.

3. Results and Discussion

Figure 3 and Figure 4 show the comparisons of the riverbed elevations at both cross-sections (cross-section 1 and cross-section 2) before and after 5 months construction of the groundsill structure in the Cimadur River. Cross section in Figure 3 is located approximately at 70 m upstream from the groundsill structure or 20 m downstream from abutment of Batan Karang bridge, while cross section in Figure 4 is located approximately at 5 m downstream from groundsill structure.

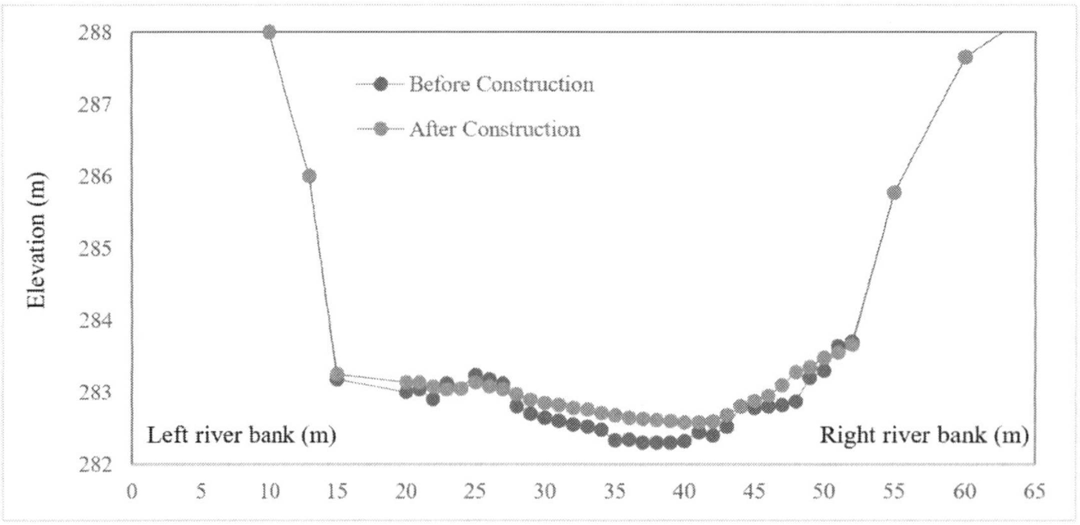

Figure 3. Riverbed elevations at cross-section 1 (upstream area of the groundsill structure).

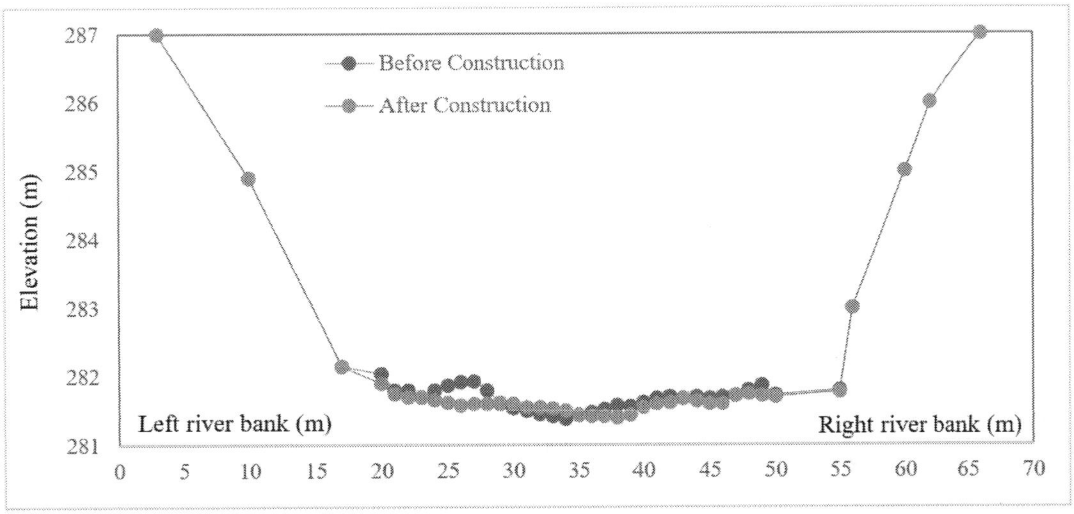

Figure 4. Riverbed elevations at cross-section 2 (downstream area of the groundsill structure).

Based on Figure 3, the result showed that the presence of groundsill structure in Cimadur River especially at the downstream of the Bantar Karang bridge has slowly help in reducing the local scouring problem around abutment of Bantar Karang bridge. In Figure 3, it has been recorded that there is increasing the riverbed elevations approximately about 25 cm in average at cross section 1 located approximately at 20 m downstream from Bantar Karang bridge structure or 70 m upstream from groundsill structure. The results are found to be in line with the research done by previous studies such as Sucipto and Tugino [21], Budi et al. [22], Permata [23] and Tungga et al. [24]. In examples, Sucipto and Tugino have recorded a returning riverbed elevations (to be normal elevations as before degradation) around abutment of one bridge in Java Province after construction of a groundsill structure at downstream area of the bridge.

Besides, based on Figure 4, there is small reduction of the riverbed elevations approximately about 10 cm in average at cross section 2 which is located approximately at 5 m downstream from the groundsill structure. It means that local scouring could be happened at downstream area of the groundsill structure and therefore more studies need to be carried in order to avoid the possibility scouring problem caused by grounsdill construction itself.

4. Conclusion

The problem of local scouring in the river body was becoming the major issue in the world since the past few decades. Groundsill structure has been constructed recently in the river as one alternative to reduce the local scouring problem in Indonesia. It is because the groundsill structure can reduce the water flow velocities and maintain the sedimentation rates in the river especially at the downstream area of the structures. Based on this study, the groundsill structure was found to be effective in slowly reducing local scouring problem in the river especially at upstream area of the groundsill structure. Unfortunately, there is a negative impact found due to the presence of groundsill structure in the river where there was a scouring recorded in the area downstream of the groundsill structure.

References

[1] Fitri A, Maulud K N A, Pratiwi D, Phelia A, Rossi F and Zuhairi N Z 2020 Trend of Water Quality Status in Kelantan River Downstream, Peninsular Malaysia *Jurnal Rekayasa Sipil (JRS-Unand)* **16**(3) pp 178-184

[2] Fitri A, Maulud K N A, Rossi F, Dewantoro F, Harsanto P and Zuhairi N Z 2021 Spatial and Temporal Distribution of Dissolved Oxygen and Suspended Sediment in Kelantan River Basin *In 4th International Conference on Sustainable Innovation 2020–Technology, Engineering and Agriculture (ICoSITEA 2020):* Atlantis Press pp 51-54

[3] Lai S H and Arniza F 2011 Application of SWAT Hydrological Model to Upper Bernam River Basin (UBRB), Malaysia *IUP Journal of Environmental Sciences* **5**(2)
[4] Maulud K N A, Fitri A, Mohtar W H M W, Jaafar W S W M, Zuhairi N Z and Kamarudin M K A 2021 A study of spatial and water quality index during dry and rainy seasons at Kelantan River Basin, Peninsular Malaysia *Arabian Journal of Geosciences* **14**(2) pp 1-19
[5] Fitri A, Hasan Z A and Ghani A A 2011 Determining the Effectiveness of Harapan Lake as Flood Retention Pond in Flood Mitigation Effort *In Proceedings of 2011 4th International Conference on Environmental and Computer Science (ICECS 2011)*
[6] Özmen B, Arch B, Gençoğlan M E & Sertkaya Z R 2019 The Neglected River: Bridging The West & The East *In International Conference on Contemporary Affairs in Architecture and Urbanism (ICCAUA-2019)* Vol. **9** p 10
[7] Fitri A, Hashim R, Song K I and Motamedi S 2015 Evaluation of morphodynamic changes in the vicinity of low-crested breakwater on cohesive shore of Carey Island, Malaysia *Coastal Engineering Journal* **57**(04) p 1550023
[8] Fitri A, Hashim R and Motamedi S 2017 Estimation and Validation of Nearshore Current at the Coast of Carey Island, Malaysia *Pertanika Journal of Science & Technology* **25**(3)
[9] Fitri A, Hashim R, Abolfathi S and Abdul Maulud K N 2019 Dynamics of sediment transport and erosion-deposition patterns in the locality of a detached low-crested breakwater on a cohesive coast *Water* **11**(8) p 1721
[10] Hashim R, Fitri A, Motamedi S, Hashim A M 2013 Modelling of Coastal Hydrodynamic Associated with Coastal Structures: A Reviews *Malaysian Journal of Science* Vol. **32** pp 149-154
[11] Fitri A, Yao L, and Sofawi B Evaluation of Mangrove Rehabilitation Project at Carey Island Coast, Peninsular Malaysia based on Long Term Geochemical Changes *In IOP Conference Series: Earth and Environmental Science* IOP Publishing Vol. **365** No. 1 p 012055
[12] Fitri A and Lao L The impact of Parameter Changes of a Detached Breakwater on Coastal Morphodynamic at Cohesive Shore; A Simulation *In IOP Conference Series: Earth and Environmental Science* IOP Publishing Vol. **365** No. 1 p 012054
[13] Hashim R, Roy C, Shamshirband S, Motamedi S, Fitri A, Petković D and Song K I 2016 Estimation of wind-driven coastal waves near a Mangrove forest using adaptive neuro-fuzzy inference system *Water resources management* **30**(7) pp 2391-2404
[14] Rasaei M, Nazari S and Eslamian S 2020 Experimental investigation of local scouring around the bridge piers located at a 90 convergent river bend *Sādhanā* 45(1) pp 1-15
[15] Kareem K A A and AL-Thamiry H A 2019 Identification of scouring zones in ungauged river by simulation: The case of Galal Badrah river, Iraq *Association of Arab Universities Journal of Engineering Sciences* **26**(3) pp 57-67
[16] Esmaeili T, Dehghani A A, Zahiri A R and Suzuki K 2009 3D Numerical simulation of scouring around bridge piers (Case Study: Bridge 524 crosses the Tanana River) *World Academy of Science: Engineering and Technology* **58**(34) pp 1028-1032
[17] Zaid M, Yazdanfar Z, Chowdhury H and Alam F 2019 A review on the methods used to reduce the scouring effect of bridge pier *Energy Procedia* Vol.160 pp 45-50
[18] Sutopo Y and Saputro A E 2020 Groundsill stability in downstream of bridge "Besi" in body of water at River Garang Semarang city, Indonesia *In IOP Conference Series: Materials Science and Engineering* IOP Publishing Vol. **830** No. 2 p 022073
[19] Wu C C, Lin J J, Lin Y C and Yeh C M 2004 The applicability of open-slot groundsills in trained river to encourage the formation of meandering *In Self-Sustaining Solutions for Streams, Wetlands, and Watersheds* American Society of Agricultural and Biological Engineers pp 35
[20] Widiyanto W 2007 Water Surface Profile in Upstream of Groundsill with Type of Broad-crested and Ogee Dinamika Rekayasa **2**(2) 71-80
[21] Sucipto and Tugino 2010 Analisis Penempatan Groundsill Sebagai Perlindungan Abutment Jembatan terhadap Gerusan Lokal Dinamika TEKNIK SIPIL **10**(2) 168-176
[22] Budi S, Dermawan V and Yuliani E 2017 Efektifitas groundsill terhadap penyebaran sedimen sungai grindulu Kabupaten Pacitan *Jurnal Teknik Pengairan: Journal of Water Resources Engineering* **8**(1) pp 112-123
[23] Permana D P 2020 Kajian Ulang Control Stabilitas Groundsill Bendung Gerak Sungai Bengawan Solo di Kabupaten Bojonegoro *De'Teksi-Jurnal Teknik Sipil Unigoro* **5**(1) pp 13-24
[24] Tungga A W, Asmaranto R and Suprijanto H 2018 Perencanaan Groundsill Pada Sungai Tinga-Tinga Desa Tukad Tinga-Tinga Kecamatan Gerokgak Kabupaten Buleleng Bali *Jurnal Mahasiswa Jurusan Teknik Pengairan* **2**(1) pp 21-30

Acknowledgments
The authors would like to thank Universitas Teknokrat Indonesia for the support in completing this study. Authors are also wishing to acknowledge assistant from Public Work and Spatial Planning (PU), Banten Province in providing the required data.

Study of the effect of climate variation on irrigated and rainfed rice productivity based on aquacrop crop modelling simulation (Case Study of Java Island)

K Aprilina[1,2*], A Susandi[3], A Sopaheluwakan[4] and H Harsa[2]

[1] Earth Science Graduate Program, Bandung Institute of Technology, Jalan Ganesha No 10, Bandung, Indonesia
[2] Research and Development Centre, Indonesian Agency for Meteorology Climatology and Geophysics, Jalan Angkasa I No 2 Kemayoran, Jakarta, Indonesia
[3]Atmospheric Research Group, Bandung Institute of Technology, Jalan Ganesha No 10, Bandung, Indonesia
[4]Center for Applied Climate Services, Indonesian Agency for Meteorology Climatology and Geophysics, Jalan Angkasa I No 2 Kemayoran, Jakarta, Indonesia

*Corresponding E-mail: kharisma.aprilina@bmkg.go.id

Abstract. Aquacrop is free-licensed Food and Agricultural Organization's crop modelling that requires minimum inputs of climate variables namely rainfall, maximum temperature, minimum temperature variables and geographic information of the area to be simulated (longitude, latitude, altitude). This study aims to measure the difference in irrigated and rainfed rice productivity from the Aquacrop crop modelling simulation to the influence of climate pattern variations in Java Island, Indonesia. The k-means clustering method applied to the rainfall, maximum, and minimum temperature variables from the bias-corrected MERRA2 data resulted in two climate regions. The principal component analysis result showed that the maximum and minimum temperature variables are the variables that most contribute to the determination of the clustering area using the k-means method compared to the rainfall variable. This study has calculated the probability of the irrigated and rainfed rice productivity resulting from the Aquacrop simulation in those climate regions during La Nina [El Nino] years that will be higher [smaller] than the mean value of rice productivity during neutral years. However, the validation between the actual irrigated and rainfed rice productivity with the Aquacrop simulation results from 2001-2014 showed low correlation values that vary between negative and positive values in all climate regions. Meanwhile, the validation on the El Nino composite years generally showed positive correlation values. In addition, the neutral and La Nina composite years resulted in varying correlation values between negative and positive correlation.

Keywords: Aquacrop, Variations of climate patterns, Rice productivity, Clustering

Track Name: Land, Water, Forest, and Food Security

1. Introduction

Java Island is the main rice-producing island in Indonesia [1], although it is not the largest island in the country. However, the population density and population on Java Island is also the highest in this country, where more than half of Indonesia's population lives on this island [2], therefore Indonesia's food security can be represented by the food security status of the people living on Java Island [3]. In order to maintain the stability of food security in Indonesia, various studies that link environmental factors to rice production and productivity have been carried out [4–9]. One of these studies produces a linear regression function that relates climate factors to rice production in Java [9]. However, this regression function only can be applied if there is an assumption of uniformity of climate patterns in all regions in Java, where it has been known that, in general, the climate pattern in Java is monsoon (annual) [10].

A recent study has shown that there is a phenomenon of variations in climate patterns in Java, such as semi-annual variations where some areas experience the peak of the rainy season other than the December–January–February (DJF) period, for example, in the March–April–May (MAM) period [11], this non-uniformity of climate patterns then can be referred as variations in climate patterns. Therefore, it can be assumed that the climate pattern in the Java Island region can also impact variations in the rice productivity in Java Island.

Generally, variations in climate patterns in Indonesia have been delineated using rainfall variables. However, the latest studies had carried out climate classification using rainfall and temperature variables [12], as well as temperature range (maximum temperature–minimum temperature) [13] or maximum temperature, minimum temperature, and relative humidity variables simultaneously [14,15]. Therefore, in this study, variations in climate patterns will be determined from clustering the variables of rainfall, maximum and minimum temperature, considering that these three variables are most often used in climate classification in previous studies.

Rice productivity as a response variable from the impact of variations in climate patterns in this study was generated from crop modelling simulations using FAO (Food and Agriculture Organization) Aquacrop. Aquacrop's crop modelling algorithm is based on water demand for plants during their growth and development period [16,17]. Aquacrop crop modelling is used in this study because it is easy to use, free license, and requires relatively few inputs of climate variables (rainfall, maximum temperature, minimum temperature) and geographic information of the area to be simulated (longitude, latitude, altitude) [18,19]. In addition, Aquacrop can simulate crop productivity using different irrigation systems, such as irrigated and rainfed systems [18]. So this study aims to determine the impact of variations in climate patterns in Java on rice crop productivity resulting from Aquacrop simulations under different irrigation systems.

2. Data and Method

2.1 Data

2.1.1 Observation Data

The data used in this study are daily rainfall data (precip), maximum temperature (tmax), and minimum temperature (tmin) from 23 stations of the Indonesian Agency for Meteorology, Climatology and Geophysics (BMKG) located on Java Island [20] for 31 years from years 1987-2017.

2.1.2 Supporting data

Supporting data used include:

a. MERRA2 data (The Modern-Era Retrospective analysis for Research and Applications, Version 2) with spatial resolution 0.5° x 0.625° which consists of daily rainfall, maximum and minimum temperature data for 31 years from 1987-2017 [21] which is cropped to Java Island between 9° S-5.5° S and 104° E-117° E.
b. Actual data on irrigated and rainfed rice productivity obtained from the BPS (Indonesia Central Bureau of Statistics) publication in Figures from the official website of the Provincial

BPS in Java, the available data are annual data for each Regency/City ranging from 13-14 years from 2001/2002- 2014.

c. Southern Oscillation Indices (SOI) data downloaded from http://www.cgd.ucar.edu/cas/catalog/climind/soi.html, for a period of 31 years from 1987-2017.

2.2 Method

Bias correction is carried out using the Quantile Mapping method from the qmap package [22] in the R program. This correction method can capture changes in the average and variability of the data as well as adjust the statistical moment of the corrected data [23]. Clustering was performed simultaneously on the predictor variables (precip, tmax, and tmin) using the k-means method from the package stats [24] in the R program. The k-means method was used because this method resulted in a more stable cluster boundary [25]. Next, the determination of the number of k-groups was conducted using the Silhouette method [26]. The results of grouping with the k-means method, then further analyzed using the Principal Component Analysis (PCA) method. PCA method is proved as an advanced solution to indicate the number of group members in the k-means clustering method [26]. In addition, the PCA method can capture important information from predictor variables and determine the grouped area's characteristics [27]. PCA was conducted on a correlation matrix because the predictor variables, namely annual rainfall, maximum and minimum temperatures, had different units. Thus data standardization was necessary. PCA was performed using the prcomp function in the R programming stats package. PCA was performed to detect the stability of the grouping results by the k-means method [28]. The loadings PC value is used to determine the contribution of each variable to the grouped area.

Aquacrop crop modelling input uses predictor variable segmentation (daily precip, tmax, and tmin from bias-corrected reanalysis data) resulting from clustering. Aquacrop is run with the assumption that apart from predictor variables, other factors such as soil type, soil structure and plant varieties are homogeneous so that the simulation output reflects the influence of the predictor variables [29]. Aquacrop is run with two modes of irrigation systems, namely regular irrigation and rainfed systems. Planting dates are assumed to be the early month of January, May, and September each year from 1987 to 2017.

Table 1. The ENSO composite years correspond to the active mode used in this study.

El Nino			La Nina		
weak	moderate	strong	weak	moderate	strong
2004	1987	1991	1996	1999	1988
2006	1994	1997	2000	2005	1998
2014	2002	2015	2008	2011	2007
	2009		2016		2010
			2017		

Then, the probabilities of irrigated and rainfed rice productivity in each group are calculated by first determining the distribution pattern of rice productivity data in the composite years experiencing climate variability mode. Determination of the composite year ENSO (El Nino Southern Oscilation) using a standardized SOI index (Table 1). While the determination of weak, medium, and strong ENSO modes refers to [30], namely weak La Nina SOI values range from 0.3 to 0.65, moderate La Nina SOI values range from 0.65 to 0.95, strong La Nina SOI values are more from 0.95, weak El Nino SOI values between -0.6 to -1.2, moderate El Nino SOI values ranged from -1.2 to -1.8, and strong El Nino SOI values less than -1.8.

3. Result and Discussion

3.1 The result of bias correction with the quantile mapping method

The MERRA2 data before being bias-corrected showed a fairly good agreement with the observation data, but for the months of January-May, the MERRA2 data was not able to reach the average monthly maximum value in those months (Figure 1a). Therefore, the bias correction on the MERRA2 rainfall data is useful for reaching the maximum rainfall value that was not previously achieved by the MERRA2 data before the correction.

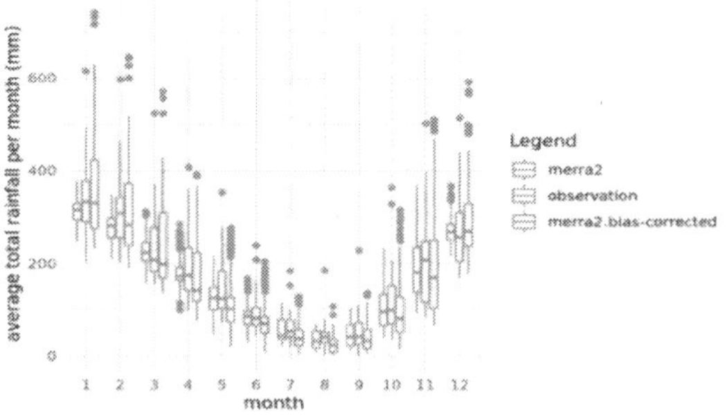

(a)

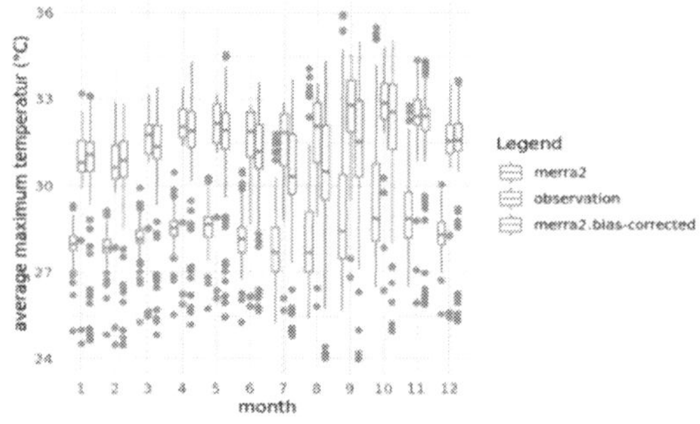

(b)

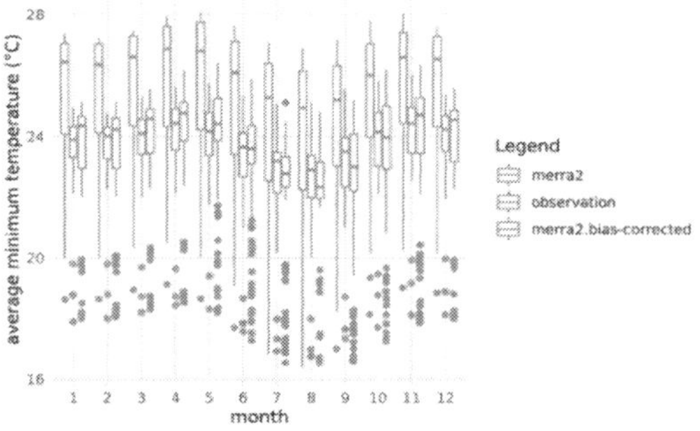

(c)

Figure 1. Boxplot of the 1987-2017 (a) aggregate monthly average rainfall, (b) maximum temperature, and (c) minimum temperature from MERRA2 data (red), observations (green), and bias-corrected MERRA2 (blue).

Meanwhile the average monthly maximum temperature data for MERRA2 (Figure 1b) before bias correction looks underestimated compared to the observation data. The corrected data tend to be able to reach the range of the average monthly maximum temperature data from the observation data. The average monthly maximum temperature data for MERRA2 in July-September has a minimum and maximum data distribution range that is longer than other months, so this also affects the distribution range of bias-corrected MERRA2 in July-September.

In contrast to the maximum temperature data, the monthly average minimum temperature on the MERRA2 data (Figure 1c) tends to overestimate the monthly minimum temperature average of the observation data. However, corrected MERRA2 data tends to be able to reach the average monthly minimum temperature range of observation data, although in several months, it still tends to be overestimated but it is better than when it has not been corrected.

3.2 Clustering of Java Island climate regions with the k-means method

The Java Island climate region clustering is conducted by grouping the variables of rainfall, maximum temperature, and minimum temperature from the bias-corrected MERRA2 data simultaneously. The determination of the optimal number of k in this k-means method uses the Silhouette method, which is a method of determining the number of groups based on the average of the Silhouette width, the wider the average, the clearer the differences between different groups and members who are in the same group will be increasingly similar [31].

The results of clustering on predictor variables data from bias-corrected MERRA2 that are aggregated on an annual scale (Figure 2a) produce two clusters or groups of climate regions in Java. The number of optimal k groups from the Silhouette method also shows k=2 (Figure 2b). Silhouette width (Si) > 0 (Figure 2c) indicates that the data group can be clustered well. The value of the Si width is closer to 1, meaning the better the grouping formed [31] [32].

Table 2. Descriptive statistic of predictor variables using annual aggregation.

Cluster	Variable	N Grid	N	Min	Q1	Mean	Median	Q3	Max	SD
1	Total rainfall (mm)	33	1023	893	1508	1852	1787	2117	4108	481
2	Total rainfall (mm)	8	248	1158	1819	2232	2111	2561	6261	653
1	Maximum temperature (°C)	33	1023	30.2	31.3	31.8	31.7	32.2	33.7	0.75
2	Maximum temperature (°C)	8	248	25.2	26.1	27.7	28.6	28.9	29.7	1.43
1	Minimum temperature (°C)	33	1023	20.4	23.3	23.7	23.9	24.4	26.2	1.1
2	Minimum temperature (°C)	8	248	17.2	18.1	18.8	18.9	19.3	20.9	0.77
1	Irrigated rice productivity (ton/ha)	33	1023	2.84	4.38	5.14	5.10	5.89	7.15	1.02
2	Irrigated rice productivity (ton/ha)	8	248	3.23	5.02	5.66	5.77	6.36	7.10	0.89
1	Rainfed rice productivity (ton/ha)	33	1023	0.77	2.38	3.55	3.18	4.43	7.14	1.38
2	Rainfed rice productivity (ton/ha)	8	248	1.28	2.93	4.25	4.22	5.37	7.11	1.49

Actually based on altitude map from [33], area of cluster 2 are dominated with highland area, meanwhile cluster 1 are dominated with lowland area. Based on descriptive statistic (Table 2), mean of total annual rainfall in cluster 2 is higher than mean of total annual rainfall in cluster 1. While mean of annual maximum and minimum temperature in cluster 2 is lower than those in cluster 1.

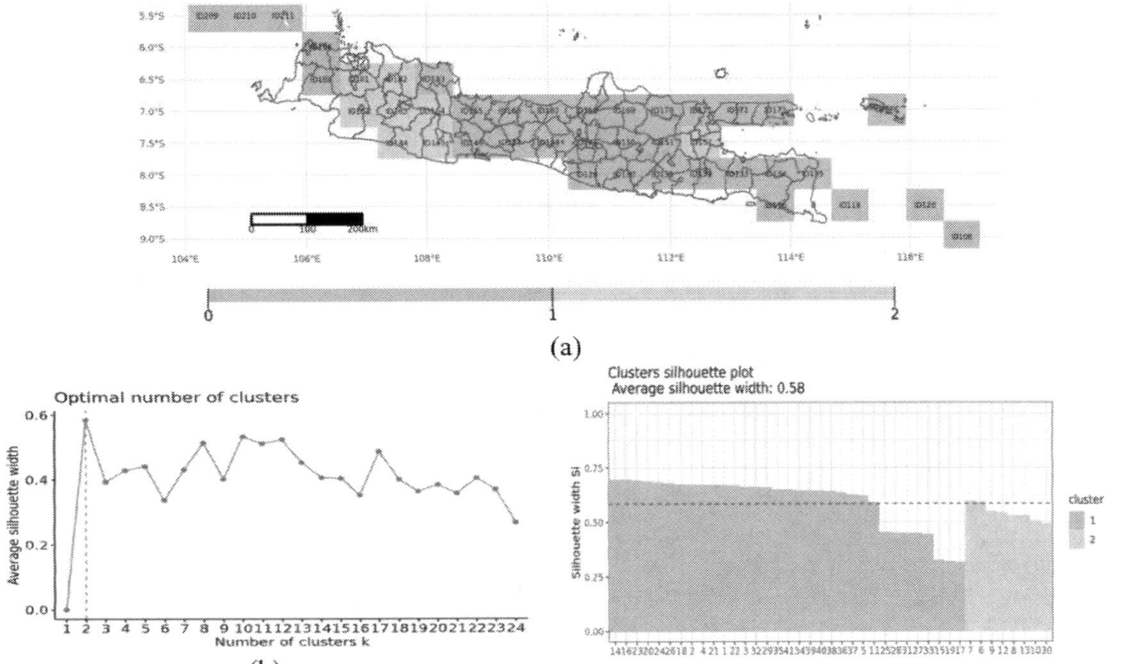

(a)

(b)

(c)

Figure 2. (a) The results of clustering predictor variables with annual aggregates using the k-means method (The numbering grid of MERRA2 only used in this study); (b) The plot of determining the optimal number of k using the Silhouette method; (c) Silhouette width histogram.

3.3 Comparison of the results of the k-means method with the Principal Component Analysis (PCA)

PCA can be used to compare the magnitude of the variance of the predictor variables that affect the clustering results of the k-means method. The loading plot (Figure 3a) can show the magnitude of the loading value (positive or negative) in dimensions 1 (PC1) and 2 (PC2) on bias-corrected MERRA2 data. Meanwhile, Table 2 provides detailed information on the loading values for each PC dimension on bias-corrected MERRA2. A positive loading value means that the variable contributes positively to the principal component (PC). A negative loading value means that the variable contributes negatively to the PC. *Loading value* is the significant contribution value given by the variable to each PC [34].

Loading plots (Figures 3a) show that in the corrected MERRA2 data, the maximum and minimum temperature variables are located at an angle of almost 90° with the rainfall variable. In contrast, the maximum temperature variable is adjacent to the minimum temperature variable. This shows that the maximum temperature and minimum temperature variables have a negative relationship with the rainfall variable, but the correlation is low; when the rainfall variable increases, the maximum temperature and minimum temperature variables tend to decrease, while the maximum temperature and minimum temperature variables are strongly correlated with one another.

PC1 on the loading plot of the corrected MERRA2 can explain the variance of 69.9 %, while PC2 can explain the variance of 24.8 %. According to Table 3, the maximum and minimum temperature variables in the clustering of MERRA2 contributed negatively and strong to PC1. In contrast, the rainfall variable contributed positively and relatively strong to PC1, then on PC2, the rainfall variable contributed negatively and strong, while the maximum and minimum temperatures contribute negatively and weak. Based on the plot and the loading value of PCA on the corrected MERRA2 data, it can be seen that the maximum temperature variable and the minimum temperature variable are variables that have more influence or contribute strongly to PC1 than the rainfall variable, and this affects the grouping results in the k-means method.

Table 3. The predictor variable loading value in each dimension of the bias-corrected MERRA2

Predictor variables	Bias-corrected MERRA2		
	Dimension 1	Dimension 2	Dimension 3
Rainfall	0.648	-0.759	0.064
Maximum temperature	-0.941	-0.170	0.293
Minimum temperature	-0.890	-0.373	-0.264

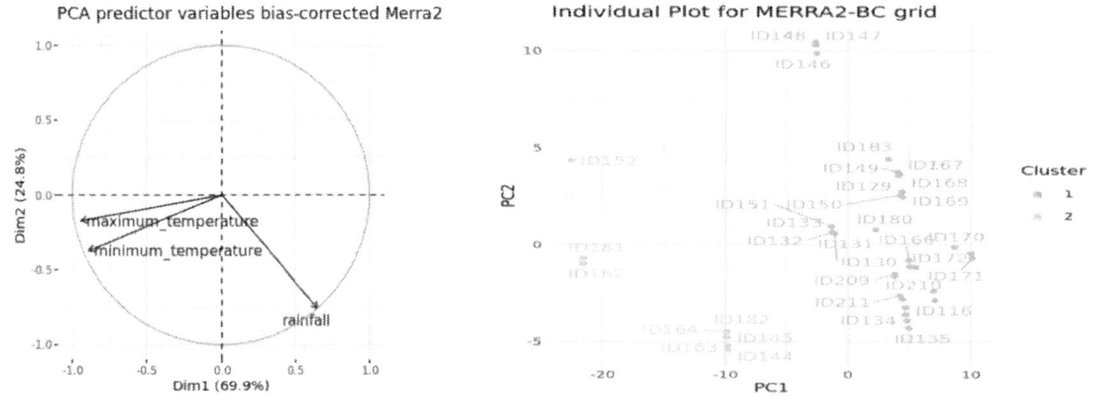

(a) (b)

Figure 3. (a) The loading plot of predictor variables on bias-corrected MERRA2, and (b) Score plot on bias-corrected MERRA2 grid.

Score is the new coordinate value on the PC axis, which is formed from PCA results. Visually (Figures 3b), the clustering result is caused by differences in the score on PC1; this means that the clustering occurs because of the influence of the predictor variable that contributes the most to PC1 (Figures 3a), namely the maximum temperature variable and minimum temperature. So the bias-corrected MERRA2 grid (Figure 3b), which has a lower average maximum and minimum temperature tend to be on the negative score side on PC1.

3.4 Probability of Irrigated and Rainfed Rice Productivity Results of Aquacrop Simulation in ENSO Year.

3.4.1 Irrigated Rice Productivity

Determination of the distribution pattern is useful for calculating the probability of the impact of ENSO on the results of the Aquacrop simulation, with a neutral year as the control year. Determination of the distribution pattern of rice productivity data is carried out by two stages of empirical data distribution tests. The first stage of the test is the Shapiro-Wilk normality test (hereinafter referred as the SW test), and if the results of the data SW test conclude that the data is not normally distributed, then a further test is carried out using the Kolmogorov-Smirnov Test (hereinafter referred as the KS test) [35,36]. Determination of the distribution of the empirical distribution of irrigated rice productivity simulation data is needed to calculate the estimated productivity probability when the ENSO mode is active (La Nina and El Nino) compared to the average productivity during a neutral year.

In this study, several data were produced whose distribution patterns could not be identified by the results of the empirical distribution test thus the determination of the distribution pattern was subjectively determined by visual of the histogram plot and the empirical cumulative density (ECDF) plot.

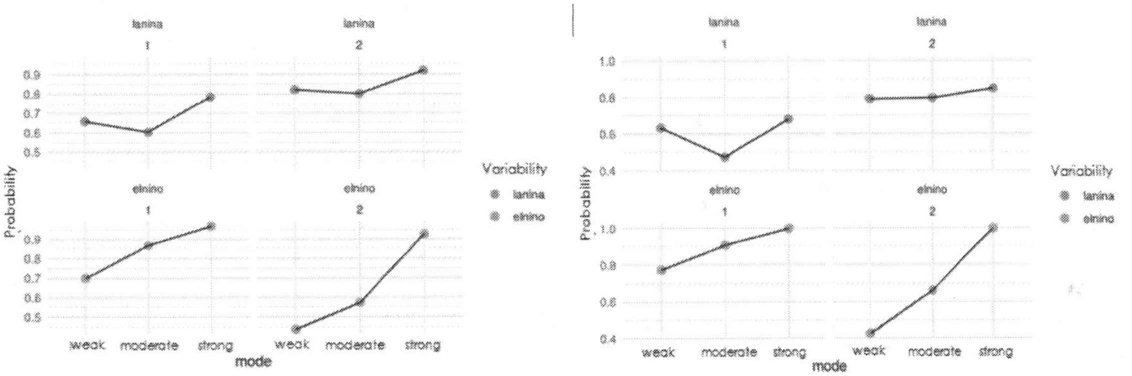

(a) (b)

Figure 4. Probability of (a) irrigated and (b) rainfed rice productivity from Aquacrop simulation (Note: in the upper panel the probability of rice producitivity during La Nina > average of rice producitivity during neutral years, while the lower panel probability of rice producitivity during El Nino < average of rice producitivity neutral years), left panel: cluster 1, right panel: cluster 2).

The distribution pattern of the simulation results of irrigated rice ranges from the lognormal, normal, gamma, and weibull distributions. The probability of irrigated rice productivity during El Nino with productivity opportunities lower than the average productivity in a neutral year (El Nino productivity probability < average of neutral years productivity) increases with the increasing El Nino mode (Figure 4a), both in cluster 1 and cluster 2. While the probability during La Nina with the opportunity for irrigated rice productivity that is greater than the average productivity in a neutral year (probability of La Nina rice > average of neutral years productivity) is different for clusters 1 and 2. In cluster 2, the probability level increases with the increase of La Nina mode, while in cluster 1 the probability in weak La Nina mode is higher than the probability of moderate La Nina mode.

3.4.2 Rainfed Rice Productivity

The distribution pattern of the simulation results of rainfed rice productivity ranges from the lognormal, normal, gamma, and weibull distributions. Similar to the results of probability calculations for irrigated rice, the probability of rainfed rice productivity during El Nino with productivity opportunities lower than the average productivity in a neutral year (probability of El Nino rice productivity < average of neutral years productivity) increases with increasing El Nino mode (Figure 4b), both in cluster 1 and cluster 2. While the probability during La Nina with rainfed rice productivity opportunities is greater than the average productivity during a neutral year (probability of La Nina productivity > average of neutral years productivity) is different for clusters 1 and 2. In cluster 2, the probability level increases with increasing La Nina mode, while in cluster 1 the probability in weak La Nina mode is higher than the probability of medium La Nina mode, this is due to during the 2016 weak La Nina accompanied by a strong negative IOD phenomenon which amplifies rainfall intensity during September-October-November 2016 [37,38]. Therefore, because the Aquacrop crop modeling is more sensitive to changes in rainfall variables, the probability of weak La Nina mode is higher than the probability of moderate La Nina mode.

The probability range (probability of La Nina productivity > average of neutral years productivity) on rainfed rice productivity from Aquacrop simulation during La Nina ranges from ~45-60 % in cluster 1 area and ~80-82 % in cluster 2. While the probability range (P El Nino < neutral year mean) in the cluster 1 region ranged from ~78-97 % and the cluster 2 region ranged from ~40-92 %.

The probability of a decrease in irrigated and non-irrigated rice productivity during El Nino in cluster 1 is higher than in cluster 2. Meanwhile, the probability of irrigated and non-irrigated rice productivity experience increasing productivity during La Nina in cluster 2 is higher than in cluster 1. It means that cluster 1 is more vulnerable to El Nino and La Nina than cluster 2. Besides that, the rainfed system is also more vulnerable to El Nino and La Nina than the irrigated yield rice system.

3.5 Validation between Actual Productivity and Simulation Productivity of Aquacrop

3.5.1 Validation for Irrigated Rice Productivity

The actual productivity data for irrigated and rainfed rice available on the official website of BPS for each province is in the form of annual productivity data per city/district within the province and the time span is only 13-14 years from 2001/2002-2014. The validation results of the correlation value of the simulation results of the Aquacrop application with the actual productivity of irrigated rice show a fairly low correlation, ranging from -0.01 to 0.58 with a data length of 13-14 years (Figure 5a) which means the Aquacrop application with default input on variables other than variables climate (daily rainfall, daily maximum and minimum temperature) and geographic location (longitude, latitude, and altitude) derived from the corrected MERRA2 grid data provide estimation results that have low reliability on the actual productivity of irrigated rice, this is as a result of input parameters used is the minimum input parameters required in running the Aquacrop application. However, the Aquacrop application which is inputted using the input parameters that have been calibrated such as the parameters of canopy cover, root depth effectiveness, and rice seed sowing rate provide a better estimate for example by [39] in Subang, West Java, Aquacrop can estimate the productivity of irrigated rice with a correlation coefficient of 0.8 (11 years of data from 2000-2010), Root Mean Square Error (RMSE) of 0.53, and a Mean Square Error (MSE) of 9%.

Because the correlation results for all available years yielded a fairly low correlation value, then the correlation value for the ENSO composite years were calculated. Calculation of the correlation value between the actual productivity and the simulated productivity of irrigated rice in the La Nina composite years resulted in a correlation value that varied between negative and positive correlations (-0.82 to 0.98) with a data length of 5 years (2005, 2007, 2008, 2010 and 2011) with the strongest negative correlation in Boyolali Regency, Central Java and the strongest positive correlation in Probolinggo City, East Java (Figure 5c).

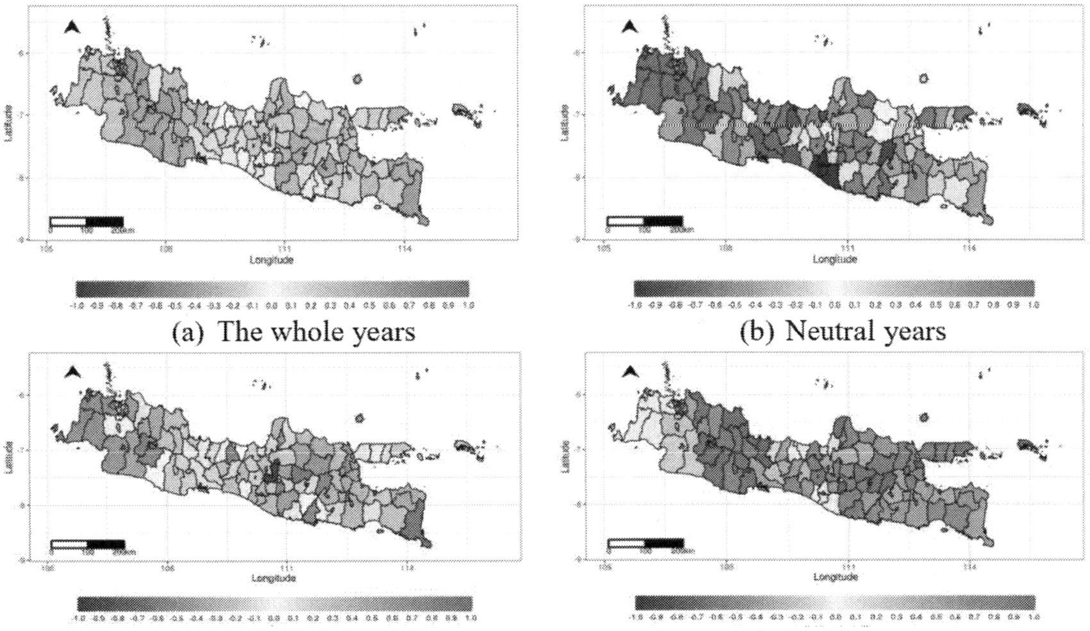

(a) The whole years (b) Neutral years

(c) La Nina years (d) El Nino years

Figure 5. The correlation plot between actual productivity and simulated productivity of irrigated rice Aquacrop per city/district in Java in (a) the whole year (2001/2002-2014); (b) composite of neutral years; (c) composite of La Nina years; and (d) composite of El Nino years.

Then the range of correlation values between actual productivity and simulated productivity of irrigated rice in the El Nino composite ranged from 0.045-0.98, with a data length of 5 years (2002, 2004, 2006, 2009 and 2014) where the highest positive correlation value was in Brebes Regency, Central Java (Figure 5b). While the neutral year composite produces a range of correlation values from negative correlation values and positive correlations (between -1 and 1) but with relatively very short data lengths between 3 and 4 years (2001, 2003, 2012, and 2013) (Figure 5d). The correlation values of the ENSO composite years are relatively higher than the correlation value in the whole years, therefore although the correlation values are relatively high, it becomes less reliable because the length of the data is inadequate since it does not have a long period of time.

3.5.2 Validation for Rainfed Rice Productivity

The correlation results for the whole year (Figure 6a) between actual productivity and non-irrigated rice productivity from Aquacrop simulation have a better correlation value than the validation results for irrigated rice, ranging from -0.40 to 0.78, with the highest correlation being in Mojokerto Regency, East Java. The correlation results which tend to be better in the validation results from the simulation of rainfed rice productivity compared to the validation of irrigated rice prove that the Aquacrop application is sensitive to rainfall variables.

In the ENSO composite years (Figure 6b, c, d), for the neutral years, the correlation range is between -1 and 1 this is because the available data range is short, resulting in a high correlation value. Then in the La Nina years, the correlation ranges from -0.88 to 0.95, each of which comes from validation in Jombang Regency and Mojokerto Regency, East Java. Correlation results for the El Nino years produce a correlation range ranging from 0 – 0.94, with the highest correlation value being in the Pekalongan Regency, Central Java.

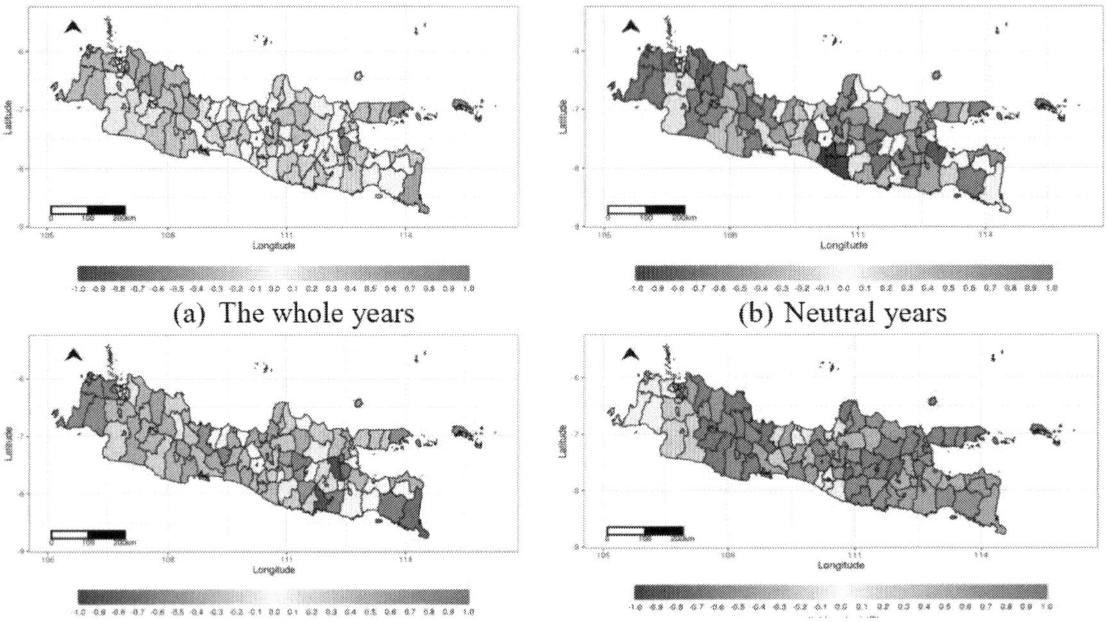

(a) The whole years (b) Neutral years

(c) La Nina years (d) El Nino years

Figure 6. The correlation plot between actual productivity and simulated productivity of rainfed rice Aquacrop per city/district in Java in (a) the whole year (2001/2002-2014); (b) composite of neutral years; (c) composite of La Nina years; and (d) composite of El Nino years.

4. Conclusion

The k-means clustering method applied to the rainfall, maximum, and minimum temperature variables from the bias-corrected MERRA2 data resulted in two climate regions. Although the loading value from the PCA results shows that the maximum and minimum temperature variables contribute the most to the clustering produced by the k-means method. However, the rainfall variable also contributes to the climate region cluster on Java Island, as indicated by the probability value of cluster 1 area to experience a decline in productivity during El Nino from the results of the Aquacrop simulation which is generally higher than the probability in cluster 2 both in irrigated and rainfed rice productivity. Likewise, the probability value for cluster 1 area to experience an increase in productivity during La Nina from the results of the Aquacrop simulation is generally lower than the probability in cluster 2 for both irrigated and rainfed rice productivity. This is because the Aquacrop simulation tends to be sensitive to rainfall variables and the cluster 2 area has a higher average rainfall than the cluster 1 area.

In general, the validation results show that between the actual productivity and the productivity of the Aquacrop simulation results, the correlation value is quite low, both in the simulation results for irrigated and rainfed rice productivity. Then it is known that the validation of the ENSO composite years show that in the El Nino years, the correlation between actual productivity and the productivity of the Aquacrop simulation results shows a positive correlation value, while in the neutral and La Nina years the correlation value varies between negative and positive values.

References

[1] Panuju D R, Mizuno K and Trisasongko B H 2013 The dynamics of rice production in Indonesia 1961–2009 *J. Saudi Soc. Agric. Sci.* **12** 27–37

[2] Sumarsono D S D and Ningsih S 2018 The Effect of Own-Source Revenue, Population, and Population Density on the Capital Expenditure of City and District Governments on Java Island: *Proceedings of the Journal of Contemporary Accounting and Economics Symposium 2018 on Special Session for Indonesian Study* Journal of Contemporary Accounting and Economics Symposium 2018 on Special Session for Indonesian Study (Bali, Indonesia: SCITEPRESS - Science and Technology Publications) pp 581–7

[3] Prasada I M Y and Masyhuri 2019 Food security in Java Island, Indonesia: Panel data of ordinary least square approach *IOP Conf. Ser. Earth Environ. Sci.* **346** 012065
[4] Naylor R L, Battisti D S, Vimont D J, Falcon W P and Burke M B 2007 Assessing risks of climate variability and climate change for Indonesian rice agriculture *Proc. Natl. Acad. Sci.* **104** 7752–7
[5] Estiningtyas W and Syakir M 2017 Pengaruh Perubahan Iklim Terhadap Produksi Padi di Lahan Tadah Hujan *J. Meteorol. Dan Geofis.* **18** 83–93
[6] Nugroho M and Nurhayanti Y 2016 Sensitivitas Produksi Padi Terhadap Perubahan Iklim Di Indonesia Tahun 1974-2015 *Agro Ekon.* **27**
[7] Amien I, Redjekiningrum P, Kartiwa B and Estiningtyas W 1999 Simulated rice yields as affected by interannual climate variability and possible climate change in Java *Clim. Res.* **12** 145–52
[8] Yuliawan T and Handoko I 2016 The Effect of Temperature Rise to Rice Crop Yield in Indonesia uses Shierary Rice Model with Geographical Information System (GIS) Feature *Procedia Environ. Sci.* **33** 214–20
[9] Naylor R L, Falcon W P, Rochberg D and Wada N 2001 Using El Niño/Southern Oscillation Climate Data to Predict Rice Production in Indonesia *Clim. Change* **50** 255–65
[10] Aldrian E and Dwi Susanto R 2003 Identification of three dominant rainfall regions within Indonesia and their relationship to sea surface temperature *Int. J. Climatol.* **23** 1435–52
[11] Fauzi M A 2014 The Effect of the Monsoon Transition on the Annual Rainfall Cycle in Java *Final Project* (Bandung, Indonesia: Meteorology Study Program, Faculty of Earth Science and Technology, Bandung Institute of Technology)
[12] Mahlstein I and Knutti R 2010 Regional climate change patterns identified by cluster analysis *Clim. Dyn.* **35** 587–600
[13] Netzel P and Stepinski T 2016 On Using a Clustering Approach for Global Climate Classification *J. Clim.* **29** 3387–401
[14] Iyigun C, Türkeş M, Batmaz İ, Yozgatligil C, Purutçuoğlu V, Koç E K and Öztürk M Z 2013 Clustering current climate regions of Turkey by using a multivariate statistical method *Theor. Appl. Climatol.* **114** 95–106
[15] Sathiaraj D, Huang X and Chen J 2019 Predicting climate types for the Continental United States using unsupervised clustering techniques *Environmetrics* **30** e2524
[16] Raes D, Steduto P, Hsiao T C and Fereres E 2009 AquaCropThe FAO Crop Model to Simulate Yield Response to Water: II. Main Algorithms and Software Description *Agron. J.* **101** 438
[17] Steduto P, Hsiao T C, Raes D and Fereres E 2009 AquaCrop—The FAO Crop Model to Simulate Yield Response to Water: I. Concepts and Underlying Principles *Agron. J.* **101** 426
[18] Foster T, Brozović N, Butler A P, Neale C M U, Raes D, Steduto P, Fereres E and Hsiao T C 2017 AquaCrop-OS: An open source version of FAO's crop water productivity model *Agric. Water Manag.* **181** 18–22
[19] Xu J, Bai W, Li Y, Wang H, Yang S and Wei Z 2019 Modeling rice development and field water balance using AquaCrop model under drying-wetting cycle condition in eastern China *Agric. Water Manag.* **213** 289–97
[20] Supari S, Ettema J and Aldrian E 2012 Spatiotemporal Characteristics of Extreme Rainfall Events Over Java Island, Indonesia *Indones. J. Geogr.* **Vol. 44** 62–86
[21] Gelaro R, McCarty W, Suárez M J, Todling R, Molod A, Takacs L, Randles C A, Darmenov A, Bosilovich M G, Reichle R, Wargan K, Coy L, Cullather R, Draper C, Akella S, Buchard V, Conaty A, da Silva A M, Gu W, Kim G-K, Koster R, Lucchesi R, Merkova D, Nielsen J E, Partyka G, Pawson S, Putman W, Rienecker M, Schubert S D, Sienkiewicz M and Zhao B 2017 The Modern-Era Retrospective Analysis for Research and Applications, Version 2 (MERRA-2) *J. Clim.* **30** 5419–54
[22] Gudmundsson L 2016 Package "qmap": Statistical Transformations for Post-Processing Climate Model Output
[23] Ringard J, Seyler F and Linguet L 2017 A Quantile Mapping Bias Correction Method Based on Hydroclimatic Classification of the Guiana Shield *Sensors* **17** 1413
[24] R Core Team 2013 *The "stats" package* (Vienna, Austria: R Foundation for Statistical Computing)
[25] Zhang Y, Moges S and Block P 2016 Optimal Cluster Analysis for Objective Regionalization of Seasonal Precipitation in Regions of High Spatial–Temporal Variability: Application to Western Ethiopia *J. Clim.* **29** 3697–717
[26] Ding C and He X 2004 K-means clustering via principal component analysis *Twenty-first international conference on Machine learning - ICML '04* Twenty-first international conference (Banff, Alberta, Canada: ACM Press) p 29
[27] Azam M, Park H, Maeng S and Kim H 2017 Regionalization of Drought across South Korea Using Multivariate Methods *Water* **10** 24
[28] Ben-Hur A and Guyon I 2003 Detecting Stable Clusters Using Principal Component Analysis *Functional Genomics* vol 224 (New Jersey: Humana Press) pp 159–82
[29] Martini L C P 2018 Sensitivity analysis of the AquaCrop parameters for rainfed corn in the South of Brazil *Pesqui. Agropecuária Bras.* **53** 934–42
[30] Hanley D E, Bourassa M A, O'Brien J J, Smith S R and Spade E R 2003 A Quantitative Evaluation of ENSO Indices *J. Clim.* **16** 10
[31] Rousseeuw P J 1987 Silhouettes: A graphical aid to the interpretation and validation of cluster analysis *J. Comput. Appl. Math.* **20** 53–65
[32] Rasid Mamat A, Susilawati Mohamed F, Afendee Mohamed M, Mohd Rawi N and Isa Awang M 2018 Silhouette index for determining optimal k-means clustering on images in different color models *Int. J. Eng. Technol.* **7** 105
[33] Yuliawan T and Handoko I 2016 The Effect of Temperature Rise to Rice Crop Yield in Indonesia uses Shierary Rice Model with Geographical Information System (GIS) Feature *Procedia Environ. Sci.* **33** 214–20
[34] Holland S M 2019 Principal Components Analysis (PCA)

[35] Razali N M and Wah Y B 2011 Power comparisons of Shapiro-Wilk, Kolmogorov-Smirnov, Lilliefors and Anderson-Darling tests *J. Stat. Model. Anal.* **2** 21–33
[36] Yap B W and Sim C H 2011 Comparisons of various types of normality tests *J. Stat. Comput. Simul.* **81** 2141–55
[37] Lim E-P and Hendon H H 2017 Causes and Predictability of the Negative Indian Ocean Dipole and Its Impact on La Niña During 2016 *Nat. Sci. Rep.* **7** 1–11
[38] Nur'utami M N and Hidayat R 2016 Influences of IOD and ENSO to Indonesian Rainfall Variability: Role of Atmosphere-ocean Interaction in the Indo-pacific Sector *Procedia Environ. Sci.* **33** 196–203
[39] Santikayasa I, Amdan W, Perdinan and Sugiarto Y 2018 Quantification the Impact of Climate Change on Paddy Field Yield Production in Subang, Indonesia *IOP Conf. Ser. Earth Environ. Sci.* **187** 012059

Acknowledgments
This paper is part of first author master thesis in Faculty of Earth Science and Technology, Bandung Institute of Technology (ITB) that funded by the scholarship from the Center of Training and Education, Indonesian Agency for Meteorology Climatology and Geophysics (BMKG). We thank to advanced comment and suggestion from anonymous reviewers.

Supplementary data

Table 4. Determination of empirical distribution with Shapiro-Wilk Test (SW) and Kolmogorov-Smirnov Test (KS) on Irrigated Rice Productivity.

Kel.	Var.	Mode	SW p-value	Decision	Kolmogorov Smirnov one sample test p-value			Distribution
1	EN	weak	0.014**)	reject H0	lnorm = 0.4	gamma= 0.5	weibull=0.2	gamma
1	EN	moderate	0.000***)	reject H0	lnorm = 1	gamma= 0.8	weibull=0.1	lognormal
1	EN	strong	0.007**)	reject H0	lnorm = 0.7	gamma= 0.6	weibull=0.6	lognormal
1	LN	weak	0.000***)	reject H0	lnorm = 0.07	gamma= 0.1	weibull=0.04	gamma
1	LN	moderate	0.021**)	reject H0	lnorm = 0.7	gamma= 0.8	weibull=0.7	gamma
1	LN	strong	0.000***)	reject H0	lnorm = 0.006	gamma= 0.01	weibull=0.05	weibull
1	N		0.001***)	reject H0	norm = 0.3	gamma= 0.1	weibull=0.2	normal
2	EN	weak	0.524	accept H0				normal
2	EN	moderate	0.305	accept H0				normal
2	EN	strong	0.879	accept H0				normal
2	LN	weak	0.002**)	reject H0	lnorm = 0.3	gamma= 0.3	weibull=0.2	lognormal[1]
2	LN	moderate	0.213	accept H0				normal
2	LN	strong	0.006**)	reject H0	lnorm = 0.1	gamma= 0.1	weibull=0.3	weibull
2	N		0.101	accept H0				normal

Notes: [1]: probability is determined subjectively by visual of the plot

Table 5. Determination of empirical distribution with Shapiro-Wilk Test (SW) and Kolmogorov-Smirnov Test (KS) on Rainfed Rice Productivity.

Kel.	Var.	Mode	SW p-value	Decision	Kolmogorov Smirnov one sample test p-value			Distribution
1	EN	weak	0.000***)	reject H0	lnorm = 0.04	gamma= 0.07	weibull=0.02	lognormal
1	EN	moderate	0.000***)	reject H0	lnorm = 0.000	gamma= 0.000	weibull=0.000	lognormal[1]
1	EN	strong	0.000***)	reject H0	norm = 0.06	lnorm= 0.08	gamma=0.02	lognormal
1	LN	weak	0.000***)	reject H0	gamma =0.05	lnorm= 0.2	weibull=0.002	lognormal
1	LN	moderate	0.000***)	reject H0	gamma =0.5	lnorm= 0.6	weibull=0.4	lognormal
1	LN	strong	0.000***)	reject H0	gamma =0.1	lnorm= 0.03	weibull=0.09	gamma
1	N		0.000***)	reject H0	gamma =0.03	lnorm= 0.01	weibull=0.01	gamma
2	EN	weak	0.015*)	reject H0	lnorm = 0.6	gamma= 0.7	weibull=07	gamma[1]
2	EN	moderate	0.52	accept H0				normal
2	EN	strong	0.047*)	reject H0	lnorm = 0.7	gamma= 0.6	weibull=0.3	lognormal
2	LN	weak	0.015*)	reject H0	lnorm = 0.4	gamma= 0.6	weibull=0.6	gamma[1]
2	LN	moderate	0.026*)	reject H0	lnorm = 0.9	gamma= 0.9	weibull=07	gamma[1]
2	LN	strong	0.002**)	reject H0	lnorm = 0.1	gamma= 0.1	weibull=0.2	weibull
2	N		0.076*)	reject H0	lnorm = 0.1	gamma= 0.2	weibull=0.7	weibull

Notes: [1]: probability is determined subjectively by visual of the plot

IOP Conf. Series: Earth and Environmental Science **880** (2021) 012028 doi:10.1088/1755-1315/880/1/012028

Numerical simulation of hydraulic performance with free overfall flow

Li Yao[1,2*], Xiao-hui Ye[3], Xiao-chong Huang [4], Ke-hong Zheng[1,2], Arniza Fitri[5] and Fera Lestari[5]

[1]School of Water Conservancy and Ecological Engineering, Nanchang Institute of Technology, Nanchang, 330099, China.
[2]National and Provincial Joint Engineering Laboratory for the Hydraulic Engineering Safety and Efficient Utilization of Water Resources of Poyang Lake Basin, Nanchang, 330099, China
[3]School of Yaohu, Nanchang Institute of Technology, Nanchang, 330099, China
[4]College of Water Conservancy and Hydropower Engineering, Hohai University, Nanjing 210098, China
[5]Department of Civil Engineering, Faculty of Engineering and Computer Science, Universitas Teknokrat Indonesia, Bandar Lampung, Indonesia

*Corresponding E-mail: yaoli0817@nit.edu.cn

Abstract. Recently, free overfall flows have attracted attention of many researchers in the world. It is because the free overfall can be used as a tool in measuring the discharges and flows in any open channel systems or water collection systems for various flow regime at different shape of channels. In this study, the hydraulic performances of free overfall flow were explored using numerical simulation for three shapes of open channels including uniform rectangular, rectangular circumscribed, and rectangular inscribed steps. The hydraulic performances discussed in this study are including the flow regime, the maximum pressure, drop length and the energy dissipation rate. Numerical simulations were performed under different discharges Q (between 0.0005 m^3/s and 0.0015 m^3/s) and excise repair densities (y/H) for 1:2, 1:3 and 1:4. Visual observations identified various flow regimes with largely identical flow patterns in different discharges and configurations. Comparisons of properties indicated that there is a better depressurization and energy dissipation performances at the open channel with the shape of rectangular inscribed steps compare to the uniform rectangular in the small charge. Comparative study also revealed that the large excise repair densities of rectangular circumscribed steps might be optimal in terms of energy dissipation performances.

Keywords: Free overfall, Flow regime, Energy dissipation, Numerical simulation

Track Name: Land, Water, Forests and Food Security

1. Introduction

Water conservancy project is a key imperative project for people's livelihood in the world [1-5], such as in China [6-8]. As an important and basic project for people's well-being, water conservancy engineering is of great significance to improve people's production and living conditions. Especially in China, those countries with abundant water resources are always considered as having a very important strategic position. Notably, in various water conservancy engineering, there is a rather common phenomenon named free overfall flow [9]. With further investigation in the function of this phenomenon like energy dissipation [10-12], flow measurement and slowdown, it is widely applied in stepped spillway dam, agricultural hydraulic engineering, municipal overfall flows, landscape design [13], etc. Take ecological restoration work in rural rivers for example, free overfall flow is utilized to accelerate aeration process in natural water, improving river quality by inhibiting the release of organic substance from the river sediment. To be more specific in the application of stepped spillway dam, like Mencesville Dam [14] in the United States, Dachaoshan Reservoir [15] of Yunnan, and Baise Reservoir [16] of Guangxi, with the aim of mitigating erosion and damage to downstream channels, multi-stage overfall flow is adopted to dissipate energy. As it is stated above, free overfall flow plays a vital role in many aspects.

In recent years, a wide range of experts and scholars in the world have conducted lots of scientific research and practice on free overfall flow, providing rich theoretical knowledge and practical engineering application. In 1995, Rand [17] fitted a formula for calculating drop length by consulting materials and experimental research. Later when critical water depth and drop height was determined, Chanson [18] obtained an empirical formula of water drop by fitting their test data with previous one. In addition, Hubert Chanson [19] studied the effect of step shape and cavity on energy dissipation of free overfall flow. In China, Zhong Tian [20] and Jianing Tian [21] also investigated the hydraulic characteristics of stepped spillway.

In terms of numerical simulation of water drop, Ramamurthy et al. [22] explored the distribution of pressure head and velocity in free overfall flow of three-dimensional open channels by establishing a VOF model, which effectiveness was verified through analyzing experimental data and the numerical results. In China, Qing Wang et al. [23] also carried out three-dimensional simulation for water flow of steep slope curves in Flow-3D. Mingjun Diao et al. [24] studied energy dissipation in two-phase flow (including water and gas) based on model tests and two-dimensional simulation, and numerical results suggest that the simulation is highly effective.

Currently, most of research work have focused on single-width flow, slope and model size of free overfall flow, but less attention is paid on body shape of the overfall flow model [25]. In the past, two-dimensional numerical simulation on overfall flow appears more frequent than the counterparts in practical engineering. Hence this paper proposes two new overfall flow models with rectangular circumscribed and rectangular inscribed steps respectively. Furthermore, by inputting different discharge volume, the influence of step shape on flow regime, maximum pressure and energy dissipation rate [26-27] are investigated in three-dimensional numerical simulation. The contribution of research here is to provide certain insights for reconstruction and expansion of some comparable engineering.

2. Materials and Methods

2.1 Water Drop Model

In this paper, the uniformed rectangular steps model is modified into the one with rectangular circumscribed and inscribed steps respectively. Based on the uniform rectangular steps, the rectangular circumscribed one is formed by filling some parts, and the other one is obtained by cutting some parts. These three models are illustrated as the following (Figure 1 and Figure 2):

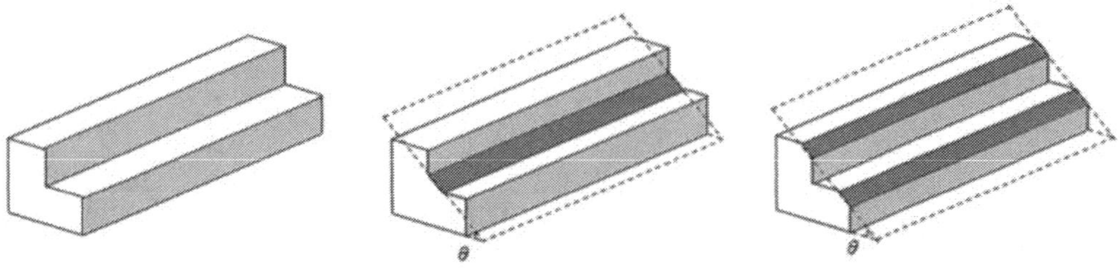

Uniform Rectangular Model (A) Rectangular Inscribed Model (B) Rectangular Circumscribed Model (C)

Figure 1. Schematic diagram of rectangular circumscribed and rectangular inscribed steps models.

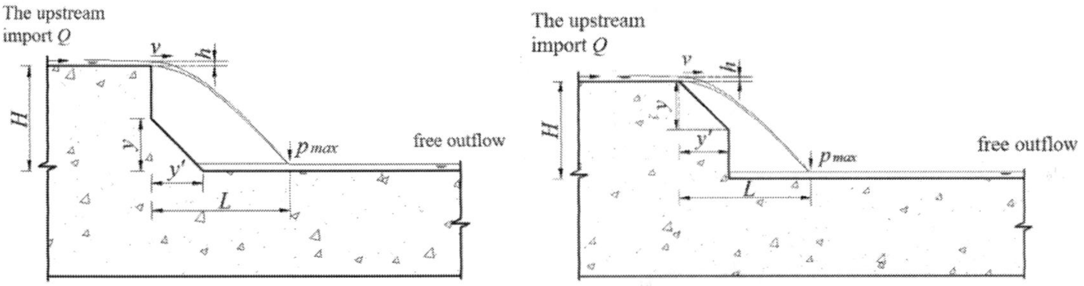

Rectangular Inscribed Model (B) Rectangular Circumscribed Model (C)

Figure 2. Schematic diagram of model parameters of rectangular circumscribed and rectangular inscribed.

In figure 2, H is model height, h is the average depth of the section at the drop point, v is the average velocity of the section at the drop point, y is the vertical height corresponding to the cut or finished part, y 'is the horizontal length corresponding to the cutting or patching part. Since the oblique Angle of the section in this simulation is 45°, corresponding y is equal to y'. P_{max} is the maximum pressure of the bottom surface obtained by numerical simulation. L is the drop length, the horizontal distance between the maximum pressure points and the drop point, y/H is the degree of cutting and patching, L/H is the drop degree, and the drop length is dimensionless. Table 1 presents the values of water discharges for various model in this study.

Table 1. Model parameters.

The model number	H(cm)	y(cm)	y/H	θ(°)	Q(m^3/s)
A	10	/	/	/	0.0005~0.0015
B_1	10	5	1:2	45	0.0005~0.0015
B_2	10	3.33	1:3	45	0.0005~0.0015
B_3	10	2.5	1:4	45	0.0005~0.0015
C_1	10	5	1:2	45	0.0005~0.0015
C_2	10	2.5	1:3	45	0.0005~0.0015
C_3	10	3.33	1:4	45	0.0005~0.0015

2.2 Mesh partitioning and boundary conditions

In order to better simulate the upstream discharge, the size of inlet rectangular block is set as 0.1 m × 0.01 m (0.1 m and 0.01 m refer to the value of initial inflow depth and model width correspondingly). The distance between flow inlet and drop point is predetermined as 0.1m to allow for full development of discharge. For grid setting, the unit size is 0.0025 m × 0.0025 m × 0.0025 m. And given that y/H = 1:2, local refinement is added into the corner part (see those solid lines in figure 3).

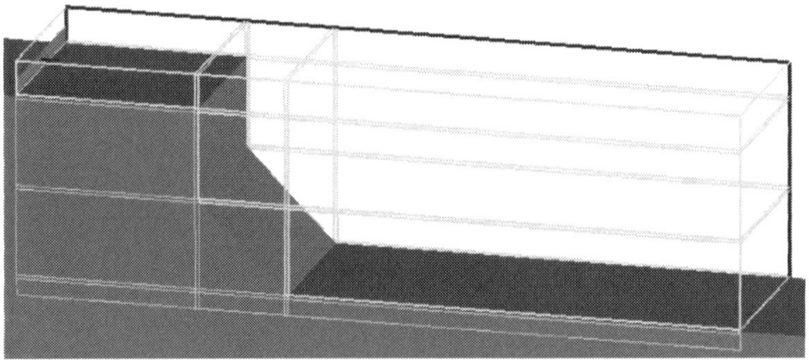

Figure 3. Schematic Diagram of Model Grid Partition.

For ensuring the stability of upstream discharge, boundary type of wall is selected at X_{min}. In case appearance of reflux affects regime of free overfall flow at X_{max}, the boundary type is determined as Outflow, and other boundaries are all set as symmetry. One reason is for the authenticity in numerical simulation, and the other one is to reduce variables regardless of the influence in other directions [22,28].

2.3 Model validation

Following the Ting Chen's research work [29], model verification process in this study is conducted based on two models in heigh of 11cm and 13cm. Under identical condition of upstream discharge, the situation of downstream discharge is monitored and recorded several times. Then data error is compared and error rate is calculated as well. The values of drop length at different flow rates when the heigh H =11 cm and H =13cm are showed in Table 2 and Table 3, respectively. Then, Figure 4 presents the model verification for upstream water depths at different water discharge (Q).

Table 2. Traditional drop length at different flow rates（H = 11cm).

Q（m^3/s）	The upstream water depth（m）	Measured values in literature	Simulated value in this paper	Error（%）
0.0004226	0.0152	0.069	0.067	-2.90%
0.0003568	0.0141	0.063	0.0659	4.60%
0.0002864	0.0130	0.058	0.059	1.72%
0.0002445	0.0123	0.053	0.0549	3.58%
0.0001975	0.0112	0.048	0.047	-2.08%

Table 3. Traditional drop length at different flow rates（H = 13cm).

Q（m^3/s）	The upstream water depth（m）	Measured values in literature	Simulated value in this paper	Error（%）
0.0001721	0.0096	0.045	0.044	-2.22 %
0.0002064	0.0102	0.051	0.0495	-2.94 %
0.0002445	0.0113	0.057	0.0555	-2.63 %
0.0002864	0.0130	0.063	0.0633	0.48 %
0.0003323	0.0135	0.069	0.067	-2.90 %

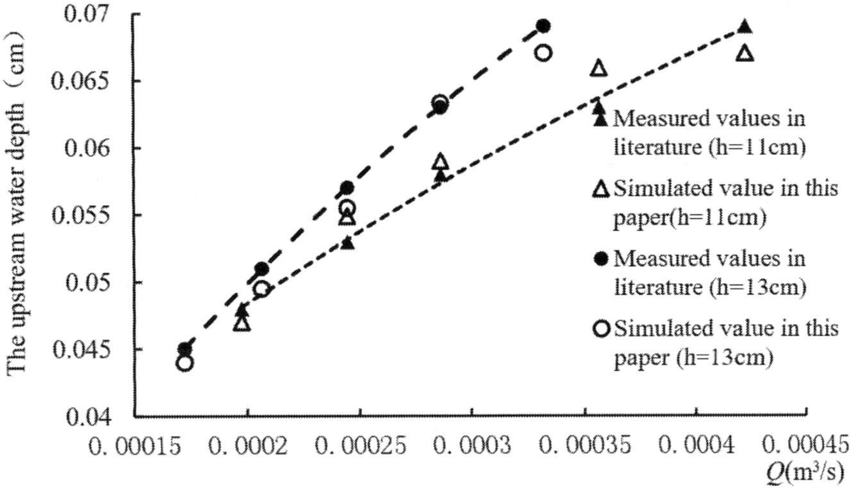

Figure 4. Model verification for upstream water depths at different Q.

From the above results, it can be detected that error between simulation data in this paper and measured one in literature is within 5%, indicating that the simulation process is effective and can accurately reflect the real-life situation of flow.

3. Results and Discussion

3.1 Influence of discharges on the flow regime

In order to study the influence of discharges on the hydraulic performances, different discharges Q (0.0005 m^3/s~0.0015 m^3/s) were set for the models. The following figure (Figure 5) shows the simulation results of the uniform rectangular, rectangular circumscribed and rectangular inscribed steps models.

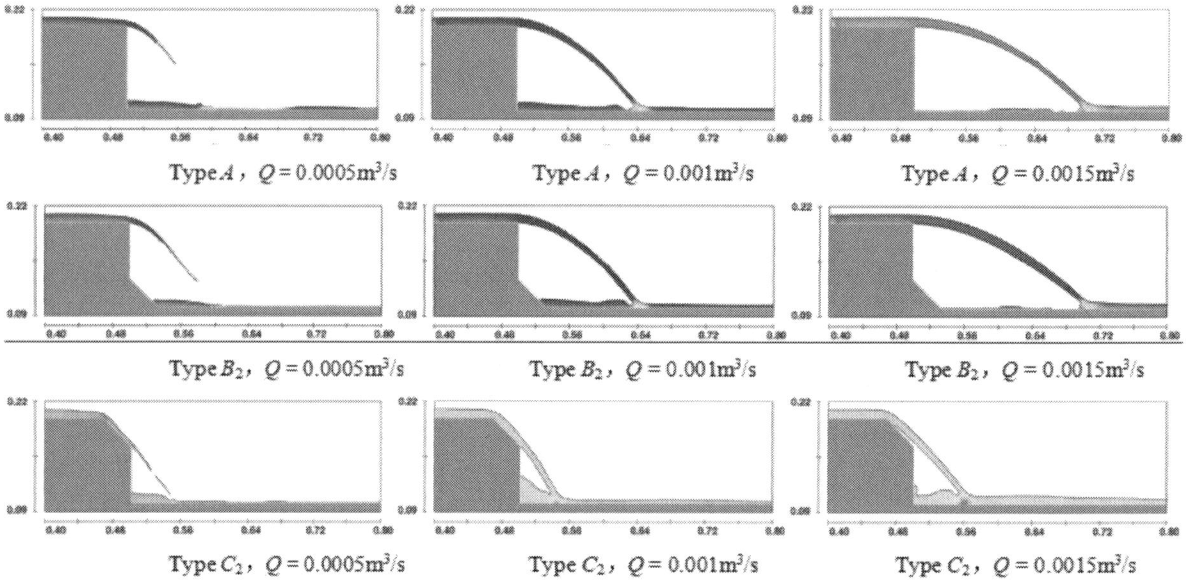

Fig. 5 Schematic diagram of different types of drop models under different discharges.

For the uniform rectangular steps, the surface of the free overfall flows is parabolic. When the water falls to the floor, part of the water flows to the left to form a back flow area, and the rest of the water continues to flow to the right due to inertia. With the increase of discharges, the water in the back flow area decreases gradually, and the falling water flows more towards the development area.

The flow regime of the rectangular inscribed steps is similar to the uniform rectangular steps. When the discharge is small, the water depth in the back flow area is slightly higher than that of the uniform rectangular steps, and the drop length is slightly smaller than that of the uniform rectangular steps. When the discharge is large, there is little difference from the uniform rectangular steps. Because the shape is rectangular inscribed, there is little connection between the flow and the cut surface in the process of falling, so the effect of the free overfall flow regime is not obvious.

There is a great difference between the flow regime of rectangular circumscribed and the uniform rectangular steps. When the discharge is small, the water drops along the section, and there is a large amount of water in the back flow area, and the vortex is formed. When the discharge increases gradually, the flow tends to break away from the section. Because the shape is rectangular circumscribed, the water flow mainly falls along the cut surface in the process of falling. When increasing the falling speed, the drop length is obviously smaller than that of the uniform rectangular steps.

3.2 Maximum Pressure characteristic results and analysis

The characteristics of maximum pressure in the rectangular circumscribed (C_2) model is shown in Figure 6 below.

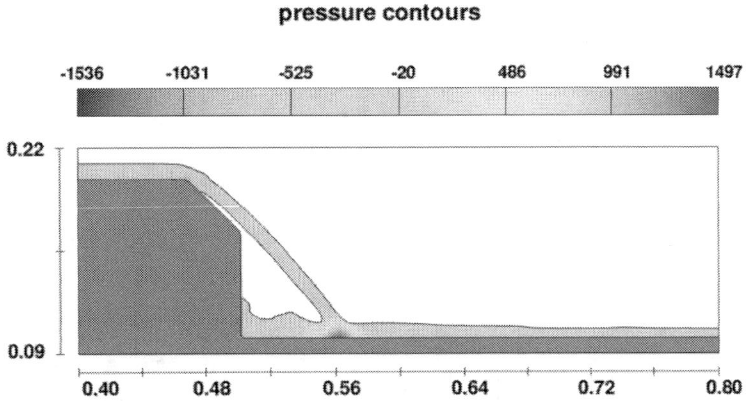

Figure 6. Pressure diagram of Q = 0.0015 m^3/s in the rectangular circumscribed (C_2) model.

Based on Figure 6, it demonstrates that for rectangular circumscribed step, it mainly has a great influence on the flow regime due to the existence of cut surface, and the overall pressure reduction effect of the rectangular circumscribed steps is more obvious than the rectangular inscribed steps. The higher of the cut and complement degree, the depressurization is more obvious in rectangular inscribed steps. In addition, at the drop point, local negative pressure may produce cavitation erosion phenomenon, so it also need to pay special attention.

3.3 Drop length results and analysis

In order to facilitate the comparison and analyses between different models, the discharge and drop length are setup to be dimensionless. Drop degree L/H is calculated to analyze the changes between the maximum pressure and the drop length degree under different Froude number $Fr = \frac{v}{\sqrt{gh}}$, where v is the average velocity of the section at the drop point, g = 9.81 m^2/s, is the gravity acceleration, and h is the average depth of the section at the drop point. Figure 7 and Figure 8 present the changes of drop length degree and Fr in uniform rectangular and rectangular inscribed steps.

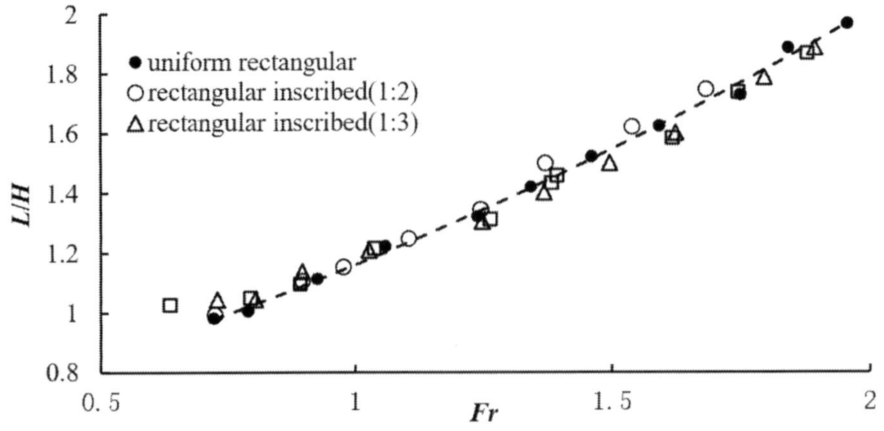

Figure 7. Changes of drop length degree and *Fr* in uniform rectangular and rectangular inscribed steps.

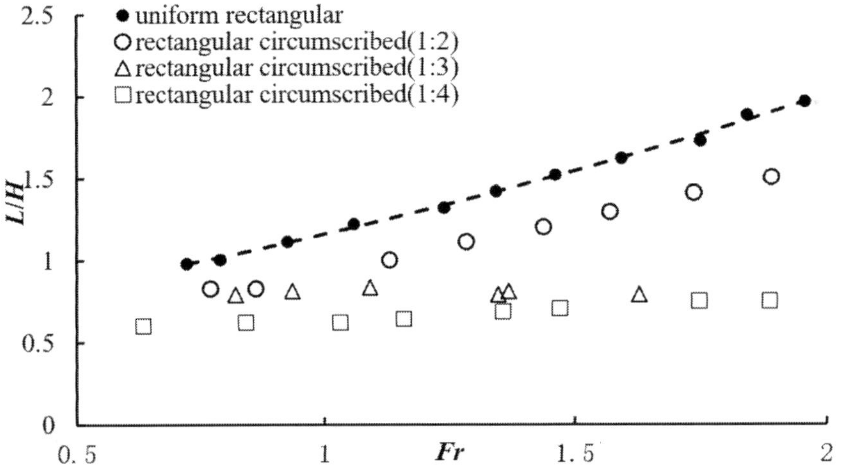

Figure 8. Changes of drop length degree and *Fr* in uniform rectangular and rectangular circumscribed steps.

The change of the rectangular circumscribed steps is more obvious than the rectangular inscribed steps, and the drop length degree is also significantly affected by the different degree of excise repair densities variable (y/H). For type C_1, in the process of the discharges gradually increasing, the water gradually breaks away from the cut surface. Under the condition of high *Fr*, the drop length degree is obviously greater than type C_2 and type C_3.

4. Conclusion

In this paper, the comparison among the uniform rectangular steps, the rectangular circumscribed and rectangular inscribed steps are analyzed under different flow conditions which are corresponding to the change of three aspects: the flow regime, the maximum pressure, and the drop length. The main conclusions are as follows:

- The rectangular inscribed steps have little effect on the flow regime. In the case of low Froude number, the pressure reduction effect on the floor is obvious, but in the case of large discharges, the maximum pressure is similar to the uniform rectangular steps. Because the existence of the cur surface has no obvious effect on the flow regime, the drop length is similar to the uniform rectangular steps.
- The rectangular circumscribed steps have obvious influence on the flow regime. The type C_1 steps do not flow along the cut surface in the fall process, and the overall drop length is between the uniform rectangular steps, type C_2 and type C_3. In the fall process of type C_2 and type C_3, the water drops along the surface, and there is a large amount of water in the back flow area, and there is a large amount of turbulent water. Therefore, the overall depressurization effect is obvious and the drop length is significantly reduced. The greater of the cut and complement degree, the greater the energy dissipation rate. However, in the case of large discharges, it should be noted that there is a local negative pressure area at the drop place, which may cause cavitation erosion.
- In general, both the rectangular circumscribed and rectangular inscribed steps have effects on free overfall flows, but the rectangular circumscribed steps have a greater effect, and have obvious effects on pressure reduction and the energy dissipation rate. However, in the actual construction, attention should be paid to the protection of the vacuum area in the drop place of the rectangular circumscribed steps under the condition of high discharges.

Reference

[1] Fitri A, Hashim R, Abolfathi S, and Abdul Maulud K N 2019 Dynamics of sediment transport and erosion-deposition patterns in the locality of a detached low-crested breakwater on a cohesive coast *Water* **11**(8) pp 1721

[2] Fitri A, Hashim R, Song K I and Motamedi S 2015 Evaluation of morphodynamic changes in the vicinity of low-crested breakwater on cohesive shore of Carey Island, Malaysia *Coastal Engineering Journal* **57**(04) pp 1550023

[3] Maulud K N A, Fitri A, Mohtar W H M W, Jaafar W S W M, Zuhairi N Z and Kamarudin M K A 2021 A study of spatial and water quality index during dry and rainy seasons at Kelantan River Basin, Peninsular Malaysia *Arabian Journal of Geosciences* **14**(2) pp 1-19

[4] Fitri A, Hashim R and Motamedi S 2017 Estimation and Validation of Nearshore Current at the Coast of Carey Island, Malaysia *Pertanika Journal of Science & Technology* **25**(3)

[5] Fitri A, Yao L, and Sofawi B 2019 Evaluation of Mangrove Rehabilitation Project at Carey Island Coast, Peninsular Malaysia based on Long Term Geochemical Changes. In *IOP Conference Series: Earth and Environmental Science* IOP Publishing Vol. **365** No. 1 p 012055

[6] Chen H, Yao L and Fitri A 2019 The Influence Mechanism Research of Inflow Temperature in Different Time Scale on the Water Temperature Structure in *IOP Conference Series: Earth and Environmental Science* IOP Publishing Vol. **365** No. 1 p 012058)

[7] Yao L, Huang X and Fitri A 2019 Influence scope of local loss for pipe flow in plane sudden expansions In *IOP Conference Series: Earth and Environmental Science* IOP Publishing Vol. **365** No. 1 p 012056

[8] Yao L, Li J, Shi S, Fitri A 2019 Simulation Take-off Angle of a Ski Jump Energy Dissipater In *IOP Conference Series: Earth and Environmental Science* IOP Proceeding Vol. **365** No. 1 p 012057

[9] Baolin J and Shirong W 2010 Review on the development of water drop projects in China Journal of Engineering *Heilongjiang University* pp 78-83 (In Chinese)

[10] Jianing T, Yoichi Y, Jianzhong L, Iwo O and Masayuki T 2004 Energy dissipation in nappe flow on stepped spillways *Chinese Journal of Hydrodynamics pp* 114-119 （In Chinese)

[11] Mingyi W 2016 Study on specific energy characteristic of stepped spillway *Northwest A&F University* (In Chinese)

[12] Nie J, Zhang J and Li S 2013 Overview of hydraulic characteristics down stepped spillways *Proceedings of the twenty-five th National Symposium on hydrodynamics and the twelve th national symposium on hydrodynamics* （In Chinese)

[13] Xinming C 2006 Numerical simulation of turbulent flow of free overfall *Zhejiang University of Technology* （In Chinese)

[14] Hansen K D 1986 Roller Compacted Concrete Developments in the USA W*ater Power and Dam Constroction* pp 9-12

[15] Jinjin Y, Hansheng L and Zongxiang L 2005 Study on the flow aeration to prevent caviation of stepped spillway for Dachaoshan Hydroelectric Station *Journal of Northwest A & F University (Natural Science Edition*) (02) pp 137-142 (In Chinese)

[16] Minan L 2004 Surface spillway tail-flaring pier joint energy dissipator of Baise RCC main dam *Guangxi Water Resources & Hydropower Engineering* pp 53-56 (In Chinese)

[17] White M P 1943 Discussion of Moore transactions *ASCE* 108:1361-1364.

[18] Chanson H 1995 Hydraulic Design of Stepped Cascades, Channels, Weirs and Spillways *Stepped Spillways* pp 1-292.

[19] Guangfu Z and Hubert C Effects of Step and Cavity Shapes on Aeration and Energy Dissipation Performances of Stepped Chutes *Journal of Hydraulic Engineering* **144** (9) 04018060

[20] Zhong T, Weilin X and Ting Y 2010 Characteristics of Energy Dissipation for "v"-type Stepped Spillway *Advanced Engineering Sciences* **42** (2) pp 21-25 (In Chinese).

[21] Jianing T, Li J and Iwo O 2004 The dissipation characteristics on stepped spillways of different slopes *Journal of Northwest A & F University (Natural Science Edition)* **32** (10) pp 119-122 (In Chinese)

[22] Ramamurthy A S, Qu J and Vo D 2014 VOF Model for Simulation of a Free Overfall in Trapezoidal Channels *Journal of Irrigation & Drainage Engineering* **132** (4) pp 425-428.

[23] Qing W and Xibin H 2016 Three-dimensional numerical simulation of flow in bend of steep slope based on Flow-3D *Journal of Water Resources and Water Engineering* **27** (05) pp 145-149 (In Chinese)

[24] Mingjun D, Yongquan Y and Yurong W 2003 Numerical simulation of water-air two-phase jet flow from flip bucket to plunge pool *Journal of Hydraulic Engineering* (09) pp 77-82 (In Chinese)

[25] Xiaoxia H, Jurui Y and Jianshu Z 2014 Study of impacts on aerated cavity length and negative pressure by Y-shape flaring gate pier combined with stepped spillway in energy dissipator *Journal of Hydroelectric Engineering* **33** (03) pp 203-209 (In Chinese)

[26] Shengcai T, Feng J, Jiahao S 2008 Experimental research on step spillway and energy dissipation ratio computation *Yangtze River* pp 43-45 (In Chinese)

[27] Jijian Y, Hansheng L and Shubing D 2015 Relation Between Relative Energy Dissipation Ratio and Relative Froude Number of Stepped Spillway *Journal of Yangtze River Scientific Research Institute* pp 86-89 (In Chinese)

[28] Lu Zou 2016 Numerical simulation of aerated flow in stepped spillway *Xi 'an University of Technology* (In Chinese)

[29] Ting Cheng 2016 Research on The Length of a Small Flow Free Overfall and Energy Dissipation *Xihua University* (In Chinese)

Acknowledgments:

This research is supported by the National Natural Science Foundation of China (Grant Nos. 51879129 and 51509128) and the Natural Science Foundation of Jiangxi Province of China (20192ACBL20041)

Evaluation of the Groundsill's stability at downstream of "Citorek" Bridge in Cimadur River, Banten Province

Arniza Fitri[1*], Hao Chen[2], Li Yao[2,3], Ke-hong Zheng[2,3], Susarman[4], Farli Rossi[4] and Yuxue Yin[1]

[1]School of Civil and Architectural Engineering, Nanchang Institute of Technology, Nanchang 330099, China
[2]School of Water Conservancy and Ecological Engineering, Nanchang Institute of Technology, Nanchang, 330099, China.
[3]National and Provincial Joint Engineering Laboratory for the Hydraulic Engineering Safety and Efficient Utilization of Water Resources of Poyang Lake Basin, Nanchang, 330099, China
[4]Faculty of Engineering and Computer Science, Universitas Teknokrat Indonesia, Bandar Lampung, Indonesia

*Corresponding E-mail: arnizafitri@gmail.com

Abstract. Scouring problems faced in the Cimadur River especially near to the Citorek bridge abutments have become the major discussion by the local researchers and the local water resources manager in Banten Province. As an effort in reducing the scouring problem around the abutment of the Citorek bridge, a groundsill structure with specific design is going to be installed in Cimadur River at downstream of Citorek bridge. To make sure the optimal function of the structure, the stability of the groundsill structure in Cimadur River need to be evaluated. This study attempt to evaluate the stability of a groundsill structure from the occurrences of rolling and sliding at both normal and flood conditions. The eccentricity of the groundsill structure is also checked during normal and flood conditions to make sure the stability of the structure. The required data (which are consisting of detail description of the groundsill structure, cross section of the river, rainfall data, topography data and sediment/soil data) are observed in the field and obtained from P.T. Saeba Konsulindo. The data are further analysed to determine: 1) design water discharges for several return period, 2) forces acting to the groundsill structure and 3) stability of the groundsill structure in the river. The results showed that the groundsill structure are stable and safety again rolling and sliding occurrences where the safety factor (SF) for rolling and sliding are higher than critical coefficients of rolling and sliding (1.5). At normal water level, safety factors (SF) for rolling and sliding are 8.07>1.5 and 2.7>1.5, respectively, while at flood water level, SF for rolling and sliding are 5.61>1.5 and 1.88>1.5, respectively. Besides, the results also found that the groundsill is safety from eccentricity at both normal and flood conditions where the calculated coefficients of eccentricities are lower than critical coefficient of eccentricity which could cause rolling and sliding.

Keywords: Citorek Bridge, Cimadur River, Groundsill Stability

Track Name: Land, Water, Forests and Food Security

IOP Conf. Series: Earth and Environmental Science **880** (2021) 012029 doi:10.1088/1755-1315/880/1/012029

1. Introduction

River is a natural water body that flows the water from downhill to downstream area due to the force of gravity [1-5]. For the transportation purposes by the local community, a bridge will be constructed to connect the areas separating by the river [6]. In water resources management, scouring problem is the most serious problem in water body which require to pay more attentions [7-13]. Scouring is an enlargement of a flow accompanied by transfer of material through the action of fluid motion. Scouring problem faced in the river can damage river structures and cause bad impact to the river environments. Since the past decades, the scouring problems in the river especially near to the bridge abutments have become the major discussion by the local researchers and the local water resources manager [14-17], while applying the groundsill structure at the downstream of the scouring area in the river is becoming one of promising alternatives [18-20].

There are scouring problems faced in the Cimadur River, Lebak distict, Banten Province since the past few years. The scouring problem in Cimadur River especially near to abutment of the bridge take more attention by the local researchers and the local water resources manager in Banten Province due to the safety consideration for the surrounding community. In 2019, a groundsill structure has been constructed at downstream of Bantar Karang bridge. The presence of the groundsill structure in downstream of Bantar Karang bridge was found to give positive impact in slowly reducing the scouring problem around the abutment of the Bantar Karang bridge. The successful of implementing a groundsill structure in reducing the scouring problem around the abutment of Bantar Karang bridge encourage to design one more groundsill structure in Cimadur River, but now, it is going to be installed near to Citorek bridge located at upstream area of Cimadur River, Lebak district, Banten Province. Before the groundsill structure is constructed at downstream of Citorek bridge, the stability of the groundsill structure in Cimadure River need to be evaluated. So, this study attempt to evaluate the stability of the groundsill structure at downstream of Citorek bridge in Cimadur River from the rolling and sliding occurrences. The evaluation is carried out at both normal water level (2 years return period of design flood discharge) and extensive flood water level (50 years return period of design flood discharge). Besides, the eccentricity of the groundsill structure is also checked during normal and flood conditions to make sure the stability of the structure. The evaluation is important to optimize the function of the groundsill structure in maintaining the condition of local sedimentation in Cimadur River.

2. Materials and Methods

2.1 Study Area

This study is carried out at the upstream area of Cimadur River. Cimadur River is located in Lebak district, Banten Province which is geographically situated at latitude of 6° S to 7° S and longitude of 105° 25" E to 106° 30" E as shown in the Figure 1 below. There are scouring problems faced around the abutments of the bridges structure in the Cimadur River since the past few years. Besides of Bantar Karang bridge, the scouring problem is also faced around the abutment of Citorek bridge.

Topographic condition in the upstream part of Cimadur River is characterized with moderately steep slope where the riverbed elevations are approximately from 400 m above MSL to 500 m above MSL. Moderately steep slope of the riverbed in this area also give potential for the local scouring problem in the river.

The climate in the Banten Province is primary subjected to two seasons during the years that are dry season (April to October) and rainy season (November to March). Average amount of yearly rainfall is about 2000 mm to 4000 mm with the temperature of between 24.5 °C and 29.9 °C.

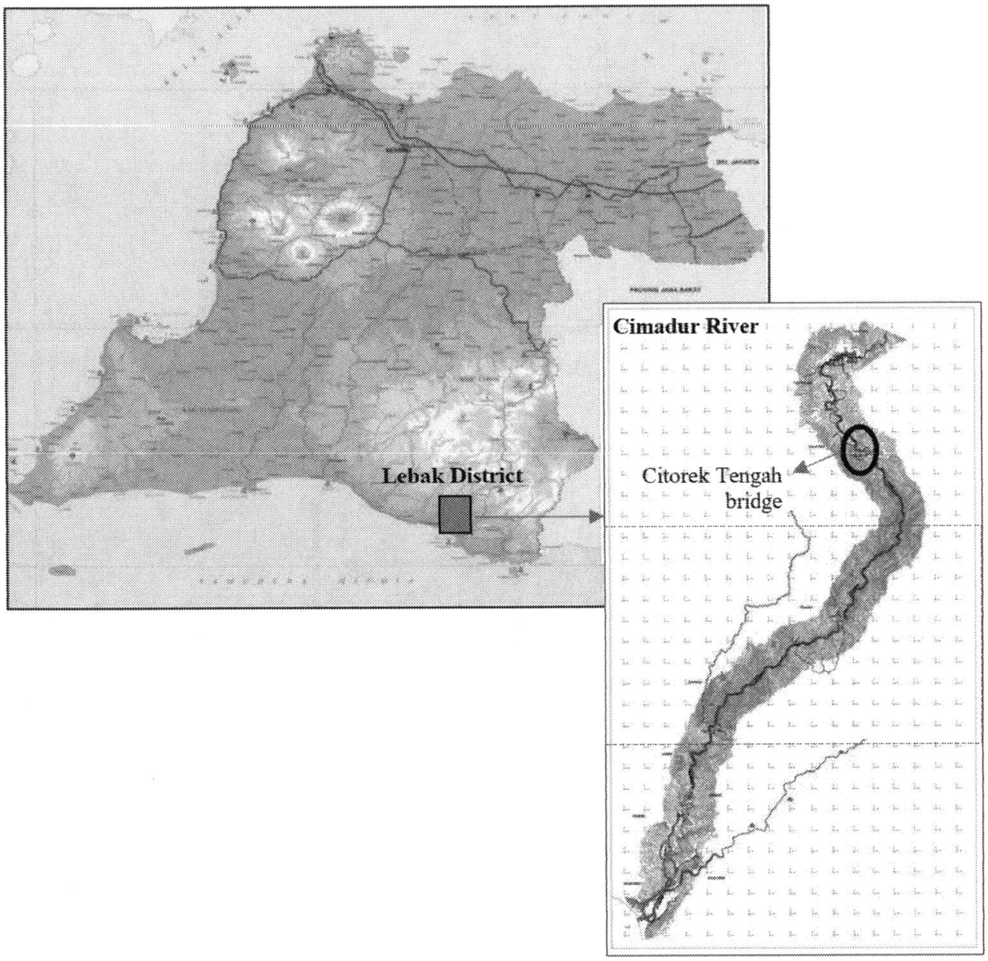

Figure 1. Location of the study area at upstream of Cimadur River, Lebak district, Banten Province.

2.2 Data Collections

In this study, the required data are consisting of detail description of the groundsill structure, rainfall data, cross section of Cimadur River, topographic data and soil/sediment data. A groundsill structure has been designed to be installed at downstream of Citorek bridge in Cimadur River. The detail description of the groundsill structure is described in Figure 2 below. The groundsill structure is made by concrete with the specific gravity (γ) is about 24 kN/m^3.

Rainfall data used in this study are obtained from P.T. Saeba Konsulindo. There are two rainfall stations in the watershed of Cimadur River that are Cibeber rainfall station and Bayah rainfall station, while there is another one rainfall station outsite of Cimadur watershed. The rainfall data for 10 years period (2011 to 2020) obtained from both rainfall stations are used in the hydrological analyses to calculate the design water discharge for several return period.

The cross sections of Cimadur River are observed in the field by P.T. Saeba Konsulindo, while the topographic data in the study area is obtained from USGS website. Further, the sediment/soil data which is consisting of depth of sediment/soil, sediment specific gravity and type of soil/sediment are obtained from P.T. Saeba Konsulindo.

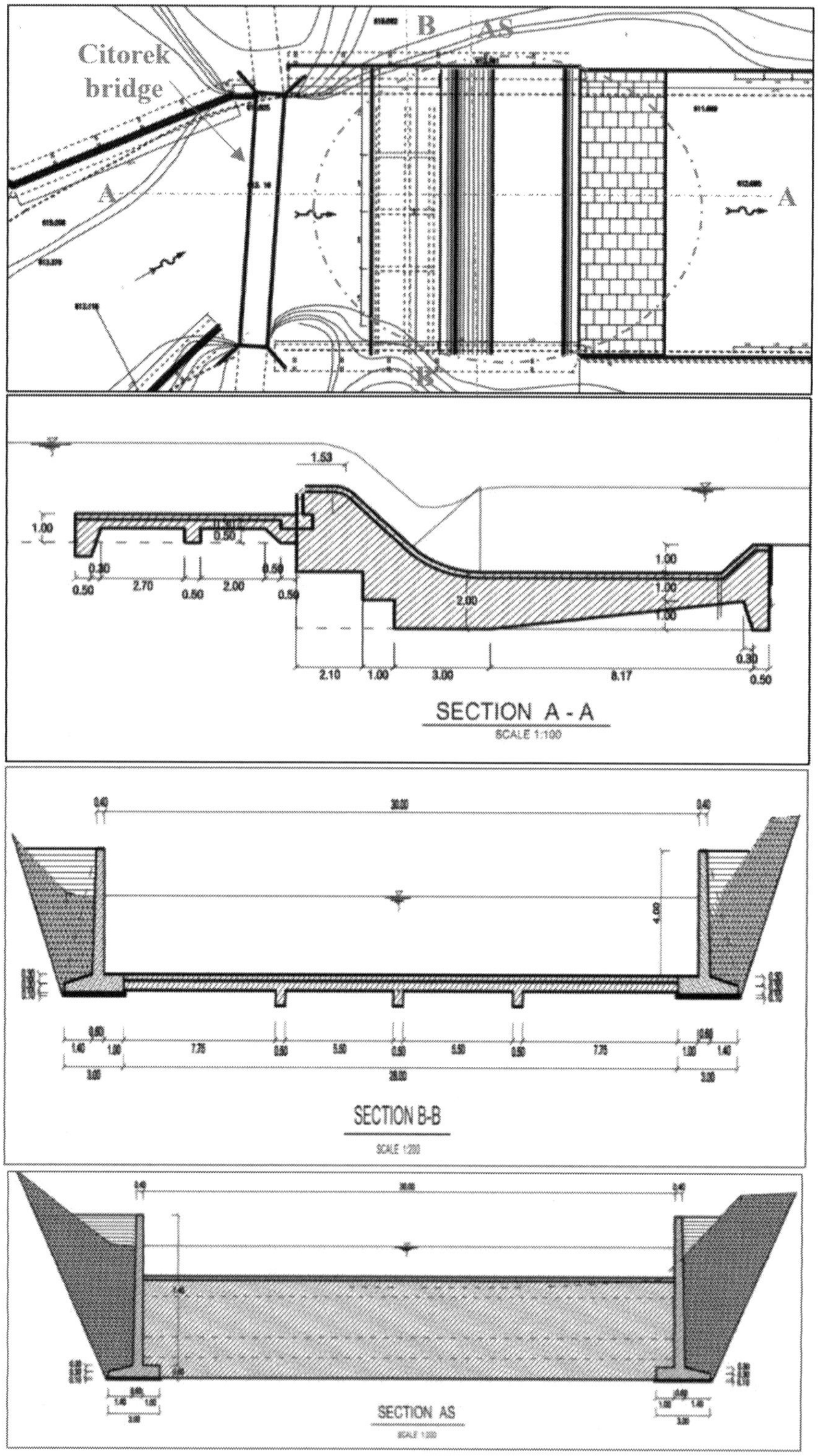

Figure 2. Detail description of the groundsill structure.

IOP Conf. Series: Earth and Environmental Science **880** (2021) 012029 doi:10.1088/1755-1315/880/1/012029

2.3 Data Analyses

In this study, available data is further analysed to determine: 1) design flood discharge for 2, 5, 10, 25, and 50 years return period; 2) forces acting on the groundsill structure and 3) the stability of the groundsill structure. Design flood discharge for 2 years return period is used to represent the condition during normal water level, while design flood for 50 years return period is used to represent the condition during extreme flood condition. HECRAS programme is applied to simulate the water level for the discharges of 2 years and 50 years return periods. The detail descriptions for the analyses are as following:

2.3.1 Analysis of Rainfall, design flood discharge and Water Level

Hydrological analyses are carried out based on maximum daily rainfall data for each year from each rainfall station. The average of maximum daily rainfall for each year is further analysed by adopting theissen method [21 - 22], while the calculation is presented in Table 1.

Table 1. Analyses Results of Maximum daily Rainfall per year.

No	Year	Cibeber Station		Bayah Station		Maximum Rainfall (mm)
		Rainfall (mm) (P1)	Catchment Area ($A1=159.6\ km^2$) (2) = (P1) * A1	Rainfall (mm) (P2)	Catchment Area ($A2=50.4\ km^2$) (4) = (P2) * A2	P = (P1*A1+P2*A2) / (A1+A2) (5) = (2+4)/(A1+A2)
1	2011	75	11970	137	6904.8	89.88
2	2012	91	14523.6	128	6451.2	99.88
3	2013	112	17875.2	185	9324	129.52
4	2014	97	15481.2	121	6098.4	102.76
5	2015	71	11331.6	101	5090.4	78.2
6	2016	93	14842.8	115	5796	98.28
7	2017	71	11331.6	120	6048	82.76
8	2018	128	20428.8	131	6602.4	128.72
9	2019	201	32079.6	112	5644.8	179.64
10	2020	184	29366.4	136	6854.4	172.48

Design rainfall is computed by adopting log pearson type III method [21-22]. Based on calculation, the values of the design rainfall for 5, 10, 25 and 50 years return periods are 128.92 mm, 153.42 mm, 191.19 mm and 224.7 mm, respectively.

Further, design flood discharges are analysed based on results of design rainfall. In this study, design flood discharge for 2 years return period is used to represent the normal condition in the field, while design flood discharge for 50 years return period is used to represent the extreme flood condition in the field. Design flood discharges are computed by adopting Nakayasu Synthetic Hydrograph Unit (SHU) method [21-22] as following:

$$Q_{max} = C * A * R_o * (1/ (3.6\ (0.3T_p+T_{0.3}))$$

Where: T_p is peak time (hour)

C is flow coefficient

Ro is single rainfall (mm)
A is the catchment area (km^2)
$T^{0.3}$ is the time of slowdown the discharge (until 30%)

Based on the calculation, the maximum discharge for 2, 5 and 10 years return periods are 398 m^3/s, 470.71 m^3/s and 560.16 m^3/s, respectively, while the maximum discharge for 25 and 50 years return periods are 698 m^3/s and 820.4 m^3/s (Figure 3).

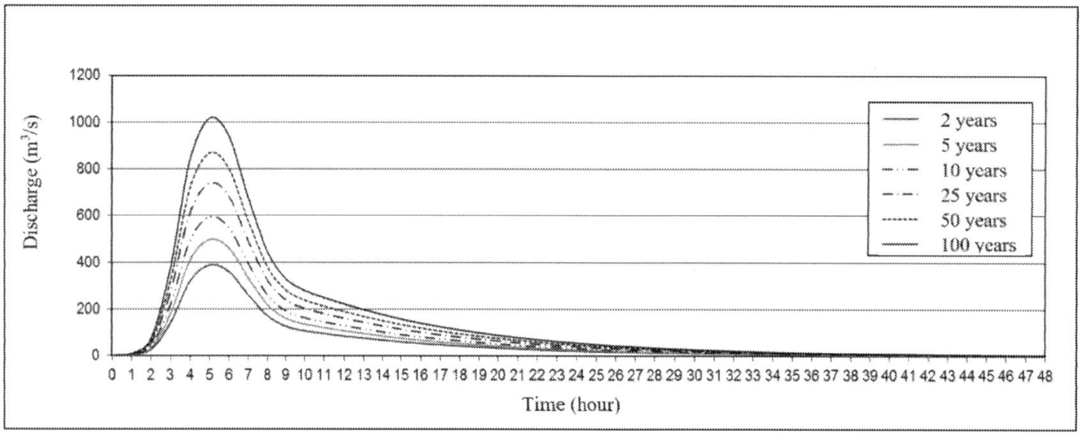

Figure 3. Design flood discharges for return periods of 2 to 50 years.

For the calculation of the forces acting on the groundsill structure, water levels during normal and flood conditions are required. In this study, water levels are simulated using HECRAS programme by inputting the cross section of the river; topographic data in the study area; 2 years return period of discharge (represent the normal condition); 50 years return period of discharge (represent the flood condition) and detail description of the groundsill structure.

2.3.2 Analyses of the Forces Acting on the Groundsill Structure under normal and flood conditions
There are several forces that can affect the stability of the groundsill structure in the field which are including the weight of the groundsill itself (dead load), hydrostatics pressure, sediment pressure, uplift pressure and soil pressure. In this study, the forces acting on groundsill structure are analyzed under two conditions (normal and flood conditions). The calculations of those forces are referred to the equations wrote in Hardiyatmo [23], as following:

- Hydrostatic pressure:

  $Pu = ((1/2 * h_u^2) - (1/2*h_1^2)) * \gamma w$

  $Pi = ((1/2 * h_i^2) - (1/2*h_2^2)) * \gamma w$

  Where:
  Pu is hydrostatics pressure at upstream (kN)
  Pi is hydrostatics pressure at downstream (kN)
  γw is specific gravity of water (kN/m^3)
  h_u is water depth at upstream (m)
  h_i is water depth at downstream (m)
  h_1 is the water level at upstream of "mercu" groundsill
  h_2 is the water level at downstream of "mercu" groundsill
- Sediment pressure:

$Ph = (1/2) * \gamma s * h^2 (1-\sin\theta) / (1+ \sin\theta)$

Where:
Ph is forces located at 2/3 depth from the top of the sediment level
θ is the sliding angle (°)
γs is the specific gravity of sediment
h is the depth of the sediment

- Uplift pressure:

$Ux = Hx - Lx/L * \Delta H$

Ux is water pressure at point x (T/m^2)
Hx is the height between point x and the level of water surface (m)
Lx is the water seepage length (m) at point x
L is total of water seepage length (m)
ΔH is the high energy difference

- Soil pressure:

$Kp = \tan 2\ (45° + \theta/2)$, where θ is the sliding angle (°)

Table 2 describes the calculation results for forces acting on the groundsill structure during normal condition and flood condition.

Table 2. Forces acting on groundsill structure.

No	Type of Forces	Normal Condition				Flood Condition			
		Force (Ton)		Moment (Ton.m)		Force (Ton)		Moment (Ton.m)	
		Vertical (R_V)	Horizontal (R_H)	Vertical (M_V)	Horizontal (M_H)	Vertical (R_V)	Horizontal (R_H)	Vertical (M_V)	Horizontal (M_H)
1	Weight (Dead load)	-74.4	11.2	-291.5	21.2	-74.4	11.2	-291.5	21.2
2	Hydrostatics Pressure		0.284		1.2		0.284		1.2
3	Sediment Pressure		0.5		2.2	8.0	1.5	26.917	6.7
4	Uplift Pressure	13.8		40.7		21.5		65.1	0
5	Soil Pressure		4.9		6.5		4.9		6.5
	Total	**-60.6**	**16.8**	**-250.8**	**31.1**	**-44.8**	**17.8**	**-199.5**	**35.6**

2.3.3 Analyses of Groundsill Stability under normal and flood conditions
Groundsill stability is analysed during two conditions that are under normal water level (2 years return period of discharge) and extreme flood water level (50 years return period of discharge). Here, we are checking the stability of the groundsill structure from the occurrence of rolling and sliding based on

Suyitno [24] and Sutopo [18]. Besides, we also check the safety and stability of the groundsill structure from the eccentricity based on Suyono [25], Sutopo [18] and Sularno [26].

In this study, we calculate the parameter of safety factor (SF) from rolling and sliding and compare with the critical coefficient of rolling and sliding (which is 1.5). The safety and stable groundsill structure must be refer to the equations below:

SF = $\frac{\Sigma M_V}{\Sigma M_H}$ (must) > 1.5, for rolling

SF = $\frac{\Sigma R_V}{\Sigma R_H}$ (must) > 1.5, for sliding

e = $(\frac{B}{2} - \sum M_V \;-\; \sum M_H)$ (must) ≤ B/6, for eccentricity

The groundsill structure is categorized as safe and stable again rolling and sliding if the safety factor (SF) for rolling and sliding are higher than value of critical coefficient for rolling and sliding (1.5). Besides, the groundsill structure also can be caterorizes as safety if the calculated eccentricities (for both normal and flood conditions) are smaller than the length of the groundsill structure divided by 6 (B/6).

3. Results and Discussion

Table 3 presents the calculation results of stability analyses for groundsill structure under occurrences of rolling, sliding and eccentricity.

Table 3. Results analyses of Groundsill's Stability from rolling, sliding and eccentricity.

No	Stability	Normal Condition		Flood Condition	
		Parameter	Checking	Parameter	Checking
1	Rolling	$\sum M_V = 250.8\ T.m$ $\sum M_H = 31.1\ T.m$ SF = $\frac{\Sigma M_V}{\Sigma M_H}$	SF $_{critical}$ = 1.5 SF $_{calculated}$=8.07 The structure is stable (safe) from rolling	$\sum M_V = 199.5\ T.m$ $\sum M_H = 35.6\ T.m$ SF = $\frac{\Sigma M_V}{\Sigma M_H}$	Sf $_{critical}$ = 1.5 Sf $_{calculated}$=5.61 The structure is stable (safe) from rolling
2	Sliding	$\sum R_V = 60.6\ T$ $\sum R_H = 16.8\ T$ SF = fx * $\frac{\Sigma R_V}{\Sigma R_H}$	SF $_{critical}$ = 1.5 SF $_{calculated}$=2.70 The structure is stable (safe) from sliding	$\sum R_V = 44.8\ T$ $\sum R_H = 17.8\ T$ SF = fx * $\frac{\Sigma R_V}{\Sigma R_H}$	Sf $_{critical}$ = 1.5 Sf $_{calculated}$=1.88 The structure is stable (safe) from sliding
3	Eccentricity	$\sum M_V = 250.8\ T.m$ $\sum M_H = 31.1\ T.m$ $\sum R_V = 60.6\ T$ B = 5.8 m	e $_{critical}$ = 0.97 e$_{calculated}$=0.725 The structure is stable	$\sum M_V = 199.5\ T.m$ $\sum M_H = 35.6\ T.m$ $\sum R_V = 44.8\ T$ B = 5.8 m	e $_{critical}$ = 0.97 e$_{calculated}$=0.755 The structure is stable

$e = (\frac{B}{2} - \sum M_V - \sum M_H) \leq$ B/6	$e = (\frac{B}{2} - \sum M_V - \sum M_H) \leq$ B/6
$0.725 \leq 0.97$	$0.755 \leq 0.97$

Based on Table 3, the safety factor (SF) for rolling which are about 8.07 during normal condition and 5.61 during flood condition are higher than the critical coefficient of rolling which is 1.5. It demonstrates that the groundsill structure at downstream of Citorek bridge is safe and stable from rolling under both normal and extreme flood water level.

The result analyses also show that the groundsill structure is stable and safe from occurrence of sliding if it is installed at downstream of Citorek bridge, Cimadur River. It is because the safety factor (SF) for sliding which are about 2.7 during normal water level and 1.88 during extreme flood water level are higher that critical coefficient of sliding (which is 1.5). Apart of that, the groundsill structure is also categorised as safe from eccentricity because coefficient eccentricities (e) which are 0.725 under normal condition and 0.755 under flood condition are found to be lower than one-sixth length of the groundsill structure divided by 6. It also demonstrates that the groundsill structure is stable from eccentricity under both normal and flood water level.

4. Conclusion

In optimizing the function of the groundsill structure at downstream of Citorek bridge in Cimadur River, the stability of the groundsill structure need to be checked before its installation in the site. Based on calculation, we can conclude that the groundsill structure is stable and safe from rolling occurrences under both condition that are during normal water level and extreme flood water level. The groundsill structure is also found to be stable and secure against sliding occurrences under normal and flood conditions if the structure is installed downstream of Citorek bridge, Cimadure River. Besides, the groundsill structure can be categorized as secure from eccentricity under both normal and flood conditions where the coefficient eccentricity values (e) is less than B/6.

References

[1] Fitri A, Maulud K N A, Pratiwi D, Phelia A, Rossi F and Zuhairi N Z 2020 Trend of Water Quality Status in Kelantan River Downstream, Peninsular Malaysia *Jurnal Rekayasa Sipil (JRS-Unand)* **16** (3) pp 178-184

[2] Fitri A, Maulud K N A, Rossi F, Dewantoro F, Harsanto P and Zuhairi N Z 2021 Spatial and Temporal Distribution of Dissolved Oxygen and Suspended Sediment in Kelantan River Basin *In 4th International Conference on Sustainable Innovation 2020–Technology, Engineering and Agriculture (ICoSITEA 2020):* Atlantis Press pp 51-54

[3] Lai S H and Amiza F 2011 Application of SWAT Hydrological Model to Upper Bernam River Basin (UBRB), Malaysia *IUP Journal of Environmental Sciences* **5**(2)

[4] Maulud K N A, Fitri A, Mohtar W H M W, Jaafar W S W M, Zuhairi N Z and Kamarudin M K A 2021 A study of spatial and water quality index during dry and rainy seasons at Kelantan River Basin, Peninsular Malaysia *Arabian Journal of Geosciences* **14**(2) pp 1-19

[5] Fitri A, Hasan Z A and Ghani A A 2011 Determining the Effectiveness of Harapan Lake as Flood Retention Pond in Flood Mitigation Effort *In Proceedings of 2011 4th International Conference on Environmental and Computer Science (ICECS 2011)*

[6] Özmen B, Arch B, Gençoğlan M E & Sertkaya Z R 2019 The Neglected River: Bridging The West & The East *In International Conference on Contemporary Affairs in Architecture and Urbanism (ICCAUA-2019)* Vol. **9** p 10

[7] Fitri A, Hashim R, Song K I and Motamedi S 2015 Evaluation of morphodynamic changes in the vicinity of low-crested breakwater on cohesive shore of Carey Island, Malaysia *Coastal Engineering Journal* **57**(04) p 1550023

[8] Fitri A, Hashim R and Motamedi S 2017 Estimation and Validation of Nearshore Current at the Coast of Carey Island, Malaysia *Pertanika Journal of Science & Technology* **25**(3)

[9] Fitri A, Hashim R, Abolfathi S and Abdul Maulud K N 2019 Dynamics of sediment transport and erosion-deposition patterns in the locality of a detached low-crested breakwater on a cohesive coast *Water* **11**(8) p 1721

[10] Hashim R, Fitri A, Motamedi S, Hashim A M 2013 Modelling of Coastal Hydrodynamic Associated with Coastal Structures: A Reviews *Malaysian Journal of Science* Vol. **32** pp 149-154

[11] Fitri A, Yao L, and Sofawi B Evaluation of Mangrove Rehabilitation Project at Carey Island Coast, Peninsular Malaysia based on Long Term Geochemical Changes *In IOP Conference Series: Earth and Environmental Science* IOP Publishing Vol. **365** No. 1 p 012055
[12] Fitri A and Lao L The impact of Parameter Changes of a Detached Breakwater on Coastal Morphodynamic at Cohesive Shore; A Simulation *In IOP Conference Series: Earth and Environmental Science* IOP Publishing Vol. **365** No. 1 p 012054
[13] Hashim R, Roy C, Shamshirband S, Motamedi S, Fitri A, Petković D and Song K I 2016 Estimation of wind-driven coastal waves near a Mangrove forest using adaptive neuro-fuzzy inference system *Water resources management* **30**(7) pp 2391-2404
[14] Rasaei M, Nazari S and Eslamian S 2020 Experimental investigation of local scouring around the bridge piers located at a 90 convergent river bend *Sādhanā* **45**(1) pp 1-15
[15] Kareem K A A and AL-Thamiry H A 2019 Identification of scouring zones in ungauged river by simulation: The case of Galal Badrah river, Iraq *Association of Arab Universities Journal of Engineering Sciences* **26**(3) pp 57-67
[16] Esmaeili T, Dehghani A A, Zahiri A R and Suzuki K 2009 3D Numerical simulation of scouring around bridge piers (Case Study: Bridge 524 crosses the Tanana River) *World Academy of Science: Engineering and Technology* **58**(34) pp 1028-1032
[17] Zaid M, Yazdanfar Z, Chowdhury H and Alam F 2019 A review on the methods used to reduce the scouring effect of bridge pier *Energy Procedia* Vol.160 pp 45-50
[18] Sutopo Y and Saputro A E 2020 Groundsill stability in downstream of bridge "Besi" in body of water at River Garang Semarang city, Indonesia *In IOP Conference Series: Materials Science and Engineering* IOP Publishing Vol. **830** No. 2 p 022073
[19] Wu C C, Lin J J, Lin Y C and Yeh C M 2004 The applicability of open-slot groundsills in trained river to encourage the formation of meandering *In Self-Sustaining Solutions for Streams, Wetlands, and Watersheds* American Society of Agricultural and Biological Engineers pp 35
[20] Widiyanto W 2007 Water Surface Profile in Upstream of Groundsill with Type of Broad-crested and Ogee Dinamika Rekayasa **2**(2) 71-80
[21] Limantara, I L M 2018 *Rekayasa Hidrologi: Edisi Revisi* Penerbit Andi
[22] Saputra A E 2019 *Stabilitas Groundsill pada Aliran Sungai Garang di Hilir Jembatan Besi Kota Semarang* Doctoral dissertation Universitas Negeri Semarang
[23] Hardiyatmo C 2010 Mekanika Tanah 2 *Gajah Mada University press* Yogyakarta
[24] Suyitno H P 1994 *Diktat mata kuliah Irigasi* Fakultas Teknik Universitas Negeri Yogyakarta Yogyakarta
[25] Suyono S 2005 Mekanika Tanah dan Teknik Pondasi PT Pradnya Paramita Jakarta
[26] Sularno 2011 *Tinjauan Analisis Bendung Tetap Studi Kasus Bendung Njaen pada Sungai Brambangan Sukoharjo* Jurusan Teknik Sipil Fakultas Teknik Universitas Sebelas Maret Surakarta

Acknowledgments
The authors would like to thanks "Public Work and Spatial Planning (PU), Banten Province" for the supporting in completing this study. Authors are also wishing to acknowledge assistant from P.T. Saeba Konsulindo in providing the required data.

Performance of fire hotspot detection system with sensor based on angle width and voltage difference monitoring for forest security

N Suryaningsih[1*], A Prasetyowati[1*] and F Busalim[1*],

[1] Electrical Engineering, Universitas Pancasila, Jl. Raya Lenteng Agung No.56-80, Srengseng Sawah, Jakarta, DKI Jakarta

Corresponding E-mail: noor.suryaningsih@univpancasila.ac.id, ane_prast@univpancasila.ac.id, fauzie.busalim@univpancasila.ac.id,

Abstract. Fires that occur naturally in the forest can cause uncontrolled fires due to the rapid spread of heat. This research is focused on the observation land in the Puncak Besar area of Pacet in Pengalengan District, Indonesia, where the area of land burned is approximately 100 hectares. The early warning system that was built in this study uses a range widening method. This method consists of 1 sector that functions to detect heat and initial gas that occurs in the remote area. The initial area of the hotspot is usually not detected by the forest operator. After testing the sensor performance on this device, the fire sensor can detect hotspots with an average range of 440 cm. The width of the detection angle using a shield can reach a range of - 45° to 45° and the wide angle of detection of a fire without a shield can reach a range of -90° to 90°. While the communication system on this device uses the GPRS/ GSM module to send data to the Android application as a monitoring system. By using this early forest fire detection system, forest operators can monitor the starting point of a fire in remote areas around the forest and minimize the spread of fire points as one solution to overcome the wider spread of forest fires.

Keywords: Android, Microcontroller, GSM/GPRS, Sensor, Fire, Gas

Track Name: Land, Water, Forests and Food Security

1. Introduction

The wide and rapid spread of forest fires in Indonesia is caused by the lack of detection of fire starting points in remote areas around the forest. The reason why the starting point of the fire in the remote area is not detected is due to the limited range of the monitoring system owned by the forest operator. Early detection systems for starting fires can accelerate the handling of larger forest fires with a wider area. Handling early detection of a rapid starting point for fire can slow the damage that occurs in the forest. The recovery of natural ecosystems from the effects of fires can also be minimized. Delay in the entry of information when a forest fire occurs makes the area that experienced fire can only be detected after the fire has become large and wide area coverage. Currently, monitoring technology in forest hotspots still relies on satellite technology [1-3]. The delay in the information provided by the satellite makes the area of forest fire spread rapidly. The factors causing these natural disasters are not only human negligence but several factors that can affect forest fires, including lightning strikes in dry forests, underground fires in peatland areas and prolonged drought. Usually, the first detected area is the occurrence of a hotspot in an area far from the range of observation. If this happens for a long time and the terrain of forest fires is difficult to reach, then large forest fires are very difficult to avoid. As a solution to the problem, we need a system that is able to analyze and monitor the indications that cause forest and land fires Previous research has made a device to monitor the situation in the forest but the tool is still using satellites as a monitoring medium [4], it is considered inefficient because it requires a very long time for the information to get back to earth. Then from the existing reference journals, similar tools were also made but without the use of satellites. The tool only uses 2 sensors namely fire sensor and gas sensor and uses computer media as monitoring media, but for computer reliability, it is considered to be less reliable because it cannot be carried anywhere and is not practical [5]. From the existing literature and reference journals, we need to make a tool that can detect the emergence of hotspots early by utilizing a faster and more practical system, which is an Android-based system The early detection system for forest fires is designed to detect hotspots early on forest land. This system works by using three Detection Sensors which are gas sensors as CO, NO and Hydrocarbon gas detectors then fire sensors as fire detectors, and temperature and humidity sensors to detect humidity and temperature in the air. The three sensors will then be processed by a microcontroller and will send data wirelessly via radio waves to the data transmission center and then will be processed using another microcontroller. Furthermore, the data that has been processed is sent by the Global Packet Radio Service (GPRS) Module to the Web Server and then the data will be sent to the Android application so the Admin can monitor it. This tool is supported by Wireless Technology which is capable of sending data without the need to use cables is expected to be one of the applicative technological developments that can support forest sustainability programs. This website-based monitoring system is able to present data in the form of indications of the condition of hot spots that have the potential for forest fires.

2. Early Detection Method of Forest Fire

The method used in this study uses a method of widening the range of hotspot detection. This method works based on certain settings of the three sensors DHT11, MQ-9 and UVtron R2868. The design of this system consists of three parts. The first system is a detection system, this system is placed in the forest. The second system is a processing system, this system is placed in a forest monitoring post and the third system is a monitoring system, this system is an application contained in a mobile phone.

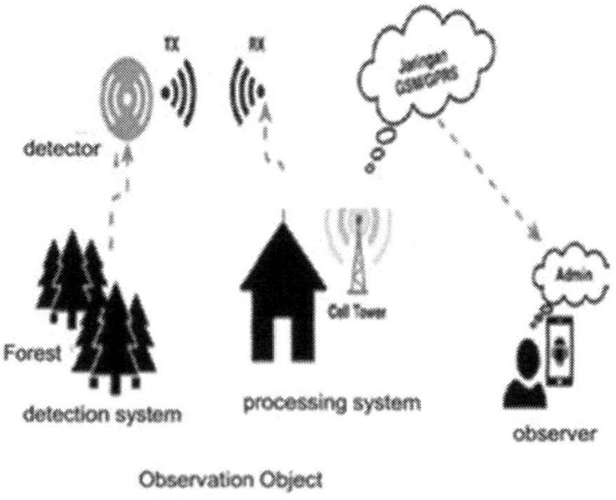

Figure 1. Overall system design diagram.

The system placed in the forest has a role to provide information such as providing data on temperature, humidity, and indications of a fire in the forest. And the second system has a role to receive data from the forest and then tasked to send data to the server to be distributed to the third system, the Android application. The following is picture 1 illustration of the overall system design. The methodology and working pattern of the application of hotspot detection in this study can be seen more clearly in the block diagram as shown in figure.2 below:

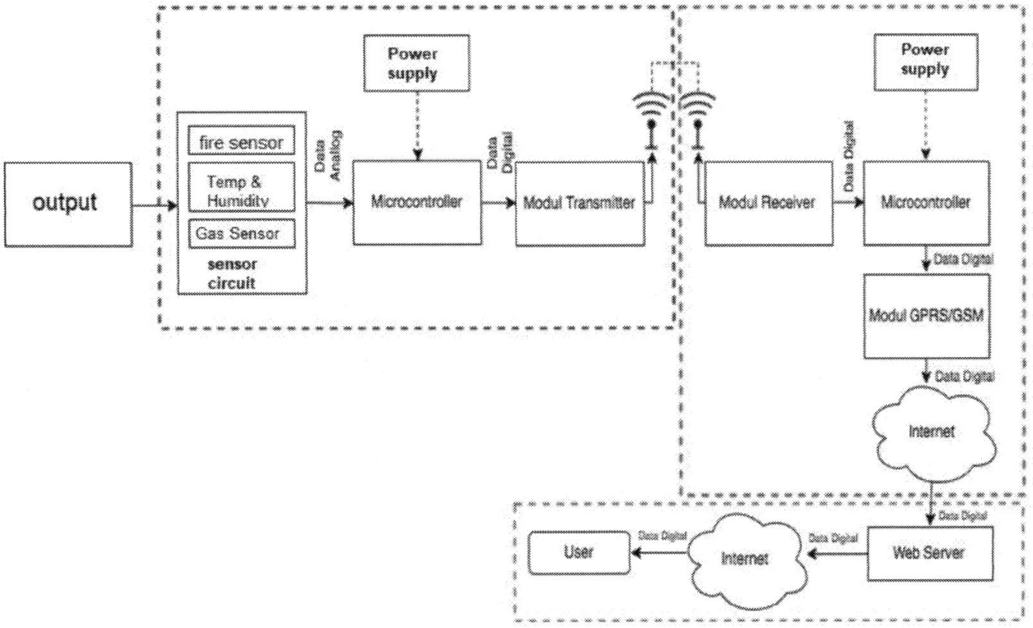

Figure 2. Diagram of the object system hardware block.

The block diagram in this study consists of a series of sensors. The sensor circuit will detect the presence of smoke or flammable gas, fire, temperature and humidity around the data system that will come out of this circuit is in the form of AO or Analog Output data. The analog data is in the form of voltage 0-5 V which represents how much the intensity of combustible gas, fire, temperature and humidity is detected by the sensor. Then the data will be entered into the microcontroller in the form of analog data and will be chopped by the Analog to Digital Converter (ADC) to 1024 the amount of data (0 Th to 1023 data). Then the data from the microcontroller in the form of digital data will be sent into the transmitter module. Radio transmitters utilize radio frequencies in digital data transmission which will then be received by the receiver module and still in digital data. Data from the receiver will enter the microcontroller. The digital data will be sent by the GPRS module to the web server as a data center before being sent to the user.

3. Result and Analysis

Testing in this study was carried out on the method of widening the range carried out by the three sensors. Tests carried out to see the performance between sensors in the detection of hot spots by setting the desired widening range. So that the detection of initial hotspots can be known with a rate of 5 seconds as the aim of this study. The temperature of 45 °C is the setpoint temperature set in the program in the design of this system. If the temperature is detected above the setpoint value, the sensor will send a notification to the user application. Temperature and humidity sensors work according to a predetermined plan where the higher the temperature the humidity will be lower. In the fire sensor testing, there are two types of testing, namely testing the distance of fire detection without the use of a protector and the use of a protector on this sensor. This test aims to determine the distance of fire detection by the UVTron sensor when it is unshielded. The test is done by placing the fire source in front of the UVTron sensor with different distances between 25 cm-450 cm, taking data based on the indicators on the serial monitor (detected or undetected). The test results are shown in Table 1.

Table 1. Testing of fire detection without protection.

Testing	Distance (cm)	Voltage (volt)	Testing	Distance (cm)	Voltage (volt)
1	25	4,93	12	300	3,92
2	50	4,81	13	325	3,60
3	75	4,70	14	350	3,50
4	100	4,69	15	375	3,39
5	125	4,63	16	400	3,39
6	150	4,57	17	425	3,31
7	175	4,48	18	430	3,30
8	200	4,42	19	435	3,28
9	225	4,39	20	440	3,27
10	250	4,04	21	425	0
11	275	3,98	22	450	0

Analysis of the testing distance fire detection using a shield which is done vertically includes the range of distances from 25 cm to 450 cm. In Table 1 it can be seen that the performance of the fire sensor without a protector can work at a distance of 440 cm. While the sensor using a protective has a performance like Table 2 below:

Table 2. Fire detection testing distance with protectors.

Testing	Distance (cm)	Voltage (volt)	Testing	Distance (cm)	Voltage (volt)
1	25	4,90	12	300	3,50
2	50	4,74	13	325	3,39
3	75	4,80	14	350	3,39
4	100	4,70	15	375	3,31
5	125	4,63	16	400	3,31
6	150	4,39	17	425	3,31
7	175	4,04	18	430	3,30
8	200	3,98	19	435	3,29
9	225	3,92	20	440	3,29
10	250	3,60	21	445	0
11	275	3,49	22	450	0

By using a shielded obtained a distance of 440 cm until the fire sensor can detect the presence of hotspots. Comparison of the performance of this sensor using a protective and without protective seen in fig. 3 below.

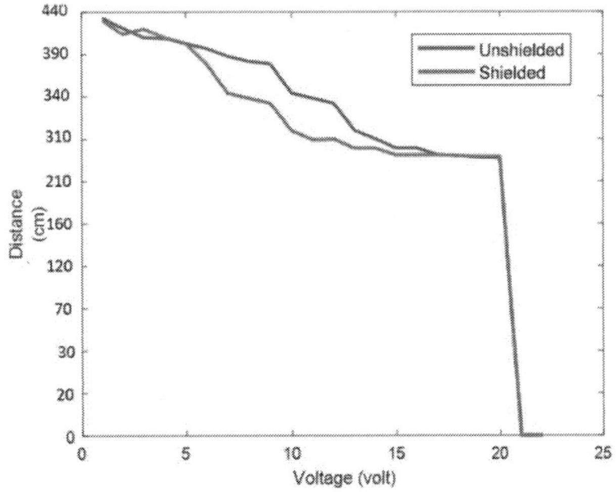

Figure 3. Comparison graph of test results for fire detection distance without and with protection.

In figure 3 we can see the comparison of the detection distance of fire without protection and with protection, in Figure 3 the response that is seen by using shielded detection has a lower detection distance compared to not using shielded with the same voltage. But at a voltage of 3.30 volts - 3.29 volts the distance to the detection with a shielded has a longer distance, this is because the voltage tends to be unstable. The voltage with this value is the minimum limit of the sensor's performance. Testing of the Wide Angle of Detection of Fire without shield is carried out to find out the maximum angle width that can be detected by the UVTron sensor without using a shield. Testing is done by placing the fire source at certain angles between -90° to 90° with a distance of 25 cm from the UVTron sensor. Figure 4 shows the results of the angular width response produced by the UVtron sensor in this study.

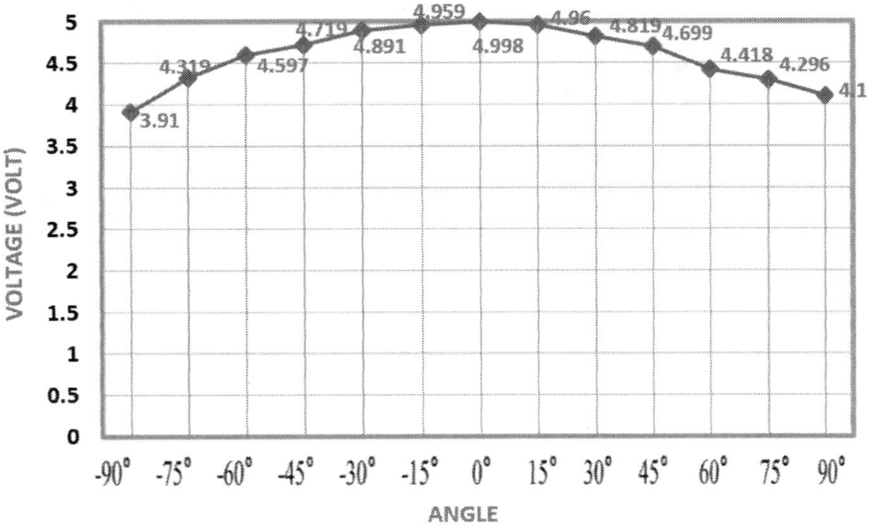

Figure 4. Test result graph of wide angle of fire detection without shield.

The response of the UVtron sensor produces a threshold voltage at 4.998 volts for an angle of 0°, the sensor range reaches - 90° at a voltage of 3.91 volts to 90° at a voltage of 4.1 volts. The maximum angle width that can be detected by the UVTron sensor using a shield. Testing is done by placing the fire source at certain angles between -90° to 90° with a distance of 25 cm from the UVTron sensor, taking data based on indicators on the serial monitor (detected or undetected). The results of comparison testing of the wide- angle of fire detection without using a shield and using a shield are shown in Figure 5 below

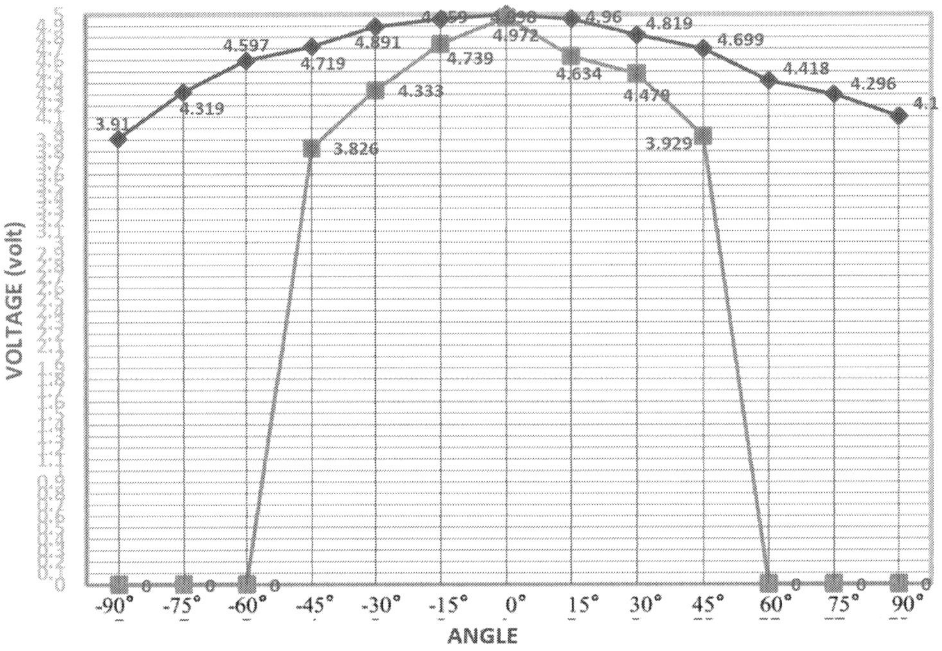

Figure 5. Graph comparison of test results for fire detection angle widths without using a shield and using a shield.

In figure 5 shows the results of a comparison of detection angle width testing if using a shield and without using a shield that can be detected. The difference in angle width between the two methods is 30 °, in the detection using the width detection angle protector is narrower than the detection using no protection. In the detection using a protector has a width of the detection angle from -90° to 90 method while the method of not using a protector has an angle width of - 60° to 60°.

Figure 6 shows a comparison of the amount of voltage generated when detecting hotspots using a shield and not using a shield. Comparison of an average of 28.77 Volts. Which indicates the voltage on the sensor will be reduced when using a protector compared to the method of detecting not using a protector.

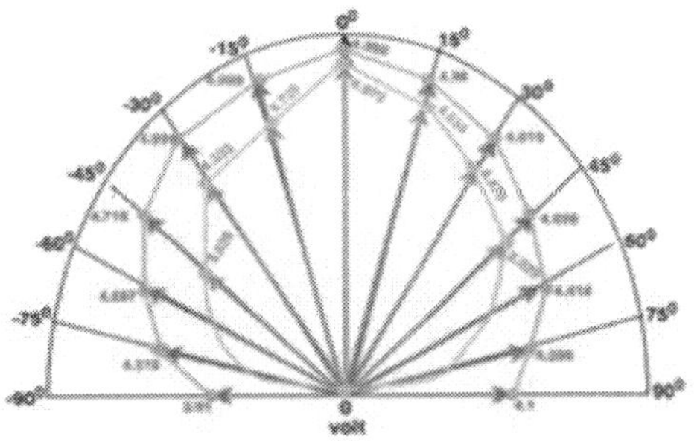

Figure 6. Comparison of voltage differences in the hotspot detection pattern.

In this study conducted communication systems using the website. Figure 7 is the output from the website for the initial hotspot detection information that will be used by the operator. The site that has been built using a specific site address with a display like figure 7 below

Figure 7. Web server main page display.

On the website menu, the history sent by the detection system is processed in the data processing system. Admin can see history data Id, Date and Time, Temperature, Humidity, CO, and Flame sent to the monitoring system. Display history on the web server can only be seen by the admin (officer) as shown in fig.8 below.

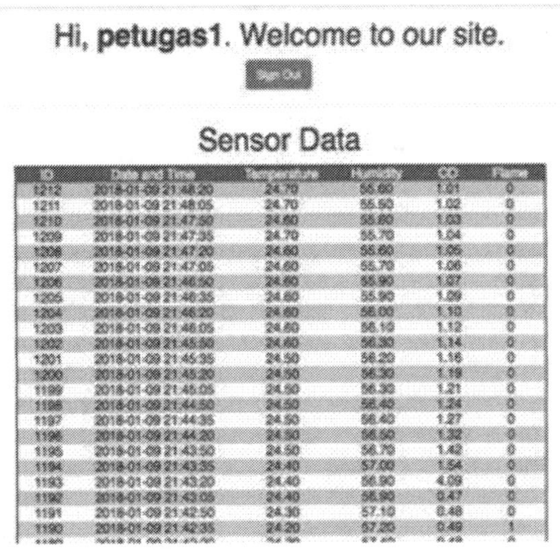

ID	Date and Time	Temperature	Humidity	CO	Flame
1212	2018-01-09 21:48:20	24.70	55.60	1.01	0
1211	2018-01-09 21:48:05	24.70	55.50	1.02	0
1210	2018-01-09 21:47:50	24.60	55.60	1.03	0
1209	2018-01-09 21:47:35	24.70	55.70	1.04	0
1208	2018-01-09 21:47:20	24.60	55.60	1.05	0
1207	2018-01-09 21:47:05	24.60	55.70	1.06	0
1206	2018-01-09 21:46:50	24.60	55.90	1.07	0
1205	2018-01-09 21:46:35	24.60	55.90	1.09	0
1204	2018-01-09 21:46:20	24.60	56.00	1.10	0
1203	2018-01-09 21:46:05	24.60	56.10	1.12	0
1202	2018-01-09 21:45:50	24.60	56.30	1.14	0
1201	2018-01-09 21:45:35	24.50	56.20	1.16	0
1200	2018-01-09 21:45:20	24.50	56.30	1.19	0
1199	2018-01-09 21:45:05	24.50	56.30	1.21	0
1198	2018-01-09 21:44:50	24.50	56.40	1.24	0
1197	2018-01-09 21:44:35	24.50	56.40	1.27	0
1196	2018-01-09 21:44:20	24.50	56.50	1.32	0
1195	2018-01-09 21:43:50	24.50	56.70	1.42	0
1194	2018-01-09 21:43:35	24.40	57.00	1.54	0
1193	2018-01-09 21:43:20	24.40	56.90	4.09	0
1192	2018-01-09 21:43:05	24.40	56.90	0.47	0
1191	2018-01-09 21:42:50	24.30	57.10	0.48	0
1190	2018-01-09 21:42:35	24.20	57.20	0.49	1

Figure 8. Display history page on web server.

Based on the web server test results, the testing process that has been carried out provides outputs according to the technical requirements that have been made. Web appearance in accordance with the design so that the admin can clearly see the information stored on the web server.

4. Conclusion

The design and realization of a site-based early forest fire detection device have been successfully carried out. Distance testing on the fire sensor is done by using a protective or cover can only detect fire as far as 440 cm and if not using a protective sensor can detect fire as far as 415 cm. Fire sensor testing based on the width of the detection angle using a shield can reach the range of -60° to 60° and the width of the angle of detection of the fire without a shield can reach the range of - 90° to 90°. The web server can store information received by the processing system to be sent back to the website application.

References

[1] Candra D and S Kustiyo 2014 Near Real Time Detection of Burned Scar Area Using Landsat-8 Imageries GOFC-*GOLD GHG Workshop Vietnam* (Vietnam)
[2] LAPAN (Lembaga Penerbangan dan Antariksa Nasional) 2015. Pedoman Pemanfaatan Data Landsat-8 untuk Deteksi Daerah Terbakar (Burned Area) *Pusat Pemanfaatan Penginderaan Jauh-Lapan* (Jakarta)
[3] Elvidge C D K and Baugh 2014 Burn Scar Mapping from Landsat 8 *Cooperative Institute for Research in Envronmental Sciences* (University of Colorado)
[4] Allison R S, Johnston J M, Craig G, Jennings S and McAlpine R S 2016 Airbone Optical And Thermal Remote Sensing For Wildfire Detection And Monitoring *International Journal Of National Fire Protection Association (Switzerland)*
[5] Enggar P 2014 Prototype Telemetri Alat Pendeteksi Dini Kebakaran Hutan *Thesis* (Universitas Negeri Yogyakarta)

Acknowledgment

Our thanks to research funding provided by the HPI-UP (Hibah Penelitian Internal Universitas Pancasila) grant and BPBD Jawa Barat for allowing us to use Puncak Besar Pacet forest in Kecamatan Pengalengan for observations ground for research.

Analyzing canopy height variations in secondary tropical forests of Malaysia using NASA GEDI

E Adrah [1], W S Wan Mohd Jaafar [2], S Bajaj [1], H Omar [3], R V Leite [4], C A Silva [5,6], A Cardil [7,8,9] and M Mohan [1,10*]

[1] United Nations Volunteering Program, Morobe Development Foundation, Lae 00411, Papua New Guinea

[2] Earth Observation Centre, Institute of Climate Change, Universiti Kebangsaan Malaysia, 43600 UKM, Bangi, Selangor, Malaysia

[3] Geoinformation Programme, Forest Research Institute Malaysia, 52109 Kepong, Selangor Darul Ehsan, Malaysia

[4] Department of Forest Engineering, Federal University of Viçosa, Viçosa 36570-900, MG, Brazil

[5] School of Forest Resources and Conservation, University of Florida, Gainesville, FL, USA

[6] Department of Geographical Sciences, University of Maryland, College Park, MD, USA

[7] School of Agrifood and Forestry Science and Engineering, University of Lleida, Av. de l'Alcalde Rovira Roure, 191, 25198 Solsona (Lleida), Spain

[8] Tecnosylva, Parque Tecnológico de León, 24009 León, Spain

[9] Forest Science and Technology Centre of Catalonia (CTFC), Ctra. Sant Llorenç de Morunys, Km 2, 25280 Solsona Lleida), Spain

[10] Department of Geography, University of California—Berkeley, Berkeley, CA 94709, USA

*Corresponding E-mail: mid_mohan@berkeley.edu

Abstract. Tropical forests play a significant role in regulating the average global atmospheric temperature encompassing 25 % of the carbon present in the terrestrial biosphere. However, the rapid change in climate, arising from unsustainable human practices, can significantly affect their carbon uptake capability in the future. For understanding these deviations, it is important to identify and quantify the large-scale canopy height variations arising from previous anthropogenic disturbances. With the advent of NASA GEDI spaceborne LiDAR (light detection and ranging), it is now possible to acquire three-dimensional vertical structural data of forests globally. In this study, we evaluate the applicability of GEDI for analyzing relative canopy height variations of secondary tropical forests of different age groups located across multiple geographical regions of peninsular Malaysia. The results for RH98 GEDI metric trends for the lowland and hill forests category across 4 different disturbance groups show a positive correlation between mean relative height and secondary forest ages. The consistency of these findings with previous studies in the region indicate the usefulness of GEDI to provide valuable insights into the patterns and drivers of forest height variation. Thus, this study contributes toward the operationalization of spaceborne LiDAR technology for monitoring forest disturbances and measuring biomass recovery rates and should help support large-scale sustainable forest management initiatives with respect to the tropical forests of Malaysia.

Keywords: Global Ecosystem Dynamics Investigation (GEDI), disturbed forests, Southeast Asia, forest canopy structures, logging

Track Name: Land, Water, Forests and Food Security

1. Introduction

Tropical forest ecosystems especially in Southeast Asian countries such as Malaysia face challenges from selective logging which has been a widespread modus operandi for commercial timber production [1-2]. Logging influences forest ecosystem dynamics - including soil greenhouse gas flux, spatial and temporal variations in light environment - as well as forest structural attributes such as canopy gap, canopy height, crown surface area, stem density to name a few [1-4]. A recent study by Okuda et al. (2019) investigated the potential impact of logging on canopy height recovery over selectively logged and primary forest regions in Pasoh Forest Reserve Peninsular Malaysia between 2003-2011. Herein, the authors employed airborne LiDAR (light detection and ranging) and found that average canopy height along with coefficient of variation and height diversity in logged forest was significantly lower than the primary forest plot suggesting that even after 53 years of logging, previously disturbed regions were not able to fully recover. Through simulation, the authors predicted that it would take another 16 years for full recovery from the previous canopy height mapping in 2011 [5]. Similar findings regarding mean canopy height were observed in one of Okuda's previous studies as well which utilized aerial triangulation and photographs to measure impact of logging in a lowland dipterocarp forest in peninsular Malaysia [2].

With increasing global deforestation and degradation along with climate change, the trend of tropical forests making an approximate net neutral contribution towards the global carbon cycle may change and tropical forests may act as net carbon sources [6]. Remote sensing technologies are widely used for monitoring tropical forest biomass over time which help predict the change in carbon sequestration capability of forests experiencing climate anomalies and/or anthropogenic disturbances [7-9]. In this regard, LiDAR sensors represent an efficient option due to its capability to directly retrieve vegetation vertical structure and have been applied extensively using aerial platforms [10-11]. Nonetheless, airborne LiDAR may be limited when it is necessary to assess large areas (e.g. regional/global scale), and hence, spaceborne data may be required. The first spaceborne LiDAR sensors onboard IceSat-1 and IceSat-2 satellites were used in several forest related studies to estimate plant traits [12-13]. However, their characteristics were not developed to optimally retrieve forest characteristics, especially over dense canopies. The recently launched Global Ecosystem Dynamics Investigation (GEDI) was designed to penetrate forest with about 98% canopy cover [14] and has been used to estimate forest biomass in different ecosystems [15-17]. Recent studies also highlighted its potential for mapping forest height and obtaining disturbance history [18-19]. For instance, Potapov et. al (2021) used GEDI data with time-series optical imagery (Landsat) for monitoring forest height and its dynamics [18]. However, its potential to characterize tropical forests' height variation of secondary forests has not been explored.

To the best of our knowledge, GEDI has yet not been employed to study forest structural attributes over the diverse tropical forests of Southeast Asia and through this pilot study we intend to explore the same. The aim of this study is to observe the trends for the chosen GEDI metric (RH98) and evaluate how post-logging forest recovery rates vary with respect to time in the various chosen geographical regions. In a prior study by Potapov et al. (2021), relative height (RH) metrics were found to be correlated with the 90th percentile of ALS-based canopy height per Landsat data pixel [18]. In this study, we compare the height distribution and growth rates of secondary forests of age groups 10-20, 20-30 and 30+ years for three different geographical regions of peninsular Malaysia and to demonstrate the applicability of GEDI for optimizing future forest management operations and supporting next generation earth system predictive models.

IOP Conf. Series: Earth and Environmental Science **880** (2021) 012031 doi:10.1088/1755-1315/880/1/012031

2. Study Area, Data and Methods

This exploratory study takes into consideration '*logged land forests, lowland forests and hill forests*' forest category as per the classifications proposed by the Forest research institute of Malaysia (FRIM). Four pre-identified disturbance groups based on logging time were sampled across the aforementioned forest category in three different geographical zones (North, South, and West) as shown in figure 1. The logging time considered was derived from the FRIM classification and the disturbance groups taken into account are (A) logged eleven to twenty (11-20) years ago, (B) logged twenty-one to thirty (21-30) years ago, (C) logged more than thirty (>30) years ago, and (D) Protection Forests. The protected forest refers to virgin forest that has never been logged (at least in the past 100 years).

The GEDI data were downloaded and subsetted from NASA EARTH DATA through the GEDI finder web tool. Level 2A data are processed waveforms providing directly measured metrics from which canopy height metrics were extracted for each 25 m diameter GEDI footprint. Level 2B products contain calculated biophysical attributes, such as plant area index (PAI) and canopy cover, while Level 1B contains geolocated corrected waveforms and geophysical correction for all shots in the GEDI beams. The GEDI instrument produces a total of eight ground tracks by three lasers, two full-power lasers dithered into four and one coverage laser diverged and dithered to produce another four beams. The produced GEDI slots are spaced 600 m apart in the cross track direction and approximately 60 m apart along the track direction.

Ground transects used in this study were filtered and considered only the full power beams to disregard and mitigate any uncertainty in the measurement due to cloud cover, day noise and scattering. Shots with low quality or incomplete information were filtered and disregarded as well. The data were then clipped, the height metrics were extracted and grouped based on the disturbance groups. Herein, RH98, the height at which 98 percentiles of energy is reached relative to the center of the ground return, was employed as a proxy for canopy height. Concerning the comparison of canopy heights between different disturbance groups, the study area was divided into a polygon grid to have a relatively similar climate and ecological zones; this helps minimize the influence of external factors such as precipitation, soil characteristics, slope, etc. that otherwise might affect the growth and recovery rates of secondary forests. The GEDI data were then overlapped over these zones individually, and subsetted for its coverage over each disturbance group after performing standard quality assessment. Three zones (North, South and West) were considered for this study, wherein a random sample of 100 GEDI footprints from each disturbance group was available after quality assessment (totalling 400 GEDI shots per zone) and the respective descriptive statistics were then calculated. A similar analysis is done considering all the zones together with 1200 GEDI footprints in total.

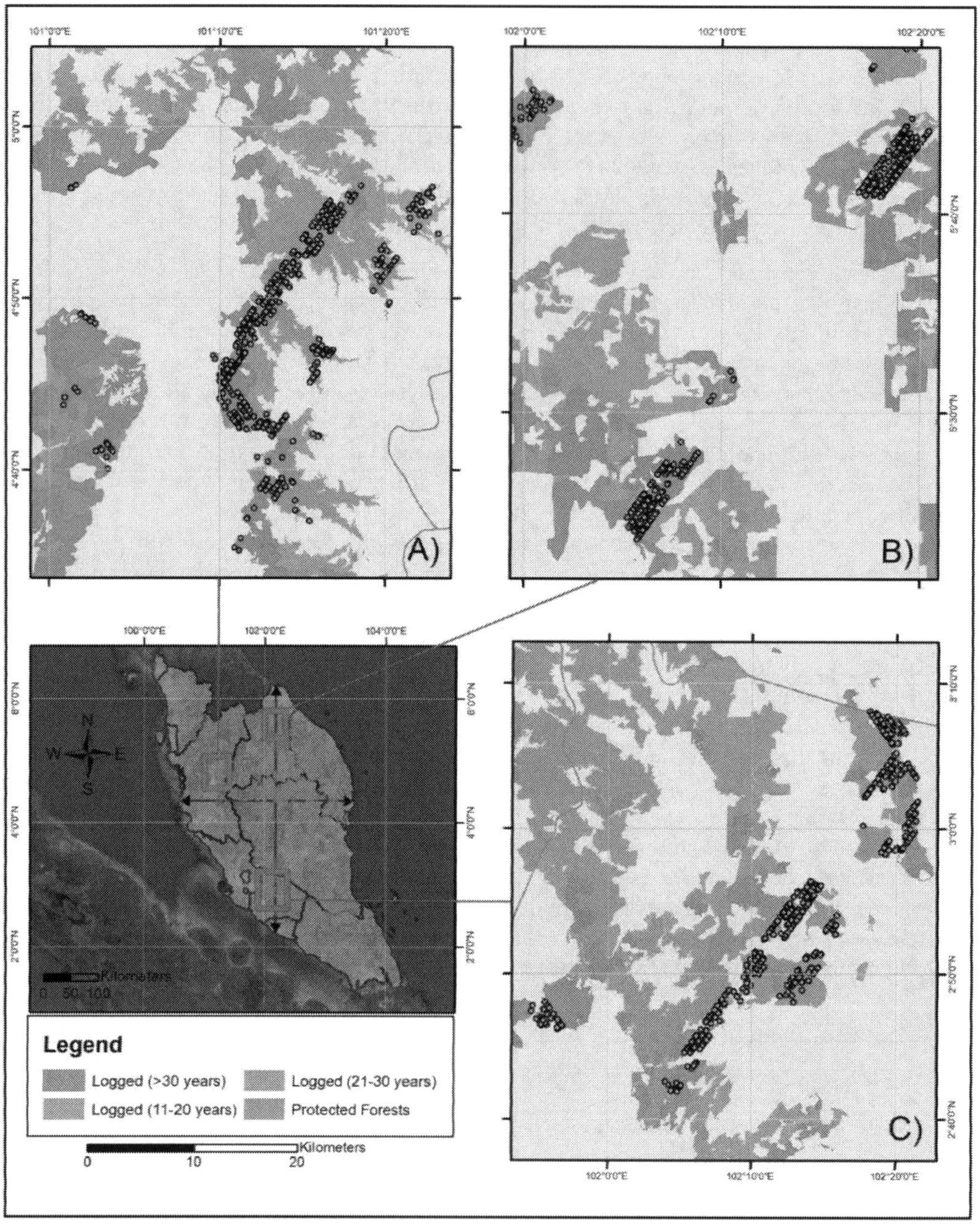

Figure 1. Study areas; a) zone A (west); b) zone B (north); c) zone C (south); the blue dots represent the GEDI footprints.

3. Results and Discussion

The results for the variation of the RH98 metric are divided into 2 sections - first, across the 4 disturbance groups (A, B, C and D) considering all the 3 chosen zones (Zone A, Zone B, Zone C) together and second considering each zone separately. The results for the former are presented in figure 2 whereas for the latter are presented in figure 3. Various parameters (min, max, median, mean, interquartile range) relevant to the box plots are enumerated in table 1 and table 2 respectively.

From Table 1, we observe a positive correlation between relative mean RH98 and secondary forest age groups, when all the zones are considered together. The upper and lower whisker extends from lowest to the highest value respectively within the interquartile range between 16.4 m to 38.83 m. As expected, protected forests had the highest average and group A (age group 11-20 years) had the lowest average for RH98. Regarding heights variations (interquartile span) within the group, it was the highest for group B (21-30 years) and lowest for group A (11-20 years).

Table 1. Summary statistics of height variations (for all the zones combined).

Disturbance group	All zones						
	Min	1st Quartile	Median	Mean	3rd Quartile	Max	Interquartile Range
A (11 - 20 Years)	1.6	16.4	23.04	23.14	29.92	52.07	13.52
B (21 - 30 Years)	2.99	13.52	22.94	23.73	31.88	65.19	18.36
C (> 30 Years)	2.54	17.95	25.77	25.66	33.05	65.95	15.1
D Protected Forests	3.17	22.2	30.22	30.59	38.83	72.2	16.63

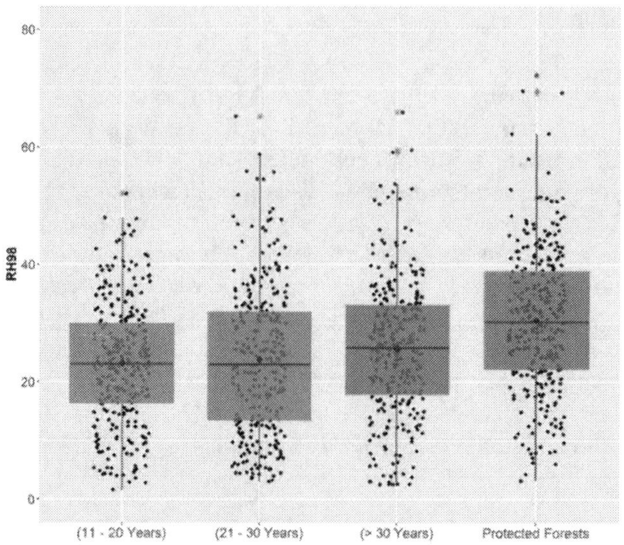

Figure 2. Boxplots representing the height variations across various disturbance groups while considering all the three zones together.

The results for the variation of RH98 metric across the different disturbance groups considering the 3 zones individually are presented in table 2 and figure 3. Positive correlation patterns are also observed between mean RH98 and secondary forest age groups and these trends persist across all zones except for the western zone (zone A) where the group A (11-20 years) have a slightly higher mean value than the group B (21-30 years) (figure 3). This can be due to differences in logging intensities experienced and/or the type of forest species predominant in these forest regions and will require further investigation before making any conclusions.

Table 2. Summary statistics of height variations (for individual zones).

Zone	Disturbance group		Min	1st Qu	Median	Mean	3rd Qu	Max	Interquartile Range
Zone A (West)	A	(11 - 20 Years)	2.77	12.43	21.64	21.26	27.57	47.76	15.14
	B	(21 - 30 Years)	2.99	11.67	20.25	20.46	26.79	54.57	15.12
	C	(> 30 Years)	4.08	16.24	20.95	23.37	29.72	52.95	13.48
	D	Protected Forests	6.06	22.06	30.46	30.01	37.02	57.99	14.96
Zone B (North)	A	(11 - 20 Years)	4.94	17.93	23.69	24.36	30.52	52.07	6.83
	B	(21 - 30 Years)	4	14.38	25.18	25.44	35.25	58.75	10.07

	C	(> 30 Years)	3.06	21.6	28.15	27.6	33.96	65.95	5.81
	D	Protected Forests	4.3	20.82	28.9	30.59	39.38	72.2	10.48
Zone C (South)	A	(11 - 20 Years)	1.6	17.67	24.25	23.79	30.67	45.22	13
	B	(21 - 30 Years)	3.77	17.52	26.26	25.31	32.6	65.19	15.08
	C	(> 30 Years)	2.54	19.38	27.04	26	33.84	53.74	14.46
	D	Protected Forests	3.17	24.54	31.85	31.17	39.3	69.55	14.76

Another explicit observation which can be inferred from the results show that the relative heights metric for the protected forest category have the highest values across all zones when compared to disturbance groups A, B and C. In other words, the RH98 metric values for all disturbance categories are well below when compared to that of protected forest when all the zones are considered, individually and together. This implies that the forests have yet not fully recovered to their original potential. In sum, the results based on the RH98 GEDI metric are consistent with previous studies in the region which took into account the actual canopy height and show a similar trend with respect to the recovery status [2,5].

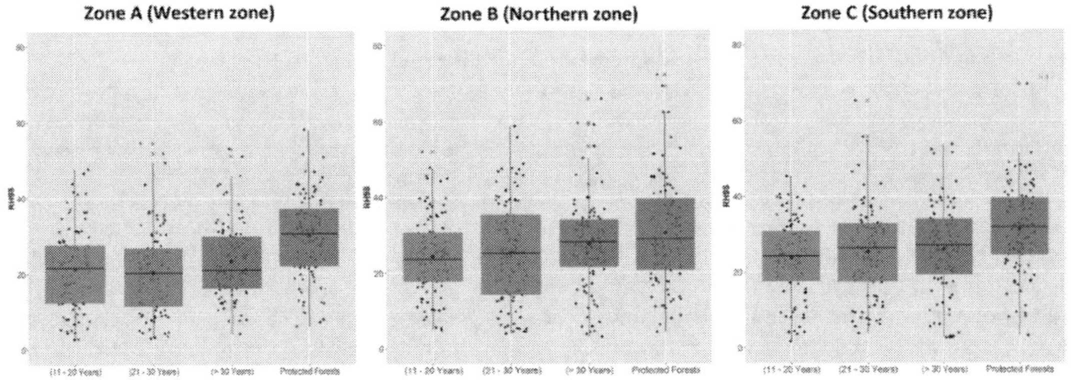

Figure 3. Boxplots representing the height variations across various disturbance groups considering all each zone individually.

Concerning heights variation across different geographical zones, as interpreted from table 2, the lowest mean values across all disturbance categories is recorded in the western zone (zone A), while the highest mean across all disturbance categories is recorded in the northern zone (zone B). Northern zone (zone B) is also the one with lowest heights variation (interquartile range) across all disturbance groups. These observations could be explored further to understand the underlying drivers of these variations with more background information about forest disturbances and ecological variants in each zone which we plan to do in our future studies.

Next, the inter-decadal difference was analyzed between the mean relative height (represented by the GEDI metric RH98) of different disturbance groups A, B and C for the 3 zones together and individually. These are presented in table 3. The range for protected forest, as shown in table 3, has little variation across the different geographical zones indicating that the protected forests are near their maturity across the 3 zones. Additionally, we observed that the difference between the inter-decadal relative height mean for disturbance group C & B (i.e. C-B) is greater for Zone A and Zone B when compared to B & A (i.e B-A). Nonetheless, the results observed are reversed for Zone C. The observed results for the zones could be possible due to a number of reasons which we intend to explore in our future studies. For instance, the intensity of logging may vary with time which may ultimately influence the mean relative height. Also, the distribution of age groups within the disturbance groups may not be uniform. For example, the year when logging occurred in a particular zone within different decades could be concentrated at extreme ends of the decade. Thus, in our future studies, we intend to utilize Landsat imagery to gain insights into intensity of logging and exact period when logging occurred. Information about the species specific for the zones could also be gathered to comprehend the results in a better way and this could help us understand how the recovery rate varies across species types.

Table 3. Height differences (in m) between secondary forests of different age groups and zones.

(a) Mean Relative Height (RH98) variations across disturbance groups in each zone

RH98 metric difference across disturbance groups	All zones together	Zone A	Zone B	Zone C
B (21-30 years) – A (11-20 years)	0.59	- 0.8	1.08	1.52
C (>30 years) – B (21-30 years)	1.93	2.91	2.16	0.69

(b) Mean Relative Height (RH98) variations across different zones for each disturbance group

Range of RH98 mean for different disturbance categories across zones	A (logged 11-20 years ago)	B (logged 21-30 years ago)	C (logged >30 years ago)	D (Protected forests)
Mean Range (max-min)	3.1	4.9	4.23	1.16

In subsequent research, calibrating GEDI data using airborne LiDAR data and furthermore, integrating it with Landsat data would be implemented for generating wall-to-wall canopy height maps. In addition, with further availability of new (version 2) and more GEDI data, a similar sampling technique could be conducted to chart the average mean heights against longitude and latitude taking vertical and horizontal tracks covering the extent of peninsular Malaysia. For instance, figure 4 shows the average heights across longitude and latitude for the entire Malaysia peninsular, taking 5000 GEDI shots over the '*logged land forests, lowland forests and hill forests*' forest category grouped per disturbance groups and sampled equally taking strip polygon grids with 5 km width in each: longitude and latitude directions. This sampling technique could be improved with the availability of more GEDI data and considering further information about the environmental gradients, climate variables, and species, to gain insights into canopy height variation across large landscapes. We also intend to extend our approach to the entire Malaysia and various other forest types and explore the patterns and drivers of canopy height variations by considering the influence of climate variables and environmental gradients.

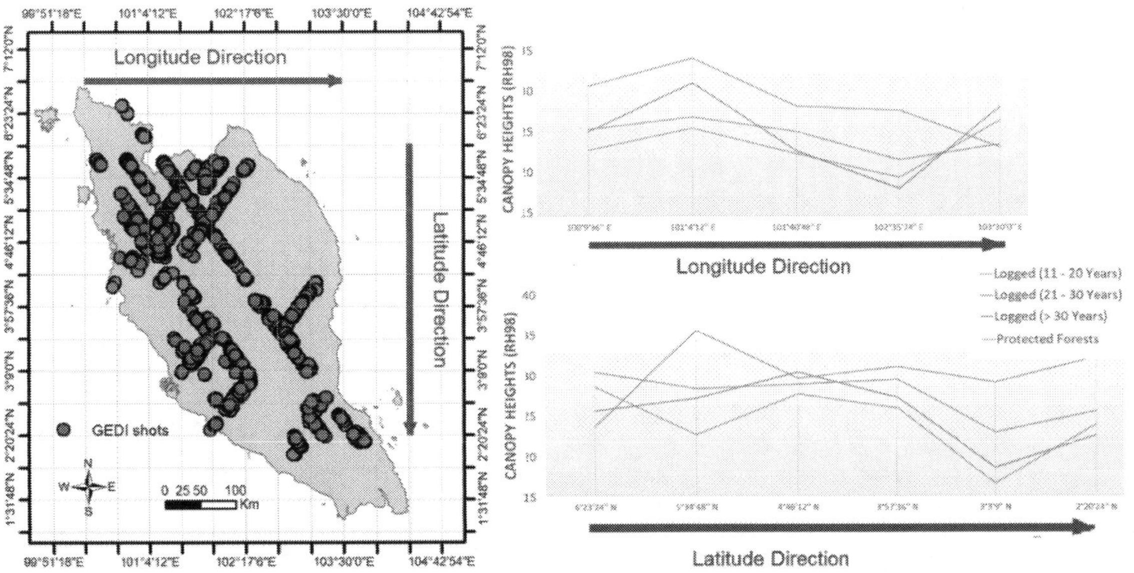

Figure 4. Relative Height (RH98) metric variation across landscape.

4. Conclusion

Through this preliminary study we have attempted to explore the relative trends in height variations using the RH98 GEDI metric for the lowland and hill forests category of peninsular Malaysia across 4 different disturbance groups (based on years after logging had happened). Our focus area is narrowed down into 3 different geographical zones after application of various steps such as quality assessment for selecting filtered highest quality GEDI data, and inclusion of a minimum of 100 shots for each type of disturbance group in each zone. The RH98 metric shows a positive correlation between mean relative RH98 and the secondary forest ages and also affirm the trend of protected forests possessing greater relative heights than logged forest categories. The consistency of these patterns with the results of previous studies in the region demonstrate the GEDI capabilities to provide insights on height variation patterns. We also find that the inter-decadal difference between the mean values of the RH98 metric varies spatially. This could be due to several reasons such as varying levels in intensity of logging across zones, non-uniformity in distribution of logging time period within each distribution group to name a few. In our future studies, we intend to explore these variations in recovery rates and height distribution as a function of environmental gradients and climate variables (such as slope and precipitation) and will be extending our analysis to other forest types. In sum, our approach demonstrates the potential of GEDI metrics to quantify the recovery and growth rates of secondary forests which can consequently be useful for estimating aboveground biomass (AGB) and understanding the impact of climate change and anthropogenic activities on tropical forest carbon sequestration capacity.

References

[1] Numata S *et al* 2006 Canopy gap dynamics of two different forest stands in a Malaysian lowland rain forest *Journal of Tropical Forest Science* **1** 109-16.
[2] Okuda T *et al* 2003 Effect of selective logging on canopy and stand structure and tree species composition in a lowland dipterocarp forest in peninsular Malaysia *Forest ecology and management* **175** (1-3) 297-320
[3] Yashiro Y *et al* 2008The effects of logging on soil greenhouse gas (CO2, CH4, N2O) flux in a tropical rain forest, Peninsular Malaysia. *Agricultural and forest meteorology* **148** (5) 799-806
[4] Yamada T *et al* 2014 Spatial and temporal variations in the light environment in a primary and selectively logged forest long after logging in Peninsular Malaysia. *Trees* **28** (5) 1355-65
[5] Okuda T *et al* 2019 Canopy height recovery after selective logging in a lowland tropical rain forest. Forest ecology and management **442** 117-23.
[6] Mitchard ET 2018 The tropical forest carbon cycle and climate change *Nature* **559** (7715):527-34.

[7] Sanchez-Azofeifa A *et al* 2017 Twenty-first century remote sensing technologies are revolutionizing the study of tropical forests *Biotropica* **49** (5) 604–19
[8] Wan Mohd Jaafar WS *et al* 2020 Carbon Emissions from Oil Palm Induced Forest and Peatland Conversion in Sabah and Sarawak, Malaysia *Forests* **11** (12) 1285
[9] Wan Mohd Jaafar WS *et al* 2020 The influence of deforestation on land surface temperature—a case study of Perak and Kedah, Malaysia *Forests* **11** (6) 670
[10] Dubayah RO et al 2010 Estimation of tropical forest height and biomass dynamics using lidar remote sensing at La Selva, Costa Rica: Forest Dynamics Using Lidar. *J Geophys Res* [Internet]. **115** (G2)
[11] Silva C *et al* 2017 Impacts of airborne lidar pulse density on estimating biomass stocks and changes in a selectively logged tropical forest. *Remote Sens (Basel)* **9** (10) 1068
[12] Duncanson LI *et al* 2010 Estimating Forest canopy height and terrain relief from GLAS waveform metrics *Remote Sens Environ* **114** (1) 138–54
[13] Narine LL *et al* 2020 Using ICESat-2 to estimate and map forest aboveground biomass: A first example *Remote Sens (Basel)* **12** (11) 1824.
[14] Dubayah R *et al* 2020 The Global Ecosystem Dynamics Investigation: High-resolution laser ranging of the Earth's forests and topography *Science of remote sensing* **1** (1) 100002
[15] Silva CA *et al* 2021 Fusing simulated GEDI, ICESat-2 and NISAR data for regional aboveground biomass mapping. *Remote Sens Environ* **253** 112234
[16] Duncanson L *et al* 2020 Biomass estimation from simulated GEDI, ICESat-2 and NISAR across environmental gradients in Sonoma County, California *Remote Sens Environ* **242** 111779
[17] Rishmawi K Huang C and Zhan X 2021 Monitoring key forest structure attributes across the conterminous United States by integrating GEDI LiDAR measurements and VIIRS data *Remote Sens (Basel)* **13** (3) 442
[18] Potapov P *et al* 2021 Mapping global forest canopy height through integration of GEDI and Landsat data *Remote Sensing of Environment* **1** (253) 112165
[19] Sanchez-Lopez N *et al* 2020 Estimating time since the last stand-replacing disturbance (TSD) from spaceborne simulated GEDI data: A feasibility study *Remote Sens (Basel)* **12** (21) 3506

Acknowledgments
We would like to thank the Forest Research Institute Malaysia (FRIM) for providing a complete dataset from field data to LiDAR data to other supplementary materials and resources. We also thank members of the Morobe Development Foundation (Papua New Guinea), UN Volunteering Program and NGEE-Tropics (LBNL) for their support and guidance. The authors gratefully acknowledge the funding for this open access publication and related proceedings of the presented work provided under UKM research grants Geran Galakan Penyelidik Muda (GGPM-2020-034) and Ganjaran Penerbitan (GP-2020-K022872).

Polyurethane modified screen – printed electrode for the electrochemical detection of histamine in fish

M A Munir[1] L Y Heng[2] and K H Badri[2, 3*]

[1]Department of Pharmcy, Faculty of Health Science, Universitas Alma Ata, 55183 Bantul, D.I. Yogyakarta, Indonesia
[2]Department of Chemical Sciences, Faculty of Science and Technology, Universiti Kebangsaan Malaysia, 43600 UKM Bangi, Selangor, Malaysia
[3]Polymer Research Center, Universiti Kebangsaan Malaysia, 43600 UKM Bangi, Selangor, Malaysia

*Corresponding E-mail: kaybadri@ukm.edu.my

Abstract. Histamine needs to be determined because of its toxicity. Histamine is commonly determined using chromatography, where not only that the instrument is expensive, the process is very tedious and require an expert. A sensor was developed using palm-based polyurethane as an electro-sensor substrate. Palm-based polyurethane (PU) was produced via condensation polymerization between palm kernel oil-based monoester polyol (PKOp) and 4,4'-diphenylmethane diisocyanate (MDI). PU offers high porosity and capability to attach onto screen–printed electrode (SPE) sturdily without being disintegrated. PU–SPE adsorbed histamine onto its pores, before being oxidized. The oxidation process was detected using cyclic voltammetry (CV) and differential pulse voltammetry (DPV). Histamine was oxidized electrochemically at +0.31 V (vs. Ag/AgCl, 1 mol·L^{-1}, pH 7.5). Differential pulse voltemmetric approach were used in order to get a satisfactory response, thus the histamine concentration was made in the range from 1×10^{-4} to 1 mmol·L^{-1}. A good sensitivity of 0.1 mmol·L^{-1} was attained with 3.07 % during intraday and 9.55 % during interday. The detection and quantification limits of histamine acquired at 0.17 mmol·L^{-1} and 0.53 mmol·L^{-1}, respectively. A wide variety of interfering compounds were also examined in order to establish their effect, if any, on the determination of histamine at the PU modified electrode. The sensor showed an excellent anti – interference property towards the other amines. The developed chemical sensor using PU – SPE has a good potential to determine histamine level in mackerel *(Rastrelliger Brachysoma)* owing to its simplicity and reproducibility.

Keywords: Histamine, polyurethane, electrode, conducting, mackerel

Track Name: Chemistry

1. Introduction

The production of food protein increases significantly owing to the high demand. Protein has an important role in human growth. Nevertheless, it will give impact to food security due to foods protein have been thoroughly investigated due to the possibility of biogenic amines accumulation, particularly histamine. The common biogenic amines obtained in foods are cadaverine, tyramine, putrescine, spermine, spermidine and histamine. However, histamine has been receiving great attention by several organization such as Food and Drug Administration (FDA) and European Food Safety Authority (EFSA) due to its toxicity [1]. Histamine is produced by the decarboxylation process of histidine, it is caused by many factors such as microbial or enzymatic process. In order to stop this microbial process, cooking or heating becomes an alternative to destroy the bacteria, nevertheless, histaminolytic bacteria found in foods, which acts as oxidizing agent, has the ability to survive inside the foods during the cooking or heating process [2,3]. Thus, the detection of histamine is imperative, in order to ensure the safety of human consumption.

Several approaches such as chromatography and electrochemical techniques have been applied and studied to detect histamine in foods. Liquid and gas chromatography are techniques that have been applied in decades. These methods are well developed, sensitive and very selective [4-9]. However, many conflicts occur during the application of these methods such as very expensive, require many chemical materials, time–consuming and only people who have background in analytical chemistry can use it. Moreover, both HPLC and GC methods need derivatization step using specific derivatizing reagent in order to increase the sensitivity of HPLC and GC. Derivatization step is needed since biogenic amines have low volatility and lack of chromophores [10].

Thus, in order to tackle these issues, electrochemical sensors are applied. Electrochemical sensors are cost-effective and have the capability to detect histamine either using biological receptors such as enzyme–modified and antibodies or using chemical receptors such as conducting polymers. They are still using chromatography methods [11,12]. Histamine biosensors coupled with a specific enzyme have attracted significant attention such as methylamine dehydrogenase, diamine oxidase and amine oxidase [13]. Nevertheless, the application of these enzymes to the sensor has some drawbacks such as life cycle and price of enzymes. Several studies have reported that immobilization enzymes causes instability. The instability of enzymes can be influenced by several factors such as pH and temperature. Some foods also containing metal that inhibit the enzyme activity [14,15].

Furthermore, biological receptors such as enzymes and antibodies are costly, complicated to produce and the selectivity is questionable. The receptors also cannot survive in high concentration of organic solvents, can be destroyed in high temperature and unstable in a specific pH. Furthermore, a synthetic receptor is strongly required so it can avoid the biological receptor issues. Chemical sensors become another alternative in order to avoid the application of antibodies, enzyme and DNA [16]. These methods use a specific receptor to capture biogenic amines physically or chemically where the electrodes must be modified chemically. The electrodes also must be both chemically stable and conductive. Carbon nanotube, graphene, gold, lithium, platinum and silicon are materials that generally applied by researchers as electrode modifiers to analyze biogenic amines. They are also friendly because they do not use organic solvents. Other advantages of chemical sensors could also be robustness, selective to small samples, tiny and need no skill on analytical chemistry to operate this sensor [17, 18].

However, the use of conducting materials such as gold, platinum and carbon are costly and the modification procedures are not straightforward. In order to solve these issues, polymers modified electrode can be applied to determine biogenic amines from samples using electrochemical technique. Conducting polymer application based composite materials as sensing elements in electrochemical sensor technology has proved to be reliable instrument based on analytical performances [19].

In this study, we have developed a bio – polymer that can be used to modify screen printed electrode (SPE) and has capability to adsorb histamine by physical adsorption. SPE modified can also operate at a lower potential. Histamine was then determined using a cyclic voltammetry (CV) and differential pulse voltammetry (DPV) methods. The chemical sensor was used to evaluate histamine in fish mackerel in order to monitor the level of histamine. Furthermore, the detection of histamine using polyurethane modified electrode has not been studied and issued elsewhere.

2. Materials and method

2.1 Instrument and chemicals

Voltammetric methods such as cyclic voltammetry and differential pulse voltammetry were performed by using Metrohm Autolab Electrochemical Workstation (UKM, Bangi, Malaysia). Three conventional electrodes used in this study such as polyurethane modified electrode as a working electrode, platinum wire as an auxiliary electrode and Ag/AgCl electrode as a reference electrode. All experiments were conducted at ambient temperature. Histamine dihydrochloride (HIS) was purchased (≥99% purity) from Sigma Aldrich Sdn. Bhd. While polyurethane was produced in UKM laboratory (Malaysia) via reaction of palm-based polyol (PKO-p, OHV of about 370 mg KOH/g and moisture content less than 0.1%) with polymeric isocyanate of 4,4-diphenylmethane diisocynate (MDI) manufactured by BASF Germany was acquired from Cosmopolyurethane (M) Sdn. Bhd., Klang, Malaysia (NCO content of 32%). Potassium chloride (KCl 0.1 M) and potassium hexacyanoferrate (III) ($K_3Fe\ (CN)_6$) were used as supporting electrolyte, HCl and NaOH were purchased from Sigma Aldrich Sdn. Bhd. and applied in order to adjust the pH solution. The sodium dihydrogen phosphate and disodium hydrogen phosphate were supplied by Sigma Aldrich Sdn. Bhd. The phosphate buffer (0.1 M) was prepared by mixing 0.1 M sodium dihydrogen phosphate and 0.1 M disodium hydrogen phosphate to the desired pH. All chemicals are of analytical grade and were used without further purification. Solutions were prepared using water purified through a Milli-Q system and kept at 4 °C until required.

2.2 Analytical procedure

10 mM of histamine stock solution was prepared using 0.1 M phosphate buffer solution (PBS) at pH 7.5. Histamine working solution was also prepared from stock solution by performing series of dilution to obtain the calibration curve. Polyurethane film was put on the electrode surface by casting method. Unmodified electrode and modified electrode were measured using cyclic voltammetry (CV) technique in the range of potential 0 to +1.6 V. After optimization of histamine standard solution using differential pulse voltammetry (DPV) method applied, calibration curve of $y = 30.917x + 1.3977$ was obtained using a series of standard histamine solution (1 - 10^{-4} mmol·L^{-1}).

2.3 Preparation of polyurethane modified screen – printed electrodes

The polyurethane modified screen-printed electrode (Figure 1) was prepared with minor modification. A screen printed electrode (SPE) was first rinsed with deionized water. The SPE was plunged in pH 7.5 PBS for 60 sec and immediately activated using electrochemical instrument by applying a scan potential started from -1.2 and +1.5 V (vs Ag/AgCl) for ten cycles and scan rate at 50 mV·s^{-1}. Afterward, 0.1 mg of polyurethane put onto the electrode surface and dried at ambient temperature for 24 h. The polyurethane electrode was flushed with the deionized water to expunge the materials on the electrode surface. Then, the polyurethane electrode stabilized in pH 7.5 PBS by performing a scan potential started at -0.5 and +1.5 V until a satisfactory response acquired. Finally, the polyurethane electrode was air-dried and ready for further experiments.

IOP Conf. Series: Earth and Environmental Science **880** (2021) 012032 doi:10.1088/1755-1315/880/1/012032

2.4 Real sample preparation
Fish and canned mackerel were purchased from a local market in Dengkil, Malaysia and were kept in a refrigerator prior to analysis. About 5 g of fish muscle was weighed and blended with 20 mL of 0.1 mol·L^{-1} PBS at pH 7.5. The mixture was centrifuged for 5 min and filtered through a Whatman filter paper. 100 µL of this solution was moved to a 10 mL volumetric flask and diluted to the mark with the PBS to make the final determination. Then, two concentrations of 0.1 and 0.01 mmol·L^{-1} of histamine standard solutions were obtained from a series of standard histamine solution. This method applied in order to validation the method and can be applied for histamine determination in real samples.

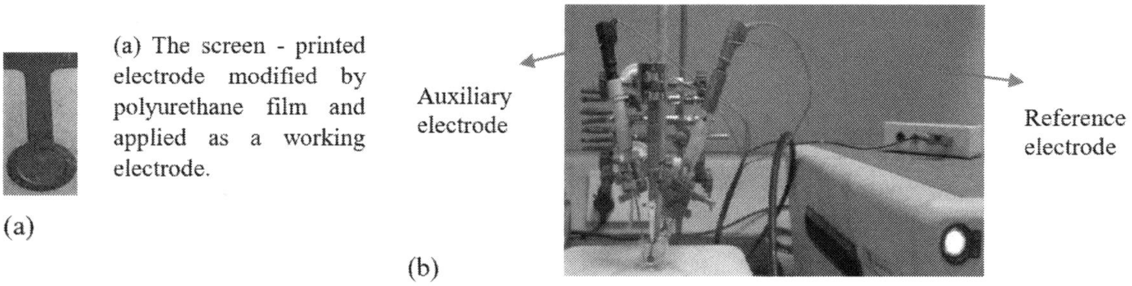

Figure 1. (a) The polyurethane modified screen – printed electrode and (b) The potentiostat instrument used to undergo the cyclic voltammetry (CV) and differential pulse voltammetry (DPV) techniques.

3. Results and discussion

3.1 Potentiostatic deposition of polyurethane on the surface of SPE
The electrochemical behavior of histamine was studied using cyclic voltammetry (CV) technique through both unmodified and modified electrodes. It was conducted at various pH values (6 – 8) in phosphate buffer solution (PBS) at 0.1 mol·L^{-1}. The studies showed that PBS was the best media in detecting the presence of histamine. Several studies reported that PBS gives the best analytical response whereas the use of hydrochloric acid (HCl) and sodium hydroxide (NaOH) were not suitable owing to the histamine can be detected by electrochemical technique [16,20]. The performance of PBS (pH 7.5) was also evaluated at concentration of 0.01, 0.05, 0.10 and 0.50 mol·L^{-1}. However, concentration of PBS did not affect the histamine analytical response when analysed using cyclic voltammetry technique. Nevertheless, 0.1 mol·L^{-1} PBS at pH 7.5 have been applied in numerous studies and found to be precise owing to a satisfactory response of histamine detection acquired. Thus, it was applied for further studies.

Following to the experimental part, polyurethane film was deposited on the electrode surface by casting method with a specific weight of polyurethane. Figure 2 shows the cyclic voltammograms of unmodified SPE (A) and polyurethane modified SPE (B) in 0.1 mM KCl containing 5 mM of ($K_3Fe(CN)_6$). The conductivity of unmodified SPE and modified SPE was compared. Solutions containing 5 mM of $K_3Fe(CN)_6$ used and determined by CV technique by applying a scan potential started at 0.0 to +1.0 V and 0.05 V·s^{-1} scan rate was performed. The voltammogram of the unmodified SPE and modified SPE exhibited similar redox couples. However, the conductivity of modified SPE was lower. The polyurethane used in this study made use of palm kernel oil – polyol as a natural polymer. It exhibited weak electrochemical signals due to its low electrochemical conductivity [21].

According to Figure 2, it shows that the anodic peaks that approximately at +0.5 V represent the oxidation process of unmodified SPE and modified SPE. The first oxidative scan of unmodified SPE and modified SPE starting from -0.2 to +1.0 V, showed a significant anodic peak at +0.5 V. Using the redox couple ($K_3Fe(CN)_6$) as the probe, the unmodified SPE showed higher current owing to its high electrical properties and active surface area, compared to the modified SPE which showed lower current. Since the conductivity of polyurethane was low, the current was also low. Smaller overpotential ($\triangle E_p$) of the $K_3Fe(CN)_6$ redox reaction on the polyurethane modified SPE compared to the SPE itself. Apparently, polyurethane can decrease the electron transfer owing to its low catalytic property. The formula of Randles – Sevcik employed in this study in order to measure the electrochemical area (A) of the polyurethane electrode [20]:

$$I_p = 2.65 \times 10^5\, n^{3/2} A v^{1/2} C D^{1/2} \text{ (Equation 1)}$$

Where, C is the concentration of analyte, D represents the diffusivity of solution (5 mM $K_3Fe(CN)_6$ diluted in 0.1 mM KCl) and n represents the total electrons transferred. The area of SPE approximately 0.2 cm^2 (0.44 cm × 0.44 cm) while 0.25 cm^2 for PU/SPE (0.5 × 0.5 cm). It shows the surface of modified electrode wider than the SPE. Furthermore, the formula below employed in order to measure the (τ) or the concentration of corresponding surface concentration:

$$I_p = (n^2 F^2 / 4RT)\, A \tau v \text{ (Equation 2)}$$

I_p represents the peak current, v is the scan rate, A is the region of electrode, F is Faraday's constant and T represents the temperature [22].

3.2 Electrochemical behaviour of histamine on PU – SPE and SPE

The sensitivity of polyurethane modified electrode and screen–printed electrode was studied and compared. 1.0 mmol·L^{-1} of histamine were investigated using these electrodes and cyclic voltammetry was applied in the potential range from 0.0 to +1.0 V with a scan rate of 0.05 V·s^{-1}. The peak currents of histamine acquired and observed on polyurethane modified electrode and screen–printed electrode. The voltammogram of the polyurethane modified electrode exhibited distinct redox couples with unmodified electrode. Figure 3 presents a well–defined anodic peak of histamine appeared approximately at +0.31 V at the modified electrode and the reverse scan found that there was no cathodic peak scanned, it was indicated that the reaction is irreversible due to no charge occurs. Whereas the peak of unmodified SPE appeared approximately at +0.58 V and there is no possibility for screen–printed electrode to react or attach with histamine whether by chemical or physical adsorption.

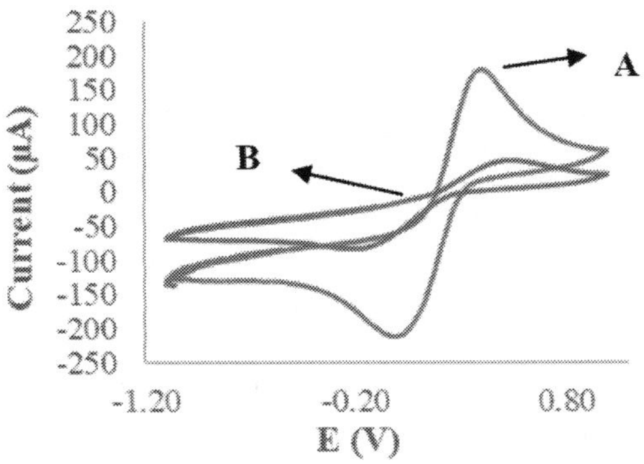

Figure 2. Cyclic voltammograms of unmodified SPE (A) and polyurethane modified SPE (B) in 0.1 mM KCl containing 5 mM of ($K_3Fe(CN)_6$).

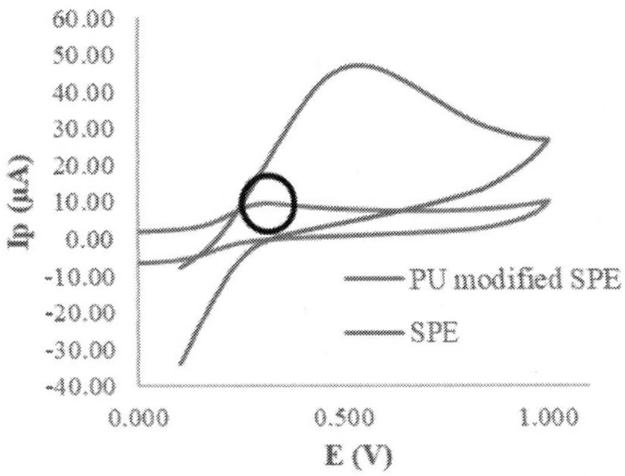

Figure 3. Comparison of the cyclic voltammograms between polyurethane modified SPE and unmodified SPE to detect 1 mmol·L^{-1} histamine in phosphate buffer, pH 7.5.

Several studies have shown that the modification of electrode biologically or chemically and the use of different methods will influence the capability of electrode in order to detect histamine and giving various potential. The use of carbon electrode and the application of amperometry method causing histamine was oxidized below +1.1 V [23]. Meanwhile, the distinct results showed when the carbon paste electrode and differential pulse voltammetry were applied. The anodic peak of histamine appeared approximately at +1.30 V whereas for carbon paste electrode modified with SWCNT, the anodic peak of histamine appeared approximately at +1.20 V [16]. Similar result reported by using lignin modified glassy carbon electrode [24]. However, the use of screen – printed electrode in this study showed different potential with others. Keow et al. (2007) [25] reported the use of SPE to determine histamine in tiger prawn, where SPE itself unable to determine histamine, thus this sensor used a photocuring method to immobilize an enzyme and entrapped in a membrane and deposited onto a SPE. A satisfactory anodic peak of histamine appeared approximately at +0.35V.

The oxidation of histamine can occur at the modified or unmodified electrode. Several studies reported that amines (-NH_2) group can be oxidized to nitro compounds (-N=O) [23]. Nevertheless, there are also several reports that histamine structure can loss $4H^+$ and alters to CN. Histamine can also be deaminated by various oxidants and enzymes such as diamine oxidase and horseradish peroxide [26]. Silver oxide has a capability to oxidize the amino acids with the implication of CO_2 loss. Nevertheless, this reaction needs to be done in high temperature and time consuming. The oxidation possibilities of histamine are illustrated in Figure 4. The oxidation of histamine on polyurethane modified screen – printed electrode.

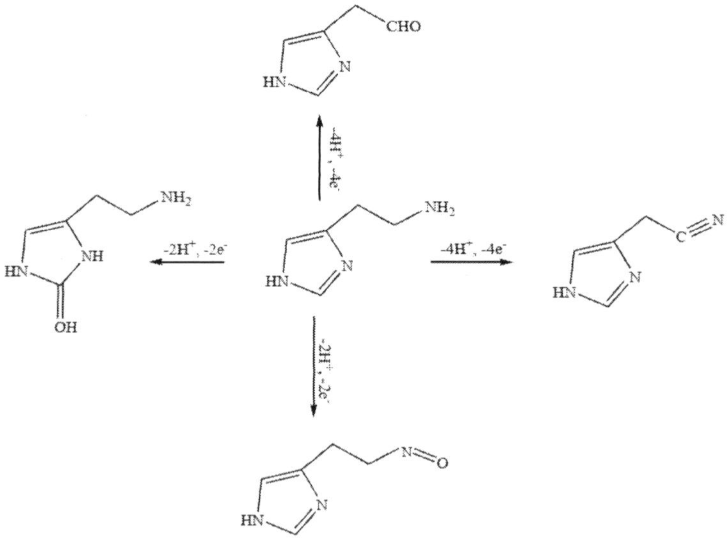

Figure 4. A possibility of histamine oxidation [20], with several modifications.

The polyurethane modified electrode showed a specific response of histamine compare to unmodified electrode. The properties of polyurethane film causing it can react with the analyte. The weigh of polyurethane film on the electrode surface was studied in order to obtain a satisfactory current response during analysis histamine using voltammetric techniques, such as CV and DPV. Approximately, 0.1, 0.3, 0.5 and 1 mg of polyurethanes investigated. The results of this study showed that the 0.1 mg of polyurethane amount giving the best response. Meanwhile, 0.3, 0.5 and 1 mg of polyurethanes reduce the current response. Furthermore, in order to obtain a satisfactory modified electrode, the thin polyurethane film suggested and the electron transfer between histamine and the electrode surface can occur.

The modification of screen – printed electrode using polyurethane is possible owing to polyurethane can be considered as a suitable material for adsorption. Many studies have noticed the advantage of polyurethane such as easy control of the pore size, stable maintenance of the quantity of cells and large-scale application at low price. A liquid material can be adsorbed to PU by a physical adsorption, by an entrapment or by a coupled chemical and physical binding [26]. In this study, histamine adsorbed into polyurethane by a physical adsorption where histamine entraps into the polyurethane. During the physical attachment of a histamine to polyurethane, the amino groups of histamine will not react with the isocyanate groups of polyurethane. This process is quite suitable for chemical sensor using CV and DPV methods owing to histamine can undergo oxidation process [23].

A voltammogram of 1 mmol·L^{-1} histamine in PBS (pH 7.5) at 0.1 mg of polyurethane film is shown in Figure 3. It can be concluded that there is no presence of histamine, no cathodic and anodic peaks appeared in unmodified screen – printed electrodes. Hence, a modified screen – printed electrode containing 0.1 mg of polyurethane was selected for the electro – oxidation of histamine in all further experiments.

3.3 Effect of pH

The shapes and heights of peak current of histamine may strongly influenced by the pH value. The pH value of PBS as a supporting electrolyte is an imperative factor that must be concerned owing to it effects the redox behavior of the histamine. Various pH of 0.1 mol·L^{-1} PBS started from 6.0 – 8.0 were investigated in order to study the correlation between pH and peak current (I_p), 1 mmol·L^{-1} of histamine was employed. The scan potential applied in range of 0.00 until +1.50 V with a scan rate 0.05 V·s^{-1} (Figure 5).

Based on this study, the pH was strongly influence the current response and peak potential of histamine. The relationship between peak potential and pH of the supporting electrolyte presented in Figure 5 and can be presented by the following equation 3:

$$E_p\ (V) = -0.1022 + 0.7316 \times pH \quad \text{(Equation 3)}$$

The slope of 0.7316 V/pH is indicating that the presence of electrons and protons influence the electrode reaction (Stojanovic et al. 2016). The histamine peak current escalated at pH 6 and reached the highest peak current at pH 7.5 yet decreased at pH 8 (Figure 6).

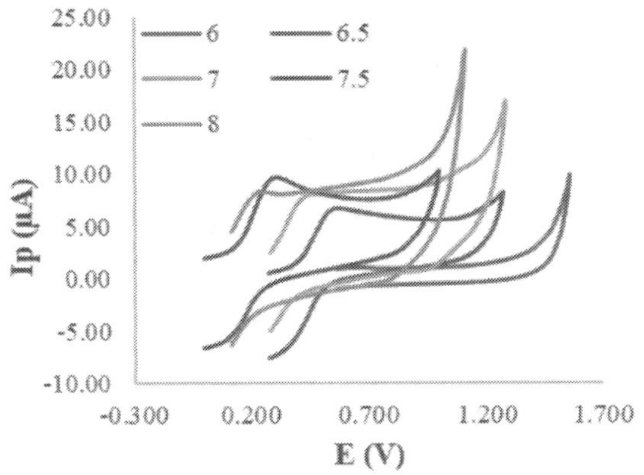

Figure 5. Cyclic voltammograms of polyurethane modified electrode in PBS at various pH (6 – 8) containing 1 mmol·L^{-1} histamine.

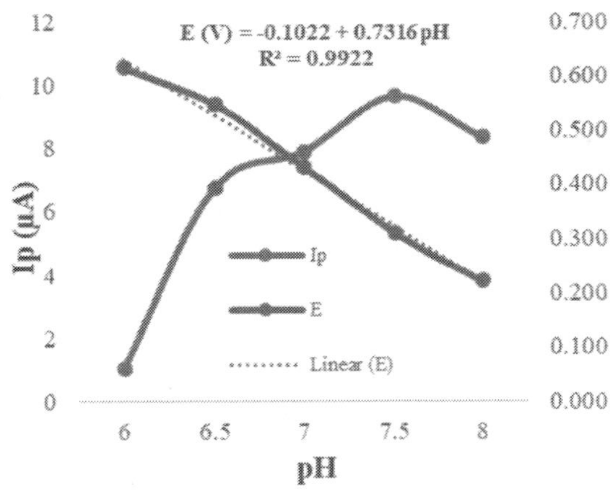

Figure 6. Effect of pH on peak potential and peak current response of 1 mmol·L^{-1} histamine in 0.1 mol·L^{-1} PBS with scan rate of 0.05 V s^{-1}.

The current signal were satisfactory developed at pH ranged 6.5 – 7.5. Nevertheless, 7.5 considered be the best pH in this study owing to giving the maximum current. A study reported, pH at 7 or 7.5 of PBS can protonate the histamine ring (pka 6.0) owing to the attraction of electrostatic, nevertheless the current can decrease when the pH 8 applied [16].

The relation between pH values and potential applied (v) of histamine at the polyurethane electrode was also studied. According to the Figure 5, it can be seen that the pH values influence the position of potential applied (v) of histamine, started to the pH 6 until 8, the potential applied moved to the positive direction indicating the participation of electrons transfer. The application of pH (6 – 8) in this study showed linear dependence on the pH with a linear regression equation and correlation coefficient (R^2) of E (V) = -0.1022 + 0.7316 pH and 0.9922, respectively. A slope of 0.7316 indicated that the electrons and protons movement influencing the histamine oxidation at the surface of polyurethane electrode.

3.4 Effect of scan rate

Several scan rates employed in the range from 0.01 to 0.10 V·s^{-1} using CV method in order to investigate the relation between potential applied, current peak and scan rate. Figure 7 shows the cyclic voltammograms of 1 mmol·L^{-1} histamine in pH 7.5 PBS at polyurethane modified electrode using various scan rates. The anodic peaks revealed in Figure 8 that the equation and correlation coefficient were i_{pa} (μA) = 0.0507 + 4.59 v (v in mV·s^{-1}) and R^2 = 0.9925, respectively (Figure 8) signifying the oxidation reaction between polyurethane electrode and histamine is adsorption controlled [24]. The obtained results indicated that current peak of histamine and scan rates employed was directly proportional and the relation between them calculated using the equation below:

$$Ip\ (\mu A) = 0.0507 v^{1/2}\ (V^{1/2} s^{-1/2}) + 4.59\ (r = 0.9925)\ \text{(Equation 4)}$$

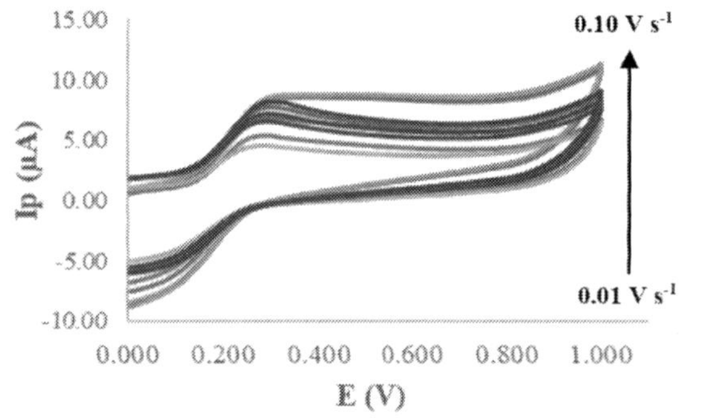

Figure 7. Cyclic voltammograms of 1 mmol·L^{-1} histamine in 0.1 mol·L^{-1} PBS at pH 7.5 on polyurethane modified electrode for series of scan rates from 0.01 to 0.10 V·s^{-1}.

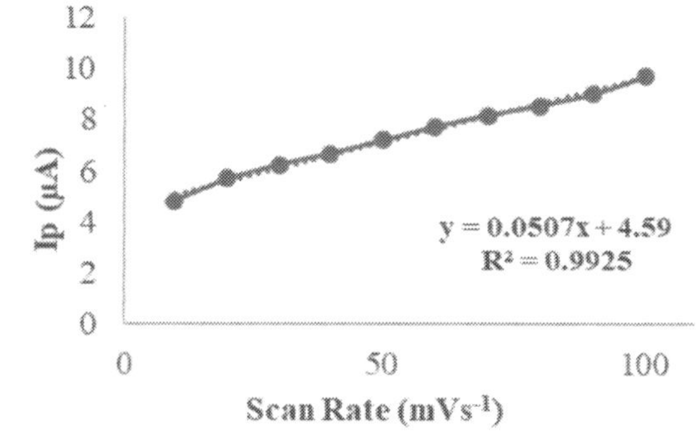

Figure 8. Calibration curve of anodic peak current of histamine versus the scan rate in the range 0.01 to 0.10 V s^{-1}.

In this study, the scan rates applied influenced the potential applied of histamine where the potential applied of histamine move to positive direction. A satisfactory calibration curve as shown in Figure 8 and the relation between them calculated using the equation below:

$$Ep\ (V) = 50.673 - 10.612\ \ln v\ (V \cdot s^{-1})\ (r = 0.9925)\ \text{(Equation 5)}$$

3.5 Analytical performance

Differential pulse voltammetry (DPV) technique has been applied in this study owing to its selectivity, simplicity and sensitivity compared to cyclic voltammetry (CV). DPV employed to verify and analyze the anodic current of histamine at polyurethane electrode. After several steps above validated and developed and in order to obtain fast analysis, the scan rate at 50 mVs^{-1} employed. The parameters of DPV technique that considered the optimal parameters chosen for further experiment such as the pulse amplitude of 100 mV and a pulse time of 25 ms.

3.5.1 Linearity

The DP voltammograms shown in Figure 9 that representing several histamine concentrations by applying the validated and developed method as mentioned before. Based on the obtained results, the signals were directly proportional to the concentration of histamine ranged from 10^{-4} to 1.0 $mmol \cdot L^{-1}$. The validation study and the analytical characteristics for the developed method are summarized in Table 1. Figure 10 presents the calibration curve of various concentrations of histamine in 0.1 $mol \cdot L^{-1}$ PBS at pH 7.5 on polyurethane modified electrode. Six replicates of every histamine concentrations analyzed and the relative standard deviation (RSD) was below 9%.

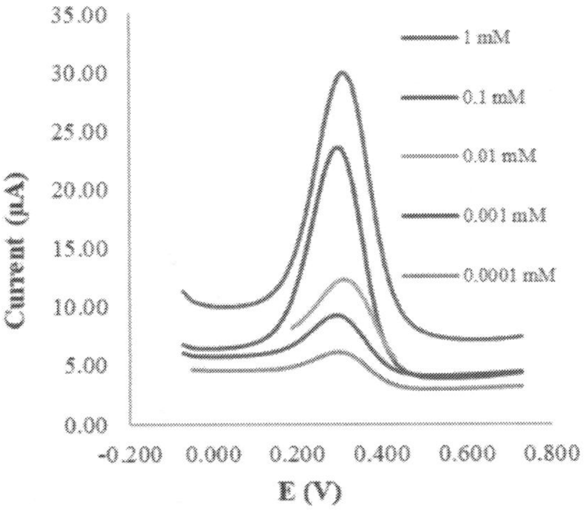

Figure 9. DPV of different concentration of histamine in 0.1 $mol \cdot L^{-1}$ PBS at pH 7.5 on SPE modified PU.

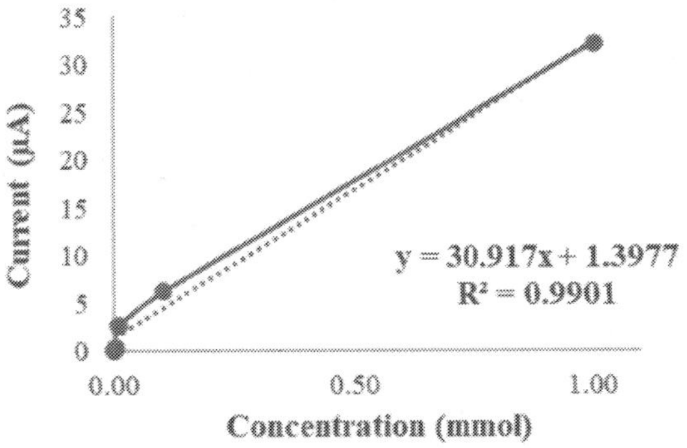

Figure 10. The calibration curve of different concentrations of histamine: 10^{-4} to 1.0 $mmol \cdot L^{-1}$.

3.5.2 Detection and Quantitation

The detection limit (DL) and quantitation limit (QL) were measured using the formula of (3.3. SD/slope) and (10. SD/slope), respectively. The detection and quantitation limits were 0.17 and 0.53 $mmol \cdot L^{-1}$ of histamine, respectively.

3.5.3 Stability and precision

The application of polyurethane modified electrode is not recommended use after one week owing to the sensitivity decreases day after day yet the selectivity is stable. The evaluation of stability of polyurethane electrode was investigated by measuring the anodic peak of histamine standard (0.1 $mmol \cdot L^{-1}$) started from day 1 until 28 with the similar electrode stored in dark and dry place in order to maintain the condition of SPE modified PU. The experimental data showed that the current response was reduced from 23.37 μA to 14.16 μA showing 39% discrepancy in current response on seventh day compared to the first day oxidation current values.

Table 1. Analytical characteristics for the determination of histamine by DPV in 0.1 mol·L^{-1} phosphate buffer at pH 7.5 on the polyurethane modified screen–printed electrode.

Analytical validated	Value
Detection potential	Anodic peak +0.31
Linear concentration range (mmol·L^{-1})	10^{-4} - 1
Coefficient of determination (R^2)	0.9901
Intraday precision (RSD, %)	3.05
Interday precision (RSD, %)	9.02
LOD (mmol·L^{-1})	0.17
LOQ (mmol·L^{-1})	0.53

Furthermore, on the 28th day, 82% reduction in current response was observed in comparison to the first day. However, the reduction of anodic peak noticed and encouraged that the polyurethane electrode do not use more than 7 days due to the sensitivity of electrode will decrease. The reliability or reproducibility of polyurethane electrode was also studied by investigating the electro – analytical response of histamine solution at 0.01 mmol·L^{-1} by performing similar conditions using six various polyurethane electrodes with the similar surface region, which were set up by the similar technique.

The intra and inter-day measurements applied in order to study the precision of the polyurethane modified electrode. The intra – day measurements were implemented at two different concentrations of histamine such as 0.1 and 0.01 mmol·L^{-1}, by six replicates analysis. The relative standard deviation (RSD) obtained below 3.05% and confirmed the satisfactory precision. Whereas, inter – day measurements were implemented the similar histamine concentration and polyurethane electrode and the investigation was reiterated every day for 7 days. The RSDs obtained at 7.54 and 9.02%, for concentration at 0.1 and 0.01 mmol·L^{-1} of histamine, respectively, showed the satisfactory reproducibility of histamine detection by the validated approach. Furthermore, this indicated the polyurethane modified electrode preparation procedure was quite easy and reliable method, including PU production, casting method and polishing step.

3.5.4. Disturbances

In order to study the selectivity of validated method, several amines were employed such as cadaverine and putrescine represent the biogenic amines, whereas aniline, xanthine and hexamine were used based on the number of carbons bonded to nitrogen atom. They were primary, secondary and tertiary amines, respectively. They were studied on similar method of histamine detection. Various amines detection at 1 mmol·L^{-1} were implemented by applying similar method of histamine detection. The obtained voltammograms showed various signal, nevertheless none of them had similar potential applied to histamine. The distinction peaks of each amines compared to histamine were negligible. Aniline as a primary amine as histamine appeared at +0.43 V, whereas putrescine and cadaverine that easily acquired in fish accompanied histamine also did not influence the histamine potential, a satisfactory potential applied of them were obtained at +0.18 and +0.19, respectively. The DPV signal height of other amines did not provoke the histamine signal. While the other amines such as xanthine and hexamine did not influence owing to appear in different potential applied compared to histamine potential.

3.6 The application of validated method
The validated and developed method of histamine standard detection employed to analyze histamine in fish mackerel and its product. The fish samples investigated in six replicates. The concentration of histamine in samples calculated using the calibration curve of histamine standard. Based on the detection using validated method, histamine detected in fish mackerel samples and canned mackerel. The content of histamine in both of them were 17 $\mu mol \cdot L^{-1}$ (1.89 ppm) and 5.3 $\mu mol \cdot L^{-1}$ (0.59 ppm), respectively. The study of recovery was performed at concentration 0.1 and 0.01 $mmol \cdot L^{-1}$ of histamine, respectively. The recovery values obtained at 94 and 103%, respectively for those concentrations. This study indicated the validated and developed method has a satisfactory accuracy. Based on the FDA regulation, the safe level of histamine concentration is below 50 $mg \cdot kg^{-1}$. Furthermore, the analyzed samples have no potential harm to human consumption. To evaluate the potential of the validated and developed method, the performance of PU modified electrode for histamine detection was compared to several reported methods as shown in Table 2 in accordance with the linear concentrations range and detection limit.

Table 2. Summary of the electrochemical performances of the validated technique compared to other studies for the determination of histamine.

Electrode	Method	LOD (μM)	Linear range (μM)	Ref.
Diamine oxidase/ carbon paste	Amperometry	3.5	0.1 – 325.9	[25]
Amine oxidase/ screen printed carbon	FIA / amperometry	3.0	10 – 300	[27]
Amine oxidase/horsedish peroxidase/ carbon	Chronoamperometry	0.18	0.4 – 2.4	[28]
MWCNT/ glassy carbon electrode	DPV	0.076	0.1 – 100	[29]
Pyrroloquinoline – Quinone modified GCE	Amperometry	341	360 – 1530	[30]
Lignin modified GCE	SWV	0.28	5 – 200	[24]
Polyurethane + SPE	CV and DPV	0.17	0.1 – 1000	This study

4. Conclusions

This work presents an innovative electrochemical sensor for histamine detection using a polyurethane modified electrode. The polyurethane film used in this study has porous surface that offers high adsorption capacity for the histamine. The reaction between PU electrode and histamine was an irreversible oxidative. A satisfactory detection limit was obtained in the linear concentration range of 10^{-4} – 1 $mmol \cdot L^{-1}$. Moreover, the simplicity of polyurethane film deposition and availability of the electrode modifier polymer, satisfactory accuracy, precision and errors that can be denied exhibited a great potential of validated and developed method for histamine detection in real samples. Although the modification of polyurethane reduced the electrochemical response of screen–printed electrode, nevertheless the capability of polyurethane film to adsorb histamine by physical reaction causing histamine can be oxidized compared to unmodified SPE. Several amines that have possibility to disturb histamine detection such as aniline, xanthine, hexamine, cadaverine and putrescine did not affect the electro – analytical response of histamine. The obtained results show that the validated and developed DPV method propose a satisfactory electro – analytical approach for qualitative and quantitative of histamine level in fish samples.

5. References

[1] Liu Y, Han F, Liu Y and Wang W 2020 *Food Anal. Methods* **13** 911
[2] Angulo M F, Flores M, Aranda M and Henriquez – Aedo K 2020 *Food Chem.* **309** 125689
[3] Alizadeh N, Kamalabadi M and Mohammadi A 2017 *Food Anal. Methods* **10** 3001
[4] Bogdanovic T, Petricevic S, Brkljaca M, Listes I and Pleadin J 2020 *Food Addit. Contam. A* **37** 815
[5] Jia W, Zhang R, Shi L, Zhang F, Chang J and Chu X 2020 *Food Chem.* **321** 126723
[6] Plakidi E S, Maragou N C, Dasenaki M E, Megoulas N C, Koupapris M A and Thomaidis N S 2020 *Foods* **9** 609
[7] Zhang W, Wang X, Yang S, Niu Q, Wu L, Li Y and Zhou J 2020 *Biomed. Chromatogr.* **34** 4740
[8] Kamankesh M, Mohammadi A, Mollahosseini A and Seidi S 2019 *Anal. Methods* **11** 1898
[9] Wojnowski W, Namiesnik J and Plotka – Wasylka J 2019 *Microchem. J.* **145** 130
[10] Papageorgiou M, Lambropoulou D, Morrison C, Namiesnik J and Plotka-Wasylka J 2018 *Talanta* **183** 276
[11] Ho L S J, Fogel R and Limson J L 2020 *Talanta* **208** 120474
[12] Kacar C, Erden P E, Dalkiran B, Inal E K and Kilic E 2020 *Anal. Bioanal. Chem.* **412** 1933
[13] Apetrei I M and Apetrei C 2016 *Sensors* **16** 422
[14] Verma N, Hooda V, Gahlaut A, Gothwal A and Hooda V 2020 *Crit. Rev. Biotechnol.* **40** 1
[15] Akhoundian M, Ruter A and Shinde S 2017 *Sensors* **17** 645
[16] Stojanovic Z S, Mehmeti E, Kalcher K, Guzsvan V and Stankovic D M 2016 *Food Analy. Methods* **9** 2701
[17] El-Nour K M A, Salam E T A, Soliman H M and Orabi A S 2017 *Nanoscale Res. Lett.* **12** 231
[18] Roales J, Pedrosa J M, Guillen M G, Lopes-Costa T, Pinto S M A, Calvete M J F and Pereire M M 2015 *Sensor Actuat. B Chem.* **210** 28
[19] Baig N, Sajid M and Saleh T A 2019 *Trend Anal. Chem.* **111** 47
[20] Butwong N, Khajonklin J, Thongbor A and Luong J H T 2019 *Microchim. Acta* **186** 714
[21] El-Raheem H A, Hassan R Y A, Khaled R, Farghali A and El-Sherbiny I M 2020 *Microchem. J.* **155** 104765
[22] Koita D, Tzedakis T, Kane C, Diaw M, Sock O and Lavedan P 2014 *Electroanalysis* **26** 2224
[23] Puthongkham P, Lee S T and Venton B J 2019 *Anal. Chem.* **91** 8366
[24] Degefu H, Amare M, Tessema M and Admassie S 2014 *Electrochim. Acta* **121** 307
[25] Keow C M, Bakar F A, Salleh A B, Heng L Y, Wagiran R and Bean L S 2007 *Food Chem.* **105** 1636
[26] Romaskevic T, Budriene S, Pielchowski K and Pielchowski J 2006 *Chemija* **17** 74
[27] Telsnig D, Kalcher K, Leitner A and Ortner A 2013 *Electroanalysis* **25** 47
[28] Alonso – Lomillo M. A, Dominguez – Renedo O, Matos P. and Arcos – Martinez M J 2010 *Anal. Chim. Acta* **665** 26
[29] Geto A, Tessema M and Admassie S 2014 *Synth. Met.* **191** 135
[30] Young J A, Jiang X and Kirchhoff J R 2013 *Electroanalysis* **25** 1589

Acknowledgments

The authors wish to thank Department of Chemical Sciences, Universiti Kebangsaan Malaysia (UKM) for providing research facilities and Universiti Kebangsaan Malaysia for the financial support through its project grant no. GGP-2019-021.

Sulfate (SO_4^{2-}) removal by electrocoagulation process under combined electrical connection of electrodes

A A Al-Raad[1,2] and M M Hanafiah[1,3*]

[1] Department of Earth Sciences and Environment, Faculty of Science and Technology, Universiti Kebangsaan Malaysia, Bangi 43600, Selangor, Malaysia
[2] Ababil School, Al-Muthanna Education Directorate, Samawa 66001, Iraq
[3] Centre for Tropical Climate Change System, Institute of Climate Change, Universiti Kebangsaan Malaysia, Bangi 43600, Selangor, Malaysia

*Corresponding E-mail: mhmarlia@ukm.edu.my

Abstract. Inorganic compounds in water can have detrimental effects on human health and the environment due to the high toxicity level of these ionic contaminants. This study assessed the efficiency of electrocoagulation process for removing sulfate (SO_4^{2-}). The technology of electrocoagulation depends mainly on electrical applied that produce coagulant species in a certain position via electro-dissolution of sacrificial anodes which are often made of iron or aluminum. EC process illustrated great potential as a vital method in eliminating numerous types of contaminants including inorganic contaminants at a lesser cost, and ecologically friendly technique. In the present study, aluminum materials were utilized in both cathode and anode electrodes. Water samples were obtained from Sawa Lake, Al-Muthanna Province located in Iraq. Electrocoagulation formations with static electrodes were used under mutual electrical connection. The effects of the different variables such as pH, current density, inter electrode distance, reaction time and stirring speed were scrutinized to obtain a higher removal of SO_4^{2-}. Preliminary outcomes exhibited the following optimal and functional conditions; pH = 8, current density = 0.8A, reaction time (RT) = 80 min, IED = 1 cm, temperature = 27 °C and agitation speed = 500 rpm. The maximum removal efficiency of SO_4^{2-} is 88 %. The present statistical rates proved the effectiveness of EC method in terms of removing salts from lake water.

Keywords: Electrocoagulation, Cathode and anode electrodes, Desalination, Water purification, Salt water

Track Name: Advanced Technology and Renewable Energy

1. Introduction

Climate change and rapid population growth which led to water paucity have a disastrous effect on the ecosystem and human health [1]. It was proved in the literature that the desalination process had a fundamental effect to yield sufficient water that may sustain large population in dry areas. In fact, seawater desalination via reverse osmosis (RO) has been established to be an effective method in removing high concentration of salts such as sulfate (SO_4^{2-}), and other salinities [2]. In the recent years, the technology of RO was considered as the most prevalent system for desalination [2]. Nevertheless, numerous operations in the system of desalination influence the performance of reverse osmosis membranes in treatment plant by destroying the membrane system. The differential pressure in RO can be noticed through abrasion processes such as degeneracy due to time or cleanings, oxidation processes, fouling or backpressure. These impacts are associated with a reduction in the salt rejection and an upsurge in the flow rate which in turn cause to growing operation budgets [3, 4]. The electrocoagulation (EC) method is also considered as one of the utilized alternative methods to treat wastewater. This method was effectively utilized in the wastewater plants and yielded promising outcomes in terms of eliminating salts, pollutants and various types of contaminants [5].

Recently, the EC process has attracted the attention of many academics. In the real situation, they could apply this process to remove different kinds of salts from wastewater. The high capacity of this technology has paved the way for numerous authors to apply this process to remove various ions, such as, heavy metals [6], bromide (Br^-) [7], chloride ion (Cl^-) [8], fluoride [9], and some elements namely lead and copper [10]. The core-mechanism of EC process depends on generating the coagulants in situ via dissolution of the electrodes-substances. The application of various electric currents on the sacrificial (anodes and cathodes) metal electrodes involves immersing them in polluted water in an EC reactor to dissolve metallic substance and induce coagulants in situ. More accurately, EC technique is utilized in delivering metallic hydroxide to water via coagulants by electro-dissolution of applied substances [5]. During EC method, particularly when electrodes have been used, the Al electrode dissolved and acted as a coagulate aid which leads to induce hydrogen gas at the cathode (denoted in Equations (1 and 2). Various aquatic aluminum species are produced, during the dissolution of Al at the anode. These kinds of Al serve as the coagulants by merging with pollutants to generate insoluble-hydroxides, which can be removed from the solution [6], as presented in Equations (1 and 2) below:

$$Al_{(s)} \rightarrow Al^{3+}_{(aq)} + 3e^- \qquad \text{anode} \qquad (1)$$

$$2H_2O + 2e^- \rightarrow H_{2(g)} + 2OH^-_{(aq)} \qquad \text{cathode} \qquad (2)$$

When Al ions produced at the anode are solvable and will thereafter react with the hydroxide ions (OH) induced at the cathode, subsequently the Al hydroxides $Al(OH)_3$ are formed as presented in Equation (3) below:

$$Al^{3+} + 3OH^- \rightarrow Al(OH)_3 \qquad (3)$$

The eco-friendly trend of the EC made it highly demanded technology since it generates a low level of waste which can simply being eradicated via purification with the aid of produced bubbles, no chemicals needed, low operating costs, as well as can be utilized with an amalgamation of renewable energies from wind or solar powers [6]. The main substances that were applied in this technique comprise iron and aluminum [11]. The purpose of the present study is to assess the effectiveness of EC technique in terms of removing the high concentration of SO_4^{2-} from saline water. The study also determines the optimal parameters for elimination of suphate (SO_4^{2-}) using EC configurations with static electrodes under combined electrical connections.

2. Materials and Method

2.1 Saline water samples

The samples of brackish water were collected from Sawa Lake, located in Al-Muthanna province in Southern Iraq between 31° 18' N (latitude) and 45° 00' E (longitude) [12]. The concentrations of SO_4^{2-} was measured using ion-chromatography (IC, ICS-2100, Dionex, Sunnyvale, CA, USA). The SO_4^{2-} concentrations was analysed using the procedures described in the standard-method [13].

2.2 Experimental design

Figure 1 illustrates the experimental setups of electrocoagulation procedure utilized in the present study. The experimental design comprises of a batch reactor made of glass. The volume of treatment water was 1 L for each run. Two aluminium plates were linked in the monopolar electrodes arrangement (MP-P) with the main power supply. Both of the electrodes were at plates configuration with equal dimensions 4x10 cm^2 (length and width) and 0.1cm thickness. The two cells of anode and cathode electrodes were connected to the positive and negative terminal with DC power supply. A magnetic field was applied to control the stirring speed. The voltage was adjusted to 6 V to each run.

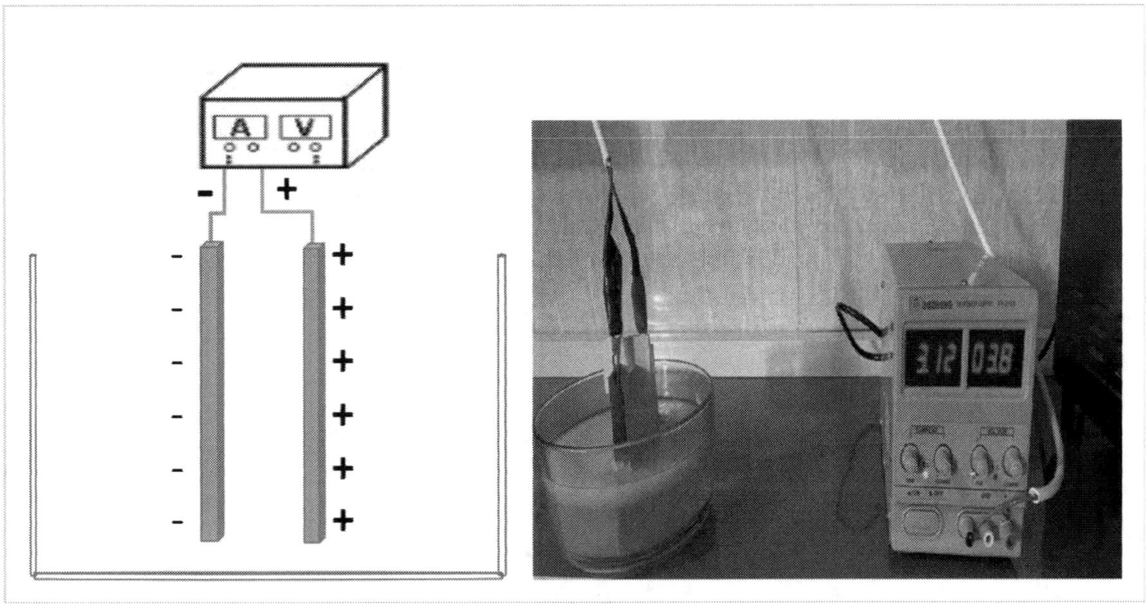

Figure 1. Experimental setup of the EC method.

Six electrical densities (0.5, 1, 1.5, 2, 2.5 and 3 mA/cm^2) were examined to gauge the effectiveness of each ratio on the coagulant method. The ammeter (Aswar DT830D, China) was employed to assess the applied current. The volt-meter was utilized to count the applied voltage by correlation of the voltmeter in parallel with the cell, and maintained at 6 V as a fixed rate for each run.

Numerous velocities (250, 350, 500, 650 and 750 rpm) were tested at baseline temperature (25-27 °C). The operating times (20, 40, 60, 80 and 100 minutes) were tested. Inter-space between cathode and the anode was fixed at three spaces 10, 15 and 20 millimeters. The experiments were also carried out at different pH values ranged from 5 to 11, with the initial pH value was 8.20. In fact, pH was steadily adjusted by adding sodium hydroxide (NaOH) and hydrochloric acid (HCl) during the experiment. Water circulation (WiseCircu ModelWCR-P6) was employed to fix the temperature between 25-45 °C during the electrocoagulation experiment. A 1-liter amount of brackish water was utilized in every process. Distilled water with a 5 % hydrochloric acid solution for a period of 10 minutes was added for internal cleaning of the reactor and electrodes. The concentrations of SO_4^{2-} was scrutinized for pre-and post-experiment. After the completion of the experiments, the best operation conditions were repeated three-times for validity and to ensure that the electrocoagulation method works competently for the treatment of brackish water. The following equation was used to compute removal efficiencies (RE %) for each of SO_4^{2-}.

$$\text{RE } (\%) = \frac{C_0 - C}{C_0} . 100 \qquad (4)$$

Where C_0 represents salts concentration in pre-treatment and C represents salts concentration in post-treatment.

3. Results and Discussion

3.1 Effect of current density and reaction time

The essential parameters that influence the performance of the EC and electrodes lifetime are current density and reaction time, thus the overall operating cost is influenced as well [5, 14]. It is worthy to mention here that the most vital parameters influencing the operating costs in the EC method are the electrode substances utilized and wasted electrical energy. The variation in removal efficiency of SO_4^{2-} was gauged over a fixed time (80 min) at numerous current densities (I = 0.5, 1.0, 1.5, 2.5, 2.5 and 3).

Figure 2 shows the outcomes of the tests of the applied current density (mA/cm^2). During the 80 min of reaction time, it was found that an upsurge in current density from 0.5 to 2.0 mA/cm^2 caused to an increase in the elimination efficiency of SO_4^{2-} from 52 % to 83 %. On the other hand, the low efficiency elimination rate was gained via the upsurge in current density from 2.0 to 3.0 mA/cm^2 during the same reaction time. It was noticeable that the increase of the current density to 2.5 and 3.0 mA/cm^2 (above the optimum current value of 2.0 mA/cm^2) resulted in producing a high number of hydroxyl radicles, which led to an increase in the pH of the solution. It is called that the coagulant breaks down in high pH, which then results in a reduction in the performance of the EC method. In this study, experiments were carried out over different reaction times ranged from 20 to 100 minutes. The samples were taken at intervals every 20 minutes. The increases of time from 20 to 80 minutes resulted in the enhancement of the elimination percentage of SO_4^{2-} from 63 % to 83 %.

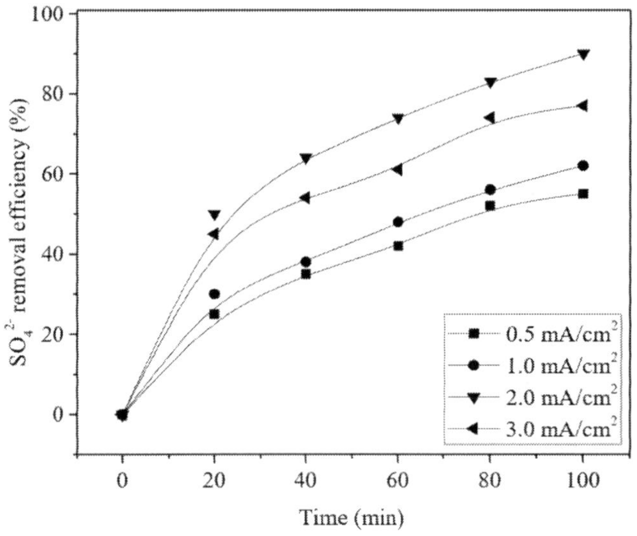

Figure 2. Effect of reaction time and current density on removal of SO_4^{2-}.

Figure 2 shown, there is a minor increase in the removal process when the reaction time is increased from 80 min to 100 minutes. Further, that increase in reaction time led to an increase in costs of electrical consumption and electrode materials. Therefore, 80 min was selected to reduce power consumption and lower operational costs. Additionally, the elimination rate was noticeably decreased when the reaction time becomes more than 100 minutes. The main cause of this decline in the efficiency is the release increase of metal ions in the solution [15]. Thus, 2.0 mA/cm^2 and 80 min were designated as the optimal operational components to decrease power consumption and lower operational costs. These results are accordance with the outcomes of that obtained in previous literatures [13, 16].

3.2 Effect of pH

The chemistry of the aqueous medium and conductivity are considered as vein of the EC mechanism. On the other hand, the impact of pH solution on electrocoagulation process has been examined in previous studies [6, 9]. In the context of the present study, the pH solution impact on SO_4^{2-} removal was examined by regulating the initial pH from 5 to 11 with the best conditions: current density = 2.0 mA/cm^2 and a RT of 80 minutes. The elimination outcomes exhibited the performance of EC method was improved as a consequence of raised pH from 6 to 8 as shown in Figure 3 with the removal efficiency of SO_4^{2-} from 84 % to 86 %. The raises observed in the pH levels may lead to water electrolysis, which resulted in generating of hydrogen and (OH) ions.

It was noticeable also that the upsurge in the value of pH solution to more than 8.5 leads to decreasing the elimination efficiencies. The indices of higher efficiency of EC technique are influenced by the optimal pH range of 6.5 to 8.4, which is close to the optimal pH for $AlOH)_{3(s)}$ formation. Flocs of $Al(OH)_{3(s)}$ can facilitate the absorption process of soluble organic molecules and trapping colloidal particles rapidly since they have large surface areas. It was proved that at low pH, ranging from 5 to 6, cationic soluble species Al^{3+} and $Al(OH)^{2+}$ are predominant [17]. When the pH is between 7 and 9, different aluminum-based polymeric and monomeric species are generated, which are finally transformed into unsolvable shapeless $Al(OH)_{3(s)}$ through complex polymerization/precipitation kinetics. When the pH is higher than 10, the soluble $Al(OH)^{4-}$ anion concentration rises at the expense of $Al(OH)_{3(s)}$ [6].

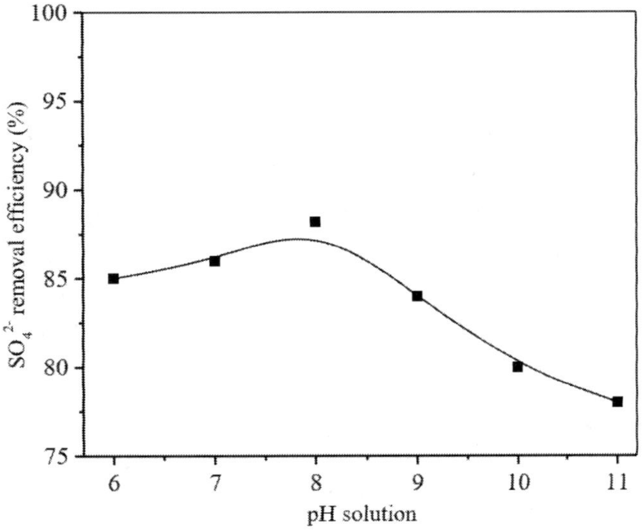

Figure 3. Impact of pH on SO_4^{2-} removal at the best conditions (I = 2.0 mA/cm^2, reaction time = 80 minutes).

3.3 Effect of inter electrode space

Many previous studies revealed the impact of the inter-electrodes space in electrocoagulation [5, 15]. The resistance between the electrodes can be increased through increasing the distance between the two electrodes. The resistance between the electrodes can be improved by increasing the distance between the two electrodes. However, this resistance requires a higher potential to be conquered and thus causing an upsurge in the treatment costs. From the theoretical perspective, the upsurge in distance between the electrodes resulting in a minimal interplay between the hydroxide polymers and ions [18].

In the current context of this study, the impact of inter space between electrodes were tested utilizing 10, 15 and 20 millimeters via employing optimal conditions of current density = 2.0 mA/cm^2, TR = 80 minutes, and pH = 8. Figure 4 indicates the optimal performance through using 1 millimeter. On the other hand, the treatment removal rate was declined via the expansion of the distance from 15 to 20 millimeters, with a drop of 86 % to 84 % for SO_4^{2-}. These outcomes are compatible with previous studies studeis [9, 19].

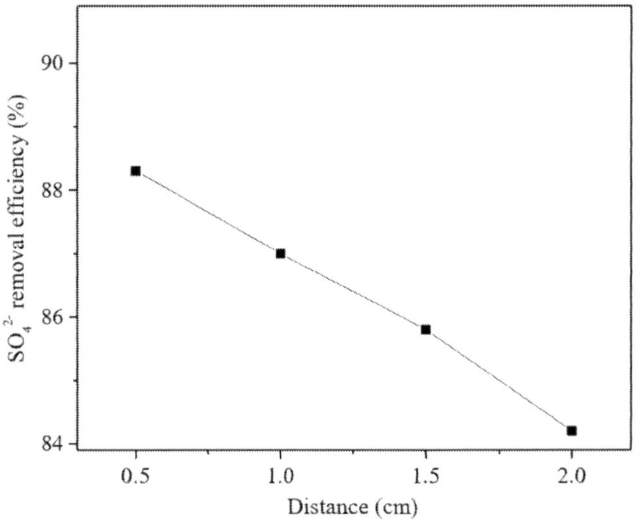

Figure 4. Influence of inter-electrode gap under best conditions (current density = 2.0 mA/cm^2, reaction time = 80 min, pH = 8).

3.4 Effect of agitation speed

As an important parameter, agitation speed plays a vital role in the electrocoagulation method [20, 21]. The upsurge in the agitation rate leads to increase mixing of the solution, subsequently conglomerating the pollutants and the precipitation as well as easier mass transfer. However, the opposite can occur particularly when the agitation rate is raised more than required. Whereas, pollutants are formed within the reactor getting degenerated inducing tiny flocks that are hard to separate from the solution. This leads to reducing the treatment efficiency [16, 20].

To scrutinize the impact of the agitation speed on SO_4^{2-} elimination efficiency in this study, agitation speed of 250, 350, 500, 650, and 750 rpm were done utilizing the optimal conditions of current density= 2.0 mA/cm^2, *RT* = 80 minutes, pH = 8, and inter electrode space = 1 millimeter. The highest elimination capacity was gained when the speed was set to 500 rpm as presented in Figure 5. The outcomes exhibit the SO_4^{2-} elimination augmented SO_4^{2-} from 83 % to 89 % for an increase in agitation speed of 250 rpm to 500 rpm. On the other hand, when the rate of agitation speed was augmented from 500 rpm to 750 rpm, the lower elimination efficiency was attained, where the SO_4^{2-} elimination reduced to 78 %. These finding were discussed in the previous literatures [16, 20].

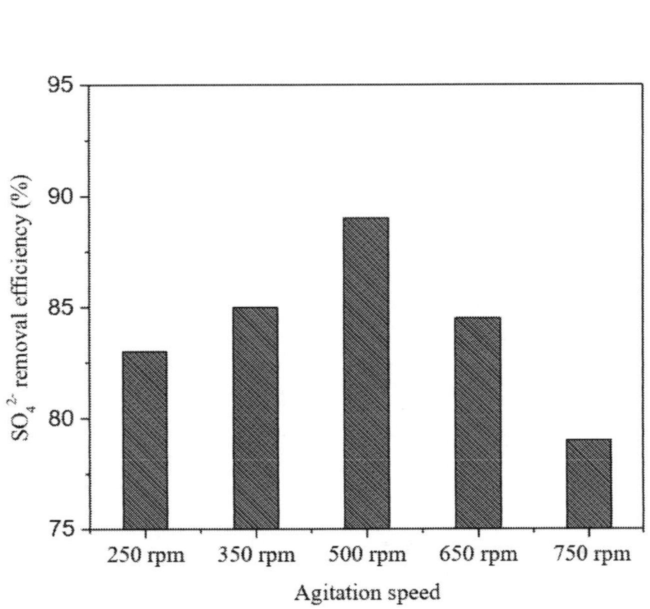

Figure 5. Influence of agitation speed under best conditions (current density = 2.0 mA/cm^2, reaction time = 80 min, pH = 8, inter electrode gap = 1 cm).

4. Conclusions

The present study assessed the feasibility of EC method in removing SO_4^{2-} from salt water utilizing aluminum electrodes. Furthermore, the effects of reaction time (minutes), current density (mA/cm^2), pH, stirrer speed (rpm), as well as the gap between the anode and cathode (millimeter) on removal of SO_4^{2-} were scrutinized. The best conditions were: I = 2.0 mA/cm^2, RT = 80 minutes, pH = 8, IED = 1 millimeter and agitation speed = 500. The EC method under the combined electrical linking of Al electrodes, was proved to be very effective for removing salts from brackish water. Utilizing optimal circumstances of the EC process, the removal efficiency was obtained at 89 % for SO_4^{2-}. The outcomes of the present study show the technical viability of electrocoagulation as a reliable method for the removal of SO_4^{2-}.

References

[1] Mohd Nizam NU, Hanafiah MM, Mahmoudi E and Halim AA 2021The removal of anionic and cationic dyes from an aqueous solution using biomass-based activated carbon *Scientific Reports* **11** 8623

[2] Elimelech M and Phillip WA 2011 The future of seawater desalination: energy, technology, and the environment J. science 333 6043 712-7

[3] Li S, Cai Y-H, Schäfer AI and Richards BS 2019 Renewable energy powered membrane technology: A review of the reliability of photovoltaic-powered membrane system components for brackish water desalination *Appl. Energy* **253** 113524

[4] Shaikh MM, Al Suhaimi AO, Hanafiah MM, Ashraf MA and Harun SN 2018 Study on migration of phenolic and volatile organic compounds from plastic pipes used in plumbing home networks into tap water *Desalin. Water Treat.* **112** 344-50

[5] Garcia-Segura S, Eiband MMS, de Melo JV and Martinez-Huitle CA 2017 Electrocoagulation and advanced electrocoagulation processes: A general review about the fundamentals, emerging applications and its association with other technologies *J. Electroanal. Chem.* **801** 267-99

[6] Al-Shannag M, Al-Qodah Z, Bani-Melhem K, Qtaishat MR and Alkasrawi M 2015 Heavy metal ions removal from metal plating wastewater using electrocoagulation: Kinetic study and process performance *Chem. Eng. J.* **260** 749-56

[7] Ramyaa T, Premkumarb P, Thanarasub A, Velayuthamb K, Dhanasekaranc A and Sivanesanb S 2019 Degradation of pesticide-contaminated wastewater (coragen) using electrocoagulation process with iron electrodes *Desalin. Water. Treat.* **165** 103-10

[8] Donneys-Victoria D, Marriaga-Cabrales N, Machuca-Martínez F, Benavides-Guerrero J and Cloutier SGJJoWPE 2020 Indigo carmine and chloride ions removal by electrocoagulation. Simultaneous production of brucite and layered double hydroxides *J. Water Process Eng.* **33** 101106

[9] Betancor-Abreu A, Mena V, González S, Delgado S, Souto R and Santana J 2019 Design and optimization of an electrocoagulation reactor for fluoride remediation in underground water sources for human consumption *J. Water Process Eng.* **31** 100865
[10] Bouguerra W, Barhoumi A, Ibrahim N, Brahmi K, Aloui L and Hamrouni B 2015 Optimization of the electrocoagulation process for the removal of lead from water using aluminium as electrode material Desalin. Water Treat. 56 2672-81
[11] Demirbas E and Kobya M 2017 Operating cost and treatment of metalworking fluid wastewater by chemical coagulation and electrocoagulation processes *Process. Saf. Environ. Prot.* **105** 79-90
[12] Al-Raad AA, Hanafiah MM, Naje AS and Ajeel MA 2020 Optimized parameters of the electrocoagulation process using a novel reactor with rotating anode for saline water treatment *Environ. Pollut.* **265** 115049
[13] Rice E, Baird R, Eaton A and Clesceri L 2012 Standard Methods Fort he Examination of Water and Wastewater *American Public Health Association, Washington* **2540** 2-66
[14] Abdel-Gawad SAB, Amin M. Omran, Kawther A. Mokhtar and Mohamed M 2012 Removal of some pesticides from the simulated waste water by electrocoagulation method using iron electrodes *Int. J. Electrochem. Sci.* **7**(8) 6654-65
[15] A Al-Raad A, Hanafiah MM, Naje AS, Ajeel MA, O Basheer A and Ali Aljayashi T, et al 2019 Treatment of Saline Water Using Electrocoagulation with Combined Electrical Connection of Electrodes *Processes* **7** (5) 242
[16] Sharma L, Prabhakar S, Tiwari V, Dhar A and Halder A 2021 Optimization of EC parameters using Fe and Al electrodes for hydrogen production and wastewater treatment *Environ. Adv.* **3** 100029
[17] Wellner DB, Couperthwaite SJ and Millar GJ 2018 Influence of operating parameters during electrocoagulation of sodium chloride and sodium bicarbonate solutions using aluminium electrodes *J. Water Process Eng.* **22** 13-26
[18] Lanzarini-Lopes M, Garcia-Segura S, Hristovski K and Westerhoff P 2017 Electrical energy per order and current efficiency for electrochemical oxidation of p-chlorobenzoic acid with boron-doped diamond anode *Chemosphere* **188** 304-11
[19] Foudhaili T, Lefebvre O, Coudert L and Neculita CM 2020 Sulfate removal from mine drainage by electrocoagulation as a stand-alone treatment or polishing step *Miner. Eng.* **152** 106337
[20] Bayar S, Yıldız YŞ, Yılmaz AE and İrdemez Ş 2011 The effect of stirring speed and current density on removal efficiency of poultry slaughterhouse wastewater by electrocoagulation method *Desalination* **280** (1-3) 103-7
[21] Manikam MK, Halim AA, Hanafiah MM and Krishnamoorthy RR 2019 Removal of ammonia nitrogen, nitrate, phosphorus and COD from sewage wastewater using palm oil boiler ash composite adsorbent *Desalin. Water Treat.* **149** 23-30.

Acknowledgement
Marlia M. Hanafiah was supported by the Universiti Kebangsaan Malaysia (DIP-2019-001 and GUP-2020-034).

Microbial fuel cells constructed from the integration of anaerobic and aerobic sanitary landfill site systems

N Abdul Harris[1] & A Abdul Halim[1]*

[1]Department of Earth Sciences and Environment, Faculty of Science and Technology, Universiti Kebangsaan Malaysia, 43600 UKM Bangi, Selangor, Malaysia

*Corresponding E-mail: azharhalim@ukm.edu.my

Abstract. The world's main source of energy now is fossil fuels but the demand for power is increasing. In addition, the burning of fossil fuels produces harmful greenhouse gases and has a significant impact on the environment. The fuel cell system in this study is based on the aerobic and anaerobic integration system are used in most waste disposal methods in Malaysia. It is a system of electrochemistry results from the oxidation of organic matter that transfers electrons to carbon graphite This research is to study the effectiveness of the method generating electricity from micro-fuel cells produced from leachate wastewater and it is also conducted to identify microbial activity using a double chamber system. The food waste obtain is divided into double chambers which is aerobic and anaerobic. Digital readings using a multimeter are performed for ten to thirteen days continuously to obtain the highest reading results for voltage and electric current. The measurement of the highest reading result on the 11th day recorded a reading as high as 146.8 mV at 2000 mV while the current reached 28 μA at 2000 μA. This study has proven that there is the production of electrical sources from the activity of organisms present in food waste using microbial fuel cell systems. The result show that food waste and cattle manure produce the highest voltage and current. This has provided an opportunity to explore alternative ways of generating electricity according to the environment and conditions of each region.

Keywords: Microbial fuel cell, aerobic landfill, anaerobic landfill, semi-aerobic landfill.

Track Name: Advanced Technology and Renewable Energy

1. Introduction

Microbial fuel cells convert chemical energy produced by the oxidation of organic or inorganic compounds by a sequential reaction in which electrons are transferred to terminal electron receivers to generate electrical current with the help of microbes. Fuel cells are used to produce electrical electrochemistry from many different chemicals such as hydrogen gas and methanol through the oxidation of fuel at the anode and chemical reduction at the cathode. Microbial fuel cells are unique because they do not require the use of a metal catalyst at the anode. Instead, they use microorganisms that oxidize organic matter and transfer electrons to the anode. The use of microbial fuel cells as an alternative source for power generation is considered a reliable, clean, efficient process, which use renewable methods and does not produce toxic by-products. Thus, recently studies on microbial fuel cells have demonstrated powerful technologies for the recovery and in-situ conversion of chemical energy into electrical energy [1].

High energy resource requirements in future are driving more research on new and more environmental approach technologies. Until now, fossil fuels have met most of the overall energy needs. This has resulted in a reduction in fossil fuel resources and leads to ecological imbalances. In addition, the burning of fossil fuels produces a lot of carbon dioxide, which is a major dangerous greenhouse gas and has shown alarming effects on the environment. According to some studies, excessive consumption of fossil fuels will lead to two unavoidable issues namely depletion of fossil fuel resources and global environmental problems [2].

Developed and developing countries face challenges in providing energy and addressing the environmental problems generated by the use of fossil fuel technology. Methods for obtaining energy sources from biomass such as ethanol, methane, and hydrogen have been created using environmental friendly technologies and some of these technologies have been applied practically. It is also very important to study new technologies that can generate energy in various forms according to the environment and conditions in each region. Apart from the above technologies, biofuel cells use microorganisms and enzymes, which can produce renewable electrical energy from organic matter contained in food waste are beginning to attract attention as a way to obtain sustainable energy [3]. It has not been practically use but if more studies are done to identify this new method thus without any problem electrical by-products can be obtained directly from this micro fuel cell method. Previous studies have been conducted and have shown success in research of microbial fuel cells constructed using graphite felt as the anode and platinum coated as the cathode [4]. Another study showed that fuel cell installations in food processing plants could potentially generate 330 kW per day of power at 7,500 kg of waste material based on 30 % efficiency [1]. Yet the two studies have one thing in common which is to have a space where it puts the anode and cathode together.

To achieve the goals of this research, a specific objective has been outlined, the first is to compare the electric current from the fuel cell generated from the leachate water of food waste by anaerobic and aerobic integration methods. The second is to determine the electrical voltage of the fuel cell resulting from the leachable water of food waste by anaerobic and aerobic integration methods.

2. Materials and Methods

2.1 Preparation of food waste and cattle manure

The sampling location for this research is located at the Faculty of Science and Technology Cafe, UKM Bangi. As in Figure 3.1, the experiment was started by collecting food waste that is already available in the waste bin and a total of five kilograms of food waste for each method is required to ensure that this research study runs smoothly. Then the sample were separated from any plastic so as not to affect the parameters when analysed later. Food waste were blended so that it becomes perfectly blended and more easily digested by bacteria and microorganisms where it will speed up the process of electricity generation by micro fuel cells. Cattle manure in this study was taken from a cattle farm near UKM Bangi.

2.2 Microbial fuel cell

There are two types of methods that were used in this study, namely semi aerobic method and full aerobic method. The semi aerobic method is a condition where there are two chamber that will form the circuit namely the anode space and the cathode space. The anode space was represented by an anaerobic chamber where no air will penetrate this area while the study is in progress. While the cathode space was represented by a semi aerobic chamber which is the area where air entered and reacted with food waste and cattle manure.

The cathode space was also replaced with a full aerobic chamber to test the efficient rate of a circuit. A full aerobic chamber is an area where each space in the container representing the cathode will be filled with air so that all food waste can react when air enters the cathode space.

Food waste were placed into both chambers representing the cathode chamber and the anode chamber as shown in Figure 1. The anode space, which is the anaerobic chamber, were closed so that no oxygen can penetrate the area. While for aerobic chamber there are two types of methods that were used, namely semi aerobic and full aerobic. The full aerobic method where airways were channelled all the way to the bottom of the surface of the chamber. While for the semi -aerobic method, the air was only channelled at the top. This is done to see the rate of oxygen efficiency if it is fully channelled or will not directly affect the rate of voltage and current generation.

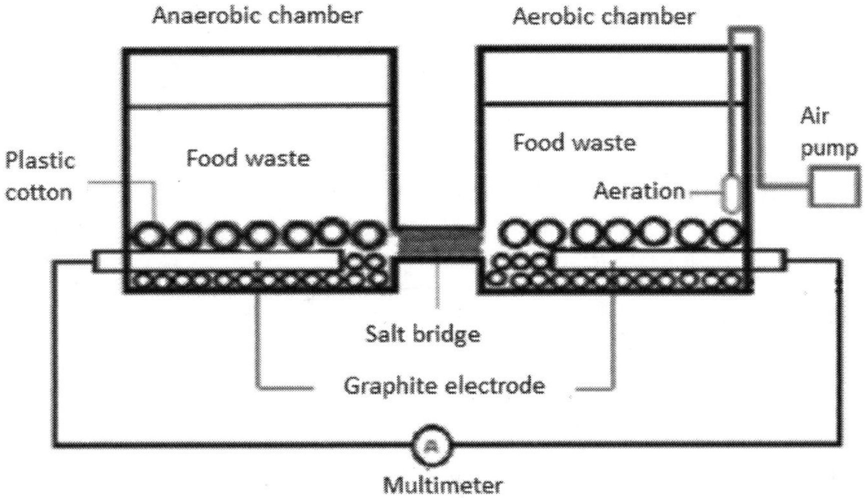

Figure 1. Microbial fuel cell diagram model represents for integration of aerobic and anaerobic landfill design.

3. Results and Discussion

Figure 2 has shown the voltage flow of semi aerobic method which uses food waste generates a voltage of 59.70 mV as the highest total voltage but the voltage generation for this method is faster yet cannot generate more voltage compared to aerobic chamber (101.2 mV) which uses food waste. This is because in a microbial fuel cell, electrons released by bacteria from substrate oxidation in the anode space are transferred to the cathode space through a conductive material. In the cathode space electrons are combined with oxygen and protons are dispersed through the proton exchange membrane or in this study the salt bridge is replaced as a channel. Microbial fuel cells require the constant release of electrons at the anode and the use of electrons at the cathode. The metabolic energy gain achieved for bacteria is directly related to the difference between the anode potential and the redox potential of the substrate. Electrode oxidation organisms use electrons from the cathode to reduce material in the cathode space. In the aerobic chamber, microorganisms can reduce oxygen to water. In an anaerobic environment, nitrate or sulfate can be reduced to nitrite, nitrogen, or sulfur ions. Another bacterial reducing potential is the conversion of carbon dioxide into methane or acetate [5]. The advantage of using this anaerobic and chemical conversion method is that it demonstrates high efficiency, stability and elimination of voltage inhibition that may occur to methanogen [6]. In terms of nutrient degradation and utilization it is clear that fuel cells can use the nutrients in the annolite efficiently and this was done in a short study period.

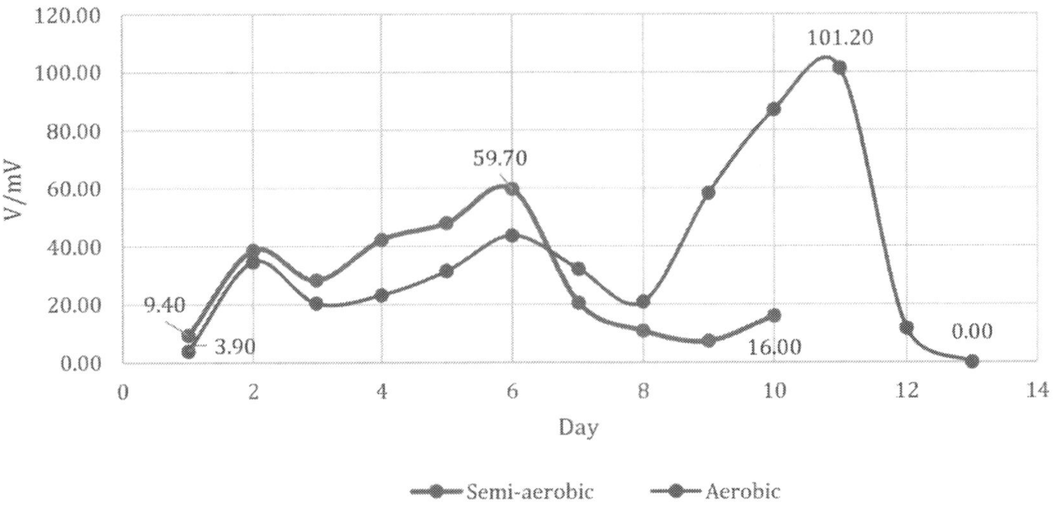

Figure 2. Voltage comparison between semi aerobics and aerobics microbial fuel cell for food waste.

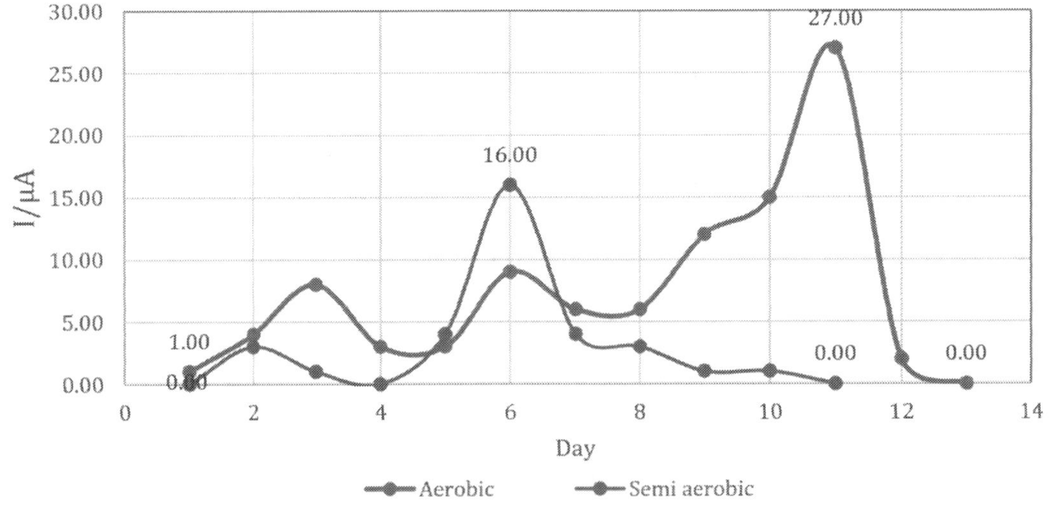

Figure 3. Current comparison between semi aerobics and aerobics microbial fuel cell for food waste.

Figure 3 shows the current flow graph of the semi aerobic method against the flow of the aerobic method. The flow of the aerobic method is almost similar to the flow of the semi aerobic method but has a different current value because at first it is the same at a current value and then both flows ascending for the second and third day, however decreased on the fourth and fifth day but the flow of semi aerobic method increased drastically until it reached the highest value on the sixth day which is 16.00 µA. The current flow of the full aerobic method increased from the sixth day and reached the highest current value on the eleventh day at a value of 27.00 µA which is consistent with the value of voltage generation 101.2 mV in Figure 2. However, after the current reached its peak value, it decreased drastically and finally recorded a value of 0.00 µA on the thirteenth day. The flow rate for both current and voltage has an internal resistance for the fuel cell but if the current flow increases the rate of voltage generation will be higher. The decrease in electron emission is due to a decrease in the microbial population when nutrients are depleted and ultimately results in lower current generation.

4. Conclusion

In this study food waste as a substrate, contains organic matter available for energy generation from microbial activity. Graphite carbon rods as electrodes and salt bridges as membranes are used to produce more microbial fuel cells generating more voltage and electric current. Using a salt bridge instead of a proton exchange membrane is more economical because it is cost effective and readily available. Microbial fuel cells that integrate anaerobic and aerobic landfill models produce more voltage and electric current compared to the coupling of anaerobic and semi-aerobic landfill models. Limitation of micro fuel cells is that they cannot operate at very low temperatures due to the fact that microbial reactions are slow at low temperatures. Optimal designs for micro fuel cells are still under research, and different materials for electrodes as well as more selective membranes for proton exchange are still being developed to improve fuel cell performance. The use of two types of methods, namely semi aerobic and full aerobic, can clearly show that oxygen plays an important role in ensuring that microbes can generate electricity.

References

[1] Logan B E*et al.* 2006 Microbial fuel cells: methodology and technology *Environ Sci Technol* **40** pp 5181-5192
[2] FAO, 2012 Towards the Future We Want: End Hunger and Make the Transition to Sustainable Agricultural and Food Systems. *Food and Agriculture Organization* (United Nations Rome)
[3] Wang Z, Lee T, Lim B, Choi C and Park J 2014 Microbial community structures differentiated in a single-chamber air-cathode microbial fuel cell fueled with rice straw hydrolysate. *Biotech Biofuels*
[4] Dhakal B, Joshi J 2015 Neutral Red Immobilized Graphite Felt Anodic Microbial Fuel Cell for Wastewater Treatment and Generation of Electricity *J Bioprocess Biotech* **5** 261
[5] David P Clark & Nanette J P 2016 Environmental Biotechnology *Biotechnology (Second Edition), Academic Cell*
[6] Goud, R K, Mohan, S V, 2011 Pre-fermentation of waste as a strategy to enhance the performance of single chambered microbial fuel cell (MFC) *Int. J. Hydrogen Energy* **36** (21) pp 13753–13762

Acknowledgement

The authors are thankful to the Faculty of Science and Technology, The National University of Malaysia for providing all the necessary facilities and research grant required to carry out this study (GUP-2019-042 and GP-2019-K007039).

Hydrogen production via methane dry reforming process over CoAl and CoMgAl-Hydrotalcite derived catalysts

Z Abdelsadek[1,2,*], P Chaudhari[3], J P Holgado[4], F Bali[2], D Halliche[2], O Cherifi[2], S Gonzalez-Cortes[5] and P J Masset[6,*]

[1]Institute of Electrical and Electronics Engineering, University of M'hamed-Bougara, Independence Avenue, 35000, Boumerdès, Algeria
[2]Laboratory of Chemistry and Natural Gas, Faculty of Chemistry, USTHB, B.P. 32 El-Alia, 16111 Bab-Ezzouar, Algiers, Algeria
[3]Dr D Y Patil School of Engineering, Lohegaon, Pune, India
[4]Instituto de Ciencia de Materials de Sevilla and Departamento de QuímicaInorgánica, CSIC University of Sevilla, Avda. AméricoVespucio, 49, 41092 Seville, Spain
[5]Inorganic Chemistry Laboratory, Department of Chemistry, University of Oxford, South Parks Road, Oxford OX1 3QR, UK.
[6]Koszalin University of Technology, Faculty of Mechanical Engineering, ul. Śniadeckich 2, 75-453 Koszalin, Poland

*Corresponding E-mail: patrick.masset@tu.koszalin.pl; zoulisabrina@yahoo.fr; z.abdelsadek@univ-boumerdes.dz

Abstract. $Co_{0.67}Al_{0.31}$ and $Co_{0.14}Mg_{0.54}Al_{0.31}$ hydrotalcite based catalysts were prepared by a co-precipitation method at a fixed pH=11, exhibiting a suitable hydrotalcite structure to be used as a catalyst in the reaction of the dry reforming of methane (DRM). Calcination at 450 °C provides the best conditions to prepare the most adapted structure and morphology to be later used in the DRM reaction. The samples were characterised by XRD, FTIR, SEM and it was shown that they exhibit a specific surface in the 30-70 g/cm² and a crystallite size of approximately 20 nm. The results of the TPR analysis showed clearly that CoAl-HT has better catalytic performances than CoMgAl-HT. This result can be explained by the presence of the Co^0 for the catalyst CoAl-HTc-R and the total absence in the sample CoMgAl-HTc-R. The solid CoMgAl- HTc-R requires high reduction temperature compared to CoAl- HTc-R due to the strong CoO-MgO interactions.

Keywords: dry reforming of methane, hydrotalcite, Co^0-catalyst, hydrogen, syngas

Track Name: Advanced Technology and Renewable Energy (Climate Change)

1. Introduction
Environmental issues and global warming resulting from the use and emission of greenhouse gases are two of the most important challenges of the 21st century. Therefore, reducing the emission of these greenhouse gases or at least converting thermo-industrial products is intensively investigated. The dry

reforming of methane (DRM) reaction to produce syngas ($CO + H_2$), a process extensively investigated in the last few decades [1]. Compared to the steam reforming reaction [2], DRM reaction produces syngas with lower H_2/CO ratio [1], which is more appropriate for the Fischer-Tropsch and methanol syntheses[1, 2]. Hence, the DRM reaction can contribute to the reduction of two greenhouse gases, what may render its utilization as an environmentally friendly process [1, 3]. Unfortunately, severe deactivation of catalysts, mostly caused by sintering of the nickel particle (active phase) and the filamentous carbon deposition which leads to the reactor blockage,[1, 3] hamper an industrialisation. The carbon formation on the catalytic active surface are owing to the inverse Boudouard's reaction (**Eq. 1**) and methane decomposition (**Eq. 2**)

$2CO \rightarrow CO_2 + C \qquad \Delta H_{298K} = -173\ kJ\ mol^{-1}$ (Eq (1))

$CH_4 \rightarrow C_{(s)} + 2H_2 \Delta H_{298K} = +75kJ\ mol^{-1}$(Eq (2))

The dry reforming of methane (DRM) by CO_2: ($CH_4 + CO_2 \leftrightarrow 2\ CO + 2\ H_2$, $\Delta H_{298K} = +247$ kJ/ mole) is an attractive reaction to produce H_2 as clean energy source and e-mobility [3]. This reaction can be carried out at high temperatures (1000-1200 °C) without catalysts[1, 3] but at the cost of the facility lifetime due to corrosion and materials issues. This can be partially alleviated by using supported noble metals (Ru, Rh and Pt) but not affordable industrially. Ni-based catalysts are among the best catalysts which exhibit comparable behaviour to noble metals. However, they undergo rapid poisoning and deactivation by carbon deposition and sintering. Confining of nanoparticles in core-shell and the core-sheath based structures can exert a spatial restriction, hinder their sintering and of carbon deposit [4]. "Solid phase crystallization (SPC)" materials ($ANiAl_{11}O_{19-x}$ (A= Ca, Sr, Ba, La) [5], pychlore $A_2B_2O_7$[6], perovskite ABO_3[7], solid solution type AO-BO [8], spinel AB_2O_4[5], ferrite spinel [9] and hydrotalcite[10-12] allow to obtain homogenous distributed metallic phases in a well-defined crystal structure. HT compounds are lamellar anion clays of general formula $[M_{1-x}^{2+} M_x^{3+}(OH)_2]^{x+}[A^{n-}]_{x/n}.mH_2O$, where M^{2+} and M^{3+} are the divalent and trivalent cations, A^{n-} is the interlamellar anion of valence n, $x = [M^{3+}]/([M^{3+}] + [M^{2+}]$ and $0.20 < x < 0.33$ [10, 11] are prepared by solid-phase crystallization method to obtain a pure lamellar and well crystallized structure. Calcined hydrotalcite materials form mixed-metal oxides having a high thermal stability and a great dispersion of fine metal active phase (after reduction) are well suited for DRM [10, 13-15]. Bhattacharyya et al.[16] compared supported HT catalysts as Ni/Al_2O_3 or $Ni/MgAl_2O_4$. The substitution of Ni for Mg leads to the higher Ni metallic particle dispersion. For Ni/CeO_2 catalysts, Ni particles are stabilized on the CeO_2 surface (reduction of Ce^{4+} to Ce^{3+} and the mobility of oxygen species) [17] . For semi-conductor perovskite $LaNiO_3$, amorphous phase NiO (not detected by classical XRD) was evidenced [18]. For Ni/ZrO_2 catalysts, XAS results have shown a lower coordination number of Ni in the sample treated with CO than that with H_2[19]. For NiAlMg reduced by H_2 at 850 °C, the $Ni_{1-x}Mg_xO$ remained partially unreduced depending onNi content [20]. Recently, we have shown that the thermal pre-treatment strongly influences the activity and selectivity of the catalysts during DRM reaction [12].The exchange of Ni for Co was investigated with the synthesis of CoAl and CoAlMg based hydrotalcite catalysts.

2. Materials and Experimental Techniques

2.1 Materials

CoAl-HT and CoMgAl-HT samples with M^{2+}/M^{3+} ratio equal to 2 were prepared by a co-precipitation method at a fixed pH (i.e.,11) from a solution containing the appropriate amounts of nitrate salts of Co^{2+}, $Mg^{2+}$and Al^{3+}. Sodium carbonate solution (1M) was added dropwise to the nitrate solution while the pH was maintained constant by adding NaOH solution (1M) under vigorous stirring at room temperature. The obtained slurry was hydrothermally treated at 70 °C for 15 h, washed several times using bi-distilled water until a neutral pH of the filtrate was reached. The recovered solid was dried overnight in an oven at 80 °C and crushed to get a homogeneous fine powder. Finally, the powder was calcined in the air in an oven at 450 °C for 6 h with a heating increment of 4 °C/min. The corresponding products are named M^{2+} M^{3+}-HTc. Temperature programmed reduction (TPR) profiles were determined using a TriStar 3000 V6.01A apparatus equipped with a TCD detector. Prior to reduction, specimens (approx. 20 mg) were flushed with argon in a quartz reactor tube, heated at 120 °C for 2 h under an argon flow with a heating rate of 5 °C/min and cooled afterwards to room temperature. Samples were submitted to a 5% H_2/Ar flow up to 950 °C with a 10 °C/min rate. The obtained solids are labelled M^{2+} M^{3+}-HTc-R.

The catalytic activity measurements were carried out under atmospheric pressure in a fixed-bed tubular reactor ("U" shaped), using approx. 100 mg of catalysts between two pompoms of quartz wool. Before reacting, the sample was reduced *in-situ* under hydrogen flow at 700 °C for 1h. Then, the reactor was cooled down to the desired temperature and the reactant mixture (CH_4/CO_2/Ar) was passed through the reactor. The CH_4/CO_2/Ar molar ratio was 20/20/60 with a total flow rate of 30 $mL.min^{-1}$. During the reaction testing, a cold-water trap containing acetone and CO_2 was used to condense the water formed during the catalytic test. The effluent gases from the reactor were analyzed by gas chromatography with a TCD detector and using porapak/molecular sieve. The catalytic properties were evaluated in terms of CH_4 and CO_2 conversions(Eq. 3), H_2/CO selectivity(Eq. 4)and carbon deposition(Eq. 5), as follows:

$$a_i = \frac{D^{in} X_i^{in} - D^{out}\ D_i^{out}}{D^{in} X_i^{in}} \quad (\text{Eq. }3)$$

$$S_{H_2/CO} = {ni'H_2}/{ni'CO} \quad (\text{Eq. }4)$$

$$X_c(\%) = \frac{D^{out}(X_{CH4}^{out} + X_{CO2}^{out} + X_{CO}^{out})}{D^{in}(X_{CH4}^{in} + X_{CO2}^{in})} \text{ x } 100 \quad (\text{Eq. }5)$$

where,

a_i: catalytic activity (conversion) while i represents CH_4 or CO_2.

D^{in}/ D^{out}: flow rate of the effluent in/out.

X^{in}_i/ X^{out}_i:CH_4 or CO_2 composition in/out.

m_{cat} : catalyst amount.

$n_{i'}$: rate of CO or H_2.

X_c : carbon balance.

2.2 Techniques

The chemical composition was determined by Atomic Absorption Spectroscopy (AAS) using Spectro-Analytical Instruments Horiba Jobin–Yvon, Ultima spectrometer. Before the analysis, the dissolution of 100 mg of the solids was affected in the presence of 20 ml of aqua regia, then further diluted to 100 ml with distilled water. Powder X-ray diffraction (XRD) patterns were recorded with Siemens D-501 equipment, using CuKα radiation in the 2θ range between 10 and 80°. The average crystallite sizes of Co^0 following reduction were estimated from Scherrer's formula. Specific surface measurements were carried out using a NOVA 2000e volumetric adsorption analyzer at -196 °C (liquid nitrogen). Before the analysis, the powder was degassed under vacuum at 150 °C for 2 h. Fourier transform infrared (FTIR) spectra were recorded using a Perkin Elmer model number spectrometer. Over 32 scans were taken to improve the signal-to-noise ratio in the wavelength range (400 - 4000 cm^{-1}). The analysis was performed on ca. 4 mg of powder mixed with 100 mg of dried spectroscopic KBr pressed into pellets. Thermogravimetry analysis (TGA) and differential thermal analysis (DTA) were performed using a Netzsch STA 409 PC/PG instrument under air as purge gas from room temperature up to 850 °C with a heating rate of 10 °C/min using alumina crucibles. Scanning electron microscopy images (SEM) of the products were recorded on a Hitachi S-4800 instrument equipped with a field emission filament using an acceleration voltage of 5 kV and a working distance of 10 mm.

3. Results and Discussion

3.1 Structural and textural properties of CoAl-HTc and CoMgAl-HTc compounds

The chemical compositions of the synthesized samples are summarized in Table 1. The values of the ratio between the divalent and trivalent metals were found to be close to 2 indicating that hydrotalcite formed. XRD analysis confirmed the presence of the crystallised hydrotalcite based phases (Table 2) in comparison to natural hydrotalcite compounds. By combining DTA/TGA, two thermal events were observed in the 100-350 °C temperature range. The dehydration of CoAl-HT and CoMgAl-HT occurs between 100 and 210 °C whereas the decarbonation/dehydroxylationat 250 and 350 °C, respectively. With the CoMgAl sample, the temperature area is much broader and included some solid phase transformation as it was shown by the presence of mixed oxides by XRD. In both cases, the sample underwent a mass loss of approximately 30 % and about 400 °C no thermal event was recorded. Once calcinated, the structure become partially amorphous with the presence of oxide MgO, Co_3O_4 or even mixed oxide such as $CoAl_2O_4$. FTIR spectra of the synthesized materials exhibit the general features often reported for the hydrotalcite like materials[10]. The presence of mixed oxide is important as it involved the presence of oxides with different oxidation states that affect the catalytic properties. The apparent bands recorded at ~ 3400 cm^{-1} were attributed to OH stretching vibrations in the brucite-like layers or to the adsorbed and interlayer water molecules. The band around 1625 cm^{-1} was assigned to H-O-H deformation mode confirming the presence of water molecules in the interlayer region. The BET specific surfaces were measured by nitrogen absorption and found close to ~ 77 and 30 $m^2 g^{-1}$ for CoMgAl-HT and CoAl-HT, respectively. The values of pore diameters for both solids non calcined and calcined lie between 4 and 7 nm. These values are typical of mesoporous materials. No significant pore size increase was observed with the addition of magnesium in the sample CoMgAl-HT. It turns out that the specific surface area and pore diameter are enhanced by the presence of Mg in hydrotalcite structure. From the SEM pictures, well dispersed agglomerates (crystallite size: approx. 20 nm from XRD measurements) are produced. After calcination, it is evident that there is a microstructural changes and reduction of the size of the agglomerates.

Table 1. Chemical composition and formula of calcined samples.

	Molar ratio $x = \frac{nM^{3+}}{n(M^{2+} + M^{3+})}$	$R = \frac{nM^{2+}}{nM^{3+}}$	Chemical formula
CoMgAl-HTc	2.20	0.31	$Co_{0.14}Mg_{0.54}Al_{0.31}$
CoAl-HTc	2.03	0.33	$Co_{0.67}Al_{0.31}$

Table 2. Crystallographic data of calcined samples * R: Hydrotalcite natural reference, MgAl-HTc[10]

	d_{003}($c=3d_{003}$) (Å)	$a = 2\ d_{110}$ (Å)	c / a
MgAl-R*	7.58 (22.74)	3.04	7.48
CoMgAl-HTc	7.74 (23,23)	3.06	7.59
CoAl-HTc	7.47 (22.41)	3.02	7.42

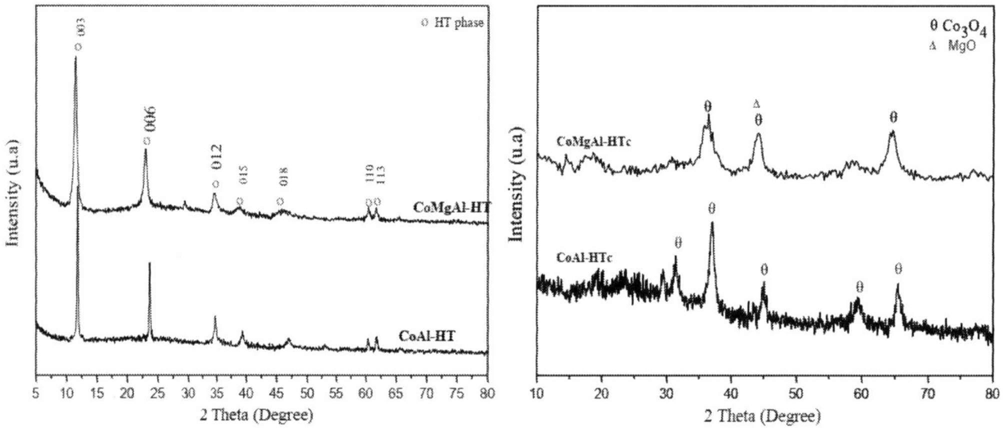

Figure 1. XRD patterns of non-calcined (left) and calcined (right) samples. Top pattern: CoMgAl-HTc and bottom pattern: CoAl-HTc.

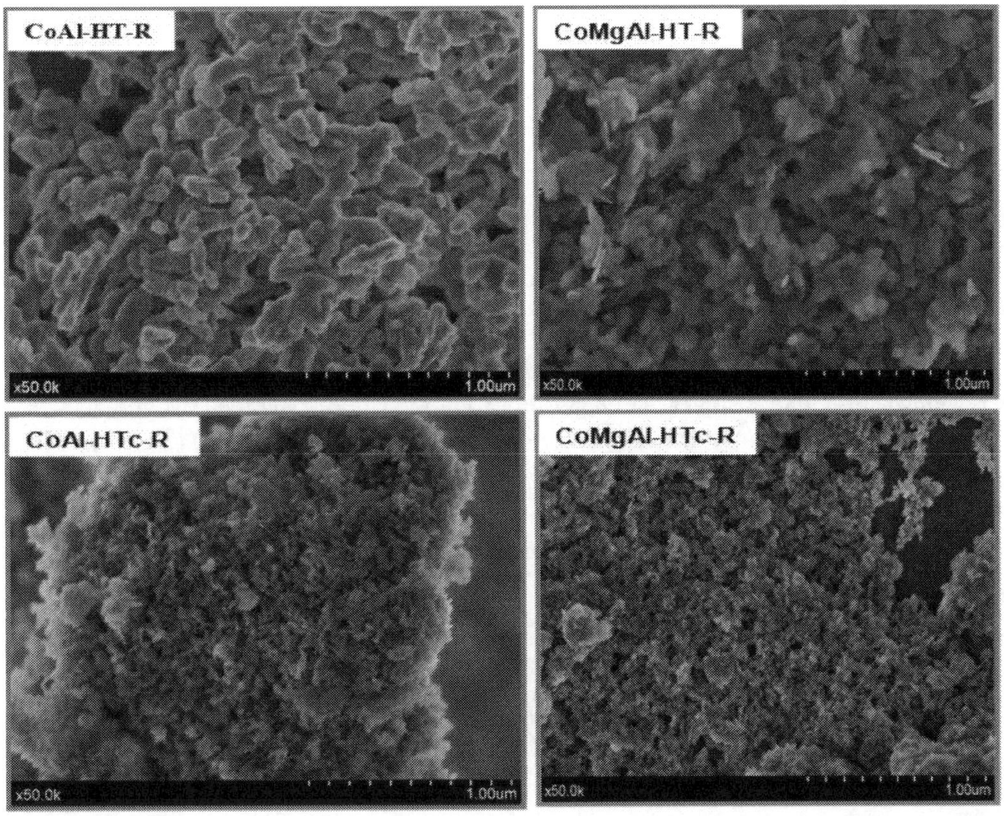

Figure 2. SEM pictures of CoAl and CoMgAl samples (c: calcinated; c-R: calcinated and reduced).

3.2 Reactivity

The stability and evolution of the reactivity of the samples over time at 700 °C was investigated. The choice of the reaction temperature was based on available data from the literature and previous laboratory investigations, which showed that interesting catalytic performance could be obtained in the temperature range 650-700 °C[1, 21]. The reduced catalytic charge was placed in the reactor and subjected to the reaction mixture (CH_4, CO_2). The reaction temperature was increased with a heating rate of 4 °C/min from ambient to 700 °C, which was maintained for this study for more than 15 hours in the frame of this work. At 700 °C, the main reaction products were H_2 and CO. The conversion rates of CH_4 and CO_2 as well as the selectivity (H_2 / CO ratio), in the presence of the catalysts CoMgAl-HTc-R and CoAl-HTc-R. The difference in catalytic behaviour between the nickel-based and the cobalt-based catalysts is mainly due to the high activity of $Ni^0$for this reaction under our operating conditions[12, 21, 22].

In the case of cobalt-based catalysts, significant catalytic inertia was observed in the case of the CoMgAl-HTc-R sample, unlike CoAl-HTc-R, which exhibits an appreciable activity and selectivity maintained beyond 11h of work (Table 3). The recorded CH_4 and CO_2 conversions reach 38% and 57 % respectively with an H_2 / CO selectivity of around 0.70. The comparison of the reduction profiles of the cobalt-based samples showed that the solid CoAl-HTc-R clearly exhibits the formation of the Co^0 species at 700 °C unlike the solid CoMgAl-HT-R whose profile of reduction only shows the formation of cobalt oxides. This is confirmed by XRD analysis after reduction at 700 °C, which confirms the absence of $Co^0$with CoMgAl-HTc-R sample (Figure 4). Figure 3 illustrates the presence of Co^0 species and cobalt oxides in the case of two samples CoAl-HTc-R and CoMgAl-HTc-R respectively. It was confirmed that Co^0 species constitute, like the Ni^0 species, the active phase for the dry reforming reaction of methane. The comparison of the catalytic performances of the solids CoMgAl-HTc-R and

CoAl-HTc-R (Figure 4, Figure 5) leads us to conclude that the Co^0 (as active phase) is formed at high reduction temperatures which can exceed 700 °C. This can be enhanced by the presence of Mg cations, which retard the reduction of cobalt and/or nickel in the catalyst.
In addition, the conversion of CO_2 is slightly higher than those of CH_4, suggesting the course of side reactions, in particular the reverse reaction of gas with water (RWGS) ($CO_2 + H_2 \rightarrow CO + H_2$) which tends to increase the conversion of CO_2.Several factors can explain the catalytic behaviour such as: textural properties (specific surface), reducibility and acid-base properties. Indeed, regarding these results, it was noticed that CoAl-HTc-R has the best catalytic performances than that of the solid CoMgAl-HTc-R. It is generally accepted that bimetallic catalysts exhibit higher performance thanthose of monometallic systems. In this same context, it has been reported in the literature [1, 3] that the addition of a low content of a given metal, such as Co, Fe, Ce, Cu, Mn, to nickel deposited on a support such as Al_2O_3 or SiO_2, could improve the catalytic performance of Ni/Al_2O_3 and Ni/Si reference solids while minimizing carbon deposition. Halliche et al. examined the doping effect of the Ni / α-Al_2O_3 and Ni/SiO_2 system with metallic additions, such as Ce, Co and Fe[23]. It was found that the effect of the dopant depended on its nature. It demonstrated the beneficial effect of the synergy caused by the addition of Co and Ce, unlike Fe, which would help deactivate the catalyst whether supported on alumina or silica.

In summary, and considering both the nature of the cations introduced as well as the presence of Mg and Co and their synergistic effect, we can classify the catalytic performances of the samples examined according to their decreasing order, as follows: CoAl-HTc-R >> > CoMgAl-HTc-R.

Table 3. Conversion and selectivity rates CoAl and CoMgAl catalysts.

	CH_4 (%) conversion rate	CO_2 (%) conversion rate	Selectivity H_2/CO
CoMgAl-HTc	0	0	0
CoAl-HTc	38	57	0.7

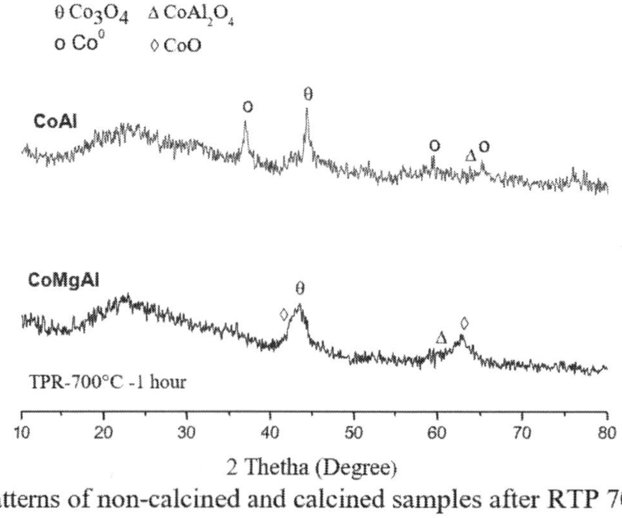

Figure 3. XRD patterns of non-calcined and calcined samples after RTP 700 °C for 1 hour

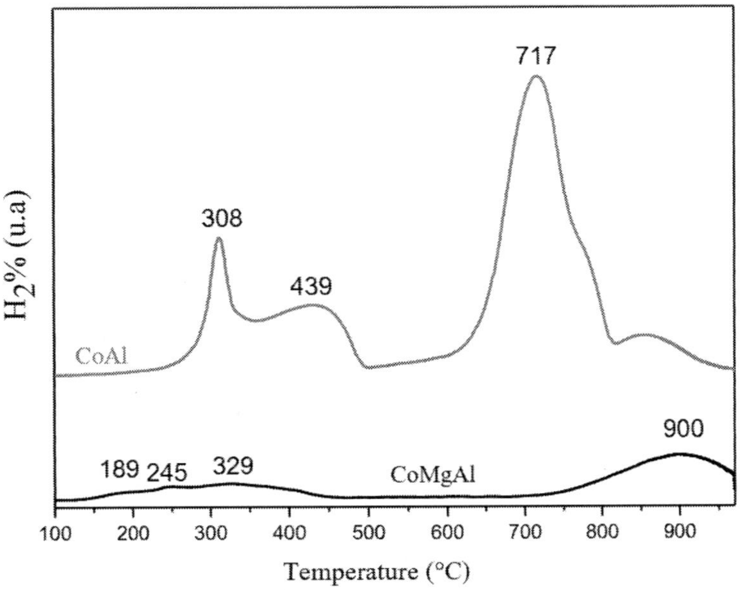

Figure 4. Reduction temperature profiles (RTP) of CoAl and CoMgAl catalysts.

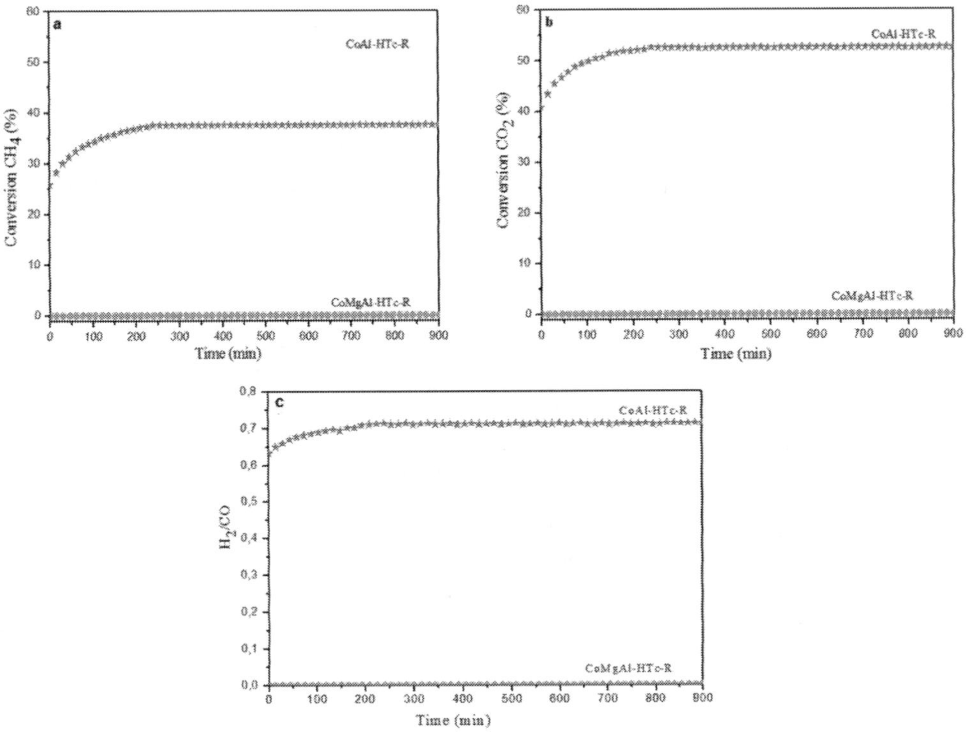

Figure 5. Evolution of the conversions rate of CH_4 (a), CO_2 (b) and H_2 / CO selectivity vs. time reaction at 700 °C for CoMgAl-HTc-R and CoAl-HTc-R.

4. Conclusions

CoAl and CoMgAl hydrotalcite based catalysts were prepared by a co-precipitation method at a fixed pH=11.Above 450 °C, no reaction occurs during the calcination step.

The two samples were tested in dry reforming of methane, the catalytic activity of catalysts can be classified as follows: CoAl-HT >>>CoMgAl-T. This result can be explained by the presence of the Co^0 for the catalyst CoAl-HTc-R and the total absence in the sample CoMgAl-HTc-R.

The solid CoMgAl-HTc-R requires high reduction temperature compared to CoAl-HTc-R due to the strong CoO-MgO interactions.

References

[1] J-M Lavoie 2014 Review on dry reforming of methane, a potentially more environmentally-friendly approach to the increasing natural gas exploitation *Frontiers in Chemistry* 2

[2] K Takehira, T. Shishido, P. Wang, T. Kosaka, K. Takaki 2003 Steam reforming of CH_4over supported Ni catalysts prepared from a Mg–Al hydrotalcite-like anionic clay *Phys. Chem. Chem. Phys* **5** 3801-3810

[3] M C J Bradford & M A Vannice 1999 CO_2 Reforming of CH_4, Catalysis Reviews **41** 1-42

[4] M. Machida, A. Sato, T. Kijima, H. Inoue, K. Eguchi, H. Arai 1995 Catalytic properties and surface modification of hexaaluminate microcrystals for combustion catalyst *Catalysis Today* **26** 239-245

[5] R H Jones, A T Ashcroft, D Waller, A K Cheetham, J M Thomas 1991 Catalytic conversion of methane to synthesis gas over europium iridate, Eu 2 Ir 2 O 7: An in situ study by x-ray diffraction and mass spectrometry *Catalysis Letters* **8** 169-174

[6] M Lang *et al* 2010 Review of A2B2O7 pyrochlore response to irradiation and pressure, Nuclear Instruments and Methods in Physics Research Section B: Beam Interactions with Materials and Atoms 268 2951-2959.

[7] S L González-Cortés 2020 Solution Combustion Synthesis of Nanostructured Solid Catalysts for Sustainable Chemistry *Sustainable Chemistry Series* **5**

[8] A. Shamsi 2004 Carbon formation on Ni–MgO catalyst during reaction of methane in the presence of CO2 and CO *Applied Catalysis A General* 277 23-30

[9] S A Rashdan & L J Hazeem 2020 Synthesis of spinel ferrites nanoparticles and investigating their effect on the growth of microalgae Picochlorum sp, Arab *Journal of Basic and Applied Sciences*, **27** 134-141

[10] F Cavani, F Trifirò, A Vaccari 1991 Hydrotalcite-type anionic clays: Preparation, properties and applications *Catalysis Today* **11** 173-301

[11] A. Vaccari 1998 Preparation and catalytic properties of cationic and anionic clays, Catalysis Today **41** 53-71

[12] Z Abdelsadek *et al* 2021 Examination of the Deactivation Cycle of NiAl- and NiMgAl-Hydrotalcite Derived Catalysts in the Dry Reforming of Methane *Catalysis Letters* 2021

[13] C Forano, T Hibino, F Leroux, C Taviot-Guého 2006 Chapter 13.1 Layered Double Hydroxides, **1** 1021-1095

[14] G Valderrama 2005 Dry reforming of methane over Ni perovskite type oxides, Catalysis Today **107** 785-791.

[15] M.-M Millet *et al* 2019 Highly Dispersed Ni0/NixMg1–xO Catalysts Derived from Solid Solutions: How Metal and Support Control the CO2 Hydrogenation *ACS Catalysis* **9** 8534-8546

[16] A. Bhattacharyya, V.W. Chang, D.J. Schumacher, CO2 reforming of methane to syngas: I: evaluation of hydrotalcite clay-derived catalysts, Applied Clay Science, 13 (1998) 317-328.

[17] V M Gonzalez-DelaCruz, J P Holgado, R Pereñíguez, A Caballero 2008 Morphology changes induced by strong metal–support interaction on a Ni–ceria catalytic system *Journal of Catalysis* 257 307-314.

[18] T-Y Cai *et al* 2017 Multiferroic Double Perovskites ScFe 1− x Cr x O 3 (1/6≤ x≤ 5/6) for Highly Efficient Photovoltaics and Spintronics *Physical Review Applied* **8** 034034

[19] V M Gonzalez-Delacruz *et al* 2011 Modifying the size of nickel metallic particles by H2/CO treatment in Ni/ZrO2 methane dry reforming catalysts *ACS Catalysis* **1** 82-88.

[20] K. Takehira 2009 "Intelligent" reforming catalysts: Trace noble metal-doped Ni/Mg (Al) O derived from hydrotalcites *Journal of Natural Gas Chemistry* **18** 237-259

[21] Z Abdelsadek *et al* 2016 In-situ hydrogasification/regeneration of NiAl-hydrotalcite derived catalyst in the reaction of CO 2 reforming of methane: A versatile approach to catalyst recycling, Journal of CO2 Utilization **14** 98-105

[22] F T Zoulikha Abdelsadek *et al* 2008 Synthesis, characterization and application of Ni/Mg/Al (La) hydrotalcites in methane dry reforming *Advanced micro-and mesoporous materials* Book Chapter **6** 60-66

[23] D Halliche *et al* 2002 Différents Cations. Application à la réaction de reformage sec du méthane, Journal of Thermal Analysis and Calorimetry **68** 997-1002

Acknowledgments
The authors would like to thank the General Directorate for Scientific Research and Technological Development (DGRSDT) of the Algerian Ministry of Higher Education and the Spanish ministry of Science for funding and technical support.

Effect of different heating rate on properties of fired brick produced from industrial waste and natural clay

N J A Hamid[1,2,3], A A Kadir[1,2,3*], N N H Hashar[1,2,3], A A Hashim[1,2,3], N A Sarani[1] and M I H Hassan[1,2]

[1] Department of Water and Environmental Engineering, Faculty of Civil Engineering and Built Environment, Universiti Tun Hussein Onn Malaysia, 86400 Parit Raja, Batu Pahat, Johor, Malaysia
[2] Micropollutant Research Centre (MPRC), Universiti Tun Hussein Onn Malaysia, 86400 Parit Raja, Batu Pahat, Johor, Malaysia
[3] Research Centre for Soft Soil (RECESS), Universiti Tun Hussein Onn Malaysia, 86400 Parit Raja, Batu Pahat, Johor, Malaysia

*Corresponding E-mail: aeslina@uthm.edu.my

Abstract. Gypsum waste is generated from the wastewater treatment plant activities from chemical industry. The environmental issues and secured landfill costs caused by this waste increasing every year. Therefore, this study is an attempted to reuse gypsum waste as a substitution material for natural clay in fired brick production. This study consisted of two stages where in the first stage, fired brick were produced from natural clay and gypsum waste with substitution of clay with 5 %, 10 %, 15 % and 20 % gypsum waste. The second stage was the investigation of the experimental when the brick samples were fired at different heating rates (1 °C/min, 2 °C/min and 3 °C/min). Next, the shrinkage, dry density, water absorption and compressive strength for each sample were examined and were compared to the control brick. Based on the results, the shrinkage and water absorption increased with the addition of gypsum waste in brick samples. In contrast, the density and compressive strength values were decreased in the brick samples containing gypsum waste. The mechanical properties met the conditions prescribed in the specified standards for all the samples and the optimization result showed that up to 10% of gypsum waste incorporation into fired brick (fired at 1°C/min heating rate) had a sufficient compressive strength over 20 MPa. The reclamation of gypsum waste was proved for the fired brick production insights reducing the environmental pollution. Gypsum waste can be used as a natural clay substitution whilst fulfilling the demand for fired brick requisition.

Keywords: Building materials, Fired brick, Heating rate, Properties, Waste reclamation

Track Name: Advanced Technology and Renewable Energy

1. Introduction

Malaysia has experienced a number of changes in its industrial development, population growth and urbanization which had increased the amount of wastes generated every day. More than 23,000 tonnes of waste are currently produced in Malaysia and only less than 5 percent of the waste is being recycled [1]. There are various industrial wastes were produced from plants and process effluents such as sludge, sawdust, chemical solvents, paint, sandpaper, paper products, manufacturing by-products, metals and toxic waste [2]. Approximate 50,000 metric tonnes of gypsum waste was generated per year from a chemical industry in Johor Bahru, Johor. Gypsum waste was resulted from the wastewater effluent activities, consequently, the amount keeps increasing and were dumped into landfill which inevitably becoming scarce in meantime. These potentially hazard industrial waste have a big impact to the health and harm to the environment as it contains a high amount of heavy metal such as zinc, chromium, lead, nickel copper and others [3, 4]. Hence the researches highlight the need for alternative methods developed to recover the wastes into new products.

Meanwhile, the utilization of industrial waste into the production of fired bricks is discovered by researchers as one of the promising way to reclaim the waste [5]. Towards striving to promote a sustainable and green building construction, the demands for fired brick incorporating with industrial residuals has also increased. In addition, the incorporation of industrial waste into fired clay brick able to contribute to the environmental protection, human health enhancement and lower operational costs for industry by minimizing the waste disposal cost [6].

Bricks is the oldest mansion that been manufactured as an important building materials that used by mankind in the world and were used as early as 4500 BC [7]. Fired clay brick can be formed when the particles in the clay are bond to each other in the high temperature [8]. In addition, it is made of natural clay that has strong tolerance to high temperatures with the optimal firing temperature ranges from 2 hours to 8 hours from 970°C to 1200°C [9]. Fired clay brick is commonly used in construction as it has good properties in term of durability, colour, texture, size, compressive strength and absorption [10]. The different shapes, colour, texture, intensity and consistency of the fired brick produced were varies depend on the materials and methods used [11]. Furthermore, fired brick tend to be more highly durable and highly strength when in cooling process. Different types of waste and sludge materials such as electroplating sludge, mosaic sludge and other waste have been incorporated successfully into fired clay brick by previous researchers [12-17]. Therefore, this paper is attempted to incorporate gypsum waste into fired brick as the alternative for disposal method. The effects of different heating rate (1 °C/min, 2 °C/min and 3 °C/min) on the properties of fired brick produced were compared with the control bricks and limitation set in the standard. Utilization of gypsum waste resolves the environmental concerns, providing alternative raw material instead of clay in bricks production and reduces dependable on natural resources, thus supported the development of sustainable goal approaches.

2. Materials and Methods

2.1 Materials

The raw materials used for the study were clay soil and gypsum waste. The clay soil was obtained from a quarry in Batu Pahat, Johor. The gypsum waste was collected in sludge form and thereafter, fired brick samples were prepared. The chemical compositions of the raw materials are given in Table 1. The analysis showed that CaO and SO_3 were significantly higher in gypsum waste with 29.57 % and 29.56 % respectively while Al_2O_3 and SiO_2 in clay soil with 38.70 % and 40.80 % respectively. The flexibility of clay minerals to be bind with other minerals at optimum temperature made the fired brick feasible to be incorporated with different type of wastes [18].

Table 1. Chemical compositions of the raw materials used in this study.

	Amount (%)	
Compounds	Clay soil	Gypsum waste
CaO	0.10	29.57
SO_3	0.27	29.56
Al_2O_3	38.70	16.13

SiO_2	40.80	8.86
MgO	0.36	7.82
Fe_2O_3	5.41	4.88
Na_2O	0.47	2.39
P_2O_5	ND	0.39
MnO	ND	0.15
K_2O	0.22	0.11
Cl	ND	418.00 mg/kg
TiO_2	1.05	344.00 mg/kg
ZnO	ND	111.00 mg/kg
Cr_2O_3	ND	56.00 mg/kg

**ND- Not Detected*

2.2 Methods

The manufacturing process of fired brick started with the preparation of raw materials. The clay soil and gypsum waste were oven dried at 105 °C for 24 hours to remove the water content as it could affect the properties of the sample. This study was conducted in two stages. In the first stage, fired brick were produced as control brick (0 % gypsum waste incorporation) and the brick incorporated with gypsum waste at the percentages at 5 %, 10 %, 15 % and 20 %. Control brick was used as a comparative during experimental testing. The raw materials were mixed together at respective percentage followed by pre-determined water using a mechanical mixer until homogenous mixture was yielded. After that, the sample was compressed at 2000 psi using a brick machine with dimension of 215 mm x 102.5 mm x 65 mm. The brick sample was dried at room temperature for 24 hours and another 24 hours in the oven at 105 °C. Next, the brick sample was put into the furnace box for firing process at firing temperature of 1050 °C for two hours, while applying different heating rates of 1 °C/min, 2 °C/min and 3 °C/min. In order to assess the properties of the fired brick production with utilization of gypsum waste, the firing shrinkage, dry density, water absorption and compressive strength were examined. The properties were tested according to BSI [19].

For the calculation of the firing shrinkage, the following equation was used:

$$L_S = L_W - L_d / L \qquad \text{(Equation 1.1)}$$

where:
L_s = Percentage of shrinkage
L_w = Length of wet
L_d = Length of dry
L = Actual length

For the calculation of the dry density, the following equation was used:

$$\rho = \frac{m}{v} \qquad \text{(Equation 1.2)}$$

where:
p = Dry density
m = Mass
v = Volume

The water absorption of the samples by volume which also represented the apparent porosity of the brick was calculated using the following equation:

$$W\Delta = \frac{M1 - M2}{M2} \times 100\%$$ (Equation 1.3)

where:
M1 = Weight of wet brick(g)
M2 = Weight of dry brick(g)
WΔ = Water absorption (%)

For the calculation of the compressive strength, the following equation was used:

$$\sigma = F / A$$ (Equation 1.4)

where,
σ = Compressive strength (MPa)
F = Applied load (kN)
A = Cross section area (mm^2)

3. Results and Discussion

The results of the tests conducted on the fired brick samples are discussed and presented as subsequent figures below.

3.1. Shrinkage

The shrinkage of bricks occurs when the water in the brick was lost during the drying and firing stages. The results of the shrinkage for the control brick and gypsum brick with different heating rates were presented as in Figure 1. The shrinkage result showed that the increment is in parallel with the addition of gypsum waste percentages. The shrinkage of the fired brick with gypsum waste was found to be not more than 2 % with respect to the heating rates of 1 °C/min, 2 °C/min and 3 °C/min. Firing temperature is one of the factors that affecting the shrinkage of brick [20]. Besides that, it was indicated that the shrinkage of brick was increased at temperature of 650 °C to 1050 °C, while was significantly affected by the clay sintering and reorganization of the bonding particles at the temperature of 900 °C [21].

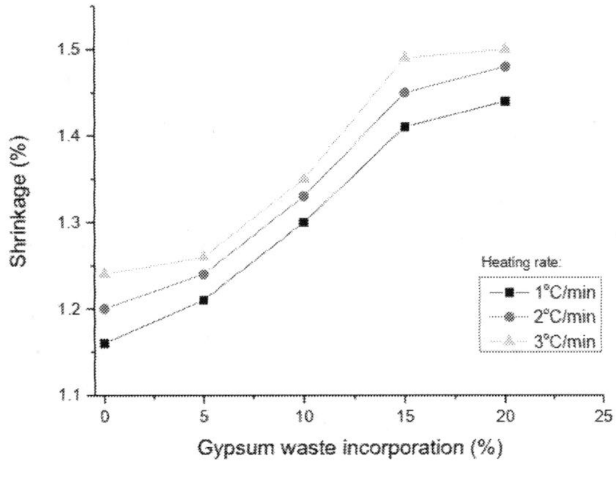

Figure 1. Shrinkage of fired brick.

3.2. Dry density

Density is the relationship between the solid mass and the overall volume of the sample which depend on the temperature and manufacturing process applied. Based on the results, the heating rate has an insignificant effect on the dry density of the fired brick [22]. The trends showed that the dry densities were quite similar within the fired brick samples. The increasing amount of gypsum waste added caused the results to decrease ranging from 1900 kg/m^3 to 1743 kg/m^3 for gypsum brick (5 %, 10 %, 15 % and 20 %) at respective heating rates (1 °C/min, 2 °C/min and 3 °C/min). This condition was related to sintering process where residual carbon phase were forming voids and eventually increased the porosity of the bricks [23]. This is supported where the waste marble powder (WMP) obtained from the marble processing factory which contained a high amount of calcium carbonate ($CaCO_3$) demonstrated the increase of density at higher temperature which is possibly caused by the incline of consolidation or vitrification within the particles in the brick bodies [24]. Meanwhile, another observation on the utilization of aluminium filter dust (AFD) into fired clay brick showed that the addition of waste did not change the mineral composition of brick but it had modified the amount and size of the pore formed after the firing process [25]. The pore-forming effect occurs at high temperature due to the increasing amount of AFD has resulted to the decrease of bulk density.

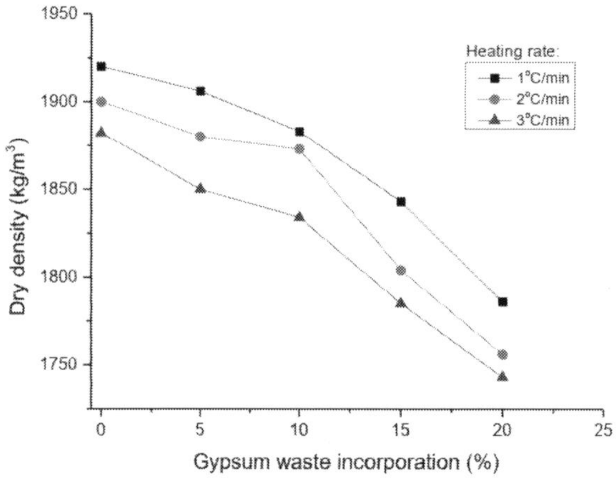

Figure 2. Dry density of fired brick.

3.3. Water absorption

Water absorption testing was performed to assess the hardness characteristics of the brick by indicated its porosity. The higher water absorption reading would cause cracks to the surface of brick samples. According to the standard limit monitored, the water absorption for severe weathering resistance bricks may not be exceed from 17 %, for moderate weathering resistance bricks may less than 22 % and for negligible weathering resistance bricks has no limit specified but must be in acceptable condition [26]. The results showed that the gypsum waste addition had caused a gradual increase of water absorption for all the gypsum brick samples for all respective heating rates. For gypsum brick 20 % fired at a heating rate 1 °C/min had water absorption at 12 % while for control brick fired at the same heating rate was recorded 3 % value which is explained the existence of higher amount of pores in the brick samples containing gypsum waste. The result is in parallel to the findings obtained by recycling waste marble sludge into fired clay bricks [27]. The apparent high porosity behaviour was formed by the dehydroxylation and combustion of calcium carbonate known as amorphous phase occurred during various reaction at higher temperature [28].

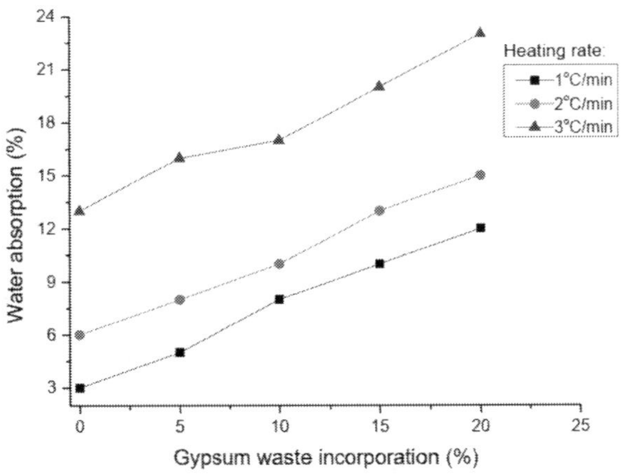

Figure 3. Water absorption of fired brick.

3.4. Compressive strength

The compressive strength test was conducted to determine the ability of the brick to hold the load under compression. This test is important in properties of fired brick classification. The values of compressive strength showed that control bricks fired at respective 1 °C/min, 2 °C/min and 3 °C/min had 31.00 MPa, 30.00 MPa and 29.68 MPa respectively. The amount of gypsum waste added had decreased the compressive strength for the gypsum bricks of 5 %, 10 % and 15 % due to high porosity compared to control bricks with ranges from 26.30 MPa to 20.63 MPa, 20.6 MPa to 18.5 MPa, 15.90 MPa to 14.60 MPa for respective heating rate of 1 °C/min, 2 °C/min and 3 °C/min. The lowest compressive strength results were from gypsum brick 20 % with 13.80 MPa, 9.06 MPa and 8.80 MPa at respective heating rate of 1 °C/min, 2 °C/min and 3 °C/min. However, the results showed that gypsum brick 10 % that fired at 1 °C/min had the sufficient requirement for compressive strength with 20.6 MPa for loading bearing application as stipulated in Malaysian standard [29]. This explained that the linear relation between density and water absorption was established and was validated that the pore amount and density have increased the amount of water penetrated into the brick bodies hence lowering the compressive strength of the brick samples [25].

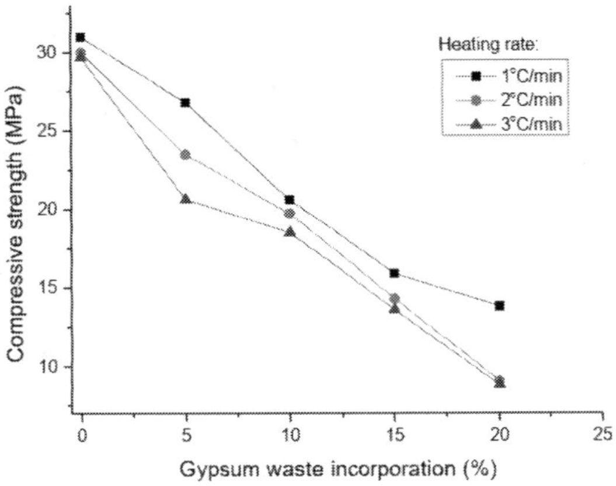

Figure 4. Compressive strength of fired brick.

IOP Conf. Series: Earth and Environmental Science **880** (2021) 012036 doi:10.1088/1755-1315/880/1/012036

4. Conclusion

This study was conducted to reclaim the gypsum waste from the chemical industry by recycling into fired brick as an alternative disposal method of landfilling. The natural clay composition in fired brick was replaced by incorporating gypsum waste at 5 %, 10 %, 15 % and 20 %. Three different heating rates was applied for firing stage which were 1 °C/min, 2 °C/min and 3 °C/min. The gypsum waste addition in brick samples had directly increased the shrinkage and water absorption readings. Meanwhile, the dry density and compressive strength were gradually decreased with the increment in gypsum waste incorporation. However, up to 10 % of gypsum waste incorporation which was fired at 1 °C/min, showed the acceptable value of compressive strength with 20 MPa and adequate requirement for other properties as well. In addition, the fired brick produced from different heating rates assemble changes in their properties due to different period of time for completing the sintering process. The results indicated that 1°C/min was the most appropriate heating rate to manufacture fired bricks as it complimented the shrinkage, dry density, water absorption and compressive strength properties. Therefore, this study demonstrated that gypsum waste could be used as a replacement for natural clay in the brick manufacturing industry insight conserving the environment and human health.

References

[1] Johari A, Alkali H, Hashim H, Ahmed S I, Mat R 2014 Municipal solid waste management and potential revenue from recycling in Malaysia. *Modern Applied Science* **8** (4) 37

[2] Renitaa A A, Kumarb P S, Srinivasb S, Priyadharshinib S, Karthikab M 2017 A review on analytical methods and treatment techniques of pharmaceutical wastewater. Desalination and Water Treatment 87 160-78.

[3] Mishra S, Bharagava R N, More N, Yadav A, Zainith S, Mani S, Chowdhary P 2019 Heavy metal contamination: an alarming threat to environment and human health. *InEnvironmental biotechnology: For sustainable future* pp. 103-125

[4] Sankhla M S, Kumari M, Nandan M, Kumar R, Agrawal P 2016 Heavy metals contamination in water and their hazardous effect on human health-a review *Int. J. Curr. Microbiol. App. Sci.* **5** (10) 759-66

[5] Sutcu M, Erdogmus E, Gencel O, Gholampour A, Atan E, Ozbakkaloglu T 2019 Recycling of bottom ash and fly ash wastes in eco-friendly clay brick production *Journal of Cleaner Production* 233 753-64

[6] Shakir A A, Mohammed A A 2013 Manufacturing of Bricks in the Past, in the Present and in the Future: A state of the Art Review. International Journal of Advances in Applied Sciences (IJAAS) **2** (3) 145-56

[7] Perkins D 2013 Encyclopedia of China: History and culture. *Routledge*

[8]. Phonphuak N, Kanyakam S, Chindaprasirt P 2016 Utilization of waste glass to enhance physical–mechanical properties of fired clay brick. *Journal of Cleaner production* **112** 3057-62

[9] Monatshebe T. Artisanal Fired Clay Bricks in Dididi and QwaQwa 2017 Their Production and Related Environmental Issues. University of Johannesburg (South Africa)

[10] Lourenço P B, Fernandes F M, Castro F 2010 Handmade clay bricks: chemical, physical and mechanical properties. *International Journal of Architectural Heritage* **4** (1) 38-58

[11] Dogan-Sağlamtimur N, Bilgil A, Szechyńska-Hebda M, Parzych S, Hebda M 2021 Eco-friendly fired brick produced from industrial ash and natural clay: A study of waste reuse *Materials* **14** (4) 877

[12] Zhang M, Chen C, Mao L, Wu Q 2018 Use of electroplating sludge in production of fired clay bricks: Characterization and environmental risk evaluation *Construction and Building Materials* **159** 27-36.

[13] Sarani NA, Kadir A A, Rahim A S, Mohajerani A 2018 Properties and environmental impact of the mosaic sludge incorporated into fired clay bricks *Construction and building materials183* 300-10.

[14] Ukwatta A, Mohajerani A, Setunge S, Eshtiaghi N 2015 Possible use of biosolids in fired-clay bricks *Construction and Building Materials* 91 86-93.

[15] Hamid N J *et al* 2021 Influence of Gypsum Waste Utilization on Properties and Leachability of Fired Clay Brick. Materials. **14** (11) 2800

[16] Chemani B and Chemani H 2012 Effect of adding sawdust on mechanical-physical properties of ceramic bricks to obtain lightweight building material. *International Journal of Mechanical, Aerospace, Industrial, Mechatronic and Manufacturing Engineering* **6** (11) 36-55

[17] Sarkar R *et al* 2017. Use of paper mill waste for brick making. *Cogent Engineering* **4** (1)1405768.

[18] Rehman M U *et al* 2020 Influence of fluxing oxides from waste on the production and physico-mechanical properties of fired clay brick: A review *Journal of Building Engineering* 27100965.

[19] B S 12 3921, B.S.I British Standards Institution. British Standard Specification for Clay Bricks; B.S.I British Standards Institution: London, UK, 1985

[20] Sahu M K *et al* 2016 Critical review on bricks. International Journal of Engineering and Management Research (IJEMR) **6** (5) 80-8

[21] Karim H M 2019 Production of lightweight thermal insulating clay bricks using papyrus *Polytechnic Journal* **9** (1)74-81

[22] Ukwatta A and Mohajerani A 2017. Characterisation of fired-clay bricks incorporating biosolids and the effect of heating rate on properties of bricks *Construction and Building Materials* 142 11-22
[23] Taurino R, Karamanova E, Barbieri L, Atanasova-Vladimirova S, Andreola F and Karamanov A 2017 New fired bricks based on municipal solid waste incinerator bottom ash *Waste Management & Research* **35**(10)1055-63
[24] Sutcu M, Alptekin H, Erdogmus E, Er Y and Gencel O 2015 Characteristics of fired clay bricks with waste marble powder addition as building materials *Construction and Building Materials* **1** 82 1-8.
[25] Bonet-Martínez E, Pérez-Villarejo L, Eliche-Quesada D and Castro E 2018 Manufacture of sustainable clay bricks using waste from secondary aluminum recycling as raw material *Materials* **11** (12) 2439
[26] ASTM C62, 2017 Edition, October 1, 2017 - Standard Specification for Building Brick (Solid Masonry Units Made From Clay or Shale)
[27] Munir MJ, Kazmi SM, Wu YF, Hanif A and Khan MU 2018 Thermally efficient fired clay bricks incorporating waste marble sludge: An industrial-scale study *Journal of cleaner production* 174 1122-35
[28] Eliche-Quesada D and Leite-Costa J 2016 Use of bottom ash from olive pomace combustion in the production of eco-friendly fired clay bricks *Waste management* 48 323-33
[29] Standard M 1972 Clay bricks. Malaysia, MS 76

Acknowledgments
This research was supported by Ministry of Higher Education (MOHE) through Fundamental Research Grant Scheme (FRGS/1/2020/WAB02/UTHM/02/9). Authors wishing to acknowledge Micropollutant Research Centre (MPRC) and Research Centre for Soft Soil (RECESS), Universiti Tun Hussein Onn Malaysia for the laboratory facilities.

Emissions of carbon dioxide and methane from dairy cattle manure

M M Hanafiah[1,2*], A J Ibraheem[1] and K K Razman[1]

[1] Department of Earth Sciences and Environment, Faculty of Science and Technology, Universiti Kebangsaan Malaysia, 43600 UKM Bangi, Selangor, Malaysia
[2] Centre for Tropical Climate Change, Institute of Climate Change, Universiti Kebangsaan Malaysia, 43600 UKM Bangi, Selangor, Malaysia

*Corresponding E-mail: mhmarlia@ukm.edu.my

Abstract. There is an increasing interest in reducing production and emissions of greenhouse gases to combat global warming. Greenhouse gases can be produced through animal production operations. One of the major sources of greenhouse gases emitted from the animal farming is dairy cattle barns. This study measured the CH_4 and CO_2 emissions from dairy cattle manure decomposition trapped inside the static chambers through anaerobic digestion process by bacteria and at regular intervals by focusing on animal age and manure storage method. Samples were analyzed using gas chromatography for the estimation of CH_4 and CO_2 emissions. Four Friesian cows were used representing two stages of cow age (3 and 10 years old) and 1 kg of fresh manure samples were collected (feces and mixture of feces with urine). It was found that CH_4 and CO_2 emissions produced by cattle at the age of 3 years were higher than age of 10 years. In addition, gases emitted from fresh slurry feces were higher than liquid manure for both ages (3 and 10 years). This is due to the fact that the organic matter degradation in the feces and amount of fresh slurry feces is twice the amount of fresh slurry feces used in the liquid manure, as well as the organic matter in the manure mass for the age of 3 years is higher than for the age of 10 years. The findings from this study can provide information for improving manure management practices in animal farms.

Keywords: Carbon dioxide, Methane, Greenhouse gas, Animal manure, Dairy cattle

Track Name: Advanced Technology and Renewable Energy

1. Introduction

Animal production farms can emit a variety of air pollutants through gas emissions, which are produced by the animal production processes such as enteric fermentation, confinement barns, manure storage, treatment systems, and manure applied to land for crop nutrients [1]. These animal breeding operations cause the greenhouse gases (GHG) to be emitted into the atmosphere and remains in the atmosphere. This in turns absorbs and emits radiation within the thermal infrared range, keeping the planet's surface warm [2-3]. The major greenhouse gases are carbon dioxide (CO_2), water vapor (H_2O), methane (CH_4), nitrous oxide (N_2O), ozone (O_3) and chlorofluorocarbons (CFCs) [4-6]. As a result of these emissions, animal production is considered as one of the most significant causes of air pollution from the agricultural sector [7]. CH_4 and N_2O emissions from enteric fermentation, manure storage and handling, and crop and pasture land are all significant sources of GHG emissions from dairy farms [8]. Air pollutants emission from these animal production facilities are influenced by many factors including type of animal, farm operation, ventilation and temperature [9]. Animal feeding operations (AFOs), which are large-scale facilities where livestock and poultry are raised for food processing, release various biological and chemical contaminants into the atmosphere [10]. Odors, gases such as hydrogen sulfide (H_2S), ammonia (NH_3), methane (CH_4), particulate matter, and bio aerosols are among the gases and materials emitted [11].

According to one report, agriculture contributes about 18% of the total global anthropogenic emissions of greenhouse gases, including CH_4 and CO_2 emissions with 9% and 34%, respectively [12] 13]. Methane and carbon dioxide are produced from manure through the anaerobic digestion process of an organic matter decomposition process. Anaerobic digestion is a microbial process that results in the decomposition of organic matter in manure in the absence of oxygen, resulting in the emission of greenhouse gases (GHGs) [14]. Such process is done through common interaction of four metabolically-linked microbes that are in hydrolytic, acidogenic, acetogenic and methanogenic stages [15-16]. Dairy cattle manure is one of the major sources of greenhouse gases emitted from dairy farming where it adversely affects the health of human, animal and the environment [17]. These gases have potential effects on changing the global climate and may especially be a concern to animal health in confined buildings [18]. Therefore, novel research to measure emissions from dairy cow manure, are needed to establish a model emission scopes for livestock farms and help in determining the impact. From the environmental and management factors of these emissions, mitigation strategies can be developed. Furthermore, measuring the amount of emitted gases from dairy manure mass is needed to reduce extreme climate change, mitigate pollution and provide a database that can be deemed useful to improve and monitor the emission impacts.

The present study aims to measure the amount of CH_4 and CO_2 emitted from manure of dairy cow barns, with a focus on the animal age and the manure storage method. This helps identify the influence of the animal's age and the manure storage method on the emission of GHG gases from a dairy cow manure. There are several factors affecting the emission on a farm level including feed, cattle, manure management and climate. Therefore, it is important to understand the representative characteristics for many dairy farms and connect the adopted manure management practices to greenhouse gas emissions for setting recommendations and policies to reduce environmental risks [19]. Hence, quantifying GHG emission from all livestock farm sources is needed for those who seeks for a solution in climate change that is related with the livestock sector [20].

1.1. Livestock manure

Dairy cows are a major source of CH_4 and CO_2 emissions, where CH_4 is produced through anaerobic digestion. As the CO_2 is produced, it is then emitted through both aerobic and anaerobic digestion due to the microbial decomposition of organic matter [21]. The decomposition of manure is done through anaerobic process leading to the decompose of organic matter that interacts with four groups of metabolically microbe-related hydrolytic, acidogenic, acetogenic and methanogenic phases leading to the output of methane and other gases [15]. Not only that, livestock manure is also a potential source of pathogens [22]. Animal production processes generate abundant amounts of animal manure, which requires appropriate management. Therefore, there is a need for appropriate disposal and management of manure in order to avoid the adverse environmental and public health affects [23].

1.1.1 Methane emission

Methane (CH_4) is a colorless, scentless and inflammable gas that exists naturally in the atmosphere and lighter than air. With a global warming potential about 25 times more than CO_2, methane affects the ozone layer degradation [24] and is the main GHG emitted from manure in storage [25]. The emission of CH_4 occurs from ruminant animals by enteric fermentation and anaerobic storage of manures at temperate conditions. In fact, enteric fermentation and manure management is responsible for 35–40 % of the total anthropogenic CH_4 emissions and 80 % of CH_4 emission from agriculture [26].

1.1.2 Carbon dioxide emission

Carbon dioxide (CO_2) is naturally present in the Earth's atmosphere through carbon cycle. It is one of the main GHGs contributing to the greenhouse effect and its anthropogenic activities may produce about 8 billion tons of CO_2 per year [27]. The major source of CO_2 emitted from solid manure is by the aerobic production of composting process. This process is affected by many factors, such as moisture content, temperature, carbon/nitrogen ratio, carbon compounds degradability, pH level and the physical structure of the organic material. It has been estimated that approximately 12.5 % of the total global GHG emission are from the livestock sector and 80 % of the total emission from agriculture is from the livestock sector [26].

1.2 Mitigation strategies of gases emission

There are many different strategies proved to be a potential in reducing GHGs emission from manure. To mitigate the impact of animal production operations on climate change there are some of the technical options such as carbon sequestration, improving diets to reduce enteric fermentation, improving manure management, and more efficient use of fertilizers [28]. In some areas, unsustainable manure management practices include a variety of disposal methods that are still commonplace with cost being one of the factors [29]. Hence, most practices involve decreasing storage duration, improving timing and application of manure, used of anaerobic digesters, covering the storage, utilization of a solid separator and changing the animal diets [30].

2. Methodology

The methodology of this study includes the measurement of CH_4 and CO_2 emissions emitted from fresh dairy cow manure samples, trapped inside static chambers by using gas chromatography. Collected manure sampling were placed in static chambers and gas emitted by the samples were collected at regular intervals of 48 hours for gas chromatography (GC) analysis. The data were analyzed using one-way ANOVA with SAS System-2009 used to run the linear regression analysis for both CH_4 and CO_2 emissions.

2.1. Samples collection

Manure samples from four Friesian cows were collected from the livestock farm located in Serdang, Selangor, Malaysia (Table 1), represented by two different age stages (3 years and 10 years) to measure gases emission from manure mass. Intact unpregnant cows were selected which were disease-free and feeding on the same diet of 4 kg concentrated feed per cow daily.

Table 1. Specifications of the selected Friesian cows.

Cow Number	Weight / kg	Age / y
T083	495	10
T092	552	10
T1404	396	3
T1405	408	3

Two tests were used to measure the gases quantity (CH_4 and CO_2) emitted from the manure mass. The first test measured gas emitted from fresh feces (fresh slurry feces) and the second test measured gas emitted from a mixture of feces with urine (fresh liquid manure) at equal amounts of 1 kg for each sample. Fresh slurry feces (FSF) and fresh liquid manure (FLM) were collected from each cow and placed in static chambers prepared for this purpose.

2.2. Data collection and instrumentation

The static chambers were transferred to a laboratory for GC analysis of gas samples and were done by collecting from the top of the chamber at regular intervals of 2 days. The static chambers were maintained at a temperature of 20 °C to 25 °C. The gas samples were collected from the top of chamber by a 25 mL syringe which were then injected into 10 mL vials prepared. From the vials, 1 mL of the gas samples were injected into the gas chromatography. Linear regression analyses were analysed using one-way ANOVA with the differences between two age stages groups and correlation between related parameters performed.

3. Results and Discussion

CH_4 and CO_2 emitted from the Friesian cows manure mass were measured based on the two age stages (3 and 10 years) and manure storage method; FSF and FLM. Due to variations in the gas sampling times and the animal age, various concentrations of CH_4 and CO_2 gas emissions were observed.

3.1 Methane emission

The results of this study showed an increasing CH_4 emission gradually from both FSF and FLM manure for both ages. Differences in the recorded concentrations of the CH_4 is shown emitted from the manure mass in the two age stages (3 and 10 years) and in the manure storage method. For the FSF storage method, the minimum and maximum values for CH_4 emissions ratio at all gas samples were between T_0 (0.00 %) to T_4 (31.41 %) for the age of 3 years. This emissions ratio was much lower for the age of 10 years with the minimum and maximum values between T_0 (0.00 %) and T_4 (12.67 %) only. As for the FLM storage method, the minimum and maximum values of CH_4 emissions ratio at all gas samples were between T_0 (0.00 %) to T_4 (10.79 %) for the age of 3 years, while the 10 years of age emission ratio being lower at only T_0 (0.00 %) to T_4 (3.02 %). Figure 1 represents a summary of the methane emission ratio with the highest value was obtained from FSF and 3 years of age and the lowest value was observed from FLM and 10 years of age.

These findings may have been contributed by the microbial community growth, which leads to organic matter decomposition and the hydrolysis process of urea in urine gradually increasing methane emission over time. The amount of CH_4 released by manure management operations is determined by three main factors which are the manure management system, the surrounding environment, and the amount and composition of manure [31]. It is also observed that CH_4 emissions from only fresh slurry feces for both ages are higher compared to the liquid manure mixture. The higher organic matter presence in the fresh slurry feces justified the finding as 1 kg of FSF was used compared to the FLM with only 0.5 kg of fresh slurry feces used and diluted with 0.5 kg of urine. These results correspond to what has been reported on the low concentration of organic matter and nutrients in liquid manure which makes it harder to recover the energy and nutrients [32].

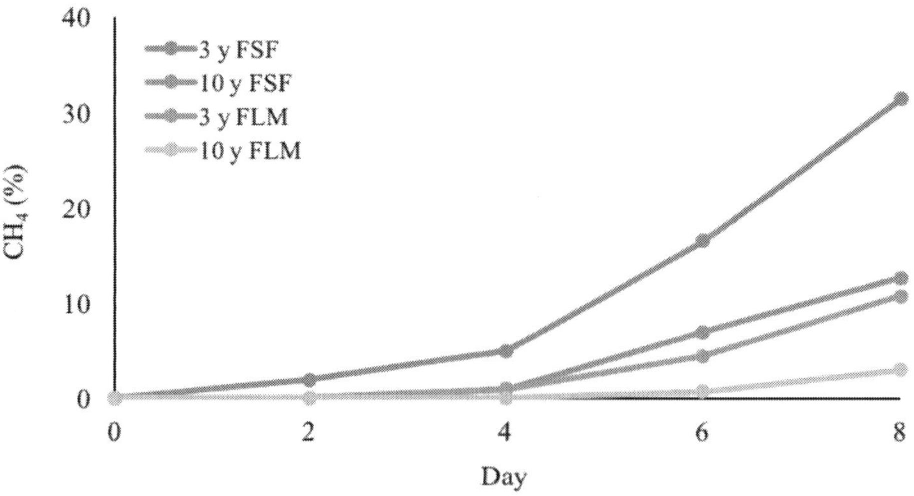

Figure 1. Methane emission ratio from manure mass for both stages.

Based on the one–way ANOVA analysis, there was significant difference in the measurement of CH_4, emitted from FSF for both age stages ($p<0.0001$). In addition to that, there was a significant difference between age stages and time ($p<0.0001$), as emissions increased over time. The statistical analysis (Table 2) showed that the standard deviation at age of 3 years was (0.00) at T_0 but showed a significant difference at T_1, T_2, T_3 and T_4.

Table 2. Statistical analysis for methane emission.

Time / d	3 y (FSF)		10 y (FSF)		3 y (FLM)		10 y (FLM)	
	Mean	Std Dev	Mean	Std Dev	Mean	Std Dev	Mean	Std Dev
T_0 / 0	0.00	0.00	0.00	0.00	0.00	0.00	0.00	0.00
T_1 / 2	1.93	0.15	0.00	0.00	0.00	0.00	0.00	0.00
T_2 / 4	4.93	1.30	0.89	0.33	0.95	0.95	0.00	0.00
T_3 / 6	16.50	3.01	6.96	1.23	4.47	0.87	0.74	0.75
T_4 / 8	31.41	0.96	12.67	3.65	10.79	2.99	3.02	3.02

3.2 Carbon dioxide emission

The recorded concentrations of CO_2, emitted from manure mass were different in the two age stages (3 and 10 years) and in the manure storage method. For the FSF storage method, the minimum and maximum of mean ratio of CO_2 emissions for the gas samples were between T_0 (1.87 %) to T_4 (50.27 %) for age of 3 years which is higher compared to the age of 10 years from T_0 (1.13 %) to T_4 (32.09 %). As for the FLM storage method, the CO_2 emissions ratio for the gas samples were between T_0 (0.92 %) to T_4 (49.94 %) for age of 3 years but much lower for age of 10 years with only between T_0 (0.90 %) and T_4 (26.1 %) (Figure 2).

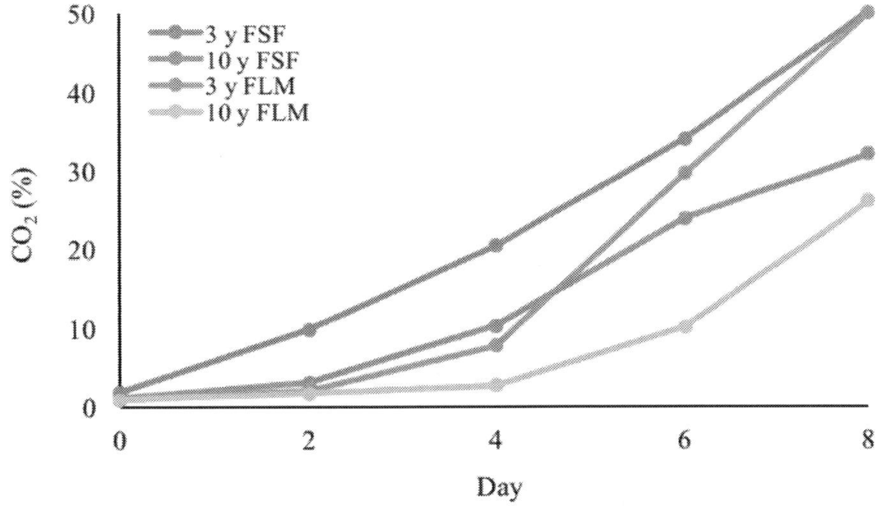

Figure 2. Carbon dioxide emission ratio from manure mass for both stages.

From the one–way ANOVA analysis, there were significant differences in the measurement of CO_2 emission from FSF at the two age stages in which the significance difference for both stages were ($p<0.0001$). In addition, there were significant differences for the time and the interaction between age stages and time which was $p<0.0001$ and $p<0.0009$, respectively. The statistical analysis in Table 3 showed that the standard deviation of CO_2 emission from FSF at age of 3 years was relatively convergent at T_0 and T_3 while it showed significant difference for T_1, T_2 and T_4.

Table 3. Statistical analysis for carbon dioxide emission.

Time / d	3 y (FSF)		10 y (FSF)		3 y (FLM)		10 y (FLM)	
	Mean	Std Dev	Mean	Std Dev	Mean	Std Dev	Mean	Std Dev
T_0 / 0	1.87	1.03	1.13	0.11	0.92	0.43	0.90	0.91
T_1 / 2	9.77	2.71	3.05	1.06	2.10	0.82	1.71	1.71
T_2 / 4	20.45	0.09	10.24	0.47	7.74	2.05	2.71	2.56
T_3 / 6	33.98	1.04	23.91	4.21	29.60	1.48	10.09	7.09
T_4 / 8	50.27	5.96	32.09	5.22	49.94	2.91	26.10	4.56

The statistical analysis showed a difference in the standard deviation of CO_2 emission from FSF for the age of 10 years for all gas samples, in which the minimum and maximum of standard deviation were 0.11 at T_0 and 5.22 at T_4. Based on the one–way ANOVA, there was a significant difference of CO_2 emission measurement from FLM between the two age stages and the time ($p<0.0001$). The standard deviation was also different at age 3 years for all gas samples, which were collected at regular intervals.

The study showed significant difference in carbon dioxide emission from dairy manure mass of FSF and FLM for both ages. This difference could be due to the major differences in pH and content of organic matter. Carbon dioxide is generated from the rapid hydrolysis of urea catalyzed by the urease enzyme and anaerobic fermentation of organic matter [33]. The results referenced the higher CO_2 concentration in the feces for both ages (3 and 10 years). It has been observed that CO_2 emission from FSF is more than that of FLM. Because of its porosity, FSF storage emits less methane than FLM as it lacks the anaerobic conditions needed for methanogenesis, reducing CH_4 emission. The amount of readily degradable carbon available for methanogenesis in FLM may be less than the amount of carbon available for methanogenesis in FSF [34]. The deterioration of organic matter in the feces could also be a factor as the amount of fresh slurry feces is twice the amount of fresh slurry feces used in the liquid manure. This was consistent with the results reported by Mathot et al. [35] showed that CO_2 emission from the solid fractions were more than the liquid fractions.

However, there was also an increase in the level of CO_2 from FLM more than the FSF. This may be attributed by the pH differences of the manure in which pH of stored liquid manure is affected by the hydrolysis process of urea in urine. The variation in physiochemical characteristics affects the release of gas remarkably as well [33].

4. Conclusions

The study found that CH_4 and CO_2 emissions for the cows at the age of 3 years were higher than those at 10 years of age. Not only that, gases emissions from FSF were also higher than FLM for both age stages of 3 and 10 years due to the fact that the organic matter degradation in the feces and amount of fresh slurry feces is twice the amount of fresh slurry feces used in the liquid manure, as well as the organic matter in the manure mass for the age of 3 years is higher than for the age of 10 years. Overall, the results from this study can help to improve the current development of manure management practices in dairy farms. However, an evaluation on the mitigation strategies for reducing emissions and improving overall sustainability of dairy farms should be performed. Detailed data of manure systems are also needed in order to propose mitigation strategies for mitigating climate change and global warming.

References

[1] Hanafiah M M, Ali MY, Aziz N I H A, Ashraf M A and Halim A A 2017 *Appl Ecol Environ Res* **15** 529-535
[2] Ali M Y, Hanafiah, M M and Khan M F 2018 *Sci. Total Environ.* **626** 1-10
[3] Hanafiah M M, Huijbregts M and Hendriks J 2010 *Sustainability* **2** 963-979
[4] Ali M Y, Hanafiah M M, Khan M F and Latif M T 2017 *Build Environ* **121** 238-246
[5] Idris S A A, Hanafiah M M, Khan M F and Hamid H H A 2020a *Chemosphere* **255** 126932
[6] Idris S A A, Hanafiah M M, Ismail M, Abdullah S and Khan M F 2020b *Arab. J. Geosci.* **13** 580
[7] Furtado D A, Amancio D, do Nascimento J W, Gomes J P and Silva R C 2012 *Engenharia Agricola* **32** 30-37
[8] Rotz C A 2018 *Int. J. Dairy Sci.* **101** 6675-6690
[9] Aziz N I H A, Hanafiah M M and Ali M Y 2019a *Renew. Energy* **132** 363-369
[10]Loftus C *et al* 2020 *Int. J. Hyg. Environ. Health* **223** 187-198
[11] Janni K 2020 *Atmosphere* **11** 453
[12] Aziz N I H A, Hanafiah M M and Gheewala S 2019b *Biomass and Bioenergy* **122** 361-374
[13] Rzeźnik W and Mielcarek P 2016 *Pol. J. Environ. Stud.* **25** 1813–1821
[14] Aguirre-Villegas H A and Larson R A 2017 *J. Clean. Prod.* **143** 169-179
[15] Aziz N I H A, Hanafiah M M, Gheewala S H and Ismail H 2020 *Sustainability* **12** 1-24
[16] Aziz N I H A, Hanafiah M M 2020 *Renew. Energy* **145** 847-857
[17] Ashraf M A and Hanafiah M M 2017 *Environ. Sci. Pollut. Res.* **24** 22753-22754
[18] Ashraf M A and Hanafiah M M 2019 *Environ. Sci. Pollut. Res.* **26** 13679-13680
[19] McCann L, Gedikoglu H, Broz B, Lory J and Massey R 2015 J. *Environ. Plan. Manag.* **58** 1346-1362
[20] Del Prado A, Mas K, Pardo G and Gallejones P 2013 *Sci. Total Environ.* **465** 156-165
[21] El-Mashad H M and Zhang R 2011 *Understanding Greenhouse Gas Emissions from Agricultural Management* (American Chemical Society)
[22] Cao J P, Huang X, Zhao X Y, Wei X Y and Takarada T 2015 *Fuel* **140** 477-483
[23] Abubakar B S U I and Ismail N 2012 *JEAS* **7** 169-172
[24] St-Pierre B and Wright A D 2013 *Animal* **7** 49-56
[25] VanderZaag A, Flesch T, Desjardins R, Baldé H and Wright T 2014 *Agric For Meteorol.* **194** 259-267
[26] Steinfeld H, Mooney H A, Schneider F and Neville L E 2013 *Livestock in a changing landscape, volume 1: drivers, consequences, and responses* (Island Press)
[27] Brand C, Goodman A, Rutter H, Song Y and Ogilvie D 2013 *Applied energy* **104** 158-169
[28] Thornton P K and Gerber P J 2010 *Mitig Adapt Strateg Glob Chang* **15** 169-184
[29] Malomo G A, Madugu A S and Bolu S A 2018 *Agricultural Waste and Residues* **119**
[30] Manale A, Hyberg S, Key N, Mooney S, Napier T and Ribaudo M 2016 *JSWC* **71** 69-81
[31] Moeletsi M E and Tongwane M I 2015 *Animals* **5** 193-205
[32] Szogi A A, Vanotti M B and Ro K S 2015 *Current Pollution Reports* **1** 47-56
[33] Dai X R, Saha C K, Ni J Q, Heber A J, Blanes-Vidal V and Dunn J L 2015 *Water research* **76** 110-119
[34] Holly M A, Larson R A, Powell J M, Ruark M D and Aguirre-Villegas H 2017 *Agric. Ecosyst. Environ.* **239** 410-419

IOP Conf. Series: Earth and Environmental Science **880** (2021) 012037 doi:10.1088/1755-1315/880/1/012037

[35] Mathot M, Decruyenaere V, Stilmant D and Lambert R 2012 *Agric. Ecosyst. Environ.* **148** 134-144

Acknowledgments
Marlia M. Hanafiah was funded by the UKM research grants (GUP-2020-034 and DIP-2019-001).

Waste-to-wealth in ASEAN countries: A case on e-waste generation from mobile phone

H Ismail[1,2] and M M Hanafiah[1,3,*]

[1] Department of Earth Sciences and Environment, Faculty of Science and Technology, Universiti Kebangsaan Malaysia, 43600 UKM Bangi, Selangor, Malaysia
[2] School of Technology Management and Logistics, College of Business, Universiti Utara Malaysia, 06010, Sintok, Kedah, Malaysia
[3] Centre for Tropical Climate Change System, Institute of Climate Change, Universiti Kebangsaan Malaysia, 43600 UKM Bangi, Selangor, Malaysia

*Corresponding E-mail: mhmarlia@ukm.edu.my

Abstract. The aim of this study was to evaluate the potential economic value of e-waste generation in ASEAN countries. The data used in this study was based on various reports. As the reliable information on e-waste generation in ASEAN countries is scarce, the study only evaluated the potential economic value of e-waste generation from mobile phone. The results showed that the potential economic value of e-waste from mobile phone around the world was valued at US$ 18,545.27 million. For ASEAN countries, it was valued at US$ 2,019.06 million, higher than 27 developed countries in European Union that valued at US$ 1,286.59 million. Currently, apart from the ineffective implementation of e-waste regulation, the lack of reliable information and data to estimate e-waste generation and its future projection was a common problem faced by ASEAN countries that impending the development of sustainable e-waste management. Therefore, the potential economic value of e-waste generation in ASEAN countries could be higher, if various other products were included in the analysis with a reliable information and data. Finally, some recommendation was made that includes the need a thorough intergovernmental cooperation and collaboration among ASEAN countries in order to reap a potential economic benefit from continuously growing e-waste in this region.

Keywords: Waste-to-wealth, E-waste generation, Mobile phone, ASEAN, Sustainable e-waste management

Track Name: Advanced Technology and Renewable Energy

1. Introduction

1.1 Production and consumption of electrical and electronic equipment (EEE)

The production an affordable and innovative electrical and electronic equipment (EEE) and the rapid growth of economic increase the consumption of electrical and electronic equipment (EEE) around the world [1, 2]. For an example, the information and communication technology (ICT) product such as computers and mobile phone – one of consumer product categorized under electrical and electronic equipment (EEE) – have been diffused at phenomenal rate into almost spheres of human activities. Apart from becoming a better and quicker way for interaction and communication in modern lifestyle, it also offers a great opportunity for economic and social development. Alongside with variety types of ICT products, the EEE also includes a wide-ranging consumer products from various types of large household appliances such as televisions, washing machines and refrigerators, to small household appliances such as rice cookers, microwave ovens and fryers.

With the variety types of EEE available in the market, the waste from EEE (e-waste or WEEE) has no general definition, but the Solving the E-waste Problem (StEP) Initiative highlighted that the "e-waste is a term used to cover items of all types of electrical and electronic equipment (EEE) and its parts that have been discarded by the owner as waste without the intention of re-use" [3]. Apart from the lack of general definition of e-waste, the classification of e-waste also varies from one country to another country [4]. For an example, the European Union countries under the recent European WEEE Directive classified the e-waste that consists of six categories types of e-waste - 1) Temperature exchange equipment, 2) Screens and monitors, 3) Lamps, 4) Large equipment, 5) Small equipment, and 6) Small IT and telecommunication equipment. On the contrary, China increased the number of product that classified as e-waste from five - television, washing machine, air-conditioner, refrigerator, personal computer to fourteen – that also includes 1) mobile phone, 2) single-machine telephone, 3) fax machine, 4) copier, 5) printer, 6) monitor, 7) range hood, 8) electric water heater, and 9) gas water heater [5]. Furthermore, the United Nation University (UNU) classified over than 900 electrical and electronic products into 54 categories of e-waste [6].

1.2 Current research on e-waste management

Due to increasing consumption of EEE around the world, the research on managing the end-of-life of EEE gained a considerable attention from various researchers in recent years [7-9]. The recent review studies showed that the number of studies in e-waste management was increasing exponentially with various aspects in e-waste management were evaluated [10-12].

According to Perez et al. [13], the research on e-waste management includes the investigation on e-waste management practices and issues, the evaluation of e-waste generation, the investigation on the characterization and composition of e-waste and its impact, the exploration on social and economic aspects of e-waste, the study on issues related to re-use and preparing for reuse of e-waste, and the study on the assessment methodologies to improve the end-of-life of EEE management. In another study, Ismail and Hanafiah [14] highlighted that various aspects in e-waste management were evaluated in previous studies that includes the evaluation of e-waste management practices in particular geographical area – at province- or district-level (e.g., Santa Elena Province in Ecuador [15]), national-level (e.g., Malaysia [16], Indonesia [17], Jordan [18]) and regional-level (e.g. European countries [19]), the investigation on the regulations and policies related to e-waste [20, 21], the design and application of extended producer responsibilities in e-waste management [22, 23], the study on consumer aspects in e-waste management such as e-waste disposal behaviour and willingness to pay [18, 24] and the investigation on performance assessment tools to improve the e-waste management such as life cycle assessment (LCA) [25-27], material flow analysis (MFA) [28] and multi-criteria decision making (MCDM) [29]. Apart from that, Ismail and Hanafiah [14] also highlighted that the quantification of e-waste also another important research area in e-waste management, where several previous studies investigated the characteristic and composition of e-waste [30, 31], the e-waste generation and its future projection amount [32-34], the international and domestic flow of e-waste [35-37] and the stock of e-waste [38, 39].

The increasing a considerable number of studies in research on e-waste management in recent years influenced by several factors. Apart from the technological advancement and the growth of economic around the world that consequently, it increases the consumption of variety types of EEE, the increasing concern on its disposal method and impact on sustainability aspects in recent years also influence the increasing number of studies in e-waste management around the world [7-9]. On one hand, the e-waste contained variety types of hazardous substances such as cadmium, chromium, lead and persistent organic pollutants that includes polybrominated diphenyl ethers (PBDEs), polychlorinated biphenyls (PCBs), and polychlorinated dibenzodioxins and dibenzofurans (PCDD/PCDFs) [40, 41], and therefore, it differs physically and chemically to other types of waste stream and it required a proper treatment and disposal method [42]. Hence, the improper disposal of e-waste will damage not only on the environment and its surrounding, but also on the human health. In another hand, the e-waste contained variety types of valuable and precious materials such as gold, silver, and platinum group of metals that includes palladium, indium and rhodium, and therefore, it provides an economic incentives and job opportunity. Therefore, the e-waste generation around the world is not only creating the environmental problems, but it also provides a potential economic and social benefits [43].

1.3 Research towards development of sustainable e-waste management system
The information on e-waste generation is very important for assisting the decision-maker in development of a proper and sustainable e-waste management [44]. This information is very useful in order to extend the knowledge in preparing the future e-waste management planning such as the infrastructure required and estimation of economic cost and revenue for proper e-waste recycling, treatment and disposal [14]. According to recent review study on e-waste management by Ismail and Hanafiah [14], various aspects in e-waste management were investigated further by using the information obtained from the work on estimation of e-waste generation. For an example, the study by Kalmykova et al. [45] and Parajuly et al. [46] investigated further the collection rate of e-waste in their studies. In addition, based on the information based on estimated e-waste generation, the study by Feszty et al. [47] and Lee et al. [48] investigated further whether the collection rate of e-waste complied with the collection rate target by the authority. In another study, Abbondanza and Souza [49] and Tan and Li [50] investigated further the distribution and density of e-waste based geographical area. While Yang et al. [51] investigated further the economic cost and benefit for e-waste recycling and disposal, Leigh et al. [52] investigated further the potential job creation based on estimated e-waste generation by using the classic economic production function. Furthermore, some studies such as Peeters et al. [53] and Kalmykova et al. [54] investigated further the availability of valuable secondary sources and its potential value.

In the literature, various methodologies to estimate e-waste generation exist [14, 32-34, 55]. The input-output method, which is consists of several variations (e.g., time step model, consumption and use model, simple delay model, etc.), is the most frequently used in analysis of e-waste generation at global- (e.g., [56]), country-level (e.g., [57]), and specific geographical areas such as district (e.g., [15]). In addition, some studies focused on e-waste generation by specific e-waste generator such as households (e.g., [9, 58]). Basically, three important data are required, namely the input data (e.g., sales), the stock data (e.g., penetration rate), and the lifetime of product, in evaluation of e-waste generation using input-output method ([16, 33, 34]).

2. Aim, Materials and Method Of The Study

The aim of this study was to evaluate the e-waste generation in ASEAN countries and its potential economic value. As a reliable information and data for ASEAN countries is scarce, the study only focused on e-waste generation from mobile phone, and the data used in this study was from various reports. The issue that related to the scarcity of information and data for estimation of e-waste generation in ASEAN countries were also mentioned by the recent United Nations Environment Programme (UNEP) report, prepared by the Basel Convention Regional Centre for South-East Asia (BCRC-SEA) [59].

To estimate the potential economic value of e-waste generation from mobile phone in ASEAN countries, two input data from two different sources were utilized, as illustrated in **Figure 1**. While the data on mobile phone subscribers was from the World Bank database [60], the value of waste mobile phone was from the United Nation University (UNU) study, in which they stated that the value of materials from waste mobile phone was € 2 per piece (or US$ 2.35), if recovered [61].

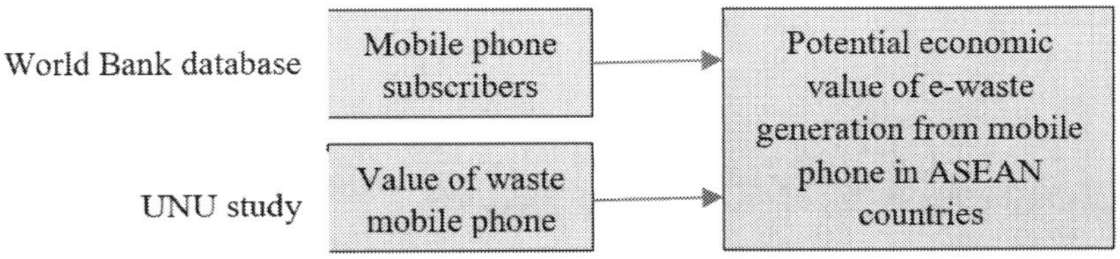

Figure 1. Methodology of the study.

3. Results

The recent World Bank reported that the total of mobile phone subscribers around the world was 7,858.27 million in 2018 [60]. Using the information from the UNU study [61], it was estimated that the potential economic value of e-waste from mobile phone was US$ 18,545.51 million, as illustrated in Figure 2 (a). In addition, it was reported that the total mobile phone subscribers in ASEAN countries was 855.54 million. Therefore, the total economic value was estimated about US$ 2,019.06 million. It is worthy to highlight that these numbers were higher compared to 27 developed countries in European Union (EU-27), in which the total of mobile phone subscribers was 545.16 million and it was valued at US$ 1,286.59 million.

Further analysis was performed to rank the potential economic value based on country in ASEAN. The results showed in Figure 2 (b), with the highest potential economic value was Indonesia (US$ 753.87 million), and the least was Brunei (US$ 1.34 million). The results also showed that the potential economic value of e-waste from mobile phone in six ASEAN countries was valued more than US$ 100 million for each country, with the top was Indonesia with the potential economic value was US$ 753.87 million, followed by Viet Nam (US$ 331.91 million), Philippines (US$ 317.64 million), Thailand (US$ 295.23 million), Myanmar (US$ 144.30 million), and Malaysia (US$ 100.10 million).

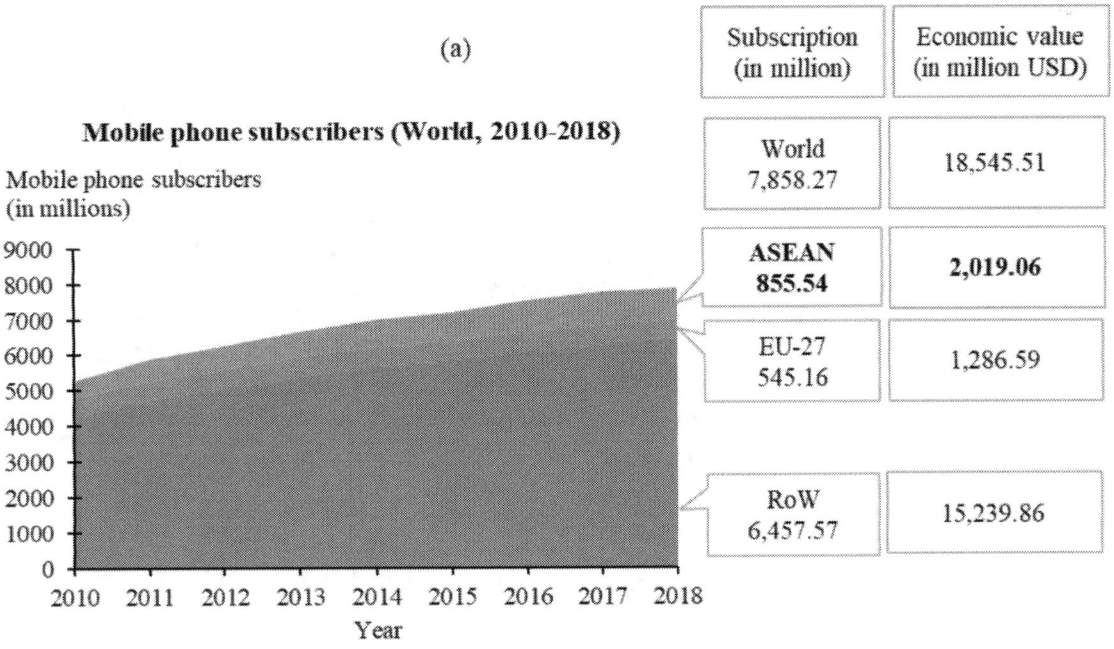

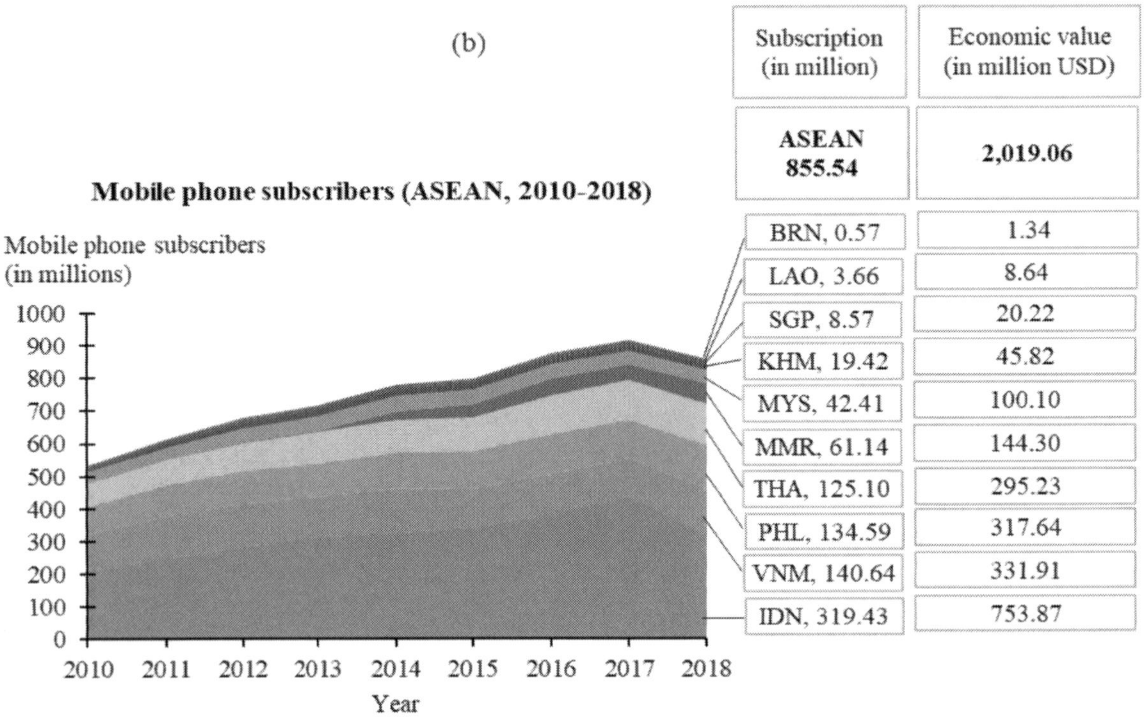

Figure 2. (a) Mobile phone subscribers from 2010-2018 and its potential economic value (World), (b) Mobile phone subscribers from 2010-2018 and its potential economic value (ASEAN countries). Sources: [60, 61].

4. Discussion

The number of e-waste generation is increasing significantly year by year, as the trend on population and economic around the world continually up growing. These situation is no difference to growing ASEAN countries. The significant increasing trend of mobile phone subscription from 2010 to 2018 in ASEAN countries could be illustrated the continually growing of population and economic of ASEAN countries.

The recent data estimated that the potential economic value of e-waste generation from mobile phone in ASEAN countries was valued at US$ 2,019.06 million. These numbers were surpassed the potential economic value of e-waste from mobile phone in 27 developed countries in European Union, in which it was valued at US$ 1,286.59 million. Out of ten ASEAN countries, the potential value of e-waste generation from mobile phone in six ASEAN countries was valued more than US$ 100 million for each country. The Indonesia topping the list with the potential economic value was US$ 753.87 million, followed by Viet Nam (US$ 331.91 million), Philippines (US$ 317.64 million), Thailand (US$ 295.23 million), Myanmar (US$ 144.30 million), and Malaysia (US$ 100.10 million).

Currently, a number of developing countries around the world, including the developing ASEAN countries facing a serious challenge in management e-waste recycling, treatment and disposal, as the regulations and policies enforcement to improve the e-waste management is still lacking. Compared to developed countries such as European Union countries, various regulations and policies were designed and implemented in order to improve the e-waste management. For an example, the European WEEE Directive were designed and implemented by providing a clear regulations and policies related to e-waste management for European Union countries [20].

Apart from the lacking on the regulations and policies enforcement to improve e-waste management in ASEAN countries, there was a numbers of challenges faced by ASEAN countries in developing a proper and sustainable e-waste management, that includes the lack of e-waste inventory data that important to estimate e-waste generation and its future projection, as reported by recent BCRC-SEA study [59]. This report also highlighted that the quantification and control on the e-waste flow – domestic and international transboundary (import and export) – is not well documented. Furthermore, the lack on public awareness program related to e-waste management and the lack of proper recycling technologies and facilities were also highlighted by the BCRC-SEA report [59].

The potential economic value of e-waste generation in ASEAN countries are expected to grow, if various other types of product are included in the analysis. However, currently the ASEAN countries faced various challenges that impending the development of sustainable e-waste management. In addition, the challenges faced by ASEAN countries was quite similar to each other such as the lack of e-waste inventory and effective e-waste regulation. Therefore, the study urged that there is a need to initiate a thorough intergovernmental cooperation and collaboration among ASEAN countries in this research area. Besides developing a proper and sustainable e-waste management, it could be sustaining the further growth of ASEAN countries by reaping the potential economic benefits from continuously growing e-waste in this region.

5. Conclusions

The present study attempted to estimate the potential economic value of e-waste generation in ASEAN countries. As a reliable information and data to estimate e-waste generation in ASEAN countries was scarce, the study only focused on potential economic value of e-waste generation from mobile phone and the study was performed using information obtained from various reports. The total potential economic value of e-waste from mobile phone in ASEAN countries was valued at US$ 2,019.06 million, but we believed that the estimated potential economic value of e-waste in ASEAN countries will further increased, if various other types of product were included in the analysis.

Currently, the ASEAN countries facing a serious challenge in developing a proper and sustainable e-waste management that includes the lack of regulations and policies, the scarcity on e-waste inventory data requires for estimation of e-waste generation and its future projection, the unwell documented information of domestic and international flows (import and export) of e-waste, the insufficient public awareness program on e-waste management, and the inadequate proper recycling technologies and facilities. As there was a numerous challenges faced by ASEAN countries in developing a proper and sustainable e-waste management, therefore, the study recommended that there is a need a thorough intergovernmental cooperation and collaboration among ASEAN countries in this research area, as it could, subsequently, provide an opportunity to enhance a proper and sustainable e-waste management from continuously growing e-waste in this region.

References

[1] Sharma M, Joshi S and Kumar A 2020 *Environ. Sci. Pollut. Res.* **27** 13325–13338
[2] Zhang S, Gu Y, Tang A, Li B, Pan D and Wu Y 2021 *J. Clean. Prod.* **283** 124575
[3] StEP Initiative 2014 *One Global Definition of E-waste* (Bonn, Germany: Solving the E-Waste Problem (StEP) White Paper)
[4] Wang F, Huisman J, Baldé K and Stevels A 2012 *2012 Electronics Goes Green 2012+ (Berlin, Germany)* p 1–6
[5] Zeng X, Gong R, Chen WQ and Li J 2016 *Environ. Sci. Technol.* **50** 1347–1358
[6] Forti V, Baldé K and Kuehr R 2018 *E-waste Statistics: Guidelines on Classifications, Reporting and Indicators* (Bonn, Germany: United Nations University, ViE – SCYCLE)
[7] Polat LO and Gungor A 2021 *Environ. Sci. Pollut. Res.* **28** 7786–7804
[8] Islam MT, Dias P, and Huda N 2020 *J. Environ. Mang.* **275** 111111
[9] Ismail H and Hanafiah MM 2019 *AIP Conference Proceedings* Vol. 2111 p 060011
[10] Andrade DF, Romanelli JP and Pereira-Filho ER 2019 *Environ. Sci. Pollut. Res.* **26** 17135–17151
[11] Gao Y, Ge L, Shi S, Sun Y, Liu M, Wang B, Shang Y, Wu J and Tian J 2019 *Environ. Sci. Pollut. Res.* **26** 17809–17820
[12] Zhang L, Geng Y, Zhong Y, Dong H and Liu Z 2019 *Environ. Sci. Pollut. Res.* **26** 21098–21108
[13] Perez-Belis V, Bovea MD and Ibanez-Fores V, 2015 *Waste Manag. Res.* **33** 3–29
[14] Ismail H and Hanafiah MM 2020 *J. Environ. Mang.* **264** 110495
[15] Campuzano MG and Crisanto T 2020 *Information and Communication Technologies. TICEC 2020. Communications in Computer and Information Science* Vol.1307 Eds. Rodriguez-Morales G, Fonseca CER, Salgado JP, Pérez-Gosende P, Orellana-Cordero M and Berrezueta S. (Springer, Cham)
[16] Ismail H and Hanafiah MM 2019 *J. Clean. Prod.* **238** 117927
[17] Mairizal AQ, Sembada AY, Tse KM and Rhamdhani MA 2021 *J. Clean. Prod.* **293** 126069

[18] Hamdan S and Saidan MN 2020 *Environ. Manag.* **66** 884–898
[19] Ibanescu D, Cailean D, Teodosiu C and Fiore S 2018 *Waste Manag.* **73** 39–53
[20] Ongondo FO, Williams ID and Cherrett TJ 2011 *Waste Manag.* **31** 714–730
[21] Nnorom IC and Osibanjo O 2008 *Resour. Conserv. Recycl.* **52** 843–858
[22] Agamuthu P and Victor D 2013 *J. Mater. Cycles Waste Manag.* **15** 411–419
[23] Wang H, Gu Y, Li L, Liu T, Wu Y and Zuo T 2017 *Resour. Conserv. Recycl.* **127** 159–167
[24] Borthakur A and Govind M 2018 *J. Clean. Prod.* **172** 1053–1066
[25] Ismail H and Hanafiah MM 2019 *J. Clean. Prod.* **232** 79-93
[26] Xue M and Xu Z 2017 *Environ. Manag.* **59** 693-707
[27] Teixeira FSM, Peres ACC, Gomes TS, Visconte LLY and Pacheco EBAV 2021 *J. Polym. Environ.* **29** 1333–1349
[28] Islam MT and Huda N 2019 *J. Environ. Manag.* **244** 344–361
[29] Marinello S and Gamberini R 2021 *Toxics* **9** 13
[30] Singh N, Duan H, Yin F Song Q and Li J 2018 *ACS Sustain. Chem. Eng.* **6** 13016–13024
[31] Singh N, Tang Y and Li J 2019 *Waste Manag. Res.* **37** 1170–1177
[32] Ikhlayel M 2016 *Resour. Conserv. Recycl.* **108** 134–139
[33] Li B, Yang J, Lu B and Song X 2015 *Waste Manag.* **35** 247–254
[34] Wang F, Huisman J, Stevels A and Balde CP 2013 *Waste Manag.* **33** 2397–2407
[35] Tong X, Wang T, Chen Y and Wang Y 2018 *Resour. Conserv. Recycl.* **135** 163–171
[36] Lepawsky J and McNabb C 2010 *Can. Geogr.* **54** 177–195
[37] Breivik K, Armitage JM, Wania F and Jones KC 2014 *Environ. Sci. Technol.* **48** 8735–8743
[38] Gusukuma M and Kahhat R 2018 *Resour. Conserv. Recycl.* **138** 142–150
[39] Jo HJ, Kang HY, Lee IS and Jun YS 2018 *J. Mater. Cycles Waste Manag.* **20** 965–974
[40] Kiddee P, Naidu R and Wong MH 2013 *Waste Manag.* **33** 1237-1250
[41] Xu X, Zeng X, Boezen HM and Huo X 2015 *Front. Med.* **9** 220–228
[42] Robinson BH 2009 *Sci. Total Environ.* **408** 183–191
[43] Herat S and Agamuthu P 2012 *Waste Manag. Res.* **30** 1113-1129
[44] Petridis NE, Stiakakis E, Petridis K and Dey P 2016 *J. Clean. Prod.* **112** 3072–3085
[45] Kalmykova Y, Berg PEO, Patrício J and Lisovskaja V 2017 *Resour. Conserv. Recycl.* **120** 65–74
[46] Parajuly K, Habib K and Liu, G 2017 *Resour. Conserv. Recycl.* **123** 85–92
[47] Feszty K, Murchison C, Baird J and Jamnejad G 2003 *Waste Manag. Res.* **21** 207–217
[48] Lee JC, Song HT and Yoo JM 2007 *Resour. Conserv. Recycl.* **50** 380–397
[49] Abbondanza MNM and Souza RG 2019 *Waste Manag.* **85** 374–384
[50] Tan Q and Li J 2014 *J. Clean. Prod.* **81** 227–233
[51] Yang J, Lu B and Xu C 2008 *Waste Manag.* **28** 1589–1597
[52] Leigh NG, Realff MJ, Ai N, French SP, Ross CL and Bras B 2007 *Resour. Conserv. Recycl.* **51** 847–869
[53] Peeters JR, Bracqguene E, Nelen D, Ueberschaar M, Van Acker K and Duflou JR 2018 *J. Clean. Prod.* **175** 96–108
[54] Kalmykova Y, Patrício J, Rosado L and Berg PEO 2015 Waste Manag. **46** 511–522
[55] Mihai F, Gnoni M, Meidiana C, Ezeah C and Elia V 2019 *Electronic Waste Management and Treatment Technology* Eds. Narasimha M, Prasad V and Vithanage M (Butterworth-Heinemann) Chapter 1 pp 1–34
[56] Yu J, Williams E, Ju M and Yang Y 2010 *Environ. Sci. Technol.* **44** 3232–3237
[57] Johnson M, Fitzpatrick C, Wagner M and Huisman J 2018 *Resour. Conserv. Recycl.* **131** 1–16
[58] Nguyen DQ, Yamasue E, Okumura H and Ishihara KN 2009 *J. Mater. Cycles Waste Manag.* **11** 358–366
[59] United Nations Environment Programme (UNEP) 2018 *Study on e-waste management in ASEAN countries* (SCRC Indonesia: Basel Convention Regional Centre for South-East Asia (BCRC-SEA))
[60] World Bank 2020 *Retrieved from* https://data.worldbank.org/indicator/IT.CEL.SETS
[61] Baldé CP, Forti V, Gray V, Kuehr R and Stegmann P 2017 *The Global E-waste Monitor – 2017* (Bonn/Geneva/Vienna: United Nations University (UNU), International Telecommunication Union (ITU) & International Solid Waste Association (ISWA))

Acknowledgments
Marlia M. Hanafiah was funded by the National University of Malaysia (GUP-2020-034 and DIP-2019-001).

Water Quality of Rooftop Rainwater Harvesting System (MyRAWAS)

N U M Nizam[1], M M Hanafiah[1,2*], M B Mokhtar[3] and N A Jalal[1]

[1] Department of Earth Sciences and Environment, Faculty of Science and Technology, Universiti Kebangsaan Malaysia, 43600 UKM Bangi, Selangor, Malaysia
[2] Centre for Tropical Climate Change System, Institute of Climate Change, Universiti Kebangsaan Malaysia, 43600 UKM Bangi, Selangor, Malaysia
[3] Research Centre for Sustainability Science & Governance (SGK), Institute for Environment and Development (LESTARI), Universiti Kebangsaan Malaysia, 43600, UKM Bangi, Selangor, Malaysia

*Corresponding E-mail: mhmarlia@ukm.edu.my

Abstract. Prolonged drought, population growth and water demand for various purposes have increased the water scarcity issue. To overcome this issue, a rainwater harvesting system can be utilized as an alternative for clean water supply. A rainwater harvesting system is a method of collecting rainwater from man-made surfaces such as rooftops and constructed surfaces and can be used for various sectors including household, agricultural and commercial. This study was conducted to determine the quality of rainwater harvested collected directly from rooftop. The quality of the rooftop rainwater was taken in three consecutive months and the water quality for before and after treatment was measured and compared. Commercial activated carbon was used to treat the rainwater obtained from the rooftop. The water quality was compared with the Water Quality Index (WQI) and the National Water Quality Standards (NWQS). The parameters involved are pH, temperature, conductivity, dissolved oxygen (DO), total suspended solids (TSS), ammoniacal nitrogen (NH_3-N), biochemical oxygen demand (BOD), chemical oxygen demand (COD), *E.coli* and total coliform bacteria. The results showed that the total value of WQI before and after treatment was 86.3 ± 8.963 and 87.6±2.081, respectively. Positive correlations were found for parameter NH_3-N, COD and pH, while paired T-test showed a significant in the COD and the presence of bacteria. Total Coliform is still at a safe level by NWQS with the average value and the standard deviation for before and after treatment were 38.11 ± 13.960 cfu/ml and 10.33 ± 6.671 cfu/ml, respectively.

Keywords: Rainwater harvesting system, Water quality, Sustainable water resource, MyRAWAS, Malaysia

Track Name: Advanced Technology and Renewable Energy

1. Introduction

Rapid urbanization causes a series of environmental problems such as clean water and water resource shortages, deteriorating urban water quality, inadequate water infrastructure, and frequent flood disaster [1]. Clean water supply crisis definitely gives a huge impact on our daily life. Therefore, the probability of occurrence of water crisis that will bring disaster to the people and the national economy should be taken more seriously. The population has increased for the last six years with a rate of 3.5 per cent per annum and this scenario requires an increase of about 10 % of water reserve capacity to meet the needs of users at peak times and to accelerate the recovery period of the water supply disruptions [2]. Some measures should be taken to overcome the water crisis that includes aspects of the present and future mitigation. In order to help overcoming the water crisis in the future, a rainwater harvesting system and its implementation has been proposed as a part of an alternative solution to the dependency of the existing water supply. In addition, a rainwater harvesting system is seen as an alternative source of which is based on green technology that is environmentally friendly and sustainable [3].

A rainwater harvesting system is a method of collecting rainwater from man-made surfaces such as rooftops and constructed area which can be used for various sectors including household, agricultural and commercial. This collection system is feasible and an educational way to reduce potable water use. Precipitation is one of the sources of water and can be obtained from the natural water cycle that continues through the process of evaporation and transpiration [3]. Rainwater harvesting system is suitable to be implemented in Malaysia because Malaysia receives high rainfall and is in the equatorial climate, which is hot and humid. Rainfall of 3000 mm per year is estimated to be equivalent to the total annual water resources of about 990 billion m^3 [4]. Therefore, the rainwater harvesting system should be implemented to resolve the water crisis in the country. However, the aspects of water quality need to be considered in the use of rainwater to ensure the quality of the water is in a class Water Quality Index (WQI) which is suitable and safe for use by humans and should be according to the standards of WQI, Department of Environment.

In addition, water quality should also be compared with the National Water Quality Standards (NWQS) drafted by the Department of Environment, Malaysia. Rainwater harvesting system provides many benefits to all aspects of the consumer and environment. Among the benefits that can be gained is the ability to reduce the use of available fresh water to a more important use like for food and beverages. Water consumption bills can also be reduced due to saving fresh water in everyday life. Furthermore, the system can slow down water runoff and may overcome the problem of clean water supply, especially in rural areas.

Currently, universities in Malaysia are experiencing clean water supply crisis especially in residential colleges [5]. Therefore, various alternatives have been introduced to address this problem and rainwater harvesting is one of the initiatives that has been introduced because the construction of rainwater harvesting systems have no negative impact and does not pollute the environment. Ungku Omar College (KUO) is a college that affected the most by the crisis among the colleges in the Universiti Kebangsaan Malaysia (UKM) and MyRAWAS project is a pilot project in which was built in KUO. High rainfall in Malaysia makes it a country appropriate for rainwater harvesting system. The quality of rainwater collected through MyRAWAS system should be monitored periodically to ensure the cleanliness and safety of the water for the use of students and residents of the college. Thus, the study of water quality including chemical, physical and biological aspects needs to be done consistently. This study aimed to determine the quality of harvested rainwater before and after the water treatment by using the WQI for the parameters of physico-chemical properties. In addition, this study was conducted to evaluate the presence of bacteria in the water based on biological parameters according to the NWQS. Field sampling was conducted in November 2015, December 2015 and January 2016.

2. Materials and Method

2.1. Sampling and study area

Rainwater harvesting system (MyRAWAS) was built between student's residence of Ungku Omar College building (K8M and K8K) as shown in Figure 1. PVC pipes were used to connect the drainage holes for rainwater collection tank, before flowing to the treatment process. Environmentally friendly method was used for rainwater harvesting system consisting of gravel, sand and activated carbon. Rainwater that has been through the treatment process was connected to the storage tank.

Sampling was done three times in November 2015, December 2015 and January 2016. The sampling was carried out in two places, drainage channel that connects the roof to the roof gutter and from the channel where rainwater harvesting treatment system was done. Each sampling was replicated three times to get a more accurate and precise reading. All parameters were analyzed in comparison with the Water Quality Index (WQI) and the National Air Quality Standards (NWQS).

Figure 1. MyRAWAS system

2.2. Sample preparation

All bottles except for the BOD and Schott bottles used prior sampling were soaked for 24 hours with 10 % nitric acid solution, rinsed with distilled water and dried. The collected water samples were preserved to prevent any biological or chemical changes that can alter the composition of real water samples. Table 1 shows the methods and the maximum preservation of water samples for each parameter.

Table 1. Preparation of samples for ex situ parameter analysis.

	Bottle	Preservation method	Preservation period
TSS	Polyethylene	Stored at 4 °C	7 days
NH_3-N	Polyethylene	Stored at 4 °C	28 days
COD	Polyethylene	Stored at 4 °C	28 days
BOD	Dark glass	Stored at 4 °C	48 days
E. coli bacteria	Schott	Stored at 4 °C	6 hours
Total coliform	Schott	Stored at 4 °C	6 hours

Source: APHA (2017)

2.3. In situ and ex situ measurement

The in-situ parameters involved for the water quality measurement are temperature, pH, conductivity and dissolved oxygen (DO). These parameters are less stable and highly dependent on environmental conditions during measurement [6]. Determination of these parameters was done using the YSI 556 multiparameter on site and the readings were recorded. Ex-situ measurements conducted in the laboratory involved physico-chemical parameters and biological parameters. Table 2 shows the methods and equipment / materials used in the analysis of ex-situ parameters.

Table 2. Methods and instruments used for ex situ parameter.

	Method	Instrument
TSS (mg/L)	Gravimetric	Vacuum pump
NH_3-N (mg/L)	Nessler	Spektrofotometer HACH DR/2010
COD (mg/L)	Digestion and calorimetry determination	Spektrofotometer HACH DR/2010
BOD (mg/L)	($DO_{initial} - DO_{final}$)	YSI 5000 dissolved oxygen meter
E. coli bacteria	Membrane filtration	Eosin methylene blue
Total coliform	Membrane filtration	Eosin methylene blue

2.4. Total suspended solid

Determination of total suspended solids (TSS) has been carried out using glass fibre filter paper with pore size of 0.45ɳm. The filter paper was first weighted (B). After filtering the water samples using vacuum pump, the filter paper was dried in the oven at 103 – 105 °C and weighted (A). Equation 1 was used for the determination of TSS.

$$\text{Total suspended solids (mg/L)} = (A-B)/V \times 1000 \text{ mL} \quad (1)$$

Where,
A = Weight of filter paper after filtration (weight of filter paper + dried residue), mg
B = Weight of filter paper before filtration (weight of filter paper), mg
V = Volume of filtered water sample, mL

2.5. Ammoniacal-nitrogen

Determination of ammonia-nitrogen (NH_3-N) concentrations was obtained using Nessler's Reagent [6]. Both the samples were shaken to make sure the samples were mixed well. 1 ml of Nessler's reagent was added to both samples and was shaken daily. Analysis was first carried out with deionised water (blank solution) followed by water samples. All sample cells were wiped clean in advance to avoid light reflection irregularities [7]. The samples were measured using wavelengths of 380 and 425 nm on a HACH DR / 2010 spectrophotometer. Water samples that turned yellow colour shows the presence of NH_3-N.

2.6. Chemical oxygen demand

Determination of chemical oxygen demand (COD) was performed with a tube in the 45 °C condition, using HACH 2021 method. The prepared sample was mixed with COD reagent and shaken. The water and blank sample solutions were heated using a Model 456000 COD Reactor with a strong oxidizing agent, such as potassium dichromate for two hours at 150 °C. Water solution and blank samples were then cooled at a temperature of 20 °C up to room temperature for 20 minutes. Then, analysis was carried out and all sample cells were wiped clean to avoid light reflection irregularities [7] before measured using wavelengths of 435 and 420 nm on a spectrophotometer HACH DR/2010.

2.7. Biochemical oxygen demand

During sampling, dissolved oxygen (DO) meter YSI 5000 was used and the initial value was recorded for $DO_{initial}$ reading (Equation 2). Then, the water samples were stored in the dark glass bottles and covered immediately after sampling. The water samples were incubated for five days at a temperature of 18°C and stored in a place that was not exposed to the sunlight in order to prevent any biological or chemical changes that can alter the composition of real water samples. Using the same method, reading on the fifth day of all samples were recorded.

$$BOD_5 = DO_{initial} \text{ (reading on the first day)} - DO_{final} \text{ (reading on the fifth day)} \quad (2)$$

2.8. E. coli bacteria and total coliform

Determination of *E. coli* and total coliform bacteria was done by using a sample of 1000 ml of water using the schott bottle, sterilized beforehand. Before the analysis, the room was made to be clean to avoid the presence of bacteria because the environment may affect the amount of bacteria in the water samples [8]. The instruments and apparatus were also sterilized. Cellulose acetate membrane filters were inserted into a petri dish containing eosin methylene blue nutrient with no air bubbles. Petri dishes were incubated for 22 hours ± 2 at a temperature of 35°C and calculation was made after completing this step. The *E. coli* bacteria was determined by counting the metallic green colonies while the total coliform colonies appeared not colored or bright purple. The colonies of both biological parameters were calculated based on Equation 3 and the unit is cfu (Colony Forming Unit) [6].

$$E. coli \text{ (100 ml)} = (100 \times \text{total of } E. coli \text{ colony})/\text{Volume of filtered water sample (mL)} \quad (3)$$

2.9. Water quality index

Classification of water quality before and after treatment of rainwater harvesting system was based on the WQI and by using Equation 4. Classification of physico-chemical parameters and biological sampling for both treatments was determined based on the National Air Quality Standards (NWQS).

$$WQI = 0.22 (siDO) + 0.19 (siBOD) + 0.16 (siCOD) + 0.15 (siNH_3\text{-}N) + 0.16 (siTSS) + 0.12 (sipH) \quad (4)$$

*si=subindex

2.10. Statistical analysis

Correlation test was used to determine the correlation between physico-chemical parameters for both sampling. T-paired test was also used to determine the difference between the water quality before and after rainwater harvesting system. The statistical analysis was run using SPSS software.

3. Results and Discussion

3.1. Total suspended solid

The TSS readings before and after treatment were 0.12 ± 0.07 mg/L and 0.07 ± 0.05 mg/L, respectively, and both parameters were in class I in WQI (Figure 3). TSS values at pre-treatment sampling were higher than after treatment. This is because the water collected directly from the roof contains sediments, dirt, dry foliage and animal faeces that are most likely to be collected and accumulated on the roof due to the uncovered roof area, and wind factors that contribute to the TSS content for sampling before the treatment. While the TSS value for the second sampling was lower due to the closed tank which does not allow any sediments and debris accumulated in the tank. Some of the factors that can influence TSS readings include wet and dry seasons, water turbulence caused by the wind and economic activities, as well as human settlement. High TSS content can cause higher turbidity that affects physical properties of water [9].

3.2. Ammoniacal-nitrogen

NH_3-N readings before and after treatment were 0.22 ± 0.28 mg/L and 0.04 ± 0.05 mg/L, respectively, and both parameters were in class I. NH_3-N consists of nitrite, nitrate, organic nitrogen, nitrogen gas and organic ammonia nitrogen which usually exists in living organisms [10]. The NH_3-N test is very important in the measurement of water quality because the presence of the nitrogen element can affect the oxygen level in the water. Water with a high content of organic nitrogen and NH_3-N is considered unsafe because it is toxic to the living organisms and may cause high irritation to human skin. The main sources that contribute to the presence of NH_3-N are animal faeces, cleaning agents, solvents or sewage materials. NH_3-N values were higher for pre-treatment sampling than after treatment, as shown in Figure 3. Compared to enclosed tanks and exposed roofs, it clearly shows that roofs are more exposed to environments that contribute to higher NH_3-N concentrations. In addition, the decomposition process by fungi and bacteria can also increase ammonia concentrations [11].

3.3. Chemical oxygen demand

The COD value before treatment was 13.67 ± 13.23 mg/L (class II), whereas after treatment was 2.0 ± 1.12 mg/L (class I). COD measurement is important to determine the amount of oxygen required by oxidizing agents in the process of oxidizing substances to water, ammonia and carbon dioxide chemically [1]. COD values increase in parallel with increasing concentrations of organic matter in water [12]. The COD is used to determine the level of water pollution caused by organic compounds that did not undergo biological decomposition [13].

3.4. Biochemical oxygen demand

The BOD parameters for before and after treatments were in class II and their values were 1.09 ± 1.16 mg/L and 1.33 ± 0.22 mg/L, respectively. BOD test was conducted to determine the level of water pollution and the presence of microorganisms involved in the process of decomposition of organic matter and the amount of dissolved oxygen to oxidize organic matter in water, which leads to an increase in the concentration of BOD [14]. The BOD parameter correlated with DO (0.63) and showed the relationship between each other. This implies that the higher the BOD, the lower the DO of the water body.

IOP Conf. Series: Earth and Environmental Science **880** (2021) 012039 doi:10.1088/1755-1315/880/1/012039

3.5. pH
The pH values before and after treatments were both in class II and their values were 6.56 ± 0.34 and 6.61 ± 0.25, respectively. Carbon dioxide found in the atmosphere dissolved in rainwater droplets and forms a very weak carbonic acid [15]. The acids then precipitate on rainwater and cause the water to be acidic. There is a relationship between NH_3-N and pH parameters, the lower the NH_3-N, the better the pH of the water body. A high pH value can cause toxicity to water due to the high ammonia content, while if it is too low, it can cause acidity to the water. The normal pH for clean water is in the range of 5-7.

3.6. Dissolved oxygen
The DO values before and after treatments were in class II with 68.18 ± 20.45 % and 54.40 ± 7.21 %, respectively. The values in unit are shown in Figure 3. Organic matters found in domestic wastes undergo an oxidation process by microorganisms and this process requires oxygen which causes the DO content in the water to be low [1,16].

The positive correlation with the value of R^2 equivalent to 1 was found for NH_3-N, COD and pH parameters. Paired T-test showed significant value ($p < 0.05$), where there was a difference in values before and after rainwater treatment for COD (0.03) with the presence of bacteria.

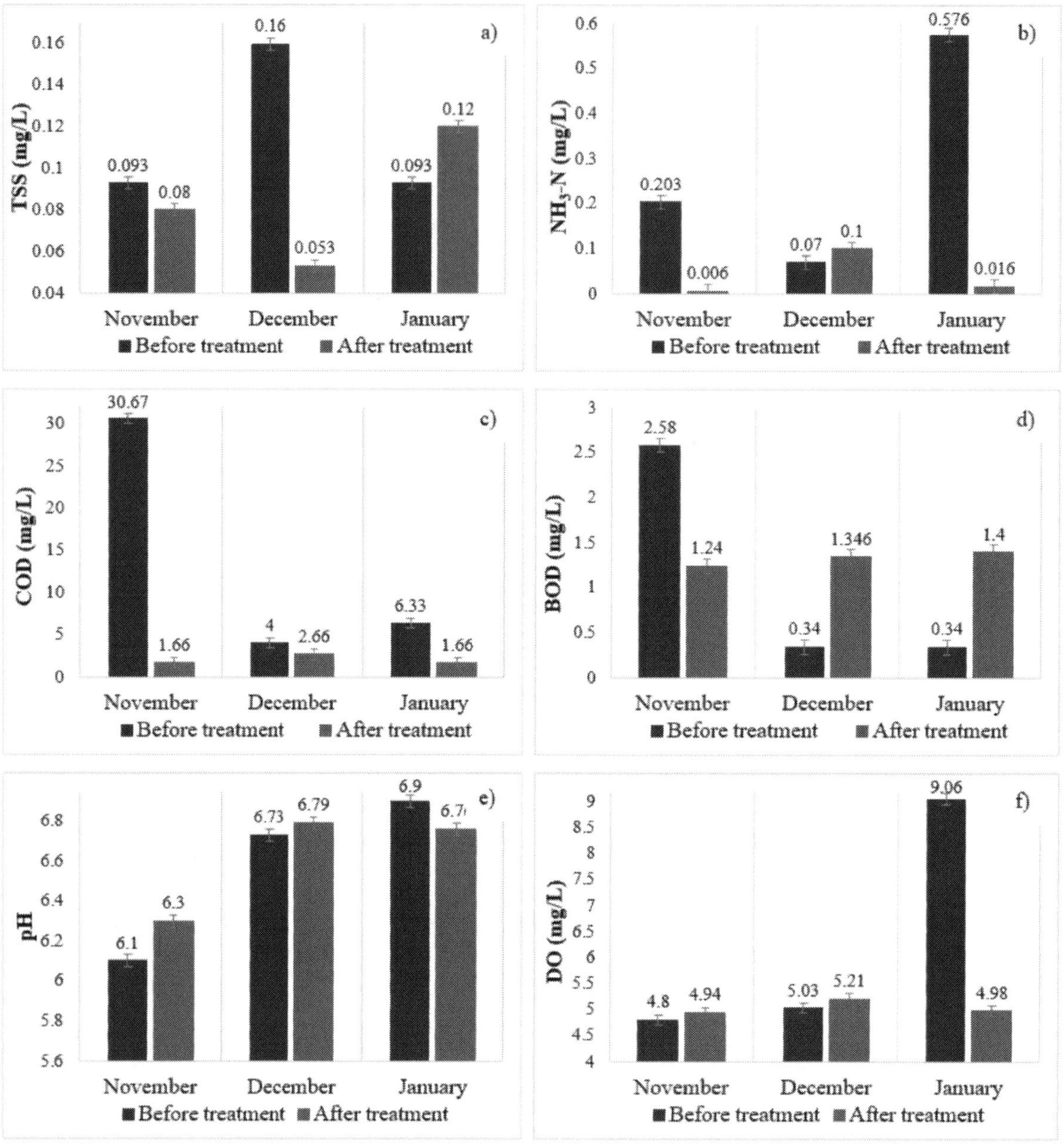

Figure 2. The parameters values of (a) TSS, (b) NH_3N, (c) COD, (d) BOD, (e) pH, and (f) DO, in before and after water treatment.

3.7. Water quality index

Based on the WQI standards in Figure 4, the rainwater quality of before and after the treatments was categorized in class II. The physico-chemical parameters for before and after treatments varies slightly due to the condition of the water that being stored in the tank for a certain duration of times. For class II, the use of water for drinking purposes requires conventional treatments. MyRAWAS is an alternative in overcoming the water supply crisis in KUO. The water collected from the rainwater harvesting system can be used for domestic use such as watering trees, washing cars and utensils and toilet flushing. However, rainwater from this system need to go for further to make it suitable for drinking water and to ensure that the rainwater is completely safe.

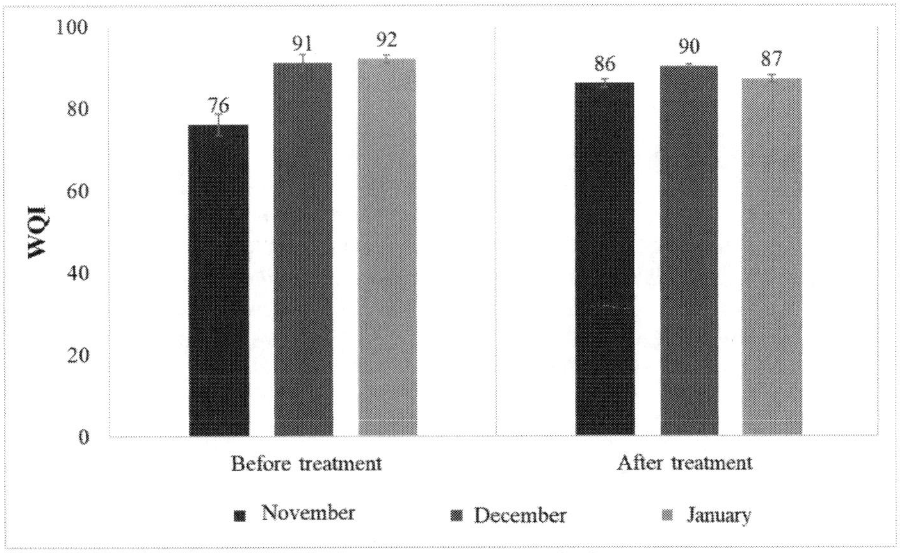

Figure 3. WQI values of before and after treatments

3.8. Biological parameters

Total coliforms were at a safe level according to NWQS (mean values and standard deviations before and after treatment were 38.11 ± 13.96 cfu/ml and 10.33 ± 6.67 cfu/ml, respectively), while the absence of *E. coli* bacteria was recorded for both treatments. This indicates that the roof was exposed to animal faeces. The total presence of these bacteria was in line with the NH_3-N value influenced by the presence of animal faeces. These biological parameters are also important in the determination of water quality. According to NWQS, the total presence of total coliform is in class I, which is at the safe level.

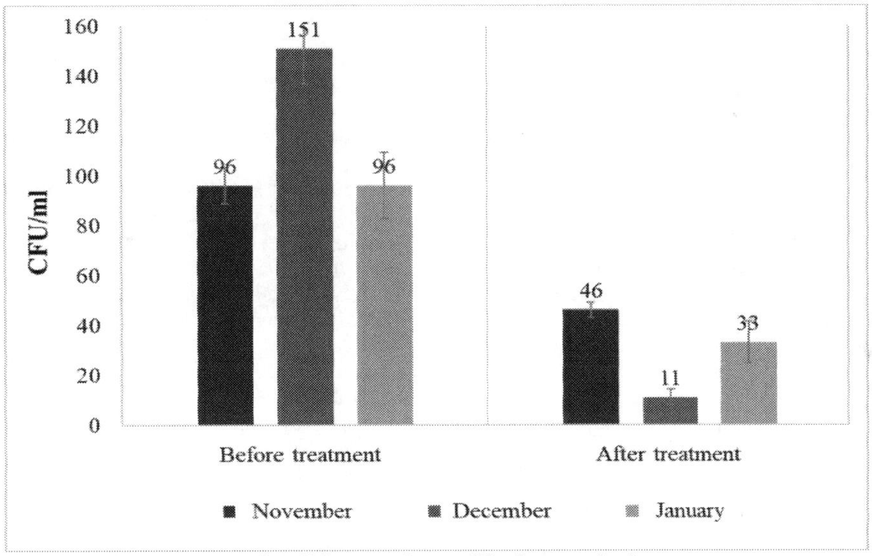

Figure 4. Total coliform values of before and after treatments

4. Conclusions

This study was done to determine the quality of rainwater harvested from roofs of residential college at the College of Ungku Omar, UKM. Rainwater quality before and after treatment were compared to observe the quality of the rainwater samples based on Water Quality Index (WQI). It was found that the quality of water before and after treatment were in class II. Based on the physico-chemical parameters, there are similarities and differences of class for both samplings. For TSS and NH_3-N, both before and after treatment are in class I, while for COD values before treatment and after treatment were in class II and class I, respectively. Other parameters like BOD, pH and DO for before and after treatment were in class II. The usage of water from rainwater harvesting system is suitable for external use and domestic, however, further treatment is needed to be used as drinking water. While for water quality based on biological parameters showed the presence of total coliform bacteria, was higher for water samples before treatment compared to after treatment. There was an absence of *E. coli* bacteria for both water samples. The presence of coliform still at a safe level according to the NWQS.

Rainwater quality studies in these universities should be done regularly and continuously to get the higher quality of harvested rainwater for students' domestic use. This study indirectly helps to increase students' awareness of the importance of rainwater harvesting as an alternative to the existing short water supply. Rainwater harvesting system is very convenient practice primarily in addressing the water crisis. With that, there are several proposals for future research in order to maintain the quality of collected rainwater. Among the recommendations for further research are good maintenance and regular cleaning of the KUO's roof to ensure the cleanliness of the water collected before the rainwater treatment process. Lastly, a more regular monitoring on the rainwater harvesting system is also necessary in order to avoid disruption of rainwater treatment.

References

[1] Xu, L., Yang, S., Zhang, Y., Jin, Z., Huang, X., Bei, K., Zhao, M., Kong, H. and Zheng, X., 2020. A hydroponic green roof system for rainwater collection and greywater treatment. *J. Clean. Prod.* **261**, 121132.

[2] Roshan, A. and Kumar, M., 2020. Water end-use estimation can support the urban water crisis management: a critical review. *J. Environ. Manag.* 268, 110663.

[3] Lancaster, B. and Lipkis, A., 2020. Rainwater harvesting for drylands and beyond: Water-harvesting earthworks (**Vol. 2**). Rainsource Press.

[4] Mayowa, O.O., Pour, S.H., Shahid, S., Mohsenipour, M., Harun, S.B., Heryansyah, A. and Ismail, T., 2015. Trends in rainfall and rainfall-related extremes in the east coast of peninsular Malaysia. *J. Earth Syst. Sci.* **124(8)**, 1609-1622.

[5] Augustine, E.E. and Mohd Hanafiah, M., 2019. Awareness level of water resource conservation of university students. *Water. Conserv. Manag.* **3(2)**, 18-21.

[6] A.P.H.A, 2017. APHA Method: Standard Methods for the Examination of Water and Wastewater, 23rd American Public Health Association 4500-F. 40 CFR 136.3(a).

[7] Hach Company World Headquarters, H. 2021. Nitrogen, Ammonia-Nessler, Hach Company/Hach Lange GmbH. Method **8038**.

[8] Tariq, M., Anayat, A., Waseem, M., Rasool, M.H., Zahoor, M.A., Ali, S., Rizwan, M., Abdel-Daim, M.M. and Alkahtani, S., 2020. Physicochemical and Bacteriological Characterization of Industrial Wastewater Being Discharged to Surface Water Bodies: Significant Threat to Environmental Pollution and Human Health. *J. Chem.* **2020**.

[9] Rashid, S.S. and Liu, Y.Q., 2020. Assessing environmental impacts of large centralized wastewater treatment plants with combined or separate sewer systems in dry/wet seasons by using LCA. *Environ. Sci. Pollut. Res.***20**, 1-17.

[10] Nizam, N.U.M., Mohd Hanafiah, M., Mohd Noor, I. and Abd Karim, H.I., 2020. Efficiency of five selected aquatic plants in phytoremediation of aquaculture wastewater. *Appl. Sci.* **10(8)**, 2712.

[11] Patil, P.B., Bhandari, V.M. and Ranade, V.V., 2021. Improving efficiency for removal of ammoniacal nitrogen from wastewaters using hydrodynamic cavitation. *Ultrason. Sonochem.* **70**, 105306.

[12] Manikam, M.K., Halim, A.A., Hanafiah, M.M. and Krishnamoorthy, R.R., 2019. Removal of ammonia nitrogen, nitrate, phosphorus and COD from sewage wastewater using palm oil boiler ash composite adsorbent. *Desalin.. Water. Treat.* **149**, 23-30.

[13] Aziz, N.I.H.A., Mohd Hanafiah, M., Halim, N.H. and Fidri, P.A.S., 2020. Phytoremediation of TSS, NH3-N and COD from Sewage Wastewater by Lemna minor L., Salvinia minima, Ipomea aquatica and Centella asiatica. *Appl. Sci.* **10(16)**, 5397.

[14] Sudiajeng, L., Wiraga, I.W., Parwita, I.G.L.M. and Budiadi, I.M., 2020, February. The effectiveness of horizontal water filtering system on deep rainwater harvesting wells. *In Journal of Physics: Conference Series* (Vol. 1450, No. 1, p. 012029). IOP Publishing.
[15] Muhammad Barzani Gasim, Ismail, B. S., Sahibin Abd. Rahim, Sujaul-Islam Mir and C. C. Tan. 2007. Hydrology and water quality assessment of the Tasik Chini's feeder rivers, Pahang, Malaysia. *America- Eurasian J. Agri. Environ. Sci.* **2(1)**, 39-47.
[16] Jamali, B., Bach, P.M. and Deletic, A., 2020. Rainwater harvesting for urban flood management–An integrated modelling framework. *Water Res.* **171**, 115372.

Acknowledgments
Marlia M. Hanafiah was funded by the National University of Malaysia (GUP-2020-034 and DIP-2019-001).

IOP Conf. Series: Earth and Environmental Science **880** (2021) 012040 doi:10.1088/1755-1315/880/1/012040

Comparative response of Malaysian weedy rice (*Oryza sativa*) initial growth towards the allelopathic potential of *Leucaena leucocephala* (Lam.) de Wit and *Dicranopteris linearis* (Burm. f.) Underw.

M S Ishak[1], Z Iberahim[1] and K Mardiana-Jansar[1*]

[1] Department of Earth Sciences and Environment, Faculty of Science and Technology, Universiti Kebangsaan Malaysia, 43600 UKM Bangi, Selangor, Malaysia

*Corresponding E-mail: mardiana@ukm.edu.my

Abstract. The infestation of weedy rice has becoming a threat which leads to competition between weedy rice and commercial rice for nutrient uptake. Weedy rice management is more dependent on the herbicides application that can contribute to various environmental problems. Thus, this study was conducted to assess the allelopathic potential of two invasive plants namely *L. leucocephala* and *D. linearis* in controlling the growth of weedy rice. The allelopathic potential was evaluated through laboratory bioassay namely the dish pack method, sandwich method and plant box method to determine the allelopathic potential on the seedling growth of two weedy rice variants collected in Peninsular Malaysia (designated as WRE and WRN) and MR220 CL2 rice variety. Highest concentration of *L. leucocephala* leaf litter (50 mg) inhibited the radicle elongation of MR220 CL2, WRN and WRE by 49 %, 51 % and 83 % compared to control, respectively. Through the dish pack method, the volatile compound from *D. linearis* leaf were found to disrupt the radicle growth of MR220 CL2, WRN and WRE by 21 %, 29 % and 25 % of the control at the nearest distance (41 mm) of the bioassay plants with the *D. linearis* leaf. Compared with *D. linearis*, the root exudate from *L. leucocephala* exhibited higher inhibitory effect towards the bioassay species. Overall, the findings from this study showed that the allelopathic potential of *L. leucocephala* was more preferable to be investigated further for sustainable weedy rice management in the future by infusing the biological approach.

Keywords: Allelopathy, rice, weedy rice, weedy rice management

Track Name: Advanced Technology and Renewable Energy

1. Introduction

Malaysian Agricultural Research and Development states that almost 80 % of paddy fields throughout Malaysia faced weedy rice problems [1]. Weedy rice resembles commercial rice in terms of morphology, growth, taxonomy and diversity which makes it difficult to distinguish and to control especially during early stage of its growth [2]. As an initiative to control weedy rice growth, MARDI introduced Clearfield® rice technology in 2010 namely MR220 CL1 and MR220 CL2 varieties that resist towards imidazolinone herbicide [3]. Unfortunately, some farmers did not follow the guidelines for the use of Clearfield® rice technology system which affects the effectiveness of this method [1].

Weedy rice seeds have spontaneous shattering characteristic during harvesting process and the ripening period is faster than commercial rice [4]. Moreover, commercial rice has to compete with weedy rice for nutrients and space to grow [5]. Allelopathy is any direct or indirect effect of one plant (including microorganisms) towards other plants through the release of chemicals into the environment (Rice 1984) [6]. This phenomenon involves the inhibition and stimulation of the growth of the recipient plant resulting from the production of chemicals from the donor plant [6]. Chemical compounds or secondary metabolites of the plants possessing allelopathic potential are known as allelochemicals [7]. Allelochemicals are released from the plants by evaporation, leaching, root exudate and decomposition of dead plants [8]. In addition, allelochemical contributes to weed control with the involvement of the allelopathic plants [9]. Hence, the allelopathic plants selected in this study are two invasive plants namely *Leucaena leucocephala* and *Dicranopteris linearis*.

Leucaena leucocephala (Lam.) de Wit (family Fabaceae) or commonly known as lead tree is a fast growing and high yielding tree which can be grown in various type of soils especially in the tropical areas [10]. Previous studies discovered that *L. leucocephala* contains secondary metabolites known as mimosine which responsible to its herbicidal activities towards some plants [11, 12]. A study by Ishak et al. [13] found that the volatilization and leaching from the foliage of *L. leucocephala* inhibited the radicle and hypocotyl length of *Ageratum conyzoides*, *Tridax procumbens* and *Emilia sonchifolia*. *Dicranopteris linearis* (Burm. f.) Underw (family Gleicheniaceae) is the most widely distributed ferns in the tropical and temperate regions [14]. According to Russell et al. [15], *D. linearis* is capable to dominate plant communities and forms enormous pure colonies, frequently. In terms of allelopathic activities, a study by Ismail & Chong [16] discovered that the extract and residues from *D. linearis* inhibited the seedlings emergence and growth of several common Malaysian weed species.

Weedy rice cannot be harvested because the seeds fall early to the ground. Although early harvesting can be done to save weedy rice seeds, but the seeds are still immature [17]. Clearfield® rice technology introduced by MARDI which is expected to be able to control the growth of weedy rice is no longer effective as the herbicide-resistant weedy rice were reported emerge from the hybridization of the herbicide-resistant rice variety with the weedy rice [18]. This causes weedy rice becomes more difficult to control and farmers have to continue the use of synthetic herbicides without following the guidelines which in turn contributes to environmental pollution. Therefore, allelopathy can be utilized as an environmentally friendly approach to control the growth of weeds especially weedy rice. Indirectly, the use of synthetic herbicides in rice fields can be reduced and subsequently minimize the impact towards the environment. In particular, this study aims to determine the allelopathic potential of leaf volatilization, foliage leaching and root exudate from *L. leucocephala* and *D. linearis* towards the growth of weedy rice.

2. Materials and Methods

In this study, the allelopathic plants (donor plants) were *L. leucocephala* and *D. linearis* while the bioassay plants (receiver plants) were weedy rice variants categorized based on its morphological features and MR220 CL2 rice variety.

2.1 Collection of the allelopathic plants and the bioassay plants

L. leucocephala were sampled around KTM UKM Station, Bandar Baru Bangi, Selangor (2.940° N, 101.787° E) while *D. linearis* were collected around Universiti Kebangsaan Malaysia, Bangi, Selangor (2.921° N, 101.784° E). The bioassay plants were two weedy rice variants and MR220 CL2 variety. The designation, sampling location and morphological features of the bioassay as shown in Table 1.

Table 1. Bioassay plants designations based on the sampling location and morphological features.

Bioassay designation	Type	Sampling location	Coordinate	Morphological features
WR1	Weedy rice	Kg. Chermai, Arau, Perlis	6.400° N 100.271° E	Awnless; brown-hulled; closed panicle
WR2	Weedy rice	Teluk Mesira, Bachok, Kelantan	6.143° N 102.343° E	Awned; straw-hulled; open/horizontal panicle
MR220 CL2	Modern rice variety	MARDI Seberang Perai	5.540° N 100.470° E	-

2.2 Dish pack method

Developed by Fujii [19], this method was utilized to identify the allelopathic potential of the donor plants towards the growth of the bioassay plants through leaf volatilization. A multi-dish plate (Thermo Scientific Inc.) with six wells (10 cm^2 each) was used. A total of 2 g of donor plant dried leaf samples were inserted into the first well (source well). The other five wells were lined with Whatman No. 1 filter paper with three replication carried out for each species. By using pipette, each lined-well was then filled with 0.7 ml distilled water. Next, three seeds of each bioassay plant were placed on the surface of the filter paper. The distance between the source well and the other neighboring wells are 41, 58, 82 and 92 mm. For control, the source well was not filled with any sample of the donor plants. To prevent the releasing of the volatile materials, the multi-dish were tightly sealed with adhesive tape. Then, each multi-dish was labeled and incubated in a growth chamber at 28 °C for seven days. The radicle and hypocotyl length of the bioassay plants were measured and recorded on the seven days after sowing.

2.3 Sandwich method

This method was developed to evaluate the allelopathic activities of the donor plants towards the growth of the bioassay plants through leaching from its foliage [20]. Two concentrations (10 and 50 mg) of oven-dried leaf of the donor plants were used in this experiment. For each multi-dish (Thermo Scientific Inc.), three wells were filled with 10 mg of oven-dried leaf of the donor plants and the remaining three wells were filled with 50 mg of the respective donor plants. To prevent fungal contamination, the multi-dish were sprayed with 70 % ethanol before use. Agar (Nacalai Tesque Inc., Kyoto, Japan) were used as the growth medium (0.75 % w/v). The agar mixture were autoclaved for 15 minutes at 121 °C and 121 kPa and the agar solution were cooled to 40-45 °C. A total of 5 ml of agar solution were poured into each wells containing 10 mg and 50 mg donor plant oven-dried leaves and left to cool until it forms first gelatin layer. After the first layer of agar gelatinized, another 5 ml of agar solution were poured to form two layers of gelatin. Three seeds of the bioassay plant seeds were added to the top layer of the gelatinized agar in each multi-dish. Multi-dish without oven-dried leaf of the donor plants was placed into the multi-dish wells and indicated as control. The multi-dish was then sealed tightly using adhesive tape and incubated in the growth chamber at 28°C for seven days. Each concentration (0. 10, 50 mg) were replicated three times in complete randomized design (CRD). The radicle and hypocotyl length of the bioassay were recorded seven days after sowing.

2.4 Plant box method

This method was developed by Fujii et al. [21] to assess the allelopathic activities of the root exudates from the donor plants towards the growth of the bioassay plants. Two-month-old donor plant seedlings were harvested and the root part was cleaned using distilled water. Next, an individual donor plant was then placed in a PVC tube equipped with nylon net (root zone separating tube) located at the corner of the magenta box (6 x 6 x 10 cm; Magenta Corporation, Chicago, USA). Each magenta box was sprayed with 70 % ethanol to prevent any fungal contamination. Autoclaved agar (Nacalai Tesque Inc.,

IOP Conf. Series: Earth and Environmental Science **880** (2021) 012040 doi:10.1088/1755-1315/880/1/012040

Kyoto, Japan) used as growth medium (0.75 % w/v; 40 °C) was poured into the magenta box and left to form gelatin. Thirty three bioassay seeds were placed on the surface of the gelatinized agar following the distances of each seed with the donor plant. Prior to root phototropism, the base of the plant box then covered with aluminium foil. The top of the plant box were sealed with transparent plastic to avoid fungal contamination. The plant boxes without the donor plants were indicated as untreated control. All plant boxes were then incubated for seven days in the growth chamber at 28 °C. Each treatments were replicated three times in complete randomized design (CRD). The radicle and hypocotyl length were recorded at seven days after sowing.

2.5 Statistical analysis

The mean ± standard error and analysis of variance from the dish pack and sandwich method were determined by using one-way ANOVA with post-hoc test using Duncan's Multiple Range Test (DMRT) at $\rho<0.05$. Data from the plant box method was analyzed by using correlation bivariate analysis (Pearson) at $\rho<0.01$ and linear regression analysis. All analysis were determined and generated using SPSS software version 23.

3. Results and Discussions

Table 2 shows the allelopathic effects of the leaf volatilization of *L. leucocephala* and *D. linearis* on root growth of the bioassay plants. The inhibitory effect was tested based on the distance of bioassay (41 mm, 58 mm, 82 mm and 92 mm) from the dry leaves of the allelopathic plants in the multi-dish. The results showed that the volatilization from the leaves of *L. leucocephala* and *D. linearis* hold allelopathic activity on the growth of the bioassay plants. The radicle growth of WR1 experienced insignificant stimulation with the presence of volatilization from *L. leucocephala* foliage at all distances while the leaf volatilization of *D. linearis* was found to inhibit radicle growth of WR1 significantly by 29 % (41 mm), 0.4 % (58 mm), 21 % (82 mm) and 19 % (92 mm) compared to control. The leaf volatilization activity of *L. leucocephala* and *D. linearis* displayed inhibitory effect on WR2 radicle growth at all distances. However, there was a growth stimulation of 3 % by the volatiles from *L. leucocephala* at distance of 82 mm. Furthermore, the volatiles from *L. leucocephala* leaves were found to inhibit the radicle growth of MR220 CL2 at all distances.

The nearest distance (41 mm) from *D. linearis* had a higher percentage of inhibition than the farthest distance (92 mm). This indicates that the presence of volatile allelochemical compounds decreases as the recipient plant moves further away from the donor plant. This study was supported by Kato-Naguchi et al. [14] who found that *D. linearis* extract had an inhibitory effect on *Echinochloa colonum* and *Avena fatua* which are two plant species that grow near *D. linearis* colonies. In addition, cinnamtannin B-1 which is a secondary metabolite of *D. linearis* was not detected in soil at a distance of 100 cm of the colony [14]. The radicle growth of MR220 CL2 were found to be inhibited at all distances by the volatiles from the leaves of both allelopathic plants. However, all bioassay plants did not display the same inhibition pattern. This is happened because the recipient plants are not sensitive enough to fully respond to the distance differences but there is still inhibitory activity by the allelopathic plant through volatilization from its foliage [22]. When comparing the efficacy between the two allelopathic plant species, volatiles from *D. linearis* leaves were found to have a significantly higher inhibitory effect rather than the volatiles from *L. leucocephala* leaves.

Table 2. Allelopathic effects of leaf volatilization from the donor plants on the radicle elongation of the bioassay by using dish pack method.

Bioassay	Distances (mm)	Radicle elongation (mm)			
		L. leucocephala	IP (%)	*D. linearis*	IP (%)
WR1	0 (control)	62.67 a	0	62.67 a	0
	41	65.50 a	+4.52	44.56 b	-28.90
	58	62.89 a	+0.35	62.44 a	-0.37
	82	64.11 a	+2.30	49.78 ab	-20.57
	92	65.44 a	+4.42	51.00 ab	-18.62
WR2	0 (control)	67.00 a	0	67.00 a	0
	41	61.56 a	-8.12	49.94 b	-25.46
	58	58.67 a	-12.43	47.22 b	-29.52
	82	69.22 a	+3.31	53.11 b	-20.73
	92	64.78 a	-3.31	54.33 b	-18.91
MR220 CL2	0 (control)	61.11 a	0	61.11 a	0
	41	59.17 a	-3.17	48.11 b	-21.27
	58	54.44 ab	-10.92	45.89 b	-24.90
	82	47.33 b	-22.55	53.89 ab	-11.81
	92	57.56 a	-5.81	51.11 b	-16.36

* Values within columns followed by different letter are significantly different at $p<0.05$ according to Duncan's multiple range test (DMRT). The symbol "-" indicates inhibitory effect, whilst the symbol "+" indicates stimulatory effect.

The sandwich method was carried out using two different total weights of allelopathic plants dry leaves namely 10 mg and 50 mg. Table 3 shows the leaching of leaves *L. leucocephala* and *D. linearis* exert an inhibitory effect on the radicle elongation of both weedy rice variants and the MR220 CL2 variety. The percentage of radicle growth inhibition of the weedy rice variants increased significantly when the total weight of the dry leaves increased from 10 mg to 50 mg. For instance, the radicle elongation of WR2 were significantly reduce by 83 % with the presence of leaching from the *L. leucocephala* foliage at highest concentration (50 mg). According to Sahid et al. [12], mimosine which is the allelochemical of *L. leucocephala* was found mostly in the young shoots and seeds of the plant. In the same study, it was found that the growth of *Ageratum conyzoides*, *Tridax procumbens* and *Emilia sonchifolia* was further inhibited as the mimosine concentration increased. Overall, the leaching from the foliage of *L. leucocephala* was found to significantly inhibit the radicle growth of both weedy rice variants compared to *D. linearis* especially at the highest total dry leaves weight (50 mg). Moreover, lower total dry leaves weight (10 mg) of *L. leucocephala* was found significantly promoted the radicle growth of MR220 CL2 rice variety and depicts that *L. leucocephala* is suitable to be investigated further for future weedy rice management.

IOP Conf. Series: Earth and Environmental Science **880** (2021) 012040 doi:10.1088/1755-1315/880/1/012040

Table 3. Allelopathic effects of foliage leaching from the donor plants on the radicle elongation of bioassay by using sandwich method.

Bioassay	Total dry leaves weight (mg)	Radicle elongation (mm)			
		L. leucocephala	IP (%)	*D. linearis*	IP (%)
WR1	0 (control)	58.67 a	0	58.67 a	0
	10	45.93 b	-21.72	41.19 b	-29.79
	50	28.93 c	-50.69	34.96 c	-40.41
WR2	0 (control)	53.72 a	0	53.72 a	0
	10	47.56 b	-11.47	51.85 a	-3.48
	50	9.19 c	-82.89	31.41 b	-41.53
MR220 CL2	0 (control)	58.72 b	0	58.72 a	0
	10	67.30 a	+14.60	50.63 b	-13.78
	50	29.93 c	-49.03	24.22 c	-58.75

* Values within columns followed by different letter are significantly different at $p<0.05$ according to Duncan's multiple range test (DMRT). The symbol "-" indicates inhibitory effect, whilst the symbol "+" indicates stimulatory effect.

Table 4 shows the allelopathic effects of the root exudates from the donor plants on the radicle growth of bioassay plants. The results discovered that the radicle growth of all bioassay plants was inhibited with the presence of exudation from the root of both donor plants. The radicle growth of WR1 was reduced by 38 % when sown with *L. leucocephala* and significantly influenced by its distance with the *L. leucocephala* plant (R^2 = 0.154). The radicle of MR220 CL2 showed less sensitivity towards the root exudate from *L. leucocephala* and only inhibited by 17 % compared to control with low R^2 value (0.006). On the other hand, the root exudates from *D. linearis* were greatly inhibited (reduced by 85 % compared to control) the radicle growth of MR220 CL2 with. According to a study conducted by Ishak et al. [13], root exudates of *L. leucocephala* cause inhibition of root growth of *T. procumbens*, *E. sonchifolia* and *A. conyzoides*. The great inhibitory effects from *D. linearis* roots were in line with the findings by Kato-Noguchi et al. [14] in turn stated that cinnamtannin B-1 may be released either by the exudate from living plants or through the decomposition of the plant residues. Some bioassay such as WR1 and WR2 were significantly influenced by its distance with the donor plant (*L. leucocephala*). This indicates that there are relationships between the distance and radicle length. The findings also suggested that distance is one of the contributing factors in the efficiency of the allelopathic plants to control other plant's growth [23]. In addition, the roots of *L. leucocephala* may secrete allelochemicals that probably high at the nearest distance and become low at the farthest distance because of the diffusion.

Table 4. Allelopathic effects of root exudates from the donor plants on the radicle growth of bioassay plants by using plant box method.

Bioassay	*L. leucocephala*		*D. linearis*	
	Radicle inhibition (%)	R^2	Radicle inhibition (%)	R^2
WR1	38**	0.154	32**	0.103
WR2	21**	0.116	74	0.005
MR220 CL2	17	0.006	85	0.001

4. Conclusions

Based on the present study, it was found that allelopathic potential was present in both *L. leucocephala* and *D. linearis* which affected the growth of both weedy rice variants. The foliage leaching and root exudates from *L. leucocephala* displayed higher inhibitory effects on the growth of weedy rice variants rather than D. linearis, whilst the volatiles from the foliage of *D. linearis* showed higher inhibitory effects on the bioassay plants compared with *L. leucocephala*. This study also revealed that *L. leucocephala* is the most promising candidate to be infused as a biological control of weedy rice and its allelopathic potential can be explored further in order to manage weedy rice or other weeds in various agricultural activities.

References

[1] MARDI 2013 80% Sawah Padi Diancam Padi Angin. https://blogmardi.wordpress.com/2013/08/01/80-sawah-diancam-padi-angin/ [22 Ogos 2020]
[2] Chauhan B S 2013 *Crop Protection* **48** 51-56
[3] Harun R *et al* 2018 *Economic and Technology Management Review* **13** 63-73
[4] Karim S R *et al* 2006 *Plant Protection Quarterly* **21**(1) 13.
[5] Olajumoke B *et al* 2016 *Chilean Journal of Agricultural Research* **76**(2) 243-252.
[6] Rice E L 1984 *Allelopathy* (Orlando, Florida: Academic Press)
[7] Kong C H *et al* 2019 *Molecules* **24**(15) 2737
[8] Dahiya S *et al* 2017 *Journal of Pharmacognosy and Phytochemistry* **6** 832-837
[9] Putnam A R 1988 *Weed Technology* **2**(4) 510-518
[10] Pandey V C and Kumar A 2013 *Genetic Resources and Crop Evolution* **60**(3) 1165-1171
[11] Xuan T D *et al* 2013 *Herbicidal activity of mimosine and its derivatives Herbicides: Advances in Research* (Rijeka, Croatia: Intech) chapter 15 pp 299-312
[12] Sahid I *et al* 2017 *Transactions on Science and Technology* **4**(2) 62-67
[13] Ishak M S *et al* 2016 *Allelopathy Journal* **37**(1) 109-122
[14] Kato-Noguchi H *et al* 2012 *Plant Ecology* **213**(12) 1937-1944
[15] Russell A E *et al* 1998 *Journal of Ecology* **86**(5) 765-779
[16] Ismail B S and Chong T V 2009 *Allelopathy Journal* **23**(2) 277-286
[17] Azmi M *et al* 2001 *Pengenalan Dan Pengawalan Padi Angin Di Malaysia* (Serdang: MARDI)
[18] Dilipkumar M *et al* 2018 *Planta Daninha* **36** 018182239
[19] Fujii Y *et al* 2005 *Thymus* **2**(3) 493-497
[20] Fujii Y *et al* 2003 *Weed Biology and Management* **3**(4) 233-241
[21] Fujii Y *et al* 2007 *Plant Box Method: A specific bioassay to evaluate allelopathy through root exudates Allelopathy: New Concepts and Methodology* eds Y Fujii and S Hiradate (Enfield: Science Publishers) chapter 3 pp 39-56
[22] Yusoff N *et al* 2018 *International Journal of Science and Applied Technology* **2**(2) 1-9
[23] Syed S et al 2014 *Pakistan Journal of Botany* **46**(5) 1693-1701

Acknowledgements

Mardiana-Jansar, K. was supported by the Ministry of Higher Education grant (FRGS/1/2020/STG02/UKM/02/3). The authors would like to thanks the staff and students at Plant Biotechnology Lab, Faculty of Science and Technology, UKM for their assistance.

Effect of mulch on glyphosate mobility in soil column

Mardiana-Jansar K[1*], Nurul Wadhihah Mazlan[1], Zainol Maznah[2] and Ishak MS[1]

[1] Department of Earth Sciences and Environment, Faculty of Science and Technology, Universiti Kebangsaan Malaysia, Bangi 43600, Selangor, Malaysia.
[2]Malaysian Palm Oil Board (MPOB), No. 6 Persiaran Institusi, Bandar Baru Bangi 43000 Kajang, Selangor, Malaysia.

*Corresponding E-mail: mardiana@ukm.edu.my

Abstract. Glyphosate is a broad-spectrum herbicide that often used in controlling weed growth in oil palm plantation. Glyphosate has a high polarity and solubility in water and is quickly absorbed by soil particles. Oil palm waste such as oil palm trunks (OPTs) and empty fruit bunches (EFBs) and the use of glyphosate in agricultural land can contribute to environmental pollution. Apart from the commercial mulch that often used in nurseries, mulch from OPT and EFB can be used to absorb glyphosate to prevent weed growth. The glyphosate concentration of soil and water samples was determined using High Performance Liquid Chromatography (HPLC). Two doses of glyphosate were used which are 5.5 mL/L (2255 mg a.i/L) and 11 mL/L (4510 mg a.i/L). Two water volumes are used which are 250 ml and 500 ml. Glyphosate residual concentration values was high in Serdang soil series at layer three while in Selangor soil series at layer one. The highest concentrations of glyphosate residues were 32.57 mg/L and 34.42 mg/L at Serdang and Selangor soil series respectively. Samples of soil in the mulch of EFB and OPT show low concentrations of glyphosate residue compared to commercial and non-mulching column. Results show that EFB and OPT indicate low residual concentration values at each layer in the soil column for both doses.

Keywords: Serdang series, Selangor series, oil palm trunks, empty fruit bunches

Track Name: Advanced Technology and Renewable Energy

1. Introduction

Malaysia is the world's second largest producer of palm oil after Indonesia. Currently, the acreage of oil palm plantations in Malaysia has increased from 5.74 million hectares in 2016 to 5.81 million hectares in 2017 [1]. Based on this increment, biomass production from oil palm processing will indirectly increase. Oil palm processing activities in the factory will produce oil palm waste which is empty fruit bunches (EFB) while oil palm trunks (OPT) are obtained from the process of replanting oil palm after 15-20 years of planting. Oil palm stems have a content of lignocellulose which serves to absorb water molecules in moist areas under hygroscopic conditions. Empty fruit bunches are also a lignocellulose material containing 25% lignin, 50% cellulose and 25% hemicellulose in the cell wall [2].

According to Sanagi [3], the use of palm oil biomass-based materials to remove environmental pollutants (effluents) is a solution to the sustainability of sustainable development. The preparation of adsorbents or mulches from empty fruit bunches and oil palm stems is able to remove heavy metal ions and organic pollutants that exist in the environment. The use of mulch on the soil is among the practices performed by farmers to retain moisture and prevent evaporation from occurring on the soil surface [4]. The study conducted by Mohamad et al. [5] showed that weed growth in oil palm plantations is able to disrupt the oil palm growth process and reduce oil palm productivity if not controlled systematically.

Glyphosate, a broad -spectrum herbicide, is often used by farmers to control the growth of weeds in oil palm plantation. Glyphosate can be applied directly to the soil surface, released from plant roots or from decomposed plant tissues, as well as involving a transport retention (adsorption-adsorption) process [6]. Therefore, this study was conducted to study the best mulch for adsorption of glyphosate on agricultural soil and reduce weed growth in oil palm plantations. This is because the use of herbicides in uncontrolled doses is one of the unsustainable environmental management and can contaminate groundwater resources.

2. Materials and Method

The sampling location was at the Faculty of Agriculture, Universiti Putra Malaysia (2° 59' 00.0" N, 101° 44' 05.7" E) for Serdang soil series and the Malaysian Oil Palm Board (2° 58' 04.2" N 101° 44' 55.0" E) for Selangor soil series. A series of concentrations of glyphosate standard solutions in the concentration ranges from 0.005 to 0.05 μg/mL were prepared for glyphosate analysis. Glyphosate herbicides was prepared at 2 different concentrations, 11 mL/L (double the recommended dose) and 5.5 mL/L (recommended dose).

2.1 Soil column preparation

30 cm soil columns were used and arranged vertically at the Greenhouse, Plant House Laboratory, Universiti Kebangsaan Malaysia (UKM). A plastic funnel and water collector cup was placed at the bottom of each soil column. A total of 48 soil columns were used according to different pesticide concentrations, mulch and watering volume. Three types of mulch were used in this study (Figure 1). The empty fruit bunch and oil palm trunk waste, the samples should be completely dry before the analysis is performed. The mulch was cut to equalize the size of the mulch for each column. Three replications were prepared at each herbicide concentration, mulch and total watering volume. Each soil column was sprayed with 10 mL of pesticide at each concentration and sprayed on the first day only. After three hours of spraying, watering simulations were performed with 250 mL and 500 mL total volume of water. For water sampling, water was taken on the first day after spraying. A total of 10 mL of collected leachate was placed in a centrifuge tube (50 mL) according to pesticide concentration, mulch and water volume for pesticide analysis in the laboratory. For soil sampling, each soil layer was taken for extraction. The soil column was cut at three different depths (27 cm, 18 cm and 9 cm) and placed in a centrifuge tube (50 mL). A total of 5g of soil was taken for each different depth.

Figure 1. Mulch; a) oil palm trunk (OPT), b) commercial mulch (K) and c) empty fruit bunch (EFB).

2.2 Glyphosate extraction method for soil and water samples

5 g of soil sample was placed in a 50 mL centrifuge tube and extracted by adding 10 mL of 0.6 M KOH and shaken using an orbital shaker for 30 min. The sample was then put into a centrifuge machine at 3500 rpm for 30 min. Only 5 mL of solution was taken for analysis. For water samples, the sample does not need to be extracted. Borate buffer solution was prepared by dissolving 1.9 g of analytical reagent disodium tetraborate decahydrate (Merck) in 100 mL of deionised water. 50 mg of fluorenylmethyloxycarbonyl chloride (FMOC-Cl) (97%, Aldrich) was dissolved in 50 mL of acetonitrile to produce a solution of FMOC-Cl reagent. Every 5 mL of solution sample extracted from the soil sample and 10 mL of water sample were placed in each 15 mL glass vial for extraction. Each sample was mixed with 0.5 mL of borate buffer solution and 0.5 mL of FMOC-Cl solution. Samples were stirred using a magnetic stirrer for 15 minutes and left at room temperature for one hour for reaction. A total of 2 mL of diethyl ether solvent was added to clean the excess FMOC-Cl reagent and a separate water phase was taken and filtered with a 0.2 µm filter to be injected directly into the HPLC (Table 1).

Table 1. HPLC condition for glyphosate analysis.

Column :	Zorbax Eclipse Plus C18
	Length: 150 mm
	Column Diameter: 3.5 µm
Column temperature :	40 °C
Flow rate:	0.5 mL/min
Flow period:	15.0 min
Detector :	Fluorescent detector
	Excitation wavelength -260 nm
Injector (volume) :	Emission wavelength - 310 nm
Mobile phase :	Automatic (50ul)
	A Phosphorus borate buffer solution (45%)
	B Acetonitrile (55%)

3. Results and Discussion

3.1 Glyphosate residue in soil column

A slow decrease of glyphosate concentration occurred in soil column from layer 1 to layer 3 for dose concentrations of 11 ml/l and 5.5 ml/l for Serdang and Selangor soil series respectively. This happened because the simulation of watering that caused the glyphosate dissolved. Referring to Figure 2, the soil column without mulch at the top of soil column (B) recorded the highest average of glyphosate concentration in layer 3 which was 21.48 mg/L when a volume of 250 mL water was poured on Serdang soil series. The absence of mulch on top of the soil column caused the glyphosate in first layer to be less adsorbed and moved to third layer along with the water. This result was supported by Aslam et al. [7] who stated that mulch influenced the presence of residues in the soil through interception and retention processes as a result of evaporation, discharge and runoff.

 IOP Publishing
doi:10.1088/1755-1315/880/1/012041

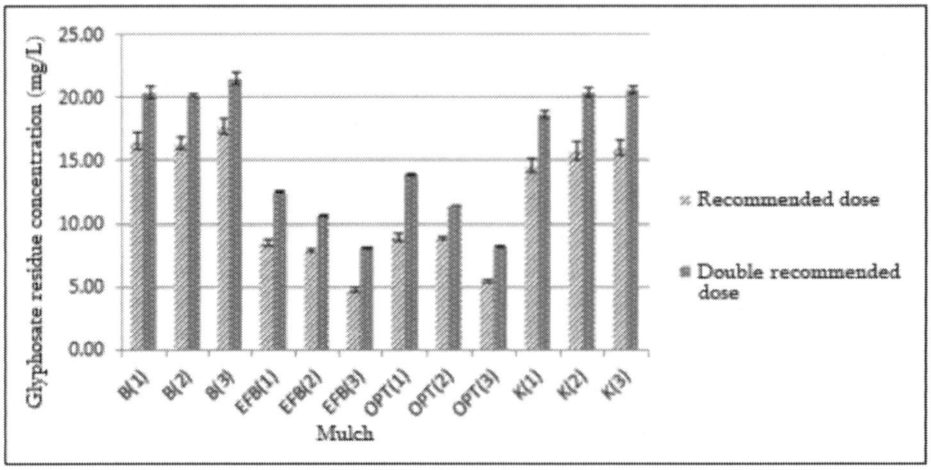

Figure 2. Mean concentration of glyphosate residual at 250ml watering volume for Serdang series soil.

Based on Figure 3, the concentration of glyphosate residue for Selangor soil series on EFB and OPT showed low residue values of 15.26 mg/L and 15.82 mg/L respectively in layer 1 compared to commercial mulch (K) and column without mulch (B) for concentration dose 11 mL/L. The high carbon and nitrogen content as well as the presence of polymers such as lignin and cellulose in empty fruit bunch and oil palm trunk would act as a natural barrier against the decomposition process [8]. The presence of high organic matter that caused higher adsorption capacity in the Selangor series soil (Table 2) also prevents the movement of glyphosate. This result was supported by Duke et al. [6] and Mardiana-Jansar [9] who stated that glyphosate is strongly bound to soil particles and has a low desorption capacity, only a small amount of glyphosate was found on the soil surface for microorganism activity, uptake by plants and interaction between metals in the soil.

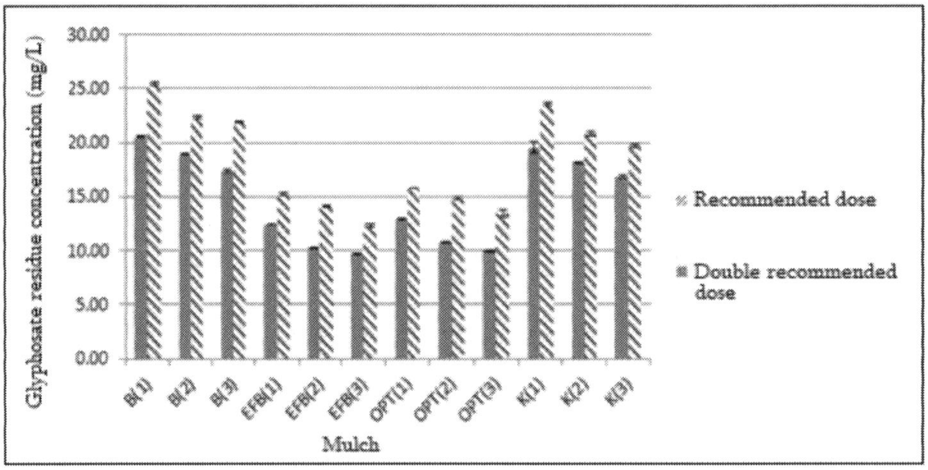

Figure 3. Mean concentration of glyphosate residual at medium watering volume for Selangor soil series.

Table 2. Physico-chemical properties of soil series.

Parameter	Selangor soil series	Serdang soil series
Organic matter (%)	26.7	0.65
Sand (%)	2.7	52.4
Clay (%)	45.4	30.5
Silt (%)	51.9	17.1
pH	4.3	5.34
CEC	24.7	5.24

At the third layer (the lowest layer), glyphosate residues were still detected at all columns explaining that glyphosate was still moved and the presence of the soil organic matter content kept the concentration of glyphosate in the soil. Based on Figure 4 and Figure 5, the lowest glyphosate residue values were recorded at 15.66 mg/L at first layer and 14.20 mg/L at third layer for the 5.5 mL/L glyphosate concentration. Higher clay content in Selangor soil series as well as its fine grain size cause less soil pores and low air space. This is supported by Okada et al. [10] who stated that the small size of soil grains makes the movement of glyphosate is limited and minimized the water to be absorbed. This explained the high concentration of glyphosate residue concentration in first layer compared to third layer when different volumes of water are applied to Selangor soil series.

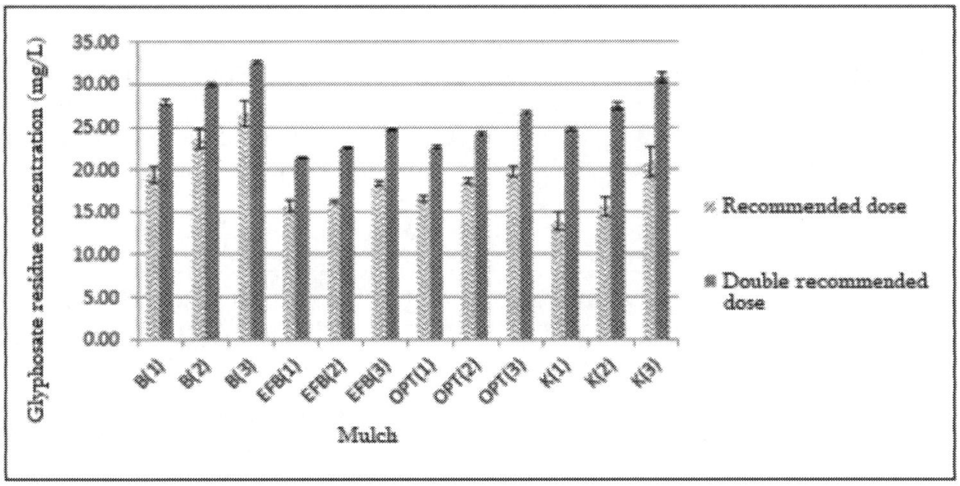

Figure 4. Mean concentration of glyphosate residual at maximum watering volume for Serdang soil series.

In contrast to the Serdang soil series, the soils were sand texture and have a larger grain size than the clay grain size. The larger grain size causes broader soil pores and faster movement of water and herbicides in the Serdang soil series compared to the Selangor soil series. In addition, a low pH value of Selangor soil series makes the adsorption of glyphosate increased especially in first layer. Soils with a pH of less than 5 will have high adsorption on soil particles. For pH above 5, glyphosate will be ionized, the negative charge is increased and the ability for glyphosate to bind to soil particles is limited [11].

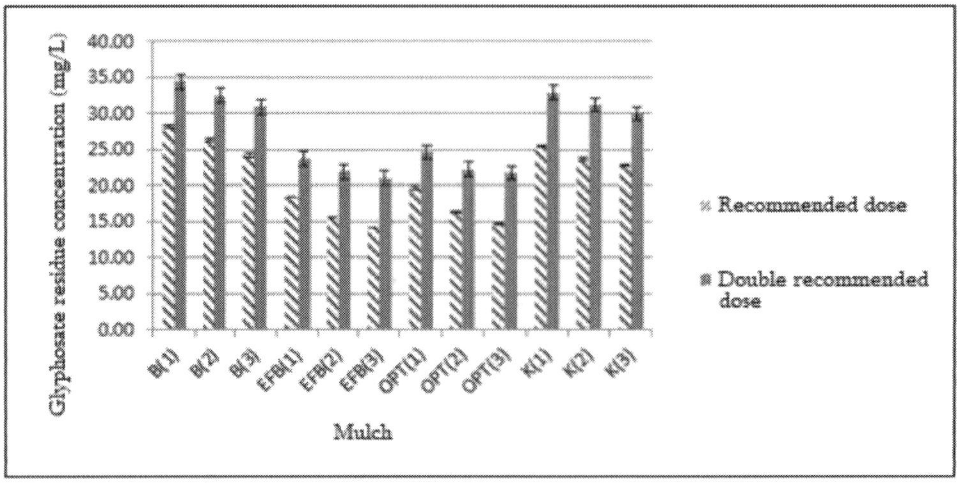

Figure 5. Mean concentration of glyphosate residual at maximum watering volume for Selangor soil series.

3.2 Glyphosate residue in water leachate

For water samples, the highest residual glyphosate concentration values detected at concentration doses of 5.5 mL/L and 11 mL/L were on Serdang soil series (Table 3). This indicates that the movement of glyphosate in water on sand-textured soils is higher compared to clay-textured soils. This is supported by Borggaard et al. [12] who stated that the concentration of glyphosate residue in water on sandy soils has low absorption capacity and slow decomposition rate compared to clay. The mobility potential of glyphosate through the soil column is influenced by the different volumes of water applied to the soil. Glyphosate leaching occurring in drains or drainage indicates a risk of contamination in groundwater but only on soils with shallow groundwater depths [12].

Table 3. Mean concentration of glyphosate residues in water at two watering volume.

Soil series	Serdang soil series		Selangor soil series	
Water volume	250 ml			
Mulch	Recommended dose	Double recommended dose	Recommended dose	Double recommended dose
Blank	3.21	3.41	3.15	3.35
EFB	2.94	2.98	2.64	2.89
OPT	2.97	3.04	2.77	2.94
K	3.17	3.38	3.11	3.26
Water volume	500 ml			
Blank	2.9	3.14	2.52	3.09
EFB	2.54	2.64	2.23	2.44
OPT	2.57	2.77	2.35	2.59
K	2.62	3.01	2.5	2.95

4. Conclusions

This study showed that the use of mulch is possible to slower the glyphosate mobility. The results showed that glyphosate adsorbed slowly in Selangor soil series conversely in Serdang soil series based on the volume of water and mulch used. The suitable mulches used to adsorb glyphosate were EFB and OPT because the lowest glyphosate residue concentrations were recorded by both mulches at both glyphosate concentration doses compared to commercial mulches and columns without mulches.

References

[1] Khusairi A, Loh SK, Azman I, Hishamuddin E, Ong-Abdullah M, Mohd Noor Izzuddin Z, Razmah G, Sundram S & Ahmad Parveez G 2018 Oil Palm Economic Performance in Malaysia and R & D Progress in 2017 *Journal of Oil Palm Research* **30** (2) 163-195

[2] Rosli NS, Harun S, Jahim JM & Othaman, R 2017 Chemical and Physical Characterization of Oil Palm Empty Fruit Bunch. *Malaysian Journal of Analytical Sciences* **21**(1) 188-196

[3] Sanagi, MM 2018 Recent Advances in the Preparation of Oil Palm Waste-Based Adsorbents for Removal of Environmental Pollutants-a Review. *Malaysian Journal of Analytical Sciences* **22**(2) 175-184

[4] Telkar SG, Singh AK, Kant K, Solanki SPS, Kumar D (2017) Types of Mulching and Their Uses for Dry-land Condition. *Biomolecule Reports* **9** 1-4

[5] Mohamad R, Mohayidin MG, Wibaya W, Juraimi AS & Lassim MM 2010 Management of Mixed Weeds in Young Oil-Palm Plantation with Selected Broad-Spectrum Herbicides *Pertanika Journal of Tropical Agricultural Science* **33**(2) 193-203

[6] Duke SO, Lydon J, Koskinen WC, Moorman TB, Chaney RL and Hammerschmidt R 2012 Glyphosate Effects on Plant Mineral Nutrition, Crop Rhizosphere Microbiota, and Plant Disease in Glyphosate-Resistant Crops. *Journal of agricultural and food chemistry* **60**(42) 10375-10397

[7] Asla S, Iqbal A, Bourdat-Deschamps M, Recous S, Garnier P and Benoi P 2015 Effect of Rainfall Regimes and Mulch Decomposition on the Dissipation and Leaching Of S-Metolachlor and Glyphosate: A Soil Column Experiment. Pest Management Science **71**(2)

[8] Siddiquee S, Shafawati SN & Naher L 2017 Effective Composting of Empty Fruit Bunches Using Potential Trichoderma Strains *Biotechnology Reports* **13**(1-7)

[9] Mardiana-Jansar K and Ismail BS 2014 Residue Determination and Levels of Glyphosate in Surface Waters, Sediments and Soils Associated with Oil Palm Plantation in Tasik Chini, Pahang, Malaysia *AIP Conf. Proc.* **1614** 795–802

[10] Okada E, Costa JL and Bedmar F 2016 Adsorption and Mobility of Glyphosate in Different Soils under No-Till and Conventional Tillage *Geoderma* **263**(78-85)

[11] Tu M, Hurd C and Randall JM 2001 *Weed Control Methods Handbook: Tools & Techniques for Use in Natural Areas*

[12] Borggaard K and Gimsing AL 2008. Fate of Glyphosate in Soil and the Possibility of Leaching to Ground and Surface Waters: A Review. *Pest Management Science* **64**(4): 441-456

Acknowledgement

Mardiana-Jansar, K. was supported by the Universiti Kebangsaan Malaysia grant (Publication Reward Grant GP-2019-K016136) for conference fees and Ministry of Science, Technology and Innovation Malaysia grant (06-01-02-SF1367) for research activities. The authors would like to thanks the officer from Malaysian Palm Oil Board for sharing the research materials (oil palm waste).

Development and mechanical properties of low fired bricks from drinking water sludge and fly ash

Z A Rahman[1*], A S M Suhaimi[1], W M R Idris[1] and T Lihan[1]

[1] Department of Earth Sciences and Environment, Faculty of Science and Technology, Universiti Kebangsaan Malaysia, 43600 UKM Bangi, Selangor, Malaysia

*Corresponding E-mail: zarah1970@ukm.edu.my

Abstract. Demand for water and energy supply has dramatically increased the amount of drinking water sludge (DW) and fly ash (FA) annually. These wastes should be properly managed and disposed to protect any potential contamination to surrounding ecosystem. Both by-products can be potentially recycled as raw material for brick development. This study aimed to examine the influence of fly ash content on mechanical properties of drinking water sludge brick at low firing temperature of 500 °C. Different ratios of FA content were added to the DWS ranged between 0 and 45%. Brick sample was moulded in 215 mm x 102.5 mm x 65 mm dimension. Samples were air-dried prior to firing at 500 °C for 3 hours in a furnace. Basic characterization of DW and FA showed pH of 5.76 and 10.1 with organic contents of 8.42% and 1.14%, respectively. Clay and silt fractions were dominant in DWS while silt more apparent than sand and clay in FA. The volume changes and water absorption of the brick samples decreased with increasing FA content. For the water absorption of the brick increased back as 40 % of FA content. The density and compressive strength dropped with the increasing amount of FA. The compressive strength of brick experienced with sulphate attack also decreased with increasing FA content. The results suggested that further study are needed to improve the compressive strength of the studied bricks.

Keywords: Sludge, fired brick, waste, compressive strength

Track Name: Advanced Technology and Renewable Energy

1. Introduction

Drinking water sludge is a waste product generated from drinking water treatment facility. Large amount of drinking water sludge has been produced annually as a result of highly demand of clean water for domestic and industrial usages. In Malaysia, it was recorded that around 0.5 million tons of waste was recorded in 2011 which involving 450 water treatment plants and its quantity is expected to rise every year [1]. Conventional treatment involves coagulation, flocculation, sedimentation, filtration and disinfection processes which ended up with large amount of residue which by-product generated during coagulation/flocculation processes [2]. Coagulation and flocculation processes by adding $Al_2(SO_3)_4.14H2O$ (alum) and poly-aluminium chloride (PAC), $Al_2Cl(OH)_5$ to separate suspended solids and the dose used depending of the quality and quantity of the raw water. It composes 1-3% of total volume of raw water during the treatment process [3]. Clean water is channelled to supply tanks while residue is discharged into lagoons which is finally dumped in designated sites for dehydration and further biological stabilization. Coal supplied 29% of energy world-wide and is expected to increase by 30% by 2035 which actual amount of coal consumed per year will increase from 3840 million tons in 2015 to 4032 million tons in 2035 [4]. It is characterized of its powdery form as a result of burning of finely ground coal in a boiler which comprised dominantly of silica content. Both wastes should be contained and properly disposed to avoid potential damage to surrounding environments. Recycling of wastes is better alternative approach instead of sending them to the landfill.

Sludge from water treatment plant contains clay, silt and sand which can substitute raw material to clay-based products [5]. Attempts to study the potential usage of drinking water sludge in development of alternative brick [6][7][8][9]. Fly ash has been widely added in cement and can improve the properties of concrete that associated with pozolanic and hydraulic reactions [10]. Range between 15% and 25% of fly ash were added in concrete application and some cases higher fly ash content of more than 45% was used to achieve better mechanical property and higher durability [11]. Conventional brick is considered not eco-friendly building product as its footprint associated with high temperature firing for length period of up to 40 hours. Therefore, this study aimed to examine the effect of fly ash on the mechanical properties of drinking water sludge brick at lower firing temperature of 500 °C for 3 hours. The mechanical characteristics of the studied brick were examined in terms of volume change, density, water absorption and compressive strength. Compressive strength of brick exposed to sulphate attack were also tested in the laboratory.

2. Materials and Methods

2.1. Wastes materials

The drinking water sludge (DW) was collected from one of raw water treatment facility in Selangor while the fly ash (FA) was provided by coal-based power plant of Sultan Salahuddin Abdul Aziz located at Kapar Selangor. Bulk samples were collected and were stored in air-tight containers. For the DWS, it was slightly wet when collected from the disposed site while FA was almost dry and very fine-grained and dark grey in colour. In the laboratory, DWS was further dried under room temperature for several days and aggregates were broken down and sieved through 2 mm. Any coarse particles or plant matters was removed from the sieve. The basic characteristics of the used materials are shown in Table 1. DW was acidic of pH 5.76 and alkaline for FA (pH 10.1). The values of specific gravity were less than 2.0. The organic content for DW higher than FA as stated by LOI test. DW displayed higher fraction of silt and clay than sand while FA was dominantly contained of silt fraction, approximately of 77%. Compaction test was performed on DW sample and resulted with maximum dry density and optimum moisture of 1.32 gcm^{-3} and 29%. XRD analysist showed the presence of silica and kaolinite in DW and silica, mullite and aluminium calcite in FA.

Table 1. Characteristics of the wastes used for making the studied bricks.

Parameters	Drinking Water Sludge (DW)	Fly Ash (FA)
pH	5.76	10.1
Specific gravity	1.99	1.82
Loss of ignition (LOI), %	8.42	1.14
Clay, %	36.31	18.05
Silt, %	48.44	76.97
Sand, %	15.25	2.11
Max. dry density, gcm^{-3}	1.32	-
Opt. moisture ct., %	29.00	-
XRD analysist	Quartz, SiO_2 Kaolinite, $Al_2SO_4H(OH)_4$	Quartz, SiO_2 Mullite, $3Al_2O_3.2SiO_2$ Aluminium calcite

2.2. Compaction tests

Compaction test was carried out to establish the value of maximum dry density, ρ_{max} and optimum moisture content, w_{opt} of the mixture of he wastes at different ratios. The ratio between the base material of DW over FA is shown in Table 2. The results from the test would be used to estimate the amount of water added to the mixture of wastes to achieve a workable mixture during preparation of brick samples. It suggested that the optimum moisture content dropped from 29.0% to approximately 26.0% for 15, 30 and 45% of FA contents.

Table 2. Compactive behaviour of DW with different ratios of added FA.

Brick sample	Ratio of mixture (DW:FA)	Max. Dry Density ρ_{max} (kgm^{-3})	Opt. Moisture Content w_{opt} (%)
DW100	100:0	1.32	29.0
DW85	85:15	1.20	26.1
DW70	75:25	1.25	26.4
DW55	55:45	1.26	26.1

2.3. Preparation of Brick Samples

Fired brick samples were prepared by dried mixing of DW and FA. The ratio of added FA was based on 0, 15, 30 and 45% of the dry mass of the DW sludge. Once the FA was added, the mixture was mixed thoroughly in the container. Then, water was added 5% higher than optimum moisture content to produce workable mixture. The mixture was transferred into mould of 215 mm x 65 mm x 102.5 mm. A total of three layers and each layer received 27 blows by compaction hammer. The samples were left dried under room temperature for 4 days. The samples were removed from the moulds before carefully transferred into furnace for 24 hours at temperature of 105°C. Then the temperature was set up to 500 °C at increment of 30 °C/minute. At this stage, the firing process would take 3 hours and then the samples were kept inside for overnight cooling down before taking out of the furnace the next day.

2.4. Testing Procedures

The brick samples were examined for linear shrinkage, density, water absorption and compressive strength. The linear shrinkage of the bricks involved measurement of each final dimension after firing process. A total of 10 samples were used and each dimension was measured at three different positions. A vernier calliper was used to measure each dimension of individual brick. The linear shrinkage, ΔL was calculated according to the following equation:

$$\Delta L = \frac{L_o - L_f}{L_o} . 100\%$$

where L_o – measurement each dimension before firing, mm; L_f – measurement of dimension after firing, mm. The volume change of brick, ΔV was calculated from the readings from the measurement of each dimension as shown by the following equation:

$$\Delta V = \frac{(V_o - V_f)}{V_o} . 100$$

where V_o – volume before dying and/or firing, mm^3; V_f– volume after dying and/or firing, mm^3. The density tests of the bricks were performed according to Archimedes' buoyancy principal [12]. The same bricks tested earlier were weighed, m_d and marked with waterproof in for labelling purpose. Samples were immersed in a tank of water for 2 hours before taking out and let to drain quickly (less than a minute). Any excess water was wiped out from the surface of the brick and its mass was recorded, m_1. Then the brick was put on the apparatus to measure its submerged mass, m_2. This apparatus consists of a steel cage to place the sample which connected to a digital weight indicator. These steps were repeated for all the bricks and the volume, V is calculated based on the following equation:

$$V = (m_1 - m_2).1000$$

where: V – volume, mm^3; m_1 – mass of wet sample (gram); m_2 – mass of submerged sample (gram). Hence, the density, D can be calculated from:

$$D = \frac{m_d}{V}.1000000$$

where: D – density of sample (kgm^{-3}); m_d – mass of dry sample in gram; V – volume of sample, mm^3. The same bricks were also used to determine the water absorption. All the bricks were placed in a tank of water for 24 hours. Then the saturated mass was recorded, m_w and water absorption, w was calculated based on the following equation:

$$w\% = \frac{m_w - m_d}{m_d} \times 100$$

where: w – percentage of water absorption; m_d – dry mass (gram); m_w – saturated mass (gram). The compressive strength test was performed by using Autocon 2000 Universal Testing Machine with 2000 kN maximum capacity. The brick sample was cleaned and a piece of hard wood was positioned on bed part of the brick. The brick was shearing by applying load at rate of 7.0 kN/s. Maximum load at failure (N_f) was recorded and would be used to determined its compressive strength, τ. Three bricks were used in this test and and the compressive strength was calculated from the following equation:

$$Compressive\ strength = \frac{Maximum\ load\ at\ failure\ (N_f)}{Area\ (m^2)}$$

Determination of water absorption and compressive strength were carried out according to the BS 3921: 1985 [13]. Meanwhile, for the assessment of the resistance of the brick against sulphate attack, the samples were exposed to 10% of potassium sulphate (K_2SO_4) by daily spraying for a minute up to 28 days. Then, the samples were tested for their compressive strength.

3. Results and Discussion

3.1. Linear shrinkage

After drying and firing processes, brick normally changes in its dimension and the change should be small. Change in dimension will end up with the overall volume of the final brick. The average and linear shrinkage of length L_P, width L_W and height L_H of the brick dimension are shown in Table 3. The linear shrinkage of each dimension dropped as FA content was increased. However, at 30% of FA content, brick showed apparent decrease of L_P, L_W and L_H represented by 28%, 45% and 66%, respectively. The shrinkage patterns in terms of volume change of the fired bricks are shown in Figure 1. As shown in Figure 1, significant change in volume was seen at 30% of FA content.

Table 3. Mean dimension and linear shrinkage of different FA contents.

FA content (%)	Length, P_f (mm)	L_P (%)	Width W_f (mm)	L_W (%)	Height H_f (mm)	L_H (%)
0	204.0±0.8	5.1	98.0±0.8	4.4	63.7±0.9	2.1
15	204.3±0.9	5.0	98.0±0.8	4.4	64.0±0.8	1.5
30	207.0±0.5	3.6	100.0±0.8	2.4	64.7±0.5	0.5
45	209.0±0.5	2.6	100.7±0.5	1.8	65.0±1.6	0.0

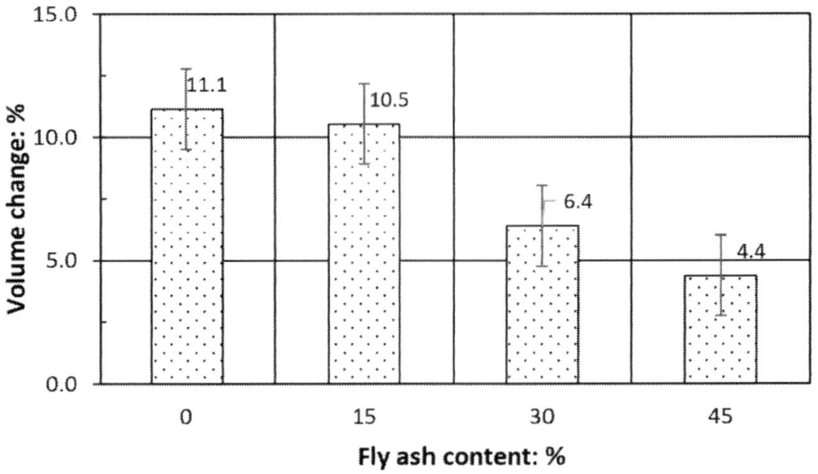

Figure 1. Volume change of the fired bricks.

3.2. Density
The results of density of the fired bricks are shown in Figure 2. The value of the density gradually changed with slightly increased as FA contents were added. Increasing of the FA content has not contributed significant change in density of the fired brick. In terms of specific gravity, FA and DWS had very close value of 1.82 and 1.99, respectively (Table 2). Thus, alteration of the ratio of brick's mixture from the addition of FA has insignificant effect on the density of the studied fired brick.

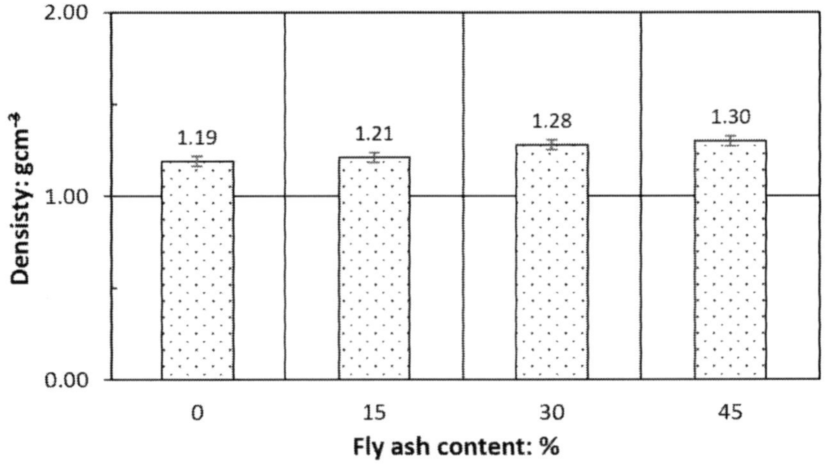

Figure 2. Density of the fired bricks.

3.3. Water Absorption
Bricks should have very small water absorption capacity and this characteristic can enhance their mechanical strength. The result of water absorption test is shown in Figure 3. The fired brick displayed a decreasing trend up to 30 % of FA content. At FA content of 45 %, the water absorption increased to

40 %. It was not very clear why the increase in water absorption occurred at this FA content. Therefore, it can be suggested that the optimum content of FA should be limited at lower than 45 % for this type of brick with similar thermal treatment.

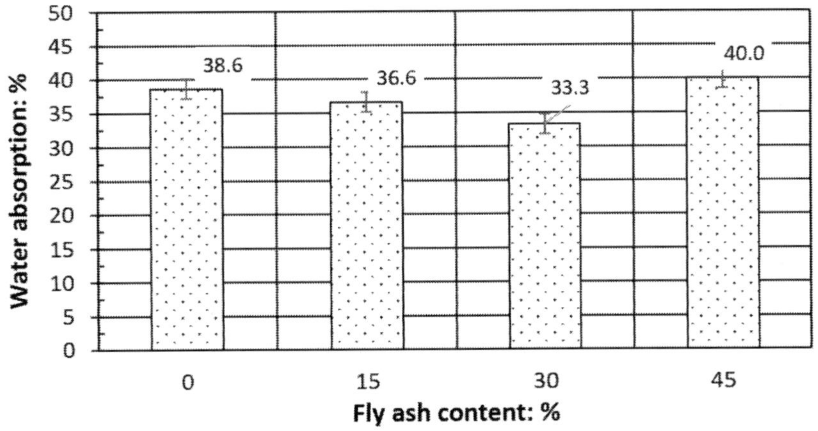

Figure 3. Water absorption of the fired bricks.

3.4. Compressive Strength

Brick is always used to support load and it load bearing capacity is measured by the compressive strength value. Different ranges of compressive strength will be intended for different engineering applications [14]. The result of compressive strength of the studied bricks is shown in Figure 4. The studied bricks displayed gradual decreasing trend as the FA contents were increased up to 30 %. However, at FA content of 45 %, a significant drop was observed from 1.83 MPa to 0.88 MPa representing of 57.4 % of loss strength. It was found that the compressive strength is limited by the amount of FA contents. Previous studies also experienced similar findings from different base materials that being added with various ranges of FA contents [15],[16].

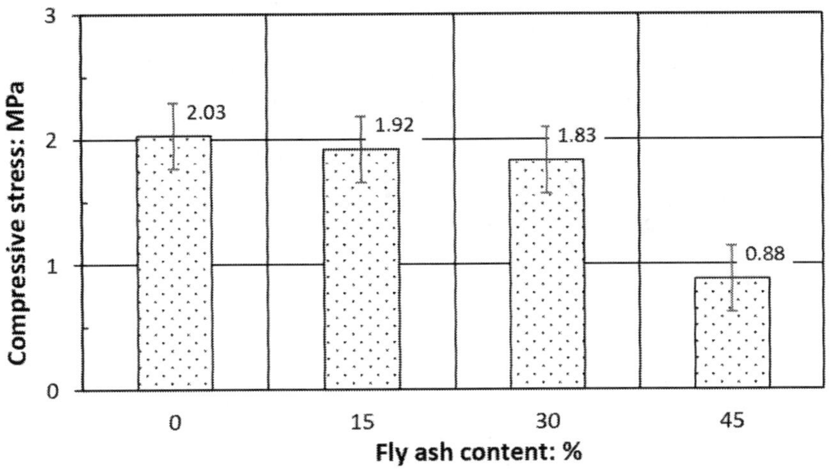

Figure 4. Compressive strength of the fired bricks.

3.5. Acid Sulphate Attack

The resistance of acid sulphate attack on the studied fired brick was accessed by the compressive strength tests. Bricks should withstand from the sulphate attack which can decay in their mechanical strength. Hot and humid nature of Malaysia pose a significant threat to brick and other masonry materials [17]. The result of compressive strength of the studied bricks against the sulphate attack is shown in Figure 5. As the content of FA increased, the compressive strength dropped from 1.91 MPa to 0.45 MPa. Similar trend was also seen where significant decreased at FA content of 45 %.

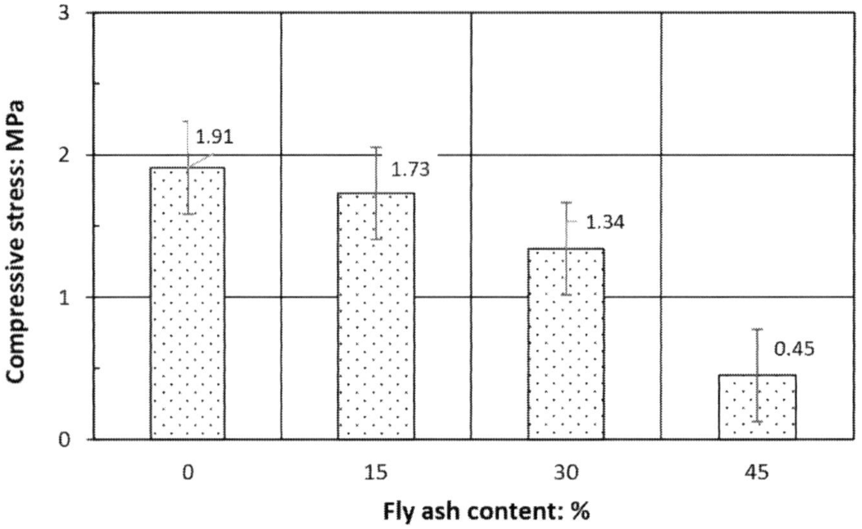

Figure 5. Compressive strength of the fired bricks against sulphate attack.

4. Conclusions

This study managed to determine the mechanical characteristics of the studied brick fired at lower temperature of 500 °C with different ratios of drinking water sludge and fly ash contents between 0% and 45% of dry weight of drinking water sludge as base material. The linear shrinkage of each dimension apparently dropped at 30% of FA content which also agreed with the volume change of the fired brick. Density of the brick was not significantly change as FA content was added the brick. For the water absorption tests, brick displayed initial decreased up to 30% of FA content which later increased as FA was 40%. The compressive strength exhibited gradual drop as FA increased to 30% and experienced drastic decrease at 45% of FA content. This trend also observed for the assessment of the compressive strength against the sulphate attack where apparently drop was seen at similar FA content. Therefore, optimum FA content was important to be known prior mixing of brick's raw materials otherwise it could deteriorate the quality of particular brick. In this study, with studied material and firing condition, it is suggested to limit the FA content of 30%.

References

[1] DOE 2013. Schedule Waste Inventory. *Malaysia Environmental Inventory Report* 2011 137-149
[2] Ren B., Zhao Y., Ji B., Wei T. & Shen C. 2020. Granulation of drinking water treatment residues: recent advances and prospects. *Water 2020*, **12** 1400. Doi:10.3390/w12051400
[3] Turner, T.; Wheeler, R.; Stone, A.; Oliver, I. 2019. Potential alternative reuse pathways for water treatment residuals: Remaining barriers and questions—A review. *Water Air Soil Pollution* 230: 227
[4] Bhatt A., Prriyadarshini S., Moanakrishnan A. A. Abri A. Sattler M. & Techapaphawit S. 2019. Physical, chemical and geotechnical properties of coal fly ash: A global review. *Case Studies in Construction Materials* **11** 1-11.

[5] Oliveira, E.M.S., Sampaio, V.G. & Holanda, J.N.F. 2006. Evaluation of the suitability of municipal waterworks waste as a raw material for red ceramic brick production. *Industrial Ceramics*, 26; 23-28
[6] Mageed et al. 2011 Utilization of water treatment plants sludge ash in brick making. *Journal of Engineering Sciences* 39(1) 195-206
[7] Anyakora, N. V., Ajinomoh, C. S., Ahmed, A. S., Mohammed-Dabo, I. A., Ibrahim, J. & Anto, J. B. 2012. Sustainable technology – based strategy for processing water works sludge for resource utilization. *World Journal of Engineering and Pure and Applied Sciences*, 2(5): 161-168.
[8] Tantawy, M.A. & Mohamed, R.S.A. 2017. Middle Eocene clay from Goset Abu Khashier: Geological assessment and utilization with drinking water treatment sludge in brick manufacture. *Applied Clay Science*, 138:114-124
[9] Haniegal A M *et al* 2020. Study on properties of clay brick incorporating sludge of water treatment plant and agriculture waste. *Case Study in Constructon Materials*, **13** 1-13
[10] Thomas M. 2007 Optimizing the use of fly ash in concrete United States: N. p., 2007. Web.
[11] Rashad A M 2017 A brief on high-volume Class F fly ash as cement replacement – A guide for Civil Engineer *International Journal of Sustainable Built Environment*, **4** 278-306
[12] AS/NZS 4456 2003 Masonry Units, Segmental Pavers and Flags: Methods of Tests, Standards Australia. Sydney
[13] BS 3921 1985 British Standard Specifications for Clay Bricks London
[14] Ramadan O M *et al* 2008. Reuse of water treatment plant sludge in brick manufacturing. *Journal of Applied Sciences Research*, 4(10): 1223-1229
[15] Koukaouzas N *et al* 2011 Synthesis of CFB-coal fly ash clay bricks and their characterisation. *Waste Biomass Valorization* **2** 87-94
[16] Ngo SH and Huynh TP 2019 Effect of f ash content on engineering properties of unfired building bricks. *Journal of Science & Technology* **4** 32-36
[17] Wan Ibrahim M H *et al* 2012 Influence of sulphate and chloride on the mechanical properties of fired clay masonry wall. *International Journal of Integrated Engineering* 4(2) 29-38

Acknowledgments
Authors would like to thank Universiti Kebangsaan Malaysia for the financial support through research grants (ETP-2015-003 & GUP-2016-068) for this project. The researchers are also grateful the technical staff for assisting in sample preparation and laboratory testing.

Parameter estimation and hypothesis testing of geographically and temporally weighted bivariate generalized Poisson regression

Setyorini Indah Purwanti[1*], Sutikno[1] and Purhadi[1]

[1] Department of Statistics, Faculty of Scientics, Institut Teknologi Sepuluh Nopember (ITS). Arief Rahman Hakim Street, Surabaya 60111 Indonesia

Corresponding E-mail: setyoriniindahpurwanti@gmail.com

Abstract. Poisson regression is used to model the data with the response variable in the form of count data. This modeling must meet the equidispersion assumption. That is, the average value is the same as the variance. However, this assumption is often violated. Violation of the equidispersion assumption in Poisson regression modeling will result in invalid conclusions. These violations are an overdispersion and an underdispersion of the response variable. Generalized Poisson Regression (GPR) is an alternative if there is a violation of the equidispersion assumption. If there are two correlated response variables, modeling will use the Bivariate Generalized Poisson Regression (BGPR). However, in the panel data with the observation unit in the form of an area, BGPR is not quite right because there is spatial and temporal heterogeneity in the data. Geographically and Temporally Weighted Bivariate Generalized Poisson Regression (GTWBGPR) is a method for modeling spatial and temporal heterogeneity data. GTWBGPR is a development of GWBGPR. In GTWBGPR, besides accommodating spatial effects, it also accommodates temporal effects. This research will discuss the parameter estimation and test statistics for the GTWBGPR model. Parameter estimation uses Maximum Likelihood Estimation (MLE), but the result is not closed-form, so it is solved by numerical iteration. The numerical iteration used is Newton-Raphson. The test statistic for simultaneous testing uses the Maximum Likelihood Ratio Test (MLRT). With large samples, then this test statistic has a chi-square distribution approximation. So the test statistic for the partial test uses the Z test statistic.

Keywords: GTWBGPR, MLE, MLRT, spatial heterogeneity, temporal heterogeneity

Track Name: Advanced Technology and Renewable Energy

1. Introduction

Poisson regression is used to see the relationship between response variables in count data with predictor variables. In modeling the data using Poisson regression, it must meet the equidispersion assumption. That is, the average value is the same as the variance. The variance value greater than the mean in the Poisson distribution is called overdispersion, while the variance value that is less than the mean is called underdispersion [1]. If this assumption is violated, the standard error will be underestimated, leading to incorrect conclusions about the significance of the independent variables [2]. The equidispersion condition is often not fulfilled. Generalized Poisson Regression (GPR) is an alternative for modeling data in case of overdispersion or underdispersion. In some studies, GPR is better at modeling data than Poisson regression [1,2,3], negative binomial regression [1,2,3], and Poisson Inverse Gaussian Regression [1]. However, if there are two correlated response variables, use the Bivariate Generalized Poisson Regression (BGPR) [4,5,6].

This regression resulted in invalid estimation results. One alternative to modeling data with spatial heterogeneity is to use Geographically Weighted Regression (GWR). GWR can accommodate spatial variations in data using geographic weighting to obtain different parameter estimates for different region points. Thus, using GWR on spatial data can accommodate Simpson's paradox. Simpson's paradox is the difference in parameter estimates in the data model in the aggregate and modeled data separately [7]. BGPR is unsuitable for spatial data because it assumes the same weights for all regions. Therefore Purhadi et al. [8] developed Geographically Weighted Bivariate Generalized Poisson Regression (GWBGPR).

Panel data is an example of data with a time dimension. Panel data is data obtained from more than one observation. This data's advantages over cross-section and time-series data are more variability, more informative, and less collinearity between variables [9]. GWBGPR is not suitable for analyzing panel data. When using GWBGPR in analyzing panel data, it is assumed that each period's parameters are the same. This assumption is certainly not realistic since data varies between periods. Geographically and Temporally Weighted Regression (GTWR) is an alternative for modeling data containing spatial and temporal heterogeneity [10,11,12]. Therefore, this time we will propose Geographically and Temporally Weighted Bivariate Generalized Poisson Regression (GTWBGPR). GTWBGPR can be used to model panel data with the observation unit in the form of an area, where the response variable is in the form of count data and correlates with each other. In modeling GTWR, it is usually done simultaneously for all periods. In GTWBGPR, parameter estimation and hypothesis testing will be carried out gradually in each period.

Parameters estimators of the regression model are obtained using the parameter estimation method. Several parameter estimation methods that often use in regression models with Generalized Poisson distributed response variables are Maximum Likelihood Estimation (MLE), Method of Moment (MM), and Generalized Method of Moment (GMM). The methods used to estimate parameters for GPR are MLE [1,2,3], MM [1], and GMM [13]. In BGPR, used MM [4] and MLE [6] to estimate parameters. GWBGPR [8] and GWMGPR [14] use MLE to estimate parameters. After getting the parameter estimator, the next step is hypothesis testing. The test statistic for testing hypotheses with a Generalized Poisson distributed response variable is obtained through the Maximum Likelihood Ratio Test (MLRT) method. Therefore, in GTWBGPR, the parameter estimator uses MLE, and hypothesis testing uses MLRT because most of the models with Generalized Poisson distributed response variables use MLE as a parameter estimation method and MLRT to obtain test statistic in hypothesis testing.

2. Materials and Method

2.1 Geographically Weighted Bivariate Generalized Poisson Regression (GWBGPR)

Suppose the response variable Y_{1i} dan Y_{2i} distributed by BGP, the probability density function of Y_{1i} and Y_{2i} from GWBGPR is as follows [8]:

$$f(y_{1i}, y_{2i}) = \lambda_0(u_i,v_i)\lambda_1(u_i,v_i)\lambda_2(u_i,v_i)$$

$$\exp\left\{-\left(\lambda_0(u_i,v_i)+\lambda_1(u_i,v_i)+\lambda_2(u_i,v_i)\right)-y_{1i}\varphi_1(u_i,v_i)-y_{2i}\varphi_2(u_i,v_i)\right\}$$

$$\left(\sum_{k=0}^{\min(y_{1i},y_{2i})} \frac{\left(\lambda_0(u_i,v_i)+k\varphi_0(u_i,v_i)\right)^{k-1}}{k!} \frac{\left(\lambda_1(u_i,v_i)+(y_{1i}-k)\varphi_1(u_i,v_i)\right)^{y_{1i}-k-1}}{(y_{1i}-k)!}\right.$$

$$\left.\frac{\left(\lambda_2(u_i,v_i)+(y_{2i}-k)\varphi_2(u_i,v_i)\right)^{y_{2i}-k-1}}{(y_{2i}-k)!}\left(\exp\left(k\left(\varphi_1(u_i,v_i)+\varphi_2(u_i,v_i)-\varphi_0(u_i,v_i)\right)\right)\right)\right) \quad (1)$$

where (u_i,v_i) are the latitude and longitude coordinates of the ith observation, and φ is the dispersion parameter with the max value $(-\lambda^{-1},-m^{-1})<\varphi<\lambda^{-1}$, m is the largest positive integer on $1+\varphi m>0$ when $\varphi<0$. If $\varphi=0$, then there is no dispersion. If $\varphi>0$, then there is an overdispersion, and if $\varphi<0$, then there is underdispersion.

The relationship between the response variables Y_{1i} and Y_{2i} and predictor variables $X_{1i},X_{2i},\ldots,X_{pi}$ stated as follows:

$$\log(\lambda_c(u_i,v_i))=\mathbf{x}_i^{\mathbf{T}}\boldsymbol{\beta}_c(u_i,v_i) \quad (2)$$

where:

$\mathbf{x}_i=\begin{bmatrix}1 & x_{1i} & x_{2i} & \ldots & x_{pi}\end{bmatrix}^T; i=1,2,\ldots,n$

$\boldsymbol{\beta}_c(u_i,v_i)=\begin{bmatrix}\beta_{c0}(u_i,v_i) & \beta_{c1}(u_i,v_i) & \beta_{c2}(u_i,v_i) & \ldots & \beta_{cp}(u_i,v_i)\end{bmatrix}^T; c=1,2$, $\boldsymbol{\beta}_c(u_i,v_i)$ is the parameter vector of the response variable Y_{ci} on the GWBGPR model.

2.2 Geographically and Temporally Weighted Bivariate Generalized Poisson Regression (GTWBGPR)

The probability density function on equation (1) does not pay attention to temporal heterogeneity. As a result, the resulting parameter estimates are the same for the exact location despite different periods. This result is, of course unrealistic, because there is temporal heterogeneity in the data. Regarding equation (1), then the probability density function of Y_{1i} and Y_{2i} on location (u_i,v_i) and period *l*, where $i=1,2,\ldots,n; l=1,2,\ldots,L$ is:

$$f(y_{1il},y_{2il})=\lambda_{0l}(u_i,v_i,t_i)\lambda_{1l}(u_i,v_i,t_i)\lambda_{2l}(u_i,v_i,t_i)$$

$$\exp\left\{-\left(\lambda_{0l}(u_i,v_i,t_i)+\lambda_{1l}(u_i,v_i,t_i)+\lambda_{2l}(u_i,v_i,t_i)\right)-y_{1il}\varphi_{1l}(u_i,v_i,t_i)-y_{2il}\varphi_{2l}(u_i,v_i,t_i)\right\}$$

$$\left(\sum_{k=0}^{\min(y_{1il},y_{2il})} \frac{\left(\lambda_{0l}(u_i,v_i,t_i)+k\varphi_{0l}(u_i,v_i,t_i)\right)^{k-1}}{k!} \frac{\left(\lambda_{1l}(u_i,v_i,t_i)+(y_{1il}-k)\varphi_{1l}(u_i,v_i,t_i)\right)^{y_{1il}-k-1}}{(y_{1il}-k)!}\right.$$

$$\left.\frac{\left(\lambda_{2l}(u_i,v_i,t_i)+(y_{2il}-k)\varphi_{2l}(u_i,v_i,t_i)\right)^{y_{2il}-k-1}}{(y_{2il}-k)!} \exp\left(k\varphi_{1l}(u_i,v_i,t_i)+k\varphi_{2l}(u_i,v_i,t_i)-k\varphi_{0l}(u_i,v_i,t_i)\right)\right) \quad (3)$$

To make writing more manageable, then (u_i,v_i,t_i) is written as a vector $(\mathbf{v}_i)$. In modeling the GTWBGPR, it is carried out in stages for each L period where $L=1,2,\ldots,L^*$. By considering equation (2), the model in the L period is:

$$\log\left(\lambda_{cL}(\mathbf{v}_i)\right)=\mathbf{x}_{iL}^T\boldsymbol{\beta}_{cL}(\mathbf{v}_i) \quad (4)$$

where:

$\mathbf{x}_{iL} = \begin{bmatrix}1 & x_{1iL} & x_{2iL} & \ldots & x_{piL}\end{bmatrix}^T; i=1,2,\ldots,n$

$\boldsymbol{\beta}_{cL}(\mathbf{v}_i)=\begin{bmatrix}\beta_{c0L}(\mathbf{v}_i) & \beta_{c1L}(\mathbf{v}_i) & \cdots & \beta_{cpL}(\mathbf{v}_i)\end{bmatrix}^T; c=1,2$

2.3 Spatiotemporal Weight

Spatial and temporal effects have been accommodated in the GTWBGPR model by assigning spatiotemporal weight $\left(w_{ii*l}\right)$ to each observation point (area). As a result, the parameter estimates at each observation point (area) will be different, which shows the local nature of the GTWBGPR model. Weight has an essential role in the spatiotemporal analysis. Spatiotemporal weight describes the relationship between observations, observations close to the ith regression point have a greater weight than observations that are further away. When forming a weighting matrix, a kernel function is needed. Commonly used kernel functions are as follows:

Fixed Gaussian Kernel Functions

$$w_{ii*l} = \exp\left(-\frac{1}{2}\left(\frac{d_{ii*l}}{q}\right)^2\right)$$

Fixed Bisquare Kernel Functions

$$w_{ii*l} = \begin{cases} \left(1-\left(\frac{d_{ii*l}}{q}\right)^2\right)^2 & \text{, if } d_{ii*l} \le q \\ 0 & \text{, if } d_{ii*l} > q \end{cases}$$

Adaptive Gaussian Kernel Functions

$$w_{ii*l} = \exp\left(-\frac{1}{2}\left(\frac{d_{ii*l}}{q_{il}}\right)^2\right)$$

Adaptive Bisquare Kernel Functions

$$w_{ii*l} = \begin{cases} \left(1-\left(\frac{d_{ii*l}}{q_{il}}\right)^2\right)^2 & \text{, if } d_{ii*l} \le q_{il} \\ 0 & \text{, if } d_{ii*l} > q_{il} \end{cases}$$

d_{ii*l} is the euclidean distance, that is $d_{ii*l} = \sqrt{\left(u_i - u_{i*}\right)^2 + \left(v_i - v_{i*}\right)^2 + \tau_L\left(t_i - t_{i*}\right)^2}$ [12], and q_{il} is spatiotemporal bandwidth. The selection of the optimum bandwidth affects the accuracy of the regression results. Bandwidth values that are too large will cause significant parameter bias. Bandwidth values that are too small will cause the model to be too coarse (under smoothing). GCV, which is one of the easy bandwidth choices in its application, is as follows [7]:

$$GCV = \frac{nL\sum_{l=1}^{L}\sum_{i=1}^{n}\left[\mathbf{y}_{il} - \hat{\mathbf{y}}_{il}\left(q_{il}\right)\right]^T\left[\mathbf{y}_{il} - \hat{\mathbf{y}}_{il}\left(q_{il}\right)\right]}{\left(nL - v_1\right)^2} \tag{5}$$

with $\left(nL - v_1\right)$ is the number of observations minus the number of the parameters.

Calculation of w_{ii*l} at L period is obtained from:

$$w_{ii*l} = A_l\left(d_{ii*l}\right); d_{ii*l} = \sqrt{\left(u_i - u_{i*}\right)^2 + \left(v_i - v_{i*}\right)^2 + \tau_L\left(L - l\right)^2}; l = 1,2,\ldots,L \tag{6}$$

So that for the random samples $Y_{1i1}, Y_{2i1}; i = 1,2,\ldots,n$ using weight w_{ii*1}. Random samples $Y_{1i2}, Y_{2i2}; i = 1,2,\ldots,n$ using weight w_{ii*2}. And so on until the random samples $Y_{1iL}, Y_{2iL}; i = 1,2,\ldots,n$ using weight w_{ii*L}.

IOP Conf. Series: Earth and Environmental Science **880** (2021) 012043 doi:10.1088/1755-1315/880/1/012043

3. Results and Discussion

3.1 Parameter Estimation

GTWBGPR model parameter estimation using MLE. In estimating the parameters in the GTWBGPR, it is carried out gradually in each period by including observations of the previous period. When estimating L period parameter using n number of data in period 1, 2, up to L or require random samples $Y_{1i1}, Y_{2i1}, Y_{1i2}, Y_{2i2}, \ldots, Y_{1iL}, Y_{2iL}; i = 1, 2, \ldots, n$, where $L = 1, 2, \ldots, L^*$. The first step in estimating the L period parameter is to form the likelihood function of the probability density function on equation (3), namely:

$$\begin{aligned}
&L\left(\lambda_{0L}(\mathbf{v}_i), \boldsymbol{\beta}_{1L}(\mathbf{v}_i), \boldsymbol{\beta}_{2L}(\mathbf{v}_i), \varphi_{0L}(\mathbf{v}_i), \varphi_{1L}(\mathbf{v}_i), \varphi_{2L}(\mathbf{v}_i); i = 1, 2, \ldots, n\right) \\
&= \prod_{l=1}^{L} \prod_{i=1}^{n} \lambda_{0L}(\mathbf{v}_i)\left[e^{\mathbf{x}_{il}^T \boldsymbol{\beta}_{1L}(\mathbf{v}_i)} - \lambda_{0L}(\mathbf{v}_i)\right]\left[e^{\mathbf{x}_{il}^T \boldsymbol{\beta}_{2L}(\mathbf{v}_i)} - \lambda_{0L}(\mathbf{v}_i)\right] \exp\{-e^{\mathbf{x}_{il}^T \boldsymbol{\beta}_{1L}(\mathbf{v}_i)} - e^{\mathbf{x}_{il}^T \boldsymbol{\beta}_{2L}(\mathbf{v}_i)} + \lambda_{0L}(\mathbf{v}_i) - y_{1il}\varphi_{1L}(\mathbf{v}_i) \\
&-y_{2il}\varphi_{2L}(\mathbf{v}_i)\} \sum_{k=0}^{\min(y_{1il}, y_{2il})} \left[\frac{\left(\lambda_{0L}(\mathbf{v}_i) + k\varphi_{0L}(\mathbf{v}_i)\right)^{k-1}}{k!} \frac{\left(\left(e^{\mathbf{x}_{il}^T \boldsymbol{\beta}_{1L}(\mathbf{v}_i)} - \lambda_{0L}(\mathbf{v}_i)\right) + (y_{1il} - k)\varphi_{1L}(\mathbf{v}_i)\right)^{y_{1il}-k-1}}{(y_{1il} - k)!} \right. \\
&\left. \frac{\left(\left(e^{\mathbf{x}_{il}^T \boldsymbol{\beta}_{2L}(\mathbf{v}_i)} - \lambda_{0L}(\mathbf{v}_i)\right) + (y_{2il} - k)\varphi_{2L}(\mathbf{v}_i)\right)^{y_{2il}-k-1}}{(y_{2il} - k)!} \exp\left(k\varphi_{1L}(\mathbf{v}_i) + k\varphi_{2L}(\mathbf{v}_i) - k\varphi_{0L}(\mathbf{v}_i)\right)\right] \qquad (7)
\end{aligned}$$

where $\mathbf{x}_{il} = \begin{bmatrix} 1 & x_{1il} & x_{2il} & \cdots & x_{pil} \end{bmatrix}^T; i = 1, 2, \ldots, n; l = 1, 2, \ldots, L$

Based on equation (7), the log-likelihood of GTWBGPR is obtained as follows:

$$\begin{aligned}
Q_L &= \log L\left(\lambda_{0L}(\mathbf{v}_i), \boldsymbol{\beta}_{1L}(\mathbf{v}_i), \boldsymbol{\beta}_{2L}(\mathbf{v}_i), \varphi_{0L}(\mathbf{v}_i), \varphi_{1L}(\mathbf{v}_i), \varphi_{2L}(\mathbf{v}_i); i = 1, 2, \ldots, n\right) \\
&= \sum_{l=1}^{L}\sum_{i=1}^{n} \log\lambda_{0L}(\mathbf{v}_i) + \sum_{l=1}^{L}\sum_{i=1}^{n} \log\left(e^{\mathbf{x}_{il}^T \boldsymbol{\beta}_{1L}(\mathbf{v}_i)} - \lambda_{0L}(\mathbf{v}_i)\right) + \sum_{l=1}^{L}\sum_{i=1}^{n} \log\left(e^{\mathbf{x}_{il}^T \boldsymbol{\beta}_{2L}(\mathbf{v}_i)} - \lambda_{0L}(\mathbf{v}_i)\right) - \sum_{l=1}^{L}\sum_{i=1}^{n} e^{\mathbf{x}_{il}^T \boldsymbol{\beta}_{1L}(\mathbf{v}_i)} \\
&- \sum_{l=1}^{L}\sum_{i=1}^{n} e^{\mathbf{x}_{il}^T \boldsymbol{\beta}_{2L}(\mathbf{v}_i)} + \sum_{l=1}^{L}\sum_{i=1}^{n} \lambda_{0L}(\mathbf{v}_i) - \sum_{l=1}^{L}\sum_{i=1}^{n} y_{1il}\varphi_{1L}(\mathbf{v}_i) - \sum_{l=1}^{L}\sum_{i=1}^{n} y_{2il}\varphi_{2L}(\mathbf{v}_i) + \sum_{l=1}^{L}\sum_{i=1}^{n} B_{il} \qquad (8)
\end{aligned}$$

where:

$$\begin{aligned}
B_{il} = &\sum_{k=0}^{\min(y_{1il}, y_{2il})} \left[(k-1)\log\left[\lambda_{0L}(\mathbf{v}_i) + k\,\varphi_{0L}(\mathbf{v}_i)\right] - \log k! + (y_{1il} - k - 1)\right. \\
&\log\left[e^{\mathbf{x}_{il}^T \boldsymbol{\beta}_{1L}(\mathbf{v}_i)} - \lambda_{0L}(\mathbf{v}_i) + (y_{1il} - k)\varphi_{1L}(\mathbf{v}_i)\right] - \log(y_{1il} - k)! + (y_{2il} - k - 1) \\
&\left.\log\left[e^{\mathbf{x}_{il}^T \boldsymbol{\beta}_{2L}(\mathbf{v}_i)} - \lambda_{0L}(\mathbf{v}_i) + (y_{2il} - k)\varphi_{2L}(\mathbf{v}_i)\right] - \log(y_{2il} - k)! + k\varphi_{1L}(\mathbf{v}_i) + k\varphi_{2L}(\mathbf{v}_i) - k\,\varphi_{0L}(\mathbf{v}_i)\right]
\end{aligned}$$

Then the log-likelihood function is weighted by spatiotemporal weight (w_{ii^*l}) for each point of observation are:

$$\begin{aligned}
Q_L^* = &\sum_{l=1}^{L}\sum_{i^*=1}^{n} w_{ii^*l} \log\lambda_{0L}(\mathbf{v}_i) + \sum_{l=1}^{L}\sum_{i^*=1}^{n} w_{ii^*l} \log\left(e^{\mathbf{x}_{i^*l}^T \boldsymbol{\beta}_{1L}(\mathbf{v}_i)} - \lambda_{0L}(\mathbf{v}_i)\right) + \sum_{l=1}^{L}\sum_{i^*=1}^{n} w_{ii^*l} \log\left(e^{\mathbf{x}_{i^*l}^T \boldsymbol{\beta}_{2L}(\mathbf{v}_i)} - \lambda_{0L}(\mathbf{v}_i)\right) \\
&- \sum_{l=1}^{L}\sum_{i^*=1}^{n} w_{ii^*l} e^{\mathbf{x}_{i^*l}^T \boldsymbol{\beta}_{1L}(\mathbf{v}_i)} - \sum_{l=1}^{L}\sum_{i^*=1}^{n} w_{ii^*l} e^{\mathbf{x}_{i^*l}^T \boldsymbol{\beta}_{2L}(\mathbf{v}_i)} + \sum_{l=1}^{L}\sum_{i^*=1}^{n} w_{ii^*l}\lambda_{0L}(\mathbf{v}_i) - \sum_{l=1}^{L}\sum_{i^*=1}^{n} w_{ii^*l} y_{1i^*l}\varphi_{1L}(\mathbf{v}_i) \\
&- \sum_{l=1}^{L}\sum_{i^*=1}^{n} w_{ii^*l} y_{2i^*l}\varphi_{2L}(\mathbf{v}_i) + \sum_{l=1}^{L}\sum_{i^*=1}^{n} w_{ii^*l} B_{i^*l} \qquad (9)
\end{aligned}$$

Furthermore, equation (9) is derived for each parameter.
Suppose that:

$$s_0 = \frac{k-1}{\left(\lambda_{0L}(\mathbf{v}_i) + k\varphi_{0L}(\mathbf{v}_i)\right)}$$

$$s_1 = \frac{\left(y_{1i^*l} - k - 1\right)}{\left(e^{\mathbf{x}_{i^*l}^T\boldsymbol{\beta}_{1L}(\mathbf{v}_i)} - \lambda_{0L}(\mathbf{v}_i)\right) + \left(y_{1i^*l} - k\right)\varphi_{1L}(\mathbf{v}_i)}$$

$$s_2 = \frac{\left(y_{2i^*l} - k - 1\right)}{\left(e^{\mathbf{x}_{i^*l}^T\boldsymbol{\beta}_{2L}(\mathbf{v}_i)} - \lambda_{0L}(\mathbf{v}_i) + \left(y_{2i^*l} - k\right)\varphi_{2L}(\mathbf{v}_i)\right)}$$

So that the following results are obtained:

$$\frac{\partial Q_L^*}{\partial \lambda_{0L}(\mathbf{v}_i)} = \sum_{l=1}^{L}\sum_{i^*=1}^{n}\frac{w_{ii^*l}}{\lambda_{0L}(\mathbf{v}_i)} - \sum_{l=1}^{L}\sum_{i^*=1}^{n}\frac{w_{ii^*l}}{\left(e^{\mathbf{x}_{i^*l}^T\boldsymbol{\beta}_{1L}(\mathbf{v}_i)} - \lambda_{0L}(\mathbf{v}_i)\right)} - \sum_{l=1}^{L}\sum_{i^*=1}^{n}\frac{w_{ii^*l}}{\left(e^{\mathbf{x}_{i^*l}^T\boldsymbol{\beta}_{2L}(\mathbf{v}_i)} - \lambda_{0L}(\mathbf{v}_i)\right)} + \sum_{l=1}^{L}\sum_{i^*=1}^{n}w_{ii^*l} + \sum_{l=1}^{L}\sum_{i^*=1}^{n}w_{ii^*l}\sum_{k=0}^{\min(y_{1i^*l},y_{2i^*l})}\left[s_0 - s_1 - s_2\right] \quad (10)$$

$$\frac{\partial Q_L^*}{\partial \boldsymbol{\beta}_{1L}(\mathbf{v}_i)} = \sum_{l=1}^{L}\sum_{i^*=1}^{n}w_{ii^*l}\frac{e^{\mathbf{x}_{i^*l}^T\boldsymbol{\beta}_{1L}(\mathbf{v}_i)}\mathbf{x}_{i^*l}}{e^{\mathbf{x}_{i^*l}^T\boldsymbol{\beta}_{1L}(\mathbf{v}_i)} - \lambda_{0L}(\mathbf{v}_i)} - \sum_{l=1}^{L}\sum_{i^*=1}^{n}w_{ii^*l}e^{\mathbf{x}_{i^*l}^T\boldsymbol{\beta}_{1L}(\mathbf{v}_i)}\mathbf{x}_{i^*l} + \sum_{l=1}^{L}\sum_{i^*=1}^{n}w_{ii^*l}\sum_{k=0}^{\min(y_{1i^*l},y_{2i^*l})}\left[s_1 e^{\mathbf{x}_{i^*l}^T\boldsymbol{\beta}_{1L}(\mathbf{v}_i)}\mathbf{x}_{i^*l}\right] \quad (11)$$

$$\frac{\partial Q_L^*}{\partial \boldsymbol{\beta}_{2L}(\mathbf{v}_i)} = \sum_{l=1}^{L}\sum_{i^*=1}^{n}w_{ii^*l}\frac{e^{\mathbf{x}_{i^*l}^T\boldsymbol{\beta}_{2L}(\mathbf{v}_i)}\mathbf{x}_{i^*l}}{e^{\mathbf{x}_{i^*l}^T\boldsymbol{\beta}_{2L}(\mathbf{v}_i)} - \lambda_{0L}(\mathbf{v}_i)} - \sum_{l=1}^{L}\sum_{i^*=1}^{n}w_{ii^*l}e^{\mathbf{x}_{i^*l}^T\boldsymbol{\beta}_{2L}(\mathbf{v}_i)}\mathbf{x}_{i^*l} + \sum_{l=1}^{L}\sum_{i^*=1}^{n}w_{ii^*l}\sum_{k=0}^{\min(y_{1i^*l},y_{2i^*l})}\left[s_2 e^{\mathbf{x}_{i^*l}^T\boldsymbol{\beta}_{2L}(\mathbf{v}_i)}\mathbf{x}_{i^*l}\right] \quad (12)$$

$$\frac{\partial Q_L^*}{\partial \varphi_{0L}(\mathbf{v}_i)} = \sum_{l=1}^{L}\sum_{i^*=1}^{n}w_{ii^*l}\sum_{k=0}^{\min(y_{1i^*l},y_{2i^*l})}\left[s_0 k - k\right] \quad (13)$$

$$\frac{\partial Q_L^*}{\partial \varphi_{1L}(\mathbf{v}_i)} = -\sum_{l=1}^{L}\sum_{i^*=1}^{n}w_{ii^*l}y_{1i^*l} + \sum_{l=1}^{L}\sum_{i^*=1}^{n}w_{ii^*l}\sum_{k=0}^{\min(y_{1i^*l},y_{2i^*l})}\left[s_1\left(y_{1i^*l} - k\right) + k\right] \quad (14)$$

$$\frac{\partial Q_L^*}{\partial \varphi_{2L}(\mathbf{v}_i)} = -\sum_{l=1}^{L}\sum_{i^*=1}^{n}w_{ii^*l}y_{2i^*l} + \sum_{l=1}^{L}\sum_{i^*=1}^{n}w_{ii^*l}\sum_{k=0}^{\min(y_{1i^*l},y_{2i^*l})}\left[s_2\left(y_{2i^*l} - k\right) + k\right] \quad (15)$$

The results of the first derivative above are not closed-form, so the Newton-Raphson iteration solves it. The Newton-Raphson iteration equation is:

$$\hat{\boldsymbol{\theta}}_L(\mathbf{v}_i)_{(m+1)} = \hat{\boldsymbol{\theta}}_L(\mathbf{v}_i)_{(m)} - \mathbf{H}^{-1}\left(\hat{\boldsymbol{\theta}}_L(\mathbf{v}_i)_{(m)}\right)\mathbf{g}\left(\hat{\boldsymbol{\theta}}_L(\mathbf{v}_i)_{(m)}\right) \quad (16)$$

$\hat{\boldsymbol{\theta}}_L(\mathbf{v}_i)_{(m)}$ is the parameter estimator vector in the mth iteration, while $\mathbf{g}\left(\hat{\boldsymbol{\theta}}_L(\mathbf{v}_i)_{(m)}\right)$ is the gradient vector with the parameters $\hat{\boldsymbol{\theta}}_L(\mathbf{v}_i)_{(m)}$, and $\mathbf{H}^{-1}\left(\hat{\boldsymbol{\theta}}_L(\mathbf{v}_i)_{(m)}\right)$ is the Hessian matrix. The iteration above will stop if $\left\|\hat{\boldsymbol{\theta}}_L(\mathbf{v}_i)_{(m+1)} - \hat{\boldsymbol{\theta}}_L(\mathbf{v}_i)_{(m)}\right\| \le \varepsilon$, and $\varepsilon > 0$ is very small. So that the parameter estimator vector is obtained, namely:

$$\hat{\boldsymbol{\theta}}_L(\mathbf{v}_i) = \left[\hat{\lambda}_{0L}(\mathbf{v}_i)\ \hat{\boldsymbol{\beta}}_{1L}^T(\mathbf{v}_i)\ \hat{\boldsymbol{\beta}}_{2L}^T(\mathbf{v}_i)\ \hat{\varphi}_{0L}(\mathbf{v}_i)\ \hat{\varphi}_{1L}(\mathbf{v}_i)\ \hat{\varphi}_{2L}(\mathbf{v}_i)\right]^T; i = 1,2,\ldots,n \quad (17)$$

3.2 Hypothesis Testing

After getting the parameter estimator for the GTWBGPR model, the next step is testing the parameters. There is simultaneous and partial parameter testing. Hypothesis testing is carried out in stages in each period L where $L=1,2,\ldots,L^*$. Simultaneous testing using the MLRT method. Simultaneous testing is carried out to measure the significance of the predictor variable to response variables simultaneously in the GTWBGPR model. Simultaneous hypothesis testing is

$H_0 : \beta_{c1L}(\mathbf{v}_i)=\beta_{c2L}(\mathbf{v}_i)=\ldots=\beta_{cpL}(\mathbf{v}_i)=0; c=1,2\ ; i=1,2,\ldots,n$

H_1 : There is at least one $\beta_{cjL}(\mathbf{v}_i)\neq 0; c=1,2; j=1,2,\ldots,p$

The test statistic in the MLRT method requires the maximum value of the log-likelihood function under H_0 and under the population. Therefore, the first step is to determine the set of parameters under H₀ (ω_L) and under the population.

The set of parameters under H₀ (ω_L):

$$\omega_L=\{\lambda_{\omega 0L}(\mathbf{v}_i),\beta_{\omega 10L}(\mathbf{v}_i),\beta_{\omega 20L}(\mathbf{v}_i),\varphi_{\omega 0L}(\mathbf{v}_i),\varphi_{\omega 1L}(\mathbf{v}_i),\varphi_{\omega 2L}(\mathbf{v}_i);i=1,2,\ldots,n\}$$

The set of parameters under the population:

$$\Omega_L=\{\lambda_{0L}(\mathbf{v}_i),\boldsymbol{\beta}_{1L}(\mathbf{v}_i),\boldsymbol{\beta}_{2L}(\mathbf{v}_i),\varphi_{0L}(\mathbf{v}_i),\varphi_{1L}(\mathbf{v}_i),\varphi_{2L}(\mathbf{v}_i);i=1,2,\ldots,n\}$$

The likelihood function under the population$(L(\Omega_L))$ is the same as equation (7), while the likelihood function under H₀ $(L(\omega_L))$ is as follows:

$$L(\omega_L)=L(\lambda_{\omega 0L}(\mathbf{v}_i),\beta_{\omega 10L}(\mathbf{v}_i),\beta_{\omega 20L}(\mathbf{v}_i),\varphi_{\omega 0L}(\mathbf{v}_i),\varphi_{\omega 1L}(\mathbf{v}_i),\varphi_{\omega 2L}(\mathbf{v}_i);i=1,2,\ldots,n)$$
$$=\prod_{l=1}^{L}\prod_{i=1}^{n}\lambda_{\omega 0L}(\mathbf{v}_i)\left[e^{\beta_{\omega 10L}(\mathbf{v}_i)}-\lambda_{\omega 0L}(\mathbf{v}_i)\right]\left[e^{\beta_{\omega 20L}(\mathbf{v}_i)}-\lambda_{\omega 0L}(\mathbf{v}_i)\right]\exp\{-e^{\beta_{\omega 10L}(\mathbf{v}_i)}-e^{\beta_{\omega 20L}(\mathbf{v}_i)}$$
$$+\lambda_{\omega 0L}(\mathbf{v}_i)-y_{1il}\varphi_{\omega 1L}(\mathbf{v}_i)-y_{2il}\varphi_{\omega 2L}(\mathbf{v}_i)\}\sum_{k=0}^{\min(y_{1il},y_{2il})}\left[\frac{\left(\lambda_{\omega 0L}(\mathbf{v}_i)+k\varphi_{\omega 0L}(\mathbf{v}_i)\right)^{k-1}}{k!}\right.$$
$$\frac{\left(\left(e^{\beta_{\omega 10L}(\mathbf{v}_i)}-\lambda_{\omega 0L}(\mathbf{v}_i)\right)+(y_{1il}-k)\varphi_{\omega 1L}(\mathbf{v}_i)\right)^{y_{1il}-k-1}}{(y_{1il}-k)!}\frac{\left(\left(e^{\beta_{\omega 20L}(\mathbf{v}_i)}-\lambda_{\omega 0L}(\mathbf{v}_i)\right)+(y_{2il}-k)\varphi_{\omega 2L}(\mathbf{v}_i)\right)^{y_{2il}-k-1}}{(y_{2il}-k)!}$$
$$\left.\exp\left(k\varphi_{\omega 1L}(\mathbf{v}_i)+k\varphi_{\omega 2L}(\mathbf{v}_i)-k\varphi_{\omega 0L}(\mathbf{v}_i)\right)\right] \tag{18}$$

$$Q_{\omega L}=\log L(\lambda_{\omega 0L}(\mathbf{v}_i),\beta_{\omega 10L}(\mathbf{v}_i),\beta_{\omega 20L}(\mathbf{v}_i),\varphi_{\omega 0L}(\mathbf{v}_i),\varphi_{\omega 1L}(\mathbf{v}_i),\varphi_{\omega 2L}(\mathbf{v}_i);i=1,2,\ldots,n)$$
$$=\sum_{l=1}^{L}\sum_{i=1}^{n}\log\lambda_{\omega 0L}(\mathbf{v}_i)+\sum_{l=1}^{L}\sum_{i=1}^{n}\log\left(e^{\beta_{\omega 10L}(\mathbf{v}_i)}-\lambda_{\omega 0L}(\mathbf{v}_i)\right)+\sum_{l=1}^{L}\sum_{i=1}^{n}\log\left(e^{\beta_{\omega 20L}(\mathbf{v}_i)}-\lambda_{\omega 0L}(\mathbf{v}_i)\right)-\sum_{l=1}^{L}\sum_{i=1}^{n}e^{\beta_{\omega 10L}(\mathbf{v}_i)}$$
$$-\sum_{l=1}^{L}\sum_{i=1}^{n}e^{\beta_{\omega 20L}(\mathbf{v}_i)}+\sum_{l=1}^{L}\sum_{i=1}^{n}\lambda_{\omega 0L}(\mathbf{v}_i)-\sum_{l=1}^{L}\sum_{i=1}^{n}y_{1il}\varphi_{\omega 1L}(\mathbf{v}_i)-\sum_{l=1}^{L}\sum_{i=1}^{n}y_{2il}\varphi_{\omega 2L}(\mathbf{v}_i)+\sum_{l=1}^{L}\sum_{i=1}^{n}B_{il} \tag{19}$$

where:

$$B_{il} = \sum_{k=0}^{\min(y_{1il},y_{2il})} \Big[(k-1)\log\left[\lambda_{\omega 0L}(\mathbf{v}_i) + k\,\varphi_{\omega 0L}(\mathbf{v}_i)\right] - \log k! + (y_{1il}-k-1)$$
$$\log\left[e^{\beta_{\omega 10L}(\mathbf{v}_i)} - \lambda_{\omega 0L}(\mathbf{v}_i) + (y_{1il}-k)\varphi_{\omega 1L}(\mathbf{v}_i)\right] - \log(y_{1il}-k)! + (y_{2il}-k-1)$$
$$\log\left[e^{\beta_{\omega 20L}(\mathbf{v}_i)} - \lambda_{\omega 0L}(\mathbf{v}_i) + (y_{2il}-k)\varphi_{\omega 2L}(\mathbf{v}_i)\right] - \log(y_{2il}-k)! + k\varphi_{\omega 1L}(\mathbf{v}_i) + k\varphi_{\omega 2L}(\mathbf{v}_i)$$
$$-k\,\varphi_{\omega 0L}(\mathbf{v}_i)\Big]$$

Next is to determine $Q^*_{\omega L}$, namely, the log-likelihood under H_0 in equation (19) multiplied by the spatiotemporal weight w_{ii^*l} to obtain the parameters estimator under H_0 at Lth period and ith location.

$$Q^*_{\omega L} = \sum_{l=1}^{L}\sum_{i^*=1}^{n} w_{ii^*l}\log\lambda_{\omega 0L}(\mathbf{v}_i) + \sum_{l=1}^{L}\sum_{i^*=1}^{n} w_{ii^*l}\log\left(e^{\beta_{\omega 10L}(\mathbf{v}_i)} - \lambda_{\omega 0L}(\mathbf{v}_i)\right) + \sum_{l=1}^{L}\sum_{i^*=1}^{n} w_{ii^*l}\log\left(e^{\beta_{\omega 20L}(\mathbf{v}_i)} - \lambda_{\omega 0L}(\mathbf{v}_i)\right)$$
$$-\sum_{l=1}^{L}\sum_{i^*=1}^{n} w_{ii^*l}e^{\beta_{\omega 10L}(\mathbf{v}_i)} - \sum_{l=1}^{L}\sum_{i^*=1}^{n} w_{ii^*l}e^{\beta_{\omega 20L}(\mathbf{v}_i)} + \sum_{l=1}^{L}\sum_{i^*=1}^{n} w_{ii^*l}\lambda_{\omega 0L}(\mathbf{v}_i) - \sum_{l=1}^{L}\sum_{i^*=1}^{n} w_{ii^*l}y_{1i^*l}\varphi_{\omega 1L}(\mathbf{v}_i)$$
$$-\sum_{l=1}^{L}\sum_{i^*=1}^{n} w_{ii^*l}y_{2i^*l}\varphi_{\omega 2L}(\mathbf{v}_i) + \sum_{l=1}^{L}\sum_{i^*=1}^{n} w_{ii^*l}B_{i^*l} \tag{20}$$

Then look for the first derivative of $Q^*_{\omega L}$ for each parameter under H_0 and equalize it to zero.
Suppose that:

$$s_{\omega 0} = \frac{k-1}{\left(\lambda_{\omega 0L}(\mathbf{v}_i) + k\varphi_{\omega 0L}(\mathbf{v}_i)\right)}$$

$$s_{\omega 1} = \frac{(y_{1i^*l}-k-1)}{\left(e^{\beta_{\omega 1L}(\mathbf{v}_i)} - \lambda_{\omega 0L}(\mathbf{v}_i)\right) + (y_{1i^*l}-k)\varphi_{\omega 1L}(\mathbf{v}_i)}$$

$$s_{2\omega} = \frac{(y_{2i^*l}-k-1)}{\left(e^{\beta_{\omega 2L}(\mathbf{v}_i)} - \lambda_{\omega 0L}(\mathbf{v}_i) + (y_{2i^*l}-k)\varphi_{\omega 2L}(\mathbf{v}_i)\right)}$$

Then the following results are obtained:

$$\frac{\partial Q^*_{\omega L}}{\partial \lambda_{\omega 0L}(\mathbf{v}_i)} = \sum_{l=1}^{L}\sum_{i^*=1}^{n}\frac{w_{ii^*l}}{\lambda_{\omega 0L}(\mathbf{v}_i)} - \sum_{l=1}^{L}\sum_{i^*=1}^{n}\frac{w_{ii^*l}}{\left(e^{\beta_{\omega 1L}(\mathbf{v}_i)} - \lambda_{\omega 0L}(\mathbf{v}_i)\right)} - \sum_{l=1}^{L}\sum_{i^*=1}^{n}\frac{w_{ii^*l}}{\left(e^{\beta_{\omega 2L}(\mathbf{v}_i)} - \lambda_{\omega 0L}(\mathbf{v}_i)\right)}$$
$$+\sum_{l=1}^{L}\sum_{i^*=1}^{n} w_{ii^*l} + \sum_{l=1}^{L}\sum_{i^*=1}^{n} w_{ii^*l}\sum_{k=0}^{\min(y_{1i^*l},y_{2i^*l})}\left[s_{\omega 0} - s_{\omega 1} - s_{\omega 2}\right] \tag{21}$$

$$\frac{\partial Q^*_{\omega L}}{\partial \beta_{\omega 1L}(\mathbf{v}_i)} = \sum_{l=1}^{L}\sum_{i^*=1}^{n} w_{ii^*l}\frac{e^{\beta_{\omega 1L}(\mathbf{v}_i)}}{e^{\mathbf{x}^T_{i^*l}\beta_{\omega 1L}(\mathbf{v}_i)} - \lambda_{\omega 0L}(\mathbf{v}_i)} - \sum_{l=1}^{L}\sum_{i^*=1}^{n} w_{ii^*l}e^{\beta_{\omega 1L}(\mathbf{v}_i)} + \sum_{l=1}^{L}\sum_{i^*=1}^{n} w_{ii^*l}\sum_{k=0}^{\min(y_{1i^*l},y_{2i^*l})}\left[s_{\omega 1}e^{\beta_{\omega 1L}(\mathbf{v}_i)}\right] \tag{22}$$

$$\frac{\partial Q^*_{\omega L}}{\partial \beta_{\omega 2L}(\mathbf{v}_i)} = \sum_{l=1}^{L}\sum_{i^*=1}^{n} w_{ii^*l}\frac{e^{\beta_{\omega 2L}(\mathbf{v}_i)}}{e^{\beta_{\omega 2L}(\mathbf{v}_i)} - \lambda_{\omega 0L}(\mathbf{v}_i)} - \sum_{l=1}^{L}\sum_{i^*=1}^{n} w_{ii^*l}e^{\beta_{\omega 2L}(\mathbf{v}_i)} + \sum_{l=1}^{L}\sum_{i^*=1}^{n} w_{ii^*l}\sum_{k=0}^{\min(y_{1i^*l},y_{2i^*l})}\left[s_{\omega 2}e^{\beta_{\omega 2L}(\mathbf{v}_i)}\right] \tag{23}$$

$$\frac{\partial Q^*_{\omega L}}{\partial \varphi_{\omega 0L}(\mathbf{v}_i)} = \sum_{l=1}^{L}\sum_{i^*=1}^{n} w_{ii^*l}\sum_{k=0}^{\min(y_{1i^*l},y_{2i^*l})}\left[s_{\omega 0}k - k\right] \tag{24}$$

$$\frac{\partial Q^*_{\omega L}}{\partial \varphi_{\omega 1L}(\mathbf{v}_i)} = -\sum_{l=1}^{L}\sum_{i^*=1}^{n} w_{ii^*l} y_{1i^*l} + \sum_{l=1}^{L}\sum_{i^*=1}^{n} w_{ii^*l} \sum_{k=0}^{\min(y_{1i^*l}, y_{2i^*l})} \left[s_{\omega 1}(y_{1i^*l} - k) + k \right] \tag{25}$$

$$\frac{\partial Q^*_{\omega L}}{\partial \varphi_{\omega 2L}(\mathbf{v}_i)} = -\sum_{l=1}^{L}\sum_{i^*=1}^{n} w_{ii^*l} y_{2i^*l} + \sum_{l=1}^{L}\sum_{i^*=1}^{n} w_{ii^*l} \sum_{k=0}^{\min(y_{1i^*l}, y_{2i^*l})} \left[s_{\omega 2}(y_{2i^*l} - k) + k \right] \tag{26}$$

The first derivative results above are not closed-form, so the Newton-Raphson iteration solves it to obtain the parameter estimator under H_0, that is:

$$\hat{\boldsymbol{\theta}}_{\omega L}(\mathbf{v}_i) = \left[\hat{\lambda}_{\omega 0L}(\mathbf{v}_i)\ \hat{\beta}_{\omega 10L}(\mathbf{v}_i)\ \hat{\beta}_{\omega 20L}(\mathbf{v}_i)\ \hat{\varphi}_{\omega 0L}(\mathbf{v}_i)\ \hat{\varphi}_{\omega 1L}(\mathbf{v}_i)\ \hat{\varphi}_{\omega 2L}(\mathbf{v}_i) \right]^T ; i = 1,2,\ldots,n. \tag{27}$$

These parameter estimators maximize the likelihood function under H_0 of the period L, namely:

$$L(\hat{\omega}_L) = \max_{\omega_L} L(\omega_L) \tag{28}$$

These parameter estimators at (17) maximize the likelihood function under the population of the period L, namely:

$$L(\hat{\Omega}_L) = \max_{\Omega_L} L(\Omega_L) \tag{29}$$

Based on equations (28) and (29) then the odds ratio is determined as follows:

$$\Lambda_L = \frac{L(\hat{\omega}_L)}{L(\hat{\Omega}_L)} < K_\alpha \tag{30}$$

where K_α is a constant that depends on α and $0 \le K_\alpha \le 1$. Furthermore, equation (30) can be written as follows:

$$\log\left(\frac{L(\hat{\omega}_L)}{L(\hat{\Omega}_L)} \right) < \log K_\alpha$$

$$-2\log\left(\frac{L(\hat{\omega}_L)}{L(\hat{\Omega}_L)} \right) > -2\log K_\alpha$$

$$G_L^2(\hat{\boldsymbol{\theta}}_L) = -2\log\left(\frac{L(\hat{\omega}_L)}{L(\hat{\Omega}_L)} \right) = 2\left(\log L(\hat{\Omega}_L) - \log L(\hat{\omega}_L) \right)$$

Based on the steps carried out by Berliana et al. [14], it was found that $G_L^2(\hat{\boldsymbol{\theta}}_L) \underset{nL\to\infty}{\sim} \chi^2_{(a-b)}$. $G_L^2(\hat{\boldsymbol{\theta}}_L)$ is the χ^2 distribution approach with degrees of freedom a-b, namely the number of parameters under the population minus the number of parameters under H_0. So that H_0 is rejected if $G_L^2(\hat{\boldsymbol{\theta}}_L) > \chi^2_{(\alpha, 2pnL)}$, which means that there are predictor variables that affect the response variable.

If the result of the simultaneous test rejects H_0, then it is followed by partial tests. According to Triyanto et al. [15], parameter estimators obtained by MLE were asymptotically normally distributed. So that it is obtained:

$$\hat{\boldsymbol{\theta}}_L(\mathbf{v}_i) \underset{nL\to\infty}{\sim} N\left(\boldsymbol{\theta}_L(\mathbf{v}_i), \left[\mathbf{I}\left(\hat{\boldsymbol{\theta}}_L(\mathbf{v}_i)\right)\right]^{-1}\right) \tag{31}$$

then it is obtained:

$$\mathbf{Z} = \left[\mathbf{I}\left(\hat{\boldsymbol{\theta}}_L(\mathbf{v}_i)\right)\right]^{-1/2}\left(\hat{\boldsymbol{\theta}}_L(\mathbf{v}_i) - \boldsymbol{\theta}_L(\mathbf{v}_i)\right) \underset{nL\to\infty}{\sim} N(\mathbf{0}, \mathbf{I}) \tag{32}$$

So that for partial testing using the Z test. The testing hypothesis for parameter β is:

$H_0 : \beta_{cjL}(\mathbf{v}_i) = 0$

$H_1 : \beta_{cjL}(\mathbf{v}_i) \neq 0$

This test is done for $c = 1,2; j = 1,2,\ldots,p; i = 1,2,\ldots,n$.

By paying attention to equation (32), the test statistic used is:

$$Z = \frac{\hat{\beta}_{cjL}(\mathbf{v}_i)}{SE\left(\hat{\beta}_{cjL}(\mathbf{v}_i)\right)} \tag{33}$$

where $SE\left(\hat{\beta}_{cjL}(\mathbf{v}_i)\right) = \sqrt{\hat{\text{var}}\left(\hat{\beta}_{cjL}(\mathbf{v}_i)\right)}$. $\hat{\text{var}}\left(\hat{\beta}_{cjL}(\mathbf{v}_i)\right)$ is obtained from the main diagonal of the covariance variance matrix $\left[\mathbf{I}\left(\hat{\boldsymbol{\theta}}_L(\mathbf{v}_i)\right)\right]^{-1}$.

Reject H_0 if $|Z| > Z_{\alpha/2}$.

The testing hypothesis for parameter λ_0 is:

$H_0 : \lambda_{0L}(\mathbf{v}_i) = 0$

$H_1 : \lambda_{0L}(\mathbf{v}_i) \neq 0$

This test is done for $i = 1,2,\ldots,n$.

By paying attention to equation (32), the test statistic used is:

$$Z = \frac{\hat{\lambda}_{0L}(\mathbf{v}_i)}{SE\left(\hat{\lambda}_{0L}(\mathbf{v}_i)\right)} \tag{34}$$

Reject H_0 if $|Z| > Z_{\alpha/2}$.

The testing hypothesis for parameter φ is:

$H_0 : \varphi_{cL}(\mathbf{v}_i) = 0$

$H_1 : \varphi_{cL}(\mathbf{v}_i) \neq 0$

This test is done for $c = 1,2; i = 1,2,\ldots,n$.

By paying attention to equation (32), the test statistic used is:

$$Z = \frac{\hat{\varphi}_{cL}(\mathbf{v}_i)}{SE\left(\hat{\varphi}_{cL}(\mathbf{v}_i)\right)} \tag{35}$$

Reject H_0 if $|Z| > Z_{\alpha/2}$.

4. Conclusions

GTWBGPR is a development of GWBGPR. GTWBGPR can accommodate spatial and temporal heterogeneity. Parameter estimation in GTWBGPR using MLE. The results obtained are not closed-form, so they are solved by Newton-Raphson iteration. Furthermore, the simultaneous hypothesis testing using MLRT and test statistic distribution was a chi-square distribution approximation. The partial testing using Z test.

References

[1] Consul P and Famoye F 1992 Generalized poisson regression model *Communications in Statistics-Theory and Methods* **21**(1) p 89-109
[2] Omzen I and Famoye F 2007 Count regression models with an application to zoological data containing structural zeros *Journal of Data Science* **5**(4) p 491-502
[3] Adams S O, Bamanga M A, Olanrewaju S O, Yahaya H U and Akano R O 2020 Modeling covid-19 cases in nigeria using some selected count data regression models *International Journal of Healthcare and Medical Sciences* **6**(4) p 64-73
[4] Vernic R 1997 On the bivariate generalized poisson distribution *ASTIN Bulletin: The Journal of the IAA* **27**(1) p 23-32
[5] Famoye F 2012 Comparisons of some bivariate regression models *Journal of Statistical Computation and Simulation* **82**(7) p 937-949
[6] Zamani H, Faroughi P and Ismail N 2016 Bivariate generalized poisson regression model: applications on health care data *Empirical Economics* **51**(4) p 1607-1621
[7] Fotheringham A S, Brunsdon C and Charlton M 2003 *Geographically weighted regression: the analysis of spatially varying relationships* John Wiley & Sons
[8] Purhadi, Sutikno, Berliana, S M and Setiawan D I 2021 Geographically weighted bivariate generalized Poisson regression: application to infant and maternal mortality data *Letters in Spatial and Resource Sciences* p 1-21
[9] Gujarati D N and Porter D C 2003 *Basic econometrics* (ed.) Singapore: McGrew Hill Book Co
[10] Fotheringham A S, Crespo R and Yao J 2015 Geographical and temporal weighted regression (GTWR) *Geographical Analysis* **47**(4) p 431-452
[11] Ma X, Zhang J, Ding C and Wang Y 2018 A geographically and temporally weighted regression model to explore the spatiotemporal influence of built environment on transit ridership *Computers, Environment and Urban Systems* **70** p 113-124.
[12] Huang B, Wu B and Barry M 2010 Geographically and temporally weighted regression for modeling spatio-temporal variation in house prices *International Journal of Geographical Information Science* **24**(3) p 383-401
[13] Allo C B G, Otok B W and Purhadi 2019 Estimation parameter of generalized poisson regression model using generalized method of moments and its application *IOP Conference Series: Materials Science and Engineering* **546**(5) p 052050
[14] Berliana S M, Purhadi, Sutikno and Rahayu S P 2020 Parameter estimation and hypothesis testing of geographically weighted multivariate generalized poisson regression *Mathematics* **8**(9) p 1523
[15] Triyanto, Purhadi, Otok B W and Purnami S W 2015 Parameter estimation of geographically weigthed multivariate poisson regression. *Appl. Math. Sci*, **9** p 4081-4093.

Acknowledgments

The authors are grateful to Statistics Indonesia (Badan Pusat Statistik), which has supported this research financially. The authors also thank lecturers and college friends at the Sepuluh Nopember Institute of Technology (Institut Teknologi Sepuluh Nopember) for their guidance and support.

Parameter estimation and hypothesis testing of geographically and temporally weighted bivariate Gamma regression model

Desy Wasani[1*], Purhadi[1] and Sutikno[1]

[1] Department of Statistics, Faculty of science and data analytics, Institut Teknologi Sepuluh Nopember (ITS). Arief Rahman Hakim Street, Surabaya 60111 Indonesia

*Corresponding E-mail: destatistik@gmail.com

Abstract. Geographically Weighted Regression (GWR) study potential relationships in regression models that distinguish geographic spaces using non-stationary parameters to overcome spatial effects. The use of gamma regression, namely regression with the dependent variable with a gamma distribution, can be an alternative if the data do not follow a normal distribution. Gamma distribution is a continuous set of non-negative values, generally skewed to the right or positive skewness. Gamma regression is developed to be Bivariate Gamma Regression (BGR) when there are two dependent variables with gamma distribution. If the observation units are location points, spatial effects may occur. The Geographically Weighted Bivariate Gamma Regression (GWBGR) model can be a solution for spatial heterogeneity. However, during its development, many cases require information from panel data. Using panel data can provide complete information because it covers several periods, but it allows for temporal effects. This study developed a Geographically and Temporally Weighted Bivariate Gamma Regression (GTWBGR) model to handle spatial and temporal heterogeneity simultaneously. The estimation of the GTWBGR model parameters uses the Maximum Likelihood Estimation (MLE) method that followed by the numerical iteration of Berndt Hall Hall Hausman (BHHH). The simultaneous testing uses the Maximum Likelihood Ratio Test (MLRT) method to get a test statistic. With a large sample size, the distribution of the test statistic approaches chi-square. Meanwhile, partial testing uses the Z test statistic.

Keywords: Gamma Distribution, GTWBGR, MLE, MLRT

Track Name: Advanced Technology and Renewable Energy

1. Introduction

Many research fields such as geography, social, economy, and epidemiology generally use geographic location data or spatial data. However, variations between locations and the relationship between variables in this spatial data can lead to spatial non-stationarity because of differences in characteristics at different locations. Geographically Weighted Regression (GWR) study potential relationships in regression models that distinguish geographic spaces using non-stationary parameters to overcome spatial effects [1].

During its development, many cases require information from panel data. In panel data, besides there are n observation units, there is also a time dimension element that shows the history of the data. Panel data has advantages because it provides a more significant number of observations, increases the degree of freedom, and reduces collinearity between predictor variables, thereby increases efficiency

[2]. The GWR model applied to panel data has two advantages in one model: reducing heteroscedasticity and collinearity. This model is called Geographically and Temporally Weighted Regression (GTWR) [3].

The gamma distribution is a continuous data distribution formed from a set of non-negative values. This distribution is widely applied in science and business, such as stochastic modeling, hydrology, reliability, and service time [4]. The gamma distribution may occur in other fields where the data are non-negative continuous values with asymmetric distribution curves. The gamma distribution forms a skewed curve, generally right-skewed. In cases where there is more than one response variable, Purhadi, Budiani, and Purnami (2018) have developed two-parameter Bivariate Gamma Regression (BGR) model [5].

Related to the spatial factors at each location, Dewi, Purhadi, and Sutikno (2019) applied the Geographically Weighted Bivariate Gamma Regression (GWBGR) model to the Maternal Mortality Rate (MMR) and Infant Mortality Rate (IMR) in North Sumatra Province in 2017 [6]. The parameter estimation result using Maximum Likelihood Estimation (MLE) was not closed-form. Therefore it was followed by numerical optimization, namely Berndt-Hall-Hall-Hausman (BHHH). Hypothesis testing used the Maximum Likelihood Ratio Test (MLRT) method. The smaller Akaike's Information Criterion corrected (AICc) and MSE values than the BGR indicate that GWBGR is more accurate than the BGR.

This study developed a two-parameter BGR model that handles the spatial and temporal heterogeneity of the panel data where the observation units are location points at several times, namely Geographically and Temporally Weighted Bivariate Gamma Regression (GTWBGR). The estimation method used in the Generalized Linear Model (GLM) is MLE because the distribution is known. Meanwhile, most of the studies with the composite hypothesis used MLRT. Therefore, this research aims to obtain parameter estimators on GTWBGR using MLE and the test statistic using MLRT.

2. Materials and Methods

2.1 Gamma Regression

Gamma regression is a non-linear regression with the response variable (Y) with a gamma distribution. The two-parameter gamma distribution has the form of a probability density function (pdf) as follows [7]:

$$f(y)=\begin{cases}\dfrac{\gamma^{\phi}y^{\phi-1}exp(-\gamma y)}{\Gamma(\phi)}, & \text{for } y>0, \phi>0, \text{and } \gamma>0\\ 0 \quad, & \text{otherwise}\end{cases} \tag{1}$$

with shape parameter ϕ, scale parameter $1/\gamma$, mean $\mu = E(Y) = \phi/\gamma$, and variance $\sigma^2 = Var(Y) = \phi/\gamma^2$. The shape parameter affects the curvature of the curve, and the scale parameter affects the curve width. Gamma functions with parameter ϕ is defined by:

$$\Gamma(\phi)=\int_0^{\infty} e^{-y}y^{\phi-1}dy \tag{2}$$

The gamma regression model has the following form [8]:

$$\mu_i = E(Y_i) = \exp(\mathbf{x}_i^T \boldsymbol{\beta}); \quad i=1,2,\ldots,n \tag{3}$$

$\mu_i = E(Y_i) = \phi/\gamma$, then $\gamma = \phi/E(Y_i) = \phi/\exp(\mathbf{x}_i^T \boldsymbol{\beta})$. So that the pdf of Y_i in the gamma regression can be written:

$$f(y_i \mid \phi, \boldsymbol{\beta}) = \begin{cases} \dfrac{\left(\dfrac{\phi}{\exp\left(\mathbf{x}_i^T \boldsymbol{\beta}\right)}\right)^{\phi} y_i^{\phi-1} exp\left(-\left(\dfrac{\phi}{\exp\left(\mathbf{x}_i^T \boldsymbol{\beta}\right)}\right) y_i\right)}{\Gamma(\phi)}, & \text{for } y_i > 0,\ \phi > 0 \\ 0 \quad , & \text{otherwise} \end{cases} \tag{4}$$

2.2 Bivariate Gamma Regression (BGR)

BGR is a form of regression with two response variables and a bivariate gamma distribution with some predictor variables (X). The bivariate gamma distribution involves the incomplete gamma function, which is defined by:

$$\Gamma(\phi, y) = \int_y^{\infty} e^{-t} t^{\phi-1} dt \tag{5}$$

The form of the BGR model is:

$$\begin{aligned} \mu_{1i} &= E(Y_{1i}) = \exp\left(\mathbf{x}_i^T \boldsymbol{\beta}_1\right) \\ \mu_{2i} &= E(Y_{2i}) = \exp\left(\mathbf{x}_i^T \boldsymbol{\beta}_2\right); \quad i = 1, 2, \ldots, n \end{aligned} \tag{6}$$

One form of the pdf of the bivariate gamma distribution can be obtained from the transformation of the gamma distribution and the beta distribution by following its characteristics, with the following form [7]:

$$f(y_1, y_2) = \begin{cases} C(y_1 y_2)^{\tau-1} \left(\dfrac{y_1}{\gamma_1} + \dfrac{y_2}{\gamma_2}\right)^{\phi-2\tau} \Gamma\left(2\tau - \phi, \dfrac{y_1}{\gamma_1} + \dfrac{y_2}{\gamma_2}\right); & \text{for } y_1 > 0,\ y_2 > 0 \\ 0 & ; \text{otherwise} \end{cases} \tag{7}$$

where $\phi > 0, \tau > 0, \gamma_1 > 0, \gamma_2 > 0$ and $C = \left((\gamma_1 \gamma_2)^{\tau} \Gamma(\tau) \Gamma(\phi)\right)^{-1}$. On the univariate gamma distribution $\mu = E(Y) = \phi/\gamma$, then the scale parameters of Y_1 and Y_2 are $1/\gamma_1 = \mu_1/\phi_1$ dan $1/\gamma_2 = \mu_2/\phi_2$. Based on Lemma 1 that $\phi_1 = \phi_2 = \phi$ [7], then $\gamma_1 = \phi/\mu_1$ and $\gamma_2 = \phi/\mu_2$ can be substituted into pdf so that it becomes:

$$f(y_{1i}, y_{2i}) = \begin{cases} C_i (y_{1i} y_{2i})^{\tau-1} \left(\dfrac{y_{1i} \exp\left(\mathbf{x}_i^{\mathbf{T}} \boldsymbol{\beta}_1\right) + y_{2i} \exp\left(\mathbf{x}_i^{\mathbf{T}} \boldsymbol{\beta}_2\right)}{\phi}\right)^{\phi-2\tau} A_i; & \text{for } y_{1i} > 0,\ y_{2i} > 0 \\ 0 & ; \text{otherwise} \end{cases} \tag{8}$$

where $\phi > 0, \tau > 0$

$$C_i = \left(\left(\frac{\phi^2}{\exp\left(\mathbf{x}_i^T \boldsymbol{\beta}_1\right) \exp\left(\mathbf{x}_i^T \boldsymbol{\beta}_2\right)}\right)^{\tau} \Gamma(\tau) \Gamma(\phi)\right)^{-1}, \quad A_i = \Gamma\left(2\tau - \phi, \frac{y_{1i} \exp\left(\mathbf{x}_i^{\mathbf{T}} \boldsymbol{\beta}_1\right) + y_{2i} \exp\left(\mathbf{x}_i^{\mathbf{T}} \boldsymbol{\beta}_2\right)}{\phi}\right)$$

2.3 Geographically and Temporally Weighted Bivariate Gamma Regression (GTWBGR)

Location and time can simultaneously be essential determinants of a variable. Previous studies have introduced location and time factors to a global regression model known as GTWR [3]. GTWR assumes the regression coefficient as a function of time and location coordinates. GTWBGR model is a development of the GTWR in the distribution of random variables and development of BGR in weighting location and time. The form of the GTWBGR model equation:

$$\begin{aligned}\mu_{1il} &= E(Y_{1il}) = \exp\left(\mathbf{x}_{il}^T \boldsymbol{\beta}_{1l}(u_i, v_i, t_i)\right) \\ \mu_{2il} &= E(Y_{2il}) = \exp\left(\mathbf{x}_{il}^T \boldsymbol{\beta}_{2l}(u_i, v_i, t_i)\right); \quad i = 1,2,\dots,n;\ l = 1,2,\dots,L\end{aligned} \tag{9}$$

Index i shows the observation at the *ith* location from the total n observations. Index l shows the lth period of the total L periods, so the total observations are $n \times L$. $\mathbf{x}_{il}$ is the vector of observation with size $(p+1)\times 1$ at the ith location of the lth period. $\boldsymbol{\beta}_{1l}(u_i, v_i, t_i)$ and $\boldsymbol{\beta}_{2l}(u_i, v_i, t_i)$ are parameter vectors of the regression coefficients with spatial and temporal effects with size $(p+1)\times 1$, where u_i, v_i is the location coordinate of the ith observation and t_i is the time coordinate of the ith observation.

The joint pdf of Y_1 and Y_2 in the GTWBGR model is as follows:

$$f(y_{1il}, y_{2il}) = \begin{cases} D_{il}(y_{1il} y_{2il})^{\tau_l - 1} \left(\dfrac{y_{1il}\exp\left(\mathbf{x}_{il}^{\mathrm{T}}\boldsymbol{\beta}_{1l}(u_i, v_i, t_i)\right) + y_{2il}\exp\left(\mathbf{x}_{il}^{\mathrm{T}}\boldsymbol{\beta}_{2l}(u_i, v_i, t_i)\right)}{\phi_l} \right)^{\phi_l - 2\tau_l} S_{il}; & \text{for } y_{1il} > 0,\ y_{2il} > 0 \\ 0 & ;\ y_{1il}, y_{2il} \text{ otherwise} \end{cases} \tag{10}$$

where $\phi_l > 0,\ \tau_l > 0$,

$$D_{il} = \left(\left(\frac{\phi_l^2}{\exp\left(\mathbf{x}_{il}^T \boldsymbol{\beta}_{1l}(u_i, v_i, t_i)\right) \exp\left(\mathbf{x}_{il}^T \boldsymbol{\beta}_{2l}(u_i, v_i, t_i)\right)} \right)^{\tau_L} \Gamma(\tau_l)\Gamma(\phi_l) \right)^{-1}$$

$$S_{il} = \Gamma\left(2\tau_l - \phi_l, \frac{y_{1il}\exp\left(\mathbf{x}_{il}^{\mathrm{T}}\boldsymbol{\beta}_{1l}(u_i, v_i, t_i)\right) + y_{2il}\exp\left(\mathbf{x}_{il}^{\mathrm{T}}\boldsymbol{\beta}_{2l}(u_i, v_i, t_i)\right)}{\phi_l} \right); i = 1,2,\dots,n;\ l = 1,2,\dots,L$$

2.4 Spatial-Temporal Weight

In the spatial-temporal regression, each location and time has a different weight. One of the methods to measure the distance between locations and times is to use Euclidean distance. If the spatial distance is d^S and the temporal distance is d^T, then the spatial-temporal distance is as follows [3].

$$d^{ST} = \eta_1 d^S + \eta_2 d^T \tag{11}$$

where η_1 and η_2 are scale factors that balance the spatial and temporal effects. These factors are needed so that spatial effects or only temporal effects do not dominate the spatial-temporal distance. If the period is expressed by l, the spatial-temporal distance function in period l can be expressed by:

$$(d_{ii^*l})^2 = \eta_1\left[(u_i - u_{i^*})^2 + (v_i - v_{i^*})^2\right] + \eta_2 (t_i - t_{i^*})^2 \tag{12}$$

Next, let δ be the ratio of the parameters η_2 / η_1. Without reducing generality, we can set $\eta_1 = 1$, so that only η_2 needs to be adjusted where $\eta_2 = \delta$. The value of δ can be optimized in several ways, including based on the smallest Akaike Information Criterion (AIC) value. Thus, the spatial-temporal Euclidean distance between observations (u_i, v_i, t_i) and $(u_{i^*}, v_{i^*}, t_{i^*})$ used in determining the spatial-temporal weight matrix becomes:

$$d_{ii^*l} = \sqrt{(u_i - u_{i^*})^2 + (v_i - v_{i^*})^2 + \delta(t_i - t_{i^*})^2} \tag{13}$$

IOP Conf. Series: Earth and Environmental Science **880** (2021) 012044 doi:10.1088/1755-1315/880/1/012044

Among the recommended weights are continuous weight values based on the Euclidean distance [9]. One of them is bisquare kernel function. The bisquare kernel weight function involving spatial and temporal effects can be formulated by:

$$w_{ii^*l} = \left[1-\left(\frac{d_{ii^*l}}{b}\right)^2\right]^2 , \text{ for } d_{ii^*l} < b \qquad (14)$$
$$= 0 \text{ , otherwise}$$

where d_{ii^*l} is the spatial-temporal distance, and b is the bandwidth. Bandwidth is an aspect of GWR that functions as a smoothing parameter. An enormous bandwidth value causes the kernel weight value to be closer to 1. So that the spatial model is getting closer to the global regression model. On the other hand, a small bandwidth causes parameter estimation to be more dependent on nearby observations. Therefore, the selection of the optimum bandwidth will determine the accuracy of the model. One of the methods to choose the optimum bandwidth is using Generalized Cross-Validation (GCV).

$$GCV = \frac{nL\sum_{l=1}^{L}\sum_{i=1}^{n}\left(\left[y_{1il}-\hat{y}_{1il}(b)\right]^2+\left[y_{2il}-\hat{y}_{2il}(b)\right]^2\right)}{\left(nL-v_1\right)^2} \qquad (15)$$

v_1 represents the number of parameters in the model. Observation i is not included in the calibration. With the minimum GCV value, the optimum bandwidth will be obtained.

2.5 Methods

The GTWBGR model is applied to panel data where the observation units are location points at several periods. The estimation of parameters in GTWBGR is carried out gradually every period. The number of locations is denoted by n, and L denotes the number of periods. The first stage estimates the parameters at n locations involving only the data for the first period. The second stage estimates the parameters at n locations involving data for the first and second periods, and so on. Finally, stage L estimates the parameters at n locations by involving data for that period and all previous periods. So, we get $n \times L$ parameter estimator vectors. The number of periods can vary where $L = 1, 2, \ldots, L^*$.

Parameters $\boldsymbol{\beta}_1$ and $\boldsymbol{\beta}_2$ are estimated for each location and time so that the location coordinates and time index are given. Meanwhile, the parameters ϕ and τ are only estimated for each period so that only a time index is given. The estimation method used is MLE with bisquare kernel weight function. The following steps carry out the MLE process:

i) Establishing a natural logarithm of the likelihood (log-likelihood) function of the GTWBGR model
ii) Multiplying the log-likelihood function by the spatial-temporal kernel weight
iii) Maximizing the function by finding the first derivative of the parameters equated to zero
iv) If the results obtained are not closed-form, it is necessary to optimize by using numerical iteration.

The numerical iteration chosen in this study is the BHHH method. The advantage of BHHH is that it only takes the first derivative of the parameters to get the Hessian matrix.

Furthermore, simultaneous hypothesis testing on parameters is carried out by MLRT by the following steps:

i) Build a simultaneous hypothesis
ii) Defining the set of parameters under the population and the null hypothesis
iii) Estimating the parameters under the population and the null hypothesis
iv) Determining the odds ratio to get a test statistic
v) Determining a test statistic, its distribution, and rejection area

Meanwhile, partial hypothesis testing uses Central Limit Theorem (CLT) to get the test statistic.

3. Results and Discussion

3.1 Parameter Estimation

The first step to estimating parameters is to form the log-likelihood function of the GTWBGR model with the following form:

$$\log L(\phi_L,\tau_L,\boldsymbol{\beta}_{1L}(u_i,v_i,t_i),\boldsymbol{\beta}_{2L}(u_i,v_i,t_i);i=1,2,\ldots,n)$$

$$=\log\left(\prod_{l=1}^{L}\prod_{i=1}^{n}\left(D_{il}(y_{1il}y_{2il})^{\tau_L-1}\left(\frac{y_{1il}\exp\left(\mathbf{x}_{il}^{\mathbf{T}}\boldsymbol{\beta}_{1L}(u_i,v_i,t_i)\right)+y_{2il}\exp\left(\mathbf{x}_{il}^{\mathbf{T}}\boldsymbol{\beta}_{2L}(u_i,v_i,t_i)\right)}{\phi_L}\right)^{\phi_L-2\tau_L}S_{il}\right)\right)$$

$$=\sum_{l=1}^{L}\sum_{i=1}^{n}\log\left(D_{il}(y_{1il}\cdot y_{2il})^{\tau_L-1}\left(\frac{y_{1il}\exp\left(\mathbf{x}_{il}^{\mathbf{T}}\boldsymbol{\beta}_{1L}(u_i,v_i,t_i)\right)+y_{2il}\exp\left(\mathbf{x}_{il}^{\mathbf{T}}\boldsymbol{\beta}_{2L}(u_i,v_i,t_i)\right)}{\phi_L}\right)^{\phi_L-2\tau_L}S_{il}\right) \quad (16)$$

where,

$$D_{il}=\left(\left(\frac{\phi_l^2}{\exp\left(\mathbf{x}_{il}^{T}\boldsymbol{\beta}_{1l}(u_i,v_i,t_i)\right)\exp\left(\mathbf{x}_{il}^{T}\boldsymbol{\beta}_{2l}(u_i,v_i,t_i)\right)}\right)^{\tau_L}\Gamma(\tau_l)\Gamma(\phi_l)\right)^{-1}$$

$$S_{il}=\Gamma\left(2\tau_l-\phi_l,\frac{y_{1il}\exp\left(\mathbf{x}_{il}^{\mathbf{T}}\boldsymbol{\beta}_{1L}(u_i,v_i,t_i)\right)+y_{2il}\exp\left(\mathbf{x}_{il}^{\mathbf{T}}\boldsymbol{\beta}_{2l}(u_i,v_i,t_i)\right)}{\phi_l}\right)$$

Estimating the parameters of location i^* at time L needs a Q_L^* function, namely the log-likelihood function multiplied by the spatial-temporal weight w_{ii^*l}, which can be shown as follows:

$$Q_L^*=\log L(\phi_L,\tau_L,\boldsymbol{\beta}_{1L}(u_{i^*},v_{i^*},t_{i^*}),\boldsymbol{\beta}_{2L}(u_{i^*},v_{i^*},t_{i^*}))$$

$$=\sum_{l=1}^{L}\sum_{i=1}^{n}w_{ii^*l}\log\left\{D_{il}^*(y_{1il}\cdot y_{2il})^{\tau_L-1}\left(\frac{y_{1il}\exp\left(\mathbf{x}_{il}^{\mathbf{T}}\boldsymbol{\beta}_{1L}(u_{i^*},v_{i^*},t_{i^*})\right)+y_{2il}\exp\left(\mathbf{x}_{il}^{\mathbf{T}}\boldsymbol{\beta}_{2L}(u_{i^*},v_{i^*},t_{i^*})\right)}{\phi_L}\right)^{\phi_L-2\tau_L}S_{il}^*\right\}$$

$$=\sum_{l=1}^{L}\sum_{i=1}^{n}w_{ii^*l}\log D_{il}^*+(\tau_L-1)\sum_{l}^{L}\sum_{i=1}^{n}w_{ii^*l}\log(y_{1il}\cdot y_{2il})+\sum_{l=1}^{L}\sum_{i=1}^{n}w_{ii^*l}\log S_{il}^*$$
$$+(\phi_L-2\tau_L)\sum_{l=1}^{L}\sum_{i=1}^{n}w_{ii^*l}\log\left(\frac{y_{1il}\exp\left(\mathbf{x}_{il}^{\mathbf{T}}\boldsymbol{\beta}_{1L}(u_{i^*},v_{i^*},t_{i^*})\right)+y_{2il}\exp\left(\mathbf{x}_{il}^{\mathbf{T}}\boldsymbol{\beta}_{2L}(u_{i^*},v_{i^*},t_{i^*})\right)}{\phi_L}\right) \quad (17)$$

where,

$$D_{il}^*=\left(\left(\frac{\phi_L^2}{\exp\left(\mathbf{x}_{il}^{T}\boldsymbol{\beta}_{1L}(u_{i^*},v_{i^*},t_{i^*})\right)\exp\left(\mathbf{x}_{il}^{T}\boldsymbol{\beta}_{2L}(u_{i^*},v_{i^*},t_{i^*})\right)}\right)^{\tau_L}\Gamma(\tau_L)\Gamma(\phi_L)\right)^{-1}$$

$$S_{il}^*=\Gamma\left(2\tau_L-\phi_L,\frac{y_{1il}\exp\left(\mathbf{x}_{il}^{\mathbf{T}}\boldsymbol{\beta}_{1L}(u_{i^*},v_{i^*},t_{i^*})\right)+y_{2il}\exp\left(\mathbf{x}_{il}^{\mathbf{T}}\boldsymbol{\beta}_{2L}(u_{i^*},v_{i^*},t_{i^*})\right)}{\phi_L}\right)$$

The GTWBGR model parameter estimators are obtained by maximizing the Q_L^* function by finding the first derivative of the parameters at location i^* in the period L and equating it to zero.

$$\frac{\partial Q_L^*}{\partial\phi_L}=0;\ \frac{\partial Q_L^*}{\partial\tau_L}=0;\ \frac{\partial Q_L^*}{\partial\boldsymbol{\beta}_{1L}(u_{i^*},v_{i^*},t_{i^*})}=0;\ \frac{\partial Q_L^*}{\partial\boldsymbol{\beta}_{2L}(u_{i^*},v_{i^*},t_{i^*})}=0$$

To simplify the derivative functions, suppose that:

$$P=\frac{y_{1il}\exp\left(\mathbf{x}_{il}^{\mathbf{T}}\boldsymbol{\beta}_{1L}(u_{i*},v_{i*},t_{i*})\right)+y_{2il}\exp\left(\mathbf{x}_{il}^{\mathbf{T}}\boldsymbol{\beta}_{2L}(u_{i*},v_{i*},t_{i*})\right)}{\phi_L}$$

Derivative of Q_L^* for ϕ_L:

$$\begin{aligned}\frac{\partial Q_L^*}{\partial \phi_L}=&-\left(\frac{2\tau_L}{\phi_L}+\psi(\phi_L)\right)\sum_{l=1}^{L}\sum_{i=1}^{n}w_{ii*l}+\sum_{l=1}^{L}\sum_{i=1}^{n}w_{ii*l}\log P-\left(\frac{\phi_L-2\tau}{\phi_L}\right)\sum_{l=1}^{L}\sum_{i=1}^{n}w_{ii*l}+\sum_{l=1}^{L}\sum_{i=1}^{n}\frac{w_{ii*l}}{\Gamma(2\tau_L-\phi_L,P)}\\&\left\{-\Gamma(2\tau_L-\phi_L,P)\log P+\frac{\pi\csc(2\pi\tau_L-\pi\phi_L)}{\Gamma(1-2\tau_L+\phi_L)}\left(-\psi(1-2\tau_L+\phi_L)+\pi\cot(2\pi\tau_L-\pi\phi_L)+\log P\right)\right.\\&\left.-\frac{1}{(2\tau_L-\phi_L)^2}P^{2\tau_L-\phi_L}\,hypergeom\left([2\tau_L-\phi_L,2\tau_L-\phi_L],[1+2\tau_L-\phi_L,1+2\tau_L-\phi_L],-P\right)+\frac{P^{2\tau_L-\phi_L}}{\phi_L\exp(-P)}\right\}\end{aligned}$$

Derivative of Q_L^* for τ_L:

$$\begin{aligned}\frac{\partial Q_L^*}{\partial \tau_L}=&-\sum_{l=1}^{L}\sum_{i=1}^{n}w_{ii*l}\log\left(\frac{\phi_L^{\,2}}{\exp\left(\mathbf{x}_{il}^{T}\boldsymbol{\beta}_{1L}(u_{i*},v_{i*},t_{i*})\right)\exp\left(\mathbf{x}_{il}^{T}\boldsymbol{\beta}_{2L}(u_{i*},v_{i*},t_{i*})\right)}\right)\\&-\psi(\tau_L)\sum_{l=1}^{L}\sum_{i=1}^{n}w_{ii*l}+\sum_{l=1}^{L}\sum_{i=1}^{n}w_{ii*l}\ln(y_{1il}y_{2il})-2\sum_{l=1}^{L}\sum_{i=1}^{n}w_{ii*l}\log P+\sum_{l=1}^{L}\sum_{i=1}^{n}\frac{w_{ii*l}}{\Gamma(2\tau_L-\phi_L,P)}\\&\left\{2\Gamma(2\tau_L-\phi_L,P)\log P-2\frac{\pi\csc(2\pi\tau_L-\pi\phi_L)}{\Gamma(1-2\tau_L+\phi_L)}\left(-\psi(1-2\tau_L+\phi_L)+\pi\cot(2\pi\tau_L-\pi\phi_L)+\log P\right)\right.\\&\left.+\frac{2}{(2\tau_L-\phi_L)^2}P^{2\tau_L-\phi_L}\,hypergeom\left([2\tau_L-\phi_L,2\tau_L-\phi_L],[1+2\tau_L-\phi_L,1+2\tau_L-\phi_L],-P\right)\right\}\end{aligned}$$

Derivative of Q_L^* for $\boldsymbol{\beta}_{1L}(u_{i*},v_{i*},t_{i*})$:

$$\begin{aligned}\frac{\partial Q_L^*}{\partial \boldsymbol{\beta}_{1L}(u_{i*},v_{i*},t_{i*})}=&(\phi_L-2\tau_L)\sum_{l=1}^{L}\sum_{i=1}^{n}w_{ii*l}\left(\frac{y_{1il}\exp(\mathbf{x}_{il}^{T}\boldsymbol{\beta}_{1L}(u_{i*},v_{i*},t_{i*}))}{y_{1il}\exp\left(\mathbf{x}_{il}^{\mathbf{T}}\boldsymbol{\beta}_{1L}(u_{i*},v_{i*},t_{i*})\right)+y_{2il}\exp\left(\mathbf{x}_{il}^{\mathbf{T}}\boldsymbol{\beta}_{2L}(u_{i*},v_{i*},t_{i*})\right)}\right)\mathbf{x}_{il}^{T}\\&+\tau_L\sum_{l=1}^{L}\sum_{i=1}^{n}w_{ii*l}\mathbf{x}_{il}^{T}-\sum_{l=1}^{L}\sum_{i=1}^{n}w_{ii*l}\left(\frac{y_{1il}\exp\left(\mathbf{x}_{il}^{\mathbf{T}}\boldsymbol{\beta}_{1L}(u_{i*},v_{i*},t_{i*})\right)P^{2\tau_L-\phi_L-1}}{\phi_L\,\Gamma(2\tau_L-\phi_L,P)\exp(P)}\right)\mathbf{x}_{il}^{T}\end{aligned}$$

Derivative of Q_L^* for $\boldsymbol{\beta}_{2L}(u_{i*},v_{i*},t_{i*})$:

$$\begin{aligned}\frac{\partial Q_L^*}{\partial \boldsymbol{\beta}_{2L}(u_{i*},v_{i*},t_{i*})}=&(\phi_L-2\tau_L)\sum_{l=1}^{L}\sum_{i=1}^{n}w_{ii*l}\left(\frac{y_{2il}\exp(\mathbf{x}_{il}^{T}\boldsymbol{\beta}_{2L}(u_{i*},v_{i*},t_{i*}))}{y_{1il}\exp\left(\mathbf{x}_{il}^{\mathbf{T}}\boldsymbol{\beta}_{1L}(u_{i*},v_{i*},t_{i*})\right)+y_{2il}\exp\left(\mathbf{x}_{il}^{\mathbf{T}}\boldsymbol{\beta}_{2L}(u_{i*},v_{i*},t_{i*})\right)}\right)\mathbf{x}_{il}^{T}\\&+\tau_L\sum_{l=1}^{L}\sum_{i=1}^{n}w_{ii*l}\mathbf{x}_{il}^{T}-\sum_{l=1}^{L}\sum_{i=1}^{n}w_{ii*l}\left(\frac{y_{2il}\exp\left(\mathbf{x}_{il}^{\mathbf{T}}\boldsymbol{\beta}_{2L}(u_{i*},v_{i*},t_{i*})\right)P^{2\tau_L-\phi_L-1}}{\phi_L\,\Gamma(2\tau_L-\phi_L,P)\exp(P)}\right)\mathbf{x}_{il}^{T}\end{aligned}$$

When the first derivatives are equal to zero, no explicit parameter estimators can be generated, so we need optimization using numerical iteration. The algorithm used for optimization is BHHH by the following steps:

1. Initialization $\boldsymbol{\theta}_{i^*L}^{(0)}$ and $m = 0$ with $\varepsilon > 0$ for convergence tolerance

   $$\hat{\boldsymbol{\theta}}_{i^*L}^{(0)} = \left[\hat{\phi}_L^{(0)} \quad \hat{\tau}_L^{(0)} \quad \hat{\boldsymbol{\beta}}_{1L}^{T(0)}(u_{i^*}, v_{i^*}, t_{i^*}) \quad \hat{\boldsymbol{\beta}}_{2L}^{T(0)}(u_{i^*}, v_{i^*}, t_{i^*})\right]^T$$

   $\hat{\phi}_L^{(0)} > 0$, $\hat{\tau}_L^{(0)} > 0$, while $\hat{\boldsymbol{\beta}}_{1L}^{(0)}(u_{i^*}, v_{i^*}, t_{i^*})$ and $\hat{\boldsymbol{\beta}}_{2L}^{(0)}(u_{i^*}, v_{i^*}, t_{i^*})$ are the estimators of univariate gamma regression coefficients.
2. Define the gradient vector $\mathbf{g}(\hat{\boldsymbol{\theta}}_{i^*L}^{(m)})$

   $$\mathbf{g}(\hat{\boldsymbol{\theta}}_{i^*L}^{(m)}) = \left[\frac{\partial \log L(\bullet)}{\partial \phi_L} \quad \frac{\partial \log L(\bullet)}{\partial \tau_L} \quad \frac{\partial \log L(\bullet)}{\partial \boldsymbol{\beta}_{1L}(u_{i^*}, v_{i^*}, t_{i^*})} \quad \frac{\partial \log L(\bullet)}{\partial \boldsymbol{\beta}_{2L}(u_{i^*}, v_{i^*}, t_{i^*})}\right]^T \Bigg|_{\boldsymbol{\theta}_{i^*L}^{(m)} = \hat{\boldsymbol{\theta}}_{i^*L}^{(m)}}$$
3. Determine the first derivative of the natural logarithm of the individual pdf in each parameter

   $$\mathbf{l}_{iL}(\hat{\boldsymbol{\theta}}_{i^*L}^{(m)}) = \left[\frac{\partial \log f(y_{1il}, y_{2il})}{\partial \phi_L} \quad \frac{\partial \log f(y_{1il}, y_{2il})}{\partial \tau_L} \quad \frac{\partial \log f(y_{1il}, y_{2il})}{\partial \boldsymbol{\beta}_{1L}(u_{i^*}, v_{i^*}, t_{i^*})} \quad \frac{\partial \log f(y_{1il}, y_{2il})}{\partial \boldsymbol{\beta}_{2L}(u_{i^*}, v_{i^*}, t_{i^*})}\right]^T \Bigg|_{\boldsymbol{\theta}_{i^*L}^{(m)} = \hat{\boldsymbol{\theta}}_{i^*L}^{(m)}}$$

   for $i = 1, 2, \ldots, n$
4. Create a hessian matrix $\mathbf{H}(\hat{\boldsymbol{\theta}}_{i^*L}^{(m)}) = -\sum_{l=1}^{L}\sum_{i=1}^{n} \mathbf{l}_{il}(\hat{\boldsymbol{\theta}}_{i^*L}^{(m)}) \mathbf{l}_{il}(\hat{\boldsymbol{\theta}}_{i^*L}^{(m)})^T$
5. Substitute value $\hat{\boldsymbol{\theta}}_{i^*L}^{(m)}$ on elements of $\mathbf{g}(\hat{\boldsymbol{\theta}}_{i^*L}^{(m)})$ and $\mathbf{H}(\hat{\boldsymbol{\theta}}_{i^*L}^{(m)})$
6. Do iteration starting from $m = 0$ with $\hat{\boldsymbol{\theta}}_{i^*L}^{(m+1)} = \hat{\boldsymbol{\theta}}_{i^*L}^{(m)} - \mathbf{H}^{-1}(\hat{\boldsymbol{\theta}}_{i^*L}^{(m)}) \mathbf{g}(\hat{\boldsymbol{\theta}}_{i^*L}^{(m)})$

   Iteration will stop if $\left\| \hat{\boldsymbol{\theta}}_{i^*L}^{(m+1)} - \hat{\boldsymbol{\theta}}_{i^*L}^{(m)} \right\| < \varepsilon$ where ε is a tiny positive number
7. If it has not converged back to step 2 and so on with $m = m + 1$ until convergent

For each L, the optimization is carried out n times to obtain the parameter estimator vector $\hat{\boldsymbol{\theta}}_{i^*L} = \left[\hat{\phi}_L \quad \hat{\tau}_L \quad \hat{\boldsymbol{\beta}}_{1L}^T(u_{i^*}, v_{i^*}, t_{i^*}) \quad \hat{\boldsymbol{\beta}}_{2L^*}^T(u_{i^*}, v_{i^*}, t_{i^*})\right]^T$ where $i^* = 1, 2, \ldots, n; L = 1, 2, \ldots, L^*$.

3.2 Hypothesis Testing

Hypothesis testing is carried out in stages every period, as in parameter estimation. The hypothesis in simultaneous testing in the period L is as follows:

$H_0 : \beta_{h1L}(u_i, v_i, t_i) = \beta_{h2L}(u_i, v_i, t_i) = \cdots = \beta_{hpL}(u_i, v_i, t_i) = 0; i = 1, 2, ..., n; h = 1, 2$

H_1 : there is at least one $\beta_{hkL}(u_i, v_i, t_i) \neq 0; k = 1, 2, \ldots, p$

p denotes the number of predictor variables.

To obtain the test statistic, the maximum value of the log-likelihood function under the H_0 and the population is required. The maximum value of the function is obtained by estimating its parameters using MLE. It is necessary to determine the set of parameters under the population and the H_0.

The set of parameters under the population:

$\Omega_L = \{\phi_L, \tau_L, \boldsymbol{\beta}_{1L}(u_i, v_i, t_i), \boldsymbol{\beta}_{2L}(u_i, v_i, t_i); i = 1, 2, ..., n;\}$

The set of parameters under the H_0:

$\omega_L = \{\phi_{\omega L}, \tau_{\omega L}, \beta_{\omega 1L0}(u_i, v_i, t_i), \beta_{\omega 2L0}(u_i, v_i, t_i); i = 1, 2, ..., n\}$.

The log-likelihood function under the population is equal to equation (16), while under the H_0 is as follows:

$$\log L(\omega_L) = \log L(\phi_{\omega L}, \tau_{\omega L}, \beta_{\omega 1L0}(u_i,v_i,t_i), \beta_{\omega 2L0}(u_i,v_i,t_i); i=1,2,\ldots,n)$$
$$= \log\left(\prod_{l=1}^{L}\prod_{i=1}^{n}\left(B_{il}(y_{1il}y_{2il})^{\tau_{\omega L}-1}\left(\frac{y_{1il}\exp(\beta_{\omega 1L0}(u_i,v_i,t_i))+y_{2il}\exp(\beta_{\omega 2L0}(u_i,v_i,t_i))}{\phi_{\omega L}}\right)^{\phi_{\omega L}-2\tau_{\omega L}} T_{il}\right)\right)$$
$$= \sum_{l=1}^{L}\sum_{i=1}^{n}\log\left(B_{il}(y_{1il}y_{2il})^{\tau_{\omega L}-1}\left(\frac{y_{1il}\exp(\beta_{\omega 1L0}(u_i,v_i,t_i))+y_{2il}\exp(\beta_{\omega 2L0}(u_i,v_i,t_i))}{\phi_{\omega L}}\right)^{\phi_{\omega L}-2\tau_{\omega L}} T_{il}\right) \quad (18)$$

where,

$$B_{il} = \left(\left(\frac{\phi_{\omega L}^{\ 2}}{\exp(\beta_{\omega 1L0}(u_i,v_i,t_i))\exp(\beta_{\omega 2L0}(u_i,v_i,t_i))}\right)^{\tau_{\omega L}}\Gamma(\tau_{\omega L})\Gamma(\phi_{\omega L})\right)^{-1}$$

$$T_{il} = \Gamma\left(2\tau_{\omega L}-\phi_{\omega L}, \frac{y_{1il}\exp(\beta_{\omega 1L0}(u_i,v_i,t_i))+y_{2il}\exp(\beta_{\omega 2L0}(u_i,v_i,t_i))}{\phi_{\omega L}}\right)$$

In obtaining parameter estimates under the H_0 at location $i*$ period L, the Q_L^{**} function is determined, namely the log-likelihood function under the H_0 given spatial-temporal weight w_{ii*l}.

$$Q_L^{**} = \sum_{l=1}^{L}\sum_{i=1}^{n} w_{ii*l}\log\left\{B_{il}^{*}(y_{1il}y_{2il})^{\tau_{\omega L}-1}\left(\frac{y_{1il}\exp(\beta_{\omega 1L0}(u_{i*},v_{i*},t_{i*}))+y_{2il}\exp(\beta_{\omega 2L0}(u_{i*},v_{i*},t_{i*}))}{\phi_{\omega L}}\right)^{\phi_{\omega L}-2\tau_{\omega L}} T_{il}^{*}\right\}$$
$$= \sum_{l=1}^{L}\sum_{i=1}^{n} w_{ii*l}\log B_{il}^{*} + (\tau_{\omega L}-1)\sum_{l=1}^{L}\sum_{i=1}^{n} w_{ii*l}\log(y_{1il}.y_{2il}) + \sum_{l=1}^{L}\sum_{i=1}^{n} w_{ii*l}\log T_{il}^{*}$$
$$+(\phi_{\omega L}-2\tau_{\omega L})\sum_{l=1}^{L}\sum_{i=1}^{n} w_{ii*l}\log\left(\frac{y_{1il}\exp(\beta_{\omega 1L0}(u_{i*},v_{i*},t_{i*}))+y_{2il}\exp(\beta_{\omega 2L0}(u_{i*},v_{i*},t_{i*}))}{\phi_{\omega L}}\right) \quad (19)$$

where:

$$B_{il}^{*} = \left(\left(\frac{\phi_{\omega L}^{\ 2}}{\exp(\beta_{\omega 1L0}(u_{i*},v_{i*},t_{i*}))\exp(\beta_{\omega 2L0}(u_{i*},v_{i*},t_{i*}))}\right)^{\tau_{\omega L}}\Gamma(\tau_{\omega L})\Gamma(\phi_{\omega L})\right)^{-1}$$

$$T_{il}^{*} = \Gamma\left(2\tau_{\omega L}-\phi_{\omega L}, \frac{y_{1il}\exp(\beta_{\omega 1L0}(u_{i*},v_{i*},t_{i*}))+y_{2il}\exp(\beta_{\omega 2L0}(u_{i*},v_{i*},t_{i*}))}{\phi_{\omega L}}\right)$$

Next, find the first derivative of the Q_L^{**} function for each parameter under the H_0 and equalize it to zero.

$$\frac{\partial Q_L^{**}}{\partial \phi_{\omega L}} = 0; \frac{\partial Q_L^{**}}{\partial \tau_{\omega L}} = 0; \frac{\partial Q_L^{**}}{\partial \beta_{\omega 1L0}(u_{i*},v_{i*},t_{i*})} = 0; \frac{\partial Q_L^{**}}{\partial \beta_{\omega 2L0}(u_{i*},v_{i*},t_{i*})} = 0$$

To simplify the derivative functions, suppose that:

$$F = \frac{y_{1il}\exp(\beta_{\omega 1L0}(u_{i*},v_{i*},t_{i*}))+y_{2il}\exp(\beta_{\omega 2L0}(u_{i*},v_{i*},t_{i*}))}{\phi_{\omega L}}$$

Derivative of Q_L^{**} for $\phi_{\omega L}$:

$$\frac{\partial Q_L^{**}}{\partial \phi_{\omega L}} = -\left(\frac{2\tau_L}{\phi_{\omega L}}+\psi(\phi_{\omega L})\right)\sum_{l=1}^{L}\sum_{i=1}^{n} w_{ii*l} + \sum_{l=1}^{L}\sum_{i=1}^{n} w_{ii*l}\log F - \left(\frac{\phi_{\omega L}-2\tau}{\phi_{\omega L}}\right)\sum_{l=1}^{L}\sum_{i=1}^{n} w_{ii*l} + \sum_{l=1}^{L}\sum_{i=1}^{n}\frac{w_{ii*l}}{\Gamma(2\tau_{\omega L}-\phi_{\omega L},F)}$$

$$\left\{ -\Gamma\left(2\tau_{\omega L}-\phi_{\omega L},F\right)\log F+\frac{\pi\csc\left(2\pi\tau_{\omega L}-\pi\phi_{\omega L}\right)}{\Gamma\left(1-2\tau_{\omega L}+\phi_{\omega L}\right)}\left[-\psi\left(1-2\tau_{\omega L}+\phi_{\omega L}\right)+\pi\cot\left(2\pi\tau_{\omega L}-\pi\phi_{\omega L}\right)+\log F\right]\right.$$
$$\left.-\frac{F^{2\tau_{\omega L}-\phi_{\omega L}}}{\left(2\tau_{\omega L}-\phi_{\omega L}\right)^2} hypergeom\left(\left[2\tau_{\omega L}-\phi_{\omega L},2\tau_{\omega L}-\phi_{\omega L}\right],\left[1+2\tau_{\omega L}-\phi_{\omega L},1+2\tau_{\omega L}-\phi_{\omega L}\right],-F\right)+\frac{F^{2\tau_{\omega L}-\phi_{\omega L}}}{\phi_{\omega L}\exp(-F)}\right\}$$

Derivative of Q_L^{**} for $\tau_{\omega L}$:

$$\frac{\partial Q_L^{**}}{\partial\tau_{\omega L}}=-\sum_{l=1}^{L}\sum_{i=1}^{n}w_{ii*l}\log\left(\frac{\phi_{\omega L}{}^2}{\exp\left(\beta_{\omega 1L0}(u_{i*},v_{i*},t_{i*})\right)\exp\left(\beta_{\omega 2L0}(u_{i*},v_{i*},t_{i*})\right)}\right)-\psi\left(\tau_{\omega L}\right)\sum_{l=1}^{L}\sum_{i=1}^{n}w_{ii*l}$$
$$+\sum_{l=1}^{L}\sum_{i=1}^{n}w_{ii*l}\log\left(y_{1il}y_{2il}\right)-2\sum_{l=1}^{L}\sum_{i=1}^{n}w_{ii*l}\log F+\sum_{l=1}^{L}\sum_{i=1}^{n}\frac{w_{ii*l}}{\Gamma\left(2\tau_{\omega L}-\phi_{\omega L},F\right)}\left\{ 2\Gamma\left(2\tau_{\omega L}-\phi_{\omega L},F\right)\log F\right.$$
$$-2\frac{\pi\csc\left(2\pi\tau_{\omega L}-\pi\phi_{\omega L}\right)}{\Gamma\left(1-2\tau_{\omega L}+\phi_{\omega L}\right)}\left(-\psi\left(1-2\tau_{\omega L}+\phi_{\omega L}\right)+\pi\cot\left(2\pi\tau_{\omega L}-\pi\phi_{\omega L}\right)+\log F\right)$$
$$\left.+\frac{2}{\left(2\tau_{\omega L}-\phi_{\omega L}\right)^2}F^{2\tau_{\omega L}-\phi_{\omega L}}hypergeom\left(\left[2\tau_{\omega L}-\phi_{\omega L},2\tau_{\omega L}-\phi_{\omega L}\right],\left[1+2\tau_{\omega L}-\phi_{\omega L},1+2\tau_{\omega L}-\phi_{\omega L}\right],-F\right)\right\}$$

Derivative of Q_L^{**} for $\beta_{\omega 1L0}\left(u_{i*},v_{i*},t_{i*}\right)$:

$$\frac{\partial Q_L^{**}}{\partial\beta_{\omega 1L0}\left(u_{i*},v_{i*},t_{i*}\right)}=\left(\phi_{\omega L}-2\tau_L\right)\sum_{l=1}^{L}\sum_{i=1}^{n}w_{ii*l}\left(\frac{y_{1il}\exp(\beta_{\omega 1L0}(u_{i*},v_{i*},t_{i*}))}{y_{1il}\exp\left(\beta_{\omega 1L0}(u_{i*},v_{i*},t_{i*})\right)+y_{2il}\exp\left(\beta_{\omega 2L0}(u_{i*},v_{i*},t_{i*})\right)}\right)$$
$$+\tau_{\omega L}\sum_{l=1}^{L}\sum_{i=1}^{n}w_{ii*l}-\sum_{l=1}^{L}\sum_{i=1}^{n}w_{ii*l}\left(\frac{y_{1il}\exp\left(\beta_{\omega 1L0}(u_{i*},v_{i*},t_{i*})\right)F^{2\tau_{\omega L}-\phi_{\omega L}-1}}{\phi_{\omega L}\,\Gamma\left(2\tau_{\omega L}-\phi_{\omega L},F\right)\exp(F)}\right)$$

Derivative of Q_L^{**} for $\beta_{\omega 2L0}(u_{i*},v_{i*},t_{i*})$:

$$\frac{\partial Q_L^{**}}{\partial\beta_{\omega 2L0}(u_{i*},v_{i*},t_{i*})}=\left(\phi_{\omega L}-2\tau_{\omega L}\right)\sum_{l=1}^{L}\sum_{i=1}^{n}w_{ii*l}\left(\frac{y_{2il}\exp(\beta_{\omega 2L0}(u_{i*},v_{i*},t_{i*}))}{y_{1il}\exp\left(\beta_{\omega 1L0}(u_{i*},v_{i*},t_{i*})\right)+y_{2il}\exp\left(\beta_{\omega 2L0}(u_{i*},v_{i*},t_{i*})\right)}\right)$$
$$+\tau_{\omega L}\sum_{l=1}^{L}\sum_{i=1}^{n}w_{ii*l}-\sum_{l=1}^{L}\sum_{i=1}^{n}w_{ii*l}\left(\frac{y_{2il}\exp\left(\beta_{\omega 2L0}(u_{i*},v_{i*},t_{i*})\right)F^{2\tau_{\omega L}-\phi_{\omega L}-1}}{\phi_{\omega L}\,\Gamma\left(2\tau_{\omega L}-\phi_{\omega L},F\right)\exp(F)}\right)$$

If the first derivative of the Q_L^{**} function for each parameter under the H_0 is equalized to zero, it does not produce an explicit parameter estimator. It is necessary to do numerical iteration using the BHHH method to obtain the parameter estimator vector under the H_0, $\hat{\boldsymbol{\theta}}_{\omega i*L}=\left[\hat{\phi}_{\omega L}\quad\hat{\tau}_{\omega L}\quad\hat{\beta}_{\omega 1L0}(u_{i*},v_{i*},t_{i*})\quad\hat{\beta}_{\omega 2L0}(u_{i*},v_{i*},t_{i*})\right]^T$ where $i*=1,2,\ldots,n;\ L=1,2,\ldots,L*$. Then, we use the following test statistic for decision-making.

$$G_L^2=-\log\Lambda_L^2=-\log\left(\frac{L(\hat{\omega}_L)}{L(\hat{\Omega}_L)}\right)^2=2\left[\log L(\hat{\Omega}_L)-\log L(\hat{\omega}_L)\right]\underset{nL\to\infty}{\sim}\chi^2_{(\alpha,df)}\qquad(20)$$

$L(\hat{\Omega}_L)$ and $L(\hat{\omega}_L)$ are likelihood functions with parameter estimators under the population and the H_0.

With a large sample size, the G_L^2 statistic can be approximated by the chi-square distribution with degrees of freedom *df,* where *df* denotes the difference between the number of parameters under the population and the number of parameters under the H_0. The area of rejection is $G_L^2 > \chi^2_{(\alpha;df)}$ [10]. If simultaneous testing gives a decision to reject the H_0, then a partial test is carried out to specify the effect of a predictor variable on the response variable. The hypothesis in the partial test is as follows:
$H_0 : \beta_{hkL}(u_i, v_i, t_i) = 0$
$H_0 : \beta_{hkL}(u_i, v_i, t_i) \neq 0$
with $h = 1,2; k = 1,2,\ldots,p; L = 1,2,\ldots,L^*$
The test statistic:

$$Z = \frac{\hat{\beta}_{hkL}(u_i, v_i, t_i)}{se\left(\hat{\beta}_{hkL}(u_i, v_i, t_i)\right)} \underset{nL \to \infty}{\sim} N(0,1) \tag{21}$$

where $se\left(\hat{\beta}_{hkL}(u_i, v_i, t_i)\right) = \sqrt{v\hat{a}r\left(\hat{\beta}_{hkL}(u_i, v_i, t_i)\right)}$, while $v\hat{a}r\left(\hat{\beta}_{hkL}(u_i, v_i, t_i)\right)$ is obtained from the k+1 diagonal element of $v\hat{a}r(\hat{\boldsymbol{\theta}}_{i*L}) = -\mathbf{H}^{-1}(\hat{\boldsymbol{\theta}}_{i*L})$. The test statistic in the partial test can be approached by a normal distribution, so the H_0 is rejected when $|Z| > Z_{\alpha/2}$ [11].

4. Conclusion

The parameter estimation of the GTWBGR model using the MLE method produces the non-explicit form of estimators. Therefore, it is optimized by numerical iteration. In this research, numerical iteration was carried out using the BHHH algorithm. With a large sample size, the test statistic in simultaneous testing approximates the chi-square distribution, and the partial test approximates the normal distribution.

References

[1] Brunsdon C, Fotheringham A S, and Charlton M E. 1996. "Geographically Weighted Regression: A Method for Exploring Spatial Nonstationarity," *Geographical Analysis*. **vol. 28.** pp. 281-298.
[2] Hsiao C. 2014. Analysis of Panel Data. New York: Cambridge University Press.
[3] Huang B, Wu B, and Barry M. 2010. "Geographically and Temporally Weighted Regression For Modeling Spatio-Temporal Variation in House Prices." *International Journal of Geographical Information Science*. **vol. 24**. pp. 383-401.
[4] Hogg R, McKean J, and Craig A. 2019. Introduction to Mathematical Statistics. Boston: Pearson Education.
[5] Purhadi, Budiani J R, and Purnami S W. 2018. "Parameter Estimation and Statistical Test in Bivariate Gamma Regression Model." in *Covergence of Basic Sciences. Toward the World's Sustainability Challenges*. Malang.
[6] Dewi D K, Purhadi and Sutikno. 2019. "Geographically Weighted Bivariate Gamma Regression in The Analysis of Maternal Mortality Rate and Infant Mortality Rate in North Sumatra Province 2017." *IOP Conference Series: Materials Science and Engineering*. pp. 1-10.
[7] Nadarajah S and Gupta A K. 2006. "Some Bivariate Gamma Distributions." *Applied Mathematics Letters 19*. pp. 767-774.
[8] Manning W G, Basu A, and Mullahy J. 2002. "Modeling Costs with Generalized Gamma Regression." The University of Chicago. Chicago.
[9] Fotheringham A, Brunsdon C, and Charlton M E. 2002. Geographically Weighted Regression. Chichester: John Wiley & Sons, Ltd.
[10] Purhadi, Rahayu A, and Wenur G H. 2021. "Geographically Weighted Three-Parameters Bivariate Gamma Regression and Its Application." *Symmetry*. **vol. 13**.
[11] Cameron C and Trivedi P K. 2005. Microeconometrics Methods and Applications. New York: Cambridge University Press.
[12] Gujarati D and Porter D. 2008. Basic Econometrics. New York: McGrawHill.
[13] Fischer M and Getis V. 2010. Handbook of Applied Spatial Analysis. Berlin Heidelberg: Springer-Verlag.

Acknowledgments
The author would like to thank BPS-Statistics Indonesia, which financially supported this research. The author also thanks the lecturers and ITS student friends for their guidance and support in completing this paper.

Parameter estimation and hypothesis testing of geographically and temporally weighted bivariate Poisson inverse Gaussian regression model

Meylita Sari[1*], Sutikno[1] and Purhadi[1]

[1]Department of Statistics, Faculty of Scientics, Institut Teknologi Sepuluh Nopember (ITS). Arief Rahman Hakim Street, Surabaya 60111 Indonesia

Corresponding E-mail: meylitasari27@gmail.com

Abstract. One of the appropriate methods used to model count data response and its corresponding predictors is Poisson regression. Poisson regression strictly assumes that the mean and variance of response variables should be equal (equidispersion). Nonetheless, some cases of the count data unsatisfied this assumption because variance can be larger than mean (over-dispersion). If overdispersion is violated, causing the underestimate standard error. Furthermore, this will lead to incorrect conclusions in the statistical test. Thus, a suitable method for modelling this kind of data needs to develop. One alternative model to outcome the overdispersion issue in bivariate response variable is the Bivariate Poisson Inverse Gaussian Regression (BPIGR) model. The BPIGR model can produce a global model for all locations. On the other hand, each location and time have different geographic conditions, social, cultural, and economical so that Geographically and Temporally Bivariate Poisson Inverse Gaussian Regression (GTWBPIGR)) is needed. The weighting function spatial-temporal in GTWBPIGR generates a different local model for each period. GTWBPIGR model solves the overdispersion case and generates global models for each period and location. The parameter estimation of the GTWBPIGR model uses the Maximum Likelihood Estimation (MLE) method, followed by Newton Raphson iteration. Meanwhile, the test statistics on the hypothesis testing is simultaneously testing of the GTWBPIGR model is obtained with the Maximum Likelihood Ratio Test (MLRT) approach, using n large samples of the statistical test is chi-square distribution. Moreover, the test statistics for partially testing used the Z-test statistic.

Keywords: Overdispersion, BPIGR, GTWBPIGR, MLE, MLRT

Track Name: Advanced Technology and Renewable Energy

1. Introduction

The Poisson distribution is a discrete distribution with a random variable value in positive integers, so it is a good choice for modelling count data. The Poisson distribution is only determined by one parameter that defines the distribution's mean and variance. Poisson regression has assumptions when the mean and variance of the response variable must be the same value (equidispersion). However, the mean and variance of count data are often not the same [1]; often, the variance is greater than the mean (overdispersion). If the overdispersion case is ignored, it can lead to an underestimate of the standard error estimation. In overcoming overdispersion cases, several models are formed, which are a combination of the Poisson distribution with several distributions, both discrete and continuous (mixed Poisson distribution). The mixed Poisson distribution is an alternative solution for the overdispersion case; one of the distributions used is the Poisson Inverse Gaussian (PIG). PIG is a combination of the Poisson distribution and Gaussian inverse [2].

The combined modelling of two or more data counts has received much attention in recent years. The bivariate count model is a case with a pair of correlated response variables and needs to be estimated together. The Poisson bivariate is the most widely used model for bivariate computations. However, it is not possible for overdispersion or under dispersion cases [3]. Thus, to overcome overdispersion, some bivariate mixing of the Poisson distribution is required. Poisson bivariate mixed distribution is an alternative solution to negotiating over-dispersion cases.

BPIGR will produce a global model that is considered valid at all data collection locations [4]. However, every location and time have different geographical, social, cultural, and economic conditions. Therefore, a regression model will be developed that considers the spatial and temporal effects, such as Geographically and Temporally Weighted Bivariate Poisson Inverse Gaussian Regression (GTWBPIGR). The GTWR parameter may be estimated anywhere in the study area given the dependent variable and where one or more independent variables have been measured in places whose location is known at any given time [5]. The spatial-temporal weighting function in the GTWBPIGR model produces a different local model for each period. The GTWBPIGR model is used to overcome over-dispersion and to generate localized models for each period. In this paper, the case of over-dispersion and developing localized models for each period can be solved using the proposed GTWBPIGR model. The parameter estimation of the GTWBPIGR model uses the Maximum Likelihood Estimation (MLE) method with Newton Raphson numerical iteration. At the same time, the test statistics on hypothesis testing are simultaneously obtained by the Maximum Likelihood Ratio Test (MLRT) method approach, using n large samples of the statistical test is chi-square distribution. Moreover, the test statistics for partially testing used the Z-test statistic approach using n large samples of the statistical test is normal (0,1) distribution.

2. Materials and Methods

2.1 Bivariate Poisson Inverse Gaussian Regression Models

Bivariate Poisson Inverse Gaussian distribution (BPIG) has two correlated count variables. Suppose two random variables independent of each other and have a Poisson distribution, Y_1 and Y_2 have mean $v\mu_1$ and $v\mu_2$. The variable v is a random variable with an Inverse Gaussian distribution with the joint probability density function as follows:

$$g(v;\tau)=(2\pi\tau^3)^{-\frac{1}{2}}e^{-\frac{(v-1)2}{2\tau v}}, v>0 \quad (1)$$

The BPIG distribution is based on the Inverse Gaussian and Poisson mixture distribution which has the following joint density function:

$$f(y_1,y_2;\mu_1,\mu_2,\tau)=\left(\frac{2z}{\pi}\right)^{\frac{1}{2}}\frac{\mu_1^{y_1}\mu_2^{y_2}e^{\frac{1}{\tau}}K_s(z)}{(z\tau)^{y_1y_2}y_1!y_2!}, y\geq 0 \quad (2)$$

with $s=y_1+y_2-\frac{1}{2}$ and $z=\sqrt{\frac{1}{\tau^2}+\frac{2(\mu_1+\mu_2)}{\tau}}$. so, $K_s(z)=K_{y_1+y_2-\frac{1}{2}}\left(\frac{1}{\tau}\sqrt{2\tau(\mu_1+\mu_2)}\right)$ is the third

type of modification of the Bessel function [6].

A bivariate regression model is a regression model with two correlated response variables and one or more predictor variables. BPIGR is used when the two response variables (Y_1, Y_2) are experiencing overdispersion. The joint probability density function for Y_1 and Y_2 is:

$$f(y_j;\boldsymbol{\beta}_j,\tau,j=1,2)=e^{\frac{1}{\tau}}K_s(z)\left(\frac{2}{\pi\tau}\right)^{\frac{1}{2}}\left(1+2\tau\sum_{j=1}^{2}\mu_{ij}\right)^{\frac{\left(2\sum_{j=1}^{2}y_{ij}-1\right)}{4}}\prod_{j=1}^{2}\frac{\mu_{ij}}{y_{ij}!} \tag{3}$$

Modelling *bivariate Poisson inverse Gaussian* regression (BPIGR) requires a logarithm (log) function. Parameters μ_{ij} are associated with predictor variables using the ln link function so that the BPIG regression model can be written as follows:

$$\log \mu_{ij} = \mathbf{x}_i^T \boldsymbol{\beta}_j$$

$$\mu_{ij} = \exp\left(\mathbf{x}_i^T \boldsymbol{\beta}_j\right)$$

with $E(Y_{ij}) = \mu_{ij} = \exp\left(\mathbf{x}_i^T \boldsymbol{\beta}_j\right)$; $\mathbf{x}_i^T = [1 \;\; x_{1i} \;\; x_{2i} \;\; \ldots \;\; x_{p_i}]$ is the vector of the predictor variable in the i-th observation and the j-th response variable $(i = 1,2, \ldots, n)$, $\boldsymbol{\beta}_j = \left[\beta_{j0} \;\; \beta_{j1} \;\; \beta_{j2} \;\; \ldots \;\; \beta_{jp}\right]^T; j = 1,2$ is the vector of the $(p + 1) \times 1$ dimensionless regression coefficient in the j-th response variable.

The parameter estimation of the BPIGR model was obtained by the MLE method with Newton Raphson numerical iteration. While the test statistics on hypothesis testing are simultaneously obtained by the MLRT method.

2.2 Geographically and Temporally Weighted Bivariate Poisson Inverse Gaussian Regression Models

The GTWBPIGR model is a development of the GWBPIGR model by considering the temporal aspect. Based on the BPIG equation (3), the joint probability function y_{1il} and y_{2il} is

$$(y_{1il}, y_{2il}) \sim GTWBPIGR(\boldsymbol{\beta}_{1l}(u_i, v_i, t_i), \boldsymbol{\beta}_{2l}(u_i, v_i, t_i), \tau_l)$$

$$f(y_{1il}, y_{2il} \,|\, \boldsymbol{\beta}_{jl}(u_i,v_i,t_i),\tau_l) = e^{\frac{1}{\tau_l}} K_{sl}(z(u_i,v_i,t_i))\left(\frac{2}{\pi\tau_l}\right)^{\frac{1}{2}}\left(1+2\tau_l\sum_{j=1}^{2}e^{\mathbf{x}_{il}^T\boldsymbol{\beta}_{jl}(u_i,v_i,t_i)}\right)^{-\frac{\left(2\sum_{j=1}^{2}y_{jil}-1\right)}{4}}\prod_{j=1}^{2}\frac{e^{y_{jil}\mathbf{x}_{il}^T\boldsymbol{\beta}_{jl}(u_i,v_i,t_i)}}{y_{jil}!} \tag{4}$$

with l-th is the period, $\mu_{jil} = \exp\left(\mathbf{x}_{il}^T \boldsymbol{\beta}_{jL}(u_i, v_i, t_i)\right)$, $\mathbf{x}_{il}^T = [1 \;\; x_{1il} \;\; x_{2il} \;\; \ldots \;\; x_{pil}]$ is the vector of the predictor variable in the i-th observation and the j-th response variable; ($i = 1,2, \ldots, n$; $l=1,2,\ldots,L$ dan j=1,2). $\boldsymbol{\beta}_{jl}(u_i, v_i, t_i) = \left[\beta_{j01}(u_i, v_i, t_i) \;\; \beta_{j12}(u_i, v_i, t_i) \;\; \ldots \;\; \beta_{jpL}(u_i, v_i, t_i)\right]^T$ are vectors of regression coefficients that are $(K + 1) \times 1$ dimensioned on the response variable in the j-th in the l-period with a temporal-spatial weighting matrix. u_i, v_i are location latitude and longitude.

The parameter estimation GTWBPIGR model using the MLE method is carried out in stages in each period by including observations of the previous period. The parameter estimation in period first uses n data in period 1. In period 2 uses n data in period 1st and n data in period 2nd. The parameter estimation in period L uses total n data in periods 1, 2 to L. The spatial-temporal weighting function in the GTWBPIGR model uses $w_{ii'l}$ [5].

For the weighting function for the period l = 1, the temporal-spatial weighting is obtained with the same Euclidean distance value as the GWR weight. During one period, weighting calculations are only performed using the data for period l total n observations to obtain the weights for $w_{i11} \ldots w_{in1}$ the i-location. When period l = 2, the weighted calculation uses data for periods 1 and 2 with a total of 2n observations. So that the weights $w_{i12} \ldots w_{in2}$ for the *i*-location are obtained. And so on for three periods to L periods.

For temporal-spatial weighting in determining the Euclidean distance between the location and time of observation (u_i, v_i, t_i) with (u_{i*}, v_{i*}, t_{i*}) using the formula in the following equation.

$$d_{ii^*l} = \sqrt{(u_i - u_{i*})^2 + (v_i - v_{i*})^2 + \gamma(t_i - t_{i*})^2}$$

So we get a temporal-spatial weighting with the Adaptive Bisquare Kernel Function, follows as

$$w_{ii^*l} = \begin{cases} \left(1 - \left(\dfrac{d_{ii^*l}}{b_{il}}\right)^2\right)^2 & ;\text{for } d_{ii^*l} \le b_{il} \\ 0; & \text{for } d_{ii^*l} > b_{il} \end{cases}$$

with, d_{ii^*l} is the Euclidean distance and b_{il} is bandwidth.
The selection of the optimum bandwidth can be made by using the Generalized Cross-Validation (GCV) method. This GCV method is defined as follows:

$$GCV = \min_h \left\{ n \sum_{i=1}^{n} \frac{\left[\mathbf{y}_i - \mathbf{y}_{\neq i}(b_{il})\right]^T \left[\mathbf{y}_i - \mathbf{y}_{\neq i}(b_{il})\right]}{(n - v_1)^2} \right\}$$

with:

$\mathbf{y}_{\neq i}(h)$: estimator value $\mathbf{y}_i$ when site observations (u_i, v_i) are not included in the estimate
$\mathbf{y}_i$: i-th response variable
b_{il} : *bandwidth*
n : number of observations
v_1 : $Tr(\mathbf{S})$
$\mathbf{S}$: a matrix that represents the estimated value of GTWBPIGR

$$S = \begin{pmatrix} \mathbf{x}_1 \left(\mathbf{X}^T \mathbf{W}(u_1, v_1) \mathbf{X}\right)^{-1} \mathbf{X}^T \mathbf{W}(u_1, v_1) \\ \mathbf{x}_2 \left(\mathbf{X}^T \mathbf{W}(u_2, v_2) \mathbf{X}\right)^{-1} \mathbf{X}^T \mathbf{W}(u_2, v_2) \\ \vdots \\ \mathbf{x}_n \left(\mathbf{X}^T \mathbf{W}(u_n, v_n) \mathbf{X}\right)^{-1} \mathbf{X}^T \mathbf{W}(u_n, v_n) \end{pmatrix}$$

The process of obtaining bandwidth that minimizes GCV value can be done using the golden section search technique [7].

Furthermore, for test statistics on the hypothesis testing simultaneously, the GTWBPIGR model is obtained with the Maximum Likelihood Ratio Test (MLRT) approach. Using n large samples of the statistical test is chi-square distribution. Moreover, the test statistics for partially testing used the Z-test statistic.

3. Results and Discussion

3.1 Parameter Estimation of GTWBPIGR Model

The parameter estimator of the GTWBPIGR model uses MLE. The presumed parameters $\boldsymbol{\beta}_{1l}(u_i,v_i,t_i)$, $\boldsymbol{\beta}_{2l}(u_i,v_i,t_i)$ and τ_l. The MLE method is carried out by taking the previous n random samples $(Y_{1il}, Y_{2il}, X_{1il}, X_{2il}, \dots, X_{pil})$ with $k = 1,2,\dots,p$ and $i = 1,2,\dots,n$. The form of the model has been mentioned in equation (4). The joint probability density function of Y_{1il} and Y_{2il} as follows:

$$f(y_{1il}, y_{2il} \mid \boldsymbol{\beta}_{1l}(u_i,v_i,t_i), \boldsymbol{\beta}_{2l}(u_i,v_i,t_i), \tau_l)$$

$$= e^{\frac{1}{\tau_l}} K_{sl}(z(u_i,v_i,t_i)) \left(\frac{2}{\pi\tau_l}\right)^{\frac{1}{2}} \left(1 + 2\tau_l \sum_{j=1}^{2} e^{\mathbf{x}_{il}^T \boldsymbol{\beta}_{jl}(u_i,v_i,t_i)}\right)^{-\frac{\left(2\sum_{j=1}^{2} y_{jil} - 1\right)}{4}} \prod_{j=1}^{2} \frac{e^{y_{jil}\mathbf{x}_{il}^T \boldsymbol{\beta}_{jl}(u_i,v_i,t_i)}}{y_{jil}!} \tag{5}$$

where:

$$sl = y_{1il} + y_{2il} - \frac{1}{2} \quad ; \quad \mu_{jil}(u_i,v_i,t_i) = e^{\mathbf{x}_{il}^T \boldsymbol{\beta}_{jL}(u_i,v_i,t_i)} \quad ; \quad z(u_i,v_i,t_i) = \sum_{l=1}^{L} \frac{1}{\tau_l} \left(1 + 2\tau_l \sum_{j=1}^{2} \mu_{jil}(u_i,v_i,t_i)\right)^{\frac{1}{2}}$$

$$K_{sl}(z(u_i,v_i,t_i)) = K_{y_{1il}+y_{2il}-\frac{1}{2}} \left[\sum_{l=1}^{L} \frac{1}{\tau_l} \left(1 + 2\tau_l \sum_{j=1}^{2} \mu_{jil}(u_i,v_i,t_i)\right)^{\frac{1}{2}} \right]$$

Parameter estimation using data from the previous period. Suppose that period L uses data information from n samples of period 1, n samples of period two until n samples of period L [8]. So that the likelihood function for the population in period L of equation (5) is

$$L(\boldsymbol{\beta}_{jL}(u_i,v_i,t_i); \tau_L; j = 1,2; i = 1,2,\dots,n; l = 1,2,\dots,L)$$

$$= \prod_{l=1}^{L} \prod_{i=1}^{n} \left(e^{\frac{1}{\tau_L}} K_{sl}(z(u_i,v_i,t_i)) \left(\frac{2}{\pi\tau_L}\right)^{\frac{1}{2}} \left(1 + 2\tau_L \sum_{j=1}^{2} e^{\mathbf{x}_{il}^T \boldsymbol{\beta}_{jL}(u_i,v_i,t_i)}\right)^{-\frac{\left(2\sum_{j=1}^{2} y_{jil} - 1\right)}{4}} \prod_{j=1}^{2} \frac{e^{y_{jil}\mathbf{x}_{il}^T \boldsymbol{\beta}_{jL}(u_i,v_i,t_i)}}{y_{jil}!} \right) \tag{6}$$

Next, the log-likelihood function for equation (6) is as follows:

$$Q_L = \log L(\boldsymbol{\beta}_{1L}(u_i,v_i,t_i); \boldsymbol{\beta}_{2L}(u_i,v_i,t_i); \tau_L; i = 1,2,\dots,n)$$

$$= \sum_{l=1}^{L} \sum_{i=1}^{n} \log \left(e^{\frac{1}{\tau_L}} K_{sl}(z(u_i,v_i,t_i)) \left(\frac{2}{\pi\tau_L}\right)^{\frac{1}{2}} \left(1 + 2\tau_L \sum_{j=1}^{2} e^{\mathbf{x}_{il}^T \boldsymbol{\beta}_{jL}(u_i,v_i,t_i)}\right)^{-\frac{\left(2\sum_{j=1}^{2} y_{jil} - 1\right)}{4}} \prod_{j=1}^{2} \frac{e^{y_{jil}\mathbf{x}_{il}^T \boldsymbol{\beta}_{jL}(u_i,v_i,t_i)}}{y_{jil}!} \right) \tag{7}$$

So we get the log-likelihood function to estimate the i^* location with w_{ii^*l} is spatial-temporal weight as follows:

$$Q_L^* = \log L(\boldsymbol{\beta}_{1L}(u_{i^*},v_{i^*},t_{i^*});\boldsymbol{\beta}_{2L}(u_{i^*},v_{i^*},t_{i^*});\tau_L;i^*=1,2,...,n)$$

$$\frac{1}{\tau_L}\sum_{l=1}^{L}\sum_{i=1}^{n} w_{ii^*l} + \sum_{l=1}^{L}\sum_{i=1}^{n} w_{ii^*l}\log K_{sl}(z(u_{i^*},v_{i^*},t_{i^*})) + \log\left(\frac{2}{\pi\tau_L}\right)^{\frac{1}{2}}\sum_{l=1}^{L}\sum_{i=1}^{n} w_{ii^*l} - \sum_{l=1}^{L}\sum_{i=1}^{n} w_{ii^*l}\log\left[\left[\frac{\left(2\sum_{j=1}^{2} y_{jil}-1\right)}{4}\right.\right.\times$$

$$\left.\left.\log\left(1+2\tau_L\sum_{l=1}^{L}\sum_{j=1}^{2} e^{\mathbf{x}_{il}^T\boldsymbol{\beta}_{jL}(u_{i^*},v_{i^*},t_{i^*})}\right)\right]\right] + \sum_{l=1}^{L}\sum_{i=1}^{n}\sum_{j=1}^{2} w_{ii^*l}\mathbf{x}_{il}^T\boldsymbol{\beta}_{jL}y_{jil}(u_{i^*},v_{i^*},t_{i^*}) - \sum_{l=1}^{L}\sum_{j=1}^{2}\sum_{i=1}^{n} w_{ii^*l}\log(y_{jil}!) \tag{8}$$

To obtain the parameter estimation of the GTWBPIGR model, the functions in the Q_L^* equation are derived respectively to $\boldsymbol{\beta}_{1L}(u_{i^*},v_{i^*},t_{i^*})$, $\boldsymbol{\beta}_{2L}(u_{i^*},v_{i^*},t_{i^*})$ and τ_L, as follows:

$$\frac{\partial Q_L^*}{\partial \boldsymbol{\beta}_{1L}(u_{i^*},v_{i^*},t_{i^*})} = \sum_{l=1}^{L}\sum_{i=1}^{n} w_{ii^*l}\frac{\partial \log K_{sl}(z(u_{i^*},v_{i^*},t_{i^*}))}{\partial \boldsymbol{\beta}_{1L}(u_{i^*},v_{i^*},t_{i^*})} + \sum_{l=1}^{L}\sum_{i=1}^{n} w_{ii^*l}2\tau_L e^{\mathbf{x}_{il}^T\boldsymbol{\beta}_{1L}(u_{i^*},v_{i^*},t_{i^*})}\mathbf{x}_{il}^T\left[-\frac{\left(2\sum_{j=1}^{2}y_{jil}-1\right)}{4}\right.$$

$$\left.\frac{1}{\left(1+2\tau_L\sum_{j=1}^{2}e^{\mathbf{x}_{il}^T\boldsymbol{\beta}_{1L}(u_{i^*},v_{i^*},t_{i^*})}\right)}\right] + \sum_{l=1}^{L}\sum_{i=1}^{n} w_{ii^*l}\mathbf{x}_{il}^T y_{1il} \tag{9}$$

with

$$\frac{\partial \log K_{sl}(z(u_{i^*},v_{i^*},t_{i^*}))}{\partial \boldsymbol{\beta}_{1L}(u_{i^*},v_{i^*},t_{i^*})} = \left[-\left(1+2\tau_L\sum_{j=1}^{2}\mu_{jlL}(u_{i^*},v_{i^*},t_{i^*})\right)^{\frac{1}{2}} M(y_{1il},y_{2il}) + \frac{\tau_L\left(y_{1il}+y_{2il}-\frac{1}{2}\right)}{\sqrt{1+2\tau_L\sum_{j=1}^{2}\mu_{jlL}(u_{i^*},v_{i^*},t_{i^*})}}\right]\frac{\mu_{2lL}(u_{i^*},v_{i^*},t_{i^*})\mathbf{x}_{il}^T}{\sqrt{1+2\tau_L\sum_{j=1}^{2}\mu_{jlL}(u_{i^*},v_{i^*},t_{i^*})}} \tag{10}$$

Then, substitution equation (10) to equation (9), and we get:

$$\frac{\partial Q_L^*}{\partial \boldsymbol{\beta}_{1L}(u_{i^*},v_{i^*},t_{i^*})} = \sum_{l=1}^{L}\sum_{i=1}^{n} w_{ii^*l}\left[-\left(1+2\tau_L\sum_{j=1}^{2}\mu_{jlL}(u_{i^*},v_{i^*},t_{i^*})\right)^{\frac{1}{2}} M(y_{1il},y_{2il}) + \frac{\tau_L\left(y_{1il}+y_{2il}-\frac{1}{2}\right)}{\sqrt{1+2\tau_L\sum_{j=1}^{2}\mu_{jlL}(u_{i^*},v_{i^*},t_{i^*})}}\right]\times$$

$$\frac{\mu_{1lL}(u_{i^*},v_{i^*},t_{i^*})\mathbf{x}_{il}^T}{\sqrt{1+2\tau_L\sum_{j=1}^{2}\mu_{jlL}(u_{i^*},v_{i^*},t_{i^*})}} - \frac{1}{2}\sum_{l=1}^{L}\sum_{i=1}^{n} w_{ii^*l}\left[\frac{\left(2\sum_{j=1}^{2}y_{jil}-1\right)\tau_L\mu_{1lL}(u_{i^*},v_{i^*},t_{i^*})\mathbf{x}_{il}^T}{\left(1+2\tau_L\sum_{j=1}^{2}\mu_{jlL}(u_{i^*},v_{i^*},t_{i^*})\right)}\right] + \sum_{l=1}^{L}\sum_{i=1}^{n} w_{ii^*l}\mathbf{x}_{il}^T y_{1il}$$

$$\frac{\partial Q_L^*}{\partial \boldsymbol{\beta}_{1L}(u_{i^*},v_{i^*},t_{i^*})} = \sum_{l=1}^{L}\sum_{i=1}^{n} w_{ii^*l}\mathbf{x}_{il}^T\left[y_{1il} - M(y_{1il},y_{2il})\mu_{1lL}(u_{i^*},v_{i^*},t_{i^*})\right] \tag{11}$$

With the same procedure, we get:

$$\frac{\partial Q_L^*}{\partial \boldsymbol{\beta}_{2L}(u_{i^*},v_{i^*},t_{i^*})} = \sum_{l=1}^{L}\sum_{i=1}^{n} w_{ii^*l}\mathbf{x}_{il}^T\left[y_{2il} - \mu_{2lL}(u_{i^*},v_{i^*},t_{i^*})M(y_{1il},y_{2il})\right] \tag{12}$$

Furthermore, the first derivative of the Q_L^* to parameter τ_L is formulated as below:

$$\frac{\partial Q_L^*}{\partial \tau_L}=\frac{\partial}{\partial \tau_L}\left[\frac{1}{\tau_L}\sum_{l=1}^{L}\sum_{i=1}^{n}w_{i_i^*l}+\sum_{l=1}^{L}\sum_{i=1}^{n}w_{i_i^*l}\log K_{sl}(z(u_{i^*},v_{i^*},t_{i^*}))+\log\left(\frac{2}{\pi\tau_L}\right)^{\frac{1}{2}}\sum_{l=1}^{L}\sum_{i=1}^{n}w_{i_i^*l}+\sum_{l=1}^{L}\sum_{i=1}^{n}w_{i_i^*l}\left[-\frac{\left(2\sum_{j=1}^{2}y_{jil}-1\right)}{4}\times\right.\right.$$

$$\left.\left.\log\left(1+2\tau_L\sum_{l=1}^{L}\sum_{j=1}^{2}e^{\mathbf{x}_{il}^T\boldsymbol{\beta}_{jL}(u_{i^*},v_{i^*},t_{i^*})}\right)\right]+\sum_{l=1}^{L}\sum_{i=1}^{n}\sum_{j=1}^{2}w_{i_i^*l}y_{jil}\mathbf{x}_{il}^T\boldsymbol{\beta}_{jL}(u_{i^*},v_{i^*},t_{i^*})-\sum_{l=1}^{L}\sum_{j=1}^{2}\sum_{i=1}^{n}w_{i_i^*l}\log(y_{jil}!)\right]$$

$$=-\frac{1}{\tau_L^2}\sum_{l=1}^{L}\sum_{i=1}^{n}w_{i_i^*l}-\frac{1}{2\tau_L}\sum_{l=1}^{L}\sum_{i=1}^{n}w_{i_i^*l}+\sum_{l=1}^{L}\sum_{i=1}^{n}w_{i_i^*l}\frac{\partial\log K_{sl}(z(u_{i^*},v_{i^*},t_{i^*}))}{\partial\tau_L}+\sum_{l=1}^{L}\sum_{i=1}^{n}w_{i_i^*l}\frac{\left(\frac{1}{2}-\sum_{j=1}^{2}y_{jil}\right)\sum_{j=1}^{2}\mu_{jiL}(u_{i^*},v_{i^*},t_{i^*})}{\left(1+2\tau_L\sum_{j=1}^{2}\mu_{jiL}(u_{i^*},v_{i^*},t_{i^*})\right)}$$

Where: $$\frac{\partial\log K_{sl}(z(u_{i^*},v_{i^*},t_{i^*}))}{\partial\tau_L}=\frac{\left(1+\tau_L\sum_{j=1}^{2}\mu_{jiL}(u_{i^*},v_{i^*},t_{i^*})\right)\left(\frac{1}{2}-\sum_{j=1}^{2}y_{jil}\right)}{\tau_L\left(1+2\tau_L\sum_{j=1}^{2}\mu_{jiL}(u_{i^*},v_{i^*},t_{i^*})\right)}+\frac{M(y_{1il},y_{2il})\left(1+\tau_L\sum_{j=1}^{2}\mu_{jiL}(u_{i^*},v_{i^*},t_{i^*})\right)}{\tau_L^2}$$

and we got:

$$\frac{\partial Q_L^*}{\partial\tau_L}=-\frac{1}{\tau_L^2}\sum_{l=1}^{L}\sum_{i=1}^{n}w_{i_i^*l}-\frac{1}{2\tau_L}\sum_{l=1}^{L}\sum_{i=1}^{n}w_{i_i^*l}+\sum_{l=1}^{L}\sum_{i=1}^{n}w_{i_i^*l}\frac{\partial\log K_{sl}(z(u_{i^*},v_{i^*},t_{i^*}))}{\partial\tau_L}$$

$$+\sum_{l=1}^{L}\sum_{i=1}^{n}w_{i_i^*l}\frac{\left(\frac{1}{2}-\sum_{j=1}^{2}y_{jil}\right)\sum_{j=1}^{2}\mu_{jiL}(u_{i^*},v_{i^*},t_{i^*})}{\left(1+2\tau_L\sum_{j=1}^{2}\mu_{jiL}(u_{i^*},v_{i^*},t_{i^*})\right)} \quad (13)$$

$$=\sum_{l=1}^{L}\sum_{i=1}^{n}w_{i_i^*l}\frac{M(y_{1il},y_{2il})\left(1+2\tau_L\sum_{j=1}^{2}\mu_{jiL}(u_{i^*},v_{i^*},t_{i^*})\right)-1}{\tau_L^2}-\sum_{l=1}^{L}\sum_{i=1}^{n}w_{i_i^*l}\frac{\left(\sum_{j=1}^{2}y_{jil}\right)}{\tau_L}$$

where: $M(y_{1il},y_{2il})=\dfrac{1}{\sqrt{1+2\tau_L\sum_{j=1}^{2}\mu_{jiL}(u_{i^*},v_{i^*},t_{i^*})}}\dfrac{K_{y_{1il}+y_{2il}+\frac{1}{2}}(z(u_{i^*},v_{i^*},t_{i^*}))}{K_{y_{1il}+y_{2il}-\frac{1}{2}}(z(u_{i^*},v_{i^*},t_{i^*}))}$, then

Equations (11)–(13) equate to zero and produce a non-explicit form. Hence, an iterative method needs to be applied for estimating parameters, so Newton Raphson's iteration will solve it. So that the Newton-Raphson iteration algorithm is as follows:

Step 1. Determine the initial value for the parameters

$\boldsymbol{\theta}_L^{(0)}(u_{i^*},v_{i^*},t_{i^*})=\left[\boldsymbol{\beta}_{1L}^{(0)}(u_{i^*},v_{i^*},t_{i^*})\quad\boldsymbol{\beta}_{2L}^{(0)}(u_{i^*},v_{i^*},t_{i^*})\quad\tau_L^{(0)}\right]^T$. The initial value of parameters $\boldsymbol{\theta}_L^{(0)}$ is using univariate Poisson regression. The initial value τ_L used the average of the observed overdispersion based on the variance of PIGD.

Step 2. Determine the gradient vector $\mathbf{g}(\boldsymbol{\theta}_L^{(m)}(u_{i^*},v_{i^*},t_{i^*}))$, which is the elements that consist of the first derivative of the log-likelihood function,

$$g(\boldsymbol{\theta}^{(m)}{}_L(u_{i^*},v_{i^*},t_{i^*}))=\begin{bmatrix}\dfrac{\partial Q_L^*}{\partial\boldsymbol{\beta}_{1L}^T(u_{i^*},v_{i^*},t_{i^*})} & \dfrac{\partial Q_L^*}{\partial\boldsymbol{\beta}_{2L}^T(u_{i^*},v_{i^*},t_{i^*})} & \dfrac{\partial Q_L^*}{\partial\tau_L^T}\end{bmatrix}^T$$

Step 3. Determine the Hessian matrix $\mathbf{H}(\boldsymbol{\theta}_L^{(m)}(u_{i^*},v_{i^*},t_{i^*}))$, which is the elements consist of the second derivative of the log-likelihood function, as follows

$$\mathbf{H}(\boldsymbol{\theta}_L^{(m)}(u_{i^*},v_{i^*},t_{i^*}))=\begin{bmatrix}\dfrac{\partial^2 Q_L^*}{\partial\boldsymbol{\beta}_{1L}(u_{i^*},v_{i^*},t_{i^*})\partial\boldsymbol{\beta}_{1L}^T(u_{i^*},v_{i^*},t_{i^*})} & \dfrac{\partial^2 Q_L^*}{\partial\boldsymbol{\beta}_{1L}(u_{i^*},v_{i^*},t_{i^*})\partial\boldsymbol{\beta}_{2L}^T(u_{i^*},v_{i^*},t_{i^*})} & \dfrac{\partial^2 Q_L^*}{\partial\boldsymbol{\beta}_{1L}(u_{i^*},v_{i^*},t_{i^*})\partial\tau_L} \\ & \dfrac{\partial^2 Q_L^*}{\partial\boldsymbol{\beta}_{2L}(u_{i^*},v_{i^*},t_{i^*})\partial\boldsymbol{\beta}_{2L}^T(u_{i^*},v_{i^*},t_{i^*})} & \dfrac{\partial^2 Q_L^*}{\partial\boldsymbol{\beta}_{2L}(u_{i^*},v_{i^*},t_{i^*})\partial\tau_L} \\ \textit{simetris} & & \dfrac{\partial^2 Q_L^*}{\partial\tau_L^2}\end{bmatrix}$$

Step 4. Start working on the Newton–Raphson iteration using the following equation:

$$\boldsymbol{\theta}_L^{(m+1)}(u_{i^*},v_{i^*},t_{i^*})=\boldsymbol{\theta}_L^{(m)}(u_{i^*},v_{i^*},t_{i^*})-\mathbf{H}^{-1}(\boldsymbol{\theta}_L^{(m)}(u_{i^*},v_{i^*},t_{i^*}))\mathbf{g}(\boldsymbol{\theta}_L^{(m)}(u_{i^*},v_{i^*},t_{i^*}))$$

with $\boldsymbol{\theta}_L(u_{i^*},v_{i^*},t_{i^*})=\begin{bmatrix}\boldsymbol{\beta}_{1L}(u_{i^*},v_{i^*},t_{i^*}) & \boldsymbol{\beta}_{2L}(u_{i^*},v_{i^*},t_{i^*}) & \tau_L\end{bmatrix}^T$ and $m=1,2,...,m^*$

Step 5. The iteration will stop when $\left\|\boldsymbol{\theta}_L^{(m+1)}(u_{i^*},v_{i^*},t_{i^*})-\boldsymbol{\theta}_L^{(m)}(u_{i^*},v_{i^*},t_{i^*})\right\|\le\varepsilon$ with ε is a very small value and will produce the estimator value for each parameter.

3.2 Hypothesis Testing of GTWBPIGR Model

The hypothesis testing of the GTWBPIGR model using the Maximum Likelihood Ratio Test (MLRT) method both simultaneously and partially [9]. Hypothesis testing is carried out in stages each period as the parameter estimates, with an L's period $L=1,2,\ldots,L^*$. This method involves two likelihood functions: $L(\Omega_L)$ is the maximum likelihood value for a model involving predictor variables and $L(\omega_L)$ is the maximum likelihood value for a simple model without involving predictor variables.

Simultaneously hypothesis testing is performed to determine the significance of the regression parameters in the model with the following hypothesis:

$H_0:\beta_{j1L}(u_i,v_i,t_i)=\beta_{j2L}(u_i,v_i,t_i)=\ldots=\beta_{jpL}(u_i,v_i,t_i)=0$; and the alternative hypothesis is at least one $H_1:\beta_{jpL}(u_i,v_i,t_i)\neq 0; j=1,2; i=1,2,...,n; k=1,2,\ldots,p$. Where k is the number of predictor variables.

Let Ω_L is a set of parameters under population with $\Omega_L=\{\boldsymbol{\beta}_{1L}(u_i,v_i,t_i);\boldsymbol{\beta}_{2L}(u_i,v_i,t_i);\tau_L;i=1,2,...,n\}$

The likelihood function $L(\Omega_L)$ for each model, as follows:

$$L(\Omega_L)=\prod_{l=1}^{L}\prod_{i=1}^{n}f(y_{1il},y_{2il}\,|\boldsymbol{\beta}_{1l}(u_i,v_i,t_i),\boldsymbol{\beta}_{2l}(u_i,v_i,t_i),\tau_l)$$

And then, let ω_L as a set of parameters under the null hypothesis $\omega_L=\left\{\beta_{10\omega_L}(u_i,v_i,t_i),\beta_{20\omega_L}(u_i,v_i,t_i),\tau_{\omega_L};i=1,2,...,n\right\}$.

The likelihood function $L(\omega_L)$ for each model, as follows:

$$L(\omega_L)=\prod_{l=1}^{L}\prod_{i=1}^{n}\left(e^{\frac{1}{\tau_{\omega_L}}}K_{sl}(z_{\omega_L}(u_i,v_i,t_i))\left(\frac{2}{\pi\tau_{\omega_L}}\right)^{\frac{1}{2}}\left(1+2\tau_{\omega_L}\sum_{j=1}^{2}e^{\beta_{j0\omega_L}(u_i,v_i,t_i)}\right)^{-\frac{\left(2\sum_{j=1}^{2}y_{jil}-1\right)}{4}}\prod_{j=1}^{2}\frac{e^{\beta_{j0\omega_L}(u_i,v_i,t_i)}}{y_{jil}!}\right)$$

So, we get the log-likelihood function under the null hypothesis, $l(\omega_L)$ with weighted spatial-temporal $w_{ii^*l\_}$ as follows:

$$l(\omega_L)=\frac{1}{\tau_{\omega_L}}\sum_{l=1}^{L}\sum_{i=1}^{n}w_{ii^*l}+\sum_{l=1}^{L}\sum_{i=1}^{n}w_{ii^*l}\log K_{sl}(z_{\omega_L}(u_{i^*},v_{i^*},t_{i^*}))+\frac{n}{2}\sum_{l=1}^{L}\sum_{i=1}^{n}w_{ii^*l}\log\left(\frac{2}{\pi\tau_{\omega_L}}\right)-\sum_{l=1}^{L}\sum_{i=1}^{n}w_{ii^*l}\frac{\left(2\sum_{j=1}^{2}y_{jil}-1\right)}{4}\times$$
$$\log\left(1+2\tau_{\omega_L}\sum_{j=1}^{2}e^{\beta_{j0\omega_L}(u_{i^*},v_{i^*},t_{i^*})}\right)+\sum_{l=1}^{L}\sum_{i=1}^{n}\sum_{j=1}^{2}w_{ii^*l}y_{jil}\beta_{j0\omega_L}(u_{i^*},v_{i^*},t_{i^*})-\sum_{l=1}^{L}\sum_{i=1}^{n}\sum_{j=1}^{2}w_{ii^*l}\log(y_{jil}!)$$

Then, derived $l(\omega_L)$ to the parameters $\beta_{10\omega_L}(u_{i^*},v_{i^*},t_{i^*})$, $\beta_{20\omega_L}(u_{i^*},v_{i^*},t_{i^*})$ and τ_{ω_L}, as below:

$$\frac{\partial l(\omega_L)}{\partial\beta_{10L_\omega}(u_{i^*},v_{i^*},t_{i^*})}=\sum_{l=1}^{L}\sum_{i=1}^{n}w_{ii^*l}\left[y_{1il}-M(y_{1il},y_{2il})e^{\beta_{10\omega_L}(u_{i^*},v_{i^*},t_{i^*})}\right]$$

$$\frac{\partial l(\omega_L)}{\partial\beta_{20\omega_L}(u_{i^*},v_{i^*},t_{i^*})}=\sum_{l=1}^{L}\sum_{i=1}^{n}w_{ii^*l}\left[y_{2il}-M(y_{1il},y_{2il})e^{\beta_{20\omega_L}(u_{i^*},v_{i^*},t_{i^*})}\right]$$

$$\frac{\partial l(\omega_L)}{\partial\tau_{\omega_L}}=\sum_{l=1}^{L}\sum_{i=1}^{n}w_{ii^*l}\left(\frac{M(y_{1il},y_{2il})\left(1+\tau_{\omega_L}\sum_{j=1}^{2}e^{\beta_{j0\omega_L}(u_{i^*},v_{i^*},t_{i^*})}\right)-1}{\tau_{\omega_L}^2}\right)-\sum_{l=1}^{L}\sum_{i=1}^{n}\sum_{j=1}^{2}w_{ii^*l}\frac{y_{jil}}{\tau_{\omega_L}}$$

Next, determine the odds ratio $\Lambda_L=\dfrac{L(\hat{\omega}_L)}{L(\hat{\Omega}_L)}$.

The decision rejects H_0 if $\Lambda_L<\Lambda_{L0}$, where $\Lambda_L<\Lambda_{L0}<1$, and $\alpha=P(\Lambda_L<\Lambda_{L0}\mid H_0\text{ is true})$.
So for decision making, test statistics are used:

$$\begin{aligned}G_L^2&=-\log\Lambda_L^2\\&=2\left[nL\left(\frac{1}{\hat{\tau}_L}-\frac{1}{\hat{\tau}_{\omega_L}}-\frac{1}{2}\ln\left(\frac{\hat{\tau}_{\omega_L}}{\hat{\tau}_L}\right)\right)+\sum_{l=1}^{L}\sum_{i=1}^{n}\left(\frac{\ln K_{sl}(z(u_{i^*},v_{i^*},t_{i^*}))}{\ln K_{sl}(z_{\omega_L}(u_i,v_i,t_i))}\right)\right.\\&\quad+\sum_{l=1}^{L}\sum_{i=1}^{n}\frac{\left(2\sum_{j=1}^{2}y_{jil}-1\right)}{4}\ln\left(\frac{1+2\hat{\tau}_{\omega_L}\sum_{j=1}^{2}e^{\beta_{j0\omega_L}(u_i,v_i,t_i)}}{1+2\hat{\tau}_L\sum_{l=1}^{L}\sum_{j=1}^{2}e^{\mathbf{x}_{il}^T\boldsymbol{\beta}_{jL}(u_{i^*},v_{i^*},t_{i^*})}}\right)\\&\quad+\sum_{l=1}^{L}\sum_{i=1}^{n}\sum_{j=1}^{2}y_{jil}\left(\mathbf{x}_{il}^T\widehat{\boldsymbol{\beta}_{jL}(u_{i^*},v_{i^*},t_{i^*})}-\hat{\beta}_{j0\omega_L}(u_i,v_i,t_i)\right)\end{aligned}\quad(14)$$

Based on steps carried out by Magdalena et al. [10], it was found that test statistic G_L^2 approximates the χ^2 distribution with v degrees of freedom where v is the number of parameters under population minus the number of parameters below H_0. The H_0 rejected criterion is $G_L^2>\chi^2_{\alpha,v}$.

Suppose the results of hypothesis testing decide to reject H_0. In that case, the conclusion obtained is that the predictor variables affect the response variable and a degree of error of 5%.

If the simultaneous testing decision is rejected H_0, the next step is to partially test the parameters to determine which parameters significantly affect the model. The test statistic Z approximates the Normal Standard distribution [11]. The hypothesis used is: the null hypothesis is $\beta_{jkL}(u_i, v_i, t_i) = 0$ vs $H_1: \beta_{jkL}(u_i, v_i, t_i) \neq 0$ with $j = 1,2$, $L = 1,2, \dots, L^*$, and $k = 1,2, \dots, p$

The test statistics used is

$$Z = \frac{\hat{\beta}_{jkL}}{se(\hat{\beta}_{jkL})} \tag{15}$$

$se(\hat{\beta}_{jkL}(u_i, v_i, t_i)) = \sqrt{\widehat{var}(\hat{\beta}_{jkL}(u_i, v_i, t_i))}$ where the $\widehat{var}(\hat{\beta}_{jkL}(u_i, v_i, t_i))$ value is obtained from the principal diagonal element of the covariance variant matrix of the following equation

$$\widehat{Var}(\hat{\theta}_L) = -\mathbf{H}^{-1}(\hat{\theta}_L)$$

.

The area of rejection H_0 is $|Z| > Z_{\frac{\alpha}{2}}$ with α is the level of significance.

Furthermore, to get partial hypothesis testing on the following τ_L parameters, the hypothesis used is:

$H_0: \tau_L = 0$

$H_1: \tau_L \neq 0$

The test statistical used is:

$$Z = \frac{\hat{\tau}_L}{se(\hat{\tau}_L)} \tag{16}$$

$se(\hat{\tau}_L) = \sqrt{\widehat{var}(\hat{\tau}_L)}$ where the $\widehat{var}(\hat{\tau}_L)$ value is obtained from the principal diagonal element of the covariance variant matrix of the following equation.

The area of rejection is the level of significance.

$$\widehat{Var}(\hat{\theta}_L) = -\mathbf{H}^{-1}(\hat{\theta}_L)$$

The area of rejection H_0 is $|Z| > Z_{\frac{\alpha}{2}}$ with α is the level of significance.

4. Conclusion

We developed a global regression to accommodate spatial and temporal aspects, called a geographically and temporally weighted bivariate Poisson inverse Gaussian regression (GTWBPIGR). We have shown the step-by-step procedure to estimate the parameters and test statistics for the GTWBPIGR model hypothesis testing. Parameter estimation obtains by using the Maximum Likelihood Estimation (MLE) method. The equation that gets is not closed-form; it continues with the Newton-Raphson algorithm. Meanwhile, hypothesis testing of GTWBPIGR model by using Maximum Likelihood Ratio Test (MLRT) method.

References

[1] A. C. Cameron and P. K. Trivedi 1998 Regression Analysis of Count Data, USA: Cambridge University Press.
[2] D Karlis, & I. Ntzoufras 2005 Bivariate Poisson and diagonal inflated bivariate Poisson regression models in R *Journal of statistical software* 14 pp. 1-36
[3] J M Hilbe 2014 Modelling count data Cambridge University Press New York
[4] M S Holla 1966 On a Poisson inverse Gaussian distribution *Metrik* 11 pp 115-121
[5] Huang, B, Wu, B, & Barry, M 2010 Geographically and temporally weighted regression for modelling Spatio-temporal variation in house prices *International Journal of Geographical Information Science* **24** (3) 383-401

[6] G E Willmot 1987 The Poisson-Inverse Gaussian Distribution as An Alternative to The Negative Binomial," Scandinavian *Actuarial Journal* **34** pp. 113-127
[7] Fotheringham A S, Brunsdon C, dan Charlton, M 2002 Geographically Weighted Regression: The Analysis of Spatially Varying Relationship *Willey* Chichester.
[8] Fotheringham, A S, Crespo, R, & Yao, J 2015 Geographical and temporal weighted regression (GTWR) *Geographical Analysis* **47** (4) 431-452
[9] N R Draper and R Smith, 1992 Applied Regression Analysis 2nd Editions New York: *John Wiley and Sons Inc*
[10] Triyanto, Purhadi, Otok, B W, & Purnami, S W 2015 Parameter estimation of geographically weighted multivariate Poisson regression. *Appl. Math. Sci* 9 4081-4093

Acknowledgements
Authors wishing to acknowledge assistance or encouragement from lecturers and college friends at the Sepuluh Nopember Institute of Technology (Institut Teknologi Sepuluh Nopember) about this research.

Mixed estimator of spline truncated, Fourier series, and kernel in biresponse semiparametric regression model

Hartina Husain[1], I N Budiantara[1]* & Ismaini Zain[1]

[1]Department of Statistics, Faculty of Science and Data Analytics, Institut Teknologi Sepuluh November, 60111 Surabaya, Indonesia

*Corresponding E-mail: nyomanbudiantara65@gmail.com

Abstract. Regression analysis is a method of analysis to determine the relationship between the response and the predictor variables. There are three approaches in regression analysis, namely the parametric, nonparametric, and semiparametric approaches. Biresponse Semiparametric regression model is a regression model that uses a combination approach between parametric and nonparametric components, where two response variables are correlated with each other. For data cases with several predictor variables, different estimation technique approaches can be used for each variable. In this study, the parametric component is assumed to be linear. At the same time, the nonparametric part is approached using a mixture of three estimation techniques, namely, spline truncated, Fourier series, and the kernel. The unknown data pattern is assumed to follow the criteria of each of these estimation techniques. The spline is used when the data pattern tends to change at certain time intervals, the Fourier series is used when the data pattern tends to repeat itself, and the kernel is used when the data does not have a specific way. This study aims to obtain parameter estimates for the mixed semiparametric regression model of spline truncated, Fourier series, and the kernel on the biresponse data using the Weighted Least Square (WLS) method. The formed model depends on the selection of knot points, oscillation parameters, and optimal bandwidth. The best model is based on the smallest Generalized Cross Validation (GCV).

Keywords: Biresponse, Fourier, Kernel, Semiparametric Regression, Spline Truncated

Track Name: Advanced Technology and Renewable Energy

1. Introduction

Regression analysis is a statistical method to determine the closeness of the relationship between the response variable and one or more predictor variables. This analysis aims to estimate the population average or the average value of the response variable based on the value of the independent variable. Three approaches are often used in regression analysis, namely the parametric, nonparametric, and semiparametric approaches. If the data pattern is known as linear, quadratic, or cubic, then the regression method used is the parametric regression method [1]. However, in reality, it is found that not all data follow certain patterns. If the relationship between the predictor variables and the response is unknown, then the nonparametric regression method is suitable for modeling the relationship between these variables [2]. Meanwhile, if the data pattern consists of parametric and nonparametric components, a semiparametric regression approach is used.

The use of semiparametric regression each year is increasingly in demand in data modeling because of its flexible and objective nature. The nonparametric component of semiparametric regression can be approached by various estimation techniques such as spline, Fourier series, and kernel [3]. Several previous studies regarding spline estimators have been conducted by other researchers [2, 4, 5], and [6]. Several studies for the Fourier series estimator were conducted [7-11]. Likewise, studies for the kernel estimators have been carried out [12-14].

Spline approach has the advantage of very good visual and statistical interpretation and has high flexibility, so it is good at handling smooth data/functions [3]. One type of basic function in spline that is often used is spline truncated. This approach can describe changes in the behavior of different curves at different intervals. The results in a better level of flexibility can adjust more effectively to local characteristics in data. Meanwhile, Fourier series is a series in the sinusoidal form (sine and cosine) used to represent periodic functions in general. One of the advantages of the nonparametric regression approach using the Fourier Series is that it can handle data that has repetitive patterns, namely the repetition of the response variable values for different predictor variables. Furthermore, kernel regression is also more flexible, easy mathematical form, and can achieve a relatively fast convergence rate [15]. The selection of smoothing parameters (bandwidth) is considered more important than selecting the kernel function.

In general, studies on semiparametric and nonparametric regression models only use the same estimation method for some or even all of the predictor variables. However, sometimes, it is found that data have different data patterns for each predictor variable. Therefore, to overcome these different data patterns, some researchers have developed mixed methods to estimate them [16-18]. Previous research on semiparametric regression on biresponse data has been carried out with the Fourier series estimator approach [19] and using the mixed estimator approach Fourier series and Spline truncated [20].

Previous studies only used a mixture of two estimators in semiparametric regression. However, there may be three or more data patterns that different estimators can approach. Therefore, this study will use a semiparametric regression model using the spline truncated estimator approach, the Fourier series, and the kernel, which will be applied to data with two responses (biresponse). Based on the formulation of the problem, this study examines the semiparametric regression estimator form of a mixture of Spline Truncated, Fourier Series, and Kernel Regression.

2. Materials and Methods

2.1 Semiparametric Regression Model

Semiparametric regression is a combination of parametric components and nonparametric components. Suppose paired data (x_i, t_i, y_i) and the relationship between x_i, y_i and t_i are assumed to follow a semiparametric regression model.

$$y_i = f(x_i) + g(t_i) + \varepsilon_i, i = 1,2,\dots,n. \quad (1)$$

y_i is the response variable, x_i and t_i are predictor variables, and ε_i is an independent random error and is normally distributed with mean zero and variance σ^2. $f(x_i)$ is a regression function a known pattern form (parametric component). In contrast, $g(t_i)$ is a regression function with no known pattern (nonparametric component). There are several nonparametric component approaches such as spline truncated, Fourier series, and kernels. This spline method is very good at modeling data whose patterns change at certain subintervals [3]. Fourier series is a trigonometric polynomial function with a degree of flexibility in handling data that has repeating patterns [21]. In contrast, the kernel estimator is used for data that does not follow a specific way [22]. Some of the advantages of kernel estimators are flexibility, easy mathematical form, and a relatively fast convergence rate [2].

If the regression curve $g(t_i)$ in equation (1) is approximated by spline truncated function of degree p with the points of knots $K_1, K_2, \dots, K_m$ $(K_1 \le K_2 \le \dots \le K_m)$ given by the equation :

IOP Conf. Series: Earth and Environmental Science **880** (2021) 012046 doi:10.1088/1755-1315/880/1/012046

$$g(t_i) = \sum_{j=0}^{p} \xi_j t_i^j + \sum_{k=1}^{m} \Phi_k (t_i - K_k)_+^p \tag{2}$$

The truncated function is given by :

$$(t_i - K_k)_+^p = \begin{cases} (t_i - K_k)_+^p & ,t \geq K_k \\ 0 & ,t < K_k \end{cases}$$

Where $\xi_0, \xi_1, \cdots, \xi_p, \Phi_1, \ldots, \Phi_m$ are unknown parameters , $i = 1,2, \ldots, n$ denotes the amount of data and K_k with $k = 1,2, \cdots, m$ is denotes k-knots.

The nonparametric component approach using Fourier series is to use the function :

$$h(z_i) = bz_i + \frac{\alpha_0}{2} + \sum_{l=1}^{L} (\alpha_l \cos lz_i) \tag{3}$$

Where , α_0, α_l , $l = 1,2, \ldots, L$ are model parameters.

Furthermore, if the kernel regression curve approaches the nonparametric component, the Nadaraya-Watson kernel estimator is used, namely.

$$\hat{q}_\phi(v) = n^{-1} \sum_{i=1}^{n} \frac{K\left(\frac{v - v_i}{\phi}\right)}{\sum_{i=1}^{n} K\left(\frac{v - v_i}{\phi}\right)} y_i,$$

$$= n^{-1} \sum_{i=1}^{n} R_{\phi i}(v) y_i \tag{4}$$

Where $\hat{q}_\phi(v)$ is the kernel regression estimation function, ϕ is bandwidth and function $R_{\phi i}(v)$ is the weighting function. One of the K kernel functions that can be used is the following Gaussian Kernel :

$$K(u) = \frac{1}{\sqrt{2\pi}} exp\left(-\frac{1}{2}u^2\right), -\infty < u < \infty \tag{5}$$

2.2 Biresponse Semiparametric Regression

Biresponse semiparametric regression is a regression model that aims to obtain a relationship between two response variables and their predictor variables. Given a biresponse semiparametric regression model for i = 1,2,…, n as follows:

$$y_{1i} = \beta_{01} + \beta_{11}x_{1i} + \cdots + \beta_{p1}x_{pi} + g_1(t_i) + \varepsilon_{1i}$$

$$y_{2i} = \beta_{02} + \beta_{12}x_{1i} + \cdots + \beta_{p2}x_{pi} + g_2(t_i) + \varepsilon_{2i} \tag{6}$$

y_{1i} and y_{2i} are the response variables , $g_1(t_i)$ and $g_2(t_i)$ are nonparametric regression functions on response one and response two, respectively.

The errors in the two regression models in equation (6), namely ε_{1i} and ε_{2i} are assumed to fulfill the following conditions :

$$E(\varepsilon_{1i}) = E(\varepsilon_{2i}) = 0, \qquad Cov(\varepsilon_{1i}, \varepsilon_{1i'}) = \begin{cases} \sigma_{11}, & i = i' \\ 0, & i \neq i' \end{cases}$$

$$Cov(\varepsilon_{2i}, \varepsilon_{2i'}) = \begin{cases} \sigma_{22}, & i = i' \\ 0, & i \neq i' \end{cases} \qquad Cov(\varepsilon_{1i}, \varepsilon_{2i'}) = \begin{cases} \sigma_{12}, & i = i' \\ 0, & i \neq i' \end{cases}$$

2.3 Methods

To get an estimation of the biresponse semiparametric regression model for the mixture of Spline Truncated, Fourier Series, and Kernel, use the following steps.

1. Given the response variable Y_1 and Y_2 with parametric component variable x , and nonparametric component variables t, z, v. Based on equation (6), it can be written as :

$$y_{ji} = f_j(x_i) + g_j(t_i) + h_j(z_i) + q_j(v_{si}) + \varepsilon_{ji} \quad (7)$$

where $i = 1,2,\cdots,n$ and $j = 1,2$.

2. Approaching the function $f_j(x_i)$ with a linear regression function
3. Approaching the function $g_j(t_i)$ with the spline truncated with M knots
4. Approaching the function $h_j(z_i)$ with the fourier series function
5. Approaching the function $q_j(v_i)$ with the kernel function
6. Expressing the $\tilde{f}(x), \tilde{g}(t), \tilde{h}(z)$, and $\tilde{q}(v)$ curves in the form of a matrix where $\tilde{\beta}$ is a parameter for linear regression, $\tilde{\xi}$ spline truncated regression parameter, and $\tilde{\alpha}$ fourier series regression parameter.
7. Converting mixed of spline truncated, fourier series, and kernel in biresponse semiparametric regression model in the form of a matrix.
8. Solving optimization using Weighted Least Square method
9. Equating the partial derivative with zero.
10. Getting the mix estimation

3. Results and Discussion

3.1 Mixed of Spline Truncated, Fourier Series, And Kernel in Biresponse Semiparametric Regression Model

Given data with response variable Y_1 and Y_2 and predictor variable consisting of a parametric component and a nonparametric component. Parametric component is x and nonparametric component variables which are approached with spline truncated, fourier series and kernel, namely t, z, v . The biresponse semiparametric regression model containing these variables can be stated by equation (7). While i states the amount of data, and j is the number of responses with i = 1, 2, ⋯, n and j = 1, 2. Meanwhile, ε_{ji} is assumed to be independent, identical, and normally distributed with mean zero and variance σ_{ji}^2.

Based on equation (7), it can be seen that this equation is a biresponse semiparametric function consisting of four constituent components. The first component, the function f, is a parametric component that is assumed to follow a linear pattern. Thus, the function is approximated by using linear regression, namely

$$f_j(x_i) = \beta_{0j} + \beta_j x_i \quad (8)$$

Meanwhile, the second component, the function g, is a nonparametric component approximated by linear spline truncated function. The relationship between the response variable y_j and the predictor variables (t_i) is assumed to fluctuate in certain sub-intervals. The linear spline truncated regression curve can be defined as follows :

$$g_j(t_i) = \xi_j t_i + \sum_{k=1}^{M} \Phi_{jk}\left(t_i - K_{jk}\right)_+ \quad (9)$$

The truncated function is given by :

$$\left(t_i - K_{jk}\right)_+ = \begin{cases} \left(t_i - K_{jk}\right)_+ & , t_i \geq K_{jk} \\ 0 & , t_i < K_{jk} \end{cases}$$

Furthermore, the third component, the function h, is a nonparametric component that is approximated by the Fourier series function. It is assumed that the relationship between the response variable y_j and the predictor variables (z_i) has a recurring pattern or forms a seasonality. The Fourier series regression curve can be defined as follows :

$$h_j(z_i) = b_j z_i + \frac{\alpha_{0j}}{2} + \sum_{l=1}^{L} \alpha_{lj} \cos l z_i \tag{10}$$

Then the fourth component, namely the q function, is a nonparametric component approached by the Kernel function. The pattern of the relationship between the response variable y_j and the predictor variables (v_i) is assumed to have no known way. In this study, the Nadaraya Watson kernel function is used, which is defined as follows:

$$q_j(v_i) = n^{-1} \sum_{i=1}^{n} \frac{K\left(\frac{v_j - v_i}{\phi_j}\right)}{n^{-1} \sum_{i=1}^{n} K\left(\frac{v_j - v_i}{\phi_j}\right)} y_{ji}$$

$$= n^{-1} \sum_{i=1}^{n} R_{\phi i}(v)\, y_{ji} \tag{11}$$

$q_j(v_i)$ is an estimator of the Nadaraya-Watson kernel regression curve, and parameter ϕ represents the bandwidth parameter, and the K function is a kernel function. As for this study, the Gaussian kernel function is used as in equation (5) so that:

$$K\left(\frac{v_j - v_i}{\phi_j}\right) = \frac{1}{\sqrt{2\pi}} exp\left(-\frac{(v_j - v_i)^2}{2\phi_j^{\ 2}}\right)$$

3.2 Mixed Estimator Of Spline Truncated, Fourier Series, And Kernel In Biresponse Semiparametric Regression Model

Before estimating the model, equation (7) components are first expressed in the form of a matrix. The first part is for the parametric component in equation (8), for response $j = 1,2$ and $i = 1,2, \dots, n$ model can be expressed in the form of a matrix to be:

$$\begin{pmatrix} f_1(x_1) \\ f_1(x_2) \\ \vdots \\ f_1(x_n) \\ \hline f_2(x_1) \\ f_2(x_2) \\ \vdots \\ f_2(x_n) \end{pmatrix} = \left(\begin{array}{cc|cc} 1 & x_1 & 0 & 0 \\ 1 & x_2 & 0 & 0 \\ \vdots & \vdots & \vdots & \vdots \\ 1 & x_n & 0 & 0 \\ \hline 0 & 0 & 1 & x_1 \\ 0 & 0 & 1 & x_2 \\ \vdots & \vdots & \vdots & \vdots \\ 0 & 0 & 1 & x_n \end{array}\right) \begin{pmatrix} \beta_{01} \\ \beta_1 \\ \hline \beta_{02} \\ \beta_2 \end{pmatrix}$$

Furthermore

$$\begin{pmatrix} \tilde{f}_1(x) \\ \tilde{f}_2(x) \end{pmatrix} = \begin{pmatrix} \boldsymbol{X_1} & \boldsymbol{0} \\ \boldsymbol{0} & \boldsymbol{X_2} \end{pmatrix} \begin{pmatrix} \tilde{\beta}_1 \\ \tilde{\beta}_2 \end{pmatrix}$$

in matrix notation can be written as:

$$\tilde{f}(x) = \boldsymbol{X}\tilde{\beta} \tag{12}$$

With vector $\tilde{f}(x)$ is sized $2n \times 1$, Matrix **X** is sized $2n \times 4$ and vector $\tilde{\beta}$ is sized 4×1.

Furthermore, the nonparametric component which is approached by Spline Truncated in equation (9), for response $j = 1,2$ and $i = 1,2,\dots,n$ model can be expressed in the form of a matrix to be:

$$\begin{pmatrix} g_1(t_1) \\ g_1(t_2) \\ \vdots \\ g_1(t_n) \\ g_2(t_1) \\ g_2(t_2) \\ \vdots \\ g_2(t_n) \end{pmatrix} = \left(\begin{array}{cccc|cccc} t_1 & (t_1 - K_{11})_+ & \cdots & (t_1 - K_{1M})_+ & 0 & 0 & \cdots & 0 \\ t_2 & (t_2 - K_{11})_+ & \cdots & (t_2 - K_{1M})_+ & 0 & 0 & \cdots & 0 \\ \vdots & \vdots & \ddots & \vdots & \vdots & \vdots & \ddots & 0 \\ t_n & (t_n - K_{11})_+ & \cdots & (t_n - K_{1M})_+ & 0 & 0 & \cdots & 0 \\ \hline 0 & 0 & \cdots & 0 & t_1 & (t_1 - K_{21})_+ & \cdots & (t_1 - K_{2M})_+ \\ 0 & 0 & \cdots & 0 & t_2 & (t_2 - K_{21})_+ & \cdots & (t_2 - K_{2M})_+ \\ \vdots & \vdots & \ddots & \vdots & \vdots & \vdots & \ddots & \vdots \\ 0 & 0 & \cdots & 0 & t_n & (t_n - K_{21})_+ & \cdots & (t_n - K_{2M})_+ \end{array}\right) \begin{pmatrix} \xi_1 \\ \Phi_{11} \\ \vdots \\ \Phi_{1M} \\ \xi_2 \\ \Phi_{21} \\ \vdots \\ \Phi_{2M} \end{pmatrix}$$

Furthermore, it can be written as:

$$\begin{pmatrix} \tilde{g}_1(t) \\ \tilde{g}_2(t) \end{pmatrix} = \begin{pmatrix} \boldsymbol{G_1} & \boldsymbol{0} \\ \boldsymbol{0} & \boldsymbol{G_2} \end{pmatrix} \begin{pmatrix} \tilde{\xi}_1 \\ \tilde{\xi}_2 \end{pmatrix}$$

in matrix notation can be written as:

$$\tilde{g}(t) = \boldsymbol{G}\tilde{\xi} \tag{13}$$

With vector $\tilde{g}(t)$ is sized $2n \times 1$, Matrix $\boldsymbol{G}$ is sized $2n \times (2M + 2)$, vector $\tilde{\xi}$ is sized $(2M + 2) \times 1$, and M is denoted number of knots.

Furthermore, the non-parametric component that is approached by the Fourier Series in equation (10), for response $j = 1,2$ and $i = 1,2,\dots,n$ model can be expressed in the form of a matrix to be:

$$\begin{pmatrix} h_1(z_1) \\ h_1(z_2) \\ \vdots \\ h_1(z_n) \\ h_2(z_1) \\ h_2(z_2) \\ \vdots \\ h_2(z_n) \end{pmatrix} = \left(\begin{array}{ccccc|ccccc} z_1 & \frac{1}{2} & \cos z_1 & \cdots & \cos Lz_1 & 0 & 0 & 0 & \cdots & 0 \\ z_2 & \frac{1}{2} & \cos z_2 & \cdots & \cos Lz_2 & 0 & 0 & 0 & \cdots & 0 \\ \vdots & \vdots & \vdots & \ddots & \vdots & \vdots & \vdots & \vdots & \ddots & \vdots \\ z_n & \frac{1}{2} & \cos z_n & \cdots & \cos Lz_n & 0 & 0 & 0 & \cdots & 0 \\ \hline 0 & 0 & 0 & \cdots & 0 & z_1 & \frac{1}{2} & \cos z_1 & \cdots & \cos Lz_1 \\ 0 & 0 & 0 & \cdots & 0 & z_2 & \frac{1}{2} & \cos z_2 & \cdots & \cos Lz_2 \\ \vdots & \vdots & \vdots & \ddots & \vdots & \vdots & \vdots & \vdots & \ddots & \vdots \\ 0 & 0 & 0 & \cdots & 0 & z_n & \frac{1}{2} & \cos z_n & \cdots & \cos Lz_n \end{array}\right) \begin{pmatrix} b_1 \\ \alpha_{01} \\ \alpha_{11} \\ \vdots \\ \alpha_{L1} \\ b_2 \\ \alpha_{02} \\ \alpha_{12} \\ \vdots \\ \alpha_{L2} \end{pmatrix}$$

Furthermore

$$\begin{pmatrix} \tilde{h}_1(z) \\ \tilde{h}_2(z) \end{pmatrix} = \begin{pmatrix} \boldsymbol{D_1} & \boldsymbol{0} \\ \boldsymbol{0} & \boldsymbol{D_2} \end{pmatrix} \begin{pmatrix} \tilde{\alpha}_1 \\ \tilde{\alpha}_2 \end{pmatrix}$$

in matrix notation can be written as:

$$\tilde{h}(z) = \boldsymbol{D}\tilde{\alpha} \tag{14}$$

With vector $\tilde{h}(z)$ is sized $2n \times 1$, Matrix $\boldsymbol{D}$ is sized $2n \times (2L + 4)$, vector $\tilde{\alpha}$ is sized $(2L + 4) \times 1$, and L is denote number of oscillations.

The fourth component is a nonparametric component which the Kernel approaches. Based on equation (11), for response $j = 1, 2$ and $i = 1, 2, \dots, n$ model can be expressed in the form of a matrix to be:

$$\begin{pmatrix} q_1(v_1) \\ q_1(v_2) \\ \vdots \\ q_1(v_n) \\ \hdashline q_2(v_1) \\ q_2(v_2) \\ \vdots \\ q_2(v_n) \end{pmatrix} = \left(\begin{array}{ccc:ccc} n^{-1}R_{\phi 1}(v_1) & \cdots & n^{-1}R_{\phi n}(v_1) & 0 & \cdots & 0 \\ n^{-1}R_{\phi 1}(v_2) & \cdots & n^{-1}R_{\phi n}(v_2) & 0 & \cdots & 0 \\ \vdots & \ddots & \vdots & \vdots & \ddots & 0 \\ n^{-1}R_{\phi 1}(v_n) & \cdots & n^{-1}R_{\phi n}(v_n) & 0 & \cdots & 0 \\ \hdashline 0 & \cdots & 0 & n^{-1}R_{\phi 1}(v_1) & \cdots & n^{-1}R_{\phi n}(v_1) \\ 0 & \cdots & 0 & n^{-1}R_{\phi 1}(v_2) & \cdots & n^{-1}R_{\phi n}(v_2) \\ \vdots & \ddots & \vdots & \vdots & \ddots & \vdots \\ 0 & \cdots & 0 & n^{-1}R_{\phi 1}(v_n) & \cdots & n^{-1}R_{\phi n}(v_n) \end{array}\right) \begin{pmatrix} y_{11} \\ \vdots \\ y_{1n} \\ \hdashline y_{21} \\ \vdots \\ y_{2n} \end{pmatrix}$$

Furthermore

$$\begin{pmatrix} \tilde{q}_1(v) \\ \tilde{q}_2(v) \end{pmatrix} = \begin{pmatrix} \boldsymbol{V}(\boldsymbol{\phi})_1 & \boldsymbol{0} \\ \boldsymbol{0} & \boldsymbol{V}(\boldsymbol{\phi})_2 \end{pmatrix} \begin{pmatrix} \tilde{y}_1 \\ \tilde{y}_2 \end{pmatrix}$$

in matrix notation can be written as:

$$\tilde{q}(v) = \boldsymbol{V}(\boldsymbol{\phi})\tilde{y} \tag{15}$$

With vector $\tilde{q}(v)$ is sized $2n \times 1$, Matrix $\boldsymbol{V}(\boldsymbol{\phi})$ is sized $2n \times 2n$ and $\tilde{y}$ is sized $2n \times 1$.

Therefore, equation (7) which is a biresponse semiparametric regression model with a truncated spline mixture estimator, Fourier series and kernel is expressed in matrix notation:

$$\begin{aligned} \tilde{y} &= \tilde{f}(x) + \tilde{g}(t) + \tilde{h}(z) + \tilde{q}(v) + \tilde{\varepsilon} \\ &= \boldsymbol{X}\tilde{\beta} + \boldsymbol{G}\tilde{\xi} + \boldsymbol{D}\tilde{\alpha} + \boldsymbol{V}(\phi)\tilde{y} + \tilde{\varepsilon} \end{aligned}$$

Or it can also be stated as:

$$\tilde{y}^* = \boldsymbol{X}\tilde{\beta} + \boldsymbol{G}\tilde{\xi} + \boldsymbol{D}\tilde{\alpha} + \tilde{\varepsilon} \tag{16}$$

Where $\tilde{y}^* = (\boldsymbol{I} - \boldsymbol{V}(\phi))\tilde{y}$

Then in equation (16), the parameter estimation is carried out using the optimization method Weighted Least Square (WLS) to be:

$$\underset{\tilde{\beta},\tilde{\xi},\tilde{\alpha}}{\text{Min}}\, Q(\tilde{\beta},\tilde{\xi},\tilde{\alpha}) = \underset{\tilde{\beta},\tilde{\xi},\tilde{\alpha}}{\text{Min}} \left\{ (\tilde{y}^* - \boldsymbol{X}\tilde{\beta} - \boldsymbol{G}\tilde{\xi} - \boldsymbol{D}\tilde{\alpha})^T \boldsymbol{W} (\tilde{y}^* - \boldsymbol{X}\tilde{\beta} - \boldsymbol{G}\tilde{\xi} - \boldsymbol{D}\tilde{\alpha}) \right\} \tag{17}$$

W is the covariance variance matrix between the two responses. Meanwhile, a simplification is made in equation (17) below :

$$\begin{aligned} Q &= (\tilde{y}^* - \boldsymbol{X}\tilde{\beta} - \boldsymbol{G}\tilde{\xi} - \boldsymbol{D}\tilde{\alpha})^T \boldsymbol{W} (\tilde{y}^* - \boldsymbol{X}\tilde{\beta} - \boldsymbol{G}\tilde{\xi} - \boldsymbol{D}\tilde{\alpha}) \\ &= \tilde{y}^{*T}\boldsymbol{W}\tilde{y}^* - 2\tilde{y}^{*T}\boldsymbol{W}\boldsymbol{X}\tilde{\beta} - 2\tilde{y}^{*T}\boldsymbol{W}\boldsymbol{G}\tilde{\xi} - 2\tilde{y}^{*T}\boldsymbol{W}\boldsymbol{D}\tilde{\alpha} + \tilde{\beta}^T\boldsymbol{X}^T\boldsymbol{W}\boldsymbol{X}\tilde{\beta} + 2\tilde{\beta}^T\boldsymbol{X}^T\boldsymbol{W}\,\boldsymbol{G}\tilde{\xi} \\ &\quad + 2\tilde{\beta}^T\boldsymbol{X}^T\boldsymbol{W}\boldsymbol{D}\tilde{\alpha} + \tilde{\xi}^T\boldsymbol{G}^T\boldsymbol{W}\,\boldsymbol{G}\tilde{\xi} + 2\tilde{\xi}^T\boldsymbol{G}^T\boldsymbol{W}\boldsymbol{D}\tilde{\alpha} + \tilde{\alpha}^T\boldsymbol{D}^T\boldsymbol{W}\,\boldsymbol{D}\tilde{\alpha} \end{aligned} \tag{18}$$

Then proceed to look for the parameter $\tilde{\beta}$, which is looking for the derivative of the function Q against $\tilde{\beta}$ obtained:

$$\frac{\partial Q(\tilde{\beta},\tilde{\xi},\tilde{\alpha})}{\partial \tilde{\beta}} = 0$$

$$\begin{aligned} &-2\tilde{y}^{*T}\boldsymbol{W}\boldsymbol{X} + 2\boldsymbol{X}^T\boldsymbol{W}\boldsymbol{X}\hat{\tilde{\beta}} + 2\boldsymbol{X}^T\boldsymbol{W}\,\boldsymbol{G}\hat{\tilde{\xi}} + 2\boldsymbol{X}^T\boldsymbol{W}\boldsymbol{D}\hat{\tilde{\alpha}} = \boldsymbol{0} \\ &\boldsymbol{X}^T\boldsymbol{W}\boldsymbol{X}\hat{\tilde{\beta}} = \tilde{y}^{*T}\boldsymbol{W}\boldsymbol{X} - \boldsymbol{X}^T\boldsymbol{W}\,\boldsymbol{G}\hat{\tilde{\xi}} - \boldsymbol{X}^T\boldsymbol{W}\boldsymbol{D}\hat{\tilde{\alpha}} \end{aligned}$$

IOP Conf. Series: Earth and Environmental Science **880** (2021) 012046 doi:10.1088/1755-1315/880/1/012046

$$\hat{\tilde{\beta}} = (\boldsymbol{X^TWX})^{-1}\boldsymbol{X^TW}\tilde{y}^* - (\boldsymbol{X^TWX})^{-1}\boldsymbol{X^TW\,G}\hat{\tilde{\xi}} - (\boldsymbol{X^TWX})^{-1}\boldsymbol{X^TWD}\hat{\tilde{\alpha}} \quad (19)$$

By assuming $\mathbf{H} = (\boldsymbol{X^TWX})^{-1}\boldsymbol{X^TW}$, then equation (19) can be simplified to form $\hat{\tilde{\beta}}$ to be :

$$\hat{\tilde{\beta}} = \boldsymbol{H}\tilde{y}^* - \boldsymbol{HG}\hat{\tilde{\xi}} - \boldsymbol{HD}\hat{\tilde{\alpha}} \quad (20)$$

Using the same method as before, derive the Q function in equation (18) against $\tilde{\xi}$ from getting the parameter $\tilde{\xi}$, namely:

$$\frac{\partial Q(\tilde{\beta},\tilde{\xi}\,,\tilde{\alpha})}{\partial \tilde{\xi}} = 0$$

$$-2\tilde{y}^{*T}\boldsymbol{WG} + 2\hat{\tilde{\beta}}^T\boldsymbol{X^TW\,G} + 2\boldsymbol{G^TW\,G}\hat{\tilde{\xi}} + 2\boldsymbol{G^TWD}\hat{\tilde{\alpha}} = \boldsymbol{0}$$

$$\boldsymbol{G^TW\,G}\hat{\tilde{\xi}} = \tilde{y}^{*T}\boldsymbol{WG} - \hat{\tilde{\beta}}^T\boldsymbol{X^TW\,G} - \boldsymbol{G^TWD}\hat{\tilde{\alpha}}$$

$$\hat{\tilde{\xi}} = (\boldsymbol{G^TW\,G})^{-1}\boldsymbol{G^TW}\tilde{y}^* - (\boldsymbol{G^TW\,G})^{-1}\boldsymbol{G^TWX}\hat{\tilde{\beta}} - (\boldsymbol{G^TW\,G})^{-1}\boldsymbol{G^TWD}\hat{\tilde{\alpha}} \quad (21)$$

By assuming $\boldsymbol{L} = (\boldsymbol{G^TW\,G})^{-1}\boldsymbol{G^TW}$, then equation (21) can be simplified to form $\hat{\tilde{\xi}}$ to be :

$$\hat{\tilde{\xi}} = \boldsymbol{L}\tilde{y}^* - \boldsymbol{LX}\hat{\tilde{\beta}} - \boldsymbol{LD}\hat{\tilde{\alpha}} \quad (22)$$

Then proceed to look for the $\tilde{\alpha}$ parameter by deriving the Q function in equation (18) against $\tilde{\alpha}$.

$$\frac{\partial Q(\tilde{\beta},\tilde{\xi}\,,\tilde{\alpha})}{\partial \tilde{\alpha}} = 0$$

$$-2\tilde{y}^{*T}\boldsymbol{WD} + 2\hat{\tilde{\beta}}^T\boldsymbol{X^TWD} + 2\hat{\tilde{\xi}}^T\boldsymbol{G^TWD} + 2\boldsymbol{D^TW\,D}\hat{\tilde{\alpha}} = \boldsymbol{0}$$

$$\boldsymbol{D^TW\,D}\hat{\tilde{\alpha}} = \tilde{y}^{*T}\boldsymbol{WD} - \hat{\tilde{\beta}}^T\boldsymbol{X^TWD} - \hat{\tilde{\xi}}^T\boldsymbol{G^TWD}$$

$$\hat{\tilde{\alpha}} = (\boldsymbol{D^TW\,D})^{-1}\boldsymbol{D^TW}\tilde{y}^* - (\boldsymbol{D^TW\,D})^{-1}\boldsymbol{D^TWX}\hat{\tilde{\beta}} - (\boldsymbol{D^TW\,D})^{-1}\boldsymbol{D^TWG}\hat{\tilde{\xi}} \quad (23)$$

By assuming $\boldsymbol{E} = (\boldsymbol{D^TW\,D})^{-1}\boldsymbol{D^TW}$, then equation (23) can be simplified to form $\hat{\tilde{\alpha}}$ to be :

$$\hat{\tilde{\alpha}} = \boldsymbol{E}\tilde{y}^* - \boldsymbol{EX}\hat{\tilde{\beta}} - \boldsymbol{EG}\hat{\tilde{\xi}} \quad (24)$$

Therefore, parameters obtained still contain other parameters, then do a process of elimination - substitution to get the parameters without loading other parameters. It can be seen in equations (20), (22), and (24) that the equation contains three parameters. Therefore, we will simplify the equation, which has only two parameters. The first step, namely the elimination process for equations (20) and (22), is obtained:

$$(\boldsymbol{I} - \boldsymbol{HGLX})\hat{\tilde{\beta}} + (\boldsymbol{HD} - \boldsymbol{HGLD})\hat{\tilde{\alpha}} = (\boldsymbol{H} - \boldsymbol{HGL})\tilde{y}^* \quad (25)$$

Furthermore, the elimination process for equations (20) and (24) is also carried out:

$$(\boldsymbol{I} - \boldsymbol{HDEX})\hat{\tilde{\beta}} + (\boldsymbol{HG} - \boldsymbol{HDEG})\hat{\tilde{\xi}} = (\boldsymbol{H} - \boldsymbol{HDE})\tilde{y}^* \quad (26)$$

Just like before, made elimination back to equation (22) and (24) were obtained:

$$(\boldsymbol{LX} - \boldsymbol{LDEX})\hat{\tilde{\beta}} + (\boldsymbol{I} - \boldsymbol{LDEG})\hat{\tilde{\xi}} = (\boldsymbol{L} - \boldsymbol{LDE})\tilde{y}^* \quad (27)$$

After obtaining an equation containing two parameters, namely equation (25), (26), and (27), the process of elimination continues again to get parameters that do not contain other parameters. The elimination process is done in equation (26) and (27) to obtain β parameters is:

$$\hat{\tilde{\beta}} = \left[(\boldsymbol{I} - \boldsymbol{HDEX}) - (\boldsymbol{HG} - \boldsymbol{HDEG})(\boldsymbol{I} - \boldsymbol{LDEG})^{-1}(\boldsymbol{LX} - \boldsymbol{LDEX})\right]^{-1}\left[(\boldsymbol{H} - \boldsymbol{HDE}) - (\boldsymbol{HG} - \boldsymbol{HDEG})(\boldsymbol{I} - \boldsymbol{LDEG})^{-1}(\boldsymbol{L} - \boldsymbol{LDE})\right]\left(\boldsymbol{I} - \boldsymbol{V}(\phi)\right)\tilde{y}$$

$$\hat{\hat{\beta}} = \boldsymbol{B}(K,l,\phi)\tilde{y} \tag{28}$$

Where $\boldsymbol{B}(K,l,\phi) = \left[(\boldsymbol{I}-\boldsymbol{HDEX}) - (\boldsymbol{HG}-\boldsymbol{HDEG})(\boldsymbol{I}-\boldsymbol{LDEG})^{-1}(\boldsymbol{LX}-\boldsymbol{LDEX})\right]^{-1}\left[(\boldsymbol{H}-\boldsymbol{HDE}) - (\boldsymbol{HG}-\boldsymbol{HDEG})(\boldsymbol{I}-\boldsymbol{LDEG})^{-1}(\boldsymbol{L}-\boldsymbol{LDE})\right]\left(\boldsymbol{I}-\boldsymbol{V}(\phi)\right)$

Furthermore, the substitution of parameter $\hat{\hat{\beta}}$ to equation (26) is obtained :

$$(\boldsymbol{I}-\boldsymbol{HDEX})\boldsymbol{B}(K,L,\phi)\tilde{y} + (\boldsymbol{HG}-\boldsymbol{HDEG})\hat{\hat{\xi}} = (\boldsymbol{H}-\boldsymbol{HDE})\tilde{y}^*$$

$$(\boldsymbol{I}-\boldsymbol{HDEX})\boldsymbol{B}(K,L,\phi)\tilde{y} + (\boldsymbol{HG}-\boldsymbol{HDEG})\hat{\hat{\xi}} = (\boldsymbol{H}-\boldsymbol{HDE})\left(\boldsymbol{I}-\boldsymbol{V}(\phi)\right)\tilde{y}$$

$$\hat{\hat{\xi}} = (\boldsymbol{HG}-\boldsymbol{HDEG})^{-1}\left[(\boldsymbol{H}-\boldsymbol{HDE})\left(\boldsymbol{I}-\boldsymbol{V}(\phi)\right) - (\boldsymbol{I}-\boldsymbol{HDEX})\boldsymbol{B}(K,L,\phi)\right]\tilde{y}$$

$$\hat{\hat{\xi}} = \boldsymbol{C}(K,l,\phi)\tilde{y} \tag{29}$$

Where $\boldsymbol{C}(K,l,\phi) = (\boldsymbol{HG}-\boldsymbol{HDEG})^{-1}\left[(\boldsymbol{H}-\boldsymbol{HDE})\left(\boldsymbol{I}-\boldsymbol{V}(\phi)\right) - (\boldsymbol{I}-\boldsymbol{HDEX})\boldsymbol{B}(K,l,\phi)\right]$

Substitute $\hat{\hat{\beta}}$ and $\hat{\hat{\xi}}$ to equation (24) to obtain the following $\hat{\hat{\alpha}}$ parameters :

$$\begin{aligned}\hat{\hat{\alpha}} &= \boldsymbol{E}\tilde{y}^* - \boldsymbol{EX}\,\boldsymbol{B}(K,l,\phi)\tilde{y} - \boldsymbol{EGC}(K,l,\phi)\tilde{y}\\ &= \left[\boldsymbol{E}\left(\boldsymbol{I}-\boldsymbol{V}(\phi)\right) - \boldsymbol{EX}\,\boldsymbol{B}(K,l,\phi) - \boldsymbol{EGC}(K,l,\phi)\right]\tilde{y}\end{aligned}$$

$$\hat{\hat{\alpha}} = \boldsymbol{F}(K,l,\phi)\tilde{y} \tag{30}$$

Where $\boldsymbol{F}(K,l,\phi) = \left(\boldsymbol{I}-\boldsymbol{V}(\phi)\right) - \boldsymbol{EX}\,\boldsymbol{B}(K,l,\phi) - \boldsymbol{EGC}(K,l,\phi)$

Based on equations (28), (29) and (30), then mixed estimator of spline truncated, fourier series, and kernel in biresponse semiparametric regression model can be written as follows:

$$\begin{aligned}\hat{\tilde{y}} &= \boldsymbol{X}\hat{\hat{\beta}} + \boldsymbol{G}\hat{\hat{\xi}} + \boldsymbol{D}\hat{\hat{\alpha}} + \boldsymbol{V}(\phi)\tilde{y}\\ &= (\boldsymbol{XB}(K,l,\phi) + \boldsymbol{GC}(K,l,\phi) + \boldsymbol{DF}(K,l,\phi) + \boldsymbol{V}(\phi))\tilde{y}\end{aligned}$$

$$\hat{\tilde{y}} = \boldsymbol{T}(K,l,\phi)\tilde{y} \tag{31}$$

Where $\boldsymbol{T}(K,l,\phi) = \boldsymbol{XB}(K,l,\phi) + \boldsymbol{GC}(K,l,\phi) + \boldsymbol{DF}(K,l,\phi) + \boldsymbol{V}(\phi)$

The model in equation (31) obtained depends on the best mix of knot point K, the oscillation parameter l, and the bandwidth ϕ using the smallest Generalized Cross Validation (GCV), namely:

$$GCV(K,l,\phi) = \frac{MSE\ (K,l,\phi)}{\left(N^{-1}trace(\boldsymbol{I}-\boldsymbol{T}(K,l,\phi))\right)^2}$$

Where $MSE\ (K,l,\phi) = N^{-1}\tilde{y}^T(\boldsymbol{I}-\boldsymbol{T}(K,l,\phi))^T(\boldsymbol{I}-\boldsymbol{T}(K,l,\phi))\tilde{y}$ dan $N = n_1 + n_2$.

4. Conclusion

Based on the analysis conducted previously, it can be concluded that mixed estimator of spline truncated, Fourier series, and kernel in biresponse semiparametric regression model can be written as follows:

$$\begin{aligned}\hat{\tilde{y}} &= \boldsymbol{X}\hat{\hat{\beta}} + \boldsymbol{G}\hat{\hat{\xi}} + \boldsymbol{D}\hat{\hat{\alpha}} + \boldsymbol{V}(\phi)\tilde{y}\\ &= \boldsymbol{XB}(K,l,\phi)\tilde{y} + \boldsymbol{GC}(K,l,\phi)\tilde{y} + \boldsymbol{DF}(K,l,\phi)\tilde{y} + \boldsymbol{V}(\phi)\tilde{y}\end{aligned}$$

Where

$$\begin{aligned}\boldsymbol{B}(K,l,\phi) = &\left[(\boldsymbol{I}-\boldsymbol{HDEX}) - (\boldsymbol{HG}-\boldsymbol{HDEG})(\boldsymbol{I}-\boldsymbol{LDEG})^{-1}(\boldsymbol{LX}-\boldsymbol{LDEX})\right]^{-1}\left[(\boldsymbol{H}-\boldsymbol{HDE})\right.\\ &\left.- (\boldsymbol{HG}-\boldsymbol{HDEG})(\boldsymbol{I}-\boldsymbol{LDEG})^{-1}(\boldsymbol{L}-\boldsymbol{LDE})\right]\left(\boldsymbol{I}-\boldsymbol{V}(\phi)\right);\end{aligned}$$

$$\boldsymbol{C}(K,l,\phi) = (\boldsymbol{HG} - \boldsymbol{HDEG})^{-1}\left[(\boldsymbol{H} - \boldsymbol{HDE})\left(\boldsymbol{I} - \boldsymbol{V}(\phi)\right) - (\boldsymbol{I} - \boldsymbol{HDEX})\boldsymbol{B}(K,l,\phi)\right]$$

$$\boldsymbol{F}(K,l,\phi) = \left(\boldsymbol{I} - \boldsymbol{V}(\phi)\right) - \boldsymbol{EX}\,\boldsymbol{B}(K,l,\phi) - \boldsymbol{EGC}(K,l,\phi)$$

And the smallest GCV based on optimum knot point K, the oscillation parameter l, and the bandwidth ϕ can be determined by the formula

$$GCV(K,l,\phi) = \frac{MSE\ (K,l,\phi)}{\left(N^{-1}trace(\boldsymbol{I} - \boldsymbol{T}(K,l,\phi))\right)^2}$$

Where $MSE\ (K,L,\phi) = N^{-1}\tilde{y}^T(\boldsymbol{I} - \boldsymbol{T}(K,l,\phi))^T(\boldsymbol{I} - \boldsymbol{T}(K,l,\phi))\tilde{y}$ dan $N = n_1 + n_2$.

References

[1] Hardle W 1990 *Applied Non-parametrik Regression* (Cambridge : Cambridge University Press)

[2] Budiantara I N, Ratna M, Zain I, and Wibowo W 2012 Modeling the Percentage of Poor People in Indonesia Using *Spline* Nonparametric Regression Approach *International Journal of Basic & Applied Sciences IJBAS-IJENS* **12** No:06 p 119-124.

[3] Eubank R L 1999 *Spline Smoothing and Nonparametric Regression* (New York: Marcel Deker)

[4] Wahba G 1990 *Spline Models for Observational Data* (Philadelphia: SIAM)

[5] Montoya E L, Ulloa N and Miler V 2014 A Simulation Study Comparing Knot Selection Methods with Equality Spaced Knots in a Penalized Regression Spline *International Journal of Statistics and Probability* **3** p 96-110

[6] Saputro D R S, Demu K R and Widyaningsih P 2018 Nonparametric Truncated Spline Regression Model on the Data of Human Development Index (HDI) in Indonesia *IOP Conf. Series: Journal of Physics* **1028** p 1-4

[7] Faber T D, Yoon D C, Service S K and White S C 2004 Fourier and Wavelet Analyses of Dental Radiographs Detect Trabucelar Changes in Osteoporosis **35** p 403-411

[8] Ratnasari V, Budiantara I N, Zain I, Ratna M and Mariati N P A M 2015. Comparison Truncated Spline and Fourier Series in Multivariable Nonparametric Regresssion Model (Application: Data of Poverty in Papua, Indonesia) *International Journal of Basic & Applied Sciences IJBASIJENS* **15** No:04 p 9-14

[9] Pane R, Sutarman, and Ampa A T 2020 Estimation of Heteroskedasticity Semiparametric Regression Curve Using Fourier Series Approach *Journal of Research in Mathematics Trends and Technology* **2** No: 1 p 14-20

[10] Bilodeau M 1992 Fourier Smoother and Additive Models *The Canadian Journal of Statistics* **3** p 257-259.

[11] Octavanny M A D, Budiantara I N, Kuswanto H, and Rahmawati D P 2021 Modelling of Children Ever Born in Indonesia Using Fourier Series Nonparametric Regression *Journal of Physics: Conference Series* **1752** 012019

[12] Aljuhani K H and Al turk L I 2014. Modification of the Adaptive NadarayaWatson Kernel Regression Estimator *AcademicJournal* **9** p 966-971

[13] Putri A P, Santoso R, and Sugito 2017Analisis Regresi Nonparametrik Kernel Menggunakan Metode *Jackknife* Sampel Terhapus-1 dan Sampel Terhapus-2 *Jurnal Gaussian* **6** No.1 p 1-10.

[14] Aydin D, Guneri O I and Fit A 2016 Choice of Bandwidth for Nonparametric Regression Models using Kernel Smoothing: A Simulation Study *International Journal of Sciences Basic and Applied Research* **26** p 47-61

[15] Dani A T R, Ratnaasari V, Budiantara I N 2021 Optimal Knots Point and Bandwith Selection in Modelling Mixed Estimator Nonparametric Regression *Proc. IOP Conference Series : Materials Science and Engineering* **1115** 012020

[16] Rahmawati D P, Budiantara I N, Prasetyo D D and Octavanny M A D 2021 Mixed Spline Smoothing and Kernel Estimator in Biresponse Nonparametric Regression *International Journal of Mathematics and Mathematical Sciences* **2021**

[17] Sudiarsa I W, Budiantara I N, Suhartono and Purnami S W 2016 Combined Estimator Fourier Series and Spline Truncated in Multivariable Nonparametric Regression *Applied Mathematical Sciences* **9** p 4997-5010

[18] Mariati N P A M, Budiantara I N, and Ratnasari V 2020 Smoothing Spline Estimator in Nonparametric Regression (Application: Poverty in Papua Province) *Proceedings of the 7th International Conference on Research, Implementation, and Education of Mathematics and Science* **528**

[19] Mardianto M F F 2015 *Model Regresi Semiparametrik Birespon dengan Pendekatan Deret Fourier* (Surabaya : ITS Repository)

[20] Dewanti P, Budiantara I N, Rumiati A T 2020 Modelling of SDG's Achievement in East Java Using Bi-responses Nonparametric Regression with Mixed Estimator Spline Truncated and Kernel *Proc. Journal of Physics : Conference Series* **1562** 012016

[21] Kuzairi, Chamidah N, and Budiantara I N 2019 Theoretical Study of Fourier Series Estimator in Semiparametric Regression for Longitudinal Data Based on Weighted Least Square Optimization *Proc. Journal of Physics: Conference Series* **410**

[22] Ardianingsih N, Budiantara I N, Purnomo J D T 2020 Mixture Model Nonparametric Regression and Its Application *Proc. International Conference on Science Education and Technology* **1842** 012044

IOP Conf. Series: Earth and Environmental Science **880** (2021) 012046 doi:10.1088/1755-1315/880/1/012046

Acknowledgments

I want to thank the second and third authors who were very helpful in completing this research. Thank you also to Lembaga Pengelola Dana Pendidikan (LPDP) for providing funds and facilities during the master study at Institut Teknologi Sepuluh Nopember

Antibacterial activity of aquatic extract of *Myrtus communis* leaves against *Periodontitis* isolated bacteria

Eman Mubdir Nayf[1] and Hamzah Abdulrhaman Salman[1*]

[1] Department of Medical Laboratory Techniques, College of Medical Sciences Techniques, The University of Mashreq, Baghdad, Iraq

[*]Corresponding Email: hamza.alayash@gmail.com; hamzah.abdulrhaman@uom.edu.iq

Abstract. *Myrtus communis* is an evergreen plant that can survive stressful environments and high-temperature seasons. Treatment using green plants was the most trended in recent years. The present study aimed to evaluate the antibacterial effects of *Myrtus communis* leaves against bacteria isolated from periodontitis. Fifty samples were collected from periodontitis subjects in both genders (female 32 % and male 68 %). The isolates were diagnosed by morphological characterization and biochemical tests. *M. communis* leaves were identified, collected, and prepared for extraction. The plant leaves were extracted using distilled water. The antibacterial susceptibility testing was performed by the well diffusion method. Antibiotics susceptibility patterns were executed using the disc diffusion method. All the isolates belonged to gram-positive bacteria. Among the isolated bacteria, 20, 18, and 12 were *Lactobacillus* spp., *Streptococcus* spp., and *Staphylococcus aureus*, respectively. The antibacterial susceptibility testing of *M. communis* extract showed a potential zone of inhibition against all the tested bacteria. Of the different concentrations, 30 mg/ml showed the highest zone of inhibition, 18.2 mm, 19.50 mm, and 30.66 mm against *Streptococcus* spp., *Staphylococcus* spp., and *Lactobacillus* spp. Among the tested antibiotics, ciprofloxacin and chloramphenicol exhibited the highest zone of inhibition against the tested bacteria. The aquatic extract of *M. communis* leaves was found to be effective against gram-positive bacteria. Further studies are warranted to investigate the active bio-compounds.

Keywords: Antibacterial activity, *Myrtus communis*, Aquatic extract, Antibiotics susceptibility testing.

Track Name: Advanced Technology and Renewable Energy

1. Introduction

Plants are not only involved in climate improvement but also involved in treating different human diseases [1-3]. *Myrtus comunis* L., is a common leafy tree plant that belongs to the family Myrtaceae. The region of this plant is generally in countries of the Mediterranean as well as Middle East countries such as Iraq, Jordan, Saudi Arabia, and Iran [4]. In Iraq, *M. communis* is widely planted in the street and the gardens of the houses. Different parts of *M. comunis* are the most effective medicinal plants that contributed to treating a wide range of diseases caused by microorganisms [4].

Moreover, plants are broadly included in cosmetics, perfume, spices, and therapeutic manufacturing [5]. The component of *M. communis* extracts can further be used as anticancer [6], antibacterial [4], antioxidants [7], and anti-inflammation [8]. Due to the sclerophyllous nature, high activity of the enzymes, and the high content of glutathione antioxidant, the *M. communis* was able to withstand climate change during the high-temperature summer seasons [9].

In the human oral cavity, gingivitis and periodontitis are considered frequently occurring diseases. In few days, gingivitis can occur due to the inflammation of the gingiva, which is influenced by plaque accumulation. The bacterial infection and untreated gingivitis lead to periodontitis [10]. Periodontitis was reported to be associated with bacteremia and may lead to infective endocarditis [11]. Ninety percent of the population, mainly developing countries, are affected by periodontitis. A large number of bacterial species (gram-positive and gram-negative) are involved in periodontal diseases [12].

The emerging antibiotic resistance and multidrug resistance against different bacterial species isolated from periodontitis have increased recently [13]. This led to the development of an alternative treatment method against various bacterial infections. The alternative treatment is herbal medicine. Many medicinal plants have been tested and recognized as antimicrobial agents [2, 14]. As per the World Health Organization, herbal medicines are the safest source for a wide range of remedies [15]. The secondary metabolites of the plant have a variety of compounds such as flavonoids, alkaloids, and phenol [16, 17]. With respect to the above facts, the present study was aimed to investigate the antibacterial activity of aquatic extract of *M. comunis* leaves and antibiotics susceptibility profile against bacteria isolated from periodontitis.

2. Materials and Methods

2.1 Isolation and identification of bacteria

Fifty samples were collected from 50 patients (male 34 and female 16) who severing from periodontitis. The samples were collected and processed as described by Salman et al. [18] using wooden toothpicks from gingiva lesions. The processed samples were inoculated into MacConkey agar, blood agar, and nutrient agar (HiMedia, India) then incubated at 37 ℃ for 24-48 h. The colonies were identified based on gram stain, morphology, and biochemical tests. The biochemical tests used were the catalase test, oxidase test, coagulase, hemolysis, lactose fermentation, mannitol fermentation, ribose fermentation, and bacitracin test. The biochemical tests were repeated in triplicate to approve the reproducibility of the results.

2.2 Collection and processing of plant material

The leaves of *M. communis* were collected and identified by a special horticulturist. The plant material was cleaned from the soil and dust using distilled water. The leaves were then dried using a hot air oven and then crushed into fine particles.

2.3 Extraction of M. communis

The plant extraction was executed using the soxhlet apparatus as described by Salman et al., [2] with few modifications. Briefly, 50 g of plant powder was added to 600 ml distilled water and kept in the apparatus. The condition of the extraction was set at 25 ℃ for 48 h. The mixture was filtered using 2 sheets of Whatman paper. The filtered extract then centrifuge at 4700 rpm for 15 min. The pellet was collected and stored in a screw cap bottle for further test.

2.4 Antibacterial susceptibility testing

The antibacterial susceptibility testing was performed using the well diffusion method described by Balouiri et al. [19]. The tested bacteria were adjusted at 0.5 Mcfarland and cultured on Mueller-Hinton agar (HiMedia, India). The well was done aseptically using a sterile cork borer with a 6-8 mm diameter. Different concentrations of plant extract solution was made (7.5 mg/ml, 15 mg/ml and 30 mg/ml). 100 µl of each concentration was inoculated in the well. Sterile distilled water was inoculated in a separate well as a control. The plates were then incubated at 37 °C for 18-24 h. The inhibition zones were measured in millimeters (mm) in diameter using a ruler. For more validation and reproducibility, the test was performed in triplicate.

2.5 Antibiotic susceptibility testing
The antibiogram profile of the tested bacteria was determined using the disc diffusion method as described by Salman and Senthilkumar [20]. The tested bacteria were inoculated on Mueller-Hinton agar (HiMedia, India). The commercial antibiotic discs employed in the present study were placed on the inoculated medium. The plates were then incubated at 37 °C for 18-24 h. The zone of inhibition was measured using a ruler (mm). The tested antibiotics were penicillin-G, gentamycin, ciprofloxacin, amikacin, neomycin, chloramphenicol, and tetracycline (HiMedia. India). For the reproducibility and confirmation of the results, the test was done in triplicate.

2.6 Statistical analysis
The standard deviation of the results in the present study was performed using Word Excel version 13.

3. Results
Of 50 collected samples, 34 (68 %) were male and 16 (32 %) were female. Among the tested samples, 20, 18, and 12 were identified as *Lactobacillus* spp., *Streptococcus* spp., and *Staphylococcus aureus*, respectively, based on biochemical tests (Table 1). The antibacterial susceptibility testing of different concentrations of *M. communis* leaves against the *Lactobacillus* spp., *Streptococcus* spp., and *Staphylococcus aureus* was demonstrated in Table 2. At concentration 30 mg/ml, *Lactobacillus* spp. showed the highest zone of inhibition (30.6 mm). The antibiotics susceptibility profile of *Lactobacillus* spp., *Streptococcus* spp., and *Staphylococcus aureus* were listed in Table 3.

Table 1. Biochemical identification of the tested bacteria.

Biochemical test	***S. aureus* (n=12)**	***Streptococcus* spp., (n=18)**	***Lactobacillus* spp., (n=20)**
Catalase test	+	-	-
Oxidase test	-	-	-
Coagulase test	+	-	-
Clamping factor	+	-	+
Hemolysis	Beta	Beta	-
Mannitol salt agar	+	-	+
Bacitracin	-	+	-
Lactose fermentation	+	+	+
Mannitol fermentation	+	-	-
Ribose fermentation	+	-	-

Table 2. Antibacterial susceptibility testing of aquatic leaves extract of *M. communis* against isolated bacteria.

***M. communis* concentrations**	**Average of zone of inhibition (mm) ± SD**		
	S. aureus	***Streptococcus* spp.,**	***Lactobacillus* spp.,**
30 mg / ml	19.50 ± 0.00	18.2 ± 5.77	30.66 ± 1.20
15 mg / ml	8.60 ± 0.00	8.5 ± 0.00	10.22 ± 1.22
7.5 mg / ml	3.33 ± 1-11	4.5 ± 0.00	7.3 ± 0.57
Sterile distilled water (control)	0.00 ± 0.00	0.00 ± 0.00	0.00 ± 0.00

mm= millimeter; SD= standard deviation

Table 3. Antibiotics susceptibility profile against isolated bacteria.

Antibiotic	**Average of zone of inhibition in (mm) ± SD.**		
	S. aureus	***Streptococcus* spp.,**	***Lactobacillus* spp.,**
Penicillin – G	19.6 ± 1.02	10.00 ± 0.57	20.66 ± 1.76
Neomycin	9.00 ± 57	11.00 ± 0.57	12.00 ± 0.00
Tetracycline	13.6 ± 1. 2	12.00 ± 0.00	15.00 ± 1.45
Chloramphenicol	19.6 ± 1.20	24.3 ± 0.3	18.00 ± 0.57
Gentamicin	16.00 ± 0.57	9.00 ± 0.00	19.00 ± 1.15
Ciprofloxacin	22.00 ± 0.00	23.00 ± 1.15	22.00 ± 0.00
Amikacin	19.6 ± 0.88	22.00± 2.88	18.00 ± 0.00

mm= millimeter; SD= Standard deviation

4. Discussion

Mother nature is still helping humans by acting as an open-source of medicine. The medicinal plants are now much in focus, especially after increasing bacterial resistance to antibiotics. The present study was aimed to investigate the antibacterial activity of aquatic extract of *M. communis* leaves against bacteria isolated from periodontitis.

In the current results, *Lactobacillus* spp. (40 %), *Streptococcus* spp. (36 %), and *Staphylococcus aureus* (24 %) were isolated from periodontitis (Table 1). *Lactobacillus* spp. is one of the main bacteria colonizing the oral cavity, which has a role in caries and periodontitis [21]. Twenty strains of *Lactobacillus* spp., were isolated in the current study from gingivitis, which is indicated the responsibility of *Lactobacillus* spp., in causing periodontitis (Table 1). However, various species of *Lactobacillus* have been associated with oral infections [22].

Streptococcus spp. was the second most prevalent bacteria isolated from periodontitis (Table 1). Arredondo et al. reported *Streptococcus* spp. as the highest prevalence bacteria isolated from periodontitis. *S. oralis* was the most common isolated species of *Streptococcus*, followed by *S. mitis*, *S. intermedius*, and *S. constellatus* [13]. The present study is consistent with Dhotre et al., which reported the isolation of different *Streptococcus* spp. from subgingival plaque [11]. Furthermore, *Streptococcus tigurinus*, a new member of viridians streptococci group, was isolated from periodontitis [23]. However, *S. mutans* and *S. sobrinus* were reported as the primary causative for dental caries [18]. Very rare species of *Streptococcus* i.e. *S. downei* and *S. dentapri* was also isolated from the human oral niche for the first time, were previously isolated only from the animal's oral cavity [24, 25].

A very recent study reported the isolation of 50 strains of *S. aureus* from periodontitis which is comparable to the current study [26]. Furthermore, *Staphylococcus aureus* and *Staphylococcus epidermidis* were also isolated from periodontitis [27].

The aquatic extract of *M. communis* leaves showed bioactivity against several bacteria and diseases [4, 28]; this is consistent with the present results. Different concentrations of aquatic extraction of *M. communis* have tested its activity against 3 different bacteria (*S. aureus, Streptococcus* spp., and *Lactobacillus* spp.,) isolated from periodontitis. Of the tested bacteria, *Lactobacillus* spp., showed the higher susceptibility among all the concentrations (Table 2). At a 30 mg/ml concentration, the zone of inhibition against *Lactobacillus* spp. was 30.6 mm. As per the literature survey, this is the first published study of the antibacterial activity of *M. communis* on *Lactobacillus* spp.

S. aureus and *Streptococcus* spp., showed almost similar susceptibility to aquatic extract of *M. communis* (Table 2). At concentration 30 mg/ml, in the present study, the zone of inhibition of the aquatic extract was found to be 19.5 mm on *S. aureus*. In contrast, a previous study reported less antibacterial activity of *M. communis* leaves against *S. aureus [29]*. This is, maybe, attributed to the solvent used or due to the variable strain of *S. aureus*. A recent study reported the bioactivity of ethanolic extract of *M. communis* leaves against gram-positive bacteria, i.e., *S. aureus* and *S. epidermidis,* with the zone of inhibition 25 mm and 15 mm, respectively [30]. A study in Iran reported the antibacterial activity of hydroalcoholic extract of *M. communis* leaves at a concentration of 80 mg/ml [31]. However, this indicates the higher activity of aquatic extraction in the present study. Zomorodian et al. reported the activity of essential oil of *M. communis* against a wide range of resistant bacteria, including methicillin-resistant *S. aureus* [32].

Among the concentrations tested, *Streptococcus* spp. demonstrated higher susceptibility to aquatic extract of *M. communis* leaves at 30 mg/ml (Table 2). This is consistent with the previously reported study, which showed the susceptibility of *S. mutans*, *S. pneumonia,* and *S. pyogenes* to the essential oil of *M. communis* [32]. Our results are similar to the report by Fani et al., who declared that the *M. communis* oil contains strong antibacterial activity against Streptococci and other oral pathogens, i.e., *S. mutans* (MIC = 31.25 μg/ml) and *S. pyogenes* (MIC = 29.68 μg/ml) [33]. Ethanolic extract of *Rhodomyrtus tomentosa* (belongs to Myrtaceae family) exhibited potent antibacterial activity against *S. pyogenes* (MIC= 3.91 - 62.5 μg/ml) [34]. Serious investigations are suggested to treat other bacteria such as *Helicobacter pylori* which is responsible for the suffering and death of a large number of human beings by causing stomach ulcers [35].

In the present study, the higher susceptibility of *Streptococcus* spp., was shown to chloramphenicol (24.3 mm) followed by ciprofloxacin (23 mm), amikacin (22 mm), tetracycline (12 mm), neomycin (11 mm), and penicillin (10 mm) (Table 3). This is consistent with a previously reported study, which demonstrated the sensitivity of *S. mutans* and *S. sobrinus* isolated from dental caries to various antibiotics [20]. However, our results showed a lower zone of inhibition in regard to penicillin and chloramphenicol (Table 3). In contrast, Arredondo et al. reported the resistant *Streptococcus* spp. to tetracycline [13]. Moreover, tetracycline was found to be resistant against *S. pyogenes* isolated from respiratory tract infection. In the same study, 95 % and 82 % of the isolated strains were sensitive to penicillin and chloramphenicol, respectively [36]. Our study was contrary to the previously reported study, which found a high rate of resistance (92 %) amikacin on *Streptococcus* spp., [37].

In the current study, ciprofloxacin was found to be the most susceptible antibiotics (22 mm) against *S. aureus*, followed by chloramphenicol (19.6 mm), penicillin (19.6 mm), amikacin (19.6 mm), gentamicin (16 mm), tetracycline (13.6 mm) and neomycin (9 mm) (Table 3). The present study is not in accordance with Gitau et al., who exhibited a high resistance rate of penicillin on *S. aureus* [38]. The same study reported the resistance of 65 % of *S. aureus* to tetracycline. However, the present study demonstrated moderate sensitivity of tetracycline on *S. aureus* (Table 3). Furthermore, one more report showed an excellent resistance rate of *S. aureus* to a penicillin (92.3 %), while a lower resistance rate was demonstrated to ciprofloxacin (4.4 %) [39]. In the previous report, amikacin was found to be susceptible against 75 % of *S. aureus* [37], which is consistent with current results.

5. Conclusion

The study concluded that the aquatic extract of *M. communis* leaves was shown potential antibacterial activity against *S. aureus*, *Streptococcus* spp., and *Lactobacillus* spp., isolated from periodontitis. In comparison to the antibiotics, the leaves extract of *M. communis* showed potent activity against the tested bacteria. Hence, further investigations have to follow to include this extract as treatment, mouth wash, and toothpaste.

References

[1] Parmesan C and Hanley M E 2015 Plants and climate change: complexities and surprises *Ann Bot* **116** 849-64
[2] Salman H A, Venkatesh S, Senthilkumar R, Gnanesh Kumar B S and Ali A M 2018 Determination of antibacterial activity and metabolite profile of Ruta graveolens against Streptococcus mutans and Streptococcus sobrinus *Journal of laboratory physicians* **10** 320-5
[3] Salman H A and Senthilkumar R J A J P C R 2015 Antibacterial activity of Annona squamosa L. and Annona reticulata L. against clinical isolates of mutans streptococci the causative agents of dental caries *Asian J Pharm Clin Res* **8** 152-5
[4] Aleksic V and Knezevic P 2014 Antimicrobial and antioxidative activity of extracts and essential oils of Myrtus communis L *Microbiological research* **169** 240-54
[5] Leri M, Scuto M, Ontario M L, Calabrese V, Calabrese E J, Bucciantini M and Stefani M 2020 Healthy Effects of Plant Polyphenols: Molecular Mechanisms *Int J Mol Sci* **21** 1250
[6] Ali Abuderman A, Syed R, A. Alyousef A, S. Alqahtani M, Shamsul Ola M and Malik A 2019 Green Synthesized Silver Nanoparticles of Myrtus communis L (AgMC) Extract Inhibits Cancer Hallmarks via Targeting Aldose Reductase (AR) and Associated Signaling Network *Processes* **7**
[7] Hennia A, Miguel M G and Nemmiche S 2018 Antioxidant Activity of Myrtus communis L. and Myrtus nivellei Batt. & Trab. Extracts: A Brief Review *Medicines (Basel)* **5** 89
[8] Rossi A, Di Paola R, Mazzon E, Genovese T, Caminiti R, Bramanti P, Pergola C, Koeberle A, Werz O, Sautebin L and Cuzzocrea S 2009 Myrtucommulone from Myrtus communis exhibits potent anti-inflammatory effectiveness in vivo *The Journal of pharmacology and experimental therapeutics* **329** 76-86
[9] Ciccarelli D, Bottega S and Spanò C 2019 Study of functional and physiological response of co-occurring shrub species to the Mediterranean climate *Saudi journal of biological sciences* **26** 1668-75
[10] Murakami S, Mealey B L, Mariotti A and Chapple I L C 2018 Dental plaque-induced gingival conditions *Journal of periodontology* **89 Suppl 1** S17-s27
[11] Dhotre S, Jahagirdar V, Suryawanshi N, Davane M, Patil R and Nagoba B 2018 Assessment of periodontitis and its role in viridans streptococcal bacteremia and infective endocarditis *Indian Heart J* **70** 225-32
[12] Curtis M A, Diaz P I and Van Dyke T E 2020 The role of the microbiota in periodontal disease *Periodontology 2000* **83** 14-25
[13] Arredondo A, Blanc V, Mor C, Nart J and León R 2020 Tetracycline and multidrug resistance in the oral microbiota: differences between healthy subjects and patients with periodontitis in Spain *J Oral Microbiol* **13** 1847431-
[14] Manandhar S, Luitel S and Dahal R K 2019 In Vitro Antimicrobial Activity of Some Medicinal Plants against Human Pathogenic Bacteria *Journal of Tropical Medicine* **2019** 1895340
[15] Organization W H 2002 *The world health report 2002: reducing risks, promoting healthy life*: World Health Organization)
[16] Salman H A and Senthilkumar Ramasamy B S M J J o C M R 2018 Detection of caffeic and chlorogenic acids from methanolic extract of Annona squamosa bark by LC-ESI-MS/MS *J Intercult Ethnopharmacol* **7** 76-81
[17] Landi M, Zivcak M, Sytar O, Brestic M and Allakhverdiev S I 2020 Plasticity of photosynthetic processes and the accumulation of secondary metabolites in plants in response to monochromatic light environments: A review *Biochimica et Biophysica Acta (BBA) - Bioenergetics* **1861** 148131
[18] Salman H A, Senthilkumar R, Imran K and Selvam K P 2017 Isolation and Typing of Streptococcus mutans and Streptococcus sobrinus from Caries-active Subjects *Contemp Clin Dent* **8** 587-93
[19] Balouiri M, Sadiki M and Ibnsouda S K 2016 Methods for in vitro evaluating antimicrobial activity: A review *J Pharm Anal* **6** 71-9
[20] Salman H A and Senthikumar R J J A P S 2015 Identification and antibiogram profile of Streptococcus mutans and Streptococcus sobrinus from dental caries subjects *J App Pharm Sci* **5** 54-7
[21] Badet C and Thebaud N B 2008 Ecology of lactobacilli in the oral cavity: a review of literature *Open Microbiol J* **2** 38-48

[22] Caufield P W, Schön C N, Saraithong P, Li Y and Argimón S 2015 Oral Lactobacilli and Dental Caries: A Model for Niche Adaptation in Humans *J Dent Res* **94** 110S-8S

[23] Dhotre S V, Mehetre G T, Dharne M S, Suryawanshi N M and Nagoba B S 2014 Isolation of Streptococcus tigurinus - a novel member of Streptococcus mitis group from a case of periodontitis *FEMS microbiology letters* **357** 131-5

[24] Salman H A, Senthilkumar R, Mahmood B S and Imran K 2019 Detection and characterization of Streptococcus downei, a rare bacterial species of mutans streptococci from caries-active patients *Indian journal of dental research : official publication of Indian Society for Dental Research* **30** 579-82

[25] Salman H A, Kumar R S, Babu N C and Imran K 2017 First Detection and Characterization of Streptococcus dentapri from Caries Active Subject *J Clin Diagn Res* **11** DM01-DM3

[26] Uribe-García A, Paniagua-Contreras G L, Monroy-Pérez E, Bustos-Martínez J, Hamdan-Partida A, Garzón J, Alanís J, Quezada R, Vaca-Paniagua F and Vaca S 2021 Frequency and expression of genes involved in adhesion and biofilm formation in Staphylococcus aureus strains isolated from periodontal lesions *Journal of microbiology, immunology, and infection = Wei mian yu gan ran za zhi* **54** 267-75

[27] O'Connor A M, McManus B A, Kinnevey P M, Brennan G I, Fleming T E, Cashin P J, O'Sullivan M, Polyzois I and Coleman D C 2018 Significant Enrichment and Diversity of the Staphylococcal Arginine Catabolic Mobile Element ACME in Staphylococcus epidermidis Isolates From Subgingival Peri-implantitis Sites and Periodontal Pockets *Frontiers in microbiology* **9** 1558-

[28] Rahimvand L, Niakan M and Naderi N J 2018 The antibacterial effect of aquatic and methanolic extract of Myrtus communis on Actinobacillus actinomycetemcomitans, Porphyromonas gingivalis and Prevotella intermedia *Iran J Microbiol* **10** 254-7

[29] Djenane D, Yangüela J, Amrouche T, Boubrit S, Boussad N and Roncalés P 2011 Chemical composition and antimicrobial effects of essential oils of Eucalyptus globulus, Myrtus communis and Satureja hortensis against Escherichia coli O157:H7 and Staphylococcus aureus in minced beef *Food science and technology international = Ciencia y tecnologia de los alimentos internacional* **17** 505-15

[30] Mir M A, Bashir N, Alfaify A and Oteef M D Y 2020 GC-MS analysis of Myrtus communis extract and its antibacterial activity against Gram-positive bacteria *BMC Complement Med Ther* **20** 86-

[31] Taheri A, Seyfan A, Jalalinezhad S and Nasery F J Z J R M S 2013 Antibacterial effect of Myrtus communis hydro-alcoholic extract on pathogenic bacteria *Zahedan J Res Med Sci* **15** 19-24

[32] Zomorodian K, Moein M, Lori Z G, Ghasemi Y, Rahimi M J, Bandegani A, Pakshir K, Bazargani A, Mirzamohammadi S and Abbasi N 2013 Chemical Composition and Antimicrobial Activities of the Essential Oil from Myrtus communis Leaves *Journal of Essential Oil Bearing Plants* **16** 76-84

[33] Fani M M, Kohanteb J and Araghizadeh A 2014 Inhibitory activity of Myrtus communis oil on some clinically isolated oral pathogens *Medical principles and practice : international journal of the Kuwait University, Health Science Centre* **23** 363-8

[34] Limsuwan S, Kayser O and Voravuthikunchai S P 2012 Antibacterial Activity of <i>Rhodomyrtus tomentosa</i> (Aiton) Hassk. Leaf Extract against Clinical Isolates of <i>Streptococcus pyogenes</i> *Evidence-Based Complementary and Alternative Medicine* **2012** 697183

[35] Salman H A, Nayif E M and Bloh A H 2021 Characterization and expression of SabA and BabA genes in Helicobacter pylori under varying pH *Korean J. Microbiol* **57** 83-90

[36] Camara M, Dieng A and Boye C S B 2013 Antibiotic susceptibility of streptococcus pyogenes isolated from respiratory tract infections in dakar, senegal *Microbiol Insights* **6** 71-5

[37] Kodati S, Eller A W and Kowalski R P 2017 The Susceptibility of Bacterial Endophthalmitis Isolates to Vancomycin, Ceftazidime, and Amikacin: a 23 Year-Review *Ophthalmol Retina* **1** 206-9

[38] Gitau W, Masika M, Musyoki M, Museve B and Mutwiri T 2018 Antimicrobial susceptibility pattern of Staphylococcus aureus isolates from clinical specimens at Kenyatta National Hospital *BMC research notes* **11** 226

[39] Iileka A E, Mukesi M, Engelbrecht F and Moyo S R J O J o M M 2016 Antimicrobial susceptibility patterns of Staphylococcus aureus strains isolated at the Namibia Institute of Pathology from 2012 to 2014 *Open Journal of Medical Microbiology* **6** 116

The use of entropy based fuzzy membership on weighted logistic regression for the unbalanced data

Ajiwasesa Harumeka[1,2*], Santi Wulan Purnami[1], Santi Puteri Rahayu[1]

[1] Department of Statistics, Faculty of Science and Data Analytics, Sepuluh Nopember Institute of Technology, Surabaya, Indonesia
[2] Statistics Indonesia, Jakarta, Indonesia

*Corresponding E-mail : aharumeka@gmail.com; ajiwasesa@bps.go.id

Abstract. Logistic regression is a popular and powerful classification method. The addition of ridge regularization and optimization using a combination of linear conjugate gradients and IRLS, called Truncated Regularized Iteratively Re-weighted Least Square (TR-IRLS), can outperform Support Vector Machine (SVM) in terms of processing speed, especially when applied to large data and have competitive accuracy. However, neither SVM nor TR-IRLS is good enough when applied to unbalanced data. Fuzzy Support Vector Machine (FSVM) is an SVM development for unbalanced data that adds fuzzy membership to each observation. The fuzzy membership makes the interest of each observation in the minority class higher than the majority class. Meanwhile, TR-IRLS developed into a Rare Event Weighted Logistic Regression (RE-WLR) by adding weight to logistic regression and bias correction. The weighting of the RE-WLR depends on the undersampling scheme. It allows an "information loss". Between FSVM and RE-WLR has a similarity, the weight based only on class differences (minority or majority). Entropy Based Fuzzy Support Vector Machine (EFSVM) is a method used to accommodate the weaknesses of FSVM by considering the class certainty of class observations. As a result, EFSVM is able to improve SVM performance for unbalanced data, even beating FSVM. For this reason, we use EF on the TR-IRLS algorithm to classify large and unbalanced data, as a proposed method. This method is called Entropy-Based Fuzzy Weighted Logistic Regression (EF-WLR). This Research shows the review of EF-WLR for unbalanced data classification.

Keywords : Classification, Entropy Based Fuzzy, Logistic Regression, Unbalanced Dataset

Track Name: Advanced Technology and Renewable Energy

1. Introduction

Logistic regression is a popular method. This method, as a linear classifier, has several advantages [1]. First, when compared to kernel-based methods, logistic regression is implemented directly on the data in the original dimension, without being transformed into a higher dimension. Second, logistic regression as a linear classifier proved to be a robust classification method by providing probability values and covering multi-class classification problems. Third, this method does not require assumptions about the distribution of the response variables. Fourth, there are no constraints to solve the optimization problem.

Komarek and Moore [2] introduced Truncated Regularized Iteratively Re-weighted Least Square (TR-IRLS) which applies Truncated Newton for large-scale data. TR-IRLS is faster than SVM. Both methods are comparable because they have similar loss functions [3]. Then, Truncated Newtons are used as an optimization method for other logistic regression problems [4]. Both TR-IRLS and SVM are designed with the assumption of balanced data conditions. When in fact, there are incidents of unbalanced data, such as tornado storms, forest covertype [1], collon and breast cancer [5], and State Failure [6].

There are two approaches to overcome the problem of unbalanced data, namely the data level approach and the algorithm level approach [7]. The algorithm-level approach has advantages over the data-level approach. It is an effective method and does not need to change the distribution of training data [7]. Then, SVM and TR-IRLS developed for unbalanced data problems using an algorithm level approach. Maalouf and Siddiqi [1], in 2014, developed the TR-IRLS to overcome unbalanced data, named Rare Event Weighted Logistic Regression (RE-WLR), by combining TR-IRLS, weighting, and bias correction [8]. RE-WLR has better accuracy than TR-IRLS. While, Lin and Wang [9], in 2002, added an element of fuzzy membership to each observation as weight and reformulated SVM, named Fuzzy Support Vector Machine (FSVM). FSVM can outperform SVM in unbalanced data. Both RE-WLR and FSVM have similar weights. The weighting value of the two methods is based on class differences only. So that, all observations in the majority class only has the same value.

Fan, et.al [10] propose Entropy Based Fuzzy Support Vector Machine (EFSVM) to handle weaknesses in FSVM by considering the class certainty of each observation. The certainty of the class is measured using entropy. The smaller the entropy value, the greater the class certainty [10]. The entropy values on the majority class observations will be grouped into subsets. Then we get the fuzzy membership value in each subset. They show that weighting using entropy based fuzzy membership (EF) can improve SVM performance when applied to unbalanced data, even beating FSVM.

Based on the success of entropy based fuzzy membership (EF) on SVM, we propose a method that applies EF to TR-IRLS algorithm, named Entropy Based Fuzzy Weighted Logistic Regression (EF-WLR). By implementing EF, the TR-IRLS algorithm is expected to be able to pay more attention to observations that have a higher class certainty value and can be implemented on large scale data.

2. Regularized Logistic Regression

Logistic regression is a regression analysis method in which the response variable is a categorical variable [11]. If the response variable has only two categories, then it called binary logistic regression. Let there are n mutually exclusive samples. From each sample there are $\vec{\boldsymbol{X}}_i$ and y_i pairs, $i = 1, 2, ..., n$. $\vec{\boldsymbol{X}}_i$ is a vector of size p which is the number of predictor variables. y_i is the value of the response variable in the form of categorical 0 or 1. Let sample with response $y_i = 0$ belong to a negative class or majority class and sample with response $y_i = 1$ belong to a positive class or minority class. Then, y_i has a Bernoulli distribution with a probability function

$$P(y_i/\pi_i) = \pi_i^{y_i}(1-\pi_i)^{1-y_i}, \qquad (1)$$

π_i is the probability of the i-th occurrence. The logistic regression equation is

$$ln\left(\frac{\pi_i}{1-\pi_i}\right) = \vec{\boldsymbol{X}}_i^{\mathrm{T}}\vec{\boldsymbol{\beta}} \qquad (2)$$

$\vec{\boldsymbol{\beta}}$ is a parameter vector of size p . Then π_i will become

$$\pi_i = \frac{e^{\vec{X}_i^T \vec{\beta}}}{1 + e^{\vec{X}_i^T \vec{\beta}}} \qquad (3)$$

The parameter estimation method that is often used is Maximum Likelihood (ML). For that, the ln-likelihood function of the variable with Bernoulli distribution is

$$ln(L(\vec{\beta})) = \sum_{i=1}^{n} ln(\pi_i^{y_i}(1-\pi_i)^{1-y_i}) \qquad (4)$$

by substituting equation (3) into equation (4), then we get ln-likelihood function of logistic regression as follows

$$ln(L(\vec{\beta})) = \sum_{i=1}^{n} \left(y_i\, ln\left(\frac{e^{\vec{X}_i^T \vec{\beta}}}{1 + e^{\vec{X}_i^T \vec{\beta}}}\right) + (1-y_i)\, ln\left(\frac{1}{1 + e^{\vec{X}_i^T \vec{\beta}}}\right) \right) \qquad (5)$$

There is an overfitting problem in modeling, the model performs well on training data but poorly on testing data. Adding regularization is one way to reduce these problems [12]. The regularization process limits the influence of the predictor variables that do not affect the response variables. Adding regularization on ln-likelihood function produce regularized ln-likelihood as follows

$$ln\left(L(\vec{\beta})\right) = \sum_{i=1}^{n} \left(ln\left(\frac{e^{y_i \vec{X}_i^T \vec{\beta}}}{1 + e^{\vec{X}_i^T \vec{\beta}}}\right) \right) - \frac{\lambda}{2}\left\|\vec{\beta}\right\|^2 \qquad (6)$$

where $\left\|\vec{\beta}\right\|^2 = \sum_{j=1}^{p} \beta_j^{\,2}$ and λ is regularization parameter. The regression model in equation (6) usually called ridge regression. Then, the objective is to maximize the likelihood function on equation (6) to get $\hat{\vec{\beta}}$, which is estimated value of $\vec{\beta}$. For binary logistic regression, there is a statistical measure called $Deviance$ [11]. $Deviance$ has the same role as Sum Square Error (SSE) in linear regression. The value of $Deviance$ is obtained by formula

$$Deviance = -2ln\left(L\left(\hat{\vec{\beta}}\right)\right) \qquad (7)$$

Maximizing equation (6) with respect to $\vec{\beta}$ results in a solution that is not closed form. Therefore, numerical method is used. One of the numeric iteration methods is Iteratively Re-weighted Least Square (IRLS) which uses the Newton-Rhapson method.

The Newton-Rhapson method requires a gradient vector and a hessian matrix which are the first and second derivatives of the likelihood function on equation (6), respectively. The gradient vector is

$$G(\vec{\beta}) = \frac{d\, ln\left(L(\vec{\beta})\right)}{d\vec{\beta}} = X^T(\vec{Y} - \vec{\pi}) - \lambda\vec{\beta} \qquad (8)$$

While the hessian matrix is

$$H(\vec{\beta}) = \frac{d\, ln\left(L(\vec{\beta})\right)}{d^2\vec{\beta}} = -X^T V X - \lambda I \qquad (9)$$

where $v_i = \pi_i(1-\pi_i)$, $V = diag(v_1, v_2, \dots, v_n)$, and I is identity matrix size $p\ X\ p$. Then the update Newton Rhapson iteration for $\vec{\beta}$ on *(c+1)*-iteration is

$$\vec{\boldsymbol{\beta}}^{(c+1)} = \vec{\boldsymbol{\beta}}^{(c)} + (\boldsymbol{X}^T\boldsymbol{V}\boldsymbol{X} + \lambda\boldsymbol{I})^{-1}\, \boldsymbol{X}^T(\vec{\boldsymbol{Y}} - \vec{\boldsymbol{\pi}}) - \lambda\vec{\boldsymbol{\beta}}^{(c)} \quad (10)$$

Since $\vec{\boldsymbol{\beta}}^{(c)} = (\boldsymbol{X}^T\boldsymbol{V}\boldsymbol{X} + \lambda\boldsymbol{I})(\boldsymbol{X}^T\boldsymbol{V}\boldsymbol{X} + \lambda\boldsymbol{I})^{-1}\vec{\boldsymbol{\beta}}^{(c)}$, then

$$\begin{aligned}\vec{\boldsymbol{\beta}}^{(c+1)} &= (\boldsymbol{X}^T\boldsymbol{V}\boldsymbol{X} + \lambda\boldsymbol{I})(\boldsymbol{X}^T\boldsymbol{V}\boldsymbol{X} + \lambda\boldsymbol{I})^{-1}\vec{\boldsymbol{\beta}}^{(c)} + (\boldsymbol{X}^T\boldsymbol{V}\boldsymbol{X} + \lambda\boldsymbol{I})^{-1}\, \boldsymbol{X}^T(\vec{\boldsymbol{Y}} - \vec{\boldsymbol{\pi}}) - \lambda\vec{\boldsymbol{\beta}}^{(c)} \\ &= (\boldsymbol{X}^T\boldsymbol{V}\boldsymbol{X} + \lambda\boldsymbol{I})^{-1}\boldsymbol{X}^T\boldsymbol{V}\vec{\boldsymbol{Z}}^{(c)}\end{aligned} \quad (11)$$

where $\vec{\boldsymbol{Z}}^{(c)} = \boldsymbol{X}\vec{\boldsymbol{\beta}}^{(c)} + \boldsymbol{V}^{-1}(\vec{\boldsymbol{Y}} - \vec{\boldsymbol{\pi}})$ is *adjusted response*.

It becomes problematic if the number of observations is large. The large unit of observation causes the Hessian matrix to be formed to be large too. This requires a long processing time to compute the inverse of the Hessian matrix. Komarek [13], first introduced Truncated Regularized Iteratively Re-weighted Least Square (TR-IRLS) in logistic regression to handle large datasets, using Truncated Newtons with linear Conjugate Gradient (CG) to obtain estimates of $\vec{\boldsymbol{\beta}}$. Truncated Newton is a flexible and powerful method for optimizing large data [4][14] .

The TR-IRLS method has two iterations. First, find the solution of the Weighted Least Square (WLS) and stop it when the relative difference of deviance statistics [11] of two consecutive iterations does not exceed the specified limit (ε_1). The WLS sub problem is

$$(\boldsymbol{X}^T\boldsymbol{V}\boldsymbol{X} + \lambda\boldsymbol{I})\vec{\boldsymbol{\beta}}^{(c+1)} = \boldsymbol{X}^T\boldsymbol{V}\vec{\boldsymbol{Z}}^{(c)} \quad (12)$$

Whereas the second one looks for a solution to the equation (12) by the CG method [15] and stops the iteration when the residual value, $\vec{\boldsymbol{r}} = \boldsymbol{X}^T\boldsymbol{V}\vec{\boldsymbol{Z}}^{(c)} - (\boldsymbol{X}^T\boldsymbol{V}\boldsymbol{X} + \lambda\boldsymbol{I})\vec{\boldsymbol{\beta}}^{(c+1)}$, does not exceed a predetermined limit (ε_2).

3. Weighted Logistic Regression for Large Scale and Unbalanced Data

In order to handle unbalanced data, Maalouf and Siddiqi [1], based on King and Zeng [8], use weighting on logistic regression. If weight is added, the regularized ln-likelihood function becomes

$$\ln\left(L_W(\vec{\boldsymbol{\beta}})\right) = \sum_{i=1}^{n}\left(w_i \ln\left(\frac{e^{y_i \vec{x}_i^T \vec{\beta}}}{1 + e^{\vec{x}_i^T \vec{\beta}}}\right)\right) - \frac{\lambda}{2}\|\vec{\boldsymbol{\beta}}\|^2 \quad (13)$$

where $w_i = \left(\frac{\tau}{\bar{y}}\right) y_i + \left(\frac{1-\tau}{1-\bar{y}}\right)(1 - y_i)$, τ is proportion of minority class on population, and $\bar{y}$ is proportion of minority class on sample. The addition of weight aims to obtain consistent estimators. So that, the minority class is given less weight than the majority [1].

Optimization processes such as those used in TR-IRLS [2] is also used . So that, we need a Gradient vector and a Hessian matrix from the likelihood function in equation (13). The gradient vector become

$$G(\vec{\boldsymbol{\beta}}) = \frac{d\ \ln\left(L(\vec{\boldsymbol{\beta}})\right)}{d\vec{\boldsymbol{\beta}}} = \boldsymbol{X}^T\boldsymbol{W}(\vec{\boldsymbol{Y}} - \vec{\boldsymbol{\pi}}) - \lambda\vec{\boldsymbol{\beta}} \quad (14)$$

where $\boldsymbol{W} = diag(w_1, w_2, \dots, w_n)$ and the Hessian Matrix become

$$H(\vec{\boldsymbol{\beta}}) = \frac{d\ \ln\left(L(\vec{\boldsymbol{\beta}})\right)}{d^2\vec{\boldsymbol{\beta}}} = -\boldsymbol{X}^T\boldsymbol{D}\boldsymbol{X} - \lambda\boldsymbol{I} \quad (15)$$

where $\boldsymbol{D} = diag\{v_1w_1, v_2w_2, \dots, v_nw_n\}$. Then the WLS sub problem is

$$(\boldsymbol{X}^T\boldsymbol{D}\boldsymbol{X}+\lambda\boldsymbol{I})\vec{\boldsymbol{\beta}}^{(c+1)}=\boldsymbol{X}^T\boldsymbol{D}\boldsymbol{z}^{(c)} \quad (16)$$

Regularization in equation (13) will produce bias [1]. So bias correction is required [17]. Maalouf and Siddiqi [1] developed the bias vector described by King and Zeng [8] and Mccullagh and Nelder [18] as follows

$$\boldsymbol{B}\left(\hat{\vec{\boldsymbol{\beta}}}\right)=(\boldsymbol{X}^T\boldsymbol{D}\boldsymbol{X}+\lambda\boldsymbol{I})^{-1}\boldsymbol{X}^T\boldsymbol{D}\vec{\xi} \quad (17)$$

where $\xi_i=0{,}5Q_{ii}((1+w_1)\hat{\pi}_i-w_1)$ and w_1 is minority class weight. While Q_{ii} is the main diagonal of the matrix $\boldsymbol{Q}$ as follows

$$\boldsymbol{Q}=\boldsymbol{X}(\boldsymbol{X}^T\boldsymbol{D}\boldsymbol{X}+\lambda\boldsymbol{I})^{-1}\boldsymbol{X}^T \quad (18)$$

The bias vector is also optimized using the Truncated IRLS method with linear CG. WLS sub problem for the bias vector is

$$(\boldsymbol{X}^T\boldsymbol{D}\boldsymbol{X}+\lambda\boldsymbol{I})\boldsymbol{B}\left(\hat{\vec{\boldsymbol{\beta}}}\right)=\boldsymbol{X}^T\boldsymbol{D}\vec{\xi} \quad (19)$$

After obtaining the bias vector, we will get $\tilde{\vec{\boldsymbol{\beta}}}$ which is the estimation of the bias corrected parameter as follows

$$\tilde{\vec{\boldsymbol{\beta}}}=\hat{\vec{\boldsymbol{\beta}}}-\boldsymbol{B}\left(\hat{\vec{\boldsymbol{\beta}}}\right) \quad (20)$$

4. Entropy Based Fuzzy Weighted Logistic Regression (EF-WLR) Algorithm

4.1 Entropy Based Fuzzy Membership

Entropy is an important measure in information theory that can be used to select an event from several events that have a probability [16]. Entropy in a case that has two chances of occurrence is written as

$$H_i=-p_{+i}\ln(p_{+i})-p_{-i}\ln(p_{-i}) \quad (21)$$

p_{+i} and p_{-i} are the probability of the i-th training sample ($\vec{\boldsymbol{X}}_i$) being the minority class and the majority class, respectively [10]. Both of these probabilities are based on it's k nearest neighbors. First, we determine the k nearest neighbors for each sample. Then, we count the number of minority class (num_{+i}) and majority class (num_{-i}) in k nearest neighbors. The probability of the i-th training sample being the minority class and the majority class is calculated by

$$p_{+i}=\frac{num_{+i}}{k}$$

$$p_{-i}=\frac{num_{-i}}{k} \quad (22)$$

Finally, the entropy of i-th training sample can be calculated by equation (21).

To get fuzzy membership value for majority class [10], first, we select entropy of majority class. Then, separated into m subset, with increasing order of entropy. Finally, we compute fuzzy membership in each subset by means of

$$FM_l=1-\varphi(l-1), l=1,2,\ldots,m \quad (23)$$

where FM_l is the fuzzy membership value of observation on l-th subset and φ is fuzzy membership parameter which has value $0<\varphi\leq\frac{1}{m-1}$.

4.2 Using Entropy Based Fuzzy Membership as a Weight on Logistic Regression

IOP Conf. Series: Earth and Environmental Science **880** (2021) 012048 doi:10.1088/1755-1315/880/1/012048

In this study, the purpose of weighing the TR-IRLS is to increase the interest of minority class observations. so that the accuracy of the minority class can be increased. It is necessary to give weight to the minority class larger than the majority [10]. Then ln likelihood function in equation (13) will have a value of w_i as follow

$$w_i = \begin{cases} 1, if\ y_i = 1 \\ \\ FM_l, if\ y_i = 0\ and\ \vec{\mathbf{X}}_i \epsilon\ l - th\ subset \end{cases} \quad (24)$$

where FM_l is the fuzzy membership value of observation on *l*-th subset which computed by equation (23) and $0 < FM_l \leq 1$.

The value of $\vec{\widetilde{\boldsymbol{\beta}}}$ is obtained by Truncated IRLS by Komarek [2]. The gradient vector and the hessian matrix have the same form with the equation (14) and (15), respectively. So the WLS sub problem still has the same form as equation (16).

The bias vector $\boldsymbol{B}\left(\vec{\widehat{\boldsymbol{\beta}}}\right)$ still have same form with equation (17). In this study, because of w_1 has value equals to 1, then ξ_i become

$$\xi_i = 0{,}5Q_{ii}(2\hat{\pi}_i - 1) \quad (25)$$

Finaly, $\vec{\widetilde{\boldsymbol{\beta}}}$ is calculated using equation (20).

Algorithm 1. WLR MLE using IRLS

Input : $\mathbf{X}, \vec{\mathbf{Y}}, \vec{\widetilde{\boldsymbol{\beta}}}^{(0)}, \boldsymbol{B}\left(\vec{\widehat{\boldsymbol{\beta}}}\right)^{(0)}, m, \varphi, k$

Output : $\vec{\widetilde{\boldsymbol{\beta}}}$

Begin

$c = 0$

$deltadev = 1$

Get $\vec{\boldsymbol{w}}$ using algorithm 2

Do While $|deltadev > \varepsilon_1|$ dan $c \leq maxIRLS$

If $c > 0$

$\vec{\widetilde{\boldsymbol{\beta}}}^{(c)} = \vec{\widetilde{\boldsymbol{\beta}}}$

End If

For *i*=1 to *n*

$$\hat{\pi}_i = \frac{\exp\left(\vec{\mathbf{X}}_i^T \vec{\widetilde{\boldsymbol{\beta}}}^{(c)}\right)}{1+\exp\left(\vec{\mathbf{X}}_i^T \vec{\widetilde{\boldsymbol{\beta}}}^{(c)}\right)}$$

$$v_i = \hat{\pi}_i(1 - \hat{\pi}_i)$$

$$z_i = \vec{\mathbf{X}}_i^T \vec{\widetilde{\boldsymbol{\beta}}}^{(c)} + \frac{(y_i - \hat{\pi}_i)}{v_i}$$

End For

$\boldsymbol{D} = diag(v_i\, w_i)$

$\boldsymbol{Q} = \mathbf{X}(\mathbf{X}^T\boldsymbol{D}\mathbf{X} + \lambda\boldsymbol{I})^{-1}\mathbf{X}^T$

For *k*=1 to *n*

$\xi_k = 0{,}5Q_{kk}(2\hat{\pi}_k - 1)$

End For

Get $\vec{\widetilde{\boldsymbol{\beta}}}^{(c+1)}$ using algorithm 3

Get $\boldsymbol{B}\left(\vec{\widehat{\boldsymbol{\beta}}}\right)^{(c+1)}$ using algorithm 4

Algorithm 1. WLR MLE using IRLS (continuation)

$$\vec{\beta} = \vec{\beta}^{(c+1)} - \boldsymbol{B}\left(\vec{\beta}\right)^{(c+1)}$$

$$deltadev = \frac{deviance^{(c)} - deviance^{(c+1)}}{deviance^{(c+1)}}$$

$c = c + 1$

End While

End

Algorithm 2. Entropy Based Fuzzy Membership to get $\vec{w}$

Input : $\boldsymbol{X}, \vec{\boldsymbol{Y}}, m, \varphi, k$

Output : $\vec{\boldsymbol{w}}$

Begin

For *i*=1 to *n*

Get k nearest neighbours of $\vec{\boldsymbol{X}}_i$

Count num_{+i} and num_{-i} in the k nearest neighbours

Calculate class probability using equation (22)

$H_i = -p_{+i}\ln(p_{+i}) - p_{-i}\ln(p_{-i})$

End For

Select entropy of majority class, H_{-}

$H_{min} = min\{H_{-}\}$

$H_{max} = max\{H_{-}\}$

For *l*=1 to *m*

$$thrUp = H_{min} + \frac{l}{m}(H_{max} - H_{min})$$

$$thrLow = H_{min} + \frac{l-1}{m}(H_{max} - H_{min})$$

For *j*=1 to *n*

If $y_i = majority\ class$

If $thrLow \leq H_i < thrUp$

$FM = 1 - \varphi(l-1)$

$w_i = FM$

End If

End If

If $\boldsymbol{y}_i = minority\ class$

$w_i = 1$

End If

End For

End For

End

Algorithm 3. Linier CG to get $\vec{\hat{\beta}}$

Input : $\boldsymbol{A} = \boldsymbol{X}^T \boldsymbol{D} \boldsymbol{X} + \lambda \boldsymbol{I}$, $\vec{\boldsymbol{b}} = \boldsymbol{X}^T \boldsymbol{D} \vec{\boldsymbol{z}}, \vec{\hat{\boldsymbol{\beta}}}^{(0)}$

Output : $\vec{\hat{\boldsymbol{\beta}}}$

Begin

$t = 0$

$\vec{\boldsymbol{r}}^{(0)} = \vec{\boldsymbol{b}} - \boldsymbol{A}\vec{\hat{\boldsymbol{\beta}}}^{(0)}$

$\vec{\boldsymbol{d}}^{(0)} = \vec{\boldsymbol{r}}^{(0)}$

Do While $\left\|\vec{\boldsymbol{r}}^{(t)}\right\|^2 > \varepsilon_2$ and $t \le maxCGBeta$

$s^{(t)} = \frac{\vec{\boldsymbol{r}}^{T(t)}\vec{\boldsymbol{r}}^{(t)}}{\vec{\boldsymbol{d}}^{T(t)}\boldsymbol{A}\vec{\boldsymbol{d}}^{(t)}}$

$\vec{\hat{\boldsymbol{\beta}}}^{(t+1)} = \vec{\hat{\boldsymbol{\beta}}}^{(t)} + s^{(t)}\vec{\boldsymbol{d}}^{(t)}$

$\vec{\boldsymbol{r}}^{(t+1)} = \vec{\boldsymbol{r}}^{(t)} - s^{(t)}\boldsymbol{A}\vec{\boldsymbol{d}}^{(t)}$

$a^{(t)} = \frac{\vec{\boldsymbol{r}}^{T(t+1)}\vec{\boldsymbol{r}}^{(t+1)}}{\vec{\boldsymbol{r}}^{T(t)}\vec{\boldsymbol{r}}^{(t)}}$

$\vec{\boldsymbol{d}}^{(t+1)} = \vec{\boldsymbol{r}}^{(t+1)} + a^{(t)}\vec{\boldsymbol{d}}^{(t)}$

$t = t + 1$

End While

End

Algorithm 4. Linier CG to get $B\left(\vec{\hat{\beta}}\right)$

Input : $\boldsymbol{A} = \boldsymbol{X}^T \boldsymbol{D} \boldsymbol{X} + \lambda \boldsymbol{I}$, $\vec{\boldsymbol{b}} = \boldsymbol{X}^T \boldsymbol{D} \vec{\xi}, \boldsymbol{B}\left(\vec{\hat{\boldsymbol{\beta}}}\right)^{(0)}$

Output : $\boldsymbol{B}\left(\vec{\hat{\boldsymbol{\beta}}}\right)$

Begin

$t = 0$

$\vec{\boldsymbol{r}}^{(0)} = \vec{\boldsymbol{b}} - \boldsymbol{A}\boldsymbol{B}\left(\vec{\hat{\boldsymbol{\beta}}}\right)^{(0)}$

$\vec{\boldsymbol{d}}^{(0)} = \vec{\boldsymbol{r}}^{(0)}$

Do While $\left\|\vec{\boldsymbol{r}}^{(t)}\right\|^2 > \varepsilon_2$ and $t \le maxCGBias$

$s^{(t)} = \frac{\vec{\boldsymbol{r}}^{T(t)}\vec{\boldsymbol{r}}^{(t)}}{\vec{\boldsymbol{d}}^{T(t)}\boldsymbol{A}\vec{\boldsymbol{d}}^{(t)}}$

$\boldsymbol{B}\left(\vec{\hat{\boldsymbol{\beta}}}\right)^{(t+1)} = \boldsymbol{B}\left(\vec{\hat{\boldsymbol{\beta}}}\right)^{(t)} + s^{(t)}\vec{\boldsymbol{d}}^{(t)}$

$\vec{\boldsymbol{r}}^{(t+1)} = \vec{\boldsymbol{r}}^{(t)} - s^{(t)}\boldsymbol{A}\vec{\boldsymbol{d}}^{(t)}$

$a^{(t)} = \frac{\vec{\boldsymbol{r}}^{T(t+1)}\vec{\boldsymbol{r}}^{(t+1)}}{\vec{\boldsymbol{r}}^{T(t)}\vec{\boldsymbol{r}}^{(t)}}$

$\vec{\boldsymbol{d}}^{(t+1)} = \vec{\boldsymbol{r}}^{(t+1)} + a^{(t)}\vec{\boldsymbol{d}}^{(t)}$

$t = t + 1$

End While

End

5. Conclusions

Entropy Based Fuzzy Weighted Logistic Regression (EF-WLR) based on Rare Event Weighted Logistic Regression. Weight on weighted logistic regression using entropy based fuzzy membership, allows sample interests not only to be seen based on class differences but also based on class certainty. Regularization process is used to reduce overfitting. Bias correction was applied to obtain unbiased parameters due to the addition of the regularization process. The computation process will be longer if the sample is large. Therefore, truncated IRLS is used to optimize the value of $\vec{\boldsymbol{\beta}}$.

Reference

[1] M Maalouf dan M Siddiqi 2014 "Weighted logistic regression for large-scale imbalanced and rare events data," *Knowledge-Based Syst.*, vol. **59**, hal. 142–148
[2] P Komarek dan A W Moore 2005 "Making logistic regression a core data mining tool with TR-IRLS," *Proc. - IEEE Int. Conf. Data Mining, ICDM*, hal. 4–7
[3] J Gareth, D Witten, T Hastie, dan R Tibshirani 2013 An Introduction to Statistical Learning. New York: *Springer Science+Business Media*
[4] S P Rahayu, J M Zain, A Embong, Juwari, dan S W Purnami 2012 "Logistic regression methods with truncated newton method" *Procedia Eng* **50** hal. 827–836
[5] F Ladaya, S W Purnami, dan Irhamah 2018 "Entropy Based Fuzzy Support Vector Machine Untuk Klasifikasi Microarray Imbalanced Data" Sepuluh Nopember Institute of Technology, Surabaya
[6] G King dan L Zeng 2001 "Improving Forecasts of State Failure," *World Polit.* **53** 4. 623–658
[7] A Ali, S M Shamsuddin, dan A L Ralescu 2015 "Classification with class imbalance problem: A review," *Int. J. Adv. Soft Comput. its Appl.* **7** 3 176–204
[8] G King dan L Zeng 2001 "Logistic regression in rare events data," *J. Stat. Softw.* **8** 137–163
[9] C F Lin dan S De Wang 2002 "Fuzzy support vector machines," *IEEE Trans. Neural Networks*, **13** 2 464–471
[10] Q Fan, Z Wang, D Li, D Gao, dan H Zha 2017 "Entropy-based fuzzy support vector machine for imbalanced datasets" *Knowledge-Based Syst.* 115 87–99
[11] D W Hoshmer dan S Lemeshow 2000 Applied Logistic Regression, Second Edition. *John Wiley & Sons , Inc*
[12] X Ying 2019 "An Overview of Overfitting and its Solutions," *J. Phys. Conf. Ser.* **1168** 2
[13] P Komarek 2004 "Logistic Regression for Data Mining and High- Dimensional Classification" Carnegie Mellon University
[14] S G Nash 2000 "A survey of truncated-Newton methods" *J. Comput. Appl. Math.* **124** 45–59
[15] M R Hestenes dan E Stiefel 1934 "Methods of conjugate gradients for solving linear systems," *J. Res. Natl. Bur. Stand.* **49** 6 409
[16] C E Shannon 1948 "A Mathematical Theory of Communication" *Bell Syst. Tech. J.* **27** 4 623–656

A study on carbon cap and trade effect to cost of electricity in accordance with the Merit Order of 300-400 MW coal power plants

A S Putra[1*], I Garniwa[1], E S Arilanggaaji[2] and S S Pradana[2]

[1] Faculty of Engineering, University of Indonesia, Depok, Indonesia
[2] Department of Operation and Maintenance, PT. Pembangkitan Jawa-Bali, Jakarta, Indonesia

*Corresponding E-mail: adityasukmanaputra@gmail.com

Abstract. Electricity is the basis of national development in a country. Power plants in Indonesia produces up to 283.8 TWh and are dominated by coal power plants which increase the amount of the greenhouse gases (GHG). In order to prevent more environmental problems, Indonesia ratified Paris Agreement by publishing the roadmap of Nationally Determined Contribution (NDC) that committed in reducing 29% of GHG emissions in 2030, which 11% of them are from the energy sector contributions. This research focuses on the implementation of the carbon cap and trade (CAT) between coal power plants having 300-400 MW capacity, which can affect their cost of electricity (Rp/kWh). It is well known that cap and trade (CAT) is a method used for reducing the mitigation cost of emission reduction in an effective way. From this research, it is found that the highest rise of incremental cost belongs to the 300 MW power plant in scenario 9 and the increase is from Rp.431.00/kWh to Rp.462.77/kWh, or approximately 7.37%. This research also shows that the most optimal carbon price is in the range of Rp. 130,165 to Rp.130,183 because the rank of the 330 MW and 400 MW power plant in merit order changes over in this condition. In the future, this research can be used as a comparison with the higher coal power plant capacity, so that an alternative way is obtained to determine the more optimal merit order.

Keywords: Greenhouse gases, Cap and Trade, Nationally Determined Contribution, Indonesia

Track Name : Human, Social, Economic and Environmental Sustainability

1. Introduction

Under Paris Agreement in 2015, Indonesia was committed in reducing greenhouse gas (GHG) emissions by publishing NDC (Nationally Determined Contribution) and pledged to reduce the emissions by 29% against the business as usual (BAU) baseline on its own effort, which 11% of them are from energy sector, and by 41% against the BAU baseline with international support, so that Indonesia could keep the average global temperature increase under 2 °C above pre-industrial level and limit the temperature growth to 1.5 °C [1]. The Electricity Supply Business Plan (RUPTL) 2019-2028 stated that in 2019 coal power plants produced 83% GHG emissions among all power plants in Indonesia in terms of mixed energy projected [2]. In 2018, based on the usage of primary energy in the power system of all over Indonesia, coal power plants dominated the system by about 60% [2]. The National Electricity Master Plan (RUKN) 2019-2038 targeted the use of coal in primary energy mix by approximately 55% in 2025 and by approximately 47% in 2038 [3]. In accordance with the Paris Agreement, Indonesia could make a policy about some methods to accelerate GHG emission reduction. Some of those methods are carbon tax and carbon emission trading. Carbon tax is a form of carbon pricing by putting tax on companies emitting carbon dioxide [4]. Carbon emission trading is also a form of carbon pricing where the government sets a carbon cap for all over the country to limit the carbon dioxide produced by companies. Companies emitting carbon dioxide under the cap can sell the allowance to the companies producing carbon dioxide over the cap [4]. Carbon emission trading prices will fluctuate based on supply and demand schemes.

Studies about emission trading in Indonesia are very rare. Ragimun has stated that Indonesia would face some challenges if it implemented carbon trading system, such as the lack of suitable system to calculate emission reduction, the absence of required carbon credit policy, no regulation tying the voluntary system, and the requisite of the big number of participants [5]. Besides, previous research is also taken from other countries. In India Surender Kumar et al. implied that if its government constituted a cap and trade system, it would obtain economic gain with the provision of banks with over a 5 year period of study [6]. Linda Meleo *et al.* presented Italian aviation case study where direct costs from EU-ETS could only slightly impact companies' revenues, airfares, and social costs [7]. E. Delarue *et al.* simulated that the cap and trade policy in EU without renewable energy generation (RES) policy could increase CO2 price and raise the quantity of emissions from the electric power sector unless there were any external factors (such as economic recession) [8].

Compared with previous studies, this paper will detail the initiation of carbon trading in Indonesia's energy sectors by analysing the changes in cost of electricity because of carbon price variations in cap and trade simulations among 25 coal power plants which produce 300 to 400 MW of electricity in Jamali power system.

This paper is organized as follows, section 2 presents briefly about Jamali (Java Madura Bali Island) power system, cost of electricity, merit order and cap & trade mechanism. Section 3 shows the methodology used for estimating cost of electricity variations based on the cap and trade scheme. Section 4 provides analysis and results of this research. Section 5 is the summary of this paper.

2. Study Literature

2.1. Jamali (Java Madura Bali Island) Power System
Jamali power system contains electric power generation, transmission, and distribution systems that are interconnected over Java, Madura, and Bali Island. It is the biggest power system built in Indonesia. Many types of power plants supply electricity into the power system, but the coal power plants dominate the system.

2.2. Cost of electricity
Based on the regulation of Ministry of Energy and Mineral Resources, electric power purchase consists of 2 main components named component A and C. Component A is the investment cost of power plant capacity which is depreciated for 20 years or more, meanwhile component C is the fuel cost of primary energy used in the power generation [2]. Calculating the primary energy needed on a thermal power plant can be done using heat rate value, as shown in equation (1) and (2). So, component C has a relationship to the heat rate of a power plant.

2.3. Merit Order
Merit order is a ranking system of all available power plants from the lowest value of marginal generation cost to the highest value of it. Because marginal generation cost relates to heat rate value [8], merit order also has a relationship to the fuel cost of a power plant (component C).

2.4. Cap and Trade Mechanism
In general, carbon cap is a carbon emission value which is set by the government to be followed by all entities which produce emissions. The gap between the emissions produced and the cap could be traded by those entities. If an entity emits more carbon dioxide than the cap, it has a carbon dioxide deficit and must buy carbon allowances to fulfil its deficit from the one which has a carbon dioxide surplus. The carbon price goes based on carbon supply and demand. Cap and trade mechanism concerns about allowance allocation and pricing. Allowance allocation could be done by grandfathering or benchmarking schemes. On the other hand, its pricing could be made by an auction system [9].

3. Methodology
This section explains sequences of the methodology used in this research. At first, describing quantitative methods is done by literary reviews from some references, such as articles/journals, books and policies that are still applicable. The next sequence is about classifying data needed which will be processed with numerical computation based on scenarios used in this research.

Technical data used in this research are power plant capacities, primary energy types used by power plants, carbon dioxide equivalent emitted by power plants that generate 300 MW to 400 MW electricity in Jamali power system. Cap and trade implementation in several countries will also be used as descriptive data of this research. In this research, the cap will be varied into 3 values and the fluctuation of carbon prices are going to be simulated also into 3 values. So, there will be 9 scenarios where each cap value will be simulated with every value of determined carbon price.

After collecting all data, numerical computation is done by putting GHG emission components, which is carbon dioxide in this case, into the cost of electricity of power plants. The calculation focuses on the coal power plants that produce 300 MW to 400 MW electricity for Jamali power system. Adding carbon cost directly to cost of electricity could affect the merit order of Jamali power system.

3.1. Jamali Power Generation System
Reference [10] provides data of 25 units of 300 MW to 400 MW coal power plants in 2019. Based on their capacity, the data are classified to 300 MW, 315 MW, 330 MW and 400 MW coal power plants as shown in Table 1.

Table 1. Coal Power Plants with 300-400 MW Capacity in Jamali Power System.

Capacity (MW)	Gross Power Generated (MWh)	Nett Power Generated (MWh)
300	7,428,025.83	7,026,816.80
315	14,900,110.29	14,127,474.42
330	7,013,188.80	6,503,017.61
350	12,067,305.53	11,429,056.58
400	17,551,572.56	16,421,882.29

Source :Data processed from the Directorate General of Electricity ESDM [10]

Heat rate tests are done by each coal power plants to measure the heat energy input of the fuel used for generating 1 kWh electricity. Table 2 shows the data taken from [11], the heat rate tests of 400 MW coal power plants based on ASME PTC PM-2010, "Performance Monitoring Guidelines for Power Plants.

Table 2. Data of Heat Rate Tests of 400 MW Coal Power Plants.

Load (MW)	Heat Rate (kCal/kWh)	Operation Cost (Rp/h)
239,260	2795	99,598,338
280,996	2760	115,507,292
325,748	2731	132,496,266
372,360	2669	148,017,061

Source: Reprocessed data of PT PJB, 2019 [11]

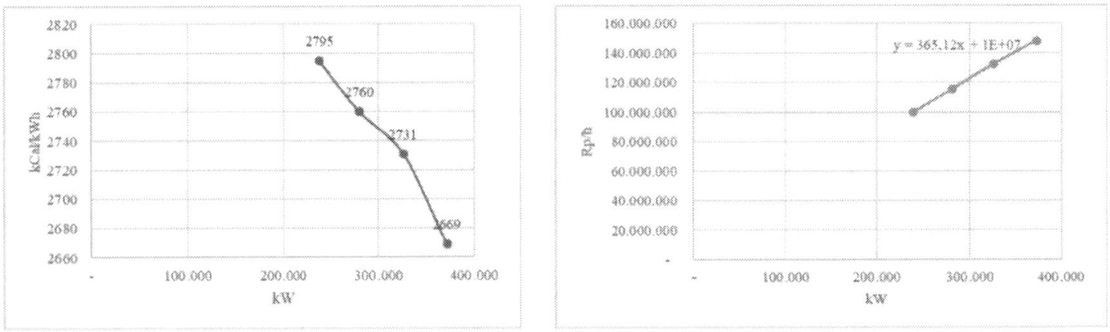

(a) (b)

Figure 1. (a) Characteristic Curve of 400 MW Coal Power Plants. (b) Incremental Cost of 400 MW Coal-Fired Power Plants.

This research uses an assumption that the coal price is Rp. 700/kg and coal heating value is 4700 kCal/kg. Those data can be processed into component C (fuel cost) used as the main variable of merit order. That cost can be formulated as follows:

$$Component\ C = E\ x \frac{NPHR}{HHV} xFP \tag{1}$$

$$Fuel\ Cost = \frac{NPHR}{HHV} x\ FP \tag{2}$$

Where E is the energy produced by a power plant, NPHR means Net Plant Heat Rate (kCal/kWh), HHV is High Heating Value (kCal/kg), FP is Fuel Price (Rp/kg) and Fuel Cost means cost of electricity made of the used fuel (Rp/kWh).

3.2. Merit Order of The Power Plants

Merit order is a method of processing technical data and economy in order to rank power plants based on their efficiency from the most efficient one. Variables that could affect merit order are fuel cost, heat rate and fuel heating value. The relation of those variables can be simplified in equation (2). From equation (1) and (2), merit order can be determined, as shown in Table 3.

3.3. Power Plant Emission Data

Based on the regulation of the Ministry of Environment and Forestry [12], power plants must report their emissions using CEMS (Continuous Emission Monitoring System) technology or using flue gas measurement through their chimney every 6 months.

Table 3. Merit Order Based on Power Plants' Capacities.

Capacity (MW)	Cost of Electricity (Rp/ kWh)	Merit Order
300	431	5
315	431	4
330	373	2
350	395	3
400	365	1

The report needs to be sent to the Directorate General of Electricity ESDM as Formula Tier-3 Method 4 [13]. In addition to the previous equations, GHG emission calculation, in accordance with [13] that referred to [14], can be obtained by using the equation of Tier-3 Method 3 of coal fuel as follows :

$$E_{CO_2} = F_{BB} x \{C_{ar} - (A_{ar} x C_{ub})\} x 44/12 \quad (3)$$

Where E_{CO_2} is total CO2 emission (ton CO2), F_{BB} is fuel consumption in a year period (ton), C_{ar} is carbon content as received (%), A_{ar} refers to ash content as received (%) and C_{ub} refers to unburned carbon content (%).

From the secondary data provided by Directorate General of Electricity ESDM [10] and data processed by manual calculation using equation (3), data of carbon dioxide emission equivalent can be made as shown in Table 4. Those data are from Jamali power system in 2019. Emission intensity is a ratio of emissions produced by power plants to the gross energy generated by them [15]. From table 4, it can be calculated that the average of 300 MW to 400 MW coal power plant emission intensity is approximately 0,98 tCO2e/MWh.

Table 4. Emission Data of the 300-400 MW Power Plants.

Capacity (MW)	Emission (ton CO2)	Gross Emission Intensity (tCO2e/MWh)
300	7,471,938.05	1.01
315	14,669,553.48	0.98
330	6,430,040.28	0.92
350	11,995,667.67	0.99
400	17,154,557.85	0.98

Source: Data processed from Directorate General of Electricity ESDM [10]

3.4. Carbon Pricing

This research calculates the cost of electricity in a cap and trade system that are varied as follows:

Table 5. Various Scenarios of Cap and Trade of the 300-400 MW Power Plants

Scenario	Cap (tCO2e/MWh)	Price (Rp/tCO2e)
1,2,3	0.98	50,000; 150,000; 300,000
4,5,6	0.95	50,000; 150,000; 300,000
7,8,9	0.90	50,000; 150,000; 300,000

3.5. Merit Order Calculation after Cap and Trade Implementation

Reference [8] implied that in order to minimize power plant emissions, emissions must be regulated into an optimal dispatch, which is alternating the high emission power plants to the lower emission power plants. In calculating merit order, power plants will be ranked by their marginal cost MC (Rp/kWh), as illustrated in equation (4), meanwhile the additional cost of electricity from carbon price can be seen in equation (5). Where $\Delta\lambda$ is the increase/decrease of the cost of electricity (%),FC means fuel cost (Rp), EC means power plant emission intensity (tCO2e/kWh), AC refers to Allowance Cost (Rp/tCO2e), and λ refers to the baseline of power plant's electricity price.

$$MC = \frac{FC}{\eta} + \frac{EC}{\eta}\text{AC} \tag{4}$$

$$\Delta\lambda = \frac{AC\ x\ EC}{\lambda}100\% \tag{5}$$

4. Analysis and Result

Based on table 5, scenario 1 – 9 have their own cap value and carbon price. Each power plant will have an obligation to buy carbon deficit if it produces emission intensity above the cap or it will have a privilege to sell its carbon surplus if it produces less emission intensity than the cap. Therefore, the buying and selling schemes will affect the power plants' cost of electricity and their merit order.

4.1. Scenario 1

In scenario 1, table 6 shows that the 300 MW power plant has the biggest deficit of carbon dioxide, which is 199,936.22 million tons of CO2e. That power plant is obliged to pay Rp 50,000.-/tCO2e for buying the carbon allowances. Therefore, its cost of electricity will be added by 1.35 Rp/kWh, which initial cost is Rp. 431.00/kWh and its final cost becomes Rp.432.35/kWh. Power plant that gets the biggest surplus is the 330 MW power plant, which surplus is 435,838.08 million tons of CO2e, and had a chance to sell the surplus. The revenue from selling its carbon dioxide surplus is able to reduce its cost of electricity by Rp.3.11/kWh where its initial cost of electricity is Rp.373.-/kWh and its final cost becomes Rp.369.89/kWh. The merit order will be the 400 MW, 330 MW, 350 MW, 315 MW and 300 MW power plant, respectively from the least cost of electricity.

Table 6. Cap of 0.98 tCO2/MWh and Carbon Price of Rp. 50,000/tCO2e Scenario.

Capacity (MW)	Surplus/Deficit (TCO2e)	Selling/ Buying (Rp/kWh)	Cost Of Electricity (Rp/kWh)	Merit Order
300	-199,936.22	-1.35	432.35	5
315	-82,416.62	-0.28	431.28	4
330	435,838.08	3.11	369.89	2
350	-181,833.15	-0.75	395.75	3
400	28,347.91	0.08	365.04	1

4.2. Scenario 2

In scenario 2, as shown in Table 7, there are no carbon dioxide surplus and deficit differences, but there is a change in the carbon price. The power plant with 300 MW capacity has the largest carbon dioxide deficit, which is 199,936.22 million tons of CO2e. That power plant is obliged to pay Rp. 150,000.-/tCO2e for buying and fulfilling its carbon deficit. That obligation makes the power plant add its cost of electricity by 4.04 Rp/kWh where its initial cost is Rp. 431.00/kWh and its final cost becomes Rp.435.04/kWh. Power plant that has the largest surplus is the 330 MW power plant, which is 435,838.08 million tons of CO2e, and it has the chance to sell its carbon surplus in order to reduce its cost of electricity. The decrease is Rp.9.32/kWh where its initial cost of electricity is Rp.373.00/kWh and its final cost is Rp.363.68/kWh. The merit order has changed. The previous most economical power plant, the 400 MW power plant, is replaced by the 330 MW power plant. The merit order becomes the 330 MW, 400 MW, 350 MW, 315 MW and 300 MW power plant, respectively.

Table 7. Cap of 0,98 tCO2/MWh and Carbon Price of Rp. 150.000/tCO2e Scenario.

Capacity (MW)	Surplus/Deficit (tCO2e)	Selling/ Buying (Rp/kWh)	Cost Of Electricity (Rp/kWh)	Merit Order
300	-199,936.22	-4.04	435.04	5
315	-82,416.62	-0.83	431.83	4
330	435,838.08	9.32	363.68	1
350	-181,833.15	-2.26	397.26	3
400	28,347.91	0.24	364.88	2

4.3. Scenario 3

In scenario 3, as shown in Table 8, there are no carbon dioxide surplus and deficit differences, but there is a change in the carbon price. The power plant with 300 MW capacity gets the largest carbon dioxide deficit, which is 199,936.22 million tons of CO2e. That power plant must pay Rp. 300,000.-/tCO2e for buying and fulfilling its carbon deficit. That obligation makes the power plant add its cost of electricity by 8.07 Rp/kWh where its initial cost is Rp. 431.-/kWh and its final cost becomes Rp. 439.07/kWh. Power plant that has the largest surplus is the 330 MW power plant, which amount is 435,838.08 million tons of CO2e, and it has the chance to sell its carbon surplus to reduce its cost of electricity. The reduction is Rp. 18.64/kWh where its initial cost of electricity is Rp.373.00/kWh and its final cost is Rp. 354.36/kWh. The merit order has changed. The initial most economical power plant, the 400 MW power plant, is still replaced by the 330 MW power plant. The merit order becomes the 330 MW, 400 MW, 350 MW, 315 MW and 300 MW power plant, respectively.

Table 8. Cap of 0.98 tCO2/MWh and Carbon Price of Rp.300,000/tCO2e Scenario.

Capacity (MW)	Surplus/Deficit (tCO2e)	Selling/ Buying (Rp/kWh)	Cost Of Electricity (Rp/kWh)	Merit Order
300	-199,936.22	-8.07	439.07	5
315	-82,416.62	-1.66	432.66	4
330	435,838.08	18.64	354.36	1
350	-181,833.15	-4.52	399.52	3
400	28,347.91	0.48	364.64	2

4.4. Scenario 4

In scenario 4, table 9 shows that the 350 MW power plant has the biggest deficit of carbon dioxide, which is 531,727.42 million tons of CO2e. That power plant is obliged to pay Rp 50,000./tCO2e for buying the carbon allowances. Therefore, its cost of electricity will be added by 1.73 Rp/kWh, which initial cost is Rp. 395.00/kWh and its final cost becomes Rp.397.20/kWh. Although it has the biggest amount of carbon deficit, the highest increase of cost of electricity occurs in the 300 MW power plant, which gets Rp.2.80/kWh increase. It happens because electric energy generated by the 300 MW power plant is significantly lower than the 315 MW power plant. Power plant that gets the biggest surplus is the 330 MW power plant, which produces 232,489.08 million tons of CO2e, and it is the only power plant that gets the chance to sell the surplus. The revenue from selling its carbon dioxide surplus can reduce its cost of electricity by Rp.1.66/kWh where its initial cost of electricity is Rp.373.-/kWh and its final cost becomes Rp.371.34/kWh. The merit order will be the 400 MW, 330 MW, 350 MW, 315 MW and 300 MW power plant, respectively from the least cost of electricity.

Table 9. Cap of 0.95 tCO2/MWh and Carbon Price of Rp.50,000/tCO2e Scenario

Capacity (MW)	Surplus/Deficit (TCO2e)	Selling/ Buying (Rp/kWh)	Cost Of Electricity (Rp/kWh)	Merit Order
300	-415,313.51	-2.80	433.80	5
315	-514,448.70	-1.73	432.73	4
330	232,489.08	1.66	371.34	2
350	-531,727.42	-2.20	397.20	3
400	-480,563.92	-1.37	366.49	1

4.5. Scenario 5

In scenario 5, as shown in Table 10, there are no carbon dioxide surplus and deficit differences from scenario 4, but there is a change in the carbon price. Power plant with 350 MW capacity has the largest carbon dioxide deficit, which is 531,727.42 million tons of CO2e. That power plant is obliged to pay Rp. 150,000.-/tCO2e for buying and fulfilling its carbon deficit. That obligation makes the power plant add its cost of electricity by 6.61 Rp/kWh where its initial cost is Rp. 395.-/kWh and its final cost becomes Rp.401.61/kWh. However, the highest increase of carbon price every kWh belongs to the 300 MW power plant because the amount of energy produced is lower than the 315 MW, 350 MW and 400 MW power plants. The only power plant that has the carbon dioxide surplus, which is the 330 MW power plant and has 232,489.08 million tons of CO2e surplus, gets the chance to sell its surplus. The sales of its surplus can decrease the cost of electricity by Rp.4.97/kWh where its initial cost of electricity is Rp.373.-/kWh and its final cost becomes Rp.368.03/kWh. Therefore, the merit order has changed. The previous most economical power plant, the 400 MW power plant, is replaced by the 330 MW power plant. The merit order becomes the 330 MW, 400 MW, 350 MW, 315 MW and 300 MW power plant, respectively.

Table 10. Cap of 0.95 tCO2/MWh and Carbon Price of Rp.150,000/tCO2e Scenario.

Capacity (MW)	Surplus/Deficit (tCO2e)	Selling/ Buying (Rp/kWh)	Cost Of Electricity (Rp/kWh)	Merit Order
300	-415,313.51	-8.39	439.39	5
315	-514,448.70	-5.18	436.18	4
330	232,489.08	4.97	368.03	1
350	-531,727.42	-6.61	401.61	3
400	-480,563.92	-4.11	369.23	2

4.6. Scenario 6

In scenario 6, as shown in Table 11, there are no carbon dioxide surplus and deficit differences from scenario 4 and 5, but there is a change in the carbon price. Power plant with 350 MW capacity had the largest carbon dioxide deficit, which is 531,727.42 million tons of CO2e. That power plant has to pay Rp. 300,000.00/tCO2e for buying and fulfilling its carbon deficit. That obligation makes the power plant add its cost of electricity by 13.22 Rp/kWh where its initial cost is Rp. 395.00/kWh and its final cost becomes Rp.408.22/kWh. However, the highest increase of carbon price every kWh belongs to the 300 MW power plant because the amount of energy produced is lower than the 315 MW, 250 MW and 400 MW power plants. The only power plant that has the carbon dioxide surplus, which amount is 232,489.08 million tons of CO2e, has the chance to sell its surplus. The sales of its carbon surplus will decrease the cost of electricity by Rp.9.95/kWh where its initial cost of electricity is Rp.373.00/kWh and its final cost becomes Rp.363.05/kWh. Therefore, the merit order is the 330 MW, 400 MW, 350 MW, 315 MW and 300 MW power plant, respectively.

Table 11. Cap of 0.95 tCO2/MWh and Carbon Price of Rp.300,000/tCO2e Scenario.

Capacity (MW)	Surplus/Deficit (tCO2e)	Selling/ Buying (Rp/kWh)	Cost Of Electricity (Rp/kWh)	Merit Order
300	-415,313.51	-16.77	447.77	5
315	-514,448.70	-10.36	441.36	4
330	232,489.08	9.95	363.05	1
350	-531,727.42	-13.22	408.22	3
400	-480,563.92	-8.21	373.33	2

4.7. Scenario 7

In scenario 7, table 12 shows that the 400 MW power plant had the biggest deficit of carbon dioxide, which was -1,358,142.55 million tons of CO2e. That power plant must pay Rp 50,000/tCO2e for fulfil its carbon deficit. Therefore, its cost of electricity will be added by 3.87 Rp/kWh, which initial cost is Rp. 365.00/kWh and its final cost becomes Rp.369.99/kWh. Although it has the biggest amount of carbon deficit, the highest increase of the cost of electricity occurs in the 300 MW power plant. It happens because electric energy generated by the 300 MW power plant is the lowest among all power plants. There is no surplus in this scenario, but the 330 MW power plant has the least deficit, which is 118,170.36 million tons of CO2e. The merit order becomes the 400 MW, 330 MW, 350 MW, 315 MW and 300 MW power plant, respectively from the least cost of electricity.

Table 12. Cap of 0.90 tCO2/MWh and Carbon Price of Rp.50,000/tCO2e.

Capacity (MW)	Surplus/Deficit (tCO2e)	Selling/ Buying (Rp/kWh)	Cost Of Electricity (Rp/kWh)	Merit Order
300	-786,714.80	-5.30	436.30	5
315	-1,259,454.22	-4.23	435.23	4
330	-118,170.36	-0.84	373.84	2
350	-1,135,092.69	-4.70	399.70	3
400	-1,358,142.55	-3.87	368.99	1

4.8. Scenario 8

In scenario 8, as shown in Table 13, the 400 MW power plant has the largest carbon dioxide deficit, which amount is 1,358,14.55 million tons of CO2e. However, the highest increase of carbon price every kWh belongs to the 300 MW power plant because it produces the least energy. There is no power plant that has the carbon dioxide surplus, but the 330 MW power plant makes the least carbon dioxide deficit, which is 118,170.36 million tons of CO2e. Therefore, the merit order is the 330 MW, 400 MW, 350 MW, 315 MW and 300 MW power plant, respectively.

Table 13. Cap of 0.90 tCO2/MWh and carbon price of Rp.150,000/tCO2e scenario.

Capacity (MW)	Surplus/Deficit (tCO2e)	Selling/ Buying (Rp/kWh)	Cost Of Electricity (Rp/kWh)	Merit Order
300	-786,714.80	-15.89	446.89	5
315	-1,259.454.22	-12.68	443.68	4
330	-118,170.36	-2.53	375.53	1
350	-1,135,092.69	-14.11	409.11	3
400	-1,358,142.55	-11.61	376.73	2

4.9. Scenario 9

In scenario 9, as shown in Table 14, the 400 MW power plant has the largest amount of carbon dioxide deficit, which is 1,358,142.55 million tons of CO2e. However, the highest increase of carbon price every kWh belongs to the 300 MW power plant because it produces the least energy. There is no power plant that has the carbon dioxide surplus, but the 330 MW power plant has the least carbon dioxide deficit, which amount is 118,170.36 million tons of CO2e. Therefore, the merit order becomes the 330 MW, 400 MW, 350 MW, 315 MW and 300 MW power plant, respectively.

Table 14. Cap of 0.90 tCO2/MWh and carbon price of Rp.300,000/tCO2e scenario.

Capacity (MW)	Surplus/Deficit (tCO2e)	Selling/ Buying (Rp/kWh)	Cost Of Electricity (Rp/kWh)	Merit Order
300	-786,714.80	-31.77	462.77	5
315	-1,259,454.22	-25.36	456.36	4
330	-118,170.36	-5.05	378.05	1
350	-1,135,092.69	-28.22	423.22	3
400	-1,358,142.55	-23.21	388.33	2

4.10. Analysis of the increases and decreases in cost of electricity based on power plants' capacities

Figure 2(a) shows that the amount of emission produced by every power plant will affect its cost of electricity, Figure 2(b) and 2(c) also illustrates the same things. Only in scenario 7, 8 and 9 shown in Figure 2(c), all power plants rise their cost of electricity because the cap is under their actual emission intensity. The changes of the power plant merit order between 330 MW and 400 MW power plant happen at the intersection of their line in Figure 2(a), 2(b) and 2(c).

IOP Conf. Series: Earth and Environmental Science **880** (2021) 012049 doi:10.1088/1755-1315/880/1/012049

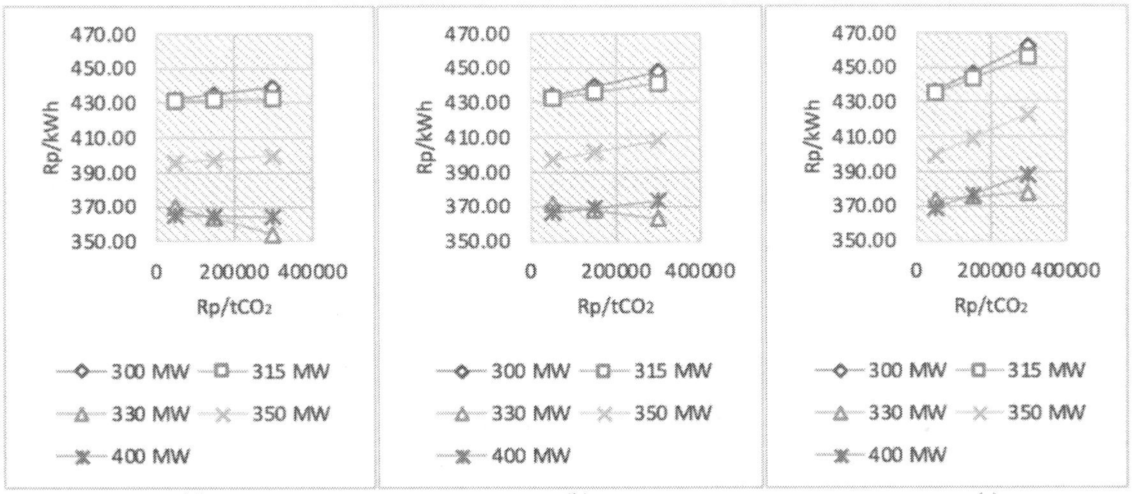

(a) (b) (c)

Figure 2. (a) Chart of the relationship between the carbon prices and the cost of electricity in scenario 1, 2 and 3. (b) Chart of the relationship between the carbon prices and the cost of electricity in scenario 4, 5 and 6. (c) Chart of the relationship between the carbon prices and the cost of electricity in scenario 7, 8 and 9.

Figure 3 gives a clear illustration about increase/decrease percentages of the power plants' cost of electricity. Only in Figure 3 (c), all power plants increase their cost of electricity. As explained before, the amount of increase/decrease percentage is affected by the cap/permitted emission limit and how many rupiahs the carbon price is.

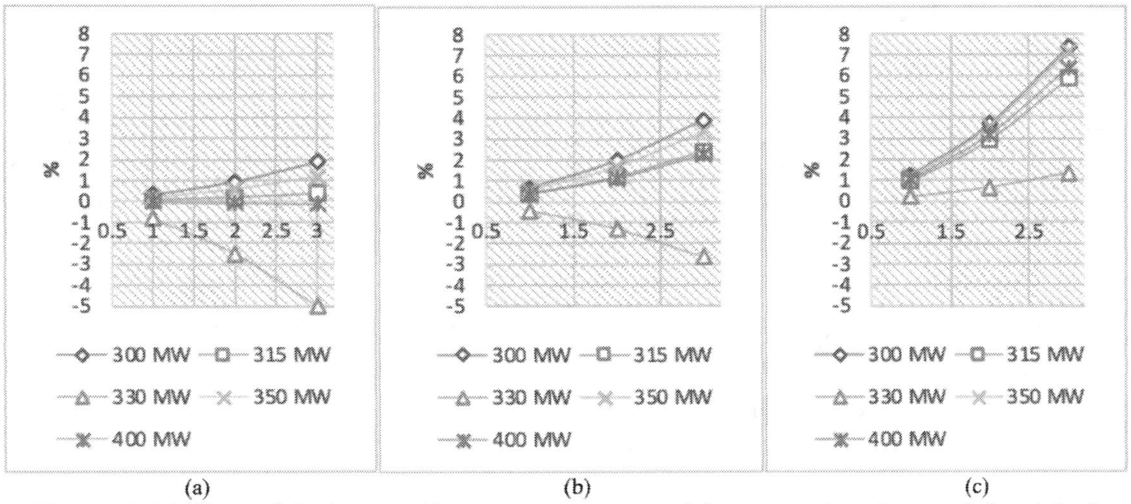

(a) (b) (c)

Figure 3. (a) Chart of the increase/decrease percentages of the power plants' cost of electricity in scenario 1, 2 and 3. (b) Chart of the increase/decrease percentages of the power plants' cost of electricity in scenario 4, 5 and 6. (c) Chart of the increase/decrease percentages of the power plants' cost of electricity in scenario 7, 8 and 9.

5. Conclusion

In this research, it can be concluded that power plants which has actual emission intensity above the cap will increase their cost of electricity, as happened for the 300 MW, 315 MW, and 350 MW power plant. The highest increase of the cost of electricity occurs in scenario 9, where the cap is 0.90

tCO2/MWh and the carbon price is Rp.300,000, That cost belongs to the 330 MW power plant which initial cost is Rp.431.00/kWh and rises into Rp.462.77/kWh or grows by 7.37%. On the other hand, the deepest decrease of the cost of electricity happens in scenario 3, where the cap is 0.98 tCO2/MWh and the carbon price is Rp.300,000.. That cost reduction belongs to the 330 MW power plant which initial cost is Rp.373.00/kWh and becomes Rp.35.76/kWh or drops by 5.00%.

The most optimal alteration of merit order occurs when the carbon price is in the range of Rp. 130,165 to Rp.130,183 where the power plant merit order changes in this range. The 400 MW power plant which earlier was in the first position of the merit order is replaced by the 330 MW power plant which is the most economical power plant.

Reference

[1] KESDM 2016 "First Nationally Determined Contribution Submitted to UNFCCC," 1–18 [Online]. Retrieved from https://www4.unfccc.int/sites/ndcstaging/PublishedDocuments/Indonesia First/First NDC
[2] K. ESDM 2019 Keputusan Menteri Energi Dan Sumber Daya Mineral Republik Indonesia Nomor 39 K/20/MEM/2019 Tentang Pengesahan Rencana Usaha Penyediaan Tenaga Listrik Pt Perusahaan Listrik Negara (Persero) tahun 2019 sampai dengan tahun 2028 *Rencana Usaha Penyediaan Tenaga List.* pp. 2019–2028
[3] K. ESDM 2019 Rencana Umum Ketenagalistrikan Nasional 2019 - 2038 p. 52
[4] O. Carbon 2020 Carbon Emission Reduction—Carbon Tax, Carbon Trading, and Carbon O ff set
[5] Ragimun 2010 Pusat Kebijakan Ekonomi Makro Badan Kebijakan Fiskal Kementerian Keuangan Republik Indonesia *Kaji. Ekon. Keuang* **14** 4 35
[6] S Kumar, S Managi, and R K Jain 2020 CO2 mitigation policy for Indian thermal power sector: Potential gains from emission trading *Energy Econ* **86** 104653 doi: 10.1016/j.eneco.2019.104653.
[7] L Meleo, C R Nava, and C Pozzi 2016 Aviation and the costs of the European Emission Trading Scheme : The case of Italy **88** 138–147 doi: 10.1016/j.enpol.2015.10.008.
[8] E. Delarue and K. Van Den Bergh 2016 Carbon mitigation in the electric power sector under cap-and-trade and renewables policies *Energy Policy* **92** 34–44 doi: 10.1016/j.enpol.2016.01.028.
[9] R. Wu, H. Dai, Y. Geng, Y. Xie, T. Masui, and X. Tian 2017 Achieving China ' s INDC through carbon cap-and-trade : Insights from Shanghai *Appl. Energy* **184** 2016 1114–1122 doi: 10.1016/j.apenergy.2016.06.011.
[10] Direktorat J. 2019 Ketenagalistrikan, "Statistik ketenaga listrikan," **53** 9 1689–1699 Retrieved from *https://Gatrik.Esdm.Go.Id/Frontend/Download_Index?Kode_Catagory=Statistik.*
[11] P. T. Cirata 2019 Facing New Business Horizon Facing New Business Horizon
[12] Indonesia environment and forestry ministry 2019 Peraturan Menteri Lingkungan Hidup Dan Kehutanan Republik Indonesia Tentang Baku Mutu Emisi Pembangkit Listrik Tenaga Termal 1–36 Retrieved from https://icel.or.id/wp-content/uploads/PERMENLHK-NO-15-TH-2019-ttg-BM-Emisi-Pembangkit-Listrik-Thermal.pdf.
[13] D J 2018 Ketenagalistrikan, "Pedoman Penghitungan dan Pelaporan Inventarisasi Gas Rumah Kaca,"
[14] IPCC 2006 "IPCC Guidelines for National Greenhouse Inventories – A primer, Prepared by the National Greenhouse Gas Inventories Programme, Eggleston H.S., Miwa K., Srivastava N. and Tanabe K.," *Iges*, p. 20
[15] W Graus and E Worrell 2011 Methods for calculating CO 2 intensity of power generation and consumption : A global perspective *Energy Policy* **39** 2 613–627 doi: 10.1016/j.enpol.2010.10.034.

Analysis of Malaysia electricity demand and generation by 2040

Azreen Harina Azman[1*], Nurul Nadrah Aqilah Tukimat[1], M.A.Malek[2] and Faizah Che Ros[3]

[1] Faculty of Civil Engineering Technology, Universiti Malaysia Pahang, Lebuhraya Tun Razak, 26300 Gambang, Pahang Darul Makmur, Malaysia

[2] Civil Engineering Deparment. College of Engineering. Universiti Tenaga Nasional, Malaysia

[3] Disaster Preparedness and Prevention Centre, Malaysia Japan International, Institute of Technology, Universiti Teknologi Malaysia, Malaysia

*Corresponding E-mail: azreenharinaazman@gmail.com

Abstract. Malaysia as an emerging country, increasing population, gross domestic product (GDP) growth and enhanced access to electricity lead to an expanding of demand. The crucial parameters to determine future energy demand and generation projections are GDP, population growth rates and weather implications due to climate change. The study aims to forecast the future trends based on the historical values and also to project the future electricity demand and generation. The electricity demand and generation growth evaluated based on 2 main elements which are population growth and weather parameters (maximum temperature and rainfall). The future trends are forecasted based on the historical values of population and weather parameters. There is 152.9% of population growth in 32 years. The population will keep on developing yet with the lower rate. The GDP trend and the population growth mirrors the pattern of emissions. The findings from Statistical Downscaling Model (SDSM) analysis shows that the rainfall distribution will diminish while the temperature will expand that depict the climate change impact as time passes by. In 2020, the most extreme temperature recorded is 31.7 °C while in 2040, the estimated greatest temperature is 32.3 °C. There will be a 0.6 °C increase in temperature in 20 years. The demand in 2040 will be expanded 50.3% more than demand in 2020. The estimated electricity demand per capita will continue expanding because of the augmentation of the populace and the significance of electricity in daily activities. The pattern shows that electricity demand and generation in Malaysia will be expanding massively year by year.

Keywords: Electricity, Demand, Generation, Gross Domestic Product

Track Name: Human, Social, Economic and Environmental Sustainability

1. Introduction

Worldwide electricity demand will extent about 38700 TWh by 2050 from 25000 TWh in 2017 [1]. There is 54.8% addition of demand in 33 years. Providing the future electricity demand will require an increment in the power generation capacity. As indicated by [2], world's net electricity generation is projected to increase 69% by 2040, from 21600TWh in 2012 to 25800TWh in 2020 and 36500TWh in 2040. In the emerging countries, population and GDP growth and improved admittance to electricity lead to a growth of demand [1]. The significant employments of electricity are residential, industrial, transportation and commercial consumption [3]. International Energy Agency (IEA) recorded the electricity demand growth in five significant activities which are space cooling (37%), residential appliances (25.5%), lighting (7.8%), heating (12.4%) and other services (17.4%) [4].

According to [5], 2019 started with a weak-to-moderate El Niño event. The average temperature recorded in 2019 was 0.95 °C past the twentieth century average of 13.9 °C, making it the second-warmest year recorded. The five (5) warmest years in the 1880–2019 history happened since 2015, while 9 of the 10 warmest years have noted since 2005. [6] expressed that the last decade has been the warmest at any point recorded where 2016 was the most sweltering year. The pinnacle demand in Malaysia was recorded at 17788 MW in April 2016 because of the El-Nino phenomena [7].

Energy demand for space cooling will triple by 2050. Cooling will lead to top electricity demand, explicitly in hot countries. By 2050, around 2/3 of the world's households might own an air conditioner. Utilizing space cooling appliances like air conditioner and electric fans accounts for about 20% of the world absolute electricity utilized in buildings [8]. As populaces increment, the utilization of climate control systems is gotten normal particularly in the sultrier regions. Rising demand for space cooling is additionally driving monstrous strain on power frameworks in a few nations. Subsequently, the objectives of the study are to estimate the future patterns dependent on the historical values and to project the future electricity demand and generation.

2. Electricity Demand and Generation Scenarios

The electricity created from different sources, for example, hydropower plants, fossil fuels, geothermal systems, solar panels, wind and biofuels including electricity-only plants and combined plants [9]. Energy demand will keep on expanding. Electricity demand is discouraged by the weather condition and economic growth. Due to humid climate, about 40% of Malaysia total energy demand in the commercial sector will be channelled for space cooling [10] while in the cold regions, the electricity demand is for heating because of the utilization of electric heaters and heat pumps [11]. Worldwide electricity demand will prompt a critical development in the electricity demand by 2050. Because of the increment in electricity creation to satisfy future interest, the impending electricity generation blend will be dependent upon spectacular switches around 2050 [12].

In view of [13], the current worldwide populace is 7.8 billion individuals and is expanding by 81 million individuals each year. The most extreme annual growth rate was 2% in the late 1960s yet the pace of increment keeps on declining in the coming years. The total populace will keep on extending in the 21st century, however with a lower rate. It is expected that it will reach 9 billion by 2037 and 10 billion persons in the year 2057. Malaysia population is 0.42% of the total world population and it is estimated to reach 32.7 million in 2020 as compared to 32.5 million in 2019 with an annual growth rate of 0.4% [14].

The high correlation between energy demand and GDP brings to use of GDP as the main factor to determine future energy demand projections. Population growth rates are also a crucial parameter. Besides that, the unpredictable and severe weather implications due to climate change become important issues because it links with energy use. [15] states that the transport sector will lead energy usage in 2040 with 39.6%, trailed by the industry sector with 30.4%.

Rainfall distribution is influenced by topography and the monsoon winds. Malaysia been blessed with abundant rainfall, with normal about 2000-4000 mm every year. Malaysia is influenced by the northeast monsoon (NEM) and southwest monsoon (SWM). NEM begins from November and keeps going till March while SWM blows from late May to September. NEM contributes more rainfall over the country compare to SWM especially in the east coast region during November or December that bring massive floods [16]. Malaysia is situated at the equator and the climate is warm and humid with the mean temperature roughly 20 °C to 32 °C [17]. The warming pattern is at 0.3 °C/decade and the country will be 2◦C warmer in the following 50 years. Both NEM and SWM essentially show expanding temperature patterns. The higher recurrence of El Niño occasions is the principal justification of the ascent of warming rates in the new decade [18].

3. Methodology

Figure 1 shows the methodology of the study. The electricity demand and generation growth assessed dependent on two (2) main elements which are population growth and weather parameters (maximum temperature and rainfall). The population historical data (1988-2019) extracted from Department of Statistics Malaysia and the population from 2020 to 2040 are projected using linear equation. The historical data for maximum temperature and rainfall (1988-2017) acquired from Malaysia Meteorological Department (MMD), the Department of Irrigation and Drainage (DID) and the Electricity Power Provider. SDSM was utilized in the weather parameters forecasting process in at regular intervals range (2020, 2030 and 2040). Electricity demand and generation historical data (1988-2019) recovered from [19] and forecasted using polynomial equation.

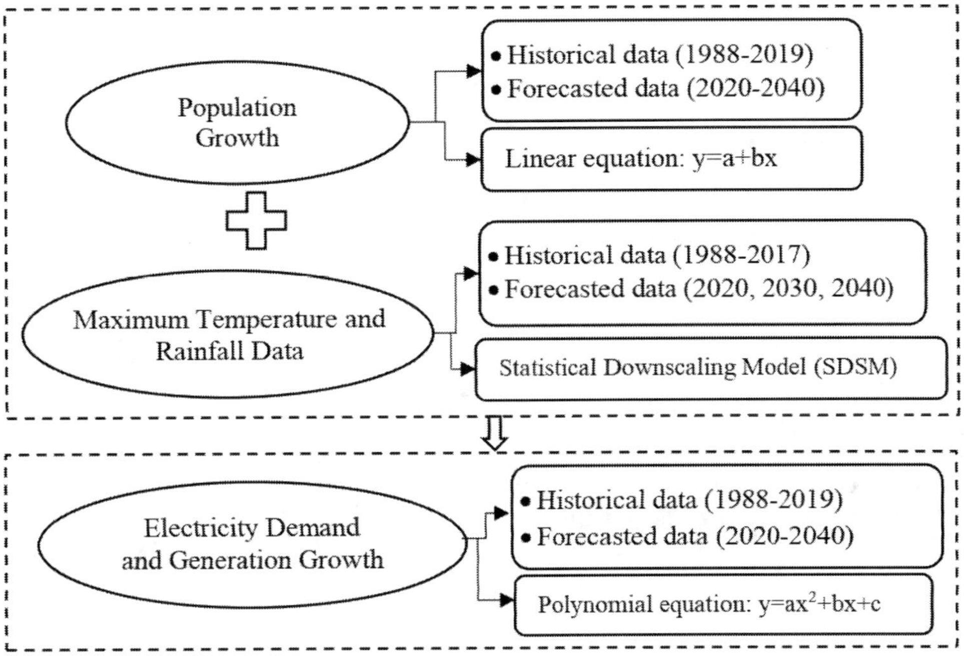

Figure 1. Methodology of the study.

4. Results and Discussions

The findings from the analysis are discussed within three (3) main elements which are population growth, weather parameters (maximum temperature and rainfall) and also electricity demand and generation scenarios. The interrelation between these elements is discussed as well.

Table 1. Malaysia population growth from 1988 to 2040.

Year	Population (million)	Year	Forecasted Population (million)
1988	17	2020	33
1992	19	2024	35
1996	21	2028	37
2000	23	2032	39
2004	25	2036	41
2008	27	2040	43
2012	29		
2016	31		

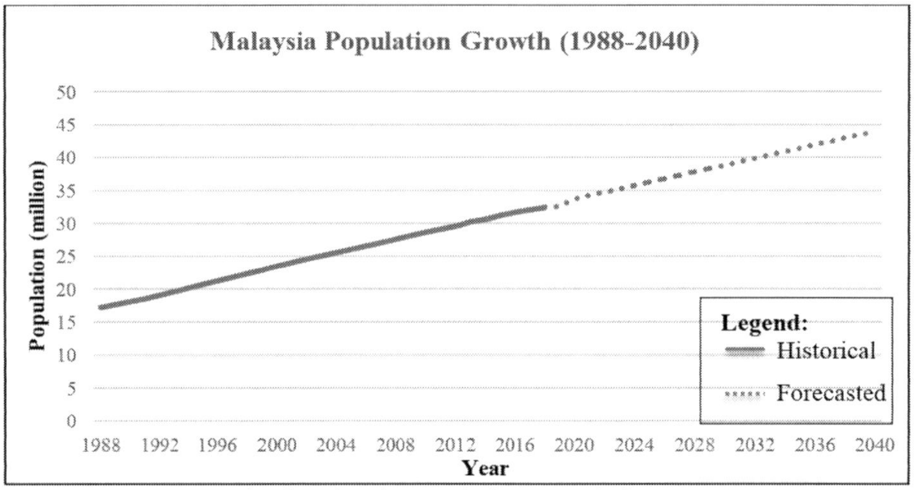

Figure 2. Graph of Malaysia population growth from 1988 to 2040.

There is 152.9% of population growth in 32 years as data in Table 1. Figure 2 clearly shows there is a constant growth from 1988 to 2019. By expecting there will be a linear growth like historical data, the forecasted population is calculated using Eq. 1 and presented in dotted line. The population is projected to increase 30.3% in 2040 compared to 2020.

$$y = 0.5154x + 16.653 \quad \text{(Equation 1)}$$

From the analysis, the population will keep on developing however with the lower rate. The growth rate for the historical data is 0.82 while the value of growth rate for forecasted data is 0.30. This finding in accordance with [13 and 20] where the pace of increment keeps on declining in the coming years and world population growth is projected to flatten in coming decades.

Table 2 recorded the estimated values from the analysis of population and weather parameters (maximum temperature and rainfall) historical data. From the projected population growth data, Malaysia population is 33 million people in 2020 and increment 5 million individuals at regular intervals wherein 2030 and 2040, Malaysia populace will accomplish 38 million and 43 million individuals separately. Figure 3 shows the forecasted population and maximum temperature for 2020, 2030 and 2040. In 2020, the most extreme temperature recorded is 31.7 °C while in 2040, the forecasted maximum temperature is 32.3 °C. There will be 0.6 °C expansion of temperature in 20 years and in accordance with [21] that expressed the temperature may increment by 0.5-1.0 °C during that period. Temperature rise is one of the aspects of climate change and its variances influence economic activities and electricity consumption [22]. As the populace develop and temperature ascend, there will be a critical expansion in electricity use for various activities in space cooling, residential appliances, lighting, water heating, and others as [23] tracked down that over 76% of United States (U.S.) electricity use or over 40% of all

U.S. energy use are utilized for space conditioning and lighting for industrial, commercial and residential buildings. Outrageous temperature occasions result in an expansion in air conditioner utilisation that influences electricity demand also [24]. Space conditioning includes gadgets to either expanding or diminishing air temperature. [25,26] additionally expressed that the pace of energy utilization assessed to increment with economic and population growth due to the usage of modern home appliances, mostly air conditioners and refrigerators.

The GDP pattern and the population growth mirrors the pattern of emissions. Human emissions of carbon dioxide and anthropogenic greenhouse gases have expanded since the pre-industrial era, driven generally by economic and population growth are a primary driver of climate change especially global warming since the mid-twentieth century [6]. Figure 4 shows the findings from SDSM analysis that reflect the climate change effect as time passes by where the rainfall distribution will diminish while the temperature will increase. The positive patterns of temperature increment have been observed since the most recent forty years. The augmentation of the greatest surface temperature is about 0.17 °C to 0.22 °C each decade [26] while a minor diminishing pattern is identified for the rainfall in Peninsular Malaysia and Sabah [21]. Outrageous climate like floods, dry spells, tempests, and heat waves are the effects of climate change [27]. Projection process for rainfall and temperature dependent on Eq. 2 and Eq. 3.

$$P_i^k = \beta_0 + \sum_{j=1}^{n} \beta_j X_{ij} + \varepsilon_i \quad \text{(Equation 2)}$$

$$\text{Ui} = \gamma_0 + \sum_{j=1}^{n} \gamma_j X_{ij} + \varepsilon_i \quad \text{(Equation 3)}$$

where P_i = amount of total precipitation, k = transformation (fourth root, inverse normal or logarithmic), U_i = daily temperature (T_{Max} and T_{Min}), X_{ij} = selected NCEP/NCAR predictors, β_j and γ_j = regression coefficients estimated for each month using least-squares regression and ε_i = model error [28, 29].

Table 2. The forecasted values from the analysis.

Year	Forecasted values		
	Population (million)	Maximum Temperature (°C)	Average Rainfall (mm/day)
2020	33	31.7	7.7
2030	38	32.2	6.9
2040	43	32.3	7

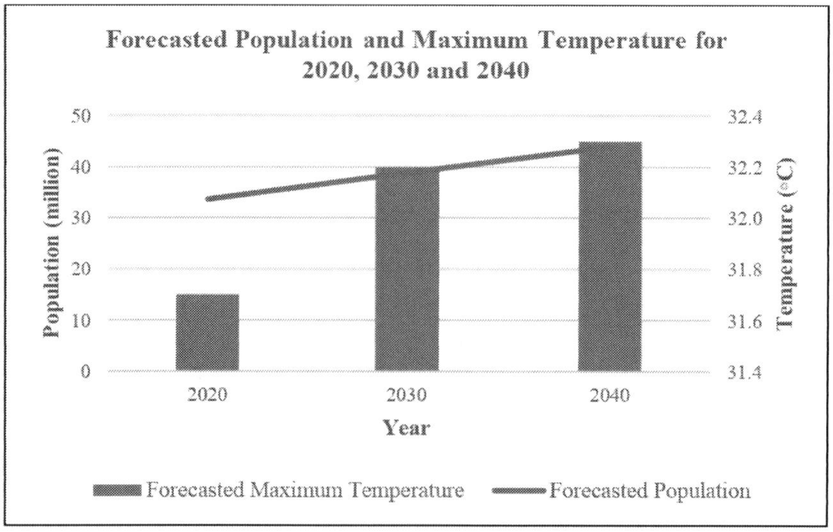

Figure 3. Graph of forecasted population and maximum temperature for 2020, 2030 and 2040.

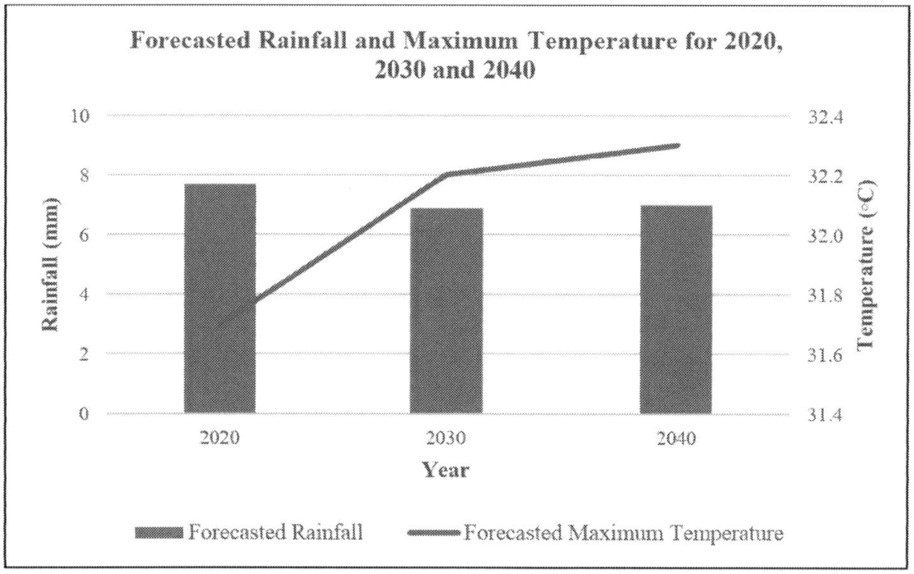

Figure 4. Graph of forecasted rainfall and maximum temperature for 2020, 2030 and 2040.

The electricity demand is essentially reliant upon the climate data [30]. In view of Table 3, electricity demand per capita for 2018 is multiple times greater than in 1988. The demand will be expanded to 7445kWh/individual in 2040 where it is 50.3% more than demand in 2020. The forecasted electricity demand per capita is determined utilizing Eq. 4 and introduced by the dotted line in Figure 5. Expansions in per capita electricity utilization impersonate potential changes in the piece of the economy, for example, shifts to more energy-intensive industries, and changes in service demand, such as growing demand for air conditioning and appliances. The forecasted electricity demand per capita will continue expanding because of the addition of the populace and the significance of electricity in everyday activities as discussed by [31], the expanding electricity consumption is driven by the improvement in GDP and population.

$$y = -0.0368x^2 + 126.48x + 828.45 \quad \text{(Equation 4)}$$

Table 3. Electricity Demand per Capita (kWh/person).

Year	Electricity Demand per capita (kWh/person)	Year	Forecasted Electricity Demand per capita (kWh/person)
1988	940	2020	4955
1992	1352	2024	5453
1996	2068	2028	5951
2000	2603	2032	6449
2004	3022	2036	6947
2008	3367	2040	7445
2012	3943		
2016	4553		

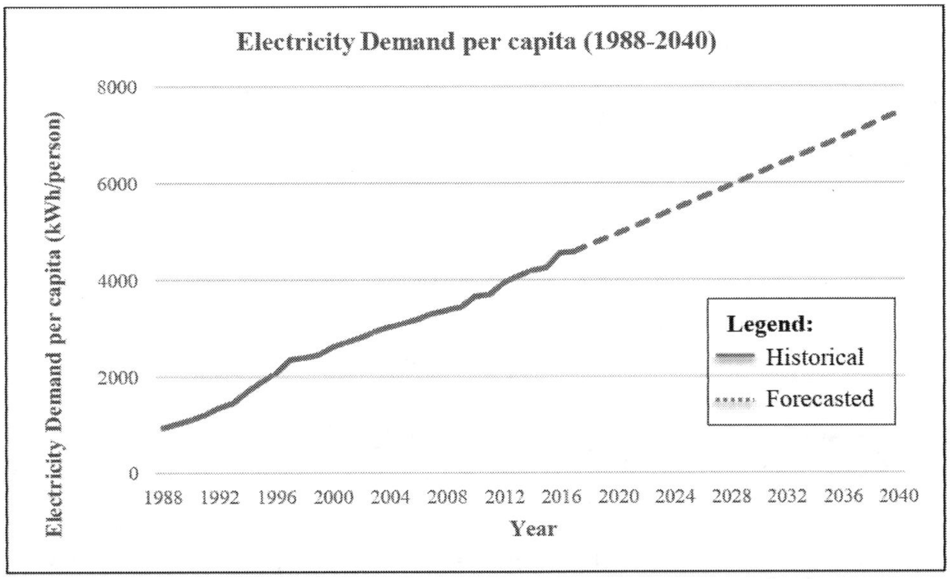

Figure 5. Graph of electricity demand per capita from 1988 to 2040.

High electricity generation prompts high power utilization. Table 4 and Figure 6 shows the values and graph of absolute electricity generation from 1988 to 2040. The anticipated absolute electricity generation are determined utilizing Equation 5. The values show the augmentation as the years progressed. The total electricity generation is relied upon to arrive at 272.8TWh in 2040, almost 60% higher than 2020. Essential energy supply and energy demand developed to pair with the population and economic growth and this pattern is relied upon to proceed [21].

$$y = 0.0071x^2 + 4.6823x + 7.1369 \quad \text{(Equation 5)}$$

Table 4. Total electricity generation from 1988 to 2040.

Year	Total Electricity Generation (TWh)	Year	Forecasted Total Electricity Generation (TWh)
1988	19.4	2020	170.7
1992	29.3	2024	191.1
1996	51.4	2028	211.5
2000	69.3	2032	232.0
2004	82.3	2036	252.4
2008	98.0	2040	272.8
2012	134.5		
2016	156.2		

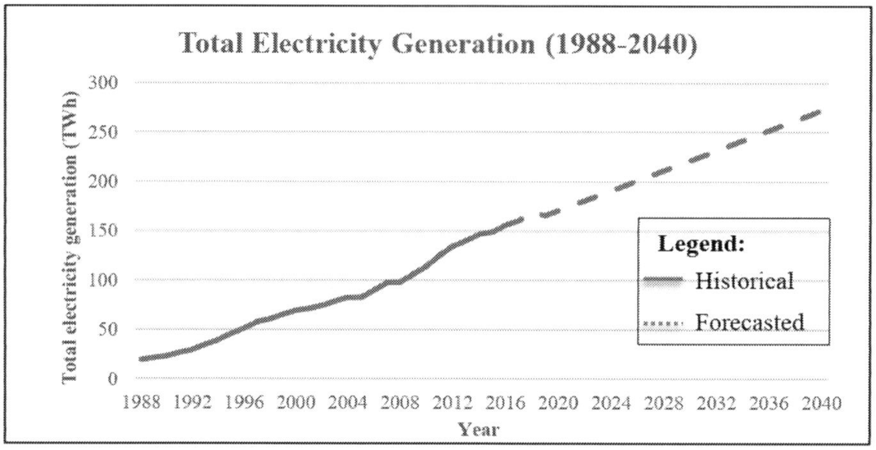

Figure 6. Graph of total electricity generation from 1988 to 2040.

5. Conclusions

The urgent boundaries to decide future energy demand and generation projections are GDP, population growth rates and weather implications due to climate change. The electricity demand and generation development assessed dependent on two (2) primary components which are population growth and weather parameters (maximum temperature and rainfall). In light of the historical values of population and weather parameters, future trends are determined. The population will keep on developing however with the lower rate. The findings from SDSM analysis mirror the climate change impact as time passes by. This tends to the significance of investigating the impacts of climatic variables on electricity demand. Rainfall distribution will diminish while the temperature will increase. The forecasted electricity demand per capita will continue expanding because of the augmentation of the populace. In a conclusion, from the analysis that has been made to crucial parameters in electricity demand and generation in Malaysia, the trending shows that both will be expanding massively year by year.

References

[1] Global Electricity Demand to Increase 57% by 2050. 2018 [Blog]. Retrieved from https://about.bnef.com/blog/global-electricity-demand-increase-57-2050/
[2] U.S. Energy Information Administration (EIA) 2016 US Energy Information Administration. International Energy Outlook 2016 (1st ed.). Washington DC.
[3] NS Energy 2018 Major uses of electricity. Retrieved on 9.10.2020.
[4] IEA, Share of global electricity demand growth to 2050, IEA, Paris. https://www.iea.org/data-and-statistics/charts/share-of-global-electricity-demand-growth-to-2050 Retrieved on 10.9.2020.
[5] NOAA National Centers for Environmental Information 2020 State of the Climate: Global Climate Report for Annual 2019. NOAA National Centers for Environmental Information.
[6] IPCC 2014 IPCC: Climate Change 2014: Synthesis Report. Contribution of Working Groups I, II and III to the Fifth

Assessment Report of the Intergovernmental Panel on Climate Change.
[7] Overview electricity Sector in Malaysia. Department of Statistics Malaysia. www.dosm.gov.my Retrieved on 21.8.2020.
[8] IEA 2018 The future of cooling, Paris https://www.iea.org/reports/the-future-of-cooling
[9] OECD 2020 Electricity generation (indicator). DOI: 10.1787/c6e6caa2-en. Retrieved on 13.10.2020.
[10] APEC Energy Demand and Supply Outlook 5th Edition 2013.
[11] Teske, S, Pregger, T, Simon, S, Negler, T, Pagenkopf, J, Adel, B, Meinshaunsen, M, Dooley, K, Briggs, C, Dominish, E, Giurco, D, Florin, N, Morris, T, and Nagrath, K 2019 Energy Scenario Results. In: Teske S. (eds) Achieving the Paris Climate Agreement Goals *Springer* 25–26
[12] World Energy Scenarios: Composing Energy Futures to 2050 World Energy Council 2013 ISBN: 978 0 946121 32 8 www.worldenergy.org/publications/2013/world-energy-scenarios-composing-en-ergy-futures-to-2050
[13] Department of Economic and Social Affairs 2019 2019 Revision of World Population Prospects. United Nations. https://population.un.org/wpp/
[14] Current Population Estimates, Malaysia 2020 Department of Statistics Malaysia. www.dosm.gov.my. Retrieved on 14.10.2020.
[15] Zulkifli Z 2016 Malaysia Country Report in Kimura S and P.Han (eds.) In Energy Outlook and Energy Saving Potential in East Asia 2016. ERIA Research Project Report 2015-5, Jakarta: ERIA, pp.215-236.
[16] Kok C T 2018 Trends of rainfall regime in Peninsular Malaysia during northeast and southwest monsoons. *J. Phys. Conf. Ser.* **995** 012122
[17] Tan K C, Tan M L, Lim H S and Mat Jafri M Z 2017 Analysis of ozone observation at atmospheric monitoring network station using Brewer ozone spectrophotometer *Proceedings Volume 10428, Earth Resources and Environmental Remote Sensing/GIS Applications VIII* 104281L. DOI: 10.1117/12.2278069
[18] Chee-Loong W, Zulkifli Y, Tarmizi I 2018 Trend of daily rainfall and temperature in Peninsular Malaysia based on gridded data set. *Int. J. GEOMATE* **14** 44 65-72
[19] Energy Commission. Malaysia Energy Information Hub. https://meih.st.gov.my/statistics. Retrieved on 21.8.2020
[20] Cilluffo, A, and G Ruiz, N 2019 World's population is projected to nearly stop growing by the end of the century. Issues Online. https://www.issuesonline.co.uk/articles/worlds-population-is-projected-to-nearly-stop-growing-by-the-end-of-the-century-
[21] Ministry of Energy, Science, Technology, Environment and Climate Change 2018 Malaysia third national communication and second biennial update report to the UNFCCC 2018. Malaysia: Ministry of Energy, Science, Technology, Environment And Climate Change.
[22] Zhang C, Liao H. and Mi Z 2019 Climate impacts: temperature and electricity consumption. Nat Hazards **99** 1259–1275. https://doi.org/10.1007/s11069-019-03653-w
[23] Quadrennial Technology Review 2015 An assessment of energy technologies and research opportunities. Increasing Efficiency of Building Systems and Technologies.
[24] Ali M, Jawed M I and Sharif M 2013 Relationship between extreme temperature and electricity demand in Pakistan *Int. J. Energy Environ. Eng.* **4** 36
[25] J S Hassan, R M Zin, M Z Abd Majid, S Balubaida, M R Hainina 2014 Building energy consumption in malaysia: an overview. *J. Teknol.* **70** 7 33–38.
[26] Ministry of Natural Resources and Environment Malaysia 2015. United Nations Framework Convention on Climate Change in December 2015. Malaysia: Ministry of Natural Resources and Environment Malaysia.
[27] Cambridge University Press 2014 Climate Change 2014: Impacts, Adaptation, and Vulnerability. Part A: Global and Sectoral Aspects. Contribution of Working Group II to the Fifth Assessment Report of the Intergovernmental Panel on Climate Change.
[28] Wilby R L and Dawson C W 2013 The statistical downscaling model: insights from one decade of application. *Int J Climatol* **33** 1707-1719
[29] Dharmaveer S, Sanjay K J and Gupta R D 2015 Statistical downscaling and projection of future temperature and precipitation change in middle catchment of Sutlej River Basin, India *J. Earth Sys. Sci.* **124** 4 843–860.
[30] D. Vu, K. M. Muttaqi & A. P. Agalgaonkar 2014 Assessing the influence of climatic variables on electricity demand in IEEE Power and Energy Society General Meeting, pp. 1-5
[31] Sharif A S S, Razman M R and Awang A 2020 The nexus of population, gross domestic product growth, electricity generation, electricity consumption and carbon emissions output in Malaysia. *Int. J. Energy Econ. Policy* **10** 3

Acknowledgment

The authors would like to thank the Ministry of Higher Education for providing financial support under Fundamental research grant No. FRGS/1/2019/TK01/UMP/02/1 (University reference RDU1901141) and Universiti Malaysia Pahang for additional financial support under Internal Research grant RDU1803156, Malaysian Meteorological Department (MMD), and Drainage and Irrigation Department (DID).

Spatiotemporal assessment of effects of leachate from landfills on groundwater in Kano Metropolis, Nigeria

H A Suleiman[1] and M M Hanafiah[1,2*]

[1] Department of Earth Sciences and Environment, Faculty of Science and Technology, Universiti Kebangsaan Malaysia, 43600 UKM Bangi, Selangor, Malaysia

[2] Centre for Tropical Climate Change System, Institute of Climate Change, Universiti Kebangsaan Malaysia, 43600 UKM Bangi, Selangor, Malaysia

*Corresponding E-mail: mhmarlia@ukm.edu.my

Abstract. Improper solid waste management in Kano Metropolis, Nigeria has become a serious threat to human health and ground water quality. Groundwater monitoring should be conducted to assess the groundwater contamination, especially originated from leachate generated from the dumpsites. Most of the dumpsites are located nearby water body that subsequently affecting the quality of the water for human purposes. This study aims to investigate the relationships between different physical-chemical and biological parameters during the two seasons (dry and wet seasons) in Kano Metropolis area as well as to identify the sources of variation during the two seasons. The level of some physical-chemical and biological parameters of the groundwater (wells) and dump sites were assessed (*i.e.* temperature, pH, conductivity, suspended solids, turbidity, hardness colour, *e-coli* and coliform). Mean concentration of some physical-chemical and biological parameters except that of temperature, colour, pH and total dissolved solids were found to be above the acceptable limit of the National and International standard of drinking water quality, NESREA and WHO. The sources and variations of the samples results were tested using statistical analysis. The water samples show a considerable level of pollution. The analysis of the groundwater and that of dump sites reveals no significant difference in the parameters measured. It is therefore recommended that the water from this source should be monitored and treated properly before consumption.

Keywords: Groundwater pollution, Water quality, Landfill, Leachate

Track Name: Human, Social, Economic and Environmental Sustainability

1. Introduction

Quality drinking water is vital for life. However, in Nigeria and many other countries around the world still facing an issue with water quality. Quality drinking water has become a scarce commodity as only a small percentage of the general population has access to treated water. Unconventional sources of water such as rainwater and ground water have become major sources of drinking water for people living in new settlements and some residents who do not have access to treated water in Nigeria. The need to assess the quality of water from some of these alternative sources has become necessary because they have a direct effect on the health of individuals. There has been an increase in the growth of the human population during the past decades in the urban cities of developing countries coupled with an increase in the number of industries and intensive agricultural activities. The increase in waste generation has been linked to the high growth of population, human behaviors and the modern life style. [1][2][3].

Furthermore, the growth and demand for energy, has adversely affected the tributaries of rivers, streams passing through major cities. Moreover, in Nigeria and developing countries more than a billion people lack access to safe water supplies; almost three billion people do not have access to adequate sanitation; five to ten million people die each year from water-related diseases or inadequate sanitation; and twenty percent of the world's irrigated lands are salt laden, thereby affecting crop production [4].

Poor treatment and inadequate waste disposal from humans, industrial waste and livestock and excessive use of limited water resources, pollutants such as heavy metals, bacteria, viruses and nitrates have infected water supplies. There is potential for natural levels of metals and other chemicals to be harmful to human health even if no sources of anthropogenic contamination existed. It was emphasized in Bangladesh that natural amounts of arsenic in groundwater were discovered to be causing toxic outcomes on the people. Unfortunately, this problem arose because the groundwater was obtained for drinking without a comprehensive water quality investigation.

Higher demand for drinking water in many arid countries which resulted from exponential growing population, primed the need for increased water supply. Many efforts were made to make water available in the area through the construction of wells which supplies water to many people living in the area. For this reason, people in the study area draw their drinking water from untreated surface water bodies like wells. Taking such untreated water directly without treatment may lead to the infestation of water borne disease among the people in the area. This is why it is important to assess the effects of waste dumps on the quality of shallow well water for drinking in the area.

A change in the taste and smell of groundwater are caused by high concentration of chemicals and threatens human health [5]. Consequently, there is a need for the prevention and control of water pollution in order to reduce the scarcity of water resources; it is of immense rational value to study how to remove the unwarranted chemicals pollutants of groundwater. Hence, it is essential to treat and remove hazardous and heavy metals pollution of groundwater using local biomass [6][7].

Water quality tests were conducted by the researchers in many places but most of which are for agricultural uses [8]. Even the ones done on drinking water were in a different environment, the few ones done in the study area were mostly on the chemical parameters of the pipe borne water and borehole water, that is why the same methodology is adopted to assess the physical-chemical and biological parameters of the open well water for drinking in the area in Kano Metropolis, Nigeria. This study also aims to compare the differences in microbial water quality between the dry and wet seasons.

2. Materials and method

2.1 Study area

Figure 1 shows the study area, Tarauni local government area that is located within Kano Metropolis, Nigeria. The metropolitan Kano lies between latitudes of 12°03'N and 8°31'E and longitudes 70°3'N and 1°35'E. The study area is among the seven local government areas that compose the metropolis city which has a lot of infrastructural facilities, utilities and services. It is well accessible from every angle in Kano Metropolis because most of its wards are along Kano-Zaria express way. The local government has ten wards and the area under study lies within the wards.

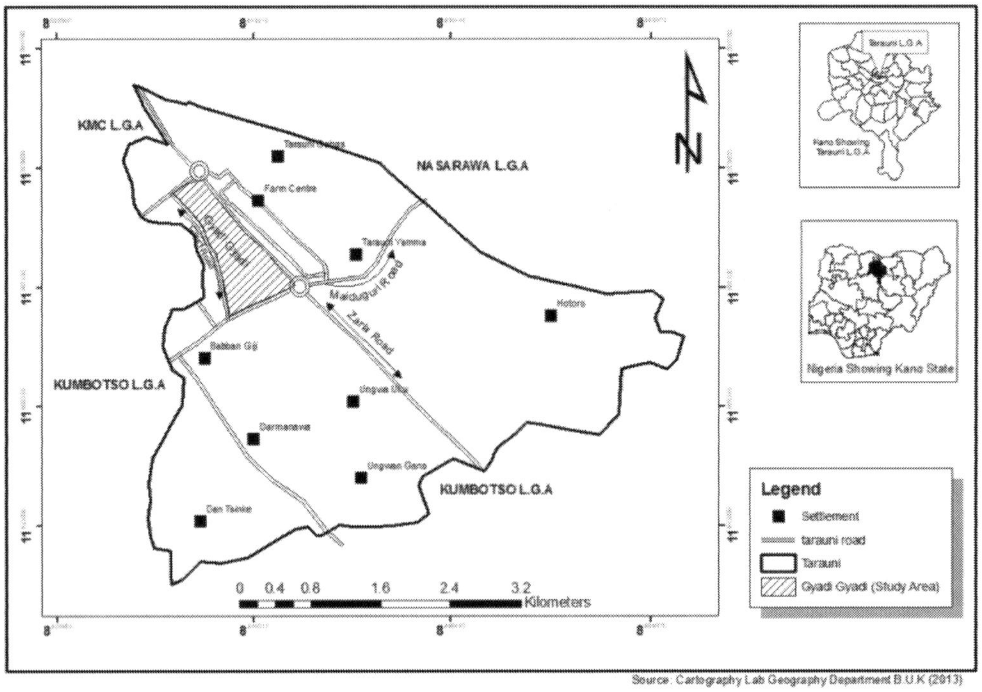

Figure 1. Study area located in Kano Metropolis, Nigeria.

The climate of the area can be regarded or categorized under AW from Köppens classification. It is also a Sudan savanna climate consists of a cool dry season from October to May, and warm rainy season from June to September. The cultivation period is conventionally during the warm rainy season. The geology of the study area belong to Eutric Gambusol in FOA UNESCO System, with soil taxonomy are moderately deep and well drained with sandy loamy texture most of these soils are underlain by iron pan depth variation between 80 -150 m. The vegetation of the study area is that of Sudan Zone in the West Africa. By cultivation, grazing, cutting and burning, the original natural vegetation of the area has been changed. With increasing setting of people, the original savanna woodland has been altered into degraded savanna in which secondary species are dominant.

2.2 Sampling techniques

To ensure statistical validity of the results, two wells were selected in each of the location of the study area and three dump sites were also selected based on 5 -15 meter radius from the open well in order to assess how much the influence at some interval (Table 1).

The detailed field work was conducted in the following way; direct observation was undertaken at the six selected open wells and three selected dumpsites were visited for identification and collection. The samples collection was done using 2 liter screw-capped bottles that have been purified to prevent cross contamination by any chemical, physical or microbial means. The collected water samples' temperatures were immediately taken before they were positioned in the ice cooler for conveyance to the laboratory. The water samples were obtained in two different seasons respectively. The seasons are in the months of April, May, and June for the dry seasons and the months of July, August and September for the wet season of the year 2019.

Table 1. Selected wells and dump sites location and distance.

Samples	Area	Location (Lat, Lon)	Distance from dumpsite per (meter)
Sample 1 (Well water)	Gyadi-Gyadi (Dangi)	11°58.644' N, 8°32.405' E	8 m
Sample 2 (Well water)		11°58.598' N, 8°32.385' E	13 m
Sample A (Waste dump)		11°58.019' N, 8°32.721' E	8 m - sample 1 &12 m - sample 2
Sample 3 (Well water)	Gyadi-Gyadi (Kasuwar Dare)	11°58.432' N, 8°32.560' E	7 m
Sample 4 (Well water)		11°58.439' N, 8°32.602' E	12 m
Sample B (Waste dump)		11°58.436' N, 8°32.569' E	5 m to sample 3 & 11 m to sample 4
Sample 5 (Well water)	Gyadi-Gyadi (Babban Layi)	11°57.998' N, 8°32.615' E	8 m
Sample 6 (Well water)		11°58.016' N, 8°32.720' E	14 m
Sample C (Waste dump)		11°58.128' N 8°32.598' E	7 m to sample 5 & 13 m to sample 6

2.3 Statistical analysis
To ensure statistical validity of the result, means, standard deviation variance and ANOVA were employed in the analysis of data. The means of physical-chemical and biological parameters in each water sample was compared with national and international standards of drinking water in order to ascertain whether they are within the acceptable limit or not. t-test was also employed to compare the wells water quality.

3. Results and discussion

3.1 Physical-chemical and biological parameters
This section analyses the physical-chemical and biological parameters of the water samples; taking water samples in the dry and wet seasons with the aim of comparing the results with that of national and international standard acceptable limits of drinking water. Table 2 and 3 shows statistical summary of the analysed parameters and their comparisons with the WHO [9] and NSDWQ [10] standards.There were variations in temperature values among the water samples ranging from 26.2 to 31 °C It was observed that high water temperature improves the growth of microorganisms and may add taste, corrosion, odour, colour, and this value falls within the limit of WHO and FEPA.

During the two seasons (dry and wet), it was noted that all the samples from wells water were marginally coloured which is either by the absorption of light or by coloured substance or due to inappropriate cover on wells. The wells water pH ranges from 6.0 to 8.0 and it remains relatively within the permissible limit. However it usually has a link with other constituents in water [11]. On its own, however, pH has no direct effect on human or animal health, but because it is so closely associated with other chemical constituents of water, it is often regarded as having an indirect effect on health [12]. It is evident that many chemical parameters generating extreme values of pH (i.e. either strong acids or bases) would impact frequently on human health [13].

The range of turbidity is from 0 to 67 NTU and WHO permissible limit is 5 NTU but from the results of the two seasons, it shows that only well 1b (Dangi) during dry season falls above the standard with 67 NTU This shows that there might be a microbial contamination which can cause substantial harm to human. Increased level of turbidity can impart an aesthetically displeasing appearance to water. The materials can also allow adsorption of microorganisms, which protect water from the effects of disinfection and can simulate bacterial growth [14]. Other wells like well 1a (dangi), well 2a (Kasuwar dare), 2b (Kasuwar dare), wells 3a and 3b (Babbanlayi) in the dry season falls below the standard with 0 NTU.

The water samples conductivity ranges from 2.36 to 2770 µs/cm, among all the water samples in the two seasons with most of the values exceeds the maximum permissible limit except well1a (dangi) in the dry season which is below the maximum limit. The effluent from the different wastes may be responsible for the intensification in the levels. WHO has not set any guideline value for EC in drinking water, it gives an indication about the level of mineralisation of the water source, which often occurs as a result of contamination [15].

The range of the total dissolved solids is from 83 to 1383 mg/L, with all the water samples falls within the acceptable limit except well 2a (Kasuwar dare) in both dry and wet seasons. Since the maximum permissible limit is 500 to 1000 mg/L, therefore the wells with high total dissolved solids will be an obstacle during treatment as it may cause filter clogging.

Hardness is one of the very important properties of ground water, from utility point of view particularly for domestic purpose. From the research results it ranges from 18.81 to 538.704 mg/L, with all the samples analyzed exceeded the maximum permissible limit except well 3a (Babban layi) which is within the permissible range and this may be due to dissolution of polyvalent metallic ions from sedimentary rocks, seepage and run off from soil.

The microbial analysis of the well water samples was carried out. Table 2 and 3 show that only well 1a (Dangi), well 1b (Dangi) and well 2a (Kasuwar dare) water samples met the national and international standard limit. Coliform bacteria must not be detectable in any 100 ml sample of all water meant for drinking. The sources of total and faecal coliform in groundwater may include agricultural runoff, effluent from septic systems or sewage discharge, and infiltration of domestic or wild animal faecal matter; poor well maintenance and construction (particularly shallow dug wells) can also increase the risk of bacteria and other harmful organisms getting into a groundwater supply [16].There is need to have uncontaminated water, since water is necessary to life. By repercussion, the wells around the selected area Gyadi Gyadi are unsafe for drinking, therefore the quality of the water needs to be improved with sufficient treatment in line with microbial standard.

Table 2. Physical-Chemical parameter during dry season.

S/n	Area	Sample	Month	Parameter									
				Temp	pH	Turb.	EC	Color	TSS	Hardness	TDS	*E.coli*	*F.coli*
1.	Dangi	1a	April	30.2	6.7	0	2.36	10	3	404.028	730	0	32
2.	Dangi	1b	April	30.4	6.8	67	717	10	5	538.704	274	0	10
3.	Kasuwar dare	2a	April	29.6	6.8	0	1940	5	0	471.366	732	0	0
4.	Kasuwar dare	2b	April	30.2	6.9	0	464	10	1	448.92	475	51	1
5.	Babban Layi	3a	April	31.0	7.2	0	840	10	2	448.92	658	1	17
6.	Babban Layi	3b	April	29.8	7.2	0	498	10	0	336.69	248	1	15
7.	Dangi	1a	May	31.2	7.23	2.06	2360	5	13	279.09	1177	179	0

8. Dangi	1b	May	31.0	7.2	27.2	1611	10	28	275.84	797	37	34
9. Kasuwar dare	2a	May	30.8	7.22	4.55	2440	5	1	296.29	1205	70	0
10. Kasuwar dare	2b	May	30.8	7.21	5.56	1552	5	1	228.29	771	50	56
11. Babban Layi	3a	May	31.1	7.32	2.11	748	5	0	18.81	471	0	33
12. Babban Layi	3b	May	30.9	7.36	2.16	1176	5	0	130.19	590	4	10
13. Dangi	1a	June	29.6	7.5	3.27	284	5	6	269.35	1056	0	5
14. Dangi	1b	June	29.2	7.3	33.6	208	10	39	255.88	945	0	76
15. Kasuwar dare	2a	June	30.4	8.0	19.7	1557	10	18	388.31	785	0	43
16. Kasuwar dare	2b	June	28.5	7.2	10.3	959	15	14	224.46	478	0	50
17. Babban Layi	3a	June	29.9	7.8	1.17	537	10	5	179.57	438	0	16
18. Babban Layi	3b	June	30.3	7.9	2.38	1154	5	5	134.68	554	0	41

Table 3. Physical-Chemical parameter during wet season.

S/n	Area	Sample	Month	Parameter									
				Temp.	pH	Turb.	EC	Color	TSS	Hardness	TDS	*E.coli*	*F.coli*
1.	Dangi	1a	July	28.7	7.5	4.1	340	5	2	350.16	1168	250	6
2.	Dangi	1b	July	28.4	7.4	7.0	162	10		269.35	809	150	37
3.	Kasuwar dare	2a	July	28.4	7.6	3.2	2390	5	2	361.38	1195	80	39
4.	Kasuwar dare	2b	July	29.2	7.6	6.1	1786	5	8	281.31	893	0	4
5.	Babban Layi	3a	July	28.5	7.7	1.2	1043	5	0	202.01	521	75	5
6.	Babban Layi	3b	July	28.8	7.6	2.13	1151	5	1	139.17	576	0	1
7.	Dangi	1a	Aug	26.2	6.2	4.33	1813	5	0	435.45	915	230	0
8.	Dangi	1b	Aug	25.9	6.3	8.44	2040	5	7	325.47	993	153	0
9.	Kasuwar dare	2a	Aug	27.6	7.6	4.38	1558	10	1	336.69	781	83	39
10.	Kasuwar dare	2b	Aug	27.6	7.6	19.4	1901	10	18	296.29	951	70	4
11.	Babban Layi	3a	Aug	27.8	7.7	6.6	1521	5	3	260.37	761	253	5
12.	Babban Layi	3b	Aug	27.9	7.6	3.2	1446	5	0	148.14	723	151	1
13.	Dangi	1a	Sept	27.3	7.0	1.35	2770	5	5	480	1383	0	0
14.	Dangi	1b	Sept	27.3	6.8	4.73	1672	5	12	583	836	0	0

15. Kasuwar dare	2a	Sept	27.3	6.	1.05	2730	5	2	763	1258	0	17
16. Kasuwar dare	2b	Sept	27.5	7.0	2.22	1967	5	6	628	83	2	12
17. Babban Layi	3a	Sept	27.4	7.2	1.2	992	5	3	448	496	2	4
18. Babban Layi	3b	Sept	27.4	7.2	3.79	1679	5	14	404	839	3	7

3.2 Recommendations

With particular reference to the findings of this research, the following recommendations are made:

- Regular monitoring of human activities in the area and public health education on water sanitation is required for all sources of water supply.
- It is recommended that people should device the means of treating water locally to reduce the hazards of biological parameters which are the most dangerous through boiling, addition of alum and local chlorination particularly those suspected to contain high concentration of coliform bacteria, and this should be done by coordination.
- The local communities should be mobilized to undertake regular maintenances and sanitation, and the government should monitor the way and manner waste is being disposed in the area.
- It would also assist in revising current strategies and involvement designed at protecting environment and improving the quality of life within urban and local societies.

4. Conclusions

The study was carried out to assess the effects of waste dump on the quality of well water in Gyadi Gyadi area of Kano Metropolis, Nigeria. The major goal of examining drinking water is to avert the spread of water borne diseases and to safeguard the health of the society. The importance of access to good quality of water cannot be over emphasized. Increase in population in Gyadi Gyadi coupled with the rise in human activity pose a great burden on provision of safe drinking water. This requires large number of people to consume water from shallow open wells which constitute a major health problem due to close proximity of wells to dump sites. This study recorded a high number of coliform counts in water samples analysed, thus making it unsafe for drinking, thus requires further treatment

This work has revealed that the wells water samples obtained within three locations from six different wells in the dry and wet seasons in the study area did not meet both the national and international standard of drinking water quality and therefore not fit for human consumption.

References

[1] Nizam M U, Hanafiah M M, Mahmoudi E and Halim A A 2021 *Scientific Reports* **11** 8623.
[2] Shaikh M M, Al Suhaimi AO, Hanafiah M M, Ashraf M A and Harun SN 2018 *Desalin. Water Treat.* **112** 344-50.
[3] A Al-Raad A, Hanafiah M M, Naje AS, Ajeel M A O Basheer A and Ali A T 2019 *Processes* **7** (5) 242.
[4] Mustapha A, Sagagi B S, Daura M M, Tanko A I, Phil-Eze P O and Isiyaka H A 2019 *Int. J. River Basin Manag.* doi.org/10.1080/15715124.2019.1606817.
[5] Guan R, Wang X, Zhao Y, Wang T, Yang Y, and Ashraf MA 2017 *Caribb. J. Sci.* **50 (2-3)** 25-28.
[6] Vystavna Y, Diadin D, Yakovlev V, Hejzlar J, Vadillo I, Huneau F and Lehmann MF 2017 *Environ.* **76** 463.
[7] Ashraf M. A and Hanafiah, M. M. 2017. *Environ. Sci. Pollut. Res.* **24** 22753-22754.
[8] Ashraf M A and Hanafiah M M 2019 *Environ. Sci. Pollut. Res.* **26** 13679-13680.
[9] WHO Guidelines for drinking water quality 2011 3rd edn. World health organization Geneva.
[10] NSDWQ Nigerian standard for drinking water quality.2007 Nigeria ind stand **554**:13-14.
[11] Mostafa M G, Uddin S M H and Haque A B M H 2017 *Appl water sci.* **7**:4663-4671.
[12] WHO *Guidelines for drinking water quality* 3rd ed vol **1**: recommendations. Geneva: WHO. Available online at: (www.who.int/water_ sanitation_ health).
[13] WHO *Guidelines for drinking-water quality*. 2011 4th ed. WHO Press Geneva, Switzerland. 541p. Available at: www.who.int/water_sanitation_health/publications.

[14] Ma J Z, Wang X S and Edmunds W M 2005 *J. Arid Environ.* **61**, 277–295.
[15] Meenakumari H R 2004 Groundwater Pollution with special reference to open wells in and around Mysore city, Karnataka state. Unpublished PhD Thesis. University of Mysore, India.
[16] Isiyaka H A, Mustapha A, Juahir H and Eze P P 2018 *MESE.* **5**, 583-593.

Acknowledgments
Hadiza Ado Suleiman was funded by the Tertiary Institutions Trust Fund Nigeria (Tetfund).
Marlia M. Hanafiah was funded by the National University of Malaysia (GUP-2020-034 and DIP-2019-001).

Solutions of contemporary problems of environmental safety as a component of national security

L M Popova[1]*, S A Skochko[2] and O V Nesterenko[3]

[1] Kharkiv National University of Civil Engineering and Architecture, Kharkiv, Ukraine;
[2] International Research and Design Center for Research and Examination of Climate Change and Community Development, Kharkiv region, Ukraine;
[3] Kharkiv National University of Civil Engineering and Architecture, Kharkiv, Ukraine

*Corresponding E-mail: image.center@kstuca.kharkov.ua; Liliya.herman01@gmail.com

Abstract. One of the priorities of the national interests of each country is to ensure environmental security as a component of national security. At the same time, among the current and projected threats to national security and national interests of almost all countries of the world are the consequences of climate change and the growth of man-made pressure on the environment. Therefore, today it is necessary to find ways to solve modern problems of environmental safety, including climate safety, which is the purpose of this scientific work. A promising way to reduce the risks of global climate threat is to ensure the environmental security of each country, which provides for the introduction of green technologies on the principle of construction of biopower plants in combination with poultry and livestock farms, as well as in combination with organic raw materials from the population, agricultural enterprises, utilities. In addition, it is planned to obtain biogas as an "alternative fuel", which will produce electricity and heat, which will be used in lighting and heating of buildings. To prevent the complication of the environmental situation and reduce the intensity of climate change, it is proposed to reduce the area of landfills for organic raw materials. Analysis of soil contamination was conducted near the landfill for solid waste in Lozova city (Ukraine), from which the results of research.

Keywords: Environment, Safety, Climate, Technologies

Track Name: Human, Social, Economic and Environmental Sustainability

1. Introduction

Climate observations confirm the fact that recent years have become the hottest for the planet in human history. This article examines the harmful factors influencing climate change and related to the use of biogas in landfills for solid household and organic waste, including livestock and poultry farms, as well as treatment plants [1-5].

Atmospheric air, soil, water and adjacent areas that affect warming and climate are studied as general indicators of pollution. The study was conducted in the city of Lozova, Kharkiv region (Ukraine), where there are large enterprises and solid household waste. Areas were surveyed for "soil respiration" and other studies were conducted. In this region, the duration of droughts has increased in recent years. In addition, in this region there are periods when it rains and hails, hurricanes occur, which leads to the effects of climate change (including the emergence of new insect pests, road problems, etc.). The purpose of this scientific work is to find ways to solve modern problems of environmental, including climatic, safety. The following research methods were used in this scientific work: bioindication method (allows to determine the level of danger of man-caused load on soil ecosystems), envelope method (by means of which soil samples were taken at a depth of up to 4 centimeters). In addition, according to regulatory methods, chemical and physiological parameters of soil samples were determined.

2. Actuality

There are currently problems with the use of fossil fuels, such as oil, carbon and gas. There is a need to replace them with renewable and environmentally friendly energy sources, while increasing energy efficiency with the introduction of new technologies. It is also advisable to reduce CO2 emissions, which will lead to a transition to a "green" energy supplier. The establishment of an environmentally friendly food production system in agriculture will help reduce hydrofluorocarbons, which are chemicals. They are used both in agricultural machines of refrigerating type, and in the conditioners established in buildings. Reducing emissions of harmful hydrofluorocarbons into the atmosphere reduces the likelihood of an irreversible climate threat, as hydrofluorocarbons have a 9000-fold stronger effect on warming than CO2 emissions.

Based on observations of the steppe zone of Ukraine, a conclusion is made about the increase in the area of such a zone and the increase in aridity. Farming there is becoming increasingly difficult, because it requires constant irrigation. We see either a lack of water in rivers that dry up in the dry summer, or severe floods. The introduction of rainwater reclamation technologies is quite expensive and inefficient, which reduces the suitability for food needs of both humans and animals. Therefore, drip root nutrition of plants is more favorable for environmental safety not only in this region but also on a global scale. After all, in recent years we have seen that with irreversible climatic processes after heavy winter snowfalls in March there is a sharp warming, and all this water does not have time to absorb into the soil, but goes into rivers, and leads to flooding of many areas.

The problem of environmental safety, which is associated with the intensity of climate change due to man-made pressures on soil ecosystems, is now becoming increasingly important. As a result of the creation of large areas of landfills for the disposal of organic raw materials in the form of solid household waste, there was a problem of catastrophic reduction of biological activity of soils. Environmental protection, rational use of natural resources, ensuring the ecological safety of human life is an integral condition for sustainable economic and social development of Ukraine. To this end, the state implements environmental policy aimed at preserving the safe for living and inanimate environment, protection of life and health from the negative effects of environmental pollution, achieving a harmonious interaction of society and nature, protection, rational use and reproduction of natural resources [1-4].

Today, Ukraine is an active participant in international relations, maintains diplomatic relations with 150 countries, and economic relations with 75 countries. External relations are diverse in form, scope, areas of implementation and are characterized by certain specifics of life and development of foreign countries [2]. Therefore, an important issue for Ukraine is to adhere to the "culture of environmental security" [3] on the path to European integration and positioning itself as a developed country in the international space.

Existing state building codes in force in Ukraine establish climatic parameters that are used in the design of buildings and structures, heating, ventilation, air conditioning, water supply, energy passport of the house, as well as in the planning and construction of urban and rural settlements [6]; aimed at ensuring the rational use of energy resources for heating, cooling and hot water supply, ensuring the normative sanitary and hygienic parameters of the microclimate of the premises, the durability of enclosing structures during the operation of buildings [7]. However, the current period of production of electricity and gas, biogas [4, 8] from fossil fuel fields is characterized by their depletion and unprecedented pollution of all geospheres, along with excessive emissions of methane, carbon dioxide, hydrogen sulfide, oxygen and other emissions into the air, heat energy through the enclosing structures of buildings and infrastructure. In addition, there are other factors that contaminate soils, groundwater, water bodies, rivers and other objects. All this affects the destruction of the ozone layer and increases the greenhouse effect, resulting in climate change.

The most common sources of methane emissions into the atmosphere are "solid waste landfills", treatment plants [9] and infrastructure facilities with low "heat transfer coefficients" [6, 7]. To implement the climate project, the authors Popova LM, Skochko SS, Nesterenko OV propose the first test version of the software product series of automated climate safety design systems, which allows to automate the design of the project to protect the environment and climate in general, and allows to prepare and implement project documentation for the construction of bioelectric power plants of secondary organic raw materials and other buildings and structures.

Thus, in order to prevent the complication of the environmental situation and reduce the intensity of climate change, it is proposed to reduce the area of landfills of the disposal of organic raw materials imported to landfills for solid household waste in the form of household waste and industrial waste. Organic raw materials will be taken from the population, agricultural enterprises, utilities and exported to the bioelectric power plant of secondary organic raw materials to produce biogas as an alternative fuel [11 - 15], from which electricity will be produced. At the same time, organic raw materials after receiving biogas from it are converted into organic biological fertilizers, which can be an alternative to inorganic fertilizers. In general, biofertilizers can displace the use of pesticides and agrochemicals in the agro-industrial complex, which are inorganic fertilizers [16, 17] and belong to the class of inorganic chemistry.

The biological fertilizers obtained in this way will reduce the area of landfills for solid waste and will lead to the elimination of landfills, as organic raw materials will not be considered household waste. Household and industrial waste, except inorganic waste, is organic raw material that should be evaluated according to a category other than household and organic waste. It can be argued that obtaining organic fertilizers at a biological power plant of secondary organic raw materials from organic raw materials will not only improve the environmental situation and reduce the effects of climate change, but also ensure rational land use and provide the agro-industrial complex with necessary biological fertilizers.

The obtained liquid biological fertilizers from liquid organic raw materials can be used for simultaneous hydraulic reclamation of lands [18] and application of organic fertilizers, which will ensure both reproduction and rational use of lands and hydraulic reclamation of lands by sub-root fertilization of plants. This will ensure the reproduction of surface water resources and will lead to the conservation, reproduction and inexhaustible use of biological diversity, reducing the likelihood of inorganic chemicals entering the soil and water bodies.

The method of bioindication allows to determine the level of danger of man-caused load on soil ecosystems. One of the most objective criteria for assessing the functional state of soil ecosystems is the enzymatic activity of soils [17]. However, as is known from the scientific literature, enzymes react differently to soil contamination by pollutants [18-20]. Experimental research was conducted on the basis of the research laboratory of the Department of Life Safety and Engineering Ecology of Kharkiv National University of Construction and Architecture. Soil samples were taken by envelope method at a depth of up to 4 centimeters. Sample preparation was performed according to regulatory methods. In soil samples were determined chemical and physiological (respiration rate - ID) indicators on MSW, 10-2 ml CO2 / g g.

Soil sampling points in the areas adjacent to the landfill for solid household waste in Lozova Lozova city (Ukraine) are shown in Figure 1. In Figure 1 section №1 is the body of the landfill for solid household waste; №2 is a site located at a distance of 1 meter from the body of the landfill for solid household waste; №3 is a site located at a distance of 5 meters from the solid household waste landfill; №4 is a site located at a distance of 20 meters from the solid household waste landfill; №5 is a site located at a distance of 50 meters from the solid household waste landfill; №6 is a site located at a distance of 200 meters from the landfill for solid household waste.

Figure 1 shows that directly near the landfill for solid household waste, the respiratory activity of soils is completely suppressed. The obtained research results are shown in Table 1.

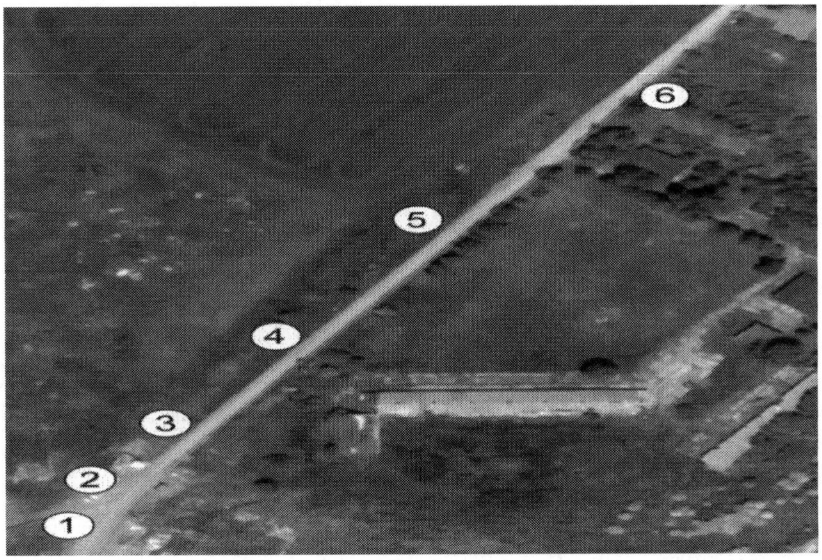

Figure 1. Soil sampling points in the territories adjacent to the solid household waste landfill in Lozova city (Ukraine).

Table 1. Intensity of soil respiration and the distance to solid household waste.

Name	Numbers of research points / Distance from the object of research (meter)					
	1/solid	2/1	3/5	4/20	5/50	6/200
Catalase activity	0,0125	0,0125	0,0125	0,01	0,012	0,02
Intensity of "soil respiration"	4,5	4,8	6,15	2,25	5,1	5,85

Soil respiration was determined by the Galstyan method, which is based on determining the intensity of CO2 release by the soil sample. Soil respiration was determined by a method based on determining the intensity of soil respiration by determining the concentration of carbon dioxide in the soil atmosphere. The intensity of carbon dioxide production was determined by the formula in mg CO2 per 100 g of soil per day.

$$SR \frac{V_2 - V_1}{m \cdot T}$$

where ***SR*** – the intensity of soil respiration;
V1 – the volume that HC_l was used for titration without soil, ml;
V2 – volume, HC_l spent on titration of the experiment, ml;
m – soil sample, g;
T – exposure time, hour.

Catalase activity was determined by measuring the rate of decomposition of hydrogen peroxide when it interacts with the soil. This indicator was determined by the permanganite method - titration of soil aqueous extract with a solution of $KMnO_4$ [4, 7, 8]. Catalase activity was calculated by the formula:

$$KA_n = \frac{V_{KMnO_4} \cdot 3}{m_n \cdot T},$$

where ***KA$_n$*** – catalase activity;
V_{KMnO4} – volume of manganese permanganate spent on titration of the sample, ml;
m – sample of soil, g;
T – exposure time, minutes.

Catalase activity was expressed in cm^3 of 0.1 normal KMn04 solution per 1 g of dry soil for 20 minutes. The results of the determination are presented in Table 1.

To determine the content of petroleum products in the soil, we used the gravimetric method, which determined the concentration of petroleum products in all soil samples. Soils were selected from the landfill according to the method based on extraction of organic matter from hexane, evaporation and removal of solvent, separation of polar compounds on a column of alumina, removal of solvent and gravimetric measurement of residual mass.

A soil sample weighing 20 g was placed in conical flasks with a volume of 250 cm3 with a ground stopper, 50 cm3 of hexane was added, closed with a glass stopper and shaken (using a special apparatus) for 15 minutes. The operation of processing soil samples with hexane was repeated three times. After settling, the extract was carefully filtered into a 250 cm3 flask, avoiding contact with soil particles. The extract was then transferred to a porcelain cup, which was previously brought to constant weight, and evaporated in a water bath in a fume hood to remove the solvent.

The content of petroleum products in the sample volume was determined as the difference between the weight of the cup with the residue after removal of the solvent and the weight of the empty cup. The concentration of petroleum products in the studied soil sample (***C_{PPg}***, mg / kg) was determined by the formula:

$$C_{PPg} = \frac{(m_2 - m_1) \cdot 1000}{M_{ss}}$$

where ***C_{PPg}*** – concentration of petroleum products in the studied soil sample;
m_1 – the mass of the porcelain cup used in the analysis, mg;
m_2 – weight of porcelain cup with extracted petroleum products, mg;
M_{SS} - mass of soil sample, g;
1000 - conversion factor, mg / kg.

The research results are presented in the Table 2.

Table 2. The content of petroleum products in the soils adjacent to the landfill of solid household waste.

Selection point / Distance from the landfill,meter	Distance from the landfill for solid household waste, meter, meter	The content of petroleum products, mg/кg
1/solid	solid	180
2/1	1	160
3/5	5	140
4/20	20	130
5/50	50	120
6/200	200	103

The tables show that directly near the solid waste landfill the respiratory activity of soils is completely suppressed, and catalase activity increased by 20% compared to point 3, selected at an environmentally friendly distance from the landfill for solid household waste. The concentration of petroleum products did not exceed 200 mg / kg. With increasing distance from the landfill for solid household waste, the concentration of petroleum products decreased by 2 times, which may have in some way affected the activity of catalase, which had a direct relationship with the concentration of petroleum products.

3. Conclusions

Thus, the intensity of soil respiration indicates the physiological state of soils, while catalase activity manifests itself as an enzyme of stress factor, which indicates an increase in the concentration of pollutants in the soil, which significantly increase their enzymatic activity. The research provides an opportunity to reduce the risks of climate threat and man-made stress, as well as to prevent the complication of the environmental situation and, accordingly, to ensure national security.

An integrated approach to solving environmental safety and rational use of organic raw materials allows to create high-quality organic biological fertilizers, the application of which into the soil both in liquid and solid or moist conditions will increase yields and protect and reproduce bioresources. This will ensure: protection of the environment and atmospheric air; reducing the likelihood of rapid development of the climate threat.

References

[1] Law of Ukraine "On environmental protection" dated 25 June 1991, №1264-XII. URL: https://zakon.rada.gov.ua/laws/show/1264-12#Text https://zakon.rada.gov.ua/laws/show/1264-12#Text
[2] Popova L M 2018 Administrative and legal Bases of Control in the Field of Entrepreneurial Activity: a monograph. Kharkiv: Maidan 536
[3] Barbashev S V 2019 On the concept of safety culture and its application in the environmental sphere. Environmental safety: problems and solutions: Coll. Science. *Articles of the XV International Scientific and Practical Conference "Environmental safety: problems and solutions"* 326 p. 18-22
[4] UN Framework Convention on Climate Change. URL: https://zakon.rada.gov.ua/laws/show/995_044#Text.
[5] National Standard of Ukraine DSTU-N B B.1.1-27: 2010 "Protection from dangerous geological processes, harmful operational influences, from fire. Construction climatology". URL: https://dbn.co.ua/load/normativy/dstu/dstu_b_v_1_1_27_2010/5-1-0-929.
[6] State Building Standards DBN B.2.6-31: 2016 "Thermal insulation of buildings" Information Bulletin of the Ministry of Regional Development, Construction and Housing of Ukraine, 2017. URL: https://www.minregion.gov.ua/wp-content/uploads /2017/02/DBN-V.2.6-31-2016-Teplova-izolyatsiya-budivel.pdf.
[7] Sherenkova I A & Nesterenko O V 2009 Aerobic-anaerobic bio-pond for biochemical treatment of wastewater contaminated with organic impurities and biogas production
[8] Yurchenko V A, Skochko S A & Nesterenko O V 2015 Possibilities of obtaining and utilizing biogas at solid household waste landfills. *Scientific Bulletin of Construction. Kharkiv: KhNUBA, KhOTV, ABU* 4 82 S.194-199
[9] Yurchenko V A, Skochko S A & Nesterenko O V 2016 Utilization of organic waste by processing in biogas plants to obtain biofertilizers *Scientific Bulletin of Construction. Kharkiv: KhNUBA, KhOTV, ABU* **84** (2) 329-332
[10] Zaitseva V G, Nesterenko O V & Onishchenko N G 2017 Greening of technologies and waste utilization *Scientific Bulletin of Construction. Kharkiv: KhNUBA, KhOTV, ABU* **1** (87) 225-228

[11] Zaitseva V G & Nesterenko O V 2017 Industrial waste management system in Ukraine *Scientific Bulletin of Construction. Kharkiv: KhNUBA, KhOTV, ABU* **2** 88 C272-276c
[12] Skochko S A, Nesterenko O V & Samokhvalova A I 2018 Investigation of activated sludge biocenosis at treatment facilities *Scientific Bulletin of Construction. Kharkiv: KhNUBA, KhOTV, ABU* **2** (92) S.274-278
[13] Zaitseva V G *et al* 2020 Environmental protection measures to address the problem of environmental protection in Ukraine *Scientific Bulletin of Construction. Kharkiv: KhNUBA, KhOTV, ABU* **2** 100 S.274-278
[14] Krasnyansky M Yu 2018 Environmental safety: a textbook. Kyiv: *Condor Publishing House* 180 p. P.46
[15] Khudoba V & Chikailo Y 2016 Ecology: teaching method. way *Lviv: LDUFK* 92 p.34 p.

Life cycle design and efficiency strategy for sustainable membrane technology

K K Razman[1], A W Mohammad[2,3] and M M Hanafiah[1,4*]

[1] Department of Earth Sciences and Environment, Faculty of Science and Technology, Universiti Kebangsaan Malaysia, 43600 UKM Bangi, Selangor, Malaysia
[2] Department of Chemical and Process Engineering, Universiti Kebangsaan Malaysia, 43600 Bangi, Selangor, Malaysia
[3] Research Centre for Sustainable Process Technology (CESPRO), Faculty of Engineering and Built Environment, Universiti Kebangsaan Malaysia, 43600 Bangi, Selangor, Malaysia
[4] Institute of Climate Change, Universiti Kebangsaan Malaysia, 43600 UKM Bangi, Selangor, Malaysia

*Corresponding E-mail: mhmarlia@ukm.edu.my

Abstract. Life Cycle Assessment (LCA) is an assessment tool to evaluate the potential environmental impact of a system or product throughout its life cycle. Results from LCA provides an insight on developing a more sustainable system while identifying its environmental hotspots. With recent advancements in membrane treatment systems especially for wastewater treatment (WWT), there is a rising interest in its sustainability aspect. Thus, it is crucial to review on previous published studies to highlight the challenges and achievements in implementing LCA for the membrane system. Hence, this paper presents a review of 33 recent papers published from year 2017-2020 conducting LCA on membrane systems for WWT. Bounded by the key frameworks of ISO standards, the analysis of papers showed variance in defining its functional unit, system boundaries, impact assessment categories and method to evaluate LCA, which makes it a challenge to do comparison. Some of the challenges discussed and highlighted in this paper include the inconsistencies in specifying the impact assessment and methodology used for some journals and the lack of LCA study in certain regional areas. A more uniform implementation of LCA should be considered to ensure the reliability and reproducibility of results to allow adoptions on novel or existing membrane systems.

Keywords: Life cycle assessment, Environmental sustainability, Wastewater treatment, Membrane technology, Reverse osmosis

Track Name: Human, Social, Economic and Environmental Sustainability

1. Introduction

Life Cycle Assessment (LCA) is a common tool utilised to analyse the environmental impacts associated with a system or product at all stages from its 'cradle-to-grave'. This includes the extraction of raw materials, product processing and manufacturing up to its final disposal. All the inputs and outputs of each process associated with the system are taken into account and its environmental impacts are assessed based on the International Standard Organization (ISO) 14040 series. The standardisation of the LCA framework begins in the late 1990s due to efforts made by certain governmental and international organisations to provide a guideline for LCA studies conducted [1][2]. The framework is guided by four fundamental phases; goal and scope definition, inventory analysis, impact assessment and interpretation [3]. With a proper framework, the comparison of studies is more objective and impartial allowing for a more reliable evaluation of the associated environmental impacts of a system. From the LCA results, improvements on the analysed system can be done and help in the decision-making process for future developments. Furthermore, LCA is a fairly versatile technique that aids in policy-making, as a marketing tool and for strategic planning [4]. The LCA approach is increasingly being used as a decision support tool in sustainable water management, providing valuable knowledge on the different environmental impacts of current or proposed water infrastructure and processes [5]. Membrane technology is significantly favoured due to the increasing complexity of wastewater contaminants and a great demand for wastewater reclamation to increase efficiency and reduce cost [5]. The technology relies on the concept of separation by removing specific contaminants from a solution or suspension through fluid passing by a membrane and retaining certain components. The scope of this analysis considers peer-reviewed journals, conference papers and chapters in books with a focus on membrane technology for wastewater treatment (WWT) system published from 2017 to 2020 as shown in Figure 1.

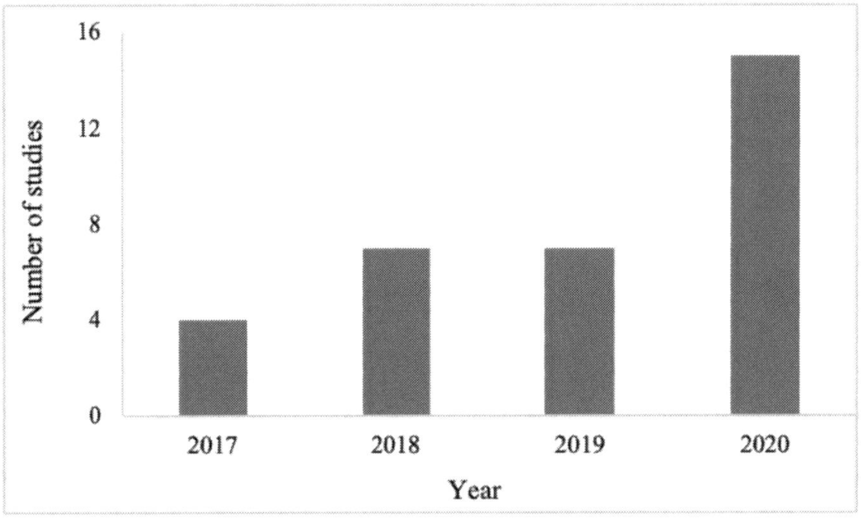

Figure 1. LCA studies on membrane technology by year (2017-2020).

LCA studies on membrane technology for WWT has seen an increasing trend generally (Figure 1) as sustainability is more incorporated into the design of membrane system. In the year 2020 alone, the LCA studies conducted are twice the number from previous years showing an increased interest in LCA as an evaluation tool. This number is projected to rise in the upcoming years as more policy-makers and organisations embrace to develop a more environmentally sustainable system for WWT and improve existing treatment systems.

2. Life cycle assessment methodology

2.1 Goal and scope definition

Various water treatment plants and systems were analysed in this paper with varying goals of studies. Some of the common goals defined were assessing the environmental impacts between different treatment systems or different scenarios implemented. The studies also managed to define the goal and

IOP Conf. Series: Earth and Environmental Science **880** (2021) 012053 doi:10.1088/1755-1315/880/1/012053

scope of the study to compare or evaluate different membrane systems or technologies that contributed to the lowest environmental impact. However, with various types of wastewater sources used for each study, the functional unit (FU) varies for different treatment systems [6] but commonly the FU is defined as 1 m^3 of treated wastewater. 91% of the studies explicitly defined the FU but 9% did not specify it. This existing variation affects making a reliable comparison among membrane treatment systems especially across the reviewed studies as different types of wastewater and FU were used for different systems. Nonetheless, the FU is unrepresentative as the quality of treated wastewater and its removal efficiency are not demonstrated [7].

In conducting an LCA study for a given system or product, there are many processes involved that are strongly interlinked together which complicates the starting and endpoint of evaluating the environmental impact. Hence, defining a system boundary is important as it sets a limit on what inputs and outputs to consider are relevant to the system. The most common system boundary defined by these LCA studies mainly concerned themselves on the construction and operation phase only. It is possible to identify other framework boundaries for LCA studies, such as cradle-to-gate studies that do not consider product distribution and use, but this depends on the study's objectives and the availability of data and/or impact assessment methodologies [3]. There is a limited number of studies doing a full LCA from 'cradle-to-grave' [6-11] as the infrastructure phase is considered negligible to the environmental impact. The final disposal phase and dismantling of a plant is also seldomly considered with only 18% taking it into account. Regardless, [12] argues that a full LCA analysis helps in determining the sustainability of the membrane system from the parameters at each stage.

2.2 Inventory analysis
The LCA software utilised in these studies was mostly Gabi as the preferred choice partly due to its ability to model each element in a system giving a broad life cycle perspective. SimaPro is followed closely as one of the highly preferred software as it has multiple databases and impact assessment tools with a powerful graphical interface that clearly displays processes having the greatest impacts [13]. As for GaBi, their database is tailored towards industrial processes and flows which can be observed from certain studies that utilise industrial wastewater [14-17]. Nevertheless, 15% of the LCA studies did not specify the software used to model their LCA which poses the inconvenience of making an objective comparison. The differences in the LCA software tool utilised also affect the result obtained due to differences in characterisation factors for each software tool [18]. From 85% of the studies that stated the software used, Gabi was chosen by 11 studies, followed by SimaPro (10), OpenLCA (5), EASETECH (1) and TOTAL (1).

2.3 Impact assessment
The related life cycle impact assessment (LCIA) methodology used for wastewater treatment systems were CML, ReCiPe and TRACI methodology.

CML method is an impact evaluation approach that limits quantitative modeling to early stages in the cause-effect chain. The results are categorized into midpoint groups based on common mechanisms (for example, climate change) or widely agreed groupings such as ecotoxicity [19].

ReCiPe method's main goal is to reduce a large number of life cycle inventory outcomes to a small number of indicator scores. The relative severity of an environmental impact group is expressed by these indicator ratings at two levels; 18 midpoint indicators and 3 endpoint indicators. The advantage of utilising ReCiPe is its comprehensive selection of effect categories at the midpoint level compared to other approaches [20].

TRACI method characterises factors for LCIA, industrial ecology and sustainability indicators. Through common equivalence units, the characterisation factors measure the possible impacts of inputs and releases on particular impact categories. Processes, goods, services, businesses, and societies all benefited from the TRACI approach [21].

From the 33 papers analysed, most studies stated the impact methodology used with 6 choosing CML, 6 ReCiPe and 8 TRACI (Table 1). As for the remaining 13 studies, mixed methodologies were incorporated or they were unspecified.

Table 1. LCIA methods used in previous LCA studies.

Method	Developer	User (Reference)
CML	Centre for Environmental Studies (CML), University of Leiden	[10] [14] [15] [22] [23] [24]
ReCiPe	RIVM, CML , PRé Consultants, Radboud Universiteit Nijmegen and CE Delft	[7] [25] [26] [27] [28] [29]
TRACI	U.S. Environmental Protection Agency (U.S. EPA)	[8] [9] [16] [30] [31] [32] [33] [34]
Mixed methodologies/Unspecified	-	[3] [6] [11] [17] [35] [36] [37] [38] [39] [40] [41] [42] [43]

As for the impact category chosen, Global warming potential (GWP), Eutrophication Potential (EP) and Acidification Potential (AP) were frequently selected. [44] stated that GWP has the most potential to improve future environmental issues especially involving agricultural-based production. Only studies by [8] [32] [27] included all 18 impact categories for evaluation. Furthermore, the number of impact categories frequently included in these LCA studies were 10 impact categories including GWP, EP and AP.

2.4 Interpretation

It has become critical to assess environmental performance using a validated and systematic approach such as the LCA [45]. Interpretation is an important step in the LCA as it provides a thorough and critical review on the analysed data. This is obtained from the result of life cycle inventory and impact assessment, where conclusions are made and stages of the systems that are contributing to the environmental hotspots are identified. From these information, the limitations of the system can be identified and suggestions are made for improvements in terms of sustainability. The key objectives of this step are to define possible areas for improvement, lend legitimacy to the LCA report, and prevent any bias arising from the practitioners' and/or organizations' personal interests and backgrounds [46]. The interpretation phase of the studies is further explored in the next section of this paper discussing the findings of the various membrane treatment system.

3. LCA of the wastewater treatment system

A comprehensive analysis was conducted on the reviewed studies to help identify and highlight the similarities and differences of LCA implementation for membrane systems in WWT along with achievements made. A total of 33 studies were reviewed within the year range of 2017 to 2020 with studies spread throughout the globe (Figure 2). While the collection of studies does not constitute a comprehensive overview of the field, they do represent the current state of the tool. The ISO framework also only provides a general methodology, the details and implementation of LCA for each membrane system differ from study to study. Through the ISO 14040/14044 guidelines, the environmental impacts were evaluated attributional as such.

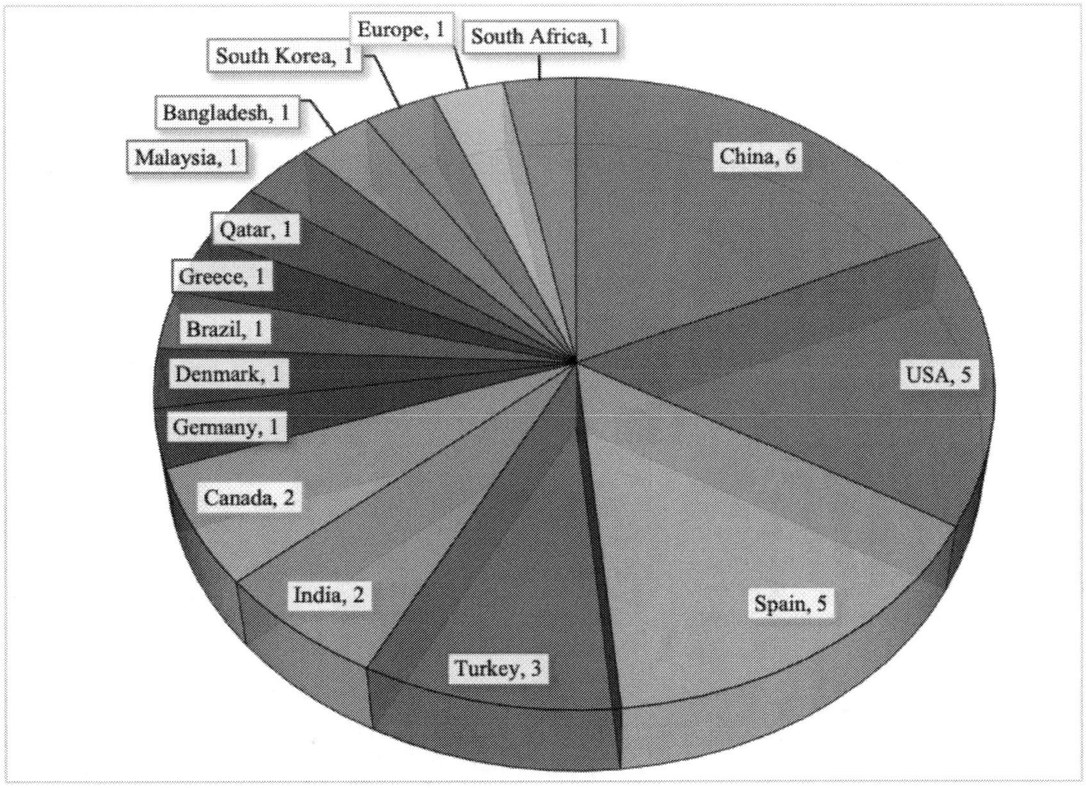

Figure 2. LCA studies on membrane technology by country of origin.

The majority of the research evaluates various technologies of membranes, to determine which treatment systems or processes have the least environmental effect in treating wastewater. Two of the most frequently studied membrane systems are the membrane bioreactor and reverse osmosis. Wastewater treatment systems have been built and operated to manage water contamination and mitigate the environmental effects of industrial and domestic wastewater discharge, but they consume energy and chemical reagents, as well as generate sludge and other emissions [4].

3.1 LCA of membrane bioreactor in treating wastewater

A membrane bioreactor (MBR) combines membrane processes such as microfiltration or ultrafiltration with a biological treatment system and is frequently analysed in these studies to identify their environmental impact in a wastewater treatment plant (WWTP). Table 2 shows an overall summary of the LCA studies focusing on MBR treatment system including anaerobic membrane bioreactor (AnMBR), anaerobic fluidised membrane bioreactor (AFMBR), submerged membrane bioreactor (SMBR) and aerobic membrane bioreactor (AeMBR). The LCA studies primarily compare the environmental impacts of membrane bioreactors (MBR) and anaerobic membrane bioreactors (AnMBR) versus other treatment systems [3] [8] [41], with an emphasis on municipal wastewater and greywater treatment.

Table 2. LCA studies on MBR treatment system.

Reference	Goal	Source of water	Functional unit	Software
[3]	Compare three decentralised treatment system	Decentralised wastewater	1 m^3 of treated wastewater for irrigation	SimaPro
[6]	Compare three biological treatment system for non-potable reuse	Greywater	Delivery of non-potable reuse water for the whole building	Not defined
[8]	Evaluate full-scale AnMBR system	Urban wastewater	Volume of treated water (m^3)	SimaPro
[9]	Evaluate greywater management system	Greywater	Annual treatment of greywater generated per person	OpenLCA
[15]	Assess two reclamation plant	Municipal wastewater	1 m^3 of reclaimed water	Gabi
[22]	Evaluate six wastewater treatment technologies	Municipal wastewater	1 m^3 of treated wastewater	Not defined
[25]	Evaluate integrated greywater treatment plant	Greywater	1 m^3 of treated greywater	Gabi
[28]	Evaluate three environmental technologies	Wastewater	10,000 m^3/d of pre-treatment, biological treatment, disinfection, wastewater discharge and sludge treatment process.	Gabi
[29]	Evaluate four water management scenarios	Greywater	Undefined	SimaPro
[31]	Compare AnMBR-based treatment system to conventional treatment system	Domestic wastewater	Treatment of 5 million gallons of medium strength wastewater with the same effluent characteristics.	Not defined
[32]	Assess three treatment processes	Textile wastewater	1 m^3 of treated effluent	SimaPro
[33]	Evaluate treatment system as sewer mining water reclamation	Municipal wastewater	1 m^3 of treated wastewater	OpenLCA
[35]	Evaluate two separate treatment plants	Municipal wastewater	1 m^3 of treated wastewater	OpenLCA
[39]	Evaluate alternative ways of MBR treatment system	Dairy factory wastewater	1000 m^3 of treated dairy wastewater	EASETECH
[40]	Conventional activated sludge versus MBR	Hostel-site wastewater	1 L of treated or processed wastewater	SimaPro
[41]	Evaluate three greywater treatment system	Greywater	The same treated volume of greywater and the same amount of sodium dodecylbenzenesulfonate (SDBS) removed	Gabi
[42]	Evaluate membrane and conventional treatment system in given scenarios	Wastewater	Flow rate of wastewater (100,000 m^3 day^{-1}) for 20 years of operation	TOTAL

29% of the wastewater treated by the MBR system is greywater and [9] found that MBR is environmentally preferred especially when the reused greywater amount is larger. The anaerobic greywater reuse system is also shown to be the most eco-efficient as the system enables energy recovery giving an environmental advantage over the other systems [29]. In decentralised wastewater treatment, MBR technology is favoured more especially when there are fewer pollutant loads compared to conventional activated sludge (CAS) [40].

It can be concluded that the MBR treatment system is more environmentally feasible when compared to conventional treatment systems such as CAS. Depending on the type of wastewater being treated and the goal of the treatment, different types of MBR should be incorporated to ensure the least impact on the environment is achieved. [33] found that AnMBR when evaluated at all scales shows a net energy benefit. In treating textile wastewater, MBR is found to be the most efficient in organic compounds and colour removal but on the industrial-scale moving bed bioreactor (MBBR) is more attractive [32]. The drawback with using the MBR treatment system is its high energy consumption contributes to the environmental impact and proves to be a major environmental hotspot.

3.2 LCA of reverse osmosis in treating wastewater

Reverse osmosis (RO) membrane treatment system is frequently used in the desalination process and reclamation of reusable water. This is mainly due to the membranes being capable of eliminating up to 90–99% of pollutants in the water supply, such as total dissolved solids (TDSs) [47]. In this analysis, the four studies that perform LCA on the RO membrane treatment system use different types of wastewater and set in achieving different goals. This is reflected in the functional unit defined by these studies with an exception from [17] (Table 3).

Table 3. LCA studies on RO treatment system.

Reference	Goal	Source water	Functional unit	Software
[10]	RO membrane versus MCDI membrane	Textile wastewater	500 m^3 of wastewater per day	SimaPro
[17]	Reverse osmosis (RO)-nanofiltration (NF) versus NF-RO	Electroplating wastewater	Not defined	Gabi
[27]	Seawater desalination versus mine water reclamation using RO membranes	Seawater and mine water	1 kL of potable water produced at a specified standard over the life cycle of each process unit	SimaPro
[38]	Compare three scenarios of tertiary wastewater treatment	Industrial wastewater	1 m^3 of water for reuse as cooling tower make up	SimaPro

From the study by [38], three scenarios were set up to identify the environmental impacts that contributed the least to the environment. It is observed that adding an intermediate softening calcite seeding is the most environmentally favoured due to maximum recovery of RO. In treating textile wastewater, [10] found that the membrane capacitive deionisation (MCDI) performs better environmentally on all of the terms compared to the RO membrane. In making a comparison between treating seawater and mine water using RO membrane, both scenarios have the operational phase as the leading environmental impact. However, [27] identified that mine water reclamation has a better overall impact on the environment than seawater desalination. In making a comparison between the stages of RO and NF, [17] found that NF-RO had more impact on the environment than RO-NF with the highest from GWP. This is justified by the principle that RO membrane requires higher pressure than NF membrane and is more suitable to lower feed concentration. By making RO the second stage, more valuable energy would be consumed and less freshwater produced.

3.3 LCA of other membrane treatment system in treating wastewater

Table 4 shows that the remaining studies incorporated various membrane systems such as fuel cell technology [11], a wood-based filtration system with different membranes [14] and a hybrid treatment system [34]. Regardless, the source of wastewater being treated are mostly municipal wastewater [26] [15, 24] and wastewater [36, 11, 23]. However, there is no consistency between defining functional units despite having similar goals of different systems such as assessing between scenarios or methods.

Table 4. LCA studies on other membrane treatment systems.

Reference	Goal	Source water	Functional unit	Software
[7]	Evaluate shale gas extraction wastewater treatment	Shale wastewater	1 m^3 of treated water	Not defined
[11]	Assess fuel cell technology	Wastewater	1 L of treated wastewater	Gabi
[14]	Assess wood-based filtration system with different membranes	Industrial wastewater	$30 \times 30 \times 5$ mm^3 of Ginkgo biloba (Gb) membrane	Gabi
[16]	Assess two scenarios of electroless nickel plating treatment system	Electroless nickel plating wastewater	Environmental impact per m^3 of treated wastewater produced by the electroless nickel plating (ENP) industry	Gabi
[23]	Evaluate two different methods on wastewater treatment case	Wastewater	10 000 m^3 sewage per day	Not defined
[24]	Assess on polishing units in treated wastewater	Domestic/municipal wastewater	1 m^3 of delivered recycled water to be used for irrigation.	Gabi
[26]	Assess SPF process for secondary effluent treatment	Municipal wastewater	1000 m^3 of treated secondary effluent per day	Gabi
[30]	Evaluate community-based sewage water use	Sewage	Annual provision of space and water heating, irrigation and water treatment per person	OpenLCA
[34]	Hybrid system versus centralised water system	Greywater	1 m^3 water used for outdoor irrigation and/or toilet flushing	SimaPro
[36]	Locally produced AC versus conventional ACs	European wastewater	Amount of AC required to reduce FAE potential to a defined level in terms of comparative toxicity unit per 1000 m^3 of wastewater	OpenLCA
[37]	Assess three reclamation scenarios from petroleum refinery	Refinery wastewater	1 m^3 refinery wastewater from API/CPI separator	SimaPro
[43]	Assess desalination technology	Wastewater/ seawater	1 L of treated water	Gabi

The studies by [7] and [43] focus on desalination of wastewater with differing desalination technology and functional units. From the study, [7] summarised that thermal-based desalination technology is the best method for desalinating water when compared to membrane distillation as it has a higher environmental impact when in comparison. However, [43] evaluated between microbial desalination cells (MDC), microbial fuel cells (MFC) and conventional treatment method to desalinate water and found that MDC has the highest environmental impact from the manufacturing and operation phase which contributes to the GWP of its whole life cycle. This is caused by the electricity consumption used to pump water for the MDC from its operation phase.

In treating municipal wastewater, [24] utilises different combination of microfiltration, ultrafiltration (UF) and ultraviolet (UV). UV treatment is deemed as the most environmental-friendly with cartridge filter and UF combination impacting the most for all impact categories evaluated [24]. [26] on the other hand uses solar photo-Fenton (SPF) and nanofiltration and identified the major environmental hotspots being electricity and chemical consumption for this type of treatment.

Studies by [16, 23, 37] were defined by assessing scenarios for wastewater reclamation with various sources of wastewater, software used and functional units. The proposed alternative scenarios from [16] proved to fare better in terms of climate change than the existing chemical treatment process for electroless nickel plating wastewater. The scenarios proposed by the study from [37] also displayed favourable results concerning the marine ecotoxicity and aquatic eutrophication impact which was reduced up to 90% and 84%, respectively. Assessing the LCA of a proposed scenario is a good way to assess the sustainability of a given system and aids in deciding which scenario provides the least environmental impact.

4. Conclusions

It can be concluded that the LCA approach is increasingly being used as a decision support tool in sustainable water management, providing valuable insights on the environmental impacts of existing membrane treatment systems and in making improvements for future novel treatment processes. With various types of software and methodology available to conduct LCA, adapting it to a specific process or product in a given system is made easier. This is especially helpful considering wastewater treatment has a variety of treatment systems depending on the types of wastewater being treated.

However, with the broad application and set up of LCA within the membrane technology itself, making a comprehensive comparison between systems poses a challenge. Some of the challenges include the inconsistencies in specifying the impact assessment and methodology used for some journals and the lack of LCA study in certain regional areas. Hence, it is suggested that a more uniform implementation of LCA should be considered to ensure the reliability and reproducibility of results to allow a better implementation on novel or existing membrane systems.

References

[1] ISO 14040 2006 *International Standards Organisation* (Geneva)
[2] ISO 14044 2006 International Standards Organisation (Geneva)
[3] Arias A Feijoo G and Moreira M T 2020 *Current Developments in Biotechnology and Bioengineering (Elsevier)*
[4] Harun S N H, Hanafiah M M and Aziz N I H A 2020 *Environmental Management* **67** 146-161
[5] Ismail H and Hanafiah M M 2020 *Journal of Environmental Management* **264** *110495*
[6] Arden S, Morelli B, Schoen M, Cashman S, Jahne M, Ma X C and Garland J 2020 *Sustainability* **12** 5459
[7] Caballero J A, Labarta J A, Quirante N, Carrero-Parreño A and Grossmann I E 2020 *Sustainability* **12** 1686
[8] Jiménez-Benítez A, Ferrer J, Rogalla F, Vázquez J R, Seco A and Robles Á 2020 *Current Developments in Biotechnology and Bioengineering (Elsevier)*
[9] Kobayashi Y, Ashbolt N J, Davies E G and Liu Y 2020 *Environ. Int.* **134** 105215
[10] Cetinkaya A Y and Bilgili L 2019 *Water Air and Soil Pollut.* **230** 149
[11] Zhang J, Yuan H, Deng Y, Abu-Reesh I M, He Z and Yuan C 2019 *Int. J. Life Cycle Assess.* **24** 1962-1975
[12] Mahmood S, Hemdi A R, Saman M Z M and Yusof N M 2013 *Re-engineering Manufacturing for Sustainability (Springer)*
[13] Liu Y, Syberfeldt A and Strand M 2019 *Prod. Manuf. Res.* **7** 490-502
[14] Niaz F, Khan Q, Ali M and Shen W 2020 *ACS Omega* **5** 4900-4906
[15] Chen Z *et al* 2018 *Bioresour. Technol.* **269** 246-254

[16] Sabeen A H, Kamaruddin S N B and Noor Z Z 2019 *IJEST* **16** 3171-3182.
[17] Zhu J, Jin Q and Dongming L 2017 *J. Water Environ. Technol.* **32** 267-275
[18] Silva D A L et al 2019 *Sustain. Prod. Consump.* **20** 304-315
[19] Guinée J B *et al* 2002 *Handbook on life cycle assessment. Operational guide to the ISO standards (Kluwer Academic Publishers)*
[20] ReCiPe 2021 *Retrieved from* https://pre-sustainability.com/articles/recipe/
[21] Bare J 2011 *Clean. Technol. Envir.* **13** 687-696
[22] Kamble S, Singh A, Kazmi A and Starkl M 2019 *Water Sci. Technol.* **79** 1102-1112
[23] Bai S, Wang X, Zhang X, Zhao X and Ren N 2017 *RSC Advances* **7** 26335-26341
[24] Büyükkamacı N and Karaca G 2017 *Water Sci. Technol.* **76** 3205-3212
[25] Yoonus H, Mannan M and Al-Ghamdi S G 2020 *World Environmental and Water Resources Congress 2020: Water, Wastewater, and Stormwater and Water Desalination and Reuse* (Reston, VA: American Society of Civil Engineers)
[26] Gallego-Schmid A and Tarpani R R Z 2019 *Water Res.* **153** 63-79
[27] Goga T, Friedrich E and Buckley C A 2019 *Water SA* **45** 700-709
[28] Su X *et al* 2019 *Chemosphere* **218** 778-792
[29] Lam C M, Leng L, Chen P C, Lee P H and Hsu S C 2017 *Appl. Energy* **202** 293-307
[30] Cabling L P B, Kobayashi Y, Davies E G, Ashbolt N J and Liu Y 2020 *Environments* **7** 36
[31] Harclerode M, Doody A, Brower A, Vila P, Ho J and Evans P J 2020 *J. Environ. Manage.* **269** 110720
[32] Yang X, López-Grimau V, Vilaseca M and Crespi M 2020 *Water* **12** 1306
[33] Cashman S, Ma X, Mosley J, Garland J, Crone B and Xue X 2018 *Bioresour. Technol.* **254** 56-66
[34] Jeong H, Broesicke O A, Drew B and Crittenden J C 2018 *J. Clean. Prod.* **174** 333-342
[35] Banti D C, Tsangas M, Samaras P and Zorpas A 2020 *Membranes* **10** 421
[36] Joseph B, Kaetzl K, Hensgen F, Schäfer B and Wachendorf M 2020 *Environ. Res. Lett* **15** 064023
[37] Muñoz, I., Aktürk, A. S., Ayyıldız, Ö., Çağlar, Ö., Meabe, E., Contreras, S., ... and Ferrer, O. (2020). J. Clean. Prod. 121967
[38] Ronquim F M *et al* 2020 *J. Clean. Prod.* **250** 119547
[39] Skrydstrup J, Larsen S L and Rygaard M 2020 *J. Ind. Ecol.* **24** 1101-1112
[40] Singh S P, Pandey A, Sharma R and Sharma M K 2019 *Indian J. Environ. Prot.* **39** 989-994
[41] Dominguez S *et al* I 2018 *Sci. Total Environ.* **621** 1047-1056
[42] Rashidi J *et al* 2018 *IJER* **12** 815-827
[43] Zhang J, Yuan H, Deng Y, Zha Y, Abu-Reesh I M, He Z and Yuan C 2018 *J. Clean. Prod.* **200** 900-910
[44] Nasution M A, Wulandari A, Ahamed T and Noguchi R 2020 *Sustainability* **12** 4101
[45] Aziz N I H A and Hanafiah M M 2020 *Environ. Pollut.* **268** 115948
[46] Martins A A, Caetano N S and Mata T M 2017 Sustainable Membrane Technology for Water and Wastewater Treatment *Springer*
[47] Maynard E and Whapham C 2020 Decontamination in Hospitals and Healthcare *Woodhead Publishing*

Acknowledgments

This project was funded by grant code KK-2019-001. Marlia M. Hanafiah was financed by the National University of Malaysia (GUP-2020-034).

IOP Conf. Series: Earth and Environmental Science **880** (2021) 012054 doi:10.1088/1755-1315/880/1/012054

Life cycle assessment of crushed glass abrasive manufacturing from recycled glass

I Zulkarnain[1], L S Lai[2], M I Syakir[2,3]*, A A Rahman[1], S Yusuff[2], M M Hanafiah[4]

[1] School of Physics, Universiti Sains Malaysia
[2] School of Industrial Technology, Universiti Sains Malaysia
[3] Centre for Global Sustainability Studies (CGSS), Universiti Sains Malaysia
[4] Research Centre for Tropical Climate Change System, Universiti Kebangsaan Malaysia

* Corresponding E-mail: misyakir@usm.my

Abstract. Sustainable Development Goals (SDGs) are the key motivation factors to determine the characteristics of green abrasives, contributing to carbon emissions reduction, reduce waste generation and build up a recycling-based sustainable blasting industry. Such aim can be evaluated through life cycle assessment as a structured basis for evaluating the performance of environmental impacts and benefits of green abrasives application in blasting industry. Crushed glass is one of zero free silica content abrasives from recycled glass, and it is widely used due to inert and safe characteristics. Database of life cycle inventory (LCI) are obtained through literature review. Production of 1 ton/year of crushed glass abrasives has been modelled at gate-to-gate boundary where the human health appears as the major impact potentials (0.71 DALY, disability-adjusted life year) at the production stage. The selection of 100% recycled glass as the raw materials in the supply chain has led to insignificant impact potentials of resources scarcity and ecosystem damages per unit production of 1 ton of crushed glass at USD 4.79 and 0.06 species. year, respectively.

Keywords: Life cycle assessment; crushed glass abrasive; glass cullet; intermediate glass processing; gate-to-gate

Track Name: Human, Social, Economic and Environmental Sustainability

1. Introduction

The first Life Cycle Assessment (LCA) has gained global attention, progressively as the awareness on the environmental protection rises. Businesses and scientists use LCA as a qualitative and comprehensive tool to analyze the environmental impacts of a product along its supply chain [1]. A life cycle assessment generally comprises four phases: goal and scope definition, life cycle inventory analysis (LCI), life cycle impact assessment (LCIA) and interpretation [2]. While it depends on the system boundary, impacts of a product can be assessed from the phase of raw material acquisition, through product manufacturing process, logistic distribution until the recycling or end-of-life disposal [3]. This systematic approach on products and services can provide a big picture on sustainability for businesses and industries to consider the health and environmental impact caused by their business activities [4,5]. LCA is useful to allow strategic planning for further environmental performance improvement contributing to effective marketing, policymaking and product development [2].

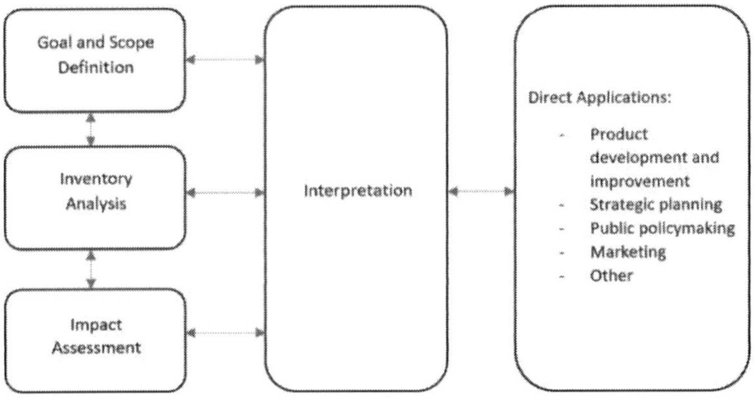

Figure 1. Phases of LCA and Applications. Source: ISO (2006a)

Glass is a material that can be found anywhere and is widely used in our everyday life. Glass can be easily sourced as it is made from silicon, the second most common element on earth [6]. It exists in the form of containers, such as bottles, food container jars, cosmetics and other products; or to be used in the industries for construction or manufacturing of electronics, automation, etc. Mass production will generate enormous amount of waste. In 2018, the United States has generated 12.3 million tons of glass waste, contributing to 4.2% of total municipal solid waste generation [7]. Therefore, recycling of glass is highly encouraged to avoid waste disposal to landfills [8].

After the realization of the glass recycling opportunity, companies such as Strategic Materials, Inc. and N.T. Ruddock Co., Inc. began to create business using recycled glass as the key product [9,10]. The recycled glass is processed into crushed glass for various uses depending on its coarseness. One of the usable end products is crushed glass abrasive. It has been recommended as alternative abrasive media due to it environmentally friendly features, generates less dust while performing higher cleaning rate [11]. However, less studies on the environmental performance of crushed glass abrasives is documented, hence, this work will serve as a precursor for evaluating the performance of environmental impacts of crushed glass abrasives application in blasting industry.

The objective of this study is to quantify the potential environmental impact associated with the production of crushed glass abrasive manufactured from recycled glass production through gate-to-gate life cycle assessment.

2. Methodology

2.1. Data Collection

Secondary data from literature or other published documents are the main source of data for Life Cycle Inventory (LCI) using the relevant keywords like “glass cullet”, “glass recycling” “glass abrasive” and “life cycle assessment”. Articles, books, journals, thesis from various web-based resource including ResearchGate, Elsevier, Google Scholar, Science Direct are reviewed. Other private company consulting report or government agency report which are relevant to glass abrasive production are studied too. Note, as well, the energy consumption and water quality impact of glass container recycling information is retrieved from Ching-Ling Tsai *et al.*, 2010 [13]. Life cycle impact assessment (LCIA) was carried out using SimaPro software. The impact potentials are quantified using ReCiPe

2016 Endpoint (E) V1.04 method, followed by simplification of the impact categories into three main areas namely, human health, ecosystem damages and natural resources depletion.

2.2. Functional Unit

The product system to be evaluated in this study is the production of 1 ton/yr of crushed glass abrasives based on the United States database.

2.3. Goal and Scope

The goal of this LCA study is to conceptualize the environmental performance of the crushed glass abrasive process for decision support at micro level of which will not cause any structural changes in the system. The result will serve as supporting knowledge towards better sustainable crushed class production management [12].

The boundary system LCA study focuses on the gate-to-gate life cycle. The assessment comprises overall intermediate cullet processing at cullet manufacturing stage encompasses processes such as crushing, screening, contaminant removal and packaging. Note, the transportation to primary or secondary consumer is beyond the scope of this LCA study. Presumably, the location of cullet processing factory is in United States based on the available inventory input data.

3. Assumption and Life Cycle Inventory

Assumption is made to scope and specify the inventory analysis. Quantity of input and output are determined based on literature database. Although in the real case scenario, the material input can be various type of glass including color sorted container glass, windshield, windows, mix container glass, borosilicate, automotive glass, etc. either from curbside or collected from suppliers, but in this study, it was narrowed down by assuming to be sorted glass cullet from general waste glass.

Figure 2 shows the conceptual of product system of 1000 kg of 3-mix-processed crushed glass abrasive comprises the material (crushed glass, kraft paper) energy (natural gas) and elementary flow (water etc) along with the product (crushed glass abrasive) as well as the impact potentials results from the Life Cycle Impact Analysis (LCIA). The unit processes are important for LCA to model the impact of crushed glass abrasive production at gate-to-gate life cycle stage.

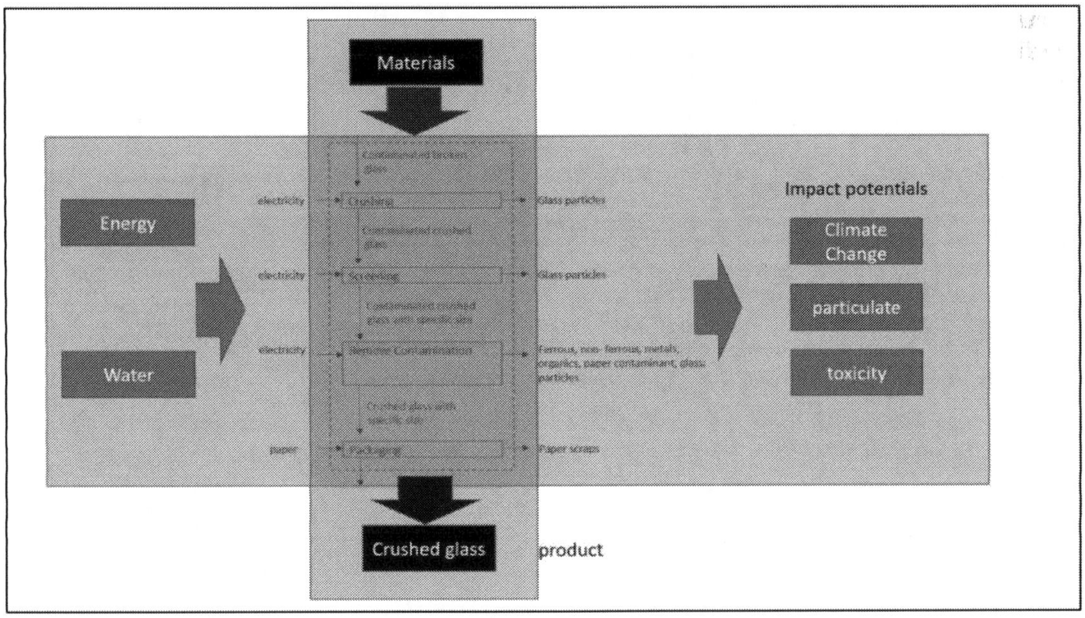

Figure 2. Product system of 1000 kg of 3-mix-processed crushed glass abrasive.

The operating process is simplified into four major processes i.e., crushing, screening, contaminant removal and packaging (Figure 3). These are the main process in intermediate cullet processing. The energy consumed for glass cullet processing is in the form of electricity considering no combustion or melting processes involved. Presumably, the type of energy for electricity generation is from natural gas considering that it is the largest energy source (38%) in United States. Note, in this study, consideration on fuel consumption is given only for skid-steer loader to transport cullet within the plant. As well, the paper used for packaging is supposed at 300g paper per standard size of 22.68 kg glass abrasive. Hence, it is estimated that about 13.2 kg of paper is used per 1000 kg of glass abrasive packaging. Furthermore, it is estimated that 26.75% of glass input will be discarded in the form of glass particles and rejected glass cullet. Another 3% of the waste is identified as non-glass waste such as paper, organics, metal contaminant and ferrous contaminants [14]. Table 1 shows the inventory input and output selected based on literature study for SimaPro life cycle impact assessment.

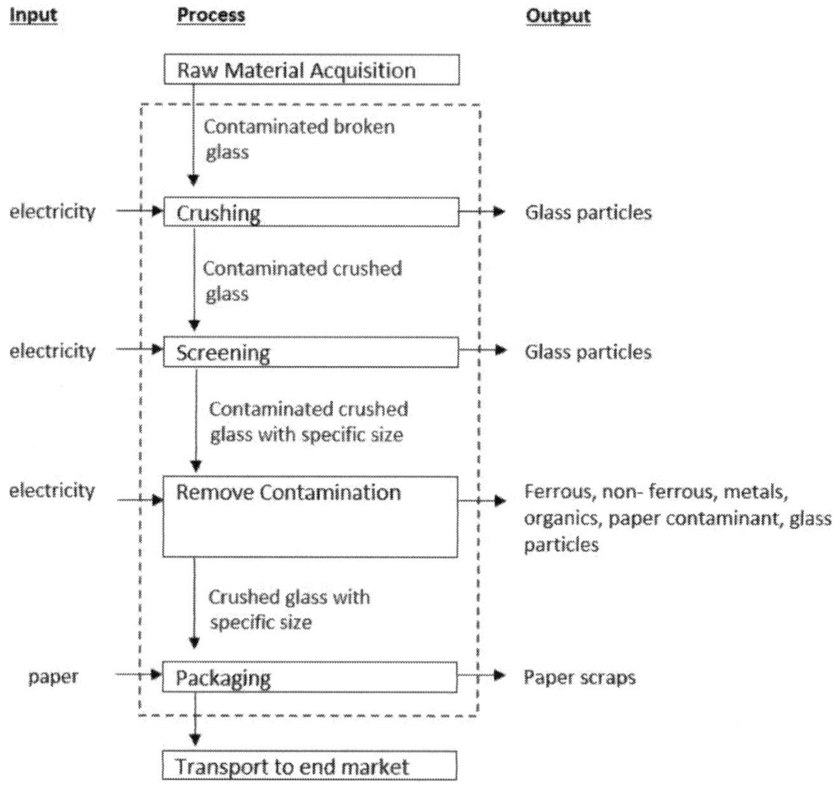

Figure 3. Life Cycle Inventory (LCI) flowchart.

Table 1. Inventory Input and Output selected in SimaPro.

	Amount
Products	
3-mixed crushed glass abrasive	1000 kg
Input	
Glass cullet, sorted {RoW}	1344.3 kg
Kraft paper, bleached	13.2 kg
Electricity sourced from natural gas	35.56 kWh
Excavation, skid-steer loader	1 m^3
Output	
Particulates < 10 um	267.5 kg
Carbon dioxide	0.062 kg
Metal waste	15 kg
Packaging waste, paper and board	2 kg
Waste, inorganic	15 kg

4. Limitation
All data sources are literature-based data. Note, the data may not be as accurate compared to primary data which measured directly from the processes without much spatial or time differences. In addition, the database in Simapro is generic, but still useful in conceptualizing the impact potentials of the product system.

5. Results and Discussion

5.1. Life Cycle Impact Assessment (LCIA)
The impact categories are selected based on total impact value of $1x10^{-4}$ or higher followed by normalized impact into the three indicators: human health (disability-adjusted life year (DALY)), ecosystem damage (species. year) and resources scarcity (USD 2013). In general, one DALY represents the loss of equivalent of one year of full health. DALY is used to measure impact towards human health categories. The ecosystem damage is described as the number of species disappear in a year (Species. year). While the USD 2013 expressed as the surplus costs of future resource production (assuming constant annual production), considering a 3% discount rate in USD unit.

The characterization of LCIA for production of 1000 kg crushed glass abrasive was observed at the mid-point level where climate change, toxicity, particulate, fossil resources and mineral resources appear as the major contributor to the product's environmental footprints (Table 2). Moreover, the selection of 100% recycled glass as the raw materials for the supply chain of crushed glass abrasives has led to impact potentials of resources scarcity per unit production of 1 ton of crushed glass, estimated at USD 4.79. In addition, surplus cost potentials of fossil resources at USD 29.30 may suggest the opportunity for the industry to switch to renewable energy-based operations, thus, potentially, the USD 29.30 may be reduced.

Table 2. Characterization of LCIA for production of 1000kg crushed glass.

Impact Category	Unit	Total	3-mixed crushed glass abrasive	Glass cullet input	Kraft paper	Natural gas	Excavation, skid-steer loader
Global warming, human health,	DALY	7.17E-3	7.75E-7	6.85E-3	2.23E-4	9.80E-5	6.28E-6
Fine particulate matter formation	DALY	8.00E-4	-	7.65E-4	3.38E-5	1.46E-6	9.07E-7
Human carcinogenic toxicity	DALY	1.67E-2	-	1.65E-2	1.99E-4	5.27E-6	2.56E-6
Human non-carcinogenic toxicity	DALY	3.87E-2	-	3.79E-2	7.53E-4	1.74E-5	3.93E-6
Mineral resource	USD2013	4.79	-	4.78	0.0128	4.24E-4	2.23E-4
Fossil resources	USD2013	29.3	-	26.4	1.69	1.15	7.14 E-2

Though, when it was normalized to the end-point results (Table 3), it showed that the health is the most pronounced impact (0.71 DALY). Given the fact that the raw material is 100% recycled glass, hence it shows almost no impact for ecosystem damage (0.06 species.year) and resource scarcity (0.00). It is depicted that the glass cullet is the main contributor to potential impact of human health, ecosystem damage and resources followed by paper (orange color) for packaging (Figure 4). Note, the packaging is considered as an ideal packing material due to its recyclable feature too. Based on the input-output determined at inventory phase, the production of crushed glass abrasive using 100% recycled glass is safe and environmentally friendly. Study has showed that at manufacturing stage, energy and emissions are reduced significantly when the recycled glass become the option as material for production [15]. Note, however, in this study, the results limit to gate-to-grave boundary.

Table 3. Normalization data of LCIA to produce 1000 kg 3-mixed crushed glass abrasive.

Damage Category	Total	3-mixed crushed glass abrasive	Glass cullet input	Kraft paper	Natural gas	Excavation, skid-steer loader
Human Health	0.710	8.68E-6	0.695	0.0136	1.37E-3	1.53E-4
Ecosystem	6.05E-2	1.84E-6	5.89E-2	1.32E-3	2.46E-4	1.94E-5
Resources	1.21E-3	-	1.11E-3	6.08E-5	4.10E-5	2.56E-6

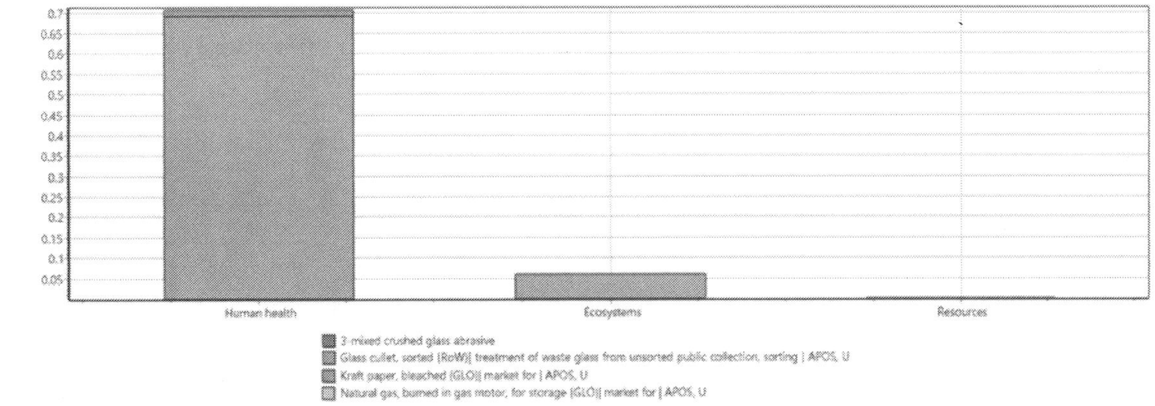

Figure 4. Normalization of LCIA for the production of 1000 kg 3-mixed crushed glass abrasive.

In addition, for every 10% of glass cullet used for crushed glass production, potentially, carbon emissions and energy consumption are reduced by 5% and 3% respectively. The used of glass cullet as raw materials will avoid the melting process, hence, reduces carbon emission [16]. In fact, if compare with the common copper slag abrasive, the crushed glass abrasive performs equally, produce a cleaner result than black colored copper slag with less risk of heavy metal and toxic pollutants. As well, NIOSH has categorized the glass dust as only "nuisance" dust because it contains less than 0.1% free silica [16]. The life cycle CO_2 emission assessed in previous study shows glass abrasive have emission of 19 kg CO_2/tonne lower than the emission for copper slag [17]. The emission of recycled glass varies depending on its application. In the case of shot blast abrasive application, the emission is relatively low compared to remanufacturing of glass containers and filtration [13].

As reported in the Circularity Gap Report 2021, in 2020, human activities have triggered global warming of 1.1 °C. The rise is expected to hit 3.2 °C this century under current trajectory. The world needs to increase the circular economy from 8.6% to 17% to halt global warming [18]. One of the main aspects of circular economy is using waste as secondary resource through reuse and recycling. That way it will maximize the uptake of circular materials in consumables that close loops and boost value in secondary markets. Crushed glass abrasive application in blasting industry can be one of the circular economy initiatives. The use of recycled-based abrasive is in line with the Sustainable Development Goals 12, responsible consumption and production. -reducing waste generation through prevention, reduction, recycling and reuse, thus, increase circulatiry of the recycled glass in value chain (circular economy).

6. Conclusion

Production of 1 ton/year of crushed glass abrasives has been modelled at gate-to-gate boundary where the human health appears as the major impact potentials (0.71 DALY) at the production stage. The selection of 100% recycled glass as the raw materials in the supply chain has led to insignificant impact potentials of resources scarcity and ecosystem damages per unit production of 1 ton of crushed glass at USD 4.79 and 0.06 species. year, respectively. This work provides first order impact potentials assessment on crushed glass abrasive manufacturing.

References

[1] Forest Product Association of Canada. Life Cycle Assessment and Forest Products: A White Paper. https://www.pwc.com/gx/en/forest-paper-packaging/pdf/fpac-lca-white-paper.pdf (accessed 12th December 2020)
[2] International Standardization Organization. ISO 14040:2006 Environmental management – Life cycle assessment – Principles and framework. https://www.iso.org/standard/37456.html (accessed 12th December 2020)
[3] Muralikrishna, I V and Manickam, V, 2017 Chapter Five - Life Cycle Assessment. Environmental Management:

Amsterdam The Netherlands .57-75
[4] Pajula T, Behm K, Vatanen S, Saarivuori E 2017 Managing the Life Cycle to Reduce Environmental Impacts. In: Grösser S., Reyes-Lecuona A., Granholm G. (eds) Dynamics of Long-Life Assets *Springer Cham*
[5] Salvador, R, Francisco, A C, Piekarski, C M and da Luz, L M. 2014 Life Cycle Assessment (LCA) as a tool for business strategy *Independent Journal of Management & Production* **5** (3) pp.733-751
[6] Royal Sociecty of Chemistry. Silicon. https://www.rsc.org/periodictable/element/14/silicon#:~:text=Silicon%20makes%20up%2027.7%25%20of,(oxygen%20is%20the%20first) (accessed 12th December 2020)
[7] United States Environmental Protection Agency. Glass: Material-Specific Data. https://www.epa.gov/facts-and-figures-about-materials-waste-and-recycling/glass-material-specific-data#:~:text=EPA%20combined%20data%20from%20the,recycling%20rate%20of%2031.3%20percent (accessed 12th December 2020)
[8] Glass Packaging Institute. Glass Recycling Facts. https://www.gpi.org/glass-recycling-facts#:~:text=Glass%20bottles%20and%20jars%20are,other%20glass%20jars%20were%20recycled. (accessed 16th January 2020)
[9] Serageldin, M, 2009 Emission Factors for Abrasive Materials
[10] European Commission - Joint Research Centre - Institute for Environment and Sustainability: International Reference Life Cycle Data System (ILCD) Handbook - General guide for Life Cycle Assessment - Detailed guidance. First edition March 2010. EUR 24708 EN. Luxembourg. Publications Office of the European Union; 2010
[11] Tsai, C L 2010 Water quality issues and life-cycle energy consumption of glass container recycling in New Jersey (Doctoral dissertation, Rutgers University-Graduate School-New Brunswick).
[12] Gaines, L L and Mintz, M M, 1994 Energy implications of glass-container recycling (No. ANL/ESD-18; NREL/TP-430-5703). Argonne National Lab., IL (United States); National Renewable Energy Lab., Golden, CO (United States)
[13] Good, N D, 2016. Documentation for Greenhouse Gas Emission and Energy Factors Used in the Waste Reduction Model (WARM).
[14] Silica dust, crystalline, in the form of quartz or cristobalite 2011 IARC Monographs on the Evaluation of Carcinogenic Risks to Humans.
[15] The European Container Glass Federation. Recycling: Why glass always has a happy CO2 ending. https://feve.org/wp-content/uploads/2016/04/FEVE-brochure-Recycling-Why-glass-always-has- a-happy-CO2-ending-.pdf (accessed 16th January 2020)
[16] Trimber, Kenneth A 1998 Evalution of Substitute Materials for Silica Sand in Abrasive Blasting 200 1–147 https://www.cdc.gov/niosh/topics/silica/pdfs/ab_p1rep.pdf
[17] A Life Cycle Analysis Report Prepared for British Glass by Enviros Consulting Ltd November 2003. Glass Recycling - Life Cycle Carbon Dioxide Emissions. http://www.packagingfedn.co.uk/images/reports/Enviros_Report.pdf
[18] Haigh, Laxmi, Marc de Wit, Caspar von Daniels, Alex Colloricchio, Jelmer Hoogzaad, Matthew Fraser, Ana Birliga Sutherland, Jim McClelland, Nanna Morgenroth, and Anna Heidtmann. 2021 The Circularity Gap Report 2021.

Acknowledgments

This paper is part of research activities under the grant (304.PTEKIND.6501034.D119).

Population trends of passage and wintering barn swallows (*Hirundo Rustica*) in Peninsular Malaysia

N A Ismail[1], U N S Daud[1], S M Nor[1], M S Mansor[1*]

[1] Department of Biological Sciences and Biotechnology, Faculty of Science and Technology, Universiti Kebangsaan Malaysia, 43600 UKM Bangi, Selangor, Malaysia

*Corresponding E-mail: msaifulmansor@gmail.com

Abstract. In Peninsular Malaysia, passage and wintering Barn Swallows (*Hirundo rustica*) often congregate at nocturnal roosts in towns. As a Holarctic migratory species, it is at risk of population collapse. Despite extensive research, little is known about this species in its wintering grounds, especially in the tropics. This study aimed to determine how climate change influences the population growth of this species across a geographical area over a decade, from 2009 to 2019. A species distribution model was constructed using data obtained from the eBird database – a citizen science project – of records from diverse locations. We used the Kernel Density tool in ArcGIS to describe Barn Swallow density and expansion pattern across Peninsular Malaysia. We demonstrate that the Barn Swallow's population has fallen by about 27.1% over the last decade. In 2009, the species was dispersed in two distinct areas: Perlis to Pulau Pinang; and the Pahang region bordering Selangor and Negeri Sembilan; in addition, some individuals chose to winter in south-eastern Johor. Within ten years, the flock has descended from Perlis to the northwest corner of Perak, while the vast wintering area in Pahang has decreased and stretched down to Melaka and Johor. The differences in Barn Swallow numbers and distribution patterns observed in this study could be a result of climate and land use changes which have caused this species to spread to parts of other wintering regions, altering the arrival and departure times of migration, thereby interfering with its breeding season.

Keywords: climate change, land-use, roosting site, swallows, wintering region

Track Name: Human, Social, Economic and Environmental Sustainability.

1. Introduction

Climatic variations can significantly affect the availability of plant and other resources for animals, resulting in population changes [1]. There is substantial evidence that the ongoing biodiversity crisis is being exacerbated primarily by human-induced global change, most notably in climate and land use [2,3]. The loss of biodiversity is one of the most severe risks to life on Earth [4]. The current species extinction rates are around a thousand times higher than the natural background rate, and the trend is moving upwards [5,6].

Birds are significant markers of overall biodiversity [7–9]. They are particularly vulnerable to climate change [10] and habitat disturbance [11]; this is especially the case for tropical species [12,13]. They are also widely used as umbrella species, as their conservation is intended to protect a vast number of naturally co-occurring species [14,15].

The Barn Swallow (*Hirundo rustica*) is a popular and ubiquitous species often observed in human settlements [16]. It is one of 83 species in the Hirundinidae, the bird family of swallows and martins (hirundines; [17]). There are six Barn Swallow subspecies: *H. r. savignii* breeding in North Africa; *H. r. transitiva* breeding primarily in the Middle East; *H. r. rustica* breeding in Europe; *H. r. tytleri* breeding in Siberia and Mongolia; *H. r. erythrogaster* breeding in North America [18]; and *H. r. gutturalis*, with a breeding range from the eastern Himalayas to northeast Russia (Siberia), China, the Korean Peninsula, and Japan [19], spending the winter in Thailand and the Malay Peninsula [20].

This study aimed to investigate the impact of climate change on the distribution and abundance of the Barn Swallow in Peninsular Malaysia between 2009 and 2019. We used data from eBird, a citizen science initiative for which data are collected by observers of all skill levels, from novices to experienced ornithologists [21]. Data from eBird have been widely used for species distribution modelling and in many recent studies (e.g., [20]). However, to our knowledge, these data have not been used to date to simulate long-term patterns in species abundance, particularly of migratory bird species, in Peninsular Malaysia. We hypothesised that the Barn Swallow numbers decrease as the pace of land use and climate change increases. This knowledge is critical for developing an adequate mitigation strategy to conserve and preserve the populations of this migratory species.

2. Methods

2.1 Study area

With a total area of 132,265 km^2, Peninsular Malaysia is divided into four distinct economic growth zones: Perlis, Kedah, Penang, and Perak are in the north; Selangor, Federal Territory of Kuala Lumpur, and Federal Territory of Putrajaya are in the west; Negeri Sembilan, Melaka, and Johor are in the south; and Kelantan, Terengganu, and Pahang are in the east. The climate of Malaysia is tropical, with hot and humid weather throughout the year. Previously dominated by tropical forests, today, this region is covered with a combination of agricultural land, woodland, wetlands, and urban–suburban–rural areas. Consequently, the loss of habitat due to deforestation, which eventually leads to climate change, has long been linked to the decline in migratory bird species, such as the Barn Swallow, in Peninsular Malaysia.

IOP Conf. Series: Earth and Environmental Science **880** (2021) 012055 doi:10.1088/1755-1315/880/1/012055

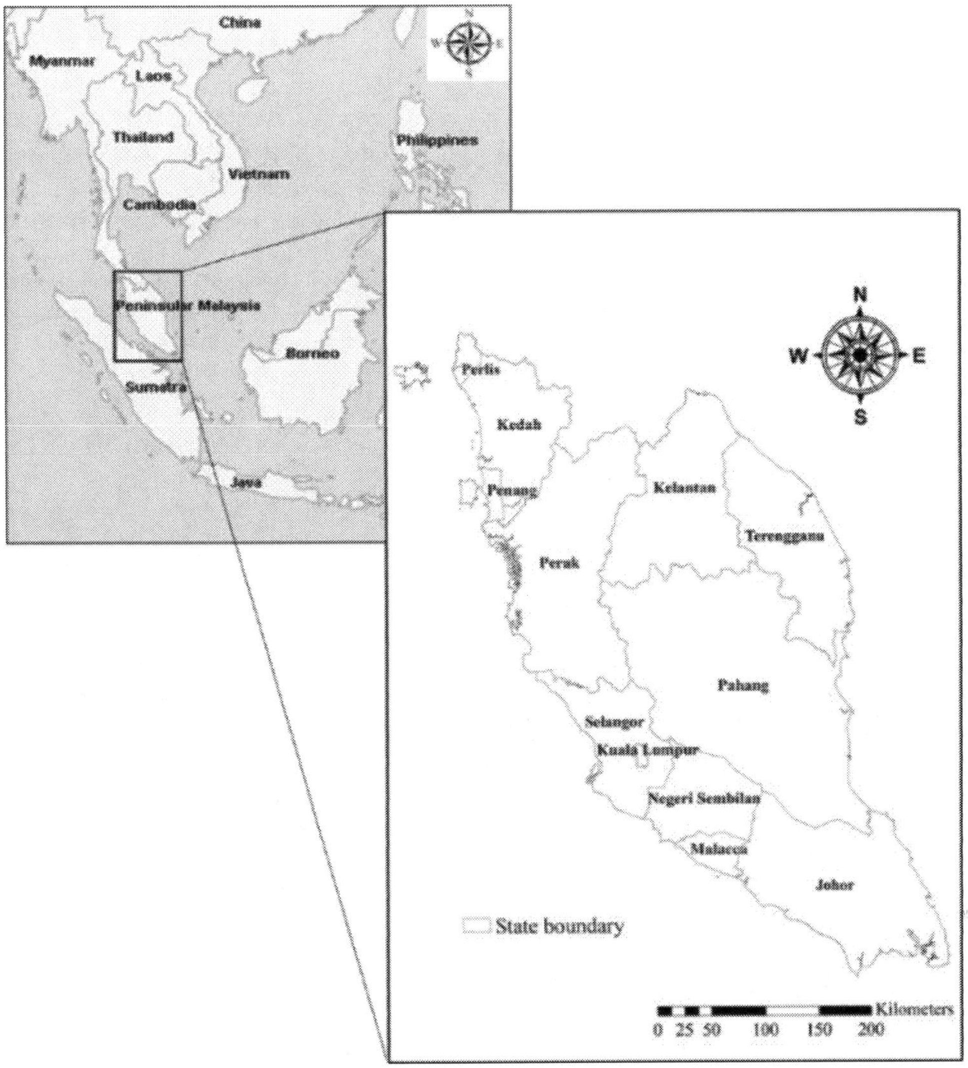

Figure 1. Map of the study sites comprising the 11 states and two federal territories of Malaysia.

2.2 eBird dataset

We estimated the distribution and range extension of the Barn Swallow using the eBird citizen science database. The dataset from eBird, which contains all checklists entered in 2009 and 2019, was obtained from the eBird website. The eBird primary dataset comprises all validated observations, including dates, observers, locations, coordinates, and observation counts. To prevent bias, we removed incomplete checklists since they did not appropriately sample the population. Checklists related to 'group identifier' were also eliminated since they were redundant checklists. We included comprehensive species checklists if the 'stationary', 'travelling', or 'exhaustive' protocols were followed [21]. For the selected years, annual data were evaluated and sorted. We examined species abundance and population size to evaluate how the distribution pattern has changed across Peninsular Malaysia.

2.3 Data analysis

The data collected spanned two separate years, 2009 and 2019, with a 10-year gap. The data were normally distributed, confirmed by inspection of quantile-quantile plots and the Shapiro–Wilk test. A t-test was used for data analysis [23] with the PAST programme (Paleontological Statistics, 2.17). Barn Swallow distribution data for 2009 and 2019 were then analysed with ArcGIS 10.7; the occurrences were set within 100 metres of the border of Peninsular Malaysia. Distributions outside the study area were omitted. Species distributions were derived based on both specified years (2009 and 2019). The Kernel Density tool was used to determine Barn Swallow density and differences in the expansion pattern across Peninsular Malaysia over the ten-year period. Kernels were determined using a Kertau RSO projection at 1-km resolution, with a 1-km search radius to prevent unnecessarily large bandwidth computed from the cell filled by points. The kernel output was stretched with a histogram equaliser and re-sampled with bilinear interpolation to improve the appearance and smoothness of the raster data.

3. Results

The eBird data yielded 26,010 numbers of swallows in 2009 (mean ± standard deviation; 2000.77 ± 1923.25) and 18,965 in 2019 (1458.85 ± 509.70) and demonstrated a 27.1% decrease in swallow abundance between the beginning and the end of the ten-year period (t = 0.27, p = 0.79) (Table. 1). Swallow numbers rose in all states except Negeri Sembilan, where they dropped by over 97.7% over the ten years (Fig. 2). The pattern of distribution revealed that the Barn Swallow observation decreased in western and eastern Peninsular Malaysia in the ten-year period (Fig. 3). Based on the data for 2009 and 2019, Negeri Sembilan and Perak were the most popular wintering spots in Peninsular Malaysia, with the highest Barn Swallow individual counts, respectively. The swallow populations in the east coast of Peninsular Malaysia have also grown slowly, particularly in Kelantan, where just ten observations were reported for 2019 in the eBird database. Based on the derived kernel values, Selangor and Pahang supported the highest density of Barn Swallow roosting spots in 2009. However, by 2019, the roosting spots had dwindled and begun to spread further north to Penang and south to Malacca.

Table 1. Abundance of Barn Swallows recorded throughout Peninsular Malaysia in the years 2009 and 2019.

States	2009	2019
Johore	74	4558
Kedah	104	829
Kelantan	0	10
Kuala Lumpur Federal Territory	0	96
Malacca	0	198
Negeri Sembilan	25077	577
Pahang	378	1244
Penang	159	3779
Perak	151	5261
Perlis	23	1267
Putrajaya Federal Territory	0	20
Selangor	44	1113
Terengganu	0	13

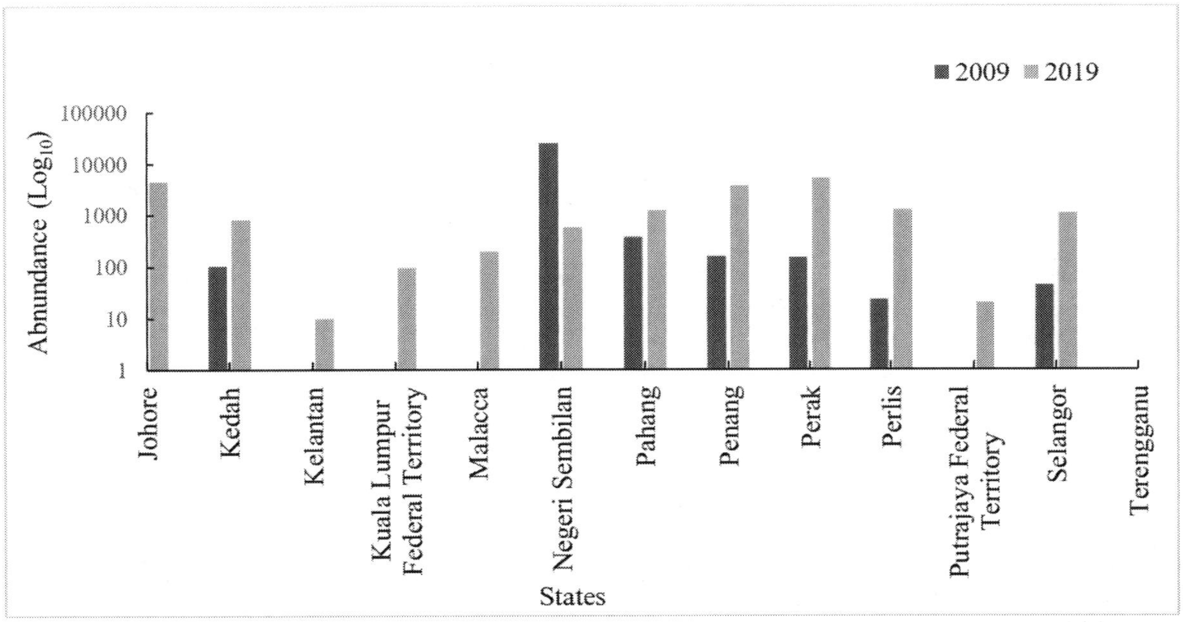

Figure 2. Occurrence of Barn Swallow in Peninsular Malaysia in 2009 and 2019.

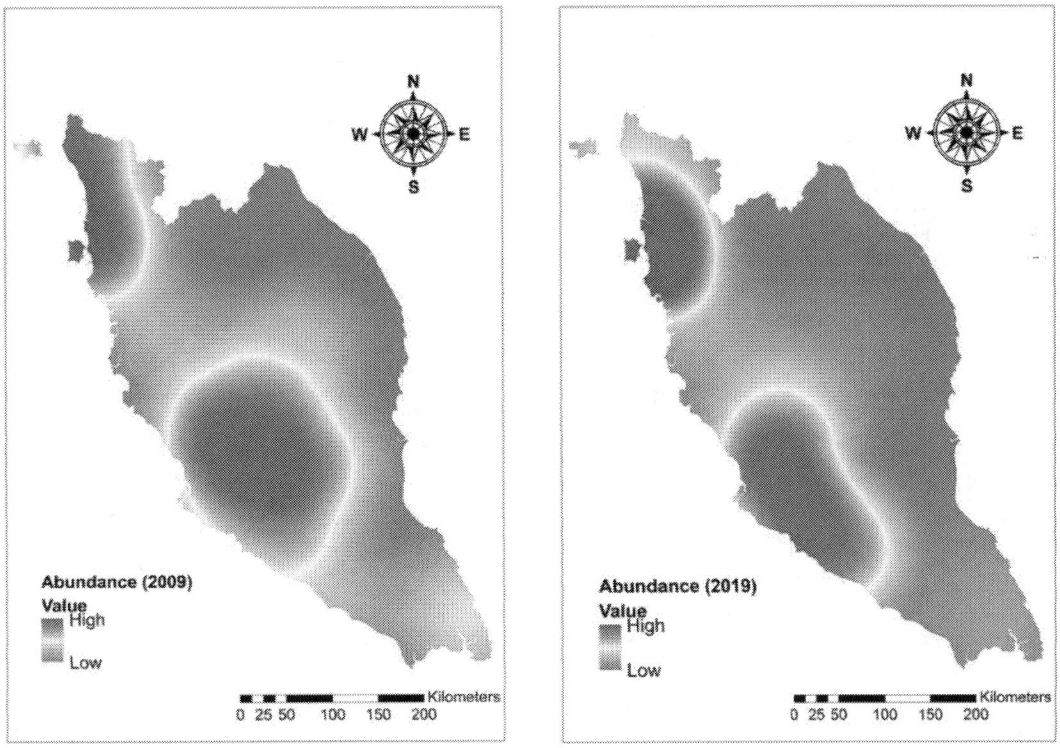

Figure 3. Spatial abundance of Barn Swallows observed throughout Peninsular Malaysia in the years 2009 and 2019. Maps were generated using ArcGIS v. 10.7.

4. Discussion

According to the eBird data for Peninsular Malaysia, Barn Swallow observations followed a rising pattern. One likely explanation is that more people began recording their sightings on eBird. Over 500,000 users have accessed eBird since its inception, and over 21 million bird records have been collected through more than 1.6 million checklist submissions. eBird has received data from over 35,000 unique users for more than 180,000 sites from the western hemisphere and New Zealand. Since eBird accepts 'historical' data, 13% of the checklists predate the November 2002 debut of eBird. Participation has increased steadily since that time [21].

However, as the number of observations has grown, the number of individual birds recorded has decreased. This was most noticeable for Negeri Sembilan, where individual records of Barn Swallows have declined by about 97.7% in the last ten years. Habitat loss is one possible factor for this decrease. Climate change has a substantial impact on habitat loss. Because of rising temperatures, flooding, and desertification, the habitats of migrating birds are at risk of changing and disappearing. According to Global Forest Watch [24], Jempol, the largest district in Negeri Sembilan, has lost 98,300 hectares (ha) of tree cover between 2001 and 2020 – a 75% loss in cover since 2000 and 36.1 megatonnes (Mt) of carbon dioxide equivalents (CO_2e) in emissions released into the atmosphere per year. During the same period, Jelebu, the second largest district in Negeri Sembilan, lost 33,300 ha of tree cover, comparable to a 26% loss in cover since 2000 and 18.4 Mt of CO_2e in emissions. Furthermore, between 2001 and 2020, Kuala Pilah district lost 21,200 ha of tree cover, a 24% reduction since 2000 and 13.9 Mt of CO_2e in emissions.

Extensive habitat loss in the tropics will continue to intensify the impact of climate change on tropical birds, particularly endemic and range-restricted species [25,26]. Coastal wetlands are an example of a habitat that migrating birds use for nesting and feeding. Regrettably, tropical coastal habitats are rapidly disappearing and are highly vulnerable to climate change and rising sea levels [27]. The loss of these coastal habitats should be of considerable concern since they provide vital ecological functions and services, such as protection against tsunami damage [28]. In addition, migrant species such as the Barn Swallow use these locations for food and rest during their journey; they can replenish their energy reserves and recuperate at these spots before continuing their arduous migrations. Unfortunately, sea-level rise caused by climate change will inundate these coastal wetlands, obliterating the habitats of birds and other animals. Without these rest stops, migrating birds' energy reserves deplete, and they struggle to complete their journeys [29].

The Barn Swallow's distribution pattern suggests that the population has migrated away from northern and eastern Peninsular Malaysia in response to habitat loss. Birds and other fauna are adjusting their ranges in reaction to climate change, with bird species moving poleward or to higher altitudes in tropical highlands [30]. The magnitude of such movements is projected to increase in the future; for example, several European birds are anticipated to experience range boundary movements of over 1,000 km [31]. However, landscapes fragmented or rendered unsuitable owing to human land use may impede birds' capacity to relocate to new, climatically acceptable habitats. Furthermore, many hotspots of bird species richness are currently found in protected areas, from which birds may migrate to unprotected areas due to climatic shifts [32].

In addition, habitat loss affects temperature fluctuations. Temperature records for Malaysia reveals a rising trend. At the same time, climate modelling employing 14 global climate models (GCMs) indicates that Malaysia might suffer temperature changes ranging from 0.7 to 2.6 °C and precipitation variation ranging from -30% to 30% [33]. These changes affect the migration routes and annual migration rhythms of many migratory birds, as the majority of these species alter their routes, reduce their journeys, or cancel them in response to changing temperatures [34].

Increasing food scarcity on wintering grounds due to decreased rainfall may impair the non-breeding performance of migrating birds and delay their departure to breeding grounds [35]. Changes in tropical wintering habitats also harm migrants' subsequent performance on their breeding grounds. Female American Redstarts (*Setophaga ruticilla*) wintering in high-quality habitat produce a higher number of offspring and fledge them earlier than females wintering in low-quality habitat [36]. Certain migratory birds that cannot match their migratory time to climate change-induced phenological changes already suffer negative consequences [37,38]. Birds that have evolved in areas with unpredictable weather are likely to cope better with climate change [39]. Unpredictable conditions affect nomadic birds, which can migrate great distances in response to droughts, floods, and other adverse weather events.

Climate change directly causes further degradation of natural resources, infrastructure, environment, and human health. In addition, some types of indirect damage are expected to be significant [40]. Climate change is likely to significantly impact bird reproduction and the entire annual cycle [41]. Numerous bird species, including swallows that migrate to higher elevations or latitudes, travel considerable distances to their nesting grounds [34]. Their migration to summering grounds is almost certainly governed primarily by photoperiod and endogenous clocks [42]. Short- and medium-distance migrants, on the other hand, can adapt to climate change more quickly, whereas long-distance migrants, such as the Barn Swallow, are disadvantaged [43]. Their migration patterns are typically more rigid, and they have difficulty adjusting to temperature changes; thus, they are more vulnerable to climate change than other birds.

In addition, mild winter temperatures help local birds to survive and thrive. As a result, they begin to make use of long-distance migrants' food supplies and breeding grounds. Long-distance migrants may find that many resident birds already occupy their breeding grounds [34] and face increased competition for space as well as food. Although there is a risk of mismatch between migrant species and their breeding grounds, it is also possible that after their arrival at the breeding grounds, the degree of snow cover and food availability of the destination fluctuates significantly over time. As a result, they could be better at responding to changes in weather and climate than other species.

5. Conclusion

Malaysia's tropical forests are diverse, and the impact of climate change on their biodiversity is a significant issue. While species composition is projected to be affected at the landscape scale by climate change, its significant variation at the local scale is expected due to soil and topographic factors. While some bird species have adapted to these changes, others have not, resulting in their extinction. The climate change we are witnessing today is different: human-induced global warming is accelerating, and many bird species are finding it increasingly difficult to adapt to the resulting shifts in climate. This study determined how climate change influences Barn Swallow numbers across an area over a decade from 2009 to 2019 and show that its community has shifted its preferred wintering region to survive. In addition, their abundance has also reduced drastically within the decade.

However, if climate change were the only new threat to birds, it is reasonable to expect that many species would adjust to the new conditions and survive, given current population and genetic variability, as well as the fact that their earlier generation survived previous climate transitions." [44]. Nevertheless, other human-made obstacles, such as habitat disruption in Negeri Sembilan, either individually or in combination, can affect the lives of birds. Birds that inhabit the harshest environments on Earth (deserts, high altitudes, and high latitudes) are already approaching the limits of their stress tolerance [34] and may be unable to deal with new difficulties. Numerous species that are already in decline due to various circumstances may be unable to cope with a multiplicity of future problems [45].

References

[1] Parmesan C, Root T L, Willig M R 2000 Impacts of Extreme Weather and Climate on Terrestrial Biota
[2] Foley JA, DeFries R, Asner GP, Barford C, Bonan G, Carpenter SR, *et al* 2005 Global consequences of land use *Science* **309** 5734 570–4.
[3] IPCC. 2013. Fifth Assessment Report - Climate Change 2013. www.ipcc.ch/report/ar5/wg1/. Retrieved 7 June 2021.
[4] Johnson C N *et al* 2017 Biodiversity losses and conservation responses in the Anthropocene. Science **356** 6335 270–5.
[5] Pimm S L, Russell G J, & Gittleman JL Brooks TM. The Future of Biodiversity 9
[6] Vos J M De *et al* 2014 Estimating the normal background rate of species extinction **29** 2 452–62
[7] Noble D G *et al* 2009 The generation and use of bird population indicators in Europe. 2008;
[8] Gregory RD, Willis SG, Huntley B, Collingham YC, Couvet D, Green RE. An Indicator of the Impact of Climatic Change

on European Bird Populations **4** 3
[9] Dan Chamberlain *et al* 2012 The altitudinal frontier in avian climate impact 205–9.
[10] Crick H Q P 2004 The impact of climate change on birds 146 48–56
[11] Fuller M R, Doyle M W &Strayer D L 2015 Causes and consequences of habitat fragmentation in river networks 1 1–21
[12] Oswald S M *et al* 2020 Urban Climate Using urban climate modelling and improved land use classifications to support climate change adaptation in urban environments : A case study for the city of Klagenfurt , Austria *Urban Clim* 31 100582
[13] Mansor M S *et al* 2018 Niche shift in three foraging insectivorous birds in lowland Malaysian forest patches. Behav Processes 157 73–9
[14] Roberge J & Angelstam PER 2004 Usefulness of the Umbrella Species Concept **18** 1 76–85
[15] Ismail N A *et al* 2020 Short communication: Roosting behavior of wintering barn swallow (*hirundo rustica*) in peninsular Malaysia *Biodiversitas* **21** 2 661–5
[16] Scordato E S C & Safran R J 2014 Geographic variation in sexual selection and implications for speciation in the Barn Swallow *Avian Research*
[17] Turner, A 2004 Family Hirundinidae (swallows and martins). In The birds of the World **9** (ed. J. Del Hoyo, A. Elliott & D. Christie) Barcelona, Spain: Lynx Edicions 602–685
[18] Del Hoyo J, Elliott A, Sargatal J, Christie D 2018 Handbook of the Birds of the World Alive. Lynx Edicions, Barcelona, Spain
[19] Dor R *et al* 2010 Phylogeny of the genus Hirundo and the Barn Swallow subspecies complex. *Mol Phylogenet Evol* **56** 1 409–18
[20] Mansor M S *et al* 2020 Cranbrook EO. Barn swallows hirundo rustica in peninsular malaysia: Urban winter roost counts after 50 years, and dietary segregation from house-farmed swiftlets aerodramus sp. *Raffles Bull Zool* **68** 238–48
[21] Sullivan B L *et al* 2014 The eBird enterprise: An integrated approach to development and application of citizen science *Biol Conserv* 169 31–40
[22] Kelling S *et al* 2015Taking a 'Big Data' approach to data quality in a citizen science project. Ambio **44** 4 601–11
[23] Hammer Ø, Harper DAT, Ryan PD 2001 Past : Paleontological statistics software package for education and data analysis **4** 1 1–9
[24] Hansen M C *et al* 2013. High-Resolution Global Maps 134 850–4
[25] Jetz W, Wilcove DS, Dobson AP 2007 Projected Impacts of Climate and Land-Use Change on the Global Diversity of Birds **5** 6
[26] Sekercioglu C H *et al* 2007 Persistence of Forest Birds in the Costa Rican Agricultural Countryside **21** 2 482–94
[27] Waycott M *et al* 2009 . Accelerating loss of seagrasses across the globe threatens coastal ecosystems **1** 19
[28] Sodhi N S *et al* 2010 Deforestation and Avian Extinction on Tropical Landbridge Islands **24** 5 1290–8.
[29] Mimura N 2013 Review Sea-level rise caused by climate change and its implications for society. **89** 7
[30] Parmesan C & Yohe G 2003 A globally coherent fingerprint of climate change impacts across natural systems
[31] Huntley B *et al* 2004 Potential impacts of climatic change upon geographical. 2006;8–28.
[32] Katrin Böhning - Gaese NL. Importance of Climate Change for the Ranges , Communities and Conservation of Birds **35** 04
[33] Malaysian Meteorological Department 2009 www.met.gov.my/documents/10661/194684/climate-scenarios.pdf/. Climate change scenarios for Malaysia (2001 – 2099).
[34] Carey C 2009 The impacts of climate change on the annual cycles of birds. Philos Trans R *Soc B Biol Sci* **364** 1534 3321–30
[35] Res C, Studds CE, Marra PP 2007 Linking fluctuations in rainfall to nonbreeding season performance in a long-distance migratory bird *Setophaga ruticilla* **35** 115–22
[36] Norris D R *et al* 2004 Tropical winter habitat limits reproductive success on the temperate breeding grounds in a migratory bird 59–64
[37] Ahola MP, Laaksonen T & Eeva T 2007 Climate change can alter competitive relationships between resident and migratory birds (Slagsvold 1975) 1045–52
[38] Both C *et al* 2006 Climate change and population declines in a long-distance migratory bird. *Nature* **441** 1 81–3.
[39] Res C, Canale CI & Henry P 2010 Adaptive phenotypic plasticity and resilience of vertebrates to increasing climatic unpredictability 43 135–47
[40] Al-amin A Q & Filho W L 2014 A return to prioritizing needs: Adaptation or mitigation alternatives? 4 359–71
[41] Coppack T, Both C. Predicting Life-Cycle Adaptation of Migratory Birds to Global Climate Change predict In G Life-Cycle Adaptation Of Migratory Birds To Global Climate Change **90** 3 369–78
[42] Kumar V *et al* 2010 Biological clocks and regulation of seasonal reproduction and migration in birds *Physiol Biochem Zool* **83** 5 827–35.
[43] Burman M S 2016 Citizen science reveals complex changes in barn swallow phenology in South Africa over three decades 245
[44] Visser M E & Both C 2005 Shifts in phenology due to global climate change : the need for a yardstick 2561–9
[45] Romain Julliard & Frédéric Jiguet D C 2003 Common birds facing global changes : what makes a species at risk? 148–54.

Acknowledgements
This study was funded by the Fundamental Research Grant Scheme (FRGS), Ministry of Higher Education Malaysia (MOHE), under grant FRGS/1/2020/STG03/UKM/02/5. We thank the numerous contributors to eBird, including the observers, project team, and data reviewers.

The adverse effect of long term intake of Monosodium Glutamate on kidney performance

Amer M. Hussin[1*], Ali A. Tala'a[2], Safa Abdul Naser Fadhil[1] and Hamzah Abdulrahman Salman[1]

[1] Medical Laboratory Techniques, College of Medical Sciences Techniques, The University of Mashreq, Baghdad, Iraq.
[2] College of Veterinary Medicine, Al-Fallujah University, Iraq

*Corresponding Email: amer17173@gmail.com; amer.muttib@uom.edu.iq

Abstract. Monosodium glutamate (MSG) is a food additive that is considered as a water and environmental pollutant and affects the tissues of the living being. This study was aimed to find the effect of long-term administration of MSG on the mass of mesangial cells of the kidneys. Forty adult male rats were divided into four groups (10 each). Control groups 1&2 were supplied orally with distilled water for 30 and 60 days, respectively. Treatment groups 1&2 were supplied orally with 15 mg/kg Bwt of MSG for 30 & 60 days, respectively. Control and treatment groups were sacrificed, specimens of kidneys were obtained, fixed with 10% neutral buffered formalin, processed by Routine histological techniques, stained by Hematoxylin and eosin, and PAS (Periodic Acid-Schiff) stains then examined under the light microscope. The result found enlargement in a mesangial mass represented by hypertrophy and hyperplasia of mesangial cells leading to mesangial proliferative glomerulonephritis. Accordingly, the study showed an increase in creatinine values, indicating a disturbance in renal function. This will lead to a decrease in the sizes of the glomeruli of renal corpuscles and a relative increase of Bowman's space. With the time of the experiment, the glomerular capillaries and gates of basement membranes will be closed, resulting in renal filtration disorders. It was concluded that the long-term intake of MSG leads to indirect narrowing of the glomerular capillary lumen, causing kidney failure.

Keywords: Monosodium glutamate, Mesangium, Kidney failure, MSG

Track Name: Human, Social, Economic and Environmental Sustainability

1. Introduction

Monosodium glutamate (MSG) is also known as a vetsin powder. It is used as a flavor enhancer in food preparation. Many studies have been referred to the effects of high dosages of MSG [1, 2]. The toxicity of MSG wastewater on the growth ability of Chinese cabbage and wheat seeds was also observed. Dong et al. [3] showed MSG manufacturing in China released ammonia into the atmosphere and maybe a potent pollutant agent. This human-made excess ammonia can cause respiratory health problems [3]. Dixit et al. reported that albino rats given 4 mg /g of body weight MSG orally for seven days showed a significant alteration in renal glomeruli with increasing the length and size of the Bowman's capsule with the widening of Bowman's space [4]. Mesangial cells are smooth muscle-like cells between the glomerular capillaries, close to the capillary lumen. Their shapes were flattened, irregular with cylindrical cell bodies and processes containing actin and myosin, giving them contractile properties. Mesangial cells act cooperatively with podocytes.

The population of mesangial cells forms about 30-40% of the total number of glomerulus cells [5]. The primary function of the mesangial cell is phagocytosis by removing the precipitates attached to the basement membrane and keeping the filtration barrier-free from debris. The mesangial cell has a contractile property in alternating the glomerulus filtration pressure [6, 7]. The anchoring filaments of mesangial cells attached to the glomerular basement membrane can decrease the flow of capillaries by altering the surface area of glomerular ultrafiltration. The functional unit of glomeruli is mesangium, podocytes, and glomerular endothelial cells through interactions of molecular signaling pathways, which are fundamental for glomerular tuft formation [8]. Mesangial cells assist the filtration by constituting part of the tuft structure of the glomerular capillary that aid in the filtration of fluids to produce urine [9].

The expansion of mesangium occurs due to an increase in extracellular matrix protein deposition [10]. The activation of metabolic pathways is a result of increased glucose levels leading to increased oxidative stress. This will finally lead to the glomerular basement membrane thickening, expansion of mesangial matrix, then fibrosis and glomerulosclerosis [11]. A previous report recommended a detailed examination to be carried out to evaluate the effects of MSG on the atmosphere, work environment, and health activities of the manufacturer's workers [12]. Furthermore, Niaz et al. reported the toxicity of MSG into community health. The results showed MSG poses harmful effects on neuron cells as well as the reproductive system [13]. In context to the above, this study aimed to focus light on the side effects of the appetizer MSG on the renal infiltration barrier.

2. Materials and methods

2.1. *Experimental setup*

Forty adult healthy male rats were collected from local markets in 2020; rats were divided into four groups (each 10). The control groups C1&C2 were supplied orally with distilled water for 30 and 60 days, respectively. Treatment groups G1&G2 were given orally with 15 mg/kg B.wt of MSG for 30 & 60 days, respectively. All rats were sacrificed, and the specimens were fixed in 10% neutral buffered formalin and prepared by Routine histological techniques. Slides were stained by Hematoxylin and eosin, and PAS then examined under the light microscope. Optica view seven image analysis software was used for measurements.

2.2. *Statistical analysis*

All data were analyzed statistically by using ANOVA one-way SPSS program version 20.

3. Results

The current results declared that the long-term intake of MSG was affected the infiltration barrier of the nephrons. The study correspondingly found a relative increase in the thickness of basement membranes, Bowman's capsule diameter, and Bowman's spaces. The study revealed hypertrophy and hyperplasia of the intraglomerular mesangial cells (Table 1). Congestion of glomerular capillaries and thickening of basement membrane was also observed in treatment groups (Figs.2-7). The values of creatinine were increased with the time of the experiment (Table 2).

Table 1. Measurement of renal corpuscles constituents in control and MSG treated groups. Different letters (A,B) denote significant ($p<0.05$) differences between groups

Groups	The thickness of Bowman's capsule basement membrane (μm) Mean± SE	Bowman's capsule diameter (μm) Mean± SE	Bowman's space (μm) Mean± SE	Number of mesangial cells per glomeruli Mean± SE
C1	0.78±0.002 B	113.9±0.15 B	24.2±0.24 B	10.85±2.43 B
G1	0.91±0.005 A	114.0±0.21 B	24.5±0.12 B	13.11±0.4 AB
C2	0.8± 0.003B	113.8± 0.17B	24.3± 0.21B	10.75±2.1B
G2	0.97±0.003 A	128.7±0.14 A	32.8±0.07 A	17.3±1.1 A

Table 2. Serum creatinine (mg/dl) levels in control and MSG treated groups. Different letters (A, B, C) denote significant ($p<0.05$) differences between groups

Groups	**Creatinine (mg/dl) Mean± SE**
C1	0.63±0.003 C
G1	1.2±0.06 B
C2	0.59 ± 0.002 C
G2	1.89±0.09 A

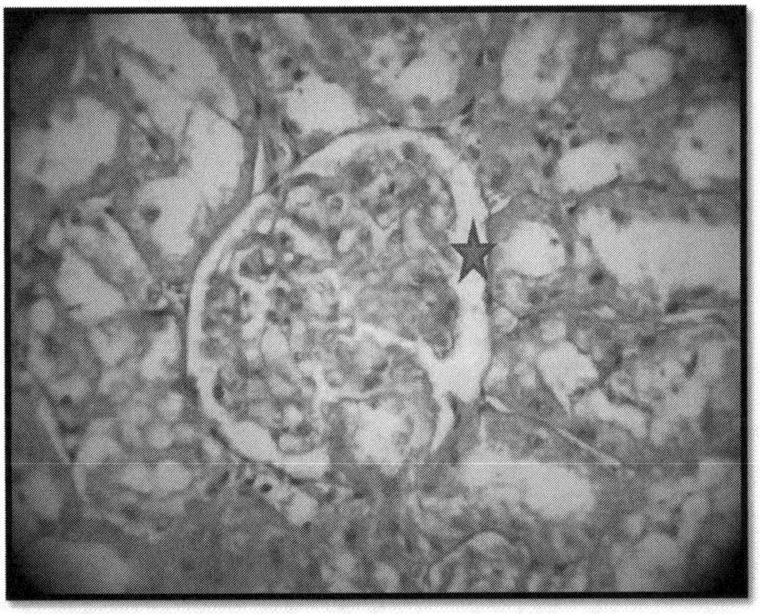

Figure 1. Histological section of kidney in control group after 30 days showing normal architecture. Star refers to normal Bowman's space. H and E stain (400X).

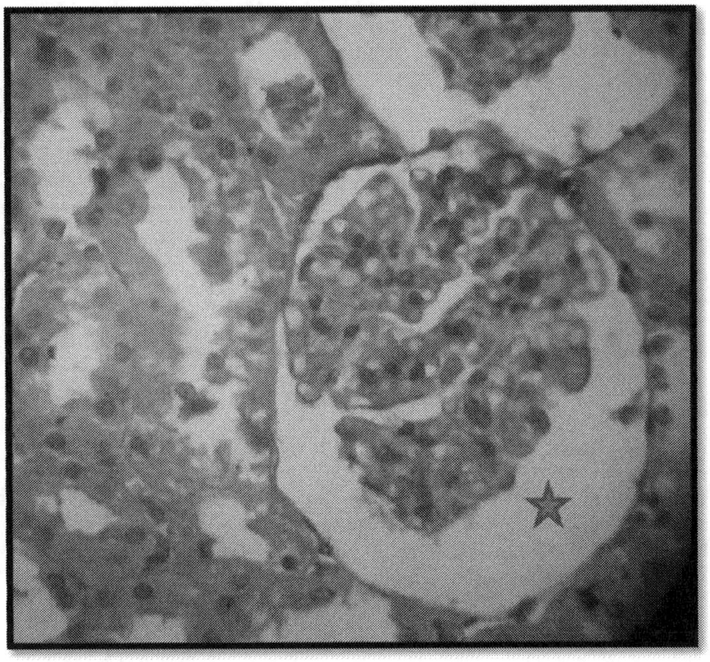

Figure 2. Histological section of kidney for MSG treated group after 30 days showing an increase in Bowman's space (star). H and E stain (400X).

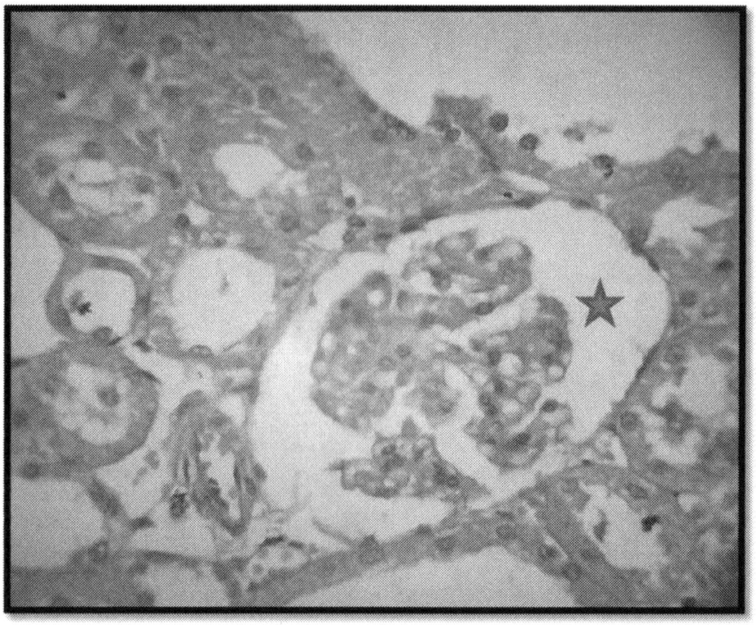

Figure 3. Histological section of kidney for MSG treated group after 60 days showing the high increase in Bowman's space (star). H and E stain (400X).

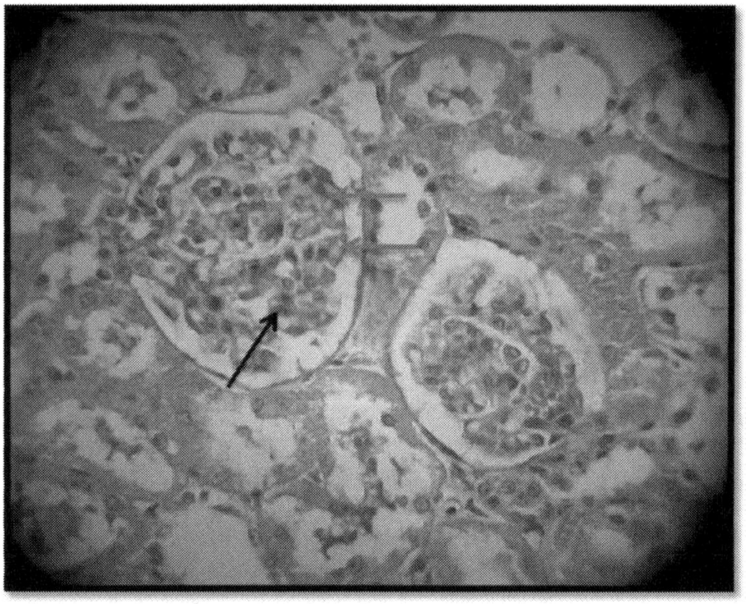

Figure 4. Histological section of kidney for MSG treated group after 30 days showing congestion of glomerular blood vessels (blue arrows) and beginning of the process of mesangial cells proliferation (black arrow) slightly. H&E stain 400X.

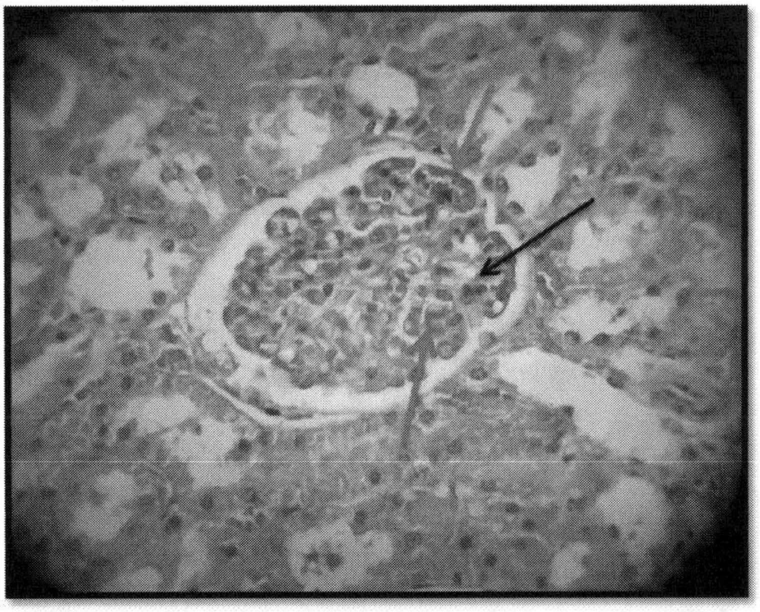

Figure 5. Histological section of kidney for MSG treated group after 60 days showing mitosis in mesangial cells (black arrow) and congested capillaries (blue arrows). H&E stain 400X.

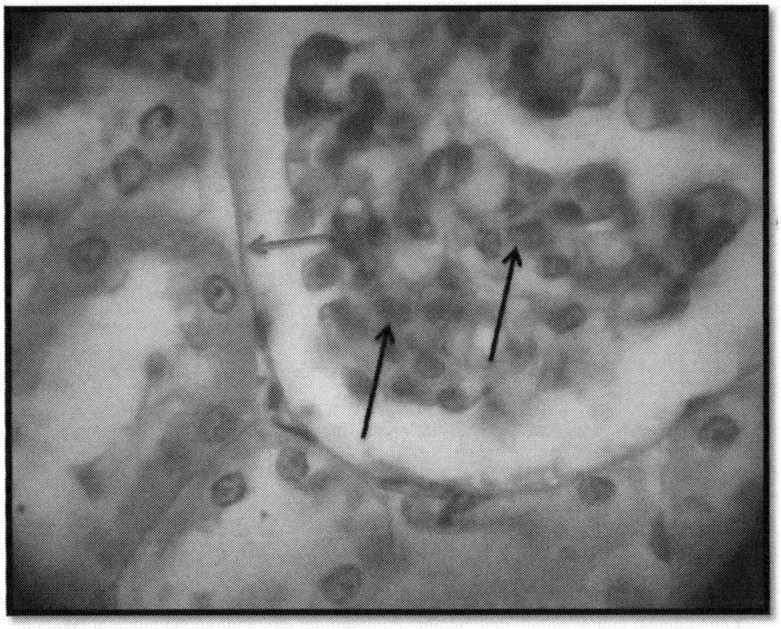

Figure 6. Histological section of kidney for MSG treated group after 60 days showing enlargement of mesangial cells (black arrows) with more apparent basement membrane (blue arrow). PAS stain. 1000X.

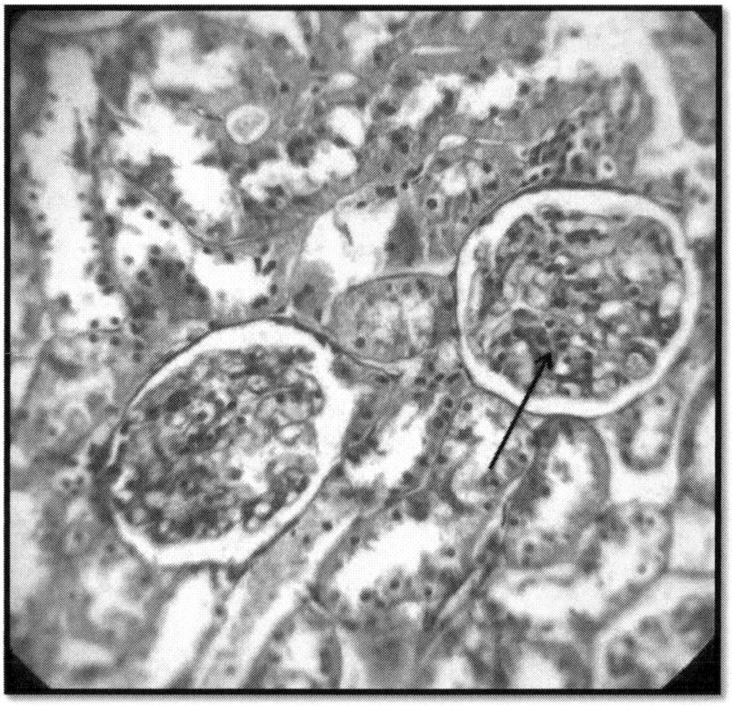

Figure 7. Histological section of kidney for MSG treated group after 60 days showing enlargement of mesangium and proliferation of mesangial cells (arrow)) PAS stain 400X.

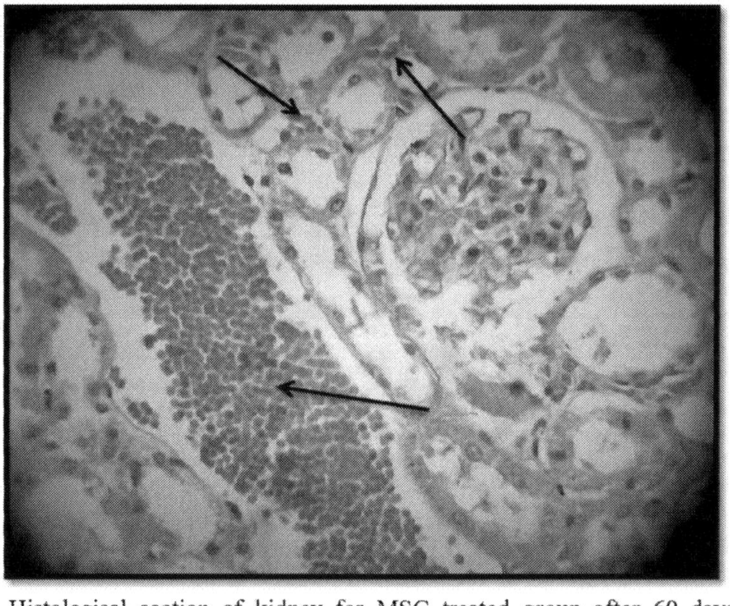

Figure 8. Histological section of kidney for MSG treated group after 60 days showing congestion and hemorrhage of blood vessels (arrows). H&E stain. 400X.

4. Discussion

The herein study stated that the renal tissue deals with MSG as a foreign body (Figure 1-8). The size of Intraglomerular mesangial cells was increased due to phagocytosis of the MSG. The mesangial enlargement sheds an external pressure on the glomerular capillaries and, in turn, on the gates of infiltration barriers of the nephrons. This was nearly similar to the findings reported earlier that the expansion of mesangium occurs due to increased extracellular matrix protein deposition [8]. The present study is consistent with the previously reported data that kidney damage may occur due to the oxidation with MSG help [14]. The increase in the thickness of the basement membrane is caused by the widening of its gates in order to get rid of this foreign body.

Table 1 showed a significant increase in the size of the basement membrane and the mesangium after 30 and 60 days, while there was a significant increase in the other constituents only after 60 days. This was because the previous components of the infiltration barrier were firstly faced the MSG present in the blood, then others come later. Despite the increase in the size and number of the intraglomerular mesangial cells (Table 1), the size of the glomeruli remains intact. Correspondingly the Bowman's space showed a relative widening that indicates the increase in the glomerular filtration rate. The mesangial cells have contractile and phagocytic characters as they are attached to the glomerular capillaries themselves. This was in agreement with the finding of [6], who mentioned that the mesangial anchoring filaments attached to the glomerular basement membrane could change the capillary flow by altering the surface area of the glomerular ultrafiltration. The current study hypothesized that this might be attributed to shifting the proliferated and enlarged mesangial cells toward the capillaries that are attached to and not toward the exterior. This will induce pressure on the glomerular capillaries leading to narrowing their lumen.

With the time of the experiment, the glomerular infiltration will be affected and finally may cease leading to kidney failure. The present hypothesis was in variance with that of [15], who reported that the decrease in the size of glomeruli was attributed to sclerosis and fibrosis of the interstitial tissue of the glomeruli. As we regarded the endothelial cells and podocytes as part of the infiltration barrier, they lose the capability to divide. It was concluded that the mitotic figure seen in the glomeruli belonged only to the mesangial cells. This is confirmed with the result of [16], who stated that the normal podocytes of humans are quiescent cells and differentiated terminally. It is not understood why the failure of podocyte proliferation in response to most injury forms. Increasing values of serum creatinine in this study indicate that we are going toward kidney disorders.

5. Conclusion

It was concluded that the long-term intake of MSG causes enlargement of the mesangium. This will adversely affect the glomerular infiltration of the kidney through pressing on their intraglomerular capillaries and narrowing their lumen. We strongly suggest further investigation be carried out to understand the effects of MSG on humans.

References

[1] Albrahim T and Binobead M A 2018 Roles of Moringa oleifera Leaf Extract in Improving the Impact of High Dietary Intake of Monosodium Glutamate-Induced Liver Toxicity, Oxidative Stress, Genotoxicity, DNA Damage, and PCNA Alterations in Male Rats *Oxidative medicine and cellular longevity* **2018** 4501097
[2] Jubaidi F F, Mathialagan R D, Noor M M, Taib I S and Budin S B 2019 Monosodium glutamate daily oral supplementation: study of its effects on male reproductive system on rat model *Systems biology in reproductive medicine* **65** 194-204
[3] Dong L, Li Y, Wang P, Feng Z and Ding N 2018 Cleaner production of monosodium glutamate in China *Journal of Cleaner Production* **190** 452-61
[4] Dixit S G, Rani P, Anand A, Khatri K, Chauhan R and Bharihoke V 2014 To study the effect of monosodium glutamate on histomorphometry of cortex of kidney in adult albino rats *Renal failure* **36** 266-70
[5] Alpers C E and Hudkins K L 2018 Pathology identifies glomerular treatment targets in diabetic nephropathy *Kidney Res Clin Pract* **37** 106-11
[6] Ryan D, Sutherland M R, Flores T J, Kent A L, Dahlstrom J E, Puelles V G, Bertram J F, McMahon A P, Little M H, Moore L and Black M J 2018 Development of the Human Fetal Kidney from Mid to Late Gestation in Male and Female Infants *EBioMedicine* **27** 275-83

[7] Takano K, Kawasaki Y, Imaizumi T, Matsuura H, Nozawa R, Tannji M, Suyama K, Isome M, Suzuki H and Hosoya M 2007 Development of glomerular endothelial cells, podocytes and mesangial cells in the human fetus and infant *The Tohoku journal of experimental medicine* **212** 81-90
[8] Schlöndorff D and Banas B 2009 The mesangial cell revisited: no cell is an island *Journal of the American Society of Nephrology : JASN* **20** 1179-87
[9] Marciano D K 2019 Mesangial Cells: The Tuft Guys of Glomerular Development *Journal of the American Society of Nephrology : JASN* **30** 1551-3
[10] Muraoka H, Hasegawa K, Sakamaki Y, Minakuchi H, Kawaguchi T, Yasuda I, Kanda T, Tokuyama H, Wakino S and Itoh H 2019 Role of Nampt-Sirt6 Axis in Renal Proximal Tubules in Extracellular Matrix Deposition in Diabetic Nephropathy *Cell Reports* **27** 199-212.e5
[11] Zhang J, Wang Y, Li L, Zhang R, Guo R, Li H, Han Q, Teng G and Liu F 2018 Diabetic retinopathy may predict the renal outcomes of patients with diabetic nephropathy *Renal failure* **40** 243-51
[12] Rim K-T 2017 Toxicological evaluation of MSG for the manufacturing workers' health: A literature review *Toxicology and Environmental Health Sciences* **9** 1-11
[13] Niaz K, Zaplatic E and Spoor J 2018 Extensive use of monosodium glutamate: A threat to public health? *EXCLI J* **17** 273-8
[14] Sharma A 2015 Monosodium glutamate-induced oxidative kidney damage and possible mechanisms: a mini-review *Journal of Biomedical Science* **22** 93
[15] Alsaad K O and Herzenberg A M 2007 Distinguishing diabetic nephropathy from other causes of glomerulosclerosis: an update *Journal of clinical pathology* **60** 18-26
[16] Shankland S J, Eitner F, Hudkins K L, Goodpaster T, D'Agati V and Alpers C E 2000 Differential expression of cyclin-dependent kinase inhibitors in human glomerular disease: role in podocyte proliferation and maturation *kidney international* **58** 674-83

AUTHOR INDEX